AF786323

Computer Assisted Radiology
Computergestützte Radiologie

Proceedings of the International Symposium
Vorträge des Internationalen Symposiums

CAR'93 Computer Assisted Radiology

Volume I

Edited by/Herausgegeben von
H.U.Lemke, K.Inamura, C.C.Jaffe, and R.Felix

Organizer/Organisator: H.U.Lemke

With 266 Figures/Mit 266 Abbildungen

Springer-Verlag Berlin Heidelberg GmbH

Prof. Heinz U.Lemke, Ph.D.
Institut für Technische Informatik
Technische Universität Berlin
Berlin, Federal Republic of Germany

Kiyonari Inamura, Ph. D.
Osaka University
College of Biomedical Technology
1 - 1 Machikaneyama, Toyonaka-shi,
Osaka 560/Japan

Prof. C. Carl Jaffe, M.D.
Department of Diagnostic Radiology
Yale University
New Haven, Connecticut, USA

Prof. Roland Felix, M.D.
Universitätsklinikum Rudolf Virchow
Freie Universität Berlin
Berlin, Federal Republic of Germany

ISBN 978-3-642-49353-9 ISBN 978-3-642-49351-5 (eBook)
DOI 10.1007/978-3-642-49351-5

This work is subject to copyright. All rights are reserved, whether the whole or part of the material is concerned, specifically the rights of translation, reprinting, re-use of illustrations, recitation, broadcasting, reproduction on microfilms or in other ways, and storage in data banks. Duplication of this publication or parts thereof is only permitted under the provisions of the German Copyright Law of September 9, 1965, in its current version, and a copyright fee must always be paid.

Springer-Verlag Berlin Heidelberg 1993
Originally published by Springer-Verlag Berlin Heidelberg in 1993

Softcover reprint of the hardcover 1st edition 1993

The use of registered names, trademarks, etc. in this publication does not imply, even in the absence of a specific statement, that such names are exempt from the relevant protective laws and regulations and therefore free for general use.

Typesetting: Camera-ready by authors

68/3020 - 5 4 3 2 1 0 - Printed on acid-free paper

Honorary Chairman:

J. Oscar M.C. Craig, MD
The Royal College of Radiologists, London (UK)

Symposium Committee

Heinz U. Lemke, PhD
Technische Universität Berlin (FRG)

Kiyonari Inamura, PhD
Osaka University (Japan)

Roland Felix, MD
Freie Universität Berlin (FRG)

C. Carl Jaffe, MD
Yale University (USA)

Symposium-Organizer

Prof. Dr. Heinz U. Lemke
c/o Institut für Technische Informatik
Technische Universität Berlin
Sekr. CG/FR 3-3
Franklinstrasse 28-29
D-1000 Berlin 10

Conference Officer
Editorial Assistant

Bibiana R. Moser
CAR'93
c/o Laser-Medizin-Zentrum
Krahmerstrasse 6-10
D-1000 Berlin 45
Tel.: +49 30 798 41 58
Fax: +49 30 216 59 73

CAR'93
in cooperation with:

TU	Technische Universität Berlin
FU	Freie Universität Berlin
acm	Association for Computing Machinery/SIGBIO
DRG	Deutsche Röntgengesellschaft
IAPR	International Association for Pattern Recognition
EuroPACS	European Society for Picture Archiving & Communication Systems
IEEE COMPUTER SOCIETY	
SIRM	Associazione Italiana di Radiologia Medica
SPIE	The International Society for Optical Engineering
GI	Gesellschaft für Informatik
NVvRd	Nederlandse Vereniging voor Radiodiagnostiek
BIR	The British Institute of Radiology
DGBMT	Deutsche Gesellschaft für Biomedizinische Technik e.V.
JRS	Japan Radiological Society
RCR	The Royal College of Radiologists
SFR	Société Française de Radiologie Médicale
IMAC	Image Management and Communication
ÖRG	Oesterreichische Röntgengesellschaft
SGMR	Schweizerische Gesellschaft für Medizinische Radiologie
SCAR	Society for Computer Applications in Radiology
WHO	World Health Organization
The Polish Medical Radiological Society	
IPMI	Information Processing in Medical Imaging
Society for Computer Assisted Radiology, P.R.China	
The Bulgarian Scientific Society of Roentgenology, Radiology and Radiobiology	
CRS	Czech Radiological Society
SRS	Slovak Radiological Society
GMDS	Ges. für Med. Dokumentation, Informatik und Statistik e.V.
ESPR	European Society of Paediatric Radiology
CEPUR	Collège d'Enseignement Post Universitaire de Radiologie
Japanese College of Radiology	

Program Committee

Paul R. Algra, MD, PhD
Medisch Ctr., Alkmaar (NL)

Yutaka Ando, MD
Keio Univ. Tokyo (Japan)

Ronald L. Arenson, MD
Hosp. of the Univ. of PA (USA)

Albert R. Bakker, PhD
BAZIS, Leiden (NL)

Roger A. Bauman, MD
Mass.Gen.Hosp.,Boston (USA)

Yves Bizais, PhD
Nantes Hôpital (France)

Antonio Chiesa, MD
Spedali Civili, Brescia (Italy)

Claus D. Claussen, MD
E.-Karl-Univ. Tübingen (FRG)

Alan C. F. Colchester, MD
Guy's Hospital, London (UK)

Jacques Cornet
General Electric - Med. Syst. (F)

Bernard L. Crowe, BA, MPH
Austral. Inst. of Health (AUS)

Hans Dahlin, PhD
HELAX, Uppsala (Sweden)

Ludovico Dalla Palma, MD
Osp. di Cattinara, Trieste (Italy)

Adrian K. Dixon, MD
Addenbr. Hosp., Cambridge (UK)

Kunio Doi, PhD
Kurt Rossmann Lab., Chicago (USA)

Samuel L. Dwyer, III, PhD
Univ. of Virginia, Charlottesville (USA)

Masahiro Endoh, PhD
Nat. Inst. of Radiol. Sc. (Japan)

Uwe Faust, PhD
Universität Stuttgart (FRG)

Eckhard Fleck, MD
Dt. Herzzentrum Berlin (FRG)

Richard M. Friedenberg, MD
Univ. of Cal., Orange (USA)

David Gilday, MD, BEE
Hosp. Sick Childr., Toronto (CDN)

Melvyn Greberman, MD, MPH
Ctr. Dev. and Rad. Health (USA)

Christian F.C. Greinacher, PhD
Siemens,UB Med., (FRG)

Atsuko Heshiki, MD
Saitama Medical School (Japan)

Karl Heinz Höhne, PhD
UKE Hamburg (FRG)

Herwig Imhof, MD
AKH Wien (Austria)

Tamon Inouye, PhD
University of Tsukuba (Japan)

Akira Itoh, PhD
Cancer Institute, Tokyo (Japan)

Herbert J. Kaufmann, MD
Freie Univ. Berlin (FRG)

Chikazumi Kuroda, MD
Osaka Adult Diseases Ctr. (J)

Philipp Lang, MD
Univ. of Cal. San Francisco (USA)

Martin J. Lipton, MD
Univ. of Chicago (USA)

Weixue Lu, PhD
Zhejiang University, Hangzhou (PR China)

Meinhard Lüning, MD
Humboldt-Univ., Charité, Berlin (FRG)

Tomoho Maeda, MD
Kyoto Univ. of Medicine (Japan)

Roman Marciniak, MD
Med. Academy Wroclaw (PL)

Dietrich Meyer-Ebrecht, PhD
RWTH Aachen (FRG)

Seong K. Mun, PhD
Georgetown Univ., Med. Ctr. (USA)

Hiromu Nishitani, MD
Tokushima Univ. (Japan)

Stelios Orphanoudakis, PhD
Inst. of Comp. Science (Greece)

Michel Osteaux, MD
PRIMIS, Free University of Brussels (B)

Helmut Oswald, PhD
Dt. Herzzentrum Berlin (FRG)

Peter E. Peters, MD
Westf. W.-Univ. Münster (FRG)

Stephen M. Pizer, PhD
Univ. of North Carolina (USA)

Osman M. Ratib, MD, PhD
Univ. Hospital of Geneva (CH)

Michael L. Rhodes, PhD
Toshiba America, Inc. (USA)

Otto Rienhoff, MD
Philipps-Univ. Marburg (FRG)

Douglas D. Robertson, MD, PhD
J.Hopkins Hosp.,Baltim. (USA)

Plinio Rossi, MD
Univ. 'La Sapienza', Rome (Italy)

Wolfgang Schlegel, PhD
DKFZ Heidelberg (FRG)

Mohamed H. Sherif, MD
El-Azhar Univ. Cairo (Egypt)

Jack Sklansky, Eng. ScD
Univ. of Cal., Irvine (USA)

H. Siegfried Stiehl, PhD
Univ. Hamburg (FRG)

Ei-ichi Takenaka, MD
Nat. Defense Med. Coll. (Japan)

Klaus D. Toennies, PhD
Technische Univ. Berlin (FRG)

Jayaram K. Udupa, PhD
Univ. of Penns., Philadelphia (USA)

Michael W. Vannier, MD
Mallinckrodt Inst. of Rad. (USA)

Erich Voegeli, MD
Kantonsspital Luzern (CH)

Gabriele v. Voigt, PhD
Technische Univ. Berlin (FRG)

Thomas Wendler, PhD
Philips Res. Lab. Hamb. (FRG)

Ian R. Young, PhD
GEC Research Ctre. (UK)

Prelude

Oscar Craig

PICTURE ARCHIVING AND COMMUNICATING SYSTEMS (PACS), MODERN TECHNOLOGY AND UNIVERSAL HEALTH NEEDS

> "Old men shall dream dreams, and young men shall see visions"
>
> Joel 2.28

It was in 1982 that Harold Glass, a regional scientific officer in London, introduced me to the concept of a filmless hospital, as we discussed the building of a new St. Mary's Hospital. When I realised the potential of this technological new world, I became obsessed that the new St. Mary's should be filmless. I was to become, in the succeeding few years, pathologically driven towards achieving the first ever filmless hospital in the world. Because of this naive and artless obsession, I was to bore my friends, terrify my radiological colleagues, con my clinical colleagues, become frustrated with my commercial associates, argue and perhaps lose the close friendships of those I thought had failed to carry me academically and underwrite me financially towards achieving this Valhalla in 1987. My disappointment, my sense of failure in not achieving this goal, was immensely intense, and now that I reflect on it, was as naive and as artless as my obsession for it in 1982.

For today in 1993, 11 years later, we may have generated a cataclysmic weight of learned papers, but in practice, despite the many PACS projects, the era of the filmless hospital is yet to come.

Heinz Lemke[1] said in October 1992 "PACS are among the most promising information technologies in health care. They are however, still in the development phase, and their implementation is taking longer than was predicted in the 1980s. Evaluation of these PACS implementations is also much more difficult than was expected."

Arenson[2] said in 1992, "Although there are several clinical PACS in operation today, there are significant problems slowing their introduction."

Racoveanu[3] said in May 1992, "PACS, a health care technology, is still in its development stage."

Seshardi[4] and Arenson said in 1992 "PACS are rapidly evolving and the total digital radiology department may be a reality within the next decade."

I, too, have learnt that there is much more to PACS than joining together developing sophisticated apparatus on a line diagram.

The clinical achievements in diagnostic radiology have been remarkable and are unequalled in any other medical specialty during the last three to four decades. The introduction of computed tomography, ultrasound, isotope imaging, digital subtraction, magnetic resonance and the diagnostic and therapeutic development of vascular and interventional radiology, all have placed radiology at the forefront of patient management, and central to the patient's clinical welfare. The volume of information and the number of images generated for each patient are enormous.

Hendee[5] pointed out that "Traditional mechanisms for managing information in radiology are outmoded and unresponsive to the needs of patients and physicians."

It is clear therefore that the high level of efficiency achieved today in clinical radiology, has to be matched by an equitable level of efficiency in handling the data that is generated. Indeed we should also be using computed science to improve the quality of the data.

To introduce PACS into any hospital will make radical changes in the work practice of radiologists and clinicians. The medical profession is traditionally conservative and resistant to radical changes. Radiologists are familiar with film, have years of experience with film, and are comfortable with film. Clinicians are anxious, that to be denied the hard copy, will disadvantage them. They fear that it might increase their dependence on their radiological colleagues. There are those radiologists, who fear that PACS, by transmitting images throughout a hospital, could by-pass the consultative process, and radiology could become totally a service specialty. However highly we regard our clinical colleagues, experience in medical practice has taught us that the trained clinical radiologist makes fewer errors in radiological interpretation than his non-trained medical colleagues. The skill of the radiologist in image interpretation must be the judge of accepted practice. One solution lies in the introduction of real time; instant reporting before transmission. This may be the aim, but it is difficult to achieve in terms of staff, staff time and cost.

Because of the perceived problems in PACS, the resistance of the medical profession, the need to change working practices and the overall cost, it is, in my view, imperative that the clinical staff of the hospital should endorse any decision to introduce a major PACS installation. I made a point of getting this in writing from each clinical division in the hospital during the planning stage of St. Mary's Hospital in 1985.

Writing in Radiographics, Choplin et al[6] said in 1992, "The goals of PACS are to improve operational efficiency, while maintaining or improving diagnostic ability."

The slow introduction of PACS throughout the Western World has been partly due to the suspicion on the part of radiologists and their clinical colleagues that PACS can actually facilitate better access to images throughout a hospital, and that the diagnostic accuracy of reporting on digital display systems can equal reporting on hard copy.

Banta[7], writing in the International Journal of Biomedical Computing in 1992 said "The effect of PACS on health outcome is not known. A PACS system has the potential to improve quality, especially if it makes the diagnostic process more efficient, but evidence that such improvement actually occurs is so far lacking".

Racoveanu[3] in 1992 said "As the technology has not been comprehensively assessed, data and facts to justify its clinical, economic and other benefits are to date insufficient." This author then goes on to call for internationally coordinated action to obtain faster and more accurate technology.

Boehme and Choplin[8] said in Radiographics - "Typically hospital information services and radiological information services and PACS have been introduced into radiological departments at various times, and are implemented without any integration strategy." Arenson[2] said "Standard interfaces are still lacking for most of the imaging equipment currently in use, making connections to these devices nearly impossible. Cost is a major issue as well."

There is no doubt that the cost of PACS is an enormous deterrent to its introduction on any large scale.

Hendee[5] wrote in 1991 in the American Journal of Roentgenology that "the enormous cost of PACS, which it is difficult to see returned, presents a major challenge to institutions struggling with financial solvency."

I can remember with extreme clarity my anxiety in trying to convince the Department of Health in the United Kingdom that to introduce a total PACS system into St. Mary's Hospital would save money. In fact we produced figures for saving money on films, porters, chemicals and the like, totally fancied and untrue, but not dishonest, because we had deceived ourselves that such calculations were truthful and reasonable.

Hilsenrath[9] et al wrote in 1991 that some technologies that may be genuinely worth while, risk being cast aside because of increasingly constrained financial environments. Bernie Huang[10] describing the implementation of a large scale PACS in Los Angeles wrote in 1993 that the two critical issues that could be studied were system performance and cost effectiveness between a digital based and a film based system. It has of course been suggested that if finance is not available then instead of a total PACS development, a mini-PACS installation would be the correct choice. McMahohn and Doi [11] wrote in 1991 "because of the great expense involved, it is likely that transition from conventional to digital radiography will be evolutionary in most medical centres with establishment of local PACS networks initially. Limited digital radiography systems can function in isolation and address specific needs successfully." I have in the past supported the phased introduction of PACS because of other factors such as the uncertainty about clinical efficacy and the uncertainty about total image traffic and retrieval times in multiple sites simultaneously. However, phased implementation involves operating a dual system and may prove more costly in the long run. It does have the advantage of less risk to the overall administration of the hospital and an ability to keep abreast with rapidly changing technology, updating piecemeal and when necessary.

Straub and Gur[12] said in 1990 that despite the considerable research and development investment and the enthusiasm for PACS technology among some radiologists, the implementation of PACS in the United States is slow because of cost/benefit concerns and the nagging belief that diagnostic accuracy with current soft display systems may not be comparable to that obtained with conventional film.

Hendee[5] in 1991 said "It is difficult to persuade those who are responsible for health service finance that PACS should be considered an up-front, non-reimbursable commitment made by an institution to improve patient care, and not a technology that will yield a direct return on investment". I am attracted to the arguments used by Straub and Gur[12] in 1990 that cost effective studies have only referred to the cost of implementing PACS, and what has largely been ignored is the substantial indirect or hidden costs that result from inefficiencies inherent in traditional archive systems.

Although frequently denied in any individual hospital, it is a fact that 20 per cent of films are lost. The need to repeat examinations is not only costly, but exposes the patient to unnecessary irradiation which, in today's health conscious climate, is totally unacceptable, and in today's litiginous climate is open to a charge of medical negligence or malpractice. It is difficult to quantify the number of repeat examinations because of technical error, but this may vary between 3 and 10 per cent. This, too, is costly in terms of finance and irradiation.

Costs such as film, space and personnel can be measured, but Straub and Gur[12] point out that indirect costs of patient care that result from delayed access to diagnostic information can be difficult to quantify. This would include repeat visits, repeat examinations, transportation costs, bed occupancy and medical staff time.

In their paper these authors said that in a hospital with 550 beds and a budget of 1.8 million dollars, 82% of physicians believed that delayed access to films frequently cost them money. If one assumed an increased length of stay in hospital of only 3 per cent, such indirect costs would be 6.4 million dollars per year. They concluded that the perceived potential impact of increased efficiency and reliability of any archival system on radiology users' practices is significant. They also noted that the elimination of repeat examinations and the reduction of patients' length of stay in hospital would be significant. I would add that as yet we have not been able to assess the increased value in terms of quality of life and indeed longevity that might result from improved diagnostic accuracy which I think is inherent in the use of digital technology.

The efficacy of VDU reporting has been a particular concern of mine since the early 1980s. When it was clear that it was not possible to install a total PACS system in St. Mary's Hospital prior to 1987, we turned our efforts to the clinical problem of diagnostic accuracy on soft copy. A team was set up to compare the accuracy of VDU reporting with film reporting. A research fellow, Dr. Richard Dawood, was appointed to plan this project. We chose three difficult pathologies, periosteal resorption of bone, pneumocystis carinii pneumonia and breast microcalcification. At the time, 1989/90, we were using digitised film, displayed lk by lk. The evaluation results were disappointing, in showing that hard copy film diagnosis was more accurate than diagnosis on digital soft copy. Our conclusions were that lk by lk display systems were unsuitable for primary diagnosis. This has been substantiated by many others, who have concluded that monitors with at least 2000 lines are necessary. However the clinical efficacy of these monitors also requires extensive evaluation. The PACS systems installed today favour 2000 line displays for primary diagnosis and l000 line displays for peripheral viewing stations. Whereas this may seem to be cost effective, I have an anxiety that a technically less efficient display system, at a peripheral site, reaches an acceptable standard of excellence. Clinicians, seeing a lk by lk image may be superficially impressed, but could ask that if the resolution on lk by lk monitors is acceptable, then why has the radiologist got 2k by 2k monitors in his department for the primary diagnosis? The clinician may think this is critical question when it is necessary for him to examine a digital image at 2 a.m. in the accident and emergency department or on the wards. The answer I am given to this anxiety is that the clinician can walk to the department of radiology to view the 2k image.

I am worried also that peripheral lk images would be medico-legally acceptable in the event of a diagnostic mishap resulting from their use. These anxieties are strong arguments in favour of immediate radiological image reporting, covering 24 hours of the day, with all its accepted difficulties.

Compared to the number of research projects on the technology of PACS, i.e. the storage and transmission issues, there are still relatively few papers on the clinical efficacy of reporting on 2000 line monitors. Lou and Huang13 in 1992 assessed a neuroradiology PACS system in clinical practice and found that the user acceptance was 3.4 on a 4 point ranking scale. Work in paediatric and intensive care units has shown that the clinical services are improved overall. Work on digital chest radiology is providing encouraging clinical results. Razavi et al14 reported a study of chest radiographs in children in the American Journal of Roentgenology in February 1992 performed in Professor Huang's department in Los Angeles. This article stated that the quality of soft copy images is still not widely accepted. However, they studied 239 images, comparing diagnosis on soft copy 2000 line images with diagnosis on digital laser printed film from computed radiography. In essence the results showed no significant difference between digital hard copy and soft copy. Nonetheless, more work including a wide range of different pathologies needs to be done by clinical radiologists on the accuracy of diagnosis from state of the art monitors. These studies should include an assessment of image compression and its effect on diagnostic accuracy.

Despite uncertainties about PACS, the growth of digital systems is progressing. Huang[15] reviewed these systems in Integrated Diagnostic Imaging in 1992.

Apart from in the USA, PACS research is in progress in eleven countries: Austria, Belgium, Switzerland, United Kingdom, Turkey, France, Germany, Netherlands, Japan, Italy and Sweden. All these countries have mini-PACS in operation. There are four large scale PACS installations developed or developing:

1. Hokkaido University
2. Vienna SMZO
3. USA MDIS four sites
4. Hammersmith Hospital (UK).

The success of these projects will surely influence the acceptance of PACS by the medical profession; and hopefully will herald a major shift towards the creation of filmless hospitals.

In the United Kingdom Dr. Ian Lavelle in Newcastle has an excellent mini-PACS system installed by Philips, linked to a R.I.S. and transmitting between two separate hospital blocks. There is a data management system with sufficient capacity to service the whole hospital. The major project in the United Kingdom is at the Hammersmith Hospital where Professor David Allison intends to convert entirely to a filmless mode in one step and is working towards this with Siemens Medical Systems.

The Hammersmith Hospital is one of London's major institutions and incorporates the Postgraduate Medical School of London. It has a large service load and a substantial teaching and research commitment. I visited David at the hospital recently and I was impressed by his determination to introduce a total PACS in one move. Such a project is very time consuming for an active clinical radiologist and the paper work itself is enough to discourage clinical radiologists from attempting to lead in this field.

I was encouraged by the improvement in the monitors, by the reduction in size of the Juke box, the working storage unit, and the total capacity. A retrieval time of 2.5 secs. for first images and less for subsequent images is achievable. There will be 138 high performance on line work stations with flicker-free viewing. There will be linkage to H.I.S. and R.I.S. There will be immediate image reporting. An enormous team has been assembled to effect this plan and includes clinical radiologists, clinical colleagues, radiographers, physicists, administrators, business managers, engineers, computer experts, commercial colleagues, accountants and others. The opening date is planned for between January and March 1994. David Allison is a man of strong will and impelling character, but for lesser mortals I would add to the team a permanent resident psychiatrist.

Physicians who are sceptical of PACS development and its total cost, question whether the enormous effort in time, finance and resources are justified in view of medical needs in a world where so much of the population is denied any health care whatever. The enormous dilemma of the challenges that face clinical radiologists in the world today came forcibly to my mind on a visit I paid to India, when as President of the Royal College of Radiologists, I accompanied a team of lecturers to hold teaching seminars for both young and established radiologists and to visit hospital centres to accredit them for teaching. I met many superb, well educated and entirely contemporary radiologists, operating very satisfactory practices, but even they would admit that they could serve only so few of the needs of that vast country. Peter Ogle16 in an editorial in Diagnostic Imaging International said "There is an uneasy relationship between imaging technology and our desire for a better world." Ogle continued "The relatively few angiographic suites, C.T. scanners and M.R. machines installed in developing nations are used to cope almost exclusively for their social elite. Sadly also the people most equipped to make a difference in their native lands often leave." In fact about 30 per cent of doctors trained in developing countries never practise there. "There are many reasons for this. Higher training is most often abroad, and having used advanced technology, many cannot accept the modest facilities found at home."

It is a fact that 70 per cent of the world's population have never had an "X-ray" and never will. Bryan Jennet[17] in his book on the Benefits and Burdens of High Technology in Medicine, says "Regarding clinical radiology, never has such a wealth of technology been made available to any specialty over such a short period of time in medical development, and this has resulted in a most costly expansion of our professional services. Scientists and high technology doctors are suspected of indulging their own curiosity and enthusiasm without sufficient thought as to either the cost or the benefit of their activities. By the early l890s it was becoming obvious on both sides of the Atlantic that the benefits that high technology promises of postponing death or reducing disabilities, sometimes exact too high a price. As only a minority of patients stand to benefit from technology, too much attention to them, diverts resources from many others who pay the price in the inadequacy of provisions for their needs. In developing countries too much attention to high technology could divert resources from more essential health needs." Professor Palmer in l984 said "The very complexity of modern radiological equipment which has been such a benefit to the art of radiology in the Western World has become a negative factor in the developing world. 80 per cent of all X-ray examinations are basic and simple. Under the initiative of Professor Chamberlain of the USA, encouraged and aided by Professor Palmer and by Professor Middlemiss of the United Kingdom, the World Health Organisation in 1980 adopted the concept of a Basic Radiology System (B.R.S.). This consisted of an apparatus guaranteed to give precise centering, precise exposure and could be powered from a battery. Individuals could be trained to operate it in 8 days and they could carry out 90 per cent of plain film examinations. Most of the pathology in remote situations is due to trauma and infection. In a survey of 400 patient films, the W.H.O. assessors found that less than 1 per cent were clinically unsatisfactory. The cost of each machine, although larger than originally intended, is approximately £20,000.

It was calculated that to serve the whole world's population would need 70,000 B.R.S. machines. Howard Middlemiss said that the B.R.S. was in fact the radiology of the future for most of the world's population. The B.R.S. is estimated to have been installed in only 500 locations worldwide (Ogle 1992). A personal communication with W.H.O. informed me that 700 such machines had been distributed. This amounts to 1 per cent or less.

Could practising doctors of conscience deny that here is a compelling challenge, to bring basic and simple diagnostic skills to those in the third world who are in poverty and need? This challenge may compete with the challenge to cross scientific frontiers, using sophisticated technology to totally alter the concept of hospital design and medical work practice as we know them today, and to radically improve medical diagnosis and treatment.

The obvious question is are these needs mutually exclusive; indeed the third world may have a great deal to gain from the sophistication of long range tele-radiology.

The developing countries and the western world require different medical resources. The developing world primarily requires the commitment of its own indigenous best medical brains, e.g. 30 per cent of doctors qualifying in the third world never practise there. The developing world needs the commitment of western doctors to provide basic medical needs and not just sophisticated needs for the privileged: the developing world needs modest finance to provide basic radiological equipment and basic radiographic training on a very wide scale for the under-privileged population.

The western world has a duty to cross scientific frontiers: it requires major resources of finance: it requires the sophisticated skills of the committed physician and the committed scientist.

It is certainly not beyond either the financial resources or the medical and scientific capabilities of the western world to fulfil the needs of both the developing world and the western world.

I am convinced that the radiology of the future in the western world lies, not only in the development of existing radiological skills, but in the creation of filmless hospitals. I believe that physicians of all specialties should now adopt a positive approach to the development and assessment of PACS.

Those who are planning a total PACS, and they are few, are taking a giant step. Their progress will be watched by anxious, hopeful eyes, and perhaps some critical eyes. When they have disappointments and difficulties, which they will have, and some of which may be major, they deserve our honest and sincere understanding and support.

We can talk and write for ever, but the time is now come for PACS to leave the womb and be born in full. As this baby grows it will develop and change, but the strength of its growth will depend on the quality of our nourishment.

I end as I began: “Old men shall dream dreams, and young men shall see visions”.

Ladies and Gentlemen,
You are all young men - and women.

References

I. Lemke, H., „European Picture Archiving and Communication Systems Projects“, Editorial Review, Current Opinion in Radiology 4, 16-22.

2. Arenson. R.L., 1992 Editorial “PACS in the United States., Current Opinion Radiology. 4. 11-15.

3. Racoveanu. N.T., 1992 “Benefits of Co-ordinated Action in PACS Evaluation”. International Journal of Biomedical Computing 30(3-4) 295-300

4. Seshadri, S.B., Arenson, R.L., 1992 “The Impact of PACS in Research and Education” International Journal of Biomedical Computing. 30(3-4) 263-6.

5. Hendee, W.R., 1991 “Information Overload and Management in Radiology” American Journal of Roentgenology 156, 1283-1285.

6. Choplin, R.H., Boehme, J.M., Maynard, C.D., 1992 “PACS : An Overview” Radiographics 12(1) 127-9.

7. Banta, H.D., 1992 “Quality Assurance Issues and PACS” International Journal of Biomedical Computing. 30(3-4) 249-33.

8. Boehme, J.M., Choplin, R.H., 1992 “Systems Integration : Requirement for a Fully Functioning Electronic Radiology Department” Radiographics 12(4) 789-94,

9. Hilsenrath, P.E., Smith, W.L., Berbaum, K.S., Owen, D.A. 1991 “Analysis of the Cost Effectiveness of PACS” American Journal of Roentgenology. 156, 177-180.

10. Huang, H.K., Taira, R.K., Lou, S.L., et al 1993 “Implementation of a Large Scale Picture Archiving and Communicating System” Computerised Medical Imaging and Graphics. 17,l.

11. MacMahon, H., Doi, K., 1991 “Digital Chest Radiography” Clinical Chest Medicine 12(1) 19-32.

12. Straub, W.H., Gur, D., 1990 “Hidden Costs of Delayed Access to Diagnostic Imaging Information. Impact on PACS Implementation” American Journal of Roentgenology. 156, 613-616.

13. Lou, S.L., Huang, H.K., 1992 “Assessment of a Neuroradiology PACS in Clinical Practice” American Journal of Roentgenology. 159, 1321-1327.

14. Razavi, M., et al, 1992 “ROC Study of Chest Radiographs in Children” American Journal of Roentgenology. 158, 443-448.

15. Huang, H.K., 1992 “PACS - A Review and Perspective” Integrated Diagnostic Imaging, Chapter 3, 39-56.

16. Ogle, P., 1992 “Our Uneasy Relationship with Imaging Technology” Diagnostic Imaging International. 8, 7, 5.

17. Jennet, B., “Benefits and Burdens of High Technology in Medicine.

Preface

Many physicians are aware that the half-life of their knowledge, in certain specific medical fields, is less than 3 years. A good illustration of this is a recent statement by Prof. Poliwoda, Tumorcentre of Hannover: "In my area of internal medicine, specifically haemotology and oncology, in particular diagnosis and therapy of malignant blood and lymphatic node deseases as well as tumors, such as carcinoma of the stomach and colon etc., the half-life of knowledge amounts to barely 2 1/2 years. This means, that physicians who do not continue with further studies would treat after 2 1/2 years only 50% of their patients according to valid standards. But how does a fulltime working physician find sufficient time to cope with the flood of new information and apply this effectively to the patient?"

Radiologists face a similar situation. One just needs to observe developments in MRI or interventional radiology. Computer information systems including digital imaging, IMAC and application systems provide a possible way to reverse the negative effects of the lack of access to up-to-date information on patients and medical knowledge.

But also in the specific medical technical developments, such as digital imaging and PACS, the half-life of knowledge can be assumed to be even less than 2 years. This makes the frequent exchange of knowledge and experience between professionals in these fields a necessity.

The biannual CAR symposia and exhibitions in Berlin, interlaced with the biannual S/CAR meetings in the USA, provide the attendees with regular opportunities to refresh their stock of knowledge on modern tools in medical imaging.

CAR'93 carries on with this tradition by carefully selecting a range of paper and poster presentations which, taken together, give an in-depth overview on digital imaging, IMAC and application systems. More than in the past, the contributions are obtained from institutions on a worldwide basis. It is this internationality and interdisciplinarity which, I hope, will provide the impetus to include increasingly also the developing world, a need so strongly enforced by Oscar Craig in his prelude.

Altogether, one can easily feel helpless facing the statistics given by Oscar Craig. Future developments of digital technology with an accompanying reduction in cost, can effectively support a better health care for a larger section of the world population. This is true not only for teleradiology, but also for image acquisition and storage. The traditional topics of the programs of the CAR events may have to be revised and augmented to also take account of the new world situation we are all living in.

Acknowledgements

If I would have had the opportunity to underwrite the financing for Oscar Craig to achieve his PACS Valhalla at St. Mary´s hospital in 1987, I surely would have signed it. His enthusiasm for taking chances with new projects and techniques is infectious.

When I first met Oscar in 1987, he encouraged me strongly to proceed with CAR at a time when the outlook for financial backing was very grim. He was then a young man from whom I have learnt that risk taking is not naive if based on visions.

Oscar has remained young in his concern and vision in the hope to assist the developing world, and we will all be fortunate if Oscar Craig, with all his visions, never retires.

Similar to previous CARs, the heavy load of administrative support is being carried by Bibiana Moser, Franziska Schweikert, and Helga Kallan. They have to deal not only with a great diversity of document preparation und distribution, but also with numerous expected and unexpected requests from speakers, attendees, media, and other interested groups.

In these difficult economic times, I am particularly greatful for any financial support. In addition to the traditional support for CAR, thanks to continuing efforts of Lutz Reinke of the Senate of Berlin., specific donations for the Poster Exhibition and CAR Introductory Seminar have come from Luis Reimer-Hevia of Schering and Ogle Burian of Nycomed, respectively.

Again there has been enthusiastic cooperation from the symposium and program committee in selecting paper and poster contributions. The Poster Exhibition in particular, has gained a high level of recognition through the enduring efforts of Roland Felix.

Apart from learning about medicine and technology from many contributors of CAR, preparing for the meeting has also taught me to recognize my own limitations in handling complex situations. In striving to overcome some of them, I have been heavily drawing on the support of my wife Jean. I hope her resourceful personality will also be assisting in further CARs.

Heinz U. Lemke
Berlin, June 1993

Contents

IMAGE MANAGEMENT AND COMMUNICATION

Picture Archiving and Communication Systems- PACS

Teleradiology - TR

Hospital Information System - HIS
Radiological Information System - RIS

Technology Assessment and Implications - TA

Standards - ST

Data Bases - DB

MEDICAL WORKSTATIONS

Computer Vision - CV

Ultrasound - US

Digital Angiography - DA

Multimodality Imaging - MMI

APPLICATION SYSTEMS - AS

Computer Assisted Radiological Diagnosis - CARD

POSTERSESSION

IMAGE MANAGEMENT AND COMMUNICATION
Picture Archiving & Communication Systems- PACS

Standards

Teleradiology

Hospital Information System - HIS
Radiological Information System - RIS

Data Bases

MEDICAL WORKSTATIONS
Computer Vision - CV

Computer Graphics - CG

Design and Application

APPLICATION SYSTEMS

Computer Assisted Radiological Diagnosis - CARD

Computer Assisted Surgical Planning - CASP

Interventional Radiology

APPENDIX

Image Management and Communication

Picture Archiving and Communication Systems

Cost Effectiveness and Patient Outcome of Modern Digital Technology

P. GRASSMANN, C.F.C. GREINACHER, H. WEISS
Siemens Medical Engineering Group, Germany, W 8520 Erlangen

Abstract:

In many countries today, politicians are faced with the difficulty of having to reduce the overproportional increase in the costs of Healthcare. In this context the contribution of medical technology to costs in the total spend are nearly always overestimated and its potential for savings underestimated.

Based on international studies, investment and subsequent running costs for so-called "major equipment" for diagnosis and therapy will be shown to account for only about 1% of the total spend for Healthcare. The benefits of medical technology will be compared to these costs.

72% of the total turnover of the Medical Engineering Group of Siemens AG for diagnostic imaging modalities in 1991/1992 , arose from digital imaging modalities, and this percentage is expected to continue to increase. Because of this we will concentrate our attention on these modern digital methods.

Impressive examples of the application of Computer Tomography, Magnetic Resonance Tomography and Digital Subtraction Angiography deliver valid statistics regarding the patient benefits and efficiency of modern imaging diagnostic systems. Studies about resulting therapeutic possibilities of Angioplasty and extra-corporal Lithotripsy demonstrate the overall economic advantages of modern medical technology.

Arising from the positive results due to the use of such equipment, the additional cost savings potential to be expected by networking these digital modalities with one another, as well as with other Information and Communications Technology systems, will be analysed. The results will be assessed and compared to experience gained from routine clinical operation of such integrated systems.

MDIS: A Large PACS and Teleradiology Project

R.G. Leckie[1], M.D. , C.S. Smith[2], D.V. Smith[2], M.D., J.Donnelly[3], M.S., M. Cawthon[4], D.O., J. Weiser[1], Ph.D., C.E.Willis[5],Ph.D, F. Goeringer[1], M.S.C.E, M.B.A.

[1]USAMMA-MDIS Project, Ft. Detrick, Fredrick, MD. 21702-5000

[2]Department of Radiology, Madigan Army Medical Center, Fort Lewis, Washington 98431

[3]Department of Radiology, Brooke Army Medical Center, Fort Sam Houston, Texas 78234

[4]Department of Radiology, U.S. Air Force Medical Center, Wright Patterson Air Force Base, Ohio 45433

[5]Department of Radiology, University of Texas Health Science Center at Houston, Houston Texas 77030

Abstract

The Medical Diagnostic Imaging Support System (MDIS) is a large project to install Picture Archival and Communications Systems(PACS) and teleradiology across several military medical sites in the U.S. and overseas. Phased implementation began at Madigan Army Medical in Tacoma, Washington in the Spring of 1992 , Wright-Patterson Air Force Medical Center in Dayton, Ohio in the Fall of 1992 and Brooke Army Medical Center in San Antonio in the Winter of 1992/93. Other sites are scheduled to follow in 1993 including teleradiology in Korea and the Azores (off the coast of Portugal) .The hardware and software components of the system will be discussed including the computed radiography units, central processing unit (CPU), working storage unit (WSU), fiber network, optical disk jukebox (ODJ), radiology information system (RIS), modality interfacing, and physician monitor workstations. The reliability of the system to date, speed of individual components, storage capacity, user acceptance ,and quality of the radiographic images will be discussed. The clinical usefulness of the system will be emphasized. Successes and concerns will be reviewed as well as future developments that are desired.

Introduction

The Medical Diagnostic Imaging Support (MDIS) System is a four-year contract to install Picture Archival and Communications Systems (PACS) and teleradiology in U.S. military medical treatment facilities. MDIS specifications were based on the results of three years of university and military tri-service research and development through the Digital Imaging Network Systems (DINS) Project and the Tactical Air Command (TAC) Teleradiology Project [1-3]. MDIS functional specifications were compiled by a composite team of radiologists, physicists, clinical engineers, hospital administrators, technologists, and computer systems specialists. As MDIS sites become operational, we are developing a better understanding of the capabilities and limitations of the technology .Presently, Madigan Army Medical Center, Brooke Army Medical Center, and Wright-Patterson Air Force Medical Center are undergoing phased implementation of this system. Several other sites are planned for 1993-94, including military medical treatment facilities in the continental U.S., Hawaii, Korea, Panama, and the Azores off the coast of Portugal .

Since we have the most accumulated experience with Madigan, this institution will be emphasized. The new Madigan Army Medical Center opened its doors in the Spring of 1992. Madigan is a 1.2 million square foot, 416-bed facility in Tacoma, Washington. Approximately 500 physicians representing nearly all subspecialties work at this tertiary care medical center and see over one million outpatient visits per year. The Department of Radiology with 12 staff and 18 residents, performs over 160,000 radiological exams per year. Presently, as part of the phased implementation, all computed radiography (CR) and fluoroscopy images are stored both softcopy and hardcopy (laser printed film). CT and MRI images are expected to be available on the PACS in the Spring/Summer of 1993. Nuclear medicine, ultrasound, angiography, and radiation therapy will be connected to the system in late 1993. A total of 9 workstations are presently in service at Madigan: one each in the orthopedics clinic, the emergency room (ER), and the intensive care unit (ICU), with the rest in the radiology department. In the Summer/Fall of 1993, 116 additional workstations will be placed throughout the clinics and wards of the hospital. Following this installation, Madigan will begin to go almost entirely filmless .

PACS components

Computed radiography

The initial phase of MDIS at Madigan includes computed radiography for all plain x-rays except mammography and scoliosis series. The use of computed radiography allows the images to be directly entered into the PACS without the labor intensive step of digitizing film-screen images into the system. The computed radiography images are processed by Siemens Digiscan 7000s and Fuji AC1 plus readers. A Kodak CR reader to be used primarily at small teleradiology sites will be introduced into our system later in 1993 . Third and fourth generation Fuji imaging plates are used. Presently the cassettes do not have leadbacking. It is our experience that using leadbacked cassettes for large body parts significantly decreases backscatter to the imaging plate thereby improving contrast. A process is underway to retrofit all of our cassettes with leadbacking.

In the initial phase of implementation of each major medical site, both a hardcopy and a softcopy of every image is maintained. This helps each site in building up a softcopy archive so that digitization of film-screen images will be required less often. A dual archival system is helpful in the initial implementation when the system is first coming on -line and problems in the hardware and software are being corrected. Also this allows for a more gradual conversion from the conventional film-screen system to a filmless environment thus decreasing anxiety among the clinician.

Central processing unit

The central processing unit in the initial MDIS site is the DEC VAX 4000-300. Future sites will have the VAX 4000- 400 which has twice the processing power. A key feature of the system is the connection between the database, workstation, and WSU. The workstation queries the database via an Ethernet line. The database provides the workstation with the necessary information and access rights to retrieve image data from the WSU. The WSU transmits images to the workstation via a direct fiber optic link to the image memory of the OPUS board, with as many images as possible stored on the OPUS board. Most of the problems with the central processing unit have been with the custom software designed for our system. Initially the processor was slower than expected and included "bugs" that corrupted the entire database. The vendor's support personnel in Germany were using the same computer address as the local PACS site, therefore the processor was occasionally getting conflicting commands and causing the system to crash. Software upgrades significantly improved the system speed and corrected previous "bugs" in the system.

Working storage unit

The Working Storage Unit (WSU) functions as the local and short-term storage for images. The WSU uses a redundant array of inexpensive disks (RAID, level 2 architecture);40 disks (magnetic media) operate in parallel; 32 disks for a 32 bit word, 7 disks for error correction , and one disk acting as a "hot spare" (single disk failure detected and corrected without loss of operation). The WSU is designed to hold inpatients for the average length of a hospital stay (4.5 days for Madigan), all outpatients for 48 hours, all exams not yet interpreted, and pertinent historical images. Images will be stored in the

WSU with approximately 2.5:1 lossless compression (currently no compression is used on the WSU). Image retrieval bandwidth is greater than 400 CR image equivalents per minute[4]. The full implementation WSU will have five times the present storage capacity (by doubling the capacity and implementing 2.5:1 compression). Because system reliability is so important, all of the major components are on an uninterrupted power supply. Most of the down-time of the WSU at the original site has been as a result of loss of the cooling capacity in the computer room. The hospital's air conditioning system has failed several times ; the temperature in the computer room rose to 34 degrees C . The WSU is designed to automatically shut off in this situation, but this has been inconsistent and on one occasion the overheating required a board in the WSU to be replaced. We plan to install a back-up cooling system in the computer rooms. When a major component fails, the computed radiography machines automatically direct two laser printed hardcopies of the image to be produced. One copy is available for the clinicians, the other is retained in the radiology department. The hardcopy image is later digitized back into the system once the problem has been corrected.

Fiber optic network

The WSU is connected to the workstations by a fiber optic network in a modified star topology. Most image data moves across the system with FDDI- like speeds (100 Mbits/sec). Patient demographic information moves over Ethernet. Images are transferred at the earliest opportunity from the WSU to the ODJ. The fiber company that does the installation and termination should be certified in the product. Laying the fiber optic network into an old hospital and between buildings can pose several problems including a lack of building floor plans, asbestosis hazards, and inaccessible conduits.

Optical disk jukebox

The ODJ holds 100 (10 Gbyte)WORM 14" optical disks. Computed radiography images are stored with 10:1 lossy compression (modified JPEG format). At 10:1 compression , the ODJ can store about one million CR images. The original interpretation of an image by the radiologist is always made on the original data from the WSU before compression is applied. ROC studies comparing images compressed verses uncompressed are planned. At Madigan, the final phase calls for two ODJs which will be able to store approximately 3 years of images on-line [4]. Initially, problems existed with archiving 2K images and the speed to dearchive an image to a workstation. Software changes corrected this, although the speed to dearchive an image to the workstation still needs improvement.

Radiology information system

The radiology information system (RIS) is an integral part of the MDIS workstation. Once the user selects a patient's study within a work list, the radiological request and dictated report (if completed) automatically show up on the bottom of the screen. Therefore, the patient's clinical history, radiological images, and dictated report are present together on the workstation monitor for the radiologist or clinician to see. Usually the radiologist dictates a report to a transcriptionist, but if he desires, he can type in the report himself. Standardized radiology reports are available on pull-down menus. These standardized reports are used almost exclusively on the Gastro-Intestinal (GI) service at Madigan, frequently with slight modifications, such that final reports are now often available to the clinician the same day as the study was done. Using the RIS, the database manager can generate a report giving the department a more exact account of workload statistics, thereby helping to locate problem areas, support QA/QC evaluations, and document man-hour requirements. The RIS is also very valuable as a research tool. Database queries can be formulated in many ways. Each report has a space to log in an ACR radiological/pathologic code and a specialty code (code designed by users) for teaching and research purposes. Most of the problems with the RIS have been due to simple errors by the users such as the mouse or keyboard not functioning because it was unplugged. The RIS has been very successful in general, but the time required for the radiologist report to be processed once approved at the workstation is too slow. The vendor is aware of this problem and a basic change in the software is planned.

Modality interfacing

Interfacing different modalities from different vendors into the PACS system has been one of our greatest problems. Although most of the vendors follow the ACR-NEMA standard, shadow groups exist in the protocols which allow vendors to place key information in areas inaccessible to the individuals attempting to make the interface. The new ACR-NEMA (DICOM 3) standard will improve the present situation, but will not solve all of the problems. We strongly support the goals of the ACR-NEMA standardization process and encourage better cooperation among vendors.

Physician workstation

The image viewing workstation is an all-important link in the PACS chain since it represents the interface between the system and the user. The basic platform for the workstation is the Macintosh IIfx computer with 8 MB RAM and 80 MB internal disk drive. Loral and Siemens then insert special image processing boards . The Macintosh-based interface is one of the keys to the user acceptance of the system. It is quite user friendly. This is critical because many of our clinicians are computer illiterate. Ease of training is directly related to the ease of use. We have also noted that the retention of knowledge on how the workstation functions is high. Both ease of initial training and retention of training are believed to be due to the mouse-driven pull-down menus. As in the normal Macintosh format, "quick keys" are available for the commonly used functions so that an experienced user can move through the functions more rapidly.

Various types of workstations are being used based on the clinical need and cost constraints. MDIS supports two basic types of workstations, a standardized and an optimized. The standardized workstation is a high volume, primary diagnostic unit whereas the optimized workstation is a lower volume unit for clinical review of images. The standardized units can have either 2K (A type) or 1K (B type) resolution portrait monitors. The optimized units have only 1K (C type) resolution landscape monitors. The primary difference between the standardized and optimized workstations is the OPUS boards with 64 MB and 32 MB of image memory respectively[5]. The primary diagnosis for CR should be made on the 2K monitors. Lower resolution modalities such as ultrasound can be read on any of the monitors. In the radiology's primary reading areas standardized workstations with four 2K monitors are used. In general, the wards and clinics will have two 1K monitors at optimized workstations, but the entire 2K data set is available by magnifying the image either by a zoom function or use of the "magic glass" tool in the region of interest.

Early System Results

The failure of a conventional film-based system on image accountability is one of the primary reasons to convert to a PACS environment. It has been stated that PACS will result in zero image loss. In reality, this has not been realized yet. Some images are still misplaced, but this time electronically. During summer of 1992, we reviewed 150 studies done in the first 3 months at Madigan after moving into the new hospital. Hardcopy laser printed images could not be found in 20% of the cases. Surprisingly, 10% of the cases could not be found on the MDIS system. Closer examination showed that some studies were located under a different folder heading, e.g., a chest x-ray listed as an abdomen. Also, in the first 3 months of operation there were a significant number of "bugs" to work out of the system. Action has been taken to address these problems. A follow-up study was done; a week (early November of 1992) was randomly selected to review image accountability. The image in question could be found in its exact heading or similar title (e.g., toe under foot heading) in greater than 98% of cases. Using both the hardcopy and softcopy formats, the images presently can be found >99% of the time[5].

A high level of reliability is critical to the success of a PACS. Our contract calls for 98% system up time. Any down time of the VAX, WSU or ODJ can affect the user at the workstation. Over the initial 10 months, the MDIS system has been up and running 98.8% of the time. Excluding air conditioning related problems, the system has been operational 99.7% of the time.

Image display speed on the workstation is primarily related to the location of the image data at the time of the request. In general, if the image is still on the WSU, it takes 5-6 seconds to display the image on the monitor. This time is fairly consistent. The vendors are working to reduce the time to less than 2 seconds. If the image is archived on the ODJ, the time to display is much longer and presently averages 1.8 minutes . The time to display an image from the ODJ is variable with several factors influencing this display speed. Only one study can be fetched at a time, therefore the time depends on the number of studies that other users are seeking at that time. Presently, if the ODJ is in the middle of archiving another case, it must complete the task before retrieving the study requested. A software change is expected shortly that will interrupt archiving to service fetch requests. A special "fetching" worklist is being developed by the vendors that will allow users to easily check on the status of an exam they are fetching while allowing them to proceed with other cases on the workstation.

As opposed to a clinician that may review a few imaging studies a day, the radiologist spends the majority of his day looking at images and making diagnoses. Even a small percentage change in the rate at which he reviews images can therefore have a major effect on his overall work efficiency. Overall, reading images on the hardcopy image is presently faster than the softcopy image. Several software features have not yet been implemented on the MDIS workstations that are expected to significantly speed up the softcopy reading times. The work lists are being improved so that a radiologist will be able to go through a specific list of studies assigned to him. A "next patient " quick key is planned. This will allow the radiologist to proceed to the next patient's images without typing in a name. The format on the MDIS workstation to compare old images to the present study is awkward; a new format for comparison images is expected soon. Also, prefetch algorithms to automatically bring the old studies from the ODJ to the WSU will be implemented once the WSU capacity is increased. We believe these features will make softcopy reading competitive with the traditional film environment.

We are generally satisfied with the image quality, but ROC studies are being initiated to more scientifically examine this question. After more than a year of clinical use and 250,000 CR images, to our knowledge, no cases have been documented in which a finding was noted on the hardcopy, but not on the softcopy image. On the other hand, several clinical findings are routinely noted on the workstation that are inconspicuous or absent on the hardcopy images. . Primary diagnosis is only made on images with lossless compression. We are using 10:1 lossy compression on CR images after a primary diagnosis has been made and before archiving the images in the ODJ.

Simple pixel replication is used on the system instead of interpolated zoom. This causes the CT and MR images to look pixelly when viewed on the workstation in the traditional 12:1 format. This is unacceptable to the radiologists and is being worked on by the vendors.

The acceptability of the MDIS PACS by clinicians is extremely high. A survey of 58 surgeons, internists, and pediatricians at Madigan revealed that 100% believe the MDIS system is useful to them, 100% believe it saves them time, and more than 98% believe it helps to improve patient care. Many clinicians have learned how to use the system before they received formal training. This is because they believe that the system saves them time and effort. With only the nine workstations currently installed, it is already better than the traditional film-based system. Clinicians now routinely go to the workstations in radiology to look for imaging studies instead of going to the fileroom. It is clear that the place where PACS is most appreciated is on the wards and clinics.

The orthopedic department has a workstation in its clinic. In the past , patients that arrived for follow-up appointments for fractures without their old x-ray films would either be canceled and told to return with the films or to retake x-rays. In their busy clinic, the orthopedic surgeon could not afford the time to go down to radiology and try to find the old film. Now he can just pull the patient's old study up on the workstation in his clinic, thus saving time, money, and/or unnecessary additional radiation exposure to the patient. Interviews with several of the orthopedic surgeons at Madigan reveal that on average the MDIS workstation saves them each approximately 1 to 1.5 hours a day. Clinician and patient satisfaction are both improved.

The GI radiology service is completely filmless now. The radiology technologists on GI believe being filmless saves them time. The radiology residents and staff also prefer the filmless environment in GI. The imaging tools, such as the magic glass, inverted gray scale and magnification, seem to be especially useful on the GI service. The general feeling of the radiologists at Madigan is that the system has great potential, but better image navigation software would enhance the value of PACS significantly. PACS is (and will remain so in the near future) most beneficial to the clinicians, not to the radiologists. Having workstations on the wards and clinics will be the pathway by which PACS will succeed and be widely accepted in the field of medicine.

Areas of Concern

Quality control, training, modality interfacing, and image navigation are the areas of greatest concern. Making the primary diagnosis directly on the workstation presents a whole new set of mandatory quality control measures to ensure adequate patient care. All the MDIS components along the pipeline of bringing images from imaging modalities to the user (e.g., CR, network connection, workstation monitor, and laser film digitizer) must be periodically checked and tested individually and collectively. The brightness of the monitors drifts gradually over time, and the luminance and spot size can vary significantly across the screen. In a traditional film-based environment, when a technologist underexposed a patient's film, it turned out white with almost no image; overexposure led to a black film. With computed radiography (CR), both the under and overexposed patient imaging plates lead to a visible image. The radiology technologist soon realizes the only risk he takes is if a patient's imaging plate is still significantly underexposed, the image looks noisy (due to quantum mottle). Therefore, there is a tendency to overexpose the patient's image plate to reduce any risk of having to repeat the study. The best way to follow the patient exposure with CR is the Sensitivity (S) number (a rough inverse correlation with x-ray dose). This number is displayed when a hardcopy laser image is printed. In a filmless environment, this number must appear somewhere with the image on the workstation monitor. Otherwise, there is a risk of routinely overexposing patients.

In the military and especially at a military training hospital, the turnover rate of personnel is high - an estimated 25% turnover rate per year is the norm. Therefore, both the initial and follow-up training are important. We strongly believe a formal training program is necessary to the success of the system. For example, with the initial conversion from film-screen to computed radiography we saw a significant increase in the repeat rate of exposures. This rate fell with time to a level below conventional film-screen. We believe the initial high repeat rate could have been avoided with better training. An improved training program has since been implemented and will hopefully avoid this problem at future sites.

As mentioned previously, improved modality interfacing by better cooperation among vendors, and improved image navigation protocols to allow the radiologist to more rapidly review large data sets are critical to the long term success of PACS.

TELERADIOLOGY

The ultimate success of teleradiology depends on the acceptance by the end users - the physicians. From a physicians' perspective in a managed health care system, several major areas need to be addressed in the ideal situation. A teleradiology system needs high image quality and a user friendly interface, qualified radiologists interpreting images, close communications between the clinician, radiology technologist, and radiologist, images rapidly accessible to the clinician with a dictated report tied to the image, a highly reliable system with the freedom to shift workloads as necessary, and the ability to utilize a large database of images for education and research purposes.

Several of these issues have been addressed above. All MDIS sites will have a radiology information system tying the spoke and hub together."Wet " readings will be available as needed using dial-up switched T1 lines. Direct phone lines and Fax communications will be used initially. With the MDIS workstation, the clinician and radiologist can be in different locations and be looking at the same patient's images at the same time. Speech recognition and teleconferencing systems are presently being evaluated and are expected to be available. Teleradiology hubs will have a designated "Chief of Remote Diagnosis " to coordinate the interactions between the hub and its' spokes. Other concepts include two-way transmission of images to balance workloads, centers of imaging excellence, and the use of time zones to maximize resources[6].

CONCLUSION

MDIS is a massive project to install PACS and teleradiology across several military medical sites in the U.S. and overseas. Early experience demonstrates that the system is very user friendly and enjoys an enthusiastic, wide acceptance from the clinicians. Clinicians using this system believe PACS can save them time, increase their overall satisfaction with practicing medicine, and improve patient care. The image quality is acceptable and the reliability of the system is remarkable. The most important area for development in the immediate future is new software for better image navigation.

References

1) Goeringer, Fred, Seong K. Mun, and Barabara D. Kerlin. Digital Medical Imaging: Implementation Strategy for the Defense Medical Establishment. SPIE. Medical Imaging III: PACS System Design and Evaluation. Vol. 1093: 429-437, 1989.

2) Mun, Seong K., Harold Benson, Steve Horii, Larry P. Elliott, Shih-Chung B. Lo, Betty Levine, Robert Braudes, Gabriel Plumlee, Brian Garra, Dieter Schellinger, Bruce Majors, Fred Goeringer, Barabara Kerlin, John Cerva, Mary-Lou Ingeholm, and Tim Gore. Completion of a Hospital-Wide Comprehensive Image Management and Communications System. SPIE. Medical Imaging III: PACS System Design and Evaluation. Vol 1093: 204-213, 1989.

3) Webb, Mark ed. TAC Teleradiology. unpublished July 1989. HQ TAC/SCUCE. 1912 Computer Systems group. Advanced Systems Division. Langley AFB, VA 23665-6349.
4) Smith,D.V. , S.Smith, F.Sauls, M.A.Cawthon and R.J.Telepak,"Design Strategy and Implementation of the Medical Diagnostic Imaging Support System at Two Large Military Medical Centers",SPIE Medical Imaging VIVol. 1654, pp. 148-159, 1992.

5) Leckie, Robert G., Fred Goeringer, Donald V. Smith, Gregory Bender, Hyung S. Choi, David R. Haynor, Yongmin Kim. Early Evaluation of MDIS Workstations at Madigan Army Medical Center, Proceedings of SPIE Medical Imaging VII, February 1993.

6) Leckie, Robert G., Robert Detreville, David Lyche, Gary Norton, Fred Goeringer, Charles E. Willis, Mike Cawthon, and Mark Hansen. The Ideal Teleradiology Configuration from a Physician's Perspective, Proceedings of SPIE Medical Imaging VII, February 1993.

Some Consideration on Hospital-wide PACS from the Experience in Tokyo Hitachi Hospital

M.Akisada MD[1], H.Okaniwa MD[2], H.Tsuneyoshi MD[2] and T.Okabe MS[3]

[1]Tama Health Management Center. Tachikawa-shi, Tokyo.
[2]Tokyo Hitachi Hospital, Hitachi Ltd.. Yushima, Bunkyo-ku, Tokyo
[3]Hitachi Medical Corp., R & D Center. Kashiwa-shi, Chiba-ken.
Japan

ABSTRACT

Tokyo Hitachi Hospital has used a hospital-wide PACS for these two years, that is probably the first experience in the world. The system runs 100Mbps optical fiber network over the hospital including outpatients clinics, wards, one treatment room, and radiology department with 11 image workstations and an image database. We have managed about 400,000 image of about 10,600 patients so far.The system has been designed for use in wards, clinics, treatment room and conference room. So physicians and engineers have made many discussions to realize these viewpoints. A complete automation for implementation of the system and the integration of Hospital Information System (HIS) with PACS has been discussed. The hospital-wide PACS has proven to be more usefull for image diagnosis and more effective for image management than that of PACS inside radiology deaprtment in the hospital. An organic integration of PACS with HIS is strongly required for the hospital-wide PACS.

keywords; PACS, hospital-wide system, image workststion, digital radiography

INTRODUCTION

Although a lot of PACS have been installed in Japan, they are the systems for image diagnosis and image management exclusively in radiology department. Japanese radiologists, estimated roughly 3,000 in qualified, in some hospital have not been responsible for all routine radiology practice such as plain chest, abdomen, bone and joint and etc., in terms of reporting system. Hence, the interpretation of images obliged to be performed by refering physicians themselves instantaneously just in front of the patients without any documents.

Tokyo Hitachi Hospital, Hitachi Ltd., has 9 clinical departments, about 430 outpatients/day and 147 beds. Untill now, the hospital has not any full-time radiologist, then physicians make primary image diagnosis in clinics. Part-time radiologists review images, such as X-CT, MRI and etc. twice a week. And the hospital has conferences concerned with interesting foci twice a week. At those times, nurses have laborious extra works to take back films, that were once scattered to the clinics and wards, to part-time radiologists and/or the conference room. These clinical situations had made image management difficult and unavailability of films had been happened sometimes.

In 1990, Dr. Okaniwa, the director, decided to introduce a hospital-wide PACS as one of the hospital modernization. Although, a lot of PACS, comprehensive and partial or modality based, have been installed in Japan so far, they are the systems for image diagnosis and management exclusively in the radiology department. The system installed in Tokyo Hitachi Hospital has been designed from viewpoint of physicians for pursuing useful image diagnosis and effective image management in the clinic, wards, and conference as well as for better reviewing by part-time radiologists. This is the major difference between this new system and already installed radiology based PAC systems. Many discussions have been made between physicians and engineers to fulfill these physicians demand. And a lot of improvements have been employed so far to already installed radiology department based PACS. The first impression with our system is that the hospital-wide PACS has been more practical and effective than that of radiology department based PACS.

Implementation, performance and discussions on this system are reported.

SYSTEM DESCRIPTION

The system runs 100Mbps optical fiber network in whole hospital including 3 wards (medicine, maternity and surgery), 6 outpatients clinics (3 medical clinics, one surgery, one orthopedic clinic and one treatment room), one conference room and radiology department with 11 image workstations. The network couples with digital imaging modalities such as X-CT, MRI, Computed Radiography(CR) for general radiology and Digital R & F Table (DR) with 2048 x 2048 matrix I.I.-TV for gastrointestinal(GI) examinations. These modalities have generated about 200 images with 50 examinations a day. The image database consists of an image data management computer(Hitachi 3050 workstation) and three libraries of 5.25" optical disk; the first is used for X-CT and MRI, the second for DR and the last for CR respectively. The system has already managed about 400,000 images of about 10,600 patients. A physician in the clinic can see one DR or CR image on the CRT from the image database in ca. 60 sec., from DR or CR in ca.13 sec. and from the hard disk in the same workstation in ca. 6 sec. on the average. The system also couples a laser image digitizer and a laser image printer. Each image workstation has two 21" CRTs of 1024 x 1280 matrix in non-interlace mode, however, one installed in the conference room has six same CRTs and a video image printer. Fig.1 shows the block diagram of the system installed in Tokyo Hitachi Hospital.

DIGITAL RADIOGRAPHY SYSTEM

One of our effort for digital imaging modalities is given below;

In Japan, R & F Table for GI exams. would often generate over 20% of the total diagnostic images in many hospitals. So high throughput of digitization and communication with these images are most important for successful PACS implementation. Both CR and a film image digitizer seem essentially being off-line system and needs much more man-power and time on operation when they are coupled with PACS. This situation has made the introduction of conventional X-ray images to PACS rather difficult so far. We have introduced an on-line real-time DR for GI exams. in our PACS in order to overcome this difficulty. DR is developed based on new optics with new components; I.I., optical lens system, Gun-diode Saticon tube and etc.. Technicians can take radiographic images with the digital format of 2k x 2k matrix at 12 bit density resolution when they found interesting foci on fluoroscopy with 1k x 1k matrix. DR has 1.2 GB hard disk and 256 MB IC memories, which can archive radiographic images as fast as one image per second and successively display 28 images of 2k x 2k matrix. Technicians can only push one button for sending GI images to clinics or wards, and images are automatically sent to the image database. If this system is not available, PACS has not been accepted by physicians.

MTF characteristics of this system at low frequency area shows the same with that of screen-film and is better than that of CR. And clinical evaluations on DR in the field of GI exams. have indicated that DR has better diagnostic capability than that of indirect radiography and almost same with that of of direct radiography.

IMPLEMENTATION OF PACS

When PACS is introduced, a complete automation and maintenance-free as well as better operability are required by physicians and the hospital administration. So various efforts have been made when implementing the system

1) Although a film image digitizer is coupled with the system, DR system plays an important role of digital archiving of conventional X-ray images in the system as reported previously.

2) A card reader is coupled with the image workstation,and patient's ID card is used to input patient's information to call images on the CRT. This trial can get a good reputation among nurses and physicians, however, touch-pannel or voice input must be tried for easier operation

3) When a patient is admitted to the ward, clerks at the front desk retrieve all his/her diagnostic images that have been filed in the image database and send them to the hard disk of the workstation installed in nurse station of the ward. These images are to be kept in the workstation as long as this patient is in the ward.

4) When a technician takes images with imaging modalities, those images are sent to the ward or clinic by pushing one button and automatically sent to the image database at night.

5) A video image printer is coupled with the workstation installed in the conference room, and any images on the CRT can be printed on a paper as physician's references.

6) Maintenance engineers are not needed for this system. Chief radiology technician examines the system for 10-20 min.in the morning. If any troubles happened, service engineers would come to restore. The frequency for servicing is very small at present.

DISCUSSIONS

A lot of discussions on how to implement PACS have been taken between physicians and engineers on the following items;

1. When making image diagnosis, doctors have necessity of comparative study on the case; present images and their past images on the CRT, which mean, when the patient registered at the reception desk, the system should automatically send a series of images to the workstation installed in the physicians clinic.

2. When technicians take diagnostic images, they compare the present images with the latest images on the CRT in the department. When physicians prescribe the patient to take images, the system is needed automatically to retrieve and to send the latest images to the department.

3. When the patients is admitted to the ward, the clerk at the front desk must retrieve all his/her images for the nurse station of the ward. The clerk wants these procedures should be automatically performed, when the patient register at the front.

4. In case of radiology department based PACS,to retrieve images on the CRT is done using patient's ID number. However,physicians have usually concerned with patients in connection not with ID number but with their name, sex, and disease or symptom. So they want to retrieve images with patient's name.

5. For displaying one DR image on the CRT from the optical disk library it takes ca. 60 sec. and from DR it takes ca. 13 sec. on the average. These throughput seem faster than those of screen-film from macroscopic viewpoint. However, when physicians diagnose patients by images in the clinics, those throughput seem too slow from microscopic viewpoint. Although prefetching system is employed, faster thoughput in PAC system is being needed.

In order to realize these demand, an organic integration of HIS with PACS is strongly required. And engineers are needed to have close negotiation with the hospital administration and doctors, while the latter two, the users, are requested to establish the philosophy and strategy on how image diagnosis and image management should be under PACS and HIS.

SUMMARY

From our experiences on hospital-wide PACS in Tokyo Hitachi Hospital, we would like to summarize as follows;

1) A hospital-wide PACS has been successfully used in Tokyo Hitachi Hospital for these two years, although it has been still under technical improvement.

2) A hospital-wide PACS has proven to be more useful for image diagnosis and more effective for image management than that of PACS inside radiology department in the hospital.

3) An organic integration of PACS with HIS is required. Then the hospital administration and physicians(includung radiologists) must establish the philosophy on how image diagnosis and image mangement in the hospital should be under PACS circumstances.

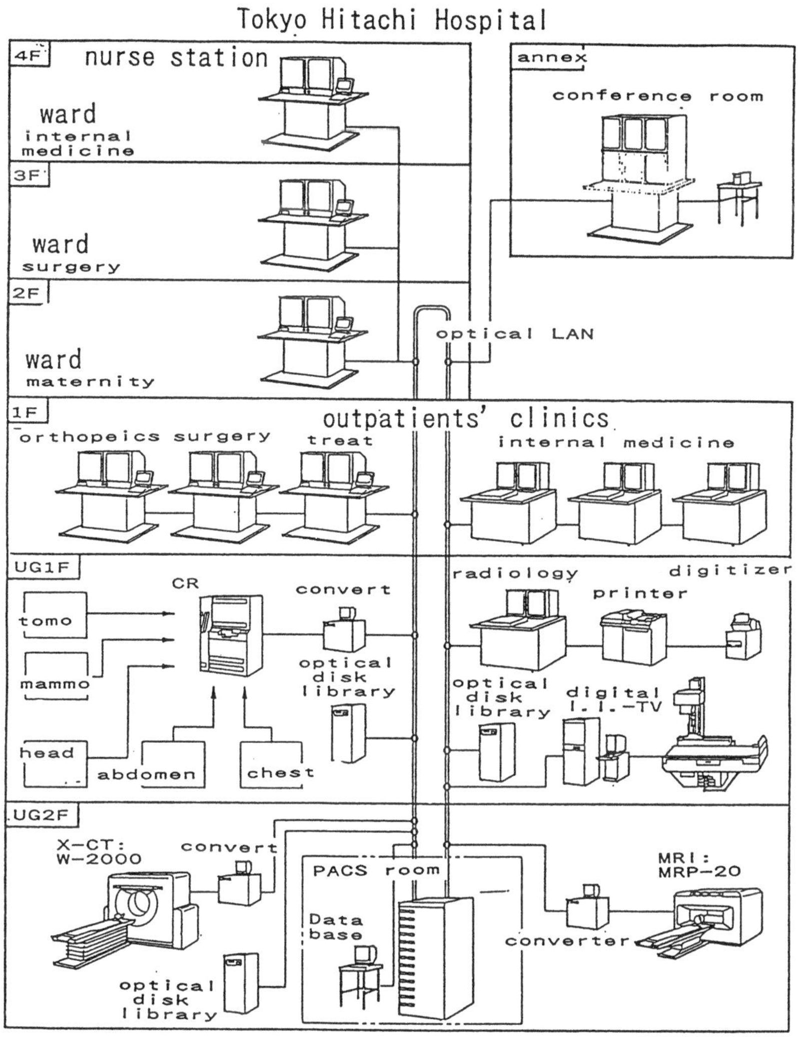

Fig. 1. Block diagram of PACS installed in Tokyo Hitachi Hospital (Tokyo)

Correspondence:
M. Akisada MD
Tama Health Management Center
Nishiki-cho, 3-10-7, Tachikawa-shi, Tokyo. 190. Japan
Tel: +81-425-28-2011
Fax: +81-425-25-7680

Designing and Implementing an „On-Demand“ IMACS

S.D. Totaro, K.M. Spicer, G.D. Frey, and R. Olinger

Department of Radiology,
Medical University of South Carolina,
Charleston, USA

Most current commercial and research Image Management, Archive, and Communication Systems (IMACS) utilize an „autorouted“ or „prefetched“ image management scheme to compensate for the limitations of network bandwidth and disk transfer time. After implementing one such commercial PACS to distribute images to a Trauma Center and several Intensive Care Units, the multidisciplinary IMACS research team at the Medical University of South Carolina (MUSC) analyzed the autorouting method to discover ways to improve the system. The resulting proposal consisted of an „on-demand“ method of image retrieval implemented with a high-speed network and a centralized storage system.

This paper describes the benefits and technical challenges of the on-demand IMACS, as well as the specific solutions developed at MUSC. Where appropriate, our experiences with the autorouting method are related to our design specifications. Some of the goals of the MUSC-IMACS were to provide the following:

1) Access to any patiens's images from any workstation
2) Access to patient images at multiple locations simultaneously
3) Protection of patient confidentiality through system security
4) Elimination of the need for „routing“ codes
5) Reduction in the time and effort required to administer the system
6) Reduction in the cost of the system

Our design approach for the logical subsystems, such as the Image Librarian and the Image security System, and their relationships is described. The technical details and their relevance to the overall system design are also discussed. These include the use of a redundant, high speed file server; RAID and SCSI-2 disk system technology; high speed networking technologies such as FDDI and Fast Ethernet; and Unix and PC workstation display systems. Software development methods and tools, such as the use of C++ and common GUI libraries, are examined.

ImAZ Emergency: a PACS Subproject of the AZ-VUB

M. OSTEAUX, R. VAN DEN BROECK, F. VERHELLE, R. MATTHEUS

A.Z.-V.U.B. Department of Radiology
Laarbeeklaan 101, B-1090 Brussels, Belgium

The context: the IMAZ Philosophy

ImAZ represents the **clinical** implementation project of the Picture Archiving and Communication System (PACS) Technology at the University Hospital of the Vrije Universiteit Brussel (AZ-VUB).

The full PACS concept (no-film option) predicted for "soon" more than 10 years ago, has failed to materialise up to now. This is due to the fact that tremendous difficulties remain in large scale applications of the technology and in its integration in the general context of the hospital informatics.
Indeed, the very large amount of data (for the AZ-VUB ± 700 beds : ± 1.5 million images per year, representing ± 3.2 terabyte) joined to the need of "instant access" from a large number of places, put the storage and retrieval capabilities as well at a network under extreme requirements. The problem of integration with the hospital information system and the radiological information system is another up to now incompletely solved problem.The lack of implemented standards in physical interface towards the radiological sources, image and transfer formats, is another very difficult problem. Last, but not least, the hospital community as a whole, and not only the radiology department, has to commute from the film environment (already **1 century** tradition !) to the digital environment, a dramatic change which can not be expected to occur in a short time.

For these reasons, according to the general philosophy of the Advanced Informatics in Medicine - AIM Program - European project HIPACS, the AZ-VUB will implement **gradually** the PACS technology in fields where it is **technically feasible, economically possible and medically contributing**.

This pragmatic approach can be characterised by different features :

- Bottom-up implementation, where decentralised islands are created around a radiological technique or a medical application
- Standardised approach, where existing industrial standards, as well as international standards for image format or transfer protocol, are followed
- Multivendor environment, the open character of the created environment will favour multiple collaborations with the industry; a link with a unique constructor has been and will be avoided in the future
- Hybrid digital-conventional environment (see fig. 2) where the frontier between the digital and conventional world will be gradually moved

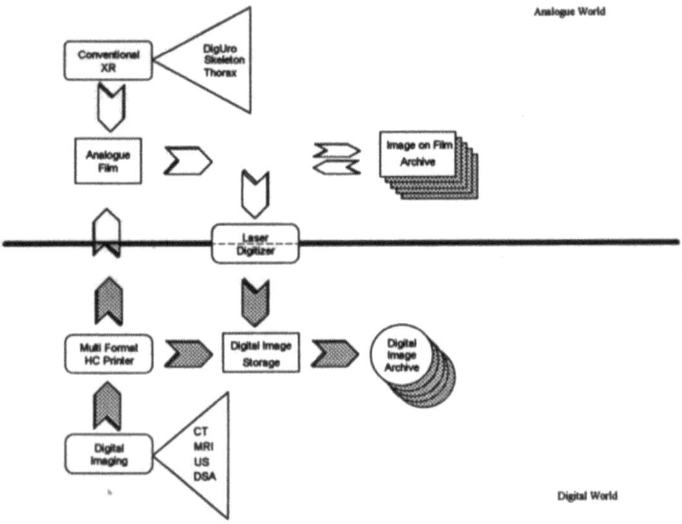

Fig 1. : Transition from analogue through hybrid digital to future full digital world

- Bridges under the form of secondary digitalisation or digital laser camera will make the medical exploitation of the hybrid environment possible.

- Integration within the hospital structure, namely via an integrated approach of the three information systems : PACS, RIS (Radiological Information System) and HIS (Hospital Information System).

As already said above, ImAZ is a clinical implementation project.

Several sub projects that encompass Echography, Intensive Care Unit, MR, Angiography, Radiotherapy, Emergency Department, can be described (fig. 2). The Emergency Department does represent an important application, developed in collaboration with KODAK Health Sciences.

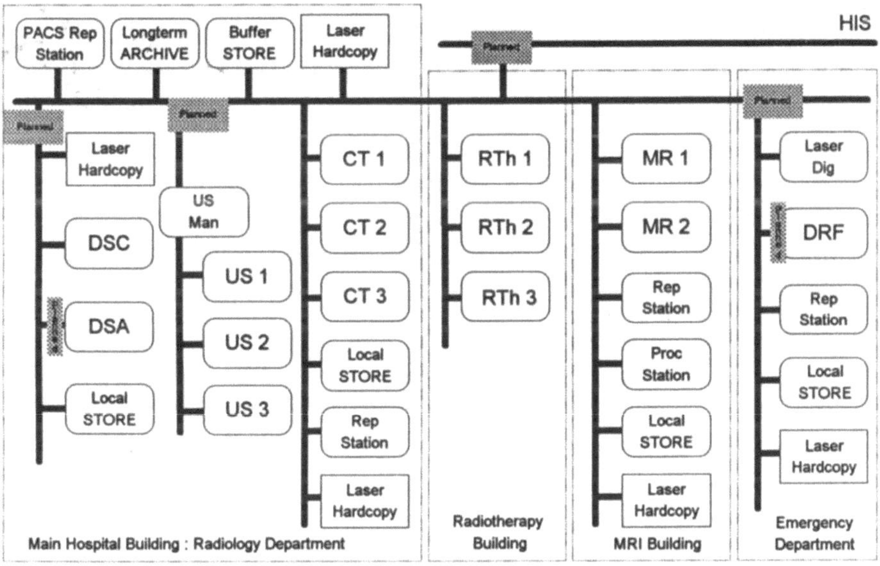

Fig. 2 represents the projected status of ImAZ for the end of '93.

Emergency-ImAZ

Medical Scope :

A correct archiving and retrieval radiological document is a crucial and very difficult point for the Emergency department : due to the nature of activities, the rate of loss of conventional documents and the subsequent medical and medico legal problems represent a real matter.

The Emergency Department of the AZ-VUB is a very busy unit : 120 patients per day in the Emergency department , of which at least 75 % will be seen at the satellite of the Radiology Department (average figures from '92).

Traumatology does represent an important proportion of those cases.

The conventional X-ray films are regularly taken directly after development, in order to be discussed with senior surgeons or as base for an urgent surgical treatment. In some instances, the documents are sent to external referring medical doctors.

The supervision of the correct diagnosis by the senior radiologist, as well as the eventual follow-up, is therefore impaired. The procedure induces also a important percentage of loss. The subsequent legal issues, very frequent in that population, where traffic and work accidents are present, put regularly serious problems.

For those reasons, in collaboration with KODAK HEALTH SCIENCES, and within the global scope of the decentralised AZ-VUB PACS project, a new digital environment will be installed, integrated and evaluated in the Emergency Department.

The objective is to digitally archive the totality of the **traumatic cases**, first for exploitation within the emergency department only, in a second step also on level of the department of Orthopaedics.

That objective will be reached by a combination between secondary digitalisation of conventional radiology (through the Kodak system) and digital fluorography.

In '92, the Kodak system has been preliminary tested and experimented in the central department of radiology. In April '93, the system has been installed in the emergency department and first used on an experimental level.

An utilisation in clinical routine is foreseen for the end of the year '93.

In '93 also, the Kodak system will be integrated with the Radiological Information System, a part of the Hospital Information System (collaboration with HIPIN, EurIPACS, and the department of Informatics of the hospital). In '94, the integration of that PACS island in the global PACS project of the AZ-VUB will be pursued.

Technical approach :

AZ-VUB - Kodak project started in March 1992 as an evaluation β-site of Kodak-Vortech.

The project consist out of different scopes :

1. Testing the hardware in a simulated clinical environment at the central radiology department.
2. Testing the organisational liability of the total environment
3. Testing the medical reliability of soft copies of secondary digitised images

The environment consist out of :

a laser digitiser
a dedicated viewing console
a data entry terminal
an single WORM 14" 10 Gb drive archive
a laser printer for hard copies

Current Functionality :

Test environment for evaluation of digitised orthopaedic cases.

Constraints :

The projected production of digital orthopaedic images at the emergency department will be about 0.5 Gb per day. The required on line availability is about forty days, in this way a jukebox with a capacity of 20 Gb will be necessary.

To provide on line availability of orthopaedic image folders produced at the radiology satellite of the emergency department. In the first phase on the viewing station at the radiology department and in the second phase at other departments, certainly the orthopaedic surgery department.
Patient data entry will be provided from HIS/RIS through an interface build by the EurIPACS project, topic HIPIN.

A PACS for Digital Video Angiography in Ophthalmology

R. Schlösser[3], S.Wolf[3], H.Toonen[2], Keizers A.[1], Schilling C.[2], M.Reim[3]

[1] Lehrstuhl für Meßtechnik, RWTH-Aachen, Templergraben 55, FRG-5100 Aachen

[2] now with 3M, Neuss

[3] Augenklinik Klinikum Aachen, Pauwelstr. FRG-5100 Aachen

Summary

We describe a PAC system for ophthalmologic application. The system has been developed at the LfM, RWTH-Aachen in cooperation with the Department of Ophthalmology, University Hospital Aachen for two purposes:

1. Validation of novel technical PACS concepts in clinical environment and

2. Support for the introduction of new approaches to diagnostic image analysis.

First results from clinical implementation show, that the ophthalmologic PACS establishes the digital infrastructure required for a joint development, a daily application and an efficient validation of novel procedures for the quantitavie and qualitative evaluation of retinal angiography image sequences. [4,6,7,8,10]

1. Introduction

Retinal fluorescence angiography is a widely applied diagnostic procedure in ophthalmology. The perfusion of the retinal blood vessel system is recorded using a scanning laser ophthalmoscope combining the advantages of video and photographic techniques [2,3]. This procedure is extremly valuable in helping to understand the diagnosis, disease process, and treatment of most of the deseases that effect the occular fundus. To develope new diagnostic tools by the use of image processing the digitizing of the analog signal is required. However, archiving and handling of digital image sequences are still cumbersome in the clinical routine. Therefore we developed a picture archiving and communication system (PACS) specially adapted to digital video angiography. From similar projects in the radiologie domain we experienced that a successful introduction of the system is dependend on the acceptance of the medical personal. So the major goal was to gain immediate advantages, measured in time saved and in an increase in diagnostic reliability.

2. System configuration

All system components described below are interconnect by the high speed opticalfiber network IMNET (IMNET-1: 1.2MB/sec, IMNET-2: 18MB/sec)[1,9]. The protocol (TCB2) implemented on this network is special designed for the transfer of large data blocks using almost the full capacity of the physical channel. So this network is exclusive used for image data transfer. For the transfer of additional managment and control data a standard coaxial network (Ethernet) is used. The full system configuration is shown by Fig 1.

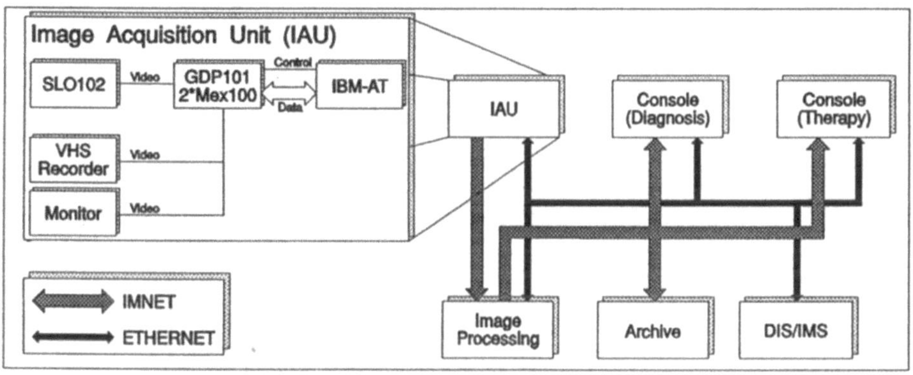

Figure 1: System configuration

2.1 Image acquisition

The scanning laser ophthalmoscope (SLO 101, Rodenstock Instr., Germany) generates a standard CCIR-video signal. The analog signal is digitized by a realtime video acquisition unit. It consists of a GDP101 framegrabber with two MEX100 memory boards (Leutron Vision) interfaced to a PC (IBM-AT486-50). This equipment is capable of capturing 22MB digital sequences. The maximum frame rate is 50 images per second with a variable definition of 256*256 up to 1024*512 pixels. For dynamic analysis of the angiography sequences the maximum frame rate is required, whereas a low definition of 256*256 is sufficient.

For morphologic analysis and documentation high resolution images are necessary. All operating modes are available on choise during a recording session. They can be interactively controlled by the software package which we have developed. The resulting data volume of a single examination is approximately 20MB. By means of state-of-the-art compression standards (MPEG or JPEG) we yielded a compression factor of 2 to 8% for a reversible (lossfree) coding.

2.2 Image presentation

For diagnosis and therapy at the moment two graphic consoles (IBM-At486-50) with a maximum resolution of 1248*1024 pixel are connected to the PACS. Low level image processing functions, e.g. zoom, contrast enhancement and filtering, are supported by this units. All image types defined by TIFF 5.0 (8 bit grayscale up to 24 bit RGB) can be displayed. To enable a prefetching of images each unit is equiped with a 200MB fixed disc for local buffering.

2.3 Image processing

Image processing is performed as a remote procedures running on a Motorola 68030-based microcomputer system. So the full functionality is available by all other system components. For image sequence analysis a true motion compensation of the eye movement is realized by a three level estimator. From the compensated sequences functional images for different parameters (aterial dye-bolus velocity, ateriovenous passage time,etc) are generated. For morphologic analyse a segmentation of the retinal blood vessel system followed by classification is performed. Parameters such as vessel diameter or length of vessel segments are extracted automaticly.

2.4 Image archiving and managment

The archive is implementated on a 68030-based microcomputer system [5]. For long time archival 2GB optical disks are used. Image data generated during the last week is also stored on fast magnetic discs to provide for a fast delivery of recent image sequences.

The image managment system (IMS) is part of the department information system (DIS). This system is realized by means of a relational object-oriented database with a SQL communication interface running on an IBM-AT486-50. The integrated security module supports various security levels for single users and user groups.

3. Quantitative requirements

All image data generated by the IAU at first is passed to the image processing workstation. The prossed image data and the additionally generated functional images are directly transfered from the workstation to the archive. For diagnosis in 70% of all cases three complete data sets are requested. For therapy mostly only the high resolution images are of interest. Fig. 2 shows the apportionment of activities for the different DTEs.

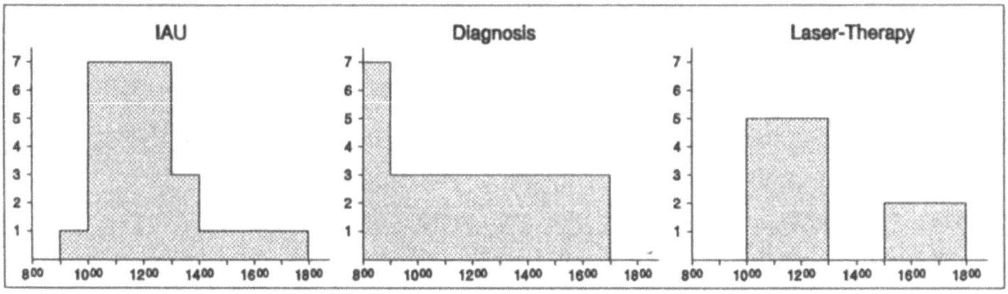

Figure 2: DTE activities per hour

The exspected peak rate of 800MB per hour is reached between 10:00 a.m. and 1:00 p.m.. So a minimum transfer rate of 0.5MB per second for image data is required to guarantee full system functionality. This transfer rate was yielded by IMNET-1 without problems. Extending the system by consoles for clinical validation and research, however, will drastically increase the data volume to transfer. Therefore an IMNET-2 interface for EISA-bus systems is under development at the LfM.

Discussion

The described system was successfully tested in clinical routine at the University Hospital Aachen. So far 700 angiograms have been recorded and stored with the system. Its capability to allow for simultaneous recording, demonstration and reporting has already proven beneficial for the daily work.

References

1. Keizers A., Meyer-Ebrecht D., Vossebürger F.: *A Fiber-Optic Line-Switching Network with a 140 Mb/s User Datarate,*, IEEE Journal on Selected Areas in Communication, Vol.10, Sept. 1992

2. Plesch A., Klingbeil U.: *Digital Laser Scanning Fundus Camera,* Applied Optics, Vol.26, p.1480-1486, April 15, 1987

3. Reim M, Wolf S.: *Videofluoreszenzangiographie zur Untersuchung der Hämodynamik des Auges,* Fortschr. Ophthalmol., No.86, p.744-750, 1989

4. Reim M, Wolf S., Toonen H.: *In Vivo measurement of Blood Flow Velocity in the Cappilaries of the Human Macula with Scanning Laser Ophthalmoscopy: A new technique,* Invest. Ophthalmol., No.31, p.382, 1990

5. Schilling C., Wein B., Meyer-Ebrecht D.: *Distributed Image Archival,* HIPACS, A second Generation PACS Concept, Ed.: Michel Osteaux, Springer Verlag Berlin, ISBN-3-540-54592-1, p.238-250, 1992.

6. Toonen H., Wolf S., Kaupp A., Reim M., Meyer-Ebrecht D.: *Digital Video Fluorescein Angiography: Quantification and Visualization of Regional Hemodynamics,* Investigative Ophthalmology & Vis. Science, No.32, p.865, 1991

7. Toonen H., Wolf S., Arend O., Schlösser R., Reim M. Meyer-Ebrecht D.: *Identification of Feeding Vessels in Subretinal Membranes by Means of Digital Picture Analysis of Indocyanine Angiograms,* Invest. Ophthalmol., No.33, p.723, 1992

8. Toonen H.: *Bewegungskompensation zur automatischen Auswertung videofluoreszenzangiographischer Bildsequenzen der Netzhaut,* Dissertation RWTH Aachen, 1992

9. Vossebürger F.: *IMNET/2 Ein lokales Breitbandnetz für die digitale Bildkommunikation,* Dissertation RWTH Aachen, 1991

10. Wolf S., Toonen H. Meyer-Ebrecht D., Reim M.: *Scanning Laser Ophthalmoscopy for the Quantification of Retinal Blood-Flow Parameters: A new Imaging Technique,* Scanning Laser Ophthalmoscopy and Tomography, München, Quintessenz, 1990.

Image Transfer within the Hospital: Developments and Experiences

M. WILTGEN[1)], G. GELL[1)], E. GRAIF[1)], R. PITZLER[1)], A. KAINZ[1)], E. SORANTIN[2)], R. FOTTER[2)]

1)Department of Medical Informatics, University of Graz, Auenbruggerplatz 9/III, 8036 Graz 2)Department of Radiology, Ped.Rad.Division, University of Graz, Auenbruggerplatz 34, 8036 Graz, Austria

Summary

At the Department of Radiology of the University of Graz a PACS has been installed and is used in daily routine. This PACS is interfaced with a RIS. Our experiences show that one of the most urgent clinical interest in PACS is not the reporting on the diagnostic console, but the image transfer to clinical departments where they are not available or only with time delays. Therefore we provide the decentralised access to the images within the PACS and activate the image transfer to the end users of the diagnostic information.

To display the images in the clinical departments, we developed software to turn every standard IBM compatible PC into a cheap image display console. To enable an easy and fast expansion of the PACS structure, a software tool has been developed that enables the definition, generation and installation of software processes which are necessary for the exchange and transfer of images via an interactive user dialogue. To import images from conventional examinations into the PACS, the user interface and software for the image acquisition by a digitizer (Konica) were developed.

Introduction

Picture Archiving and Communication Systems (PACS) have been developed to enable an easy and fast management of the digital images [1-4]. PACS offers advantages like the reduced cost for film and archive space, the faster access to the images and the image processing capabilities of the viewing consoles. One of the main tasks that a PACS should fulfil is the image transfer within the hospital and the direct access of the end users of the diagnostic information (like surgery, radio therapy etc.) to the images in the radiology. We present the development of software tools for image display, image acquisition by an digitise and for an easy expansion of the PACS structure, in that sense that new processes for the images transfer in clinical departments can simply be added to the existing configuration.

System configuration

At the Department of Radiology of the University of Graz an integrated PACS-RIS is used in daily routine. To the advantages of the RIS like sharing data for patient care, fast access to the reports etc. the PACS adds the advantages of a compact and easy to handle digital image archive. The PACS is interfaced with the RIS in that way that the optical archive is managed exclusively by the RIS.

PACS: The PACS at the Department of Radiology of the University of Graz [5,6] was installed in collaboration with SIEMENS Erlangen and includes 4 CT-scanners (one of them is located in the radiotherapy at a distance of 1 km away from the Department of Radiology), a DSA, an MRI-scanner, a radiotherapy planning system, an archive with 2 drives for optical disks, a diagnostic console with 3 high resolution monitors and several image display consoles. The configuration consists of two Ethernet-based networks: a local network which is restricted to the department of Radiology and a hospital wide network which connects a lot of departments inside the hospital. A special software process (PACS-MONITOR) enables the user to survey and handle the image and data flow in the PACS. To eliminate troubles which occurs in routine operation a build-in expert system (OPERAS) is used for maintenance.

The PACS is running 24 hours a day and all the examinations from the 4 CT-scanners are archived on optical disk. One optical disk (12 inch, 2 GB) can store up to 26000 256x256 images and 22 to 25 optical disks are needed per year.

RIS: The RIS (AURA) was developed in our department [7]. AURA is based on free text and contains a flexible record structure and index-sequential files. All the examinations from CT, US, DSA, conventional radiography and MRI are stored in AURA. At the moment the RIS includes 60 terminals for patient admission, report writing, retrieval of previous examinations and scientific studies. The management of the optical archive is performed by AURA and the images are selected for retrieval by using patient identification or medical criteria.

Software structures

The PACS is organised as a group of independent software processes which are running in the background. The processes communicate which each other by using REQUEST, GET and SUBMIT commands according to the ACR-NEMA standard. The processes, which are distributed over several computers, operate simultaneously and execute tasks like image transfer, conversion of images, submission of data into databases etc.

Emphasis of Development

Our experiences show that the most urgent clinical interest in PACS is not the primary diagnosis in radiology, where only a small additional benefit is to be expected by the replacement of films, but in the wards, where images are presently not available or only with time delays. For example: In the conventional film archive of the Department of Radiology 25- 30% of the films are not available when requested. Therefore we provide the decentralised access to the images within the PACS and activate the image transfer to the end users of the diagnostic information. To enable the access to the images, software processes for image transfer and management and a PACS terminal for image display are necessary.

Process Configuration Editor

The Process Configuration Editor (PCE) is used to add the software processes which are needed for the image distribution to the PACS software structure. This tool mainly includes:
-a dialogue oriented installation program
-a set of standard processes for given applications which are necessary for image distribution like image transfer, management of local databases
-the Process Configuration Structure Database with the addresses, functions and interactive connections of all the processes in the process structure.

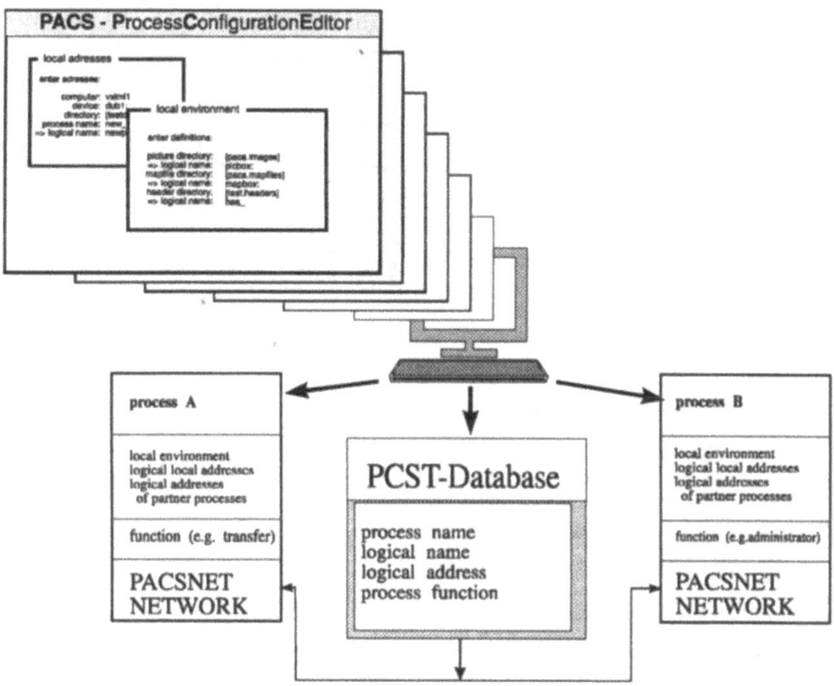

figure 1:
processes for image transfer are connected to the PACS via the PCE user dialogue

Via the installation program the standard process for the requested application is selected and named (fig. 1). Then the needed addresses (network address, process logical address) and the local environment of the process (image directories, device name, node name) are defined in an interactive dialogue. The procedures which are necessary for a successful startup and running of the process are generated automatically. Then after the process is logically linked to the PACS, the process, the start-up procedures and the process environment are installed on the target computer.

PACS-VIEW

PACS-View is an image display console based on PC. With the PACS-View software every standard PC can be turned into an image display console allowing image processing capabilities like windowing and pixel lens. The PC can be linked with a high resolution and high density monitor (IMLOGIX). The images themselves are stored on a VAX and the PC has access to the images via the network operating software PCSA.

Integration of Conventional Images

To enable the integration of conventional X-ray images into the PACS structure, a radiographic film digitizer (manufactured by Konica) has been connected to the PACS network via a µVAX-II-computer. The appropriate software interface to acquire and manage the (patient- and image-oriented) data has been developed in our department. At the moment, this device is used mainly to collect images for scientific purposes, but not yet in daily routine.

Image transfer handling

The image transfer is initiated by selecting an examination with patient data at a RIS terminal. Then after the user has entered the destination the images are retrieved from the archive and transferred to the desired workstation (fig. 2). At the workstation the user can select the examination in the local database and display the images on the PC.

Initial Clinical experiences:

At the moment mainly CT and MRI images are transferred within the hospital. In daily clinical practice this application is used for:

-Image display of CT examinations is used for synoptic reports.

-Demonstration of CT examinations in conference-rooms using existing video-devices within clinical, interdisciplinary conferences.

For this purposes screen resolution is sufficient.

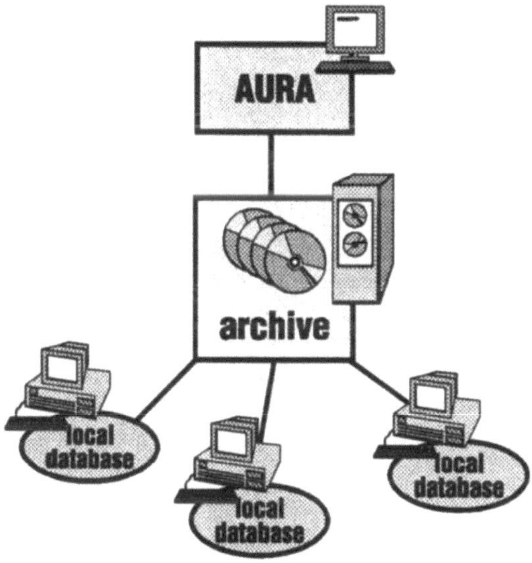

Figure 2
The images are selected via an AURA terminal and transferred to the PACS-VIEW station.

Outlook

At the moment we expand the PACS-View software in that way that the user can select images which are of special interest for his purposes out of the examination and collect them in a special folder. One application will be that the surgeon can collect all the images which are important for him before the operation and display them on an IMLOGIX monitor in the operating room.

Further we plan the combination of the PC with a IMLOGIX Monitor for radiological reporting.

Furthermore, the installation of a jukebox is planned to enable a fully automatic image distribution. At the moment the needed optical disks have to be inserted manually into a single disk drive. With a jukebox the images of at least one years production can be held online.

Acknowledgement

We would like to thank the Österreichische Nationalbank for supporting this work.

References

1. van Poppel, B.M; Lodder, H; Elsakkers, P.H.; Koens, M L.; Bakker, A.R.:
Prefetching offers prospects for a quick responding PACS
In Computer Assisted Radiolgy, Proceedings of the International Symposium CAR91 (EDS: H U Lemke, M L Rhodes, C C Jaffe, R Felix)
Springer, Berlin, Heidelberg, New York, (1991) 464-470

2. Glass, H.:
The Impact of PACS on Hospital Information and Practice.
In Medical Imaging 3: PACS System Design and Evaluation R. (EDS: H Schneider, S J Dwyer, R G Jost), SPIE Vol. 1093 SPIE- The International Society for Optical Engineering. P. O. Box 10, Bellingham, Washington 98227-0010 USA, (1989) 354-361

3. Hruby, W.; Mosser, H.; Urban, M.; Rüger, W.:
The Vienna SMZO- PACS-Project:
The Totally Digital Hospital
In Computer Assisted Radiolgy, Proceedings of the International Symposium CAR91 (EDS: H U Lemke, M L Rhodes, C C Jaffe, R Felix)
Springer, Berlin, Heidelberg, New York, (1991) 436-441

4. Knots, M.J.J:; van Erning, L.J.Th.; Scaf, J.; Guijt, W.; Ruijs, J.H.J.:
PACS in practice: on-line communications in daily routine
Med. Inform. (1990), vol. 15, no.1, 11-14

5. Wiltgen, M.; Gell, G.; Schneider, G.H.; Gnoyke, H.:
PACS: A pilot system in routine operation
Electromedica (1990), vol. 58, 4, 124-129

6. Wiltgen, M.; Gell, G.; Graif, E.; Stubler, S.; Kainz, A.; Pitzler, R.:
An Integrated PACS-RIS in a Radiological Department
Journal of Digital Imaging, Vol 6, No 1 (February), 1993: pp16-24

7. Gell, G.:
Routine documentation of medical texts
Methods Inf Med 14: 63-68, 1983

Multi-user Applications for Cooperative Work in Cardiology and Radiology

Lutz Kleinholz, Martin Ohly, Joachim Steffens, Roland Felix, Eckard Fleck, Bernd Mahr

Universitätsklinikum Rudolf Virchow, Strahlenklinik und Poliklinik, Projekt BERMED, Augustenburger Platz 1, 1000 Berlin 65, FRG

Abstract

Existing and exciting cooperation between physicians in radiology and cardiology may be enhanced through the introduction of computer support in the form of multi-user applications within desktop conferencing (DTC). An understanding of the way the existing media of images and documents are used is important for the design of such facilities. An analysis has been carried out of the shared workspace activities of physicians. The findings have been used to derive the functional requirements and design recommendations for DTC. This paper summarizes the analysis, the findings, an architecture for the integration of multimedia medical information systems with desktop conferencing, and briefly describes a prototype that has been implemented.

1 Introduction

Computer Supported Cooperative Work (CSCW) is the label given to a recently formed research field which is concerned with the understanding of the way people work in groups and the role of computer support [1]. It is an interdisciplinary research field, involving computer scientists, psychologists, sociologists, and others. Examples of medical activities that we shall term cooperative are joint diagnosis, therapy planning, consulting, presentation of findings, tuition, and conferences. The success of other projects[1] in telemedicine has clearly shown the requirement in radiology and cardiology for conferencing facilities to support the existing cooperation with external physicians. Technological support for these activities can bring many benefits to patients, physicians, and hospitals. For example, teleconferencing facilities could lead to faster and more precise diagnosis, improve diagnostic information and the communications between clinics, and bring new opportunities to consult experts.

The term synchronous, or real-time, desktop conferencing is applied to systems that aim to integrate audio/visual teleconferencing within existing desktop workstation environments [2]. In

[1] This work has been motivated by the very positive results of the DeTeBerkom funded BERMED project and EEC funded projects concerned with telemedicine that have been or are still being carried out in Berlin in conjunction with the Deutsches Herzzentrum, the Department of Radiology at the Universitätsklinikum Rudolf Virchow (UKRV), and the Department of Computing Science at the Technische Universität Berlin.

contrast to analog based video conferencing, desktop conferencing systems are relatively inexpensive, may be located in any office, can use available broadband communication links, and allow the seamless integration with existing PACS and HIS/RIS. One aspect of DTC is application sharing. A multi-user application can be considered as providing a shared workspace that allows all conference participants to view and manipulate a common set of objects. The workspace activities of radiologists usually utilize a lightbox and a range of documents which include text and images. A computer based shared workspace would allow physicians, who may be located in different hospitals, to view and work on the same set of multimedia[2] documents. Our aim is to develop multi-user applications that, using the BERMED developed distributed information management system (HDMS) [3], integrate existing RIS, HIS, and PACS services, together with image processing capabilities.

Recent CSCW research has led to several important findings which may improve the design of multi-user applications. An understanding of the way existing media are used is considered important for success [4]. In this respect, we have analysed the shared workspace activities of physicians and used the results to derive a set of requirements and recommendations for DTC. Our findings have shown the importance of adequate support for gesture[3] using telepointers[4], lightweight conference management, adaptable floor control[5], and system performance. This paper presents the preliminary results of our analysis, the derived requirements, design recommendations, and an overview of a prototype implementation.

2 Analysis

Certain cooperative activities of physicians at the Universitätsklinikum Rudolf Virchow (UKRV) have been analysed. Repeated observation has led to the identification of the form of the activity (consultation, presentation, or collaboration), the objects that constitute the shared workspaces, the actions performed on these, and the problems associated with the existing meeting media. An example activity is the fortnightly meeting between the orthopaedists of the Oscar-Helene Heim (OHH), Orthopädische Klinik, and the radiologists of the UKRV. The OHH physicians refer on average five patients each week to the UKRV for MRI and CT examinations. A radiologist travels the 10km to the OHH to present the findings. The meeting serves several purposes: the presentation of findings; consulting, the orthopaedists consult the expert radiologist concerning third party examinations, image interpretation, or choice of suitable examination technique; further education in radiology, the orthopaedists learn about new examination possibilities; and, not least, to maintain good relations between the clinics. Figure 1 illustrates the objects of the shared and private workspaces and actions performed during the presentation of the findings of one case. The actions are grouped into those that involve the

[2] Multimedia is a term given to the integration of text, graphics, still image, video, and audio.
[3] Gesture is "purposeful hand movements which communicate information" [4].
[4] A telepointer is a graphic marker visible to all participants that can be moved to point at details.
[5] Floor control is necessary to ensure that only one person at a time can provide input to a shared application.

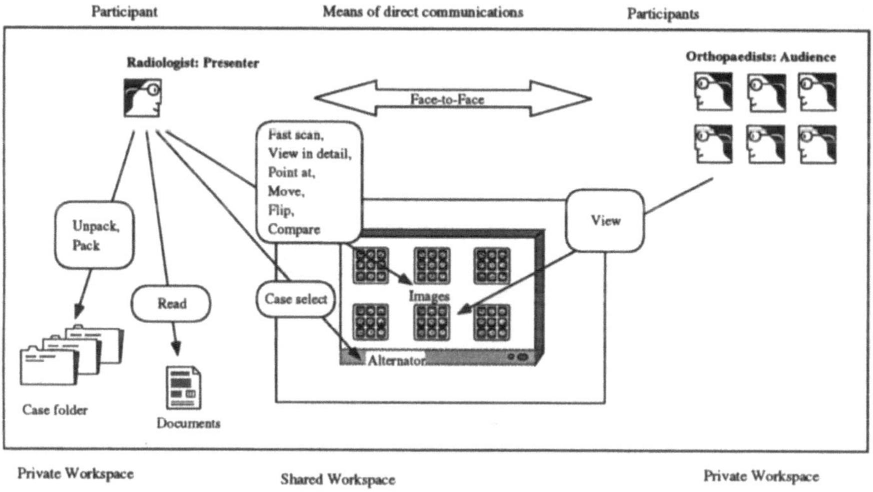

Figure 1: Objects and Actions for the Presentation of Findings

shared objects and those that involve the *private objects*. Actions involving *shared objects* include the mounting and selection of hardcopies on the alternator, the rapid scan of images, the movement and *flipping* of hardcopies, the view in detail and comparison of images, the selection of representative images to highlight the findings[6], the joint viewing in detail of images, and the review of the request and findings documents. Actions involving *private objects* include the unpacking and packing of case folders, and the writing of notes.

Several problems have been identified with the existing meeting format and media. It occasionally occurs that the patient has already been treated before the meeting takes place. Sometimes not all relevant documents are available and the time spent by the radiologist in travelling is not negligible. Due to the small format of the individual images, it is difficult for all those present to see sufficient details and it is not possible to perform any image processing.

These problems may be alleviated through the introduction of DTC. The findings may be presented directly after the examination. A standard video projector can be used to display the contents of the workspace so that all members of a remote audience can clearly see the details of images.

3 Requirements

Requirements and design recommendations for multi-user applications and DTC systems have been derived from the results of the analysis. These are divided into basic and advanced requirements.

[6] This action may be aided through pointing with a finger, pen, or other instrument.

3.1 Basic Requirements

The core of the DTC is a shared workspace that allows all participants to view and manipulate a common set of objects. These objects may include medical images, text documents, sketches, applications programs, etc. Telepointing is a prerequisite for DTC in medical applications, careful design of the facilities is essential. There are two possible implementations: either the pointers move continuously with the mouse, such as a cursor, or they are *markers* that must be explicitly dragged. Experience has shown that the latter implementation is very effective for medical applications. The pointer should move in real-time, i.e. with minimal latency, with the mouse. It must be possible to simply change the size, colour, orientation, and label on the pointer.

An audio channel is a prerequisite for DTC in medical applications. High quality audio is essential to avoid possible misunderstandings during conversations [5]. Basic image manipulation facilities, such as zooming, panning, rotation, flipping (about vertical and horizontal axes), and magnifying glass and basic image processing facilities, such as contrast and intensity adjustments are essential. Fast access to all documents from RIS/HIS, image management and archiving systems is important, in particular to review the findings of an emergency examination. Arbitrary queries should be allowed. On-line access to other RIS services such as appointment scheduling and current examination status reports may also be beneficial.

Since all documents relevant for a conference cannot be displayed simultaneously in high resolution on existing monitors, the DTC should allow the arbitrary iconification, stacking, movement, and selection of these. After certain medical conferences, a report containing the agreed findings or results must be written. A facility to aid this preparation should be included in the DTC system. The report may include the manipulated and annotated images.

Simplicity is essential. In emergency situations, time pressed physicians do not want to navigate superfluous menu hierarchies. The conference setup or registration procedure should be automated. Performance and robustness, together with ease of use, are likely to be requirements that strongly influence the acceptance of the final system. These qualities are essential for any system that is to be relied upon in emergency situations.

The suitable design of the user interface is seen as particularly critical for the success of multi-user applications [6]. Consideration of the security and data protection aspects are essential for medical applications. These topics are being studied in the BERMED project.

3.2 Advanced Requirements

Advanced requirements include further image processing capabilities, such as filtering, adaptive histogram equalization, edge enhancement, 3-D visualization, and image analysis facilities, such as defining regions of interest (ROI), segmentation, and quantification of volumes, flows, etc. All observations have shown the importance of speed and fluency for presentations.

The DTC system should offer a *slideshow* facility, allowing the preparation and replay of a presentation. A video channel allowing the viewing of the partners in a conference is essential when discussing complex cases. Inclusion will also lead to better *group awareness* [7], and hence improve user acceptance.

Since many medical documents are not stored digitally, the DTC system must include some form of high resolution scanner, or document camera with lightbox. Textual and graphical (e.g. with highly visible arrows) annotation of images and other documents should be allowed. These should be erasable without destroying the underlying image or text. Floor control, which is needed to ensure that only one person at a time can provide input to a multi-user application, should be adaptable. The DTC should include both a free and a controlled floor.

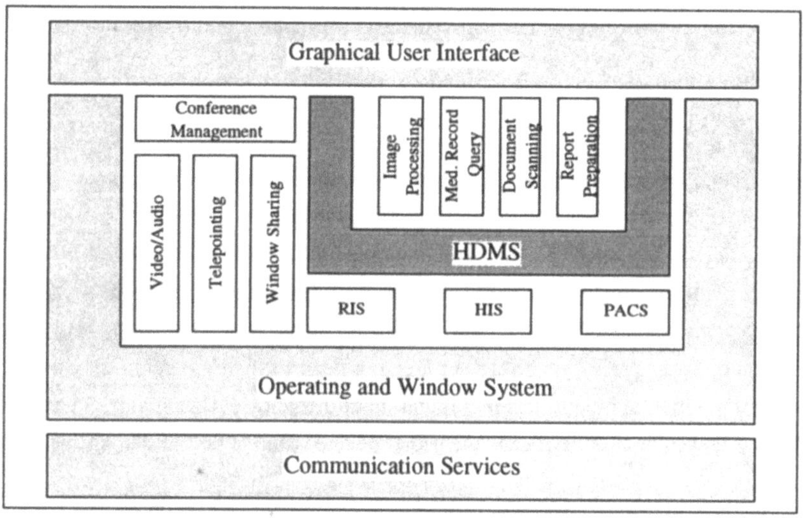

Figure 2: Architecture for a Medical DTC System

4 Design and Prototype Implementation

Multi-user applications can have a centralized or replicated architecture. Replication requires data objects to be copied to each remote site and introduces the need for synchronization to avoid state divergence. A replicated architecture requires custom built multi-user applications The advantage of replication is that performance (after copying data objects) is much better since only input events need be distributed. This may be important when sites are connected by low speed data communications. Since the Bermed project uses metropolitan and wide area networks with between 2 and 140 Mbit/s we have chosen a centralized architecture. Data objects are *remotely visualized*, hence patient data is not replicated. The output of the application is distributed to all conference sites using a *window sharing* tool [2]. Floor control ensures that only one person at a time can provide input to the application. Different control policies will be allowed, including an explicit chalk passing mechanism for strictly controlled, formal conferences, and a free floor (or an implicit request protocol with timeout release) for informal interactions. The telepointer component provides clearly identifiable coloured markers for each participant. The label on the marker, giving the name of the participant name, can be simply changed.

The architecture of our implemented DTC is illustrated in figure 2. The hospital PACS, RIS/HIS are integrated within this using the Bermed HDMS platform. The complete functionality of the system is presented to the user through a uniform interface. The conference management component is responsible for the connections to remote sites, the window sharing, floor control, telepointing, the video, and audio. The prototype has been implemented on Sun SparcStations under Unix, X11, and OSF/Motif. The video transmission uses the Video Extension for X Windows (VEX) and JPEG compression standard. The duplex audio uses the standard Unix vtalk protocol.

5 Discussion

Initial experience with the prototype using window sharing and the telepointer markers has proven encouraging. Several aspects require further investigation. One issue is whether all participants should have exactly the same screen contents (termed *What You See Is What I See, WYSIWIS*). Unless identical graphics hardware is used and window managers are disabled this is difficult to achieve. The effect of delays introduced by the communications network need to be investigated. When transmission delays are significant, synchronization may be necessary to ensure that pointing at objects and manipulation of the workspace can only continue when it is guaranteed that each participant has the same view. Extensive field trials will be carried out in the near future to investigate these and other aspects.

References

1. Irene Greif. Computer-Supported Cooperative Work, Morgan Kaufmann, 1988
2. J. Chris Lauwers. Collaboration Transparency in Desktop Teleconferencing Environments. Phd Thesis, Stanford University, July 1990
3. H. Hansen, R.-D. Kutsche, J. Steffens. The PADKOM System Model - an Open Platform for Medical Applications in a Distributed Environment, Proc. of the IFIP TC6/WG6.4 Int. Workshop on Open Distributed Processing, Berlin, October, 1991
4. John C. Tang. Findings from Observational studies of collaborative work *Int. J. Man Machine Studies*, Vol.34, No.3, March 1991
5. Christian Heath and Paul Luff. Disembodied Conduct: Communication through video in a multimedia office environment, Proc. of *ACM CHI*, 1991
6. C. A. Ellis, S. J. Gibbs, and G. L. Rein. Groupware: Some issues and experience, *Communications of the ACM*, Vol.34, No.1, Jan. 1991
7. Stephen Gale. Desktop video conferencing: technical advances and evaluation issues. *Computer Communications*, Vol.15, No.8, October 1992

Software Development and Management for a Picture Archiving and Communication System

LU J. HUANG, DANIEL J. VALENTINO, R. K. TAIRA

Clinical PACS - 172115
Department of Radiological Sciences
University of California, Los Angeles
Los Angeles CA 90024-1721

ABSTRACT

With the gradual maturation of Picture Archiving and Communications Systems (PACS), software development and management has become one of the focal points of PACS research. We have analyzed the current characteristics of PACS software, identified the most important factors for effective PACS software development and management, and we developed a PACS software engineering environment that enforces PACS software integrity while increasing PACS programmer productivity. The use of such an environment is the first step toward the production of reliable, verifiable, reusable, and portable PACS software modules.

INTRODUCTION

Picture Archiving and Communication Systems (PACS) are electronic systems that support the systematic acquisition, transmission, archival and presentation of radiological image and text data. To meet the diverse needs of PACS users (radiolologists, clinicians and others), the objectives, and thus the capabilities, of different PACS subsystems may vary significantly. They range from closed, dedicated imaging modality PACS, to complex PACS for total digital radiology departments and total digital hospital image management systems. Our main interest is in fully functional, and thus complex, PACS such as the UCLA Clinical PACS.

PACS software, built on top of the underlying physical hardware, is an essential and indispensible part of PACS and directly affects the clinical viabilityof the PACS. With the gradual maturation of PACS implementations [1,2], PACS software development will become one of the focal points in future PACS research and development. In this report we discuss the major issues concerning PACS software, identify the requirements for productive PACS software development and management, and describe our implementation of the UCLA Clinical PACS.

BASIC CHARACTERISTICS OF PACS

A fundamental characteristic of PACS is distributed processing in a heterogeneous computing environment. The heterogeneity of PACS computing arises as a result of three different degrees of granularity for PACS components:

1. System Heterogeneity.

A PACS is composed of several different sub-systems, each of which performs a particular and distinct function. Each sub-system is optimized for the greatest performance within commercial and economic constraints. Thus PACS are intrinsically heterogeneous because each sub-system is very different from other sub-systems.

2. Component Heterogeneity.

Each sub-system may comprise a number of hardware components, each of which has its own computing capability. The component heterogeneity comes from the fact that the host computers, consisting of a single major computing group, may have different hardware platforms, or run under different operating systems and have different communication protocols and capabilities.

3. Software Heterogeneity.

Some PACS software development environments have converged to the UNIX operating system, the C programming language, the TCP/IP networking protocol, the X Windows graphical user interface(GUI), and the Structured Query Language (SQL) for database operations. However, significant differences still exist due to hardware and software component heterogeneity and PACS software development without standardization. Software heterogeneity can be identified by various factors, for example, different operational modes for fundamentally similar operations such as radiological image acqisition [3]. Additional factors include the need to consider the underlying hardware architecture and software development tools used in the normal software development cycle, and the syntactical and semantical differences between computer commands when executed on different host computers with different architectures.

A great deal of effort has gone towards increasing PACS connectivity through industry standardization. The recently proposed ACR-NEMA standard, Digital Imaging and Communication in Medicine (DICOM) version 3, compliant with the ISO model, makes widespread networking protocols, such as TCP/IP and Ethernet, become standard communication methods for PACS. Another significant feature of the new standard is its use of object-oriented data modelling for radiological data [4,5]. The use of an object-oriented data modelling method and the extension of ACR-NEMA network protocols substantially broadens the scope of the new ACR-NEMA standard, making it much more desirable for PACS than previous versions of the standard.

Radiological data, in the form of image, graphics, text, voice, etc., are processed by distributing the processing tasks among different computers in different groups. The primary function of PACS software is to make different PACS components work in a harmonious manner, so that PACS operations such as image data acquisition, image data archival, database updates, and image retrieval can be carried out in an automatic, reliable, non-interruptible and time-efficient fashion. Equally important are the following design objectives: (a) reliable and efficient operation modes with adequate fault-tolerant schemes for automatic or semi-automatic error recovery and dynamic data routing, (b) facilitating and enforcing the open architecture of the underlying physical infrastructure for easy adaptation of more heterogeneous PACS components and integration with other information systems, (c) conformance to industry standards, such as the ACR-NEMA standard, to increase PACS connectivity and portability and to decrease the software engineering efforts at one PACS site.

MAJOR PACS SOFTWARE ISSUES AND REQUIREMENTS

The major issues impacting PACS software management are: (a) functionality, (b) evolution, (c) development environment, and (d) management and maintenance.

1. Functionality

A typical PACS software system can be devided, according to its functionality, into certain subsystems or modules:

1. Modality-to-network interface.
2. Image and text data transfer.

3. Data archival and the central database management.
4. Image display and the presentation database management.
5. Special image display and printing softwares.

Each of these sub-systems should incorporate adequate fault-tolerant schema to perform automatic or semiautomatic systems error recovery and dynamic data routing.

2. Evolution

PACS software sub-systems are developed and operated in an environment that is ususally under constant revision. Thus PACS software subsystems evolve when:
1. Hardware is added or updated.
2. New end-user functionality is added.
3. Other information systems (HIS/RIS) are integrated into the existing system.
4. Formats are changed for conformance with industry standards.

These events can serve as the indicators that show that a major PACS software version update is in order.

3. Development Environment

The typical development environment for PACS software has these characteristics:
1. Networked environment.
2. Multiple graphic user interface(GUI) systems, (Open Look, Motif).
3. Multiple operating systems, and multiple computer programming languages, (C, C++, FORTRAN77).

4. Management and Maintenance

The characteristics of PACS software and their dynamic nature mandate careful consideration and planning of a PACS software development and management environment to ensure programming productivity and software reliability and integrity.

5. Requirements

Summarizing the discussion in the previous sections, the major requirements for PACS systems programming, testing and management are:
1. Centralized planning and coordination of software development activities.
2. Centralized software management and maintenance. The centralized control is emphasized by controlling the code between the systems manager and systems programmers.
3. Clear identification and division of PACS software subsystems, and independent development of each one of them.
4. Well-defined procedures and facilities for PACS software testing and customer software integration without any interruption of the normal PACS clinical operation.
5. Transparency in network, language and operating system for both software engineering and testing.
6. Well-defined and timely-updated common programming paradigm, emphasizing good and persistent programming styles.
7. A well-established documentation system, both on-line and in hard copy.

By meeting these requirements, PACS system software is expected to be reliable and portable, at source or binary level, among PACS computers.

IMPLEMENTATION

According to the requirements we identified in the previous section, we have implemented a PACS software development and management environment for the UCLA Clinical PACS.

1. Centralized Planning and Coordination of Software Development.

The software development environment is divided into two categories: (a) PACS infrastructure, and (b) PACS display. The infrastructure involves acquisition, archival, and transmission (distribution). The display environment involves the presentation of images and ancillary information to the PACS user.

2. Centralized Software Management and Maintenance.

The control of software is emphasized at the source code level. The programmer views source code as belonging to a particular software sub-system, regardless of the physical host computer on which the current version is being developed. After initial testing, the programmer submits the source to the PACS system manager for integration and execution.

3. Testing Procedures and Facilities.

Well defined procedures and facilities for PACS software testing are required to avoid interruption of daily Clinical PACS operation. We defined procedures and developed facilities to test new PACS modules with no or minimal impact on daily operations.

4. Documentation and Support.

Documentation of PACS software is enforced in three steps: (a) design, (b) implementation, and (c) operation. Flow-of-control charts are used to develop programs and documentation. Implementation documentation takes place both during design and coding. It is essential to document commands and software libraries accurately during the design and implementation phases; documentation includes the data structures and files related to the module.

CONCLUSION

By systematic analysis of the major issues impacting PACS software development and management, we have identified the basic requirements for a productive and consistent PACS software engineering environment. Our implementation enforces the standards of the UCLA Clinical PACS in the Department of Radiological Sciences. Since many system configuration related details are now hidden, the environment makes it more convenient to access, develop, and test the PACS software for the PACS manager, programmer and operator.

REFERENCES

[1] Huang HK, Kangarloo H, Cho PS, Taira RK, Ho BKT, and Chan KK. Planning a Total Digital Radiology Department. AJR, 154:635-639, 1990.

[2] Huang HK and Taira RK. Infrastructure Design of a Picture Archiving and Communication System. AJR, 158:743-749, April 1992.

[3] Weinberg WS, Loloyan M, Taira RK, and Chan KK. Automatic Acquisition of CT, MT, and US Images for PACS. In Huang HK, Ratib O, Bakker AR, and Witte G, editors, Picture Archiving and Communication Systems in Medicine. Berlin: Springer-Verlag, 1991.

[4] Horii SC and Bidgood WD Jr. Network and ACR-NEMA DICOM Protocols. Radio-Graphics, 12:537-548, 1992.

[5] ACR-NEMA Committee. Digital Imaging and Communications in Medicine(DICOM) version 3.0. draft 0.1. Washington DC: National Electrical Manufacturers Association, 1991.

ISAC Developments and Applications

Nagaaki OHYAMA

Imaging Science and Engineering Laboratory,
Tokyo Institute of Technology
4259 Nagatsuta, Midori-ku, Yokohama, 227 Japan

Abstract

ISAC (Image Save And Carry) is an off-line system, where medical information including images can be easily interchanged among computer systems and modalities. Because ISAC system requires a full compatibility, a lot of efforts have been paid for standardization works. This paper reviews the developments of ISAC system and the security mechanism indispensable for medical applications. As an application of ISAC system in medicine, ISAC project in National Cancer Center will be also introduced.

1. Introduction

In contrast to PACS (Picture Archiving and Communication System) [1], which is essentially on-line, ISAC (Image Save And Carry) [2] is an off-line system using portable recording media with high capacity. Because ISAC system is used for digital data storage and interchange with ease, it should be fully compatible in terms of recording media. The concept of this system was first proposed in 1989 [2,3], and all necessary works concerning development and standardization were begun by the ISAC committee, which was organized as a joint committee of JPACS (Japanese PACS Society) and MEDIS-DC (Medical Information System Developing Center). JPACS and MEDIS-DC, respectively, are an industry organization having studied standards for PACS and a government foundation developing new medical information systems. Because ISAC system is expected to be useful for both health care and preventive medicine, which are quite important for an aging society coming soon in Japan, the development of ISAC system has been supported by Japanese ministries: MITI (Ministry of International Trade and Industry) and MHW (Ministry of Health and Welfare).

This paper first describes a short review of ISAC system and explains the ISAC experimental project, which has been carried out since 1991 in National Cancer

Center Hospitals with financial support from MITI and MHW. Because ISAC for medical use should have a strong file protection, the security and file protection mechanism developed for Japanese domestic use will be also explained.

2. Review of ISAC system

For an easy handling of digital data including images of large size, ISAC was proposed as an off-line system using portable and full compatible recording media of high capacity. From the users' requirement, recording media for ISAC system have to be portable, rewritable, of high capacity and randomly accessible. These requirements except for the rewritability are definitely clear, but there are some users who oppose the rewritability and insist on the usage of WORM (Write Once Read Many) type disk. Their claims are based on the fear of falsification of official documents or diagnostic reports, and the file protection mechanism for ISAC system was developed so as to protect files safely. Of course, serious discussion about the security of recorded files has been made with taking into consideration the system operation and we concluded that ISAC system with file protection mechanism was good for practical use in medicine. Through these considerations, magneto-optical disks (MODs) [4] were selected as portable recording media.

The basic concept of ISAC system is illustrated in fig.1, where all stations have their own MOD drives, and we can store, for example, images in a MOD as soon as they are reconstructed by CT, MRI etc. or digitized by a film digitizer. These images as well as other data, thus stored, can be commonly accessed by the display stations. Because ISAC system is off-line, stations in fig.1 are not necessarily in a hospital; we can easily move a huge amount of stored data among hospitals and clinics. It is needless to say that digital data can be copied without any degradation, which can not be done by analog technology.

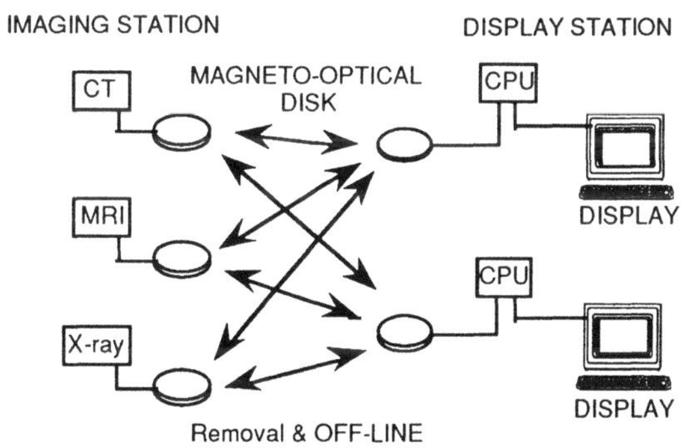

Fig. 1 A Schematic diagram of the ISAC system

MOD of our current concern has the size of 130mm in diameter and its capacity is about 640 Mbytes for both sides. This is because the physical specification of MOD medium was standardized by ISO SC23, and also because the inspection to confirm the compatibility of MOD media and drives is being carried out by ISAC committee. In order to avoid any confusion for users, MOD media and drives which pass the compatibility conformance test are explicitly shown to be ISAC products by the trade mark of "IS&C". Smaller MOD with 90mm in diameter is under discussion and standardization works will be begun. Therefore, in a few years ISAC system will be able to use the small MOD, whose capacity will be more than 250 Mbytes for single side.

3. The organization of ISAC committee

Fig.3 shows the organization of ISAC committee in 1993, which has two management comittees; one for business and the other for standardization. The business committee is in charge of business management necessary for a wide use of ISAC system such as a decision of the trade mark of ISAC and planning of ISAC exhibition, etc. The standardization committee has 6 working groups (WGs); WG1-6 are, respectively, standardizing or studying the format of volume and file structure, the data format specifying image size or codes etc., system design especially focusing on the security and file protection, the technology assessment based on the results obtained in the field tests which were carried out from 1989 to 1991, the specifications to confirm compatibility of MOD medium and drive, and the formats of numerals and text. WG2 has two sub WGs, WG2-1 and WG2-2, which are, respectively, standardizing data formats for monochromatic images used mainly in radiology and color images in endoscopy and pathology, etc. All of the committee members from industry belong to JPACS and now 60 companies are collaborating the standardization works.

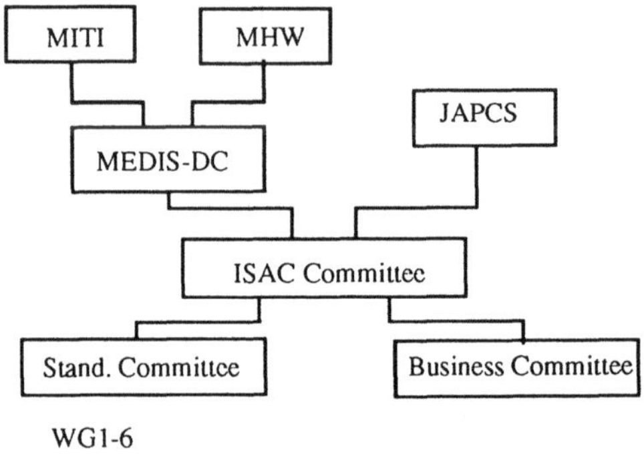

Fig.2 The organization of ISAC committee

ISAC committee has issued drafts of ISAC standards written by WG1, WG2 and WG5 in English so far. Because the ISAC format of volume and file structure is quite original, sample software, written in C language, were made and installed into several kinds of computer system such as Macintosh, Next, Sun work station etc. Several manufacturers have modified the sample software so as to optimize it in their own products while keeping compatibility. The ISAC data format for monochromatic images is subject to the ACR/NEMA standards vs. 2.0, but some modifications were made in order to make the formats applicable to a filing system as well as a communication system. The drafts written by WG5 show the specifications of both MOD medium and drive, which are necessary for the compatibility.

4. An application of ISAC system

In April 1991 MITI and MHW jointly organized the ISAC project as an practical application of ISAC system to medicine. This project will be funded until March 1994 by the ministries and is searching the effectiveness of ISAC system in medicine; PACS is mainly said to be useful for data management in hospital, but its effectiveness in terms of diagnosis and possibility to provide new information to doctors have been hardly discussed. Therefore, in this project ISAC system is not designed for the purpose of rationalization of hospitals, but as an experimental system to search merits with ISAC in practice. A field test using ISAC system is under preparation in National Cancer Center Hospital in Tokyo, and all patients' records will be stored in their personal MODs [5,6], which enable the doctors to examine the records very easily. The organization of the research group for this project is shown in fig.3. While the ISAC committee is in charge of standardization works, the research group is performing the ISAC project and 22 companies are collaborating. Departments of radiology, endoscopy and pathology in National Cancer Center Hospitals are participating in the project. The research subjects are, for example, ISAC/HIS coupling, storage of patient records in MODs and data interchange between two hospitals of National Cancer Center: one located in Tokyo and the other in Chiba prefecture. Each company has almost developed an ISAC apparatus such as CT and viewing station based on ISAC format. Because this project is the first trial of ISAC system operation in medicine, all stations support the ISAC security mechanism.

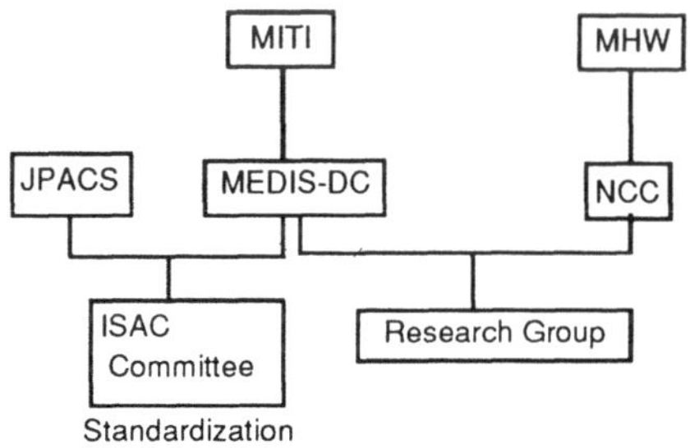

Fig.3 ISAC Project

5. Practical use of ISAC system in medicine

Before a practical use of ISAC system in medicine, digital images should be thoroughly reviewed in terms of image quality, prevention of falsification, security of privacy and life time in comparison with film images. These four items have been reviewed by a research group directly supported by MHW, and will be briefly explained here.

5-1. Image quality
One of the requirements for digital images is that the image quality never becomes worse than that of film images. Review tells that images originally digital such as CT and MRI never lose any quality and that images originally taken by film should be digitized with high resolutions. Because films are routinely examined by naked eyes, the digitization of film images may be said to keep good quality if films are digitized by 100 mm pitch with 12 bits, which is the resolution limit given by the definition of the eye power of 0.9.

5-2. Prevention of falsification
Because medical images are well secured and falsification should be prevented, ISAC system shall have a file protection mechanism for security. The mechanism developed for ISAC system will be explained in the next section.

5-3. Privacy problem
The question concerning privacy is whether privacy of a patient may be in danger because of digital data. The answer made by the review is that all MODs storing medical images should be used under doctors' responsibility just like film images so that the privacy is well secured as it is.

5-4. Life time
Because film images are stored for more than 5 years, the life time of digital images becomes of our concern. MOD's life time is estimated to be more than 20 years through the accelerated experiments, so digital data can be safely stored in MODs. Furthermore, we can have backups or store data in other MODs, because digital data are easily copied without degradation.

From these considerations it was concluded that usage of ISAC system for image storage would be ready to be authorized by MHW as soon as the security mechanism for ISAC system would be appropriately developed.

6. Security and file protection for ISAC system

6-1. File status
As is clear in the precious section, ISAC system needs to have a file protection mechanism for the security of medical data stored in MODs. ISAC is, of course, a computer system , so we can protect files by software with a definition of a specific attribute denoting a permanent file, for example. The file protection mechanism ,

explained here, should have higher security than that of conventional operating system. Therefore, ISAC system must have an additional mechanism to protect data beyond users' control. This file protection mechanism supports four kinds of file status as shown in table 1. Because medical images to be protected are those which are used by doctors for diagnostic purpose, their status will be changed as shown in fig.4; images generated by medical apparatus are first assigned to the original status, which allows read and erase operations. As soon as the original images are used in diagnosis, their status is changed into the authorized original, which is declared by a doctor. Once authorized, images can be only read and never changed. ISAC system also supports two more status; one is the authorized copy, indicating that the image is a copied image of the authorized original. The other is the not-specified status, which is given to general files.

File status	Access control
Original	Read/Erase
Auth. Org.	Read
Auth. Copy	Read
Not Specified	Read/Erase/Write

Table 1. File Status

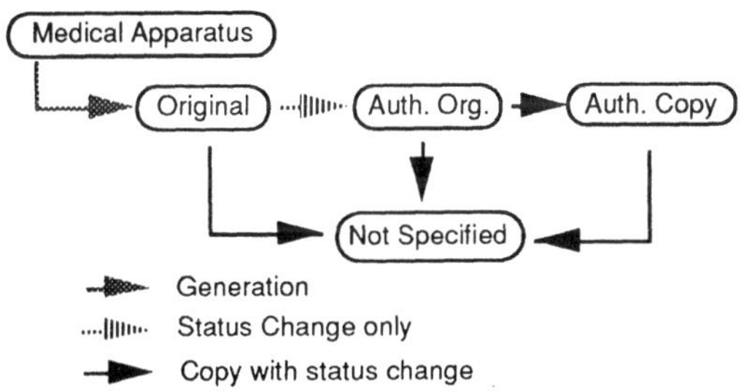

Fig4. File status in ISAC system.

The four kinds of status are managed within ISAC system, and should be safely protected from any access made by other systems. For this purpose ISAC system with security uses the file manager, MOD drive and MOD (see fig.5) which are all different from those used in ISAC system for general purpose.

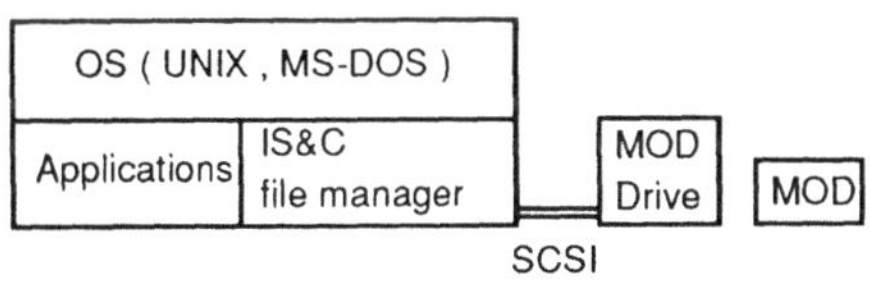

Fig5. Example of IS&C station

6-2. Compatibility of drive and media

Because MOD used in ISAC system has been standardized by ISO-SC23 and may have compatibility, any protection mechanism managed by software will be easily violated, unless MODs with security lose compatibility to the other MODs without security. Consequently, MODs with security are differently formatted so that general MOD drives cannot make any access to them. As a result, MOD drives and media have or lose compatibility summarized in table 2.

7. Summary

This paper introduces the standardization activities of ISAC system and the ISAC project, that is being carried out in National Cancer Center Hospitals. Because data filing systems for medical use require file protection or access control, the security mechanism has been developed for the domestic use of ISAC system in medicine. For an international data exchange, data should be copied from Japanese medical ISAC to general purpose ISAC, and copied files are out of Japanese medical security.

		Media		
		Medical IS&C	IS&C	ISO
DRIVE	Medical IS&C	◎	×	×
	IS&C	×	◎	○
	Others	×	○	△

◎: Compatibly is confirmed ○: Maybe usable
△: Risky ×: No compatibility

Table2. Compatibility of drive and media

4 years have passed since ISAC system was first proposed, and ISAC becomes ready to be authorized by the Japanese government. Authorization of the use of ISAC in medicine will provide ISAC system a good chance of wide use in Japan.

References

1) S.J.Dwyer, A.W.Templeton, N.L.Martin et al., "The Cost of Managing Digital Diagnostic Images" Radiology vol.144 No.2 (1982) 313-318

2) N.Ohyama "Transportable Image Recording Media - A Proposal of ISAC system" Proceeding of IMAC'89 (1989) 250-255

3) N.Ohyama "ISAC (Image Save And Carry) strandardization" Proceeding of IMAC'91 (1991) 60-65

4) There are many papers. For example, M.Ojima, A.Saito, T.Kaku, M.Ito, Y.Tsunoda, S.Takayama and Y.Sugita "Compact magnetooptical disk for coded data storage" Appl. Opt., vol.25, No.4, (1986) 483-489

5) K.Kaneko, S.Ikeda and N.Ohyama, "Research on Personal Health Data Recording System" The proceeding of IMAC '89 (1989) 72-76

6) N.Ohyama, M.Kaneko, S.Ikeda and T.Maeda, "A proposal of an off-line PACS based on personal database" Medical Informatics, vol.16 No.1 (1991) 7-13

ImNet/2: The High-Speed Network

N. Söderman 1), F. Vosseburger 1), A. Keizers 2)

1) IMTEC Imaging Network AB
Glunten
S-75183 Uppsala; Sweden
Phone: 46 18 55 58 00

2) Lehrstuhl für Messtechnik
Aachen University of Technology
D- 5100 Aachen; Germany
Phone: 49 241 807860

Abstract

ImNet/2 is a fiber-optic local area network which was developed for high speed image communication in Picture Archiving and Communication Systems (PACS). ImNet/2 is optimized for the transmission of large data files. The topology of the network is multi-star i. e. an arbitrary tree structure of stars. This topology allows multiple transmissions to proceed simultaneously. Therefore the overall throughput of ImNet/2 can be a multiple of the data rate per link.

ImNet/2 is constructed in accordance with the ISO/OSI reference model. For the physical layer the MIC connector is used, which is also standardized for FDDI. Up to layer 3, all protocols are handled by hardware or firmware. At layer 4 a special high speed protocol makes the full network speed available on top of the network-specific OSI layers. Layer 4 allows for several different network-independent protocols to be implemented e.g. DICOM, ACR-NEMA or MEDICOM. Interfaces are available for VMEbus (UNIX, OS/9), ISAbus (UNIX, OS/2, MS-DOS) and SCSI II (under development).

ImNet/2 is commercially available and is currently being installed at several hospitals in Scandinavia. At the Klinikum in Aachen, Germany, the construction of a prototyp network is supported by the AIM project EURIPACS.

1. PACS Calls for Innovative Image Networks

Picture Archiving and Communication Systems for Medical Application (PACS) are being developed to replace diagnostic films with digital media. The backbone of PACS will be the communication system which will link the various image sources with digital image archives and with "digital lightboxes". One step towards a simplification of the complicated communication tasks in a PACS is to seperate the different data flows into pure image data flows and management data flows [10].

Information systems in the hospital (HIS) or in the radiological department (RIS) and image management systems (IMS) communicate small sets of alphanumerical data with various imaging modalities and one or more image workstations. General purpose LANs are used for this tasks. A strictly separated network communicates only images. Networks which are designed for this purpose can be optimally adapted to the special communication requirements of image equipment and thus be much more efficient than general purpose LANs.

2. ImNet is designed for this purpose.
ImNet is such a specialized network [6], [7], [8]. It is based on fiber optic data transmission technologies and operates at a 175 MHz carrier frequency. Its data transport mechanism permits asynchronous data transfers between Image Equipment (IE) without packeting data sets into small data frames. The distributed routing and control mechanism supports a simple and efficient protocol on the network dependent ISO/OSI layers. This results in effective data rates of 140 Mb/s which can be simultaneously obtained on all independent network routes.

ImNet consists of two elements : *Switch Modules* which autonomously establish individual message data paths between any two arbitrary nodes (IE units connected to the network) for the duration of a message transfer, and *Link Modules* which connect IE units to switch modules or switch modules among each other. Figure 1 shows the schematical drawing of the Link Module between an IE unit and a Switch Module.

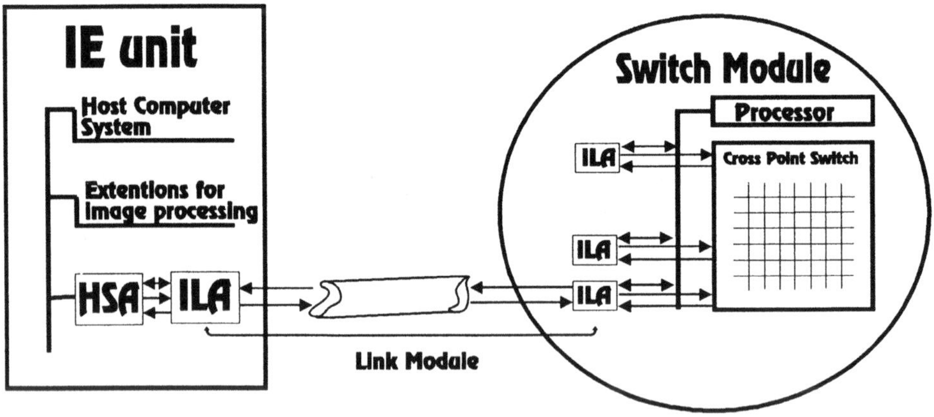

Figure 1. The ImNet Link Module consists of 2 ILAs, connected with an optical fiber. The Link Module shown connects an IE unit to one of the 14 ImNet-Link-Adapter-ports (ILA) of a Switch Module. A Host-System-Adapter (HSA) connects the bus system of the IE unit to the command port and data ports of the ILA.

The Link Module can be up to 4 km by using fiber optic cable. For short istances (< 100 m) a less expensive module using shielded twisted pair cable can be used. Because ImNet uses two cables for each point-to-point link, the link module works bi-directionaly and handles all protocols up to OSI-Level 3. The data rate per link is up to 140 MBit/s (which is equivalent to a clock rate 175 MHz). The Switch Module consists of the control unit and the cross-point-switch array. The array has up to fourteen interfaces for Link Modules and is able to handle up to fourteen data transfers with a maximal transfer rate of 400 MBit/s simultaneously. Therefore the maximal throughput of a Switch Module is 5.6 GBit/s. Out of these modules a multi-star network can be build i. e. an arbitrary tree structure of stars. This topology makes it possible to perform multiple transmission at the same time as long as they do not require identical links. Therefore the overall throughput of ImNet can be a multiple of the data rate per link.

Two types of information are transferred via the network: *messages* (data blocks of arbitrary length) which carry images and related data and *control commands* (32-bit data words) which initialize or terminate the message transmission. The network operates in different modes for those two types of data. Control commands are processed by the Switch Modules. Depending on their control function they are either broadcasted to other Switch Modules or forwarded to the addressed destination node. For message data blocks, in contrast, the network applies the line switching principle in order to connect two IE units as closely as possible during the message transfer.

ImNet offers a connection oriented service to the transport protocol. To make ImNet's speed available for the network independent protocol layer a special streamlined protocol on the transport layer is implemented. Before transferring an image, the transmitter and the receiver determine the maximal usable blocksize e.g. 16 MBytes. At ISO/OSI layers 5 to 7 different network independent protocols are implemented e.g. ACR/NEMA, DICOM or MEDICOM.

3. ImNet's interfaces to IE units.

To interface an IE unit,a standard ImNet-Link-Adapter unit (ILA) and a specific Host-System-Adapter (HSA) are needed. The ILA handles the low-level protocols (up to ISO/OSI layer 3) by means of on-board hardware and firmware. It offers the choice of an 8-, 16- or 32-bit parallel interface for message data and a separate 16-bit parallel interface for control commands. In order to make effective use of the data transfer rate of the network, the ILA message data port should be coupled to the bus of the IE unit via a DMA port implemented on the HSA. The HSA should also contain a dedicated microprocessor to execute the network dependent protocols (up to ISO/OSI layer 4) in order to prevent frequent transactions of the host CPU. Then the host CPU has to process nothing but the higher level protocol (ISO/OSI layers 5 to 7). Such a task distribution makes it possible even for small-scale computer systems to fully utilize the network capacity.

An HSA has been developed for a VME/VSbus host computer. Message data are transferred via the VSBus while the VMEbus port is only used for control data . The HSA works in both a Master and a Slave mode on the VMEbus. Transfer tasks to and from the network are operated by a local microprocessor. The microprocessor also controls the HSA hardware, sets up network connections and the DMA for the specific transfer tasks. The optional VSbus, which is a standardized subbus to the VMEbus, can be used to keep the VMEbus free for host activities during network transfers. This HSA is optimized for application in image workstations. The host processor is then free to operate the graphical user interface and to control image processing hardware, while the HSA simultaneously transfers images to and from the network. It also works very efficiently in image file server systems, where the host processor transfers the data between the disks and RAM, while the HSA operates the RAM to network I/O.

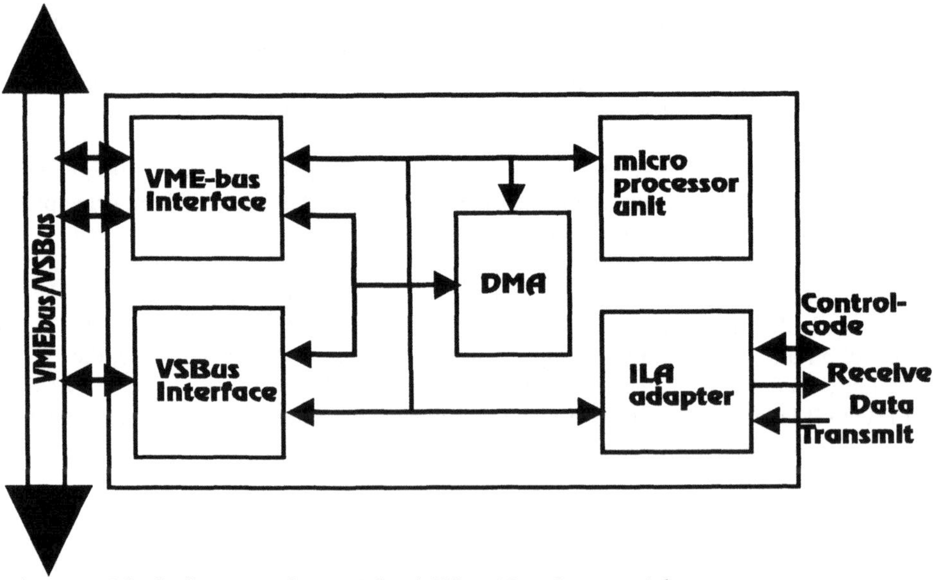

Figure 2. Block diagram of a VMEbus/VSbus Host System Adapter

There are two additional HSAs available from IMTEC. One is a low cost *ISAbus interface* with up to 16 MByte buffer memory on board for ATbus based PCs. The buffer memory is necessary because the ISAbus is not able to handle ImNet's data rate. There is no dedicated microprocessor on board. This interface together with the AT is used as a gateway to other communication environments like ISDN (64 kb/s- 2Mbit/s) or Ethernet and to very specialized interfaces like DEC's DR11-W.

The other HSA is ImNet's interface to *SCSI 2.* The SCSI bus is well known as an interface to large storage capacity devices like hard discs, optical discs or tape streamers. The new standard SCSI 2 is able to handle data rates up to 20 MByte/s and defines additional commands used not only to interface mass storage devices but also clear-cut communication devices. This interface is able to work as a target device at a host computer as a communication device or as an initiator with archiving storage capabilities.

4. Current State and installations.

Network components of ImNet are industrially available [9]. Interface hardware and driver software exist for VMEbus computers with OS/9 and UNIX and ATbus PCs with MSDOS and OS/2. IMTEC is working on several installations at hospitals in Sweden and Denmark.

An ImNet installation at the Aachen University Hospital in Germany is in progress which will replace the preceding network step by step. Additionally ImNet is intended for the digital reporting and archiving of chest X-rays and for image communication between the Radiology Department and the Anesthesiology Department under routine conditions.

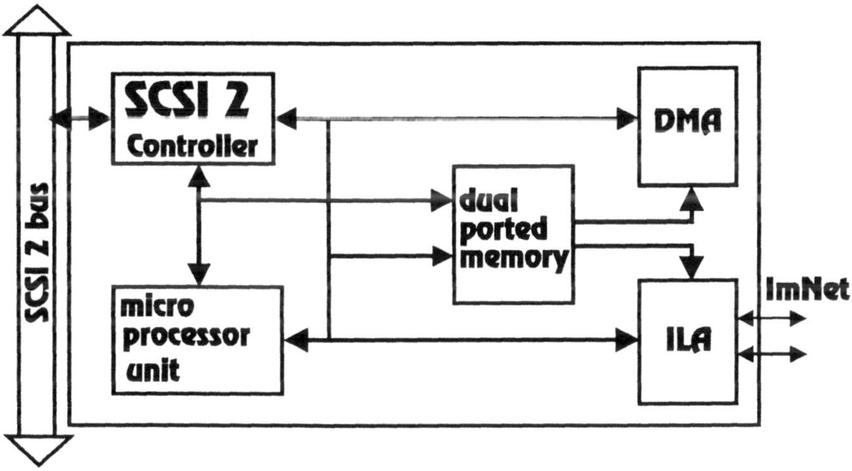

Figure 3. ImNet's SCSI 2 interface

Another installation is in one of the five towns (Lund, Ystad, Trelleborg, Helsingborg, Landskrona) in southern Sweden : Ystad. The installation is not one of the more complex among the five, but will give an indication of the possible complexity of the included modalities. The following types and manufacturers are included:
Workstation IMAGE 1200 from IMTEC. Resolution 1280 x 1024 pixels.
Teleradiology interface for ISDN connections to other hospitals in Sweden.
Printerserver for Agfa Laser Hardcopy. 600 MByte harddisk etc.
ImNet switch unit equipped with 5 of the 14 possible connector boards (ILA).
Interface for Somatom CR from Siemens AG.
Interface for DSA from Philips.
Interface for Ultrasound SAL-100 from Toshiba.

Conclusion.
ImNet is designed for high speed network connections between different manufacturer's modalities giving the user ease of us

Acknowledgements.
We wish to thank Prof. Rolf Günther, who is in charge of the Radiology Department of the Aachen University Hospital, for supporting the installation of the Aachen pilot network.

The work is financially supported in part by the European Commission under grant A2009 (AIM project EurIPACS) and by the Swedish National Board for Technical Development, NUTEK.

References.

[1] Meyer-Ebrecht D, Keizers A, Stut WJJ, Vossebürger F, Image Networking in Technology, Chapter 5.1 in: Osteaux M (ed.) HIPACS, Springer 1991

[2] Dahm M et al, Radiologists start designing their digital workplaces, in: Lemke HU et al, Computer Assisted Radiology Proc CAR'91, Springer, Berlin 1991

[3] Newsfront, OSI throughput performance: Breaktrough or Bottleneck?, Data Communications, May 87

[4] Wittgen HG et al, PACS: ein Pilotsystem im Routinebetrieb, electromedica 4/90 pp 124

[5] Chotas HG, Ravin CE, A computed radiography imaging network, medicamundi 34/3 pp 114

[6] Fasel B, Vossebürger F, Meyer-Ebrecht D, Image Network for PACS, in: Lemke HU et al, Computer Assisted Radiology, Proc. CAe and no frustrating wait for image transfers. New interfaces for old and new modalities are under continuos developement.R'87, Springer, Berlin 1987: pp 558

[7] Fasel B, ImageNet - Ein neues Datennetz für Bildinformationssysteme in der Medizin, Fortschritt-Berichte VDI Reihe 10 Nr. 123 VDI-Verlag, Düsseldorf 1990

[8] Vossebürger F, ImageNet/2: ein lokales Breitbandnetz für die digitale Bildkommunikation; Thesis RWTH Aachen 1991

[9] ImNet Specification, IMTEC Imaging Network AB, Uppsala, Sweden

[10] Meyer-Ebrecht D, Wendler T, An Architectural Route through PACS, IEEE Computer Vol. 16, No. 8, Aug. 1983: pp 19

[11] ACR-NEMA Standards Publication No. 300-1985, Digital Imaging and Communication, Washington 1985

[12] Stut WJJ, van Steen MR, Groenewegen LPJ, Bakker AR, Picture Archiving and Communication System Design Issues: The Importance of Modelling and Simulation, J. Digital Imaging, Vol. 3, No. 4 (November), 1990: pp 246-263

[13] Keizers A, Meyer-Ebrecht D, Vossebürger F, A Fiber-Optic Line-switching Network with 140 Mb/s User Data Rate, IEEE JSAC Vol 10, No. 7, Sept. '92: pp 1197

Maximizing End-to-End Throughput in Image Communication Systems

W. J. Dallas, R. Vercillo, K. M. McNeill, W. P. Klein, K. Maloney, M. Parra

Department of Radiology, The University of Arizona Health Sciences Center Tucson, Arizona 85724, USA

Summary

We describe two systems designed and constructed in the Radiology Department of the University of Arizona.

The first system is a high-speed experimental network that is optimized for transmitting radiologic images. The XFT (eXperimental Fiberoptic Token ring) is a fiberoptic-based network that provides a raw band width of 500 Mbits/sec. Various design considerations have allowed us to project a buffer-to-buffer transfer speed of over 300 Mbits/second. We are presently testing transfers in loop-back mode. Using simplified protocol processing we have attained a transfer rate of 400 Mb/s. The XFT project includes the protocol specification and implementation of both the network interface hardware and software.

The second project is a 64 MByte memory board of novel design which we have installed as an upgrade to the Arizona Viewing Console (AVC). This memory board borrows several techniques generally applied to Central Processing Unit (CPU) design to speed operation. For instance, the triple-ported image memory is supplied with caching for each access port. Write-through data cache and preemptive data pre-fetching are also realized.

Introduction

Maximizing end-to-end throughput is a complicated task that requires the individual components to preform well and to be designed to fit together. The research network performance motivated us to look at the factors that contribute to the efficiency of network systems and the impact of the operating environment on that efficiency. After careful consideration of questions, we decided that the only reliable way to answer them was to construct a network system that we hypothesized to be efficient for managing medical images. This network is the XFT (eXperimental Fiberoptic Token ring) network .

The end-to-end throughput in a network system is dependent not only on the network but also on the nodes. As a particular example of a node we have chosen a display station that we have constructed in the University of Arizona Radiology Department. We have chosen this device because it illustrates both efficiency in coupling to the network, and an internal efficiency that comes from incorporating advanced memory management techniques. The name we have given this display station is the AVC (Arizona Viewing Console).

The XFT

The XFT electronics is composed of three boards: transceiver, memory, and interface-logic. The transceiver board is purchased, the memory and interface-logic boards we designed using our workstation-based Cadence Computer Aided Design, CAD, software. The only board that is connected to the host computer system of the network node is the memory board. This design decision was made in order to allow easy movement from one bus to another. The present memory board includes an interface to a VMEbus. We are constructing three XFT station interfaces.

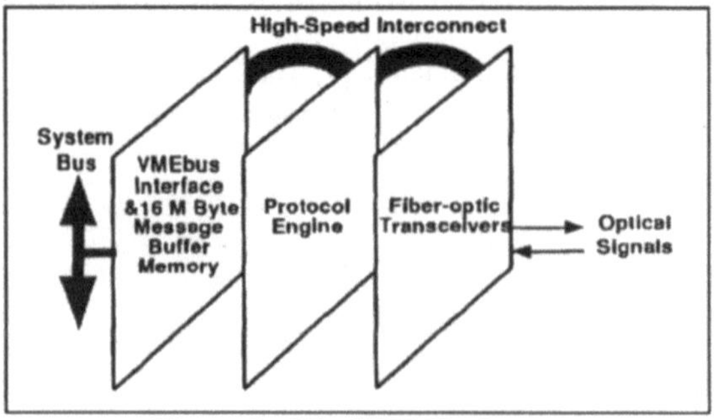

Figure 1: The XFT Hardware

The transceiver board

In designing the XFT we made use of fast serial link hardware manufactured by the firm IPItek. The IPItek links gave us the 500 Mb/s electro-optic modulators together with parallel-to-serial and serial-to-parallel modules that relieved us of designing electronics that required exotic technologies. We purchased the IPItek components already mounded on their transceiver boards. We substituted super-stable clocks, 5 part per million, for those supplied by IPItek.

The memory board

The memory board has a 9U format. The capacity is sixteen megabytes .The design incorporates several features that work to maintain the board's high throughput. One of the novel features is a page allocation table that allows fragmented memory to appear contiguous to the application from both the high-speed port and from the VMEbus..

This memory board with its page allocation table and fifo's is able to dynamically reallocate memory for both send and receive so that the same board can be used with a source node or a sink node . Communication with the interface-logic board is

Figure 2: The Memory Board

through a mailbox for interrogating status of the network, active stations, and other network parameters. In contrast, image data is transported directly over the high-speed link. The memory board has first-in first-out, fifo, queues that provide small elastic buffers. The memory board supports eighteen different queues for various functions plus addressing. The queues for both transmit and receive operate independently of the message memory but concurrently with it. There is a great deal of parallelism built in. Thanks to this care in design the high-speed port is able to handle 50 MByte/sec (400Mbit/sec) in sustained data transfers.

The interface-logic board

The interface-logic board consists of several state machines that implement the handling of the parallelized data entering and exiting the transceiver boards.. That state machines that comprise the intelligence of this board make a decisions every 40 ns. This handling includes hardware protocol processing according to the ISO/OSI stack model up through the middle of the data-link layer. The entire token ring management protocol is realized in hardware, all token ring responses to events on the ring are managed by the hardware. The actions of:a station joining the ring (which is traumatic), data handling , and error recovery are all handled by the hardware at the full system speed, i.e., in 40 ns.

The protocol

The token ring architecture of the XFT provides fairness in access to the communications channel. Each station is guaranteed to have access at a fixed maximum interval even when all the stations on the ring have traffic. In order to make the transfer of image data efficient, the XFT protocol allows for a maximum transmission of 1 MBytes each time a station acquires the token. This transmission may be a single 1 MB packet (efficient for image data) or multiple smaller packets (efficient for non-image traffic). When the ring is fully utilized, time between token arrivals at a specific station depends upon the number of stations on the ring. In order to make this delay acceptable in digital radiology, the protocol currently specifies a maximum of thirty-one stations on a single ring. This will allow a 2Kx2Kx16bit image to be transferred between any two stations within about 5 seconds when the ring is fully loaded.

The XFT protocol specifies several components which comprise the sub-network layers for a local area network. The protocol components are illustrated in Figure 5. The operation of the components and the interface primitives which provide communication between them are all detailed in the protocol specification.

The software

The following is a very brief functional overview of the lower layer XFT software. The material is intended to give only a flavor of the considerations involved in designing the software.

- XFT Interface Device Driver/Memory Manager: Provides 1) communications between software components of the UA_NSA executing on the host system and the physical hardware components of the network interface, and 2) management of the message buffer memory associated with the network interface, including allocation/deallocation of message buffers on behalf of higher layer processes. Function 1 is protocol specific and hardware specific while function 2 is protocol independent but hardware specific.

- Link Control Sub-layer (LCS) Process: This software implements the upper sub-layer of the XFT Data Link layer. (See Specification Document UA XFT.2.1990 for detailed description).

- Server Process: This process is specific to the OS/9 implementation environment. It handles asynchronous signals from the XFT interface device driver which indicate the availability of message(s) to be passed to the LCS process. This process isolates some of the operating system dependent aspects of receiving messages from the network interface.

- Station Management Function at Layers 1 & 2: This software implements XFT specific portions of the Station Management Function. These functions include loading and checking of parameters associated with the network protocol functions implemented in hardware, and error handling and reporting.

Tx. Min Optical Power:	**-3.5 dBm**
Rx. Min. Optical Power:	**-18 dBm**
Max. Tx. distance (Multimode fiber):	**2 Km**
Max. Tx. distance (Singlemode fiber):	**25 Km**
Bit Error Rate (Min. Optical Power):	**10^{-12}**
Fiber Baud Rate:	**500 MHz**
Byte Period:	**20 ns**
Fiber Symbol Size:	**10 Bits**
Fiber Framing Overhead:	**1 Bit**
XFT Symbol Overhead:	**1 Bit**
Tx. Time (2048 Symbols):	**40.96 µsec**
Tx Time (1 MByte Packet):	**20.97 msec**
Token Rotation Time (Loaded Ring):	**650 msec**
Tx Time (8 MB Image - Loaded Ring):	**5.2 sec.**

Figure 3: Performance Summary

The AVC

Overview of the AVC

We will give only a very brief description of the general system and move rapidly to concentrate on the new base memory board. The AVC hardware from the user's point of view consist of a color command monitor, a desk with a keyboard, trackball and six shaft encoder knobs and two 19 inch monochrome high resolution image display monitors each capable of displaying 1024 by 1536 pixel images. Each display monitors has been fitted with a scanned infrared touch screen which allows the user to manipulate the images just by touching them. The color command monitor is used to display functional menus for image manipulation and to access the image files stored on the local hard disk. The color monitor is also fitted with a pressure sensitive touch screen which enables the user to quickly invoke the commands on the menu currently being displayed.

Image manipulation features are pan, move, place, horizontal flip, vertical flip, and delete. The image processing functions are window, level, high and low pass filtering, image zoom and minify. Images can be activated by touching them on the display monitors. Once an image has been activated, any image manipulation or image processing only affects the active image even though other images may be displayed on the same display monitor. This allows each image to be independently controlled without disturbing any other image being displayed.

The Review Console architecture is shown in Figure #3. Each of the major units, the disk controller, base image memory, image processor, display memories and video controllers contain many special features. However, we will concentrate our attention on the new base image memory.

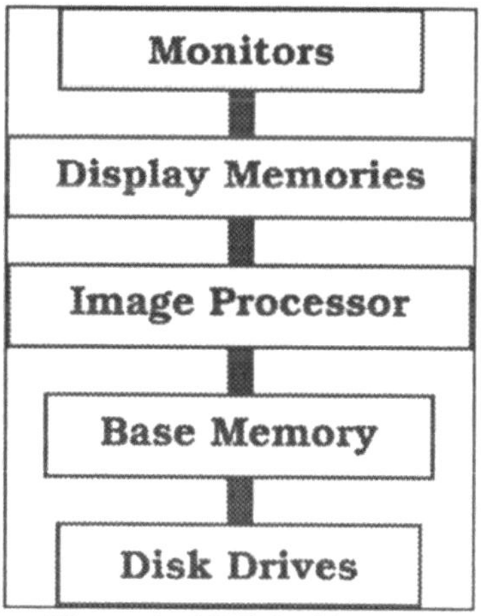

Figure 4 AVC Architecture

The New Base Memory

Results of experiments with the AVC have shown a need to increase the original 8 megabyte image memory's in order to accommodate high resolution images. The new 64 megabyte base memory stores the raw image data for images up to a 5000 pixel by 4000 line image with sixteen bits of storage reserved for each pixel. The image data and graphics bits are combined in the sixteen-bit pixel field. The size of the image data field and the number of graphics bits will vary according to the type of imaging modality. In addition, the image data will be signed and unsigned variable length data. Several techniques used to speed up CPU based systems are being applied to a VMEbus based peripheral memory board. In the new design a cache memory for each access port has been inserted between the DRAM array and the access port's interface. If the data accesses are sequential through memory, a significant improvement in the memory's response time can be achieved. With additional intelligence in each port's cache controller, write-through data caching and preemptive data pre-fetching protocols can also be realized. These techniques, coupled with the increased memory storage reduce the image memory's latency time to provide a faster overall system response to the radiologist.

Due to the fact that medical images can be extremely large, and require equally large memory arrays, these memory arrays are designed using DRAMs. Unfortunately, DRAMS are not as fast as their SRAM counterparts. Compound this problem by placing a large DRAM array to be shared in a multi-ported environment and the performance of your system will be degraded.

By using IDT's 7050 four port memory as a high speed cache with each it's port dedicated to the three access ports in the system, namely the VMEbus, VSBbus and the image processor, the 64 megabyte DRAM array starts to look like 64 megabytes of fast static RAM.The address space is divided in half, separate read and write caches are obtained. The read cache is further divided into two logically separate caches to support data prefetch thereby allowing for sustained data accesses of the DRAM array at near SRAM speeds.

Three access ports, each with a unique set of performance requirements, are required to interface the network interface module, image processor and the VMEbus to the base image memory. These three access ports make the base image memory the focal point for all the image data being used by the AVC. The first access port is interfaced to the VMEbus and allows the VMEbus to directly access the entire memory array. In keeping with the high speed design goal the data bus on the base image memory is a full 32 bits wide as specified by the VMEbus specification.The second access port is dedicated to a VSBbus interface. This port can support data transfers at 100 megabits per second. The third access port is a read-only port dedicated to the image processor, it provides the image and graphics data at a sustained rate of 20 megapixels per second. Through address mapping, the base image memory's organization appears differently to each of the three independent access ports. Both the VSBbus port and the image processor port accesses to the base image memory are serviced with page mode memory operations supported by the DRAMs. The VMEbus accesses the base image memory in a more conventional random access mode.

One of the more difficult design aspects of a multiported memory is the implementation of the arbitration controller to service simultaneous memory requests. We have implemented a controller that is a hybrid of time-division multiplexing and prioritizing strategies.

A memory allocation table (MAT) is used to re-map the addresses to the base memory array. The MAT is essentially a look-up-table that is configured under software control and is transparent to all the access ports, namely the VMEbus, VMSbus and the image processor port.

A dynamic port sizing feature is included. Under software control, the addressing range for individual access ports can be limited from a small work space up to the entire base memory array. This feature protects reserved areas in the base memory from being corrupted by data being written from another port and provide security by not allowing data to be read outside the port's addressing range.

Summary

Maintaining end-to-end throughput in networked medical imaging systems requires a novel approaches in system design and integration. The design The XFT network and AVC display console are illustrative of the design considerations.

Acknowledgements

This work was supported in part by National Institutes of Health grant R01CA49261 and in part by the Toshiba Corporation.

Data Flow Analysis for Transition from Film to Electronic Management

M.W. Hedgcock Jr., M.D., T.S. Levitt, Ph.D., V.M. Shadle, M.S.

Department of Veterans Affairs Medical Center/Radiology (114B)
4150 Clement Street, San Francisco, Ca. 94121-1598
U.S.A.
(415) 750-6901 FAX: (415) 750-6925

Summary

The transition from analog film-based imagery management to electronic imagery management in radiology departments and clinics requires good planning. Most importantly, it necessitates making accurate projections of storage and network needs and capacities prior to equipment selection. We offer below some simple mechanical calculation tools which can be used to estimate needs for electronic image handling when starting from film-based operations.

Introduction

Most readers are familiar with the units used for computer storage, but for review purposes, 1 byte = 1 "computer word," usually 8 bits for text data. For image data, the number of bits in a byte varies depending on the "depth" of the image. Most formatted data is 12 bits deep. Other terms include:

1 kilobyte (Kbyte) = 1,000 bytes;
1 megabyte (MByte) = 1,000 KBytes;
1 gigabyte (GByte) = 1,000 MBytes;
1 terabyte (TByte) = 1,000 GBytes.

Calculating the Size of the Total Archive

In a medical facility, the central medical image archive usually consists of a remote file area for inactive patient records and a smaller file area for active patient image files.

In most medical settings, an image archive which allows rapid access to the last two years of medical imagery is adequate. We have shown previously that one

14" X 17" equals 6 MBytes of electronic storage, assuming that one uses computed radiography or film digitization devices which acquire imagery with an approximate 2000 X 2000-byte matrix[1]. Using this conversion figure, there are two methods by which one can rapidly calculate the on-line electronic storage equivalent for a film-based operation.

One method involves estimating the number of stored radiographs by multiplying the linear feet of storage by 1000. This provides an average figure which assumes an even mix of patients incurring many medical images and patients incurring only a few. The resulting range of films per linear foot goes from 800 (when patients incur only a few images) to 1,200 (if most patients have a lot of images). This range occurs because film jackets occupy more space than a film. In the file rooms of institutions where patients incur only a few films each, more space is occupied by jackets. The conversion unit to electronic storage is therefore 6,000 MBytes/linear foot, or 6 Gigabytes/linear foot. Assuming the intermediate value of 1,000 films per linear foot, the equation to convert film storage to electronic storage is as follows:

Electronic Storage = Linear Feet of Film Storage X 6 GByte/linear foot

In our own institution, for example, the active film-based archive is 1,080 linear feet, or the equivalent of 6.480 GBytes (6.48 TBytes) electronic uncompressed storage.

The remote, or inactive, archive is 4,867 linear feet, which equals 29.2 TBytes electronic uncompressed storage.

Estimating Additions to the Archive

To estimate additions to the total archive per year, the total number of sheets of film used per year is multiplied by 6 MBytes to get the total electronic storage added per year. In our case, we calculate:

130,000 sheets film X 6 MByte/sheet = 780,000 MBytes (780 GBytes).

Predicting Data Flows

To predict data flows in the network, one must determine patterns of medical imagery usage. In a film-based system, all of the patient's imagery (the film jacket) is usually transported physically to where the imagery is needed. All of this imagery is not used, however.

Example:

Mr. Green is seen in the clinic. His film jacket sent to the clinic contains:

98 films	chest PA and lateral
25 films	portable chest
18 films	chest CT scans (3 scans, 6 films each)
4 films	lumbar spine (2 PA's and laterals)
8 films	cervical spine (24-view series)
16 films	ankle (from a previous fracture)
169 films	Total films transported

Films used

8 films	chest PA and lateral (4 sets)
12 films	chest CT scans (2)
20 films	Actual films used

Clearly, the actual images used represent only a fraction of the films transported in a film-based system.

Developing an Algorithm for Images to be Sent

In a previous survey of three large medical centers, physicians were satisfied if the imagery transported included the most recent PA and lateral chest radiograph and the three most recent comparisons, as well as the most recent exams of other types and the most recent comparison of each type. Using this algorithm with a week's worth of imagery requests, the average number of images needed is 12 per patient, or 72 MBytes of uncompressed electronic data equivalent.

Distribution Differences: Film-based vs. Electronic

With film-based distribution, blocks of films are moved to reading or treatment areas according to patient scheduling. At our institution, for example, approximately 200 jackets are moved to the clinics in the morning and returned to Radiology Service in the afternoon. Over a period of two weeks, we counted the number of films that this represented and found that, on average, 8400 films were being moved. In electronic equivalent, this represents 50.4 Gigabits of data. We also counted the films that were actually used and discovered an average per day of only 2400 films, or 14.4 Gigabytes of electronic equivalent. The rate of use equalled 1 to 3 Gigabytes per hour.

Conclusion

A working estimate of electronic archive sizes needed for active and inactive imagery archives can be computed easily by assuming that each film used, stored, or transported represents 6 MBytes of electronic equivalent storage. Since it is impractical and unnecessary to send a patient's entire imaging file electronically to each location, it is easy to develop algorithms which anticipate imagery most likely to be needed at a given medical center location.

References

1. Hedgcock, M. W.; Levitt, T.S.; Smith, S. Predicting Optical Storage Needs for a Medical Image Archive, SPIE Proceedings,1093 (1989) 551-557.

2. Hedgcock, M.W., Clinical Requirements for an Electronic Image Archive, Electronic Imaging West 90 Advance Printing of Paper Summaries, Vol. 2 (1990) 377-380.

ITACA: An Irreversible Threshold-based Angio-cardiographic Image Compressor and Archiver Based on New Error Limitation Criteria

F. PINCIROLI[a], C. COMBI[a], G. POZZI[a], M. NEGRETTO, L. PORTONI, G. INVERNIZZI[b]

a Dipartimento di Bioingegneria, Politecnico di Milano, Italy
b Divisione di Cardiologia, Ospedali Riuniti di Bergamo, Bergamo, Italy

Summary
We defined, implemented and experimented on angiocardiographic static images three new irreversible compression techniques: Brightness Error Limitation (BEL), Pseudo-gradient Adaptive Brightness Error Limitation (PABEL), Pseudo-gradient Adaptive Brightness and Contrast Error Limitation (PABCEL). We applied these techniques to some 168 static images selected from 35mm angiocardiographic films. Considering threshold values limited to maintain all significant clinical informations, we achieved best compression results applying PABCEL method: we obtained a mean compression ratio of about 8:1.Consulted cardiologists did not find significant diagnostic differences between original images and reconstructed ones. The three methods implemented using a C compiler are now collected in a package named ITACA (Irreversible Threshold-based Angiocardiographic images Compressor and Archiver).

1 Introduction

Recently, diagnosis by images has been attracting great interest in the clinical environment. The procedures for storing images are also interesting for the purpose of substituting digital archives for the present analog archives, from which rapid image-retrieval is often difficult. Large-capacity storage devices, typically based on optical technologies, are needed for this purpose but are unfortunately not yet adequately available. This stimulates reseach for new, more efficient data-compression techniques.

2 Source of the Images and Instrumentation

The angiocardiographic images we used were acquired in the Cath Labs, Cardiology Division, Ospedali Riuniti, Bergamo, and refer to angiocardiographic examinations of patients that had undergone Percutaneous Transluminal Coronary artery Angioplasty (PTCA).
The instrumentation used for angiocardiographic examinations is a Philips X-ray system. The instrumentation used to acquire digital-format images is a 80386-based Kontron Instruments Cardio 500 workstation dedicated to angiocardiographic and echocardiographic image processing: it is also equipped with image-processing specific hardware. Images are converted into a 512*512-pixel matrix with 8-bit per pixel quantization (256 grey levels); images used for compression are acquired by a moviola with cinevideocamera Tagarno. Resolution after digital conversion is 35 μm/pixel minimum. Storing is performed by writing a file in the typical Cardio 500 format (bit-map format with a 128-byte heading at file start). The digital-format images can then be transferred by diskette in MS-DOS format.

3 Compression Algorithms

We defined three new methods for the irreversible compression of angiocardiographic images: Brightness Error Limitation (BEL), Pseudo-gradient Adaptive Brightness Error Limitation (PABEL), Pseudo-gradient Adaptive Brightness and Contrast Error Limitation (PABCEL). We followed the basic principle of limiting, for each pixel, the brightness error incurred by approximating the original image to the reconstructed one. Our compression techniques are made of three subsequent steps : a) image scanning along Peano-Hilbert's path; b) irreversible compression; c) encoding.

3.1 Peano-Hilbert's Scan Path

When applying the irreversible-compression techniques we devised, we wanted to use an image-scanning path without preferential directions and to proceed in such a way that, at least in general, any contiguous pixels in the image are consecutively scanned.

For this purpose we implemented an image-scanning algorithm based on a function known as Peano-Hilbert's curve [1], [2]. This is a recursive algorithm that generates a Peano's curve having an initial and an end point beginning from four different elementary movements. The standard scan path for an image having dimensions 2^k*2^k (k=1,2,3..) is shown in detail in Fig. 1. A characteristic of Peano-Hilbert's path is that it never leaves an area of the image, quadrant by quadrant, before having scanned all its pixels.

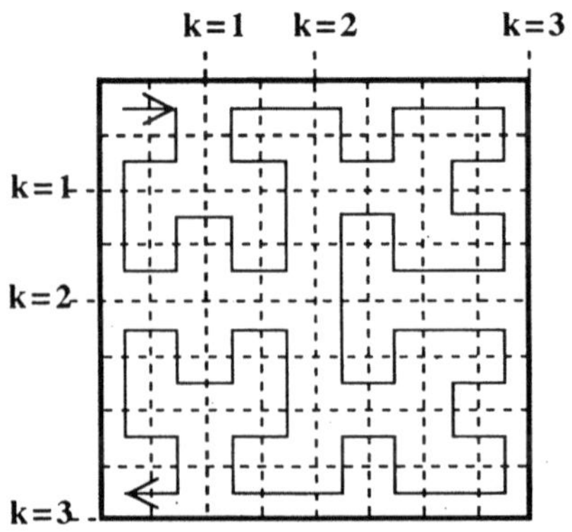

Fig.1. Scan path based on Peano-Hilbert's plane-filling curve for a matrix sized 8*8 pixels.

3.2 Irreversible Compression Algorithms

The irreversible-compression algorithms we defined are meant to achieve a high compression level by storing the brightness value only of those pixels that are regarded as indispensable to save all the clinically relevant information present.

BEL method consists in limiting the maximum brightness difference that may occur between corresponding pixels in the original image and in the reconstructed one, to a preset, constant value. Fig. 2 illustrates three different graphs related to BEL, PABEL and PABCEL methods.

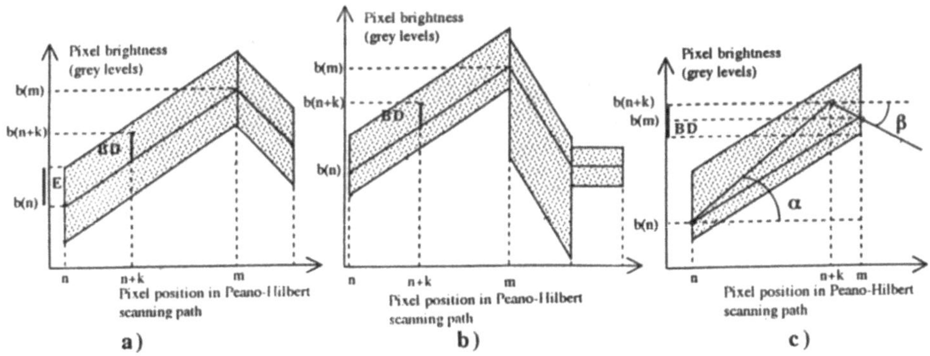

Figure 2. a) BEL method, b) PABEL method, c) PABCEL method.

The X-axis shows the position of the pixels along Peano-Hilbert's path; the Y-axis represents the pixels' brightness value in grey levels. Pixels $p_1(n,b(n))$, $p_2(n+k,b(n+k))$ and $p_3(m,b(m))$ are also shown. Consider the line from p_1 to p_3: evaluate the brightness error, in grey levels, introduced by approximating the actual brightness of p_2 to the value obtainable by linear interpolation of the brightness of p_1 and p_3. This brightness difference, BD, is expressed by:

$$BD = [b(n+k)-b(n)]-\{[b(m)-b(n)]*k\}/(m-n) \tag{1}$$

BEL compression method is based on limiting the brightness difference BD as follows:

$$|BD| \leq E$$

where E is a positive, constant threshold value in grey levels, experimentally determined. A confidence corridor, which includes all the allowed values of BD according to considered method, is highlighted in fig. 2(a).

Consider now fig. 2(b), related to **PABEL** method. A typical property of PABEL algorithm is that the threshold value for the maximum allowable brightness-difference between corresponding pixels of the original image and the reconstructed one varies according to the brightness contrast existing among the pixels of different areas in the image being scanned along Peano-Hilbert's path: greater brightness-errors are allowed for those pixels whose brightness value could hardly be distinguished by our eyes. The confidence corridor highlighted in fig.

2(b) is now asymmetrical; the allowed values of BD are set up by the following inequalities:

a) if sign{BD} = sign{b(m)-b(n)} and |(b(m)-b(n))/(m-n)|> E/H, then

$$|BD| < |H * [(m-n)/(m-n-H)] * [(b(m)-b(n))/(m-n)]|$$

b) if sign{BD} = sign{b(m)-b(n)} and |(b(m)-b(n))/(m-n)| > E/H, then

$$|BD| < |H * [(m-n)/(m-n+H)] * [(b(m)-b(n))/(m-n)]|$$

c) finally, if |(b(m)-b(n))/(m-n)| ≤ E/H, then

$$|BD| < E$$

where the values of H (0<H<1) are experimentally determined. The threshold values E were experimentally determined for both the above described methods. We implemented two new versions of the BEL and PABEL algorithms, in which the E values are expressed as a percentage of the difference between the highest and lowest grey level in the original image. **PABCEL** method may be regarded as a development of the PABEL method, aiming at adaptively limiting the brightness error but also the contrast error that occurs when considering corresponding pixels from the original and reconstructed images along Peano-Hilbert's scan path. This should leave the existing contrast in the original image's pixels in the high-contrast areas, such as those around the blood vessels, as unaltered as possible. A new value, contrast difference, is defined and indicated with CD:

$$CD = tg\alpha - tg\beta$$

where $tg\alpha = [b(n+k)-b(n)]/k$; $tg\beta = [b(m)-b(n+k)]/(m-n-k)$.
The confidence corridor highlighted in fig. 2(c) related to PABCEL method includes all the allowed values of BD according to PABEL method, but we now must consider also the limitation for the constrast difference CD:

$$|CD| < SLOPE\_DIF$$

where SLOPE_DIF assume empirically defined values.

3.3 Encoding Technique

Output message from the irreversible compression goes through a final encoding by an adaptive reversible compression method based on Lempel-Ziv-Welch's algorithms [3].

3.4 Algorithms testing

We applied our compression techniques to some 168 static images from 14 different PTCA patients. Each method was applied using four different threshold values. Considering threshold values limited to preserve all the diagnostically significant information of original images, the best results were obtained by the PABCEL method. The mean compression ratio was 8:1 approximately; consulted cardiologists found no clinically significant difference between the original and reconstructed images. To test our compression algorithms also on images coming from public files, 11 images coming from a collection of teaching videotapes [4] were compressed.

In order to highlight the brightness error due to the applied compression method in each pixel, we obtained the difference between the original image and the corresponding reconstructed one. For the cases that had been evaluated by the cardiologists, we calculated, for each original and reconstructed image, brightness-difference histograms between consecutive pixels along Peano-Hilbert's scan path, in grey levels.

4 The software package ITACA

Compression techniques we defined were implemented using a C compiler running on a 386/33 PC. These new irreversible techniques are now collected in a software package named ITACA (Irreversible Threshold-based Angiocardio-graphic images Compressor and Archiver). ITACA offers three main working modalities: a) compression; b) decompression; c) visualization of compression-related information.

Consider first the compression modality: in this case the user is requested to specificy following items:

- irreversible compression method to be applied and related threshold values;
- name of the file containing the image to compress;
- name to assigne to the file, where compressed image will be stored, after coding by Lempel-Ziv-Welch's method.

Following steps are performed by the Compressor:

1) parameters specified by the user are read and threshold values used by the chosen method are initialized;

2) size of original image, as pixels' number, is read in the header of the file containing image to compress in Cardio-500 format;

3) image to be compressed is loaded in main memory in a matrix having suitable size;

4) image is scanned according to the Peano-Hilbert path;

5) the chosen irreversible compression method is applied during image scan; image compressed is stored in a temporary file with the format illustrated in fig. 3;

6) this temporary file is coded according to Lempel-Ziv-Welch's method: file containing the compressed image is created. This file has a header containing the following information: compression method, threshold values, compression ratio obtained, size of the original image (pixel*pixel), bits per pixel. The next part of the file contains compressed image.

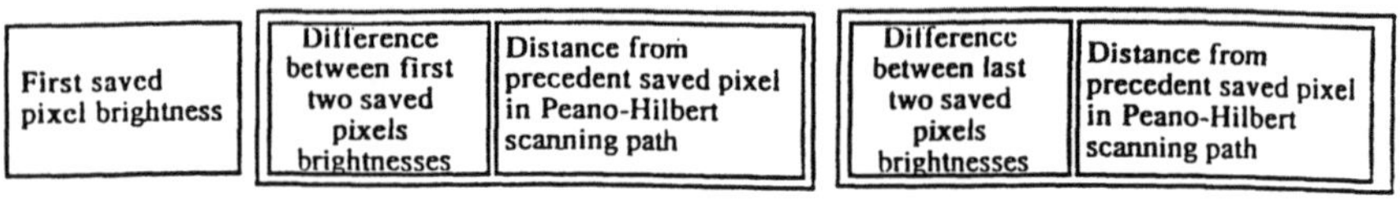

Fig.3. Message format after irreversible compression

Consider now the decompression modality. The user is requested to specify the file containing the compressed image and the name to be assigned to the file where the reconstructed image will be stored. Following steps are performed by the Decompressor:

1) the header of the file containing the compressed image is read; in this way ITACA-files are identified;

2) next part of the file is decoded according to Lempel, Ziv and Welch's algorithms and stored in a temporary file with the format illustrated in fig.3;

3) the image is reconstructed in bitmap format, by linear interpolation starting from temporary file at step 2);

4) image is stored in a file having the name specified from the user.

Finally consider visualization of compression-related information modality: users can access to information stored in the header of the file containing compressed image.

5 Conclusion

We created a software package named ITACA (Irreversible Threshold-based Angiocardiographic images Compressor and Archiver) to compress and archive angiocardiographic static images: it collects the new irreversible compression techniques BEL, PABEL and PABCEL. Because of the non-standard performances of PCs in displaying high-resolution images (e.g. CGA, EGA, VGA) we decided not to release ITACA for the MS-DOS environment. Our purpose is to develop the ITACA software package in Sun UNIX environment, too. The Sun version of ITACA will also use images in PAPYRUS format [5], which is based on the standard image format ACR-NEMA[6].

References

1. Lempel A., Ziv J., Compression of Two-Dimensional Data, IEEE Trans. Inform. Theory, vol. IT-32, No. 1, January 1986, pp. 2-8.
2. Gardner M., Mathematical Games, Scientific American, Decembre 1976, pp. 124-133.
3. Welch T.A., A Technique for High Performance Data Compression, IEEE Computer, June 1984, pp. 8-19.
4. Douglas J., PTCA Video Demonstration Series, Cardiac catetherization lab, Emory Univ. Hospital, Atlanta.
5. Ratib O., PAPYRUS, Image file format, version 2.3, UIN/HCUG, 1990, 1991.
6. ACR-NEMA standard publication, Digital Imaging and Communications, #300-1988.

Superimposition of MR-Images and Scintigrams as an Application of PACS

R.NICOLETTI, M.WILTGEN, L.KRONBERGER jr, G.RANNER, R.PITZLER, K.M.SIMONIC, R.STOLLBERGER, G.F.FUEGER, G.GELL, F.EBNER

Karl-Franzens-University Graz: Department of Radiology, Department of Surgery, Institute for Medical Informatics and Documentation, Institute for Magnetic Resonance

Summary
Picture archiving and communication systems (PACS) were originally designed to enable widespread use of radiographic images at various locations utilizing central archives and network communication. We consider one further application of a PACS even more important, i.e. the superimposition of images of different imaging modalities for synoptic interpretation. We integrated the equipment of the Nuclear Medicine Department in our existing PACS to facilitate the superimposition of magnetic resonance images and scintigrams for the diagnosis of suspect recurrencies after rectal cancer. The synoptic interpretation by this image fusion enhanced the specificity of the investigation to the value of 1.

Background

The combination of different imaging modalities, for example digital subtraction angiography (DSA), computer tomography (CT), magnetic resonance imaging (MR) or nuclear medicine imaging (single photon emission computed tomography, SPECT), may enhance the diagnostic results since the different nature of their respective information content complements each other [1]. A comprehensive picture archiving and communication system (PACS) may be used to facilitate the complementation of images of one kind with those of a different kind. Such a system has been developed at the Department of Radiology of the University Hospital Graz during the last ten years, which includes four CTs, MR-, DSA- and radiotherapy equipment, diagnostic consoles and the archive [2].

Method

The nuclear medicine equipment was joined to the PACS to facilitate the superimposition of MR tomograms and scintigraphic anti-

CEA immuno images obtained by SPECT. Three steps were necessary for accomplishing the fusion of those images:

1) The transfer of the image files into the same computer, which included the file-transfer itself and the translation of at least one of the image studies into a common format, including all ancillary information (header, acquisition parameters etc).

2) The superimposition of the two images based upon the selection of corresponding slices and the alignment of the anatomical structures.

3) A suitable presentation of the fused image.

In our system the first step is accomplished in the following way: The MR-studies (Philips Gyroscan S15) are sent via ethernet to a gateway-computer (DEC MicroVax 2000), where an automatically started routine, developed at the Institute for Medical Informatics Graz, translates the studies from ACR/NEMA-format into the Elscint-format. From this gateway-computer the studies are transferred via ethernet to the nuclear medicine image processor (Elscint SP-1), where the superimposition of MR-images and scintigrams is realized.

For the identification of corresponding slices appropriate markers on the body-surface of the patients are used in MR-imaging as well as with scintigraphy. These markers, 57-Co-point sources in scintigraphy and thin tubes filled with $CuSO_4$ in MR-imaging, define a reference-slice. As the slice distances are known in both modalities, all corresponding slices can be defined in relation to this reference slice.

If the pair of corresponding slices is defined, the scintigraphic slice is automatically zoomed to the scale of the MR-slice. For anatomical correct alignment we use two methods: the landmark method which allows the definition of up to four landmarks on the scintigram as well as on the MR-image, and a method of interactive shift and rotation of the scintigram corresponding to the body contour of the MR-slice.

The fused image showes the unmodified MR-slice in black and white with a coloured overlay of the contours of the scintigraphic slice.

Clinical Application

Up to February 1993 we have applied this superimposition of scintigrams and MR-images on studies of 34 patients, which were admitted for routine follow-up after operation of rectal cancer in order to search for suspected recurrencies. The mean time gap between operation and follow-up investigation was 25 months and ranged from 6 to 72 months.

MR- and SPECT-investigations were performed on the same day within a maximum of 2 hours. Since the tracer (anti-CEA moAB) is excreted into urine and collects in the urinary bladder where it causes untoward imaging effects, the bladder is continuously emptied by catheter prior to and during MR imaging and SPECT.

The MR image series consisted of transversal T1-weighted, contrast enhanced spin-echo images (repetition time (TR) 750 ms, echo time (TE) 30 ms, slice thickness (THK) 5 mm, field of view (FOV) 310 mm, scan matrix 230x256, 0.2 ml/kg Gd-DTPA), and T2-weighted, transversal slices (TR = 2500 ms, TE = 30/90 ms, THK = 6 mm, FOV = 310 mm, scan matrix = 154x256). 5 mg Glucagon were administered intravenously to minimize artefacts caused by peristaltic movements.

Immuno-scintigraphy was performed with 700 to 900 MBq 99m-Tc labelled anti-CEA (Behring MAB 431/26). An Elscint 409 ECT-scintillation camera was used for acquiring 60 frames (40 s acquisition time) over 360 degrees, the image-matrix was 64x64. Both investigations were analysed separately first, followed by the evaluation of the computed fusion.

Results

Evaluating the MR-investigations 11 patients revealed various kinds of abnormalities, four of them were diagnosed as suspect recurrancies. 22 patients had no significant scar tissue or suspicion of recurrance. One patient had to be excluded because of artefacts caused by an unexpected shell splinter. In SPECT 9 patients (six different from those with MR findings) showed abnormal tracer uptake, whereas 25 patients had no pathological scintigraphic findings.

Evaluation of the superimposition showed only three cases of malignant recurrance, all of whom were histologically verified.

The abnormal MR findings of the other eight patients were caused by various scar formations.
In SPECT seemingly pathological tracer uptake was caused in three cases by the post-surgical dorsal extension of the urinary bladder in patients with abdomino-perineal resections, in the other three cases by activity in the ureter, in the sacral marrow and in the colostomy respectively.
The above results correspond to a specificity of 0.97 of the MR-investigation and a specificity of 0.80 of the SPECT-investigation. The superimposition of MR-images and scintigrams improved the specificity to the value of 1.

Discussion

Superimposition of images gained by different modalities has been described for many applications, e.g. brain [3,4], chest [5,6], or pelvis [6,7,8], but most of this work was done with CT-images. The differentiation of malignant recurrences from post-surgical scar tissue is limited in CT because of the small differences in density [9,10,11]. For this reason we used MR-images because of the better sensitivity for the differentiation between scar tissue and malignant recurrencies. In contrast to the work described bei Liehn et al [8], our MR- and SPECT investigations were performed on the same day with a maximum delay of 2 hours.
The registration problems are the same with SPECT-MR- and SPECT-CT-superimposition and require special attention in the abdominal area: There are no anatomical structures in this area, which can be used as landmarks in the SPECT-slices, even the bifurcation of the great vessels is not reliable. External landmarks seem to be very accurate, but we found that even they can change their position between the two studies, probably because of a shift of the skin especially in thick patients. An interesting solution to this problem is described by Liehn et al. who acquires a 99m-Tc-bone scan simultaneously to the 111-In-Immunoscintigraphy for visualizing the bone structure [8]. But this method is not applicable when using 99m-Tc-labelled antibodies.
The lack of suitable anatomical structures in the SPECT-slices

causes problems in finding the anatomical correct alignment using the landmark method. It seems better to interactivly move the SPECT-slice into the overlayed contours of the MR-slice after automatically correction to the same scale. Phantom measurements showed that this method worked well in spite of the slight geometric distortion of the MR-slices.
If the technical problems of image transfer and registration can be solved in such a way, that images of different modalities can be gathered by simple menu selection, the image fusion should become a routine tool for many applications.

Conclusion

Image fusion definitly improved the diagnostic value of tomographic investigations by combining the morphological information of MR and the functional information of SPECT. Specifically it enabled accurate anatomical delineation, particularly of the post-surgical anatomy, diagnosis and localisation of malignant recurrance, the interpretation of tracer distribution (SPECT) by anatomical correlates of MR, and that of anatomical structures by radioactive tracer uptake. Computed image fusion (MR and SPECT) was of help in follow-up of curatively operated patients with colo-rectal carcinoma, since it improved early detection. It further reduced false positive findings and increased specificity.

References

1. Correia, J.A.: Registration of Nuclear Medicine Images. J Nucl Med 31 (1990) 1227-1229.
2. Wiltgen, M.; Gell, G.; Schneider, G.H.: Some software requirements for a PACS: Lessons from experiences in clinical routine. Int J Biomed Comput 28 (1991) 61-70.
3. Stollberger, R.; Fazekas, F.; Payer, F.; Flooh, E.; Valetitsch, H.; Offenbacher, H: Superposition of morphological MR information and functional SPECT images. MedicaMundi 36 (1991) 117-120.
4. Levin, D.N.; Pelizzari, C.A.; Chen, G.T.Y.; Chin-Tu Chen; Malcolm D. Cooper: Retrospective Geometric Correlation of MR, CT, and PET Images. Radiology 169 (1988) 817-823.
5. Pitcher, E.M.; Stevens, P.H.; Davies, E.R.; Goddard, M.D.; Jackson, F.R.C.R.; Jackson, P.C.: Transfer and combination of digital image data. Brit J Radiol 58 (1985) 701-703.
6. Kaplan, I.L.; Swayne, L.C.: Composite SPECT-CT Images: Technique and Potential Application in Chest and Abdominal Imaging. AJR 152 (1989) 865-866.

7. Kramer, E.L.; Noz, M.E.; Sanger, J.J.; Megibow, A.J.; Maguire, G.Q.: CT-SPECT Fusion to Correlate Radiolabeled Monoclonal Antibody Uptake with Abdominal CT Findings. Radiology 172 (1989) 861-865.
8. Liehn, J.C.; Loboguerrero, A.; Perault, C.; Demange, L.: Superimposition of computed tomography and single photon emission tomography immunoscintigraphic images in the pelvis: validation in patients with colorectal or ovarian carcinoma recurrence. Eur J Nucl Med 19 (1992) 186-194.
9. Lange EE, RE Fechner, HJ Wanebo: Suspected recurrent rectosigmoid carcinoma after abdominoperineal resection: MR imaging and histopathologic findings. Radiology 170 (1989) 323.
10. Krestin GP, W Steinbrich, G Friedmann: Recurrent rectal cancer: Diagnosis with MR imaging versus CT. Radiology 168 (1988) 307.
11. Barzen, G.; Zwicker, K.; Neumann, K.; Löhde, E.; Raakow, R.; Böse-Landgraf, W.; Richter, W.; Hierholzer, J.; Langer, M.; Felix, R.: Wertigkeit der Radioimmunszintigraphie im Vergleich zur Computertomographie in der Diagnostik und Rezidivdiagnostik kolorektaler Tumoren. Fortschr.Röntgenstr. 156,1 (1992) 3-10.

(Acknowledgement: This work was supported as project no. 7037 by the "Fond zur Förderung der wissenschaftlichen Forschung".)

Digital Image Quality: A Contrast-Detail Study Using the Leeds Phantom

G.D. FREY[1], C.W. STARR[1,2] and L.B. USHER[1]

[1]Department of Radiology
Medical University of South Carolina
Charleston, SC

[2]Department of Computer Science
College of Charleston
Charleston, SC

Abstract
A contrast-detail study using the Leeds phantom was conducted to evaluate five viewing areas in a multi-station clinical PACS system. The study showed that conventional film viewed on a viewbox was superior to all CR images when noise rather than resolution is dominant. However, the 18 x 24 CR plates were better than film for the smallest objects. For softcopy viewing the ambient lighting conditions were more important that the data matrix size or bit depth used on the viewing station.

Introduction

The introduction of multi-station picture archiving and communication systems (PACS) has raised the question of how image quality is measured across a PACS system. We chose to use a contrast-detail phantom to measure multi-station image quality [1,2]. Contrast-detail phantoms are test objects that have embedded within them disks of varying size and contrast [3,4]. The threshold contrast for each object size was determined by having readers determine the object with the least contrast that can be separated from the background. The use of the contrast-detail diagram allows one to compare observer performance under a wide variety of viewing conditions. This work was undertaken to accomplish two goals. The first was to develop a method for measuring contrast-detail curves for both hard and soft copy viewing of images on a PACS system. The second was to compare the various hard and soft copy viewing technologies and to determine the critical features that effect observer performance when viewing PACS images.

Materials and Methods

Images for this study were obtained by radiographing a contrast-detail phantom[1]. This phantom

---

[1]Leeds T0.10 Phantom, Leeds University, Leeds, UK

was placed on a 15 cm stack of 20 cm x 20 cm plates of acrylic. Imaging was done at 70 kVp using a conventional 80 kW radiographic generator[2]. The x-ray generator was checked to confirm the accuracy of the kilovoltage and the stability of the phototiming system. Images were taken at 100 cm using the conventional Bucky arrangement with a 12/1 focused grid. Film images were obtained by phototiming. For film images a 24 x 30 cm film size was used[3] in a regular speed cassette system[4]. Films were developed using a daylight film processing system[5]. Daily sensitometry was done on the film processor. After processing the films were digitized using a laser digitizer[6].

Computed radiography (CR) images were obtained in the same way as conventional film except that the same exposure time was used as for the conventional films. Images were obtained on 18x24 cm, 24x30 cm and 35x43 cm CR plates[7]. CR images were processed in the plate reader using the technique used for AP abdominal radiographs[8]. CR films were printed on film[9]. Conventional imaging was done using a single cassette. CR films used the same plate holder but different CR plates. Films were done in a single session. Digitized CR and Film images were transmitted by ethernet to viewing stations for interpretation.

Images were scored by seven readers; one radiologist, two medical physicists, three computer scientists and one medical student. In addition one medical physicist, one medical student and two computer scientists re-scored the data at a later time. Image observers were trained on a set on images and asked to determine the number of hole of lowest contrast of each size visible on the image. They were to score a positive if they were 75% certain that they could detect the hole above the background contrast.

Five viewing modalities were evaluated in this study. Films were viewed on a conventional four

---

[2] Super 80CP, Philips Medical Systems, Shelton, CN, USA

[3] TMG, Eastmann Kodak, Rochester, NY, USA

[4] Curix Ortho Regular, Miles-Agfa, Ridgeway Park, NJ, USA

[5] Curix Capacity, Miles-Agfa, Ridgeway Park, NJ, USA

[6] Lumesis, Sunnyvale, CA, USA

[7] Fuji, Burbank, CA, USA

[8] CR 7000, Philips Medical Systems, Shelton, CN, USA

[9] Fuji, Burbank, CA, USA

bulb viewbox. The viewbox was in a darkened room and masks were used to exclude extraneous light. CR images were viewed on four viewing stations of two types. The diagnostic viewing stations(DVS1 & DVS2) processed a 2048 x 2048 by 10 bit matrix for display on a 1024 x 1024 matrix of 8 bits[10]. The DVS would zoom and window so the full range of data could be used. The remote viewing stations(RVS1 & RVS2) displayed data as a 1024x1024 matrix of 8 bits[11]. Viewing stations had different ambient light levels which were subjectively evaluated.

Images were interpreted in groups called packets. Each packed contained a conventional film image, and one 18x24 cm, one 24x30 cm and one 35x43 cm CR image. The packets were distributed to readers in a random order for viewing on the five modalities. Each observer was presented with the four images in each of the five packets on each modality. Image observers were freely allowed to used the window, level and zoom and rove controls on the viewing stations. Data analysis was performed using SAS software[12]. A type I level of 0.05 was chosen for the per-experiment confidence level. We found no interaction effect significant using three or more variables at a 0.05 level. These interactions were subsequently dropped from the model. Mean responses were calculated for all main effect variables of reader (seven levels), image type (four levels), image samples (five levels), display modality (five levels), and reader replications (four levels, i.e., four readers reread two film samples), and one interaction (film type and display modality). A one-way analysis of variance (ANOVA) was performed on the disk observations to identify significant mean differences. Tukey's (HSD) test was performed on each significant factor (alpha = 0.05) and the per experiment error rate was controlled.

Results

The seven observers are described in Table I. The light levels at the viewing stations ranged from normal room brightness for RVS2 to quite dark for the viewbox and RVS1 with little variation. Both of the DVS's were at intermediate brightness but varied somewhat at different times of the day. The sizes of the disks in the Leeds T0.10 phantom are presented in Table II. The contrast

---

[10] Comm-View Turbo, Philips Medical Systems, Shelton, CN, USA

[11] Critical Care Workstation, Philips Medical Systems, Shelton, CN, USA

[12] SAS Version II

levels are presented in Refs. [5] . The statistical analysis of the data produced a number of significant results. There was a significant difference found between readers for all disk sizes (p(disk) < 0.001, disk=1..12). Four reader groups were found: readers 1, 2, 4 and 7 formed a group, and as did reader 3, reader 5 and reader 6. There were significant differences between modalities and packets for all images (Table III).

Viewing modalities were analyzed by film type (Table IV). There were significant differences between film types for the different modalities (Table V). A contrast-detail diagram showing the minimum detectable contrast as a function diameter for the five modalities is shown in Fig. 1. Fig. 2 shows the same data for the film and CR plate sizes.

Table I: Observer Characteristics

Number	Observer
1	Medical Student
2	Computer Scientist
3	Medical Physicist
4	Computer Scientist
5	Medical Physicist
6	Computer Scientist
7	Radiologist

Table II: Leeds T0.10 Disk Sizes

Disk Group	Disk Size (mm)
1	11.10
2	7.90
3	5.60
4	4.00
5	2.80
6	2.00
7	1.40
8	1.00
9	0.70
10	0.50
11	0.35
12	0.25

Table III: Significant Differences Over All Images

Significant Groupings	Hole Group					
	1	2	3	4	5	6
Modality Groupings	5>2,3,4 1>3	5>2,3,4 1>4	5,1>2,3 ,4	5>4	5,3,1,2 >4	-
Packet Groupings	4,5,3,1 >2	5,3>2	3,1>4,2	1>2,3	1>4	1,4>2,3 ,5
Significant Groupings	Hole Group					
	7	8	9	10	11	12
Modality Groupings	1>4	1>5,3,4 2>3,4	2,1>4	-	-	3,2>4
Packet Groupings	1,2>5,3 4>3	1>3,2	3>2,4	4>1,2 5,3,1>2	3,4,1,5 >2	5,1,3,4 >2

For modality groupings 5>2,3,4 means modality 5 was significantly better than modalities 2, 3 and 4. Numbers occur in order of decreasing mean contrasts from left to right. Modalities: 1=View Box, 2=DVS1, 3=DVS2, 4=RVS1, and 5=RVS2. Packets: 1-5 represent the five packets under study.

Discussion

Although a training session was held for the readers to reduce inconsistent readings, the readers still fell into three groups. This is likely due to the subjective nature of the task. Since the observers had a priori knowledge of the disk locations it is likely that they would set differing contrast detection thresholds. Readers were self-consistent. Reread data showed no significant difference between readers at the 0.01 level for all disk sizes.

When conventional film was digitized and viewed on the various viewing stations, it was not better than the CR plates and for the smaller disk sizes it was inferior to the 18 x 24 CR plates. For the larger disk sizes (disks 2 mm and greater) the best results were achieved on the viewbox and on the RVS that was in a darkened room. The two DVS's were intermediate in performance and the RVS in a brightly lighted room was inferior to the other modalities. For disks in midrange (1.4 to 0.7 mm), the dimly lighted RVS remained significantly better than the brightly lighted RVS. For these disks , the viewbox was the superior performer only to the brightly lighted RVS. As disk size decreased, the two DVS's performed significantly better than the brightly lighted RVS. Thus for disks 0.7 mm and greater the lighting conditions in the room were a larger factor than the nature of the viewing station (DVS or RVS). For the three smallest disks (less than 0.7 mm), no differences were found between any viewing stations. These disks are not well resolved by any of the modalities.

Table IV: Significant Modality Groupings by Film Type

	Hole Group					
Film Type	**1**	**2**	**3**	**4**	**5**	**6**
Conventional	1,5>3,4	1,5>4	1,5>4	1,5>4	1,5>4	-
18x24 cm CR	-	-	-	-	-	-
24x30 cm CR	5>3	-	-	5>4	-	-
35x43 CR	-	5>2	-	-	-	-
	Hole Group					
Film Type	**7**	**8**	**9**	**10**	**11**	**12**
Conventional	1>4,5	1>2-5	1>4,5	1>4	4>1,2,3	4>3
18x24 cm CR	-	1>4,3	-	-	-	-
24x30 cm CR	-	1>3,4	2>4	-	-	-
35x43 CR	-	-	-	-	-	-

1,5>3,4 means modalities 1 and 5 were significantly better than modalities 3 and 4. Numbers occur in order of decreasing mean contrasts from left to right. Modalities: 1=View Box, 2=DVS1, 3=DVS2, 4=RVS1, and 5=RVS2.

Table V: Significant Film Type Groupings by Viewing Modality

	Hole Group					
Modality	**1**	**2**	**3**	**4**	**5**	**6**
View Box	1>3,4,2	-	1>2	-	-	1>4
DVS1	-	-	-	-	-	-
DVS2	-	-	-	-	-	-
RVS1	-	4,3,2>1	-	-	-	4,3,2>1
RVS2	1>2	-	1>2	1>4	-	-
	Hole Group					
Modality	**7**	**8**	**9**	**10**	**11**	**12**
View Box	1>4	1,2>4	1,2,3>4	3>4	3>4	2>1
DVS1	-	-	3>1	3>4,1	3>4	3,2>1
DVS2	-	-	3>4	3>2,1	3>2,1,4	3>2,1,4
RVS1	-	-	3,2,4>1	3,2,4>1	3,2,4>1	3,2,4>1
RVS2	-	-	-	3>1	3>4,1	3>1

1>3,4,2 means film type 1 was significantly better than film types 3, 4 and 2. Numbers occur in order of decreasing mean contrasts from left to right. Film types: 1=conventional, 2=18x24 cm CR, 3=24x30 cm CR and 4=35x43 cm CR.

For the detection of disks 0.7 mm through 11.1 mm on the viewbox, the conventional film provided the greatest responses and was significantly superior than at least one CR plate size. The performance of the small CR plates exceeded that of conventional film when resolution demands were highest. Conventional film also

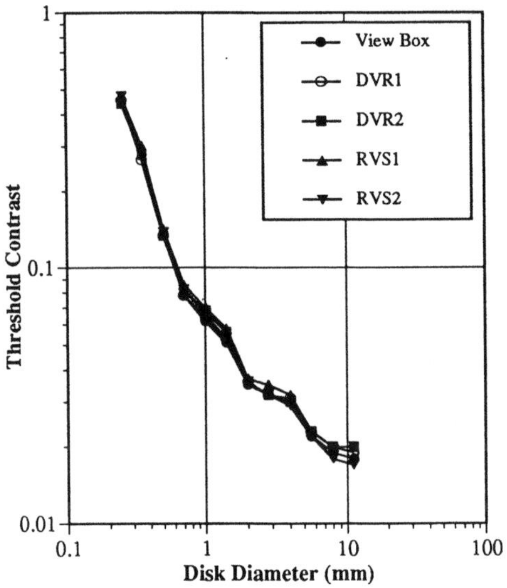

Fig. 1: Contrast-Detail Curves for Modalities

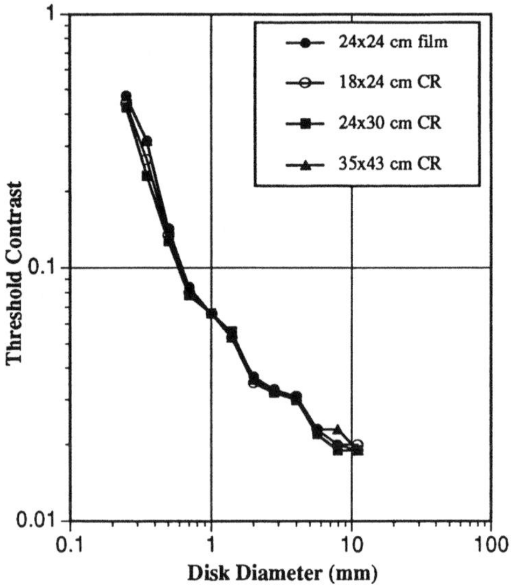

Fig. 2: Contrast-Detail Curves for Image Types

provided superior performance over small and large CR plates for the larger disks on the RVS2 modality where the ambient lighting conditions were similar.

When viewing conditions were not optimal as was the case with DVS1, DVS2 and RVS1, the differences in spatial and contrast resolution represented by conventional film, scanned film, and CR images, are masked and not appreciated in this contrast-detail study using the Leeds T0.10 phantom.

Conventional film observed on the viewbox is superior to all CR images observed in this study when noise rather than resolution is dominant. However for high resolution, the small CR plates were better than the conventional film (0.25 mm) on the viewbox. Also evident was the inferiority of the large CR plates for resolving detail over all other film types except for the scanned conventional film. Interestingly, the medium CR plates were significantly better than all other film types for the smallest three disk sizes across all digital modalities.

References

1. Hay et al, A set of X-ray test objects for quality control in television fluoroscopy. BJR 58, 335-344, 1985
2. Hay and Chesters, Threshold mechanisms in the presence of visual noise. In: "Medical Images; Formation, Perception and Measurement, Institute of Physics, Bristol, 1976
3. Webster and Wipfelder, J. Soc. Mot. Pict. Tel. Eng. 73, 617-624, 1964
4. Cohen, Contrast-Detail Analysis of Imaging Systems: Caveats and Kudos. In: Recent Developments in Digital Imaging, AAPM Monograph 12, AIP, New York, 1985
5. Cowen et al., Leeds X-ray Test Objects, The University of Leeds, Leeds, 1987

Assessment of DCT-Based Method for X-Ray Image Compression

D.TRIPPI (*), D.CARAMELLA (*), G.BRACCINI (*),R. DE DOMINICIS (**), G.RUSSO (***), M.SALUSTRI (***), P.TALONE(***).

(*) Department of Radiology, University of Pisa
(**) Department of Radiology, University of Florence
(***) Fondazione Bordoni, Rome

Summary

This work deals with the optimum compression rates that can be obtained by means of two DCT coding techniques in order to preserve the radiograms "diagnostic information".

The results of a quality analysis of compressed digitized radiographics images of incomplete fractures are presented. A consistent meaningful set of radiograms was coded by Bordoni Foundation in Rome and quality tested by a team of radiologists working at the Radiology Department of the University of Pisa.

ROC curves [7] were obtained from the results of the observations of trained radiologists who analyzed the compressed images without any previous knowledge of the algorithms used for coding.

Image acquisition

The radiographic images used in this work were acquired by means of a high resolution CCD scanner (Rita!, Vision Ten, USA) and coded and displayed on a Quadra 900 computer (Apple Computers, Cupertino, USA). Digitized images had a size of 4096 x 4096 pixels; they were acquired with 12 bit precision, and stored with 8 bit/pixel after a remapping of the grey levels [6]. For each image the remapping function was specifically selected by the radiologist, among a small allowable set implemented in the Rita! scanner. The analysis was carried out on a set of 12 meaningful portions of the digitized images having a size of about 1024x1024 pixels. They were "cut" from the previous 4096 x 4096 digitized radiograms.

In the experiment two original versions of each radiogram were compressed:

- the original image without any pathology;
- the original with a non displaced (incomplete) fracture, simulated by means of a thin pencil line.

Statistical analysis

A statistical analysis of 8x8 and 64x64 DCT transformed images [1], [4] was carried out in order to establish the optimum quantization matrices to be used in the DCT-based compression techniques.

---

* This work has been carried out in the framework of the agreement between Fondazione Bordoni and the Italian PT Administration

Amplitude distributions of radiograms' DCT coefficients were compared with the homologous of generic photographic image luminance components for which the quantization steps are well established [2].

For both 8x8 and 64x64 DCT, coefficient distributions presented Laplacian like shapes with zero mean and variances decreasing with the coefficient order. However radiogram coefficients had variances smaller than the homologous of photographic images due to the slower contrast variations presented by X-ray images.

In Fig.1 two instances of the statistical analysis results are showed. On the left a Laplacian-like distribution of an 8x8 DCT coefficient (3,0) is reported; on the right some radiograph 64x64 DCT coefficients' ranges are showed.

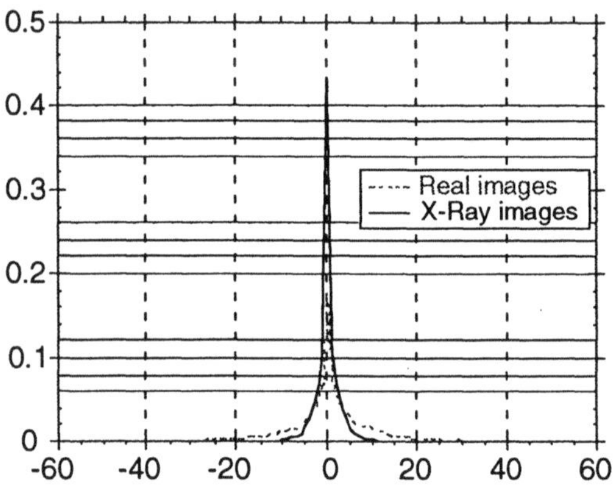

Fig.1 - 8x8 DCT: (3,0) coefficient distribution

DCT 64 Coefficient	Ranges: Real images	Ranges: Radiograms
(0, 8)	-396÷672	-177÷145
(8, 0)	-240÷209	-170÷353
(8, 8)	-117÷127	-73 ÷ 57
(0, 32)	-139÷99	-13 ÷ 19
(32, 32)	-11÷15	-1 ÷ 3
(56, 56)	-5 ÷ 9	-1 ÷ 3

Statistical analysis clearly showed that value ranges of radiograms' DCT coefficients were contained in the photographic ones.

Consequently, in order to have, for radiograms, quality levels similar to those usually obtained for photographic images, the quantization of the radiograms' DCT coefficients had to be performed using smaller quantization steps than the photographic ones.

JPEG Standard

JPEG image compression algorithm is an international standard supported by ISO (International Organisation for Standardization) and CCITT (the International Telegraph and Telephone Consultative Committee)[2].

The main benefits of using JPEG technique are the availability of multi-vendor, multi-system, low cost specialized hardware and software and the use of a common interchange format that guarantees image portability on different systems.

The block diagram of the JPEG procedure is reported in Fig.2.

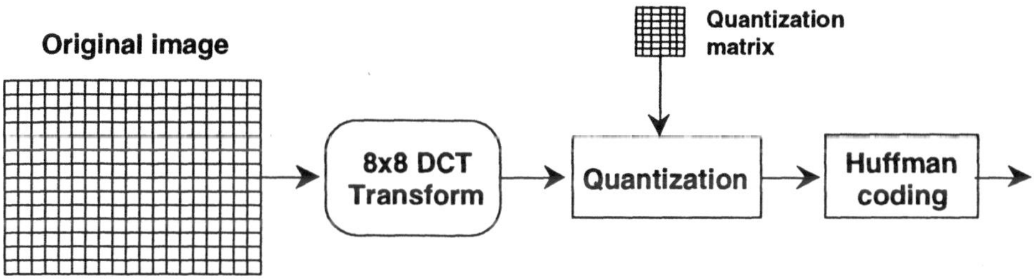

Fig.2 - JPEG compression procedure

The compression scheme implies information loss due to the quantization of the DCT coefficients in each 8x8 block.

The choice of the quantization matrix determines the compression ratio ($\frac{\text{original image cost} = 8\text{bit/pixel}}{\text{compressed image cost}}$) and, at the same time, the reconstructed image quality.

The resulting effect of the quantization on the reconstructed image is twofold:

- a general contrast decrease similar to that produced by a low-pass filtering operation;
- the introduction of spurious frequencies that produce artefacts.

The most annoying artefact is the "blocking" effect that is the border enhancement of the adjacent blocks.

JPEG standard suggests a quantization matrix for the luminance component of photographic images. This matrix was obtained as result of a huge amount of psychovisual tests.

The (1,1) element (DC coefficient) of each transformed block represents the spatial block mean value and it is generally coded with a quantization step (16) that preserves the integer value of the block's spatial mean component. Dividing this matrix by 16 (its (1,1) element) we obtain values indicating the ratios of each quantization step in respect to that used for the DC one:

	0.6875	0.625	1	1.5	2.5	3.1875	3.8125
0.75	0.75	0.875	1.1875	1.625	3.625	3.75	3.4375
0.875	0.8125	1	1.5	2.5	3.5625	4.3125	3.5
0.875	1.0625	1.375	1.8125	3.1875	5.4375	5	3.875
1.125	1.375	2.3125	3.5	4.25	6.8125	6.4375	4.8125
1.5	2.1875	3.4375	4	5.0625	6.5	7.0625	0.5625
3.0625	4	4.875	5.4375	6.4375	7.5625	7.5	6.3125
4.5	5.75	5.9375	6.125	7	6.25	6.4375	6.1875

Fig.3 - Normalized JPEG luminance quantization matrix

Values less or greater than one indicate that the coefficients have to be quantized finer or ruder than the DC one respectively. The inverses of these values can represent the psychovisual importance of the 8x8 DCT frequencies in respect to the mean component.

If we assume that the quantization ratios are those reported in Fig.3, different quantization matrices are obtained multiplicating the above matrix by a scaling factor k.

The scaling factor sets the coding precision of the DCT coefficients thus determining the compression ratio and the reconstructed image quality.

The present paper aims to establish the optimum value of JPEG compression ratio in order to obtain the target image quality (requested by the radiologist team).

The statistical analysis reported in the previous Section points out that the JPEG luminance quantization matrix is too rude for X-ray images.

Consequently quantization scaling factors smaller than that proposed in the JPEG Standard (k=16) have to be used in order to reduce the blocking effect.

Using scaling factors values in the range from 3.5 to 15, several compression ratios depending on the scaling factor and the particular image were obtained. Assuming that the main factor that determines the image quality is the compression ratio, the compressed images were divided in four classes identified as:

Method	JPEG – A	JPEG – B	JPEG – C	JPEG – D
Cost (bit/pixel)	0.7 ÷ 0.6	0.45 ÷ 0.35	0.3 ÷ 0.25	0.25 ÷ 0.2
Compression ratio	11÷13	17 ÷ 23	26 ÷ 32	32 ÷ 40

The compressed radiographic images were proposed to the radiologists. The quality analysis is reported in the appropriate Section.

Overlapped DCT-64

The compression rate achievable by the JPEG procedure is mainly limited by the "blocking" effect that appears at the blocks' boundaries in the reconstructed images.

The use of large blocks is a way to obtain higher compression rates and to reduce the length of the block boundaries.

A DCT-based method using 64x64 overlapped blocks, has been ad-hoc developed to achieve the coding benefit of a large transform avoiding the drawbacks of the full frame DCT.

In fact full frame DCT is a powerful method to avoid the blocking effects that appear at block boundaries. However it presents a number of drawbacks as the computational complexity, the variable size DCT implementation and the full frame quantization matrix, definition and sending.

In order to overcome the above problems, preserving the advantage of a large DCT, a method that uses 64x64 overlapped blocks was developed.

The use of large dimension blocks preserves several of the full frame technique advantages and minimizes its drawbacks. In addition, the reduction of the boundaries' blocks length, allows to use an overlapping technique to minimize the residual blocking effects with an acceptable coding data overhead [3], [4].

Block dimensions and overlapping size are determined as a compromise between the DCT transform computational cost and the coding data overhead.

In Fig.4 and Fig 5, the overlapping structure and the interpolation scheme to evaluate the boundary pixels are showed.

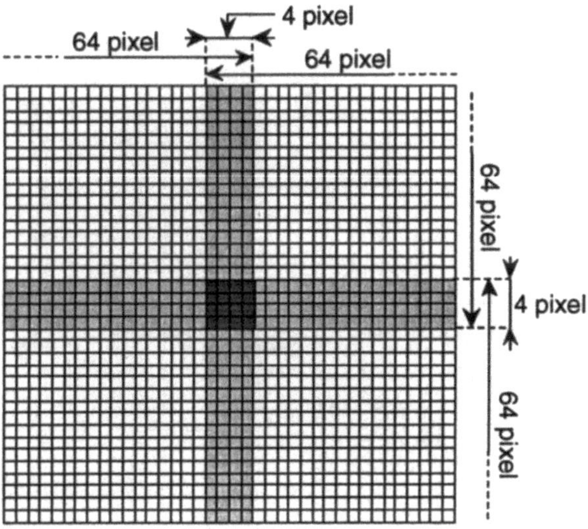

Fig.4 - Overlapping structure

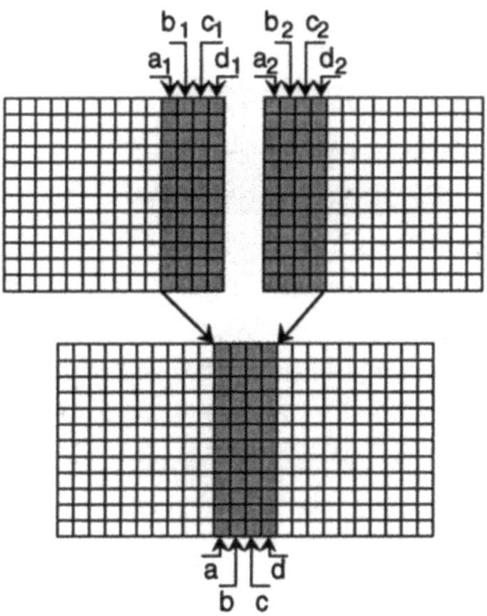

Fig.5
Boundary pixels interpolation for rows or columns

Interpolator:

$$\mathbf{a} = \frac{2}{3}\mathbf{a_1} + \frac{1}{3}\mathbf{a_2}$$

$$\mathbf{b} = \frac{1}{2}\mathbf{b_1} + \frac{1}{2}\mathbf{b_2}$$

$$\mathbf{c} = \frac{1}{2}\mathbf{c_1} + \frac{1}{2}\mathbf{c_2}$$

$$\mathbf{d} = \frac{1}{3}\mathbf{d_1} + \frac{2}{3}\mathbf{d_2}$$

The interpolation is performed by rows and by columns indipendently using the same interpolator. The interpolator weights were selected among a small set in order to minimize the reconstructed images MSE (Mean Square Error). The quantization of the transformed blocks is performed likewise the JPEG method. The quantization matrix was synthesised from the 8x8 JPEG one on the basis of the 8x8 ↔ 64x64 DCT frequencies' mapping. The frequencies mapping scheme is showed in Fig.6. Lacking values were obtained through a bilinear interpolation of the existing ones.

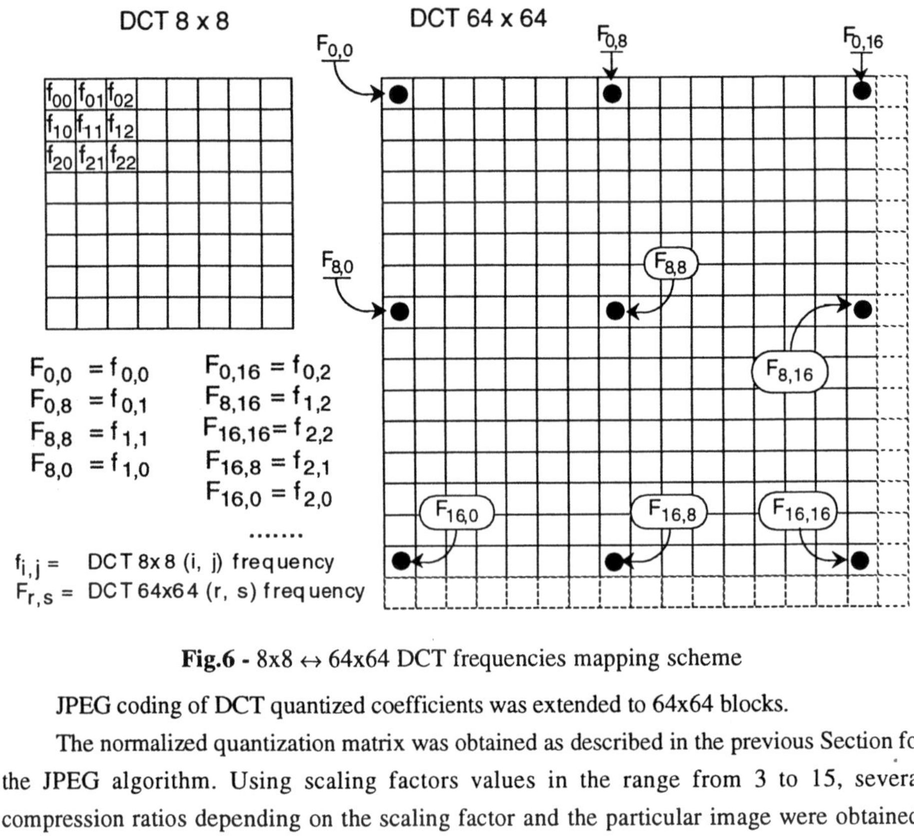

Fig.6 - 8x8 ↔ 64x64 DCT frequencies mapping scheme

JPEG coding of DCT quantized coefficients was extended to 64x64 blocks.

The normalized quantization matrix was obtained as described in the previous Section for the JPEG algorithm. Using scaling factors values in the range from 3 to 15, several compression ratios depending on the scaling factor and the particular image were obtained. Assuming that the main factor that determines the image quality is the compression ratio, the compressed images were divided in three classes identified as:

Method	DCT64 – A	DCT64 – B	DCT64 – C
Cost (bit/pixel)	0.59 ÷ 0.35	0.33 ÷ 0.25	0.24 ÷ 0.14
Compression ratio	14÷24	24 ÷ 32	33÷ 57

Also these compressed images were proposed to the radiologists. The quality analysis is reported in the following Section.

Quality analysis

The quality test was carried out by visually comparing a total of 240 images, either uncompressed or decompressed after the application of different coding methods, concerning 12 cases.

Twenty images of each case, consisting of uncompressed and compressed versions of the original image with and without the presence of simulated fracture were presented to 4 radiologists separately, in a random sequence.

For each image the radiologists indicated the presence or the absence of the pathology. The answer's scale was the following:

1- not present;
2- probably not present;
3- don't know;
4- probably present;
5- present.

The results of the observations of 4 trained radiologists who analyzed the compressed images without any previous knowledge of the algorithms used for coding are summarized in the following Tables.

Answer type	JPEG – A		JPEG – B		JPEG – C		JPEG – D	
	N	N+S	N	N+S	N	N+S	N	N+S
1	42	0	41	0	40	0	39	1
2	3	0	3	0	5	0	6	0
3	0	0	0	0	0	0	0	2
4	1	4	2	0	1	10	1	10
5	0	42	0	46	0	36	0	33
	Data match the perfect decision Fig.7		Data match the perfect decision Fig.7		Data match the perfect decision Fig.7		ROC-curve in Fig.8	

Answer type	DCT64 – A		DCT64 – B		DCT64 – C		Originals (uncompressed)	
	N	N+S	N	N+S	N	N+S	N	N+S
1	54	0	37	2	67	7	87	0
2	8	0	4	2	7	4	4	0
3	0	0	0	1	0	2	0	0
4	2	6	2	9	2	11	0	4
5	0	60	0	37	0	44	1	88
	Data match the perfect decision Fig.7		ROC-curve in Fig.8		ROC-curve in Fig.8		ROC-curve vertical at FPF=0.0109	

N: radiograms without pathology; N+S: radiograms with pathology.

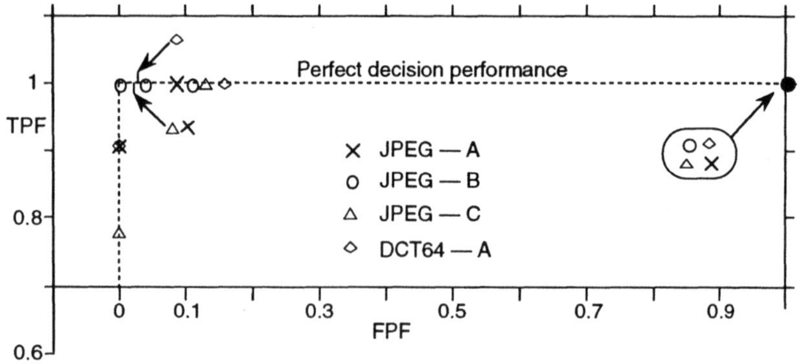

Fig.7 - ROC-curves for data matching the perfect decision performance.

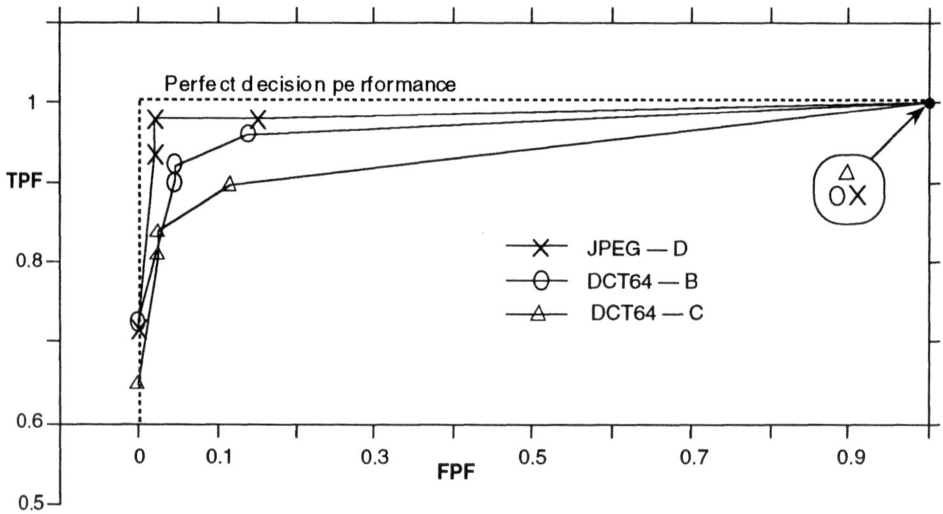

Fig.8 - ROC-curves for data not matching the perfect decision performance.

The analysis points out that, in the cases taken into account, the DCT techniques have excellent performance for low-medium compression ratios ($10 \div 30$).

However we believe that two main factors have to be more investigated in order to validate the preliminary results above reported:

- the number of observations needed to have reliable statistical data;
- the design of the experiment structure in order to minimize the polarization of the trials.

Conclusions

The problem of information loss is one of the main items in X-Ray image storing and it is strictly connected to the forensic medicine aspects.

Therefore, for data-bank storage purposes, radiographic images are usually coded by means of lossless compression algorithms [5].

However, there are a number of practical cases in image retrieval procedures where the compression factors achievable by lossless techniques are not compatible with the needs of a remote rescue service. For instance, the remote access to data-banks via radio links makes available only low bit-rate channels.

DCT coding techniques seems to be a good candidate for the purpose previously illustrated especially when the required compression ratio do not exceed the value of 30.

Moreover image databases and information systems, based on DCT techniques, allow to retrieve images with a large number of compression rates starting from a high quality (lossless or semi-lossless) transformed and stored image, simply tuning an ad-hoc quantization process before sending.

References

[1] **R. J. Clarke**, "Transform Coding of Images", Academic Press, 1985.

[2] **ISO/IEC IS 10918-1 — CCITT Rec. T.81:** Digital compression and coding of continuous-tone still images — Part 1: Requirements and guidelines, 1992.

[3] **K. N. Ngan, R. J. Clarke**, "Low-pass filtering in the cosine transform domain", Proceedings of ICC 1990, pp.31.7.1 - 5 Seattle.

[4] **K.R.Rao, P.Yip**, "Discrete Cosine Transform — Algorithms, Advantages, Applications", Academic Press, San Diego, 1990.

[5] **G.R.Kuduvalli, R.M.Rangayyan**, "Performance Analysis of Reversible Compression Techniques for High-Resolution Digital Teleradiology", IEEE Transaction on Medical Imaging, vol. 11, No.3, pp.430 - 445, September 1992.

[6] **M. Ibrahim Sezan, Kwok-Leung Yip, Scott J. Daly,** "Uniform Perceptual Quantization: Applications to Digital Radiography", IEEE Trans. on Systems, Man, and Cybernetics, vol. SMC-17, No.4, pp. 622 - 634, July/August 1987.

[7] **Gur D, Rockette HE, Good WF, Slasky BS, Cooperstein LA, Straub WH, Obuchowski NA, Metz CE.** Effect of observer instruction on ROC study of chest images. Investigative Radiology: 25, 230-234, 1990.

Teleradiology

Teleradiology in a Large Radiology Practice

RICHARD L. MORIN, DANIEL F. BRODERICK, BYRN WILLIAMSON, GLENN S. FORBES

Department of Radiology
Mayo Clinic Jacksonville
4500 San Pablo Road
Jacksonville, FL 32224

Teleradiology Practice Considerations

The large number of teleradiology products and features may confuse radiologists who are considering the incorporation of this technology into their practice. It is important that practice patterns and needs be evaluated prior to specification and acquisition of a teleradiology system. An appropriate beginning for this decision concerns whether the system will be utilized for primary diagnosis rather than consultation or image review. For instance, teaching or consultation at a remote location may demand real-time interactions to be of use while on-call or overreading coverage from a remote location could be performed with batch image transmission and remote interpretation without person-to-person interaction. The needs of the radiology practice determine the hardware and software requirements of a teleradiology system. We shall outline three general modes of practice before summarizing our clinical evaluations.

Batch Image Transmission with Remote Hard-Copy Review

In this method, image acquisition and display stations are typically based on personal computer platforms and images are typically transmitted over standard telephone lines by means of modems. A technologist or clerical staff member either digitizes a conventional image with a video or charge-coupled-device (CCD) camera or laser image digitizer or directly digitizes video images for transmission. The images are then transmitted in batch mode to the remote station, where hard-copy images are produced (e.g., multiformat camera or laser printer) by a technologist or clerical staff member. The hard-copy images are then transported to the radiologist who makes an interpretation and consults with the referring physician by telephone or facsimile.

An advantage of this method is that two radiologists or the radiologist and referring physician are not required to be in specified locations at a prearranged time. This is of course not unlike communication with a colleague while one reads an outside film. Other advantages of this technique are the minimal time commitments for the radiologist and retention of the convenience and desirable aspects of conventional film viewing. Disadvantages of this method include the possibly long image transmission times due to data transfer rates over conventional phone lines and the degradation and image quality which may occur because of digitization and conversion into hard-copy.

Image Transmission with Remote Soft-Copy Review

The systems used in this mode of practice are configured as described above, with the exception of the hard-copy imaging device. In practice because of the length of transmission time, typically all images for a case are often prefetched and transmitted to the remote location. Consultation may then occur either during image review or after interpretation of the entire case. An advantage of this scenario is the familiarity of the consultative setting without the necessity for both individuals to be physically present at a specific location and time. This system represents a compromise between hard-copy and real-time image review and is widely used by radiologists responsible for evening and weekend on-call support.

Image Transmission with Real-Time Review

Typically the systems utilized in this setting are video-based and images are transmitted by means of dedicated coaxial cable or fiberoptics, satellite, and most recently telephone lines that incorporate Integrated Services Digital Network (ISDN). In this scenario, a radiologist or referring physician occupies one review station and the consultant radiologist the other. Both parties view the same image at the same time with simultaneous audio communication. With some systems, it is also possible for each person to manipulate an electronic pointer and thus identify and indicate relative image structures during the consultation session. The obvious advantage of this scenario is its close resemblance to the natural process of on-demand consultation with colleagues and referring physicians. Hence this type of communication can mimic the common local modes of practice even though individuals are separated by several floors within a hospital, several blocks within a campus, or even several thousand miles within one country or between countries.

Teleradiology at Mayo Clinic

Introduction

During the past seven years various technologies have been used to transmit images about the Rochester, Minnesota, campus as well as between the Rochester campus and group practice locations in Jacksonville, Florida, and Scottsdale, Arizona. Many different technologies have been evaluated and utilized, and this compilation summarizes the clinical experience with these teleradiology activities. For any system, specifications and performance parameters are important. In this discussion primary emphasis is placed on the pragmatic, clinical aspects of each system. From a technical standpoint all systems demonstrated approximately 512 x 512 display and could resolve mesh patterns of approximately 30 lines per inch.

PC-Based Workstations

Image capture transmission and display by means of PC-based workstations (KEETS; Eastman Kodak, Rochester, New York, USA) with modem transmission over telephone lines was used for teleradiology applications for approximately five years on the Rochester campus and approximately three years at the group practice locations in Jacksonville and Scottsdale. Typically images were transmitted in batch-mode and hard-copy was generated at the receiving site. At Rochester, the systems captured image data by means of a video camera and digitization of the video signal. In Jacksonville, a different system (RVS; American Telephone and Telegraph, Parsippany, New Jersey, USA) was used for communication between Mayo Clinic Jacksonville and St. Luke's Hospital (approximately 10 miles apart). This system used a CCD camera for analog images. Several observations were made about this form of teleradiology. Initially, it was determined that the most crucial link in rendering the original image with high fidelity was the image capture process. In particular masking extraneous light to avoid vignetting and a proper degree of magnification were found to be important. Thus proper instruction of personnel who performed these duties was given high priority and ongoing review. At first, to reduce the necessary transmission time, only portions of an imaging examination were transmitted. The consulting radiologists considered this practice unsatisfactory. Often, it became necessary to view other parts of the examination (particularly in cross-sectional imaging studies) to make a confident interpretation. These systems were similar to those presently commercially available and were considered practical for their intended use. They were economical since all data were transmitted over common telephone lines without specialty use charges. Overall such systems appeared to be useful when interactive communication was not felt to be of the highest priority.

Image Transmission by Satellite

Facilities for teleconferencing (ku band) by means of satellite between the Rochester campus and the group practices at Jacksonville and Scottsdale have been in place for approximately six years. These facilities are primarily used for two-way or three-way video teleconferences. Portions of these conferences sometimes include images transmitted by means of a ceiling-mounted video camera with rear illumination of the xray film. Additionally, teleconferencing has been utilized within the Radiology Department for resident teaching conferences. At both locations residents and staff have participated in much the same way they do at the usual teaching conferences conducted within the department. This system has proved useful in both settings currently in use; however, the necessity of gathering personnel at specific locations at a specified time (which can be confusing across time zones) can be somewhat rigid. Additionally, this is a costly solution if these facilities are not provided as an institutional resource which is shared for purposes other than teleradiology.

Real-Time Video Transmission

This aspect of teleradiology has been in place for approximately five years on the Rochester campus. The system consists of 19 portable stations (Proview, G.E. Medical Systems, Milwaukee, WI, USA), each of which has a video camera for image capture and monitor for image display. Additionally, features of masking, magnification, and two-way electronic pointing are provided. The system is typically used for consultation among staff radiologists for both academic and clinical purposes. Since a two-way audio link is provided, the qualitative feel of a session is similar to that of a reading room consultation. The various display stations are interconnected by means of a video network provided by the institution and composed of coaxial and fiberoptic cabling. These connections could be established over any medium capable of video transmission (e.g., satellite or microwave). It is noteworthy that the system was infrequently used until all units were installed within easy access of the reading areas. This experience seems to reinforce the fundamental observation that access must be convenient for the system to have significant clinical utility. Currently, the majority of teleradiology at Mayo Clinic Rochester is performed with this system.

Real-Time Digital Transmission

A recent investigation concerned the use of teleradiology over a wide-area network using ISDN capability. This system was installed and evaluated on the Rochester campus as well as the group practice locations at Mayo Clinic Jacksonville and St. Luke's Hospital. Evaluations were performed using compressed video over the ISDN facilities. This service is requested from the Telephone Company and the bandwidth is achieved by multiplexing a number of independent lines to achieve sufficient data rates. Charges are based upon the time and number of lines which are required during the consultation session. Although the video image demonstrates increased image lag compared with satellite or closed-circuit television, this system provides not only image display but two-way audio and video communication for those who participate in the consultation. Because it is economical over long distances and becoming widely available, this system may be useful and competitive for the future as either a stand-alone system or as an adjunct to an existing wide-area PACS installation. Initial use of the usual manipulation features (e.g., magnification, contrast, etc.) demonstrates that contrast and spatial resolution can be maintained with this technology.

Summary

Our observations and evaluations have led to implementation of different teleradiology devices depending upon clinical requirements and practice needs. Our experience suggests that the nature and specification of a teleradiology system is highly dependent upon the clinical environment and care must be taken to fully evaluate the radiology practice needs and desires prior to implementation of teleradiology.

The Use of Teleradiology System Linking a Regional Center of Radiology to its District Hospitals and Clinics

Abdalla S.A.Mohamed

Department of Biophysics,Faculty of Medicine & Health Sciences,
United Arab Emirates University,
P.O.Box 17666, Al-Ain,
United Arab Emirates.

Summary

United Arab Emirates university has applied telecommunication in medicine to link a diagnostic medical center located in Faculty of Medicine and Health Sciences with primary health care centers and hospitals in the district of Al-Ain city. X-ray films,ECG charts, and patient records are transferred across the communication network.Two workstations are involved in involved for patient data management , and image processing and analysis.

Introduction

Most of the population in the United Arab Emirates [UAE] lives in large urban area where towns are widely spread.Great number of local inhabitants live in rural areas where delivery of specialist health care needs travel to tertiary hospitals. For example, in Abu Dhabi Emirate and specially the city of Al-Ain (150 Km from the capital),there are about eight primary health care centers [PHCC]and three hospitals (300-500 beds).The PHCC's are distributed over distances between 20 to 130 Km from the hospital assigned to deliver the main health care.The PHCC staff includes two doctors and four nurses working at two shifts,five hours each.The main problems associated with the delivery of primary and specialist services in PHCC are: 1)reluctance of local population to travel to the main hospital to achieve medical services and their loyalty to their own doctor and nurses; 2)lack of all specialist services where provision of them is possible but it its cost is inefficient; 3)PHCC staff suffers from lack of continuing medical education sessions or seeking the advice of colleagues where new recruits go to the rural areas first; and 4)resulting from the above situations only least qualified doctors are attracted to work at these locations.

Aims of the Project

Providing the PHCC with facilities (equipment and staff) to use x-ray, ECG,and laboratory basic tests(blood,urine,and faces)is possible, but assigning qualified specialist with high experience to interpret these findings is really cost inefficient. Therefore,this project aims at: 1-Measure the effects of linking such PHCC's with a diagnostic medical center [DMC] on such objective measurements as consultation,admission rate,and subjective opinions such as patient satisfaction, and the desire of doctors and nurses to remain in rural areas.2-Identify the needs for diagnostic imaging in individual PHCC.3-Evaluate the financial aspects of such services, staffing/equipment requirements, cost/ benefit ,and the potential of continuing medical education benefits.

Preparatory Study

This project started at the beginning of 1991 as a pilot project between departments of Biophysics and Radiology in the Faculty of Medicine and Health Sciences[FMHS]. The goal of that project was to investigate the possibilities of linking existing information systems in FMHS by means of the available telecommunication networks and standardizing types of messages.The study was primary focused at the exchange of medical imaging and related patient care data. To reach this goal ,we started with determining the available communication networks and protocols assigned for using them to control data transfer.Local telephone network was tested to determine the efficient baud rate for data transmission and modem protocol with different types of digital formats(ASCII,binary,...,etc.)[1].Moreover,the attitude of medical staff (doctors,and nurses) was investigated and compared with the current system(faxing documents,and examining x-ray films on light screen)The main bottleneck in communication process was the lower baud rate of transmission where an image(binary format) with.512x512 pixels and 256 gray scale takes about 17 minutes for transmission.Therefore, data compression was included before transmission. Another communication network assigned for electronic mail [EMNET]was also tested but telephone network was preferred because of its lower cost where all local telephone calls are free of charge.

Teleradiology System

Currently,the proposed system consists of a link between a diagnostic health center[DHC] located in FMHS and individual center that provides health services (PHCC's and hospitals) to guarantee clinical information exchange via telephone lines as shown in Fig.1. At the PHCC,the system components are an IBM-AT or compatible computer with 4- Mbyte of RAM and 40-Mbyte hard disk linked with a very high resolution scanning camera type AS1600 (3840x2560),video monitor to control the focus of the camera.The AS1600 can scan both x-ray films, charts, and text (patient records).The supporting hardware of AS1600 includes data compression/decompression, and modem interface to minimize the time of transferring information over telephone line as shown in Fig.2. At the DMC, a storage server system with an interface to modem and data compression/decompression is connected to a

24" high resolution monitor (1600x1200). Also,an ETHERNET interface is avaiable to communicate with Apollo workstations HP 9000/433 to provide information processing and archiving. A supporting software is designed to permit displaying the received information on different window.

The diagnostic dialogue between PHCC/hospital and DHC starts from PHCC by sending a message (telephone call) requesting case study followed by information transfer.This information may contain x-ray image,ECG trace,and patient file.ACR-NEMA format with an error correcting mode are used during image transfer.Faculty members from departments of radiology,cardiology,and internal medicine with up to date experience are available in DHC for examining ,diagnosing,and deciding the patient treatment plan.A feedback from DHC to doctor of PHCC requesting case study will be issued by sending all information displayed on the DHC's monitor (windows) to the PHCC Video monitor A voice conversation via telephone line will start to investigate the case and show the reasons of deciding the treatment plan. Meanwhile,the received information at DHC will be branched into two basic paths.The first path will include patient data file and clinical reports to the department of Family Medicine where the medical database will be updated and patient records management will take place.The second path is directed to another workstation located at the department of Biophysics where medical imaging and charts will be processed and analyzed using Picture Station PS1000 software(Krig Research product) before archiving and the results with helpful features will be sent to DMC.

Feasibility Study

The expenditure of the current phase of teleradiology project would come under three main headings:1-Installation of system components at both PHCC/DHC and interface with informatic workstations.2-Data collection :(i)A radiological technologist is needed to liase with the rural PHCC for processing x-ray films. (ii)Special training programs were needed for preparing nurses to be familiar with the new system especially scanning image and charts,and patient data entry.3-Running costs:i-The main additional items would be the rental of telephone lines,modems from telecommunication department (Etisalat).ii-Maintenance of equipment.The savings of the proposed project becomes significant where the project covers its costs over the first year of operation.

Discussions

A computer-based system for medical informatics (communication and archiving)can provide assistance for both patients and medical staff(clinicians,nurses,and technologists). This project has a tremendous impact on UAE population especially patient management and admission policy and bed use in hospitals.The high quality of received images after transmission and the ease of system use led to modify the level of health care provision as well as confidence and quality of clinical advice that may be given. On-line discussion between doctor at PHCC and expertise at DHC reduces the costs of patient transfer and improves the background of inexperienced physician.Comparing the proposed system with others [2,3],three basic features would be found:1)suitability to the society for providing health services;2)simultaneous decision-making and reasoning for patient treatment plan;and;3)Experience transfer to rural PHCC which makes servicing in these areas convenient especially for new recruiting doctors. The primary results encourage to develop the proposed system through two phases (Fig.1):i)including ISDN digital network and video phone for voice conversation instead of the direct telephone lines;and ii)using mobile communication via satellite communication channels to offer health services to areas where telephone lines or digital networks are not available especially at oil fields in the desert and Gulf water. Simultaneously, another collaborative work for organ reconstruction using Picture Center to offer 3-D visualization of an organ before surgery is iniated.

References

1. Lee,T.Latham,J.,and et.al . A new CT image transfer system:How does this effect the management of referrals to a regional neurosurgical services?, Lancet, II(1990)101-103.
2. Hasman,A.,Telecommunication in medicine- The 3I Project, Int. J. Biomed. Compu. 26 (1990)229-236.
3. Griffith,H.B. The IMTRAN revolution telephone transmission of x-rays and scans. Clinical Radiology,41(1986)229-231.

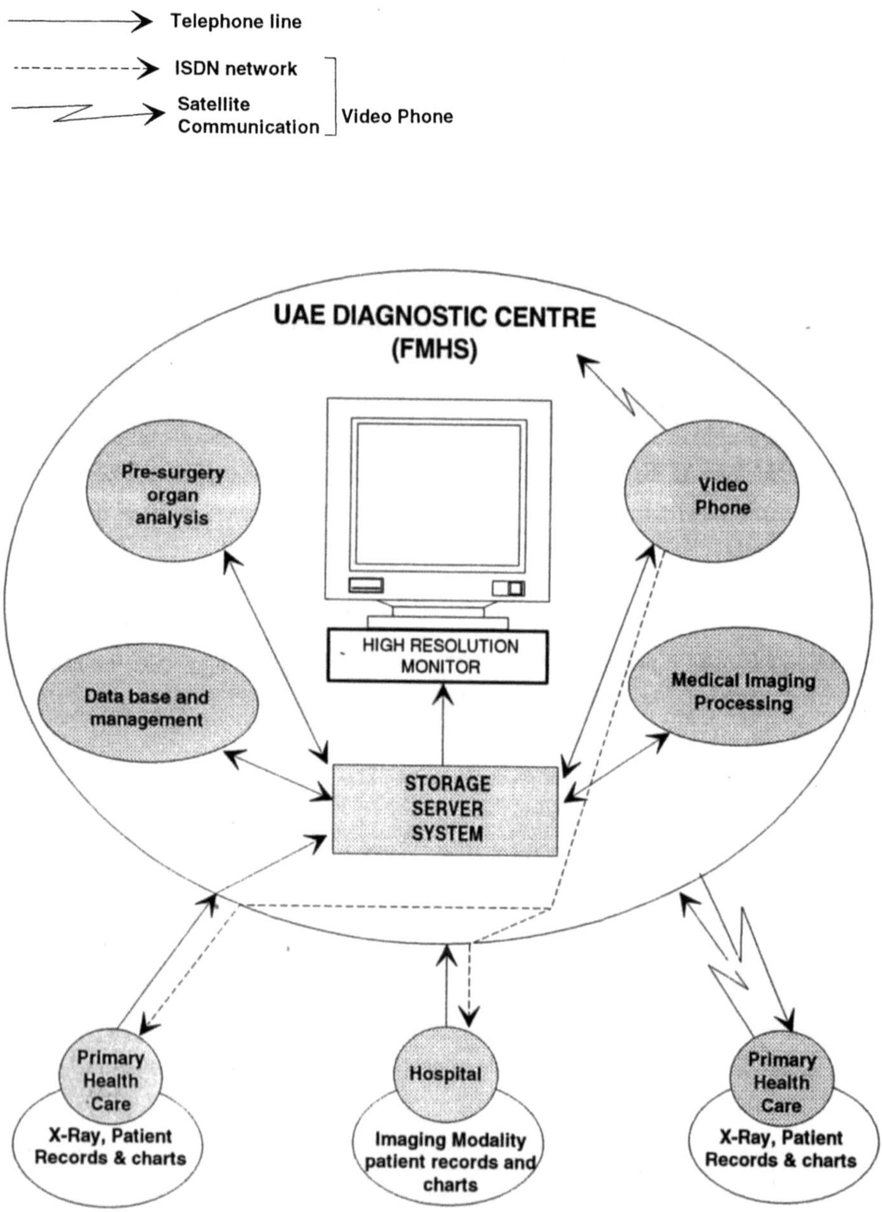

Fig. 1 Communication links among teleradiology system components

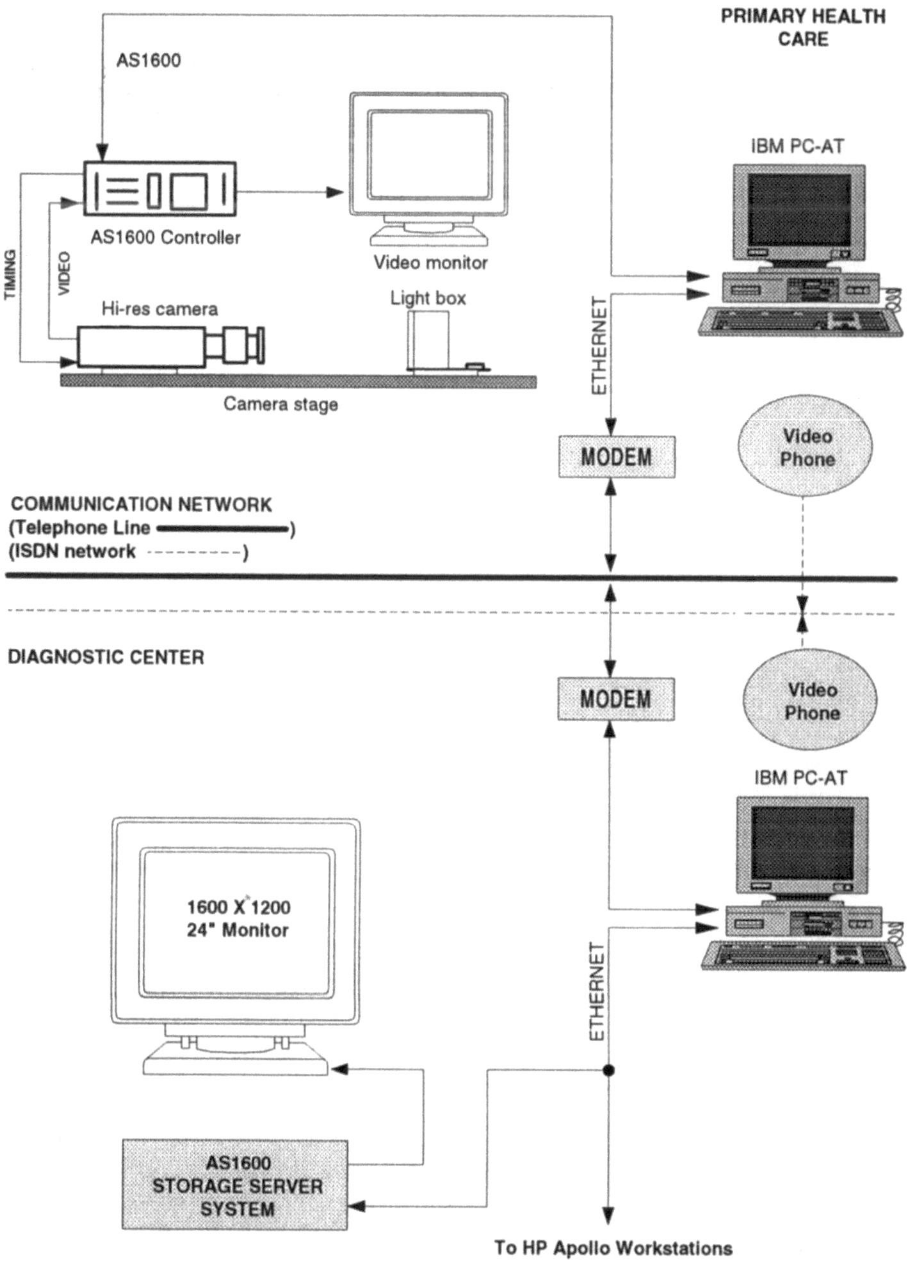

Fig.2 Information flow between MDC and PHCC .

A Metropolitan Area Network for Teleradiology and Remote Expert Consultation Based on ISDN

Y. LIGIER, O. RATIB, C. GIRARD, P. RUBIN, M. REJMER

Digital Imaging Unit, CIH, University Hospital of Geneva

and Telia Research, Malmö, Sweden

1211 Geneva 14, Switzerland, tel +41 22 37 26 267, fax +41 22 37 26 198

ABSTRACT

The purpose of this project is to evaluate image communication over metropolitan ISDN. Although the speed of 64 Kb/s of standard ISDN is insufficient for image transmission on a routine basis, it can still be used for occasional consultations. Provided that the images are transmitted in advance between two locations, a remote expert consultation conference can take place. This project is an extension of a hospital-wide PACS project that is currently under development at the University Hospital of Geneva (HCUG). Several peripheral clinics have an urgent need to access the images and to transmit their own images for consultation. The HCUG has therefore elected to explore the feasibility of using ISDN for transmission of medical images and to develop synchronization protocols for cooperative consultation.

A multimodality image manipulation and analysis software called OSIRIS was developed at the University Hospital of Geneva. This software is currently used on different hardware platforms. An extension of this software was recently designed to allow cooperative work on remotely located workstation. The images are exchanged between the different workstations in an image file format called PAPYRUS. A special communication protocol was developed to transmit the different actions performed on one station to the other. The message passing protocol is based on the ACR-NEMA standard. The lower level communication protocol is TCP-IP.

This project was carried out first with a geriatric hospital (HOGER) located at 7 Km. from the HCUG. The radiology of HOGER is run by radiologists from the HCUG. They require to be able to easily show studies to other colleagues in the hospital for second opinions and discussions.

1. INTRODUCTION

To improve the hospital-wide information system that has been developed at the University Hospital of Geneva (Switzerland), a general infrastructure has been designed and developed for the management of medical images. This Picture Archiving and Communication System (PACS) allows the management of all the life cycle phases of images. It mainly consists of the possibilities to acquire images, to archive them, to print them, to display them, and to transmit them. Images can be provided by different imaging modalities. Radiological images as well as non-radiological images can be managed. Automatically, following their acquisition, images are archived in an archive server called a juke-box. The transmission and display systems are certainly the main components of a PACS and of a teleradiology system. They deserve to be explained in more details in the next paragraphs.

The Geneva PACS [1] is based on an open architecture with heterogeneous systems and multi-vendor equipment. This design raises the problem of integration of the different components. One needs to find the glue that can make such a system work properly. In our design, the glue is the file format. A standard file format was developed based on the ACR-NEMA standard of image communication [2]. This file format, called PAPYRUS [3], allows to store sets of images as a sequence of standard ACR-NEMA messages in an encapsulated file format. A folder structure allows to keep the list of images present in the file. In our PACS system, all the images acquired are automatically converted to the PAPYRUS format. Then all the other components of the PACS deal only with one file format: PAPYRUS. This is the key of modularity and integration of our PACS components.

As a PACS allows an easy access to images, new functionalities are possible. In the Geneva PACS context, an important one is the remote expert consultation. The geriatric hospital is located seven kilometers away from the main hospital but its radiology department is managed by radiologists of the main hospital. It was important to provide them with a possibility to exchange images and to work on them cooperatively. The transmission of images is based on commercially available standard networks. Within the hospital an Ethernet network allows to transmit images between different consultation workstations. Outside of the hospital we use ISDN as well as broadband MEGACOM lines, whose bandwidth is 2Mbits/s. ISDN lines, called Swissnet in Switzerland, can be characterized by an easy access, a low cost, a good reliability and a relatively limited bandwidth (64 Kbits/s). They are sufficient for occasional usage. In our configuration ISDN lines are used to connect different regional hospitals while MEGACOM lines have been used in the frame of the European teleradiology project called TELEMED [4].

A very important component of the PACS system is the facility to display the images. As we need to display them on a variety of workstations, we elected to develop a modular and portable software called OSIRIS [5]. This software is designed to handle images from different imaging modalities as individual images or as sets of images.

2. OSIRIS AND THE SYNCHRONIZATION

The main choice criteria for the development platforms was the keyword standard where standard means widely available, manufacturer independent and officially supported. In this optic two environments have been selected for the development of the OSIRIS software: the UNIX based workstations with X11 and OSF/Motif windowing system because it can be considered as being a standard; and the Macintosh computer family as it tends to become the office desktop system in the hospital (at least in Geneva) and there are lots of users' demands to use it also as a medical station.These two environments are very different, thus making it impossible to carry out the development of OSIRIS for one platform and then install it on the other one. We decided to develop OSIRIS for both platforms simultaneously while trying to minimize system-specific developments. We opted for an object oriented approach which allowed us to structure the software into different layers of modules in a relatively independent manner. The choosen object oriented programming language is C^{++}.

The OSIRIS software is designed as a general digital medical image manipulation and analysis software. It can serve for image review as well as for image processing and more quantitative image analysis. It can be used in different clinical implementations with some degree of customization while maintaining consistency between different setups. The graphical user interface, based on windows, icons, buttons and menus, is adapted to the needs of physicians and does not require extensive skills in operating a computer. The user interface remains the same whatever kind of workstation you are working on. In a typical OSIRIS session, several windows can be opened, and a window allows the display of one single image or a set of images specific to a patient examination. To manage this capability, OSIRIS has different modes of image presentation : the stack mode, the movie mode and the tile mode. The images are displayed in overlapping resizable windows. The figure 1 shows the user interface of the OSIRIS software.

OSIRIS provides some tools that allow the basic manipulation as well as the processing or the analysis of images such as the zoom functionalities, the inverse video mode, the magnifying glass, the rotations, the flipping, the image reordering, the color manipulation, the cross section, the histogram, the text annotations, the regions of interest, the angles and local density measurements, the coordinates, the filters, the isocontours, and so on.

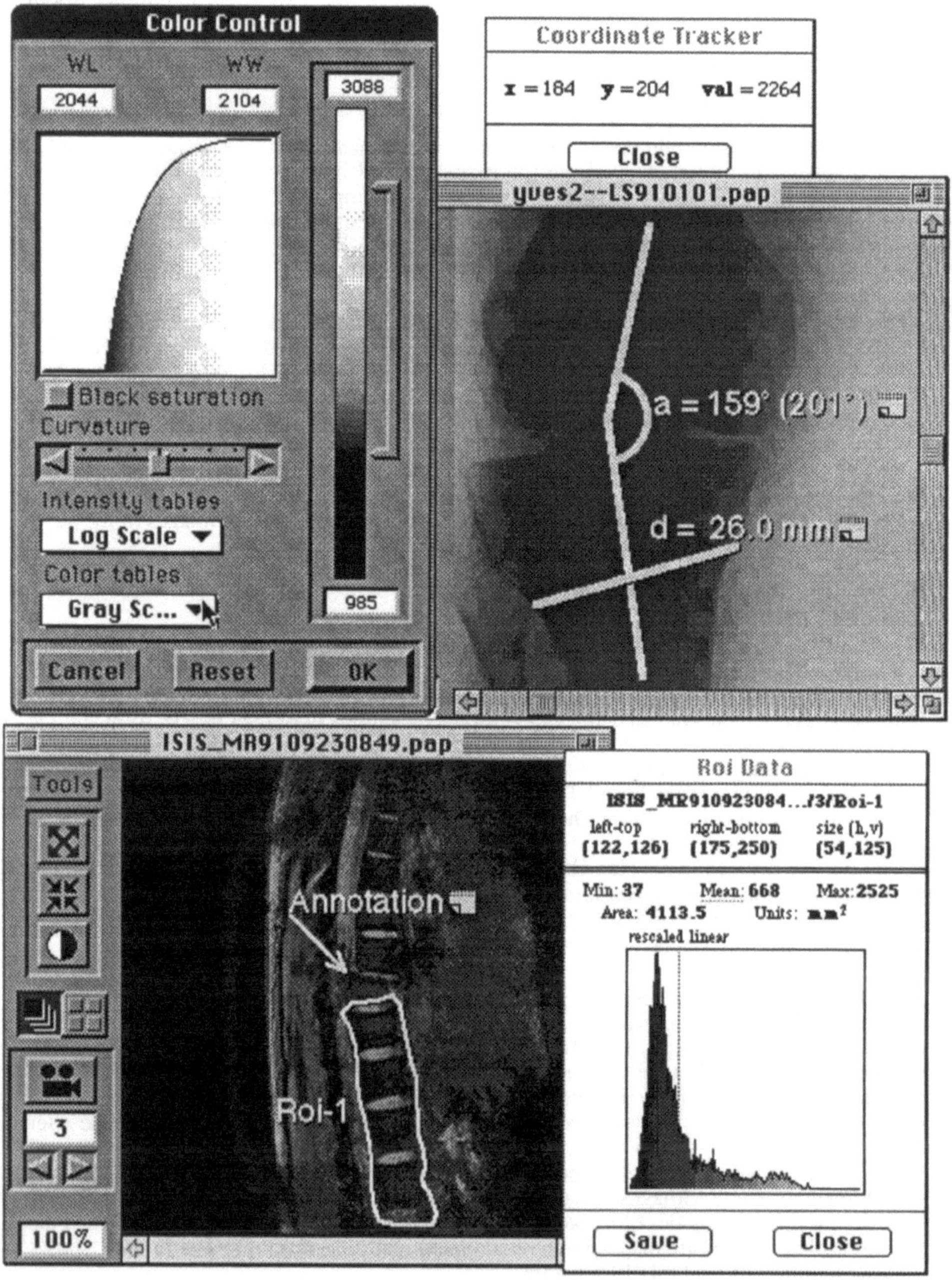

Figure 1: Typical view of the OSIRIS user interface

To allow for teleradiology and remote consultations a special synchronization feature was added to OSIRIS. This facility allows remote consultations and communications between physicians over metropolitan and wide area networks. The images must be first exchanged through the network. The interactive session can then take place. The communication protocol between the two remote stations allows for simultaneous manipulation of the same images and a live conversation through a vocal link. When establishing the communication link, it must be decided which station is the master and which is the slave. It is possible to switch from one mode to the other during the session. The transfer of messages between both stations is managed by a specific software called HERMES. It has been developed by the TELIA Research company which is a Swedish partner of the TELEMED European teleradiology project.

The figure 2 shows how the synchronization facility has been implemented and what needs to be done to have it running between different platforms. One important feature was to define a concept of synchronization event completely independent from the underlying system. All the objects into OSIRIS which are concerned with synchronization (sending or receiving a synchronization event) are system independent. Only the implementation of the synchronization object is system dependent. Concerning the Macintosh version, this object exchanges Apple events with the HERMES process who exchanges messages with another HERMES running on another machine.

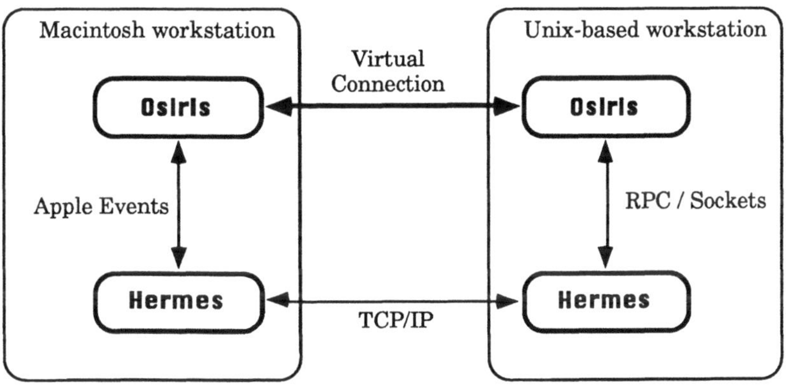

Figure 2: Implementation of the synchronization facility over different operating systems

3. THE HELIOPOLIS PROJECT

The HELIOPOLIS project is a local extension of one of the work carried in the TELEMED project. Unlike TELEMED which was developed around expensive international broadband networks, HELIOPOLIS is based on more accessible and affordable ISDN networks. It demonstrates the clinical feasibility of teleradiology on widely available public networks. The current implementation is based on Macintosh personal computers (Apple Computer, Cupertino, CA) equipped with an ISDN interface card. This setting represents an affordable solution which can be adopted by other small peripheral clinics or by private physicians. Several local clinics in Geneva are currently studying a similar system. The figure 3 shows the configuration used beetween the main hospital and the geriatric clinic.

This project was carried out first with the geriatric hospital (HOGER) located at 7 kms. from the HCUG. The radiology department of HOGER is run by radiologists from the HCUG.

The patients from the geriatric hospital are transferred to the main hospital for specialized examinations such as CT, MRI or angiographic studies. The radiologist on duty at the geriatric hospital as well as the referring physician benefit from a faster turnaround of the images where the results of the radiological studies can be accessed directly through the ISDN network rather than waiting for the images to be sent by regular mail. Besides, the radiologists require to be able to easily show radiological studies to other colleagues in the hospital for second opinions and discussions. This becomes feasible in a more convenient way than having to physically go to the main hospital or send the images to the experts in the main hospital. For this project a special software and hardware was implemented to allow the communication between the two remote workstations. The synchronization tool of OSIRIS allows them to work cooperatively.

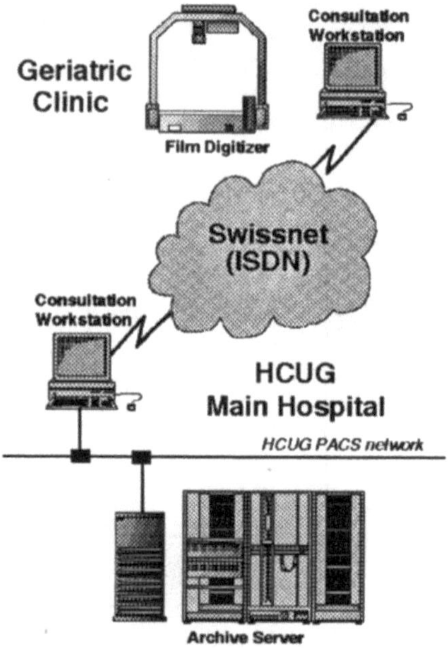

Figure 3: The HELIOPOLIS project configuration

4. CONCLUSION

The ISDN network offers an affordable and convenient solution for teleradiology. The very low throughput makes it unsuitable for the communication of large amounts of image data in a routine practice. It is however appropriate and cost effective for occasional usage of a few consultations per week. It is important to notice that the extra cost of teleradiology and wide area network access does not generate any additional income. The cost of such implementations are then usually supported by the institutions which will try to maintain it as low as possible. The ISDN solution remains one of the most affordable solution despite it requires a much longer transmission time. It should be used by more and more local clinics for teleradiology beetween them and our main hospital.

5. REFERENCES

1. Ratib O., Ligier Y., Hochstrasser D., Scherrer J.R.: *Hospital Integrated Picture Archiving and Communication System (HIPACS) at the University Hospital of Geneva*, Proc. of SPIE conf. (Medical Imaging IV), San Jose, California, vol. 1446, Feb. 91.

2. NEMA: *Digital Imaging and Communications*, ACR-NEMA standard publication, #300-1988.

3. Ratib O., Ligier Y., Appel R., Scherrer J.R.: *PAPYRUS: a Portable Image File Format*, NATO ASI Series, Vol F74, Picture Archiving and Communication Systems (PACS) in Medicine, Edited by H.K. Huang et al., Springer-Verlag Berlin, 1991, pp 91-94.

4. Ratib O.: *TELEMED Project*, NATO ASI Series, Vol F74, Picture Archiving and Communication Systems (PACS) in Medicine, Edited by H.K. Huang et al., Springer-Verlag Berlin, 1991, pp 335-336.

5. Ligier Y., Ratib O., Funk M., Perrier R., Girard C., Logean M.: *Portable Image Manipulation Software: What is the Extra Development Cost?*, Journal of Digital Imaging, W.B. Saunders Company, 1992, Vol 5, No 3, pp 176-184.

Results of a Remote Expert Consultation Project

F.-R. BARTSCH, M. GERNETH, M. KRAUSE*, M. BRADO*, R. SCHOSSER

Department of Experimental Surgery, University of Heidelberg,
Im Neuenheimer Feld 347, D-6900 Heidelberg

*Department of Radiodiagnostics, University of Heidelberg,
Im Neuenheimer Feld 110, D-6900 Heidelberg

Summary

The application pilot "Remote Expert Consultation in Radiology" has been implemented between 8 university hospitals in 7 European countries as part of the EC-sponsored TELEMED project (R1086). Advanced telecommunication services like videoconference (VC) and digital image communication (DIC) have been evaluated regarding their usability for remote consultations. For this purpose ISDN and satellite or terrestrial networks at 2 Mbit/s or 140 Mbit/s line speed were utilized. In a prospective multicenter study, videoconference proved to be a reliable tool for diagnosing digitally recorded images while some classes of conventional images are subject to a significant loss of information and diagnostic accuracy. Digital image communication promises image transmission free-of-loss but still needs more development effort before being fully fit for use in clinical routine.

1. Introduction

The development of ever more specialized medical techniques for diagnosis and therapy leads to a situation where experts in certain fields can only be encountered in highly specialized centers of excellence on a national or even international scale. At the same time, consultation of those experts for diagnosis and treatment is increasingly desired, particularly from physicians in smaller hospitals or general practice. Advanced telecommunication services that become available in many countries can be utilized to satisfy such demand. It was the aim of the EC-sponsored TELEMED project (R1086) to encourage and evaluate such medical applications on a European scale.

Since the early 1950ies, teleradiology has been subject of multiple research projects [3]. In most cases, the use of teleradiology systems for educational or consultative purposes was confined to applications within the respective hospitals. Some authors report on teleradiology systems that were employed to provide specialized radiological knowledge to small hospital in remote areas [9]. Special research effort was put into teleradiology by the US Department of Defense [1] promoting the installation of teleradiology systems at military bases to overcome a lack of radiologists in the Armed Forces.

Today, advanced telecommunications technology and digital high speed networks provide the means for international teleradiological consultations on demand. Our remote expert consultation (REC) trial was designed and implemented as a demonstrator of the benefits that public broadband networks can provide to health care.

2. Concept and implementation

To establish a consultation environment that easily adapts to daily routine, three modes of consultation were derived from radiologist's daily routine:

1. Batch mode comprises a normal, asynchronous consultation. A radiologist requesting advice or a second opinion sends images and accompanying information to the expert's electronic mailbox. The expert reviews the case at a time convenient to him and returns annotated images and a report into the requestant's mailbox.
2. Interactive mode portrays an emergency consultation. The requestant calls the remote expert and asks for an immediate consultation session to jointly review the case.
3. Mixed mode describes the teaching scenario. The requestant sends images and textual information into the expert's mailbox. At a time convenient to him, the expert initiates an interactive consultation and jointly with the requestant reviews the case.

Based on these consultation scenarios, requirements for radiological consultations were elaborated and translated into technical specifications. The implementation comprised the use of public videoconference (VC) services and digital image communication (DIC) via public broadband networks. The videoconference demonstrator was implemented between the university hospitals of Berlin (D), Geneva (CH), Heidelberg (D), Lund (S), and Tromsø (N). For the second stage of implementation, university hospitals of Barcelona (E), Florence (I), and Montpellier (F) were additionally connected to the consultation network.

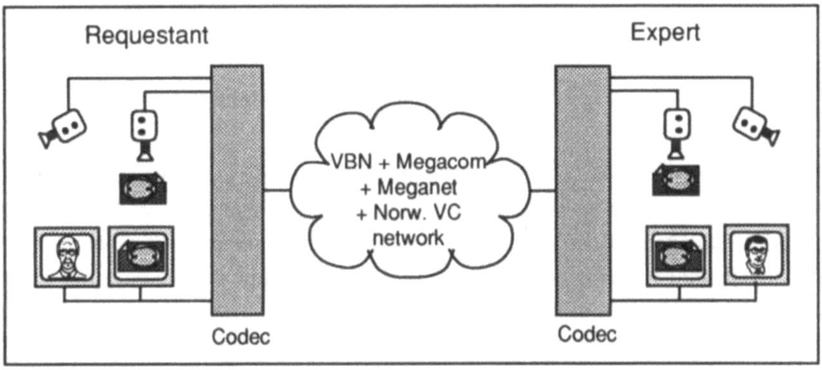

Fig.1. Standard videoconference setting

2.1. Videoconference

The standard VC setting for REC comprises person camera and high resolution document camera for recording (Fig. 1). Receive and preview monitors allow for viewing of images and for controlling outgoing information. Audio equipment encompasses microphones, loudspeaker, and echocanceller. To convert analogue to digital signals, a codec is employed. If the studio is linked to a 2 Mbit/s network, the codec additionally compresses the video signal.

German participants subscribed to the 140 Mbit/s VBN network that allows for full resolution transmission of the CCVS signal. Swiss and Scandinavian participants utilize interlinked national 2 Mbit/s videoconference networks allowing for automatically dialled conferences. To serve German radiologists for self-dialled international VCs, a semi-automatic gateway was installed between VBN and MEGACOM in Stuttgart (D) in a close cooperation between German Telekom and Swiss Telecom. The gateway acts as a passive subscriber to both networks and passes through the signal as soon as it is dialled up from both sides.

2.2. Digital Image Communication

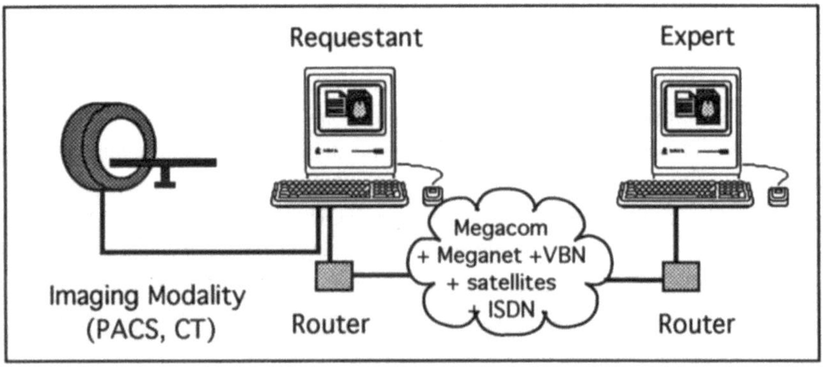

Fig.2. Setting for digital image communication

In the general implementation of DIC digital images are imported to Apple Macintosh based communication workstation either from PACS (Fig. 2) or directly from the modality. The workstation is connected to a router via Ethernet. The routers transfer data traffic using public broadband networks, satellite links, or ISDN. In addition to the networks utilized for VC, hospitals in Barcelona and Florence are connected by 2 Mbit/s satellite links on demand. Heidelberg and Montpellier use up to six automatically dialled up ISDN channels adding to a 384 kbit/s coupling.

Several software modules have been developed to support remote consultations of expert radiologists using DIC (Fig. 3). The image file format PAPYRUS [8] is based on the ACR-NEMA standard 2.0 and was developed by University of Geneva. In the TELEMED project it is used to exchange and store radiological images in digital format. For the exchange of image data and text information, the TCP/IP based communication module RECPHONE [2] was developed by University of Patras. The imaging software OSIRIS [6] is employed to display and manipulate radiological images. It was developed by University of Geneva. Finally, to synchronize the workstations for interactive manipulation of images during consultation sessions the HERMES protocol [7] a development of Telia Research of Malmö is employed.

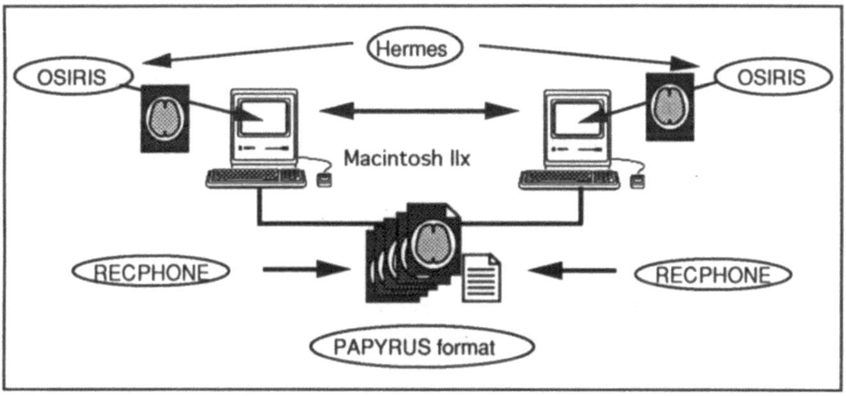

Fig.3. Software modules for DIC

3. Experiences

We investigated the diagnostic performance of standard VC technology in a prospective multi-center study.

3.1. Study design for evaluating VC

Five sets of different images and pathology were investigated, three of them consisting of conventional (analog) chest X-ray (signs of interstitial disease, e.g. cardiac lung, silicosis, lymphangiosis carcinomatosa etc.), mammography (microcalcifications), and hand-skeleton radiographs (hyperparathyroidism), and two of digital images namely liver CT (hypodense areas, signs of a focal lesions in the liver e.g. metastases of abdominal malignant tumors, hemangioma or cysts) and liver MRI (focal lesions, Gadolinium as contrast medium).

Each set of films consisted of 100 images from different cases, 50 positive and 50 negative, except for hand skeleton 78 films (19 positive, 59 negative) and for liver MRI 68 cases (33 positive, 35 negative) were selected. The films were randomized within each set, placed on a conventional light box assembly, recorded with a studio quality camera mounted over the light box and transmitted via VC networks from the study center (Heidelberg) to Berlin, Geneva and Lund. Three to four months after the VC sessions, each observer had to reexamine the original films in a conventional light box setting. The 12 participating radiologists were asked to rank their confidence of having detected lesions compatible with the given pathology, using a five score scale.

To achieve independence from decision threshold, the data was subjected to receiver operating characteristic (ROC) analysis according to individual performance, comparing VC to conventional film reading by a modified maximum likelihood estimation algorithm of Dorfman and Alf [5].

3.2. Study results

The study results clearly demonstrate that remote expert consultation using standard VC equipment and public networks is feasible. There is no loss of diagnostic accuracy with images from digital modalities. Limitations become obvious from analog images showing very subtle lesions and requiring high spatial and contrast resolution for being detected.

3.3. Digital image communication

Since no information loss due to media conversion is expected for DIC and earlier studies already investigated the accuracy of reporting on screen [4], evaluation of the second phase concentrated on feasibility studies and measurements of effective transmission speed on terrestrial broadband networks, satellite links and ISDN.

While consultations in normal mode only require conclusive textual information accompanying the images, the situation is more complex for interactive consultations as in emergency and mixed mode. A minimal communication configuration for interactive manipulation of radiological images in a consultation requires the synchronization of the viewing software and an audio link using hands-free telephone.

Measurements of the effectively used bandwidth revealed that for terrestrial 2 Mbit/s connections, image data were transferred at up to 500 kbit/s, while consultations via satellite only achieved a speed of 130 kbit/s. The same effective speed was measured via the ISDN connection. Data traffic for interactive consultations does not look like a continuous data stream. It resembles bursts when image data are transferred and low data traffic during interactive manipulation of the images.

4. Discussion

4.1. Videoconference

Videoconference is a communication tool that is available off-the-shelf. With minor modifications like the customized document unit, publicly available videoconference services at 2-140 Mbit/s can be utilized for teleconsultations in radiology. Since standardized solutions for interfacing of the equipment and public networks are available, installation is rather simple. Self-dialling of consultations provides the flexibility in scheduling required for medical applications. Since most radiology departments still work with film radiographs, videoconference offers a simple way to record and transmit a multitude of formats. For presenting a case using VC, radiologists have to adapt their daily routine procedures only marginally.

However, resolution of transmitted images is restricted if VC is used. No degradation in diagnostic reliability of images of digital modalities was found, while conventional radiographs with a high demand for spatial or contrast resolution must be treated carefully. Fortunately, the critical cases constitute only a small fraction of the film material to be evaluated in routine practice. Yet there must be considered, that no further clinical information nor history of the patient was given in our study. This interaction is a paramount advantage of VC technology. Therefore this study presents the "worst case" for remote expert consultation.

To guide through demanding cases, even the requestant needs a minimum training. He should be able to point out the relevant images or regions of interest. Another limitation of videoconference is that only one image is displayed at a time. To overcome this problem, freeze frame facilities and a second monitor were introduced in our study places.

4.2. Digital image communication

The main advantage of digital image communication is the transmission of image data without loss of information. Transmission and display fits optimal for images from digital modalities. Conventional radiographs are scanned once and can be handled as image files. Other than with videoconference, were images on film are recorded into an analogue signal, converted to digital for transmission on the network and decoded to analogue again for display, DIC requires no media transitions for the image transmission.

It is a major drawback for DIC that no collectively accepted image standard exists. Digital modalities from different vendors generate images in different formats. For a truly digital image handling chain, many individual conversion software modules have to be created firsthand. Since reporting on screen is not yet commonplace, many radiologists still feel uncomfortable. For efficient interaction during consultation, a voice or even videoconference link is needed in addition to the synchronization of actions on both screens.

Although their capacity was not fully exploited in our setting, broadband networks provide the line speed necessary for an immediate image file transfer for consultation purposes. However, tariffing has to be oriented on the actually used transport capacity for acceptance of the service.

5. Conclusion

We succeeded in implementing a remote expert consultation demonstrator on a European scale using integrated broadband communication networks (IBCN). Videoconference technology is commonly available and proved to be useful for teleconsultations in radiology if limitations in resolution are obeyed. Digital image communication has the potential for future integrated applications. However, more research and development work is needed to bring DIC into clinical routine. The ultimate solution has VC implemented into DIC.

6. Acknowledgement

The authors thank all members of TELEMED's working group 1 for their enthusiasm and active participation.

7. References

1. Allman, R.M., Curtis, D.J., Smith, J.P., Brahman, S.L. and Maso, E.C., Potential contribution of teleradiology to the management of military radiologist resources, Milit Med, 148 (1983) 897-900.
2. Basiroglou, P., Bouras, C., Garofalakis, J., Papoutsopoulos, G. and Spirakis, P., RECPHONE User Manual Version 2.2, Computer Technology Institute and University of Patras, Patras, 1991,
3. Batnitzky, S., Rosenthal, S.J., Siegel, E.L., Wetzel, L.H., Murphey, M.D., Cox, G.G., McMillan, J.H., Templeton, A.W. and Dwyer, S.J.,III., Teleradiology: an assessment, Radiology, 177 (1990) 11-17.
4. Cox, G.G., Cook, L.T., McMillan, J.H., Rosenthal, S.J. and Dwyer, S.J.,III., Chest Radiography: Comparison of High-Resolution Digital Displays with Conventional and Digital Film, Radiology, 176 (1990) 771-776.
5. Dorfman, D.D. and Alf, E.,Jr., Maximum likelihood estimation of parameters of signal detection theory and determination of confidence intervals data, J math Psychology, 6 (1969) 487-496.
6. Ligier, Y., Ratib, O., Girard, C., Perrier, R. and Logean, M., Portable and Extensible Software for the Manipulation of Medical Images. In J.P. Morucci, R. Plonsey, J.L. Coatrieux and S. Laxminarayan (Eds.), Proc. 14th A. Int. Conf. of the IEEE Eng. in Med. and Biol. Soc, IEEE Service Center, Piscatawa, NJ, 1992, pp. 1209-1210.
7. Ligier, Y., Ratib, O., Rejmer, M. and Hultman, J., Interactive software for cooperative medicine and remote consultations. In SPIE Medical Imaging Conference 1899, SPIE, Bellingham, 1993,
8. Ratib, O., Appel, R. and Scherrer, J.R., PAPYRUS - A multimodality image file format for PACS and teleradiology, Radiology, 177(p) (1990) 320.
9. Souhrada, L., Teleradiology may ease rural staffing problems, Hospitals, 63 (1989) 36.

Generics of a Media-Integrating System for Remote Consultations in Radiology

MARLENE GERNETH, FRANK-REINHARD BARTSCH, RUDOLF SCHLOSSER

University of Heidelberg, Department of Experimental Surgery
Im Neuenheimer Feld 347, D-6900 Heidelberg

Summary

For efficient consultation of remote experts in radiology, comprehensive media-integrating (tele)communication systems are desirable. Such systems should combine different ways of communication and should thus allow to communicate the broad range of information necessary for consultation.
In this paper, possible ways of communication are analyzed regarding their applicability for consultations, and characteristics of a media-integrating (tele)communication system are derived.

1. Introduction

In recent years, the increasing specialization in medicine and the resulting geographic scattering of experts in highly specialized centers of excellence on a national or even international scale has lead to a growing need for consultations, particularly remote consultations. In multiple research projects, modern telecommunication technologies have been investigated and proposed as an appropriate means to support such communication [2-5].

The information to be communicated in radiology is not homogeneous, but rather complex and heterogeneous: still images (e.g. CT or x-ray images) as well as motion images (e.g. radiological procedures like fluoroscopy or the patient and/or colleague) are relevant besides text and graphics. Since all these different information categories form the basis for medical decision-making, they must be integrated in the communication with colleagues, i.e. exchanged during a consultation.

This, however, is not possible with conventional communication means like mail or telephone. Instead, a comprehensive media-integrating communication system is required for effective consultation. In the following, possible ways of communication are analyzed regarding their applicability for consultations. Based on the results of this analysis, requirements and generic features of such a media-integrating system for remote consultations in radiology are derived.

2. Consultation of experts in radiology

The general scenario underlying the consultation of experts can be depicted as follows: A (perhaps quite unexperienced) radiologists faces a problem when preparing a report, making a diagnosis or proposing the method of treatment for a particular case. In order to receive help, he decides to consult an expert. A successful consultation and discussion of the case at question will then lead to an optimal decision on the patient's treatment.

Three modes of consultation can be derived from this general scenario:

- In normal consultation (batch mode), a requestant sends images and related textual information to the expert. At a time convenient to him, the expert reviews the case and prepares a report with possibly annotating the images. Subsequently, he returns his report (and the images, if annotated) to the requesting physician.
- In emergency mode, a highly interactive scenario, the requestant calls the expert and asks him for an instant consultation session. Images are exchanged and discussed interactively by the requestant and the expert.
- In the teaching scenario, finally, the requestant sends images and related textual information to the expert, like in normal mode. The expert reviews the case and, at a time convenient to him, calls the requestant for an interactive discussion to provide expertise.

Communication means to be employed for consultating experts have to cover the following three major aspects that can be deduced from the different consultation modes:

- exchange of radiological images (possibly also motion images from e.g. interventional procedures) and accompanying textual information,
- review of images and accompanying information, including viewing and manipulation,
- and joint discussion of the case at question.

2.1. Personal contact

Relying upon personal contact, exchange of images (and accompanying textual information) is done by the requestant himself, carrying the images and other relevant material with him. Joint reviewing, reporting, and manipulating is accomplished using films and light boxes or viewing stations. Regarding information exchange and discussion, this way of communicating is most effective for consultation, since joint viewing and cooperative working are possible and work faster than e.g. using mail or telephone. However, in case of large distances, the efforts for travelling (in time and money) are considerable, so that consultations relying upon personal contact are not always possible. Thus, a surrogate has to be looked for that comes as close as possible to personal contact.

2.2. Conventional ways of communication: mail and telephone

By mail (usually postal services or courier), images and accompanying textual information (letters or notes) can be exchanged, however, quite slowly. Reviewing, reporting and manipulating are done by the expert on light boxes in his department in general. Information exchange and discussion are performed in written form. Thus, no interactive discussions are possible.

Using telephone, no exchange and reviewing of images (and accompanying textual information) is possible. For this, the communication partners have to rely on additional means like e.g. mail. Regarding information exchange and discussion, telephone offers valuable opportunities for interactive discussion of the case at question that result in a more efficient communication than e.g. writing notes or letters. However, both partners have to be present for discussion at the same time. Comparable to mail, images and their contents have to be described verbally, no pointing to interesting regions is possible.

2.3. Advanced communication technologies

As examples for advanced communication technologies, videoconferencing and workstation-based communication have been investigated within the frame of the TELEMED project regarding their usability for remote expert consultation in radiology.

Using videoconferencing [1], exchange of images (and accompanying textual information) is performed using document transmission units comprising of high-resolution cameras mounted on top of a horizontal light box. Like in every-day practice, radiological films are handled which makes this way of exchanging images very easy-to-use. Reviewing, reporting and manipulating of images is done closely connected to the image exchange, using document transmission units and video monitors. In an international multicenter study, the diagnostic reliability of such a communication system could be demonstrated. Interactive information exchange and discussion are feasible. Exceeding the possibilities of e.g. pure audio communication via telephone, videoconferencing supports non-verbal communication, too. Hence, videoconferencing as a means for consultation comes very close to personal contact.

Using workstation-to-workstation-communication for exchanging images (and accompanying textual information) [1] is far more abstract than e.g. personal contact or even videoconferencing. On the other hand, the full range of information can be exchanged, when communicating digital images which are directly imported from the respective modality. Reviewing, reporting and manipulating are done on the workstations, offering a wide range of viewing and manipulations functionalities. For effective cooperative working, synchronization of image display on both workstations is required, enabling also information exchange and discussion. Furthermore, exchange of text messages is possible, however, medical users are not used to this way of communicating. Hence, pure workstation-based communication constitutes a rather timely and inefficient way of communicating.

3. Characteristics of a media-integrating system for consultations

As it becomes obvious from the above analysis, all the various communication alternatives reveal specific advantages and disadvantages for transmitting the broad range of heterogeneous information needed for (diagnostic and/or therapeutic) consultations. An optimal consultation system should thus combine the advantages of the different communication means, eliminating as far as possible the respective disadvantages. For information exchange and discussion, surrogates as close as possible to personal contact are most appropriate.

3.1. System components

As necessary components, such media-integrating (tele)communication systems should comprise interpersonal communication facilities and still video image transmission as well as workstation-based image communication.

Interpersonal communication facilities should include both audio and video communication. Audio communication is implemented via microphones (relevant for the sender) and loudspeakers (receiver). For video communication, person camera(s) are employed on the sending side. Alternatively to receiving motion images from the communication partner on a separate video monitor, computer screen display is possible via video frame capture and display facilities. This is, however, not very appropriate as soon as the screen is covered with several simultaneously displayed image files.

Still image transmission can be realized using a document transmission unit (like in videoconferencing), including a high-resolution camera with adjustable iris function (sender) resp. high-resolution video monitor (receiver) for displaying radiological images recorded with a document camera.

Workstation-based communication must include the communication of digitized or primarily digital images (e.g. from CT or MR) and their synchronized display on high-resolution screens using appropriate software, as well as the exchange of text documents and messages.

3.2. System requirements

The consultation system should be

- integrated as one piece of equipment (instead of multiple ones; however, several monitors may be necessary) into the environment of a radiological department,
- modular, i.e. adaptable to different communication needs by combination of individual components as well as offering comprehensive viewing and reporting functionalities,
- flexible, i.e. offering various bandwidths according to the actual communication needs, and
- compatible, i.e. providing standardized interfaces to other communication systems and radiological modalities as well as compatible to other systems and procedures used in the working routine of a radiology department.

Establishing communication links should be automated. Traffic-adaptive networks (with corresponding tariffing) should be used to take into account the inhomogeneities of data flow during consultation sessions (bursts before, at the beginning or the end of a session; smaller data volumes, e.g. command sequences, during consultation).

Concerning the handling of the application and user acceptance aspects, establishing consultation sessions on demand according to the actual working schedule and workload is most essential. Thus, self-dialling and switched networks are required.

3.3. Introduction of such integrated communication services

According to our experiences, such integrated communication services should be introduced on a regional level, following existing communication needs and paths. Extensions will eventually evolve on a national and even international level, according to communication needs. However, establishing regional self-contained islands without respect to established standards must be avoided.

A first introduction of such advanced communication services should be performed in selected medical disciplines and application fields regarding the demand and the relevance of the specific topics of communication for further refinement and improvement of the service. We regard radiology an appropriate area for future research on these issues.

4. Conclusion

Enabling remote consultations via a media-integrating (tele)communication system combines the advantages of different forms of communication. However, current developments are not yet able to level all existing disadvantages of either communication means. Furthermore, technical limitations like storage capacity required for the large data volumes are not yet overcome.

Today, the basic components of such a communication system as proposed here are already available. Thus, media-integrating communication in radiology can be considered realistic. However, considerable further developments are necessary in order to reach the level of maturity needed for routine usage, in particular regarding integration and adaption into the radiological environment.

Integrated communication is supposed to accelerate the medical decision-making process by making expert knowledge widely available, to improve health care (particularly in rural areas without expertise near-by), and to lower health care costs (e.g. by avoiding double examinations through more effective communication).

Economic considerations as well as the future potential for reduplication and evolution into routine usage should stimulate the development of such communication systems.

5. Acknowledgements

This work was partly sponsored by the RACE research programme of the Commission of the European Communities under contract R-1086. The authors would like to thank the participants of TELEMED's working group 1 for their enthusiasm and active participation in implementing and evaluating videoconferencing and digital image communication for international remote expert consultation.

6. References

1. Bartsch F-R, Gerneth M, Schosser R, Krause M, Brado M (1993) Results of a Remote Expert Consultation Project. In: Computer Assisted Radiology - Proceedings of the International Symposium CAR'93, Springer, Berlin Heidelberg
2. Carey LS (1985) Teleradiology: part of a comprehensive telehealth system. Radiol Clin North Am 23:357-362
3. Jutras A (1959) Teleroentgen diagnosis by means of video-tape recording. Am J Roentgenol 82:1099-1102
4. Sund T, Rinde E, Stormer J (1991) Full-Scale Replacement of a Visiting Radiologist Service with Teleradiology. In: Computer Assisted Radiology; Proceedings of the International Symposium CAR'91, edited by Lemke HU, Rhodes ML, Jaffe CC, Felix R. Springer, Berlin Heidelberg
5. Webber MM, Wilk S, Pirruccello R, Aiken J (1973) Telecommunication of Images in the Practice of Diagnostic Radiology. Radiology 109:71-74

MIMACS: Results of a Study to Evaluate the Clinical Relevance of Teleradiology

E. List-Hellwig[1], R. Rechid[1], R. Michalik-Himmelmann[1], M. Gartenschläger[1], K.P. Stocker[1], W. Gelmetti[1], A. Lütcke[2], G. Spalke[2], M. Koller[3], O. Rienhoff[4], K.J. Klose[1]

University Hospital Marburg
1. Department of Diagnostic Radiology.
2. Department of Neuroradiology
3. Department of Theoretical Surgery
4. Institute for Medical Informatics
Baldingerstraße
DW-3550 Marburg/Lahn

Summary

The purpose of the teleradiology project within the Marburg Image Management and Communication System (MIMACS) was to evaluate the teleconsultation process between the Departments of Diagnostic Radiology and Neuroradiology at the University Hospital of Marburg. The project was divided into three phases. In the first phase, completed in June 1992, the technical and organisational framework was developed. In the second phase, completed in October 1992, radiologists and neuroradiologists evaluated the functions of Siemens workstations DRC40 and Litebox by means of a questionnaire. It turned out that workstation functions most often used at both sites were: magnifying and rearranging images, displaying image data, using window/level presets and modifying the grayscale. Users rarely took advantage of the magic glass, zoom and density measurement functions. In the third phase of the study, the impact of teleconsultation on diagnoses and clinical recommendations was analysed. The study design included three radiologists who specified their diagnoses and clinical recommendations before and after consulting with the neuroradiologists. The degree of difference between these two points of judgement was evaluated by three radiologists who were not involved in the consultation process. The results of phase three show that teleconsultation changed the initial diagnostic findings in 34% of the cases and changed the clinical recommendation in 22% of the cases.

Introduction

Teleradiology is defined as the transmission of radiologic images from one place to another. The development of teleradiology systems took place in many countries, starting in 1950 with Gershon-Cohen and Cooley in Philadelphia ([5]). Since then there have been numerous trials of clinical teleradiology for the purpose of diagnosing radiologic images from remote sites where radiologists are not available (telediagnosis), for consulting remote experts (teleconsultation), and for training people at distant sites (tele-education). Different techniques for image aquisition have been used: facsimile ([5], analog video modulation ([22],[1],[8],[17]), image digitization ([6],[7],[13],[15],[20]) and direct digital data acquisition from modalities ([4],[12],[11]) and for image transmission telephone lines ([5],[22],[6],[7],[13],[15],[20]), coaxial cables ([9],[19]), terrestrial microwave ([14],[1]) and satellite microwave transmission ([10],[16],[3],[21]), and fiber optics ([2],[18],[4],[12],[11]). IMACS are being used more frequently to improve image communication within and between hospitals.

Marburg Teleradiology Project

The diagnostic accuracy and clinical acceptance of teleradiology systems have been evaluated in many studies. The purpose of our study was to evaluate the teleconsultation process between the Departments of Diagnostic Radiology and Neuroradiology at the University Hospital of Marburg with respect to the impact of teleconsultation on diagnoses and clinical recommendations in neuroradiology cases.

The hardware and software platform of the Marburg teleradiology project is SIENET which has been built up since 1989. An inter-hospital fiber optic Ethernet link was established between the Departments of Radiology and Neuroradiology (distance 3,5 km) for teleconsultation purposes. For Teleconsultation, one DRC40 in the Department of Radiology and a two screen Litebox in the Department of Neuroradiology were used.

Experimental Design and Results

The teleradiology study was divided into three phases:

Phase 1

In the first phase of the study, completed in June 1992, the technical and organisational framework was established. Teleconsultation between the radiologists and the neuroradiologists (expert) took place four times a week with a maximum of four actual routine cases per consultation. Teleconsultation within the study was limited to cranial CT examinations. Every examination consisted of approximately 23 or 46 images, depending on the application of contrast media or the calculation of high resolution images of the skull. The digital matrix size of every CT image was 512x512 pixels with a grayscale coding of 16 bits per pixel. The transmission speed from the PACS-host (Microvax II) to the Siemens-Litebox (Apple Macintosh IIx) was 20 seconds per image.

A questionaire designed to evaluate the DRC40 and Litebox image manipulation functions was tested with 25 consultations.

A special telephone with the capability of free speaking was installed at both sites. After testing the telephone the users demanded a louder tone. Therefore, an additional loudspeaker was connected, but because of overlaying echos and unfiltered noises in the environment the loudspeaker was not accepted and the consultation partners communicated by normal telephone. An additional microphone could have improved the free speaking function.

Phase 2

In the second phase, completed in October 1992, the questionaire was applied to 100 consultations. Depending on the workstation functionality, different ways of workstation handling were established. At the Litebox, the diagnostic view (mosaic view) was preferred because of its ability to display 24 scans simultaneously on two monitors. Thus the CT images were reduced to 256x256 pixels. At the DRC40, the stack view of maginified images (1024x1024 pixels) was preferred because only 16 images could be displayed in the diagnostic view.

It turned out that workstation functions most often used at both sites were: magnifying and rearranging images, displaying image data, using window/level presets and modifying the grayscale. Users rarely took advantage of the magic glass, zoom and density measurement functions.

In 33% of the cases, users wished to have synchonization of the workstations for simultaneous image display and manipulation and the ability to point to relevant findings at the remote workstation using the mouse.

In 49% of the cases, neuroradiologists demanded to access previous examinations in order to achieve greater diagnostic reliability.

Phase 3

In the third phase, completed in February 1993, the influence of teleconsultation on diagnoses and clinical recommendations was analysed. The study design included three radiologists who specified their diagnoses and clinical recommendations before and after consulting with the neuroradiologists. The degree of difference between these two points of judgement was evaluated by the consultation partners (internal raters) and three radiologists (external raters) not involved in the consultation process.

The CTs were categorized according to the following clinical questions: (a) bleeding/ischemia, (b) inflammation, (c) trauma, (d) neoplasma, (e) anomaly and (f) metabolic disorder.

In order to assess the difference between the diagnoses, the following six categories were defined: (1) no differences in diagnoses, (2) differences in secondary diagnoses, (3) slight differences in primary diagnoses, (4) slight differences in primary diagnoses and differences in secondary diagnoses, (5) distinct differences in primary diagnoses, (6) distinct differences in primary diagnoses and differences in secondary diagnoses.

Modification of the clinical recommendation was jugded by the two categories (I) yes and (II) no.

During the third phase of the study, consultations were done on 50 cranial CT examinations. The distribution of cases based on the clinical question is presented in Table 1.

Table 1: Distribution of cases based on the clinical question.

Teleconsultation problem	N (= 50)	percent
Bleeding/Ischemia	21	42 %
Neoplasma	18	36 %
Trauma	9	18 %
Inflammation	1	2 %
Anomaly	1	2 %
Metabolic Disorder	0	0 %

The internal raters (IR) assessed changes in the primary diagnosis (categories 3 to 6) in 32% of the 50 cases consulted; whereas, external raters (ER) assessed a difference in 36% (mean value of three raters) in the primary diagnosis (see Table 2).
The ratings were analysed based on the clinical question. A difference in the primary diagnosis (assessment categories 3 to 6) was found for **bleeding/ischemia** in 33% (IR) / 38% (ER) of 21 cases, for **neoplasma** in 39% (IR) / 35% (ER) of 18 cases and for **trauma** in 22% (IR) / 37% (ER) of 9 cases (see Table 2). The raw data for each rater group can be found in Table 3.

Table 2: Difference in primary diagnosis according to the clinical question after teleconsultation. (IR = internal raters, ER = mean of external raters)

Clinical question	IR	ER
Bleeding/Ischemia	33 %	38 %
Neoplasma	39 %	35 %
Trauma	22 %	37 %
Overall	32 %	36 %

Table 3: Assessment of the difference in diagnoses after teleconsultation. (IR = Internal raters, ER1..ER3 = External raters, diagnostic categories are defined above)

diag-	bleeding/ischemia				neoplasma				trauma			
nosis	IR	ER1	ER2	ER3	IR	ER1	ER2	ER3	IR	ER1	ER2	ER3
1.-2.	14	13	13	13	11	10	12	13	7	4	5	8
3.-6.	7	8	8	8	7	8	6	5	2	5	4	1
Sum	21	21	21	21	18	18	18	18	9	9	9	9

diag-	inflammation				anomaly				total			
nosis	IR	ER1	ER2	ER3	IR	ER1	ER2	ER3	IR	ER1	ER2	ER3
1.-2.	1	1	1	1	1	0	1	1	34	28	32	36
3.-6.	0	0	0	0	0	1	0	0	16	22	18	14
Sum	1	1	1	1	1	1	1	1	50	50	50	50

The clinical recommendations were modified in 16% (IR) / 27% (ER) of the 50 cases after consultation (see Table 4). Again, based on the clinical question, the recommendations were changed for **bleeding/ischemia** in 24% (IR) / 37% (ER) of 21 cases, for **neoplasma** in 11% (IR) / 17% (ER) of 18 cases, and for **trauma** in 11% (IR) / 30% (ER) of 9 cases (see Table 4). The raw data for each rater group can be found in Table 5.

Table 4: Difference in clinical recommendation according to the clinical question after teleconsultation. (IR = internal raters, ER = mean of external raters)

Clinical question	IR	ER
Bleeding/Ischemia	24 %	37 %
Neoplasma	11 %	17 %
Trauma	11 %	30 %
Overall	16 %	27 %

Table 5: Assessment of clinical recommendations after teleconsultation. (IR = Internal raters, ER1..ER3 = External raters)

	bleeding/ischemia				neoplasma				trauma			
	CP	R1	R2	R3	CP	R1	R2	R3	CP	R1	R2	R3
yes	5	8	9	6	2	3	2	4	1	3	2	3
no	16	13	12	15	16	15	16	14	8	6	7	6
Sum	21	21	21	21	18	18	18	18	9	9	9	9

	inflammation				anomaly				total			
	IR	ER1	ER2	ER3	IR	ER1	ER2	ER3	IR	ER1	ER2	ER3
yes	0	0	0	0	0	0	0	0	8	14	13	13
no	1	1	1	1	1	1	1	1	42	36	37	37
Sum	1	1	1	1	1	1	1	1	50	50	50	50

Conclusions

Teleconsultation was implemented in order to provide radiologists with neuroradiological expert advice more efficiently in their clinical routines. Digital image transfer to the Neuroradiology Department resulted in a considerable amount of time being saved for the radiologists since they were no longer required to commute between the distant departments. However, the actual transmission time of twenty seconds per CT image has to be improved.

The results have shown that teleconsultation can be practiced with relatively basic workstation functionality. In 34% of the cases, teleconsultation changed the initial diagnostic findings. Teleconsultation also changed the clinical recommendation in 22% of the cases. Thus, it makes sense to apply teleconsultation in the daily practice of a radiology department in the case that the expert is at a distant site and direct communication is not possible.

References

1. Andrus W.S., Dreyfuss J.R., Jaffer F., Bird K.T.: Interpretation of Roentgenograms via Interactive Television. Radiology, 116, 25-31, 1975
2. Carey L.S., O'Connor B.D., Bach D.B., Hobbs B.B., Hutton L.C., Lefcoe M.S., Lyons R.O., Munro T.G., Paterson R.G., Rankin R.N., Rutt B.K.: Digital Teleradiology: Seaforth-London Network. Journal of the Canadian Association of Radiologists, 40 (2), 71-74, 1989
3. Desai A.B., Bhargava S.K., Mehta D.K., Deshpanda A.S., Ramji, S.: Telemedicine - A Feasibility Experiment for National Application. Indian Pediatrics, 21 (10), 773-775, 1984
4. Felix R., Langer M., Astinet F., Zendel W.: Das Berliner Radiologie-Kommunikations-Projekt. Erste klinische Erfahrungen. Fortschr. Röntgenstr. 152 (6), 667-672, 1990
5. Gershon-Cohen J. and Cooley A.G.: Telognosis. Radiology, 55, 582-587, 1950
6. Gitlin J.N.: Teleradiology. Radiologic Clinics of North America, 24 (1), 55-68, 1986
7. Heshiki A., Amanuma M., Tomioka K., Ohta H., Inamura K.: Teleradiology at Gunma University Hospital. Med. Inform., 13 (4), 295-301, 1988
8. Jelaso D.V., Southworth G., Purcell L.H.: Telephone Transmission of Radiographic Images. Radiology, 127, 147-149, 1978
9. Jutras A.: Teleroentgen Diagnosis by Means of Video-tape Recording (Editorial). American Journal of Roentgenology, 82, 1099-1102, 1959

10. Lester R.G., O'Foghludha F., Porter F., Friedman D.S. and Pedolsky H.R.: Transmission of Radiologie Information by Satellite. Radiology, 109, 731-732, 1973
11. List-Hellwig E. , Rechid R., Kleinöder T.: Einsatz des Marburger PACS in der Neurochirurgie und Neuroradiologie, 6. Frankfurter Gespräche über Digitale Radiographie, Bad Nauheim, 8.-10.10.92, Schnetztor Verlag, Konstanz
12. List-Hellwig E., Rienhoff O., Klose K.J.: Requirements for teleradiology system. ECR´91, Vienna, 16.-20.09.91
13. Markivee C.R., Mahanta B., Savci S., Abbas S.A., Chenoweth J.L., Luther R.K., Patel B.K., Vas W.G., Hill T.G.: Diagnostic Accuracy of a Teleradiology Image Transmission System. MD Computing, 6 (2), 88-93, 1989
14. Murphy R.L.H., Barber D., Broadhurst A. and Bird K.T.: Microwave Transmission of Chest Roentgenograms. American Review of Respiratory Disease, 102, 771-777, 1970
15. Paakkala T., Aalto J., Kahara V., Seppanen S.: Diagnostic performance of a teleradiology system in primary health care. Comput Methods Programs Biomed. 36(2-3), 157-160, 1991
16. Page G., Sylvestre J., Roberge F.A., Gregoire A., Chalaoui J., Galand C., Laperriere J., Seguin R.: Narrowband Teleradiology. Journal of the Canadian Association of Radiologists, 33 (4), 221-226, 1982
17. Roberge F.A., Page G., Sylvestre J., Chahlaoui J.: Telemedicine in northern Quebec. Canadian Medical Association Journal, 127 (8), 707-709, 1982
18. Souhadra L.: Strategy: Communication Links with Physicians. Hospitals, 64 (10), 55-62, 1990
19. Steckel R.J.: Daily x-ray rounds in a large Teaching Hospital using High-Resolution Closed Circuit Television. Radiology 105, 319-321, 1972
20. Viitanen J., Sund T., Rinde E., Stoermer J., Kormano M., Heinilä J., Yliaho, Ahonen J.: Nordic teleradiology development. Comput Methods Programs Biomed, 37 (4), 273-277, 1992
21. Watson D.S.: Telemedicine. The Medical Journal of Australia, 151, 62-73, 1989
22. Webber M.M. and Corbus H.F.: Image Communication by Telephone. Journal of Nuclear Medicine, 13, 379-381, 1972

Expert Consultation for Mobile MRI Examinations by Teleradiology

F.H. BARNEVELD BINKHUYSEN (1), F.P. OTTES (2), J. VALK (3).

1. Eemland Hospital, location de Lichtenberg, Amersfoort, dept. of radiology. the Netherlands.
2. BAZIS, Central Development and Support Group Hospital Information System, Leiden, the Netherlands.
3. Academic Hospital of the Free University Amsterdam, dept of radiology, the Netherlands (AZVU).

INTRODUCTION

At the end of 1990 the first mobile MRI-scanner (Fysiologic B.V.) started to function in the Netherlands. To support the function of MRI in general hospitals in terms of diagnostic efficacy and efficiency, this service was extended by teleradiology (Rogan B.V.).
This teleradiologyproject is a joint effort of Fysiologic B.V., Rogan B.V., BAZIS and the Academic Hospital of the Free University of Amsterdam (AZVU). It is financially supported by the Dutch Ministery of Economic Affairs.

PURPOSE OF THE TELERADIOLOGYPROJECT

The goals of the project were:
1. To support the mobile MRI service in general hospitals by using the knowledge of a recognized MRI expertise center (AZVU).
2. To estimate the required teleradiology equipment (hardware and software) and the required organizational support to provide teleradiology in daily clinical routine.
3. To get more information on the diagnostic efficacy of a teleradiology consulting system for a mobile MRI-unit?

APPLICATION SCENARIO

The mobile MRI-unit was used by different general hospitals every week, with a frequency ranging from two days a week through one day per four weeks. During the day when the mobile MRI-unit was situated at a general hospital, the present radiologist had the possibility to request additional information about a MRI investigation by means of teleradiology. The most common reason for the radiologist to use the concultancy function by means of teleradiology was that he was not (completely) sure about his interpretation of the MRI images.
When the radiologist decided to use the expert consultation service, he asked the technician to send the complete MRI procedure to the expert center (AZVU). At the same time he had to complete a form which was sent by fax together with the images. At the additional fax request the next data and questions had to be filled in: patient data, medical indication for the MRI procedure, preliminary findings of the radiologist who performed the MRI procedure, preliminary conclusion and eventual additional questions.
The MRI expert responded by fax. This whole procedure had been accomplished in 43 cases during 1992.

TELERADIOLOGY CONFIGURATION

The teleradiology network consists of two systems of which one is built in the mobile MRI-unit (the sending system) and the other one is installed in the department of radiology of the AZVU. Both systems use a PC with a 486 processor, a colour screen, which is driven by a special graphical chart (ATI wonder) and with a 600 MByte magnetic disc. Both systems work with HyperPACS ROGAN software, which uses the MS-DOS operating system. The systems are coupled with a modem to tranfer the MRI images to the AZVU via ordinary telephone lines using data-compression with the JPEG algorithm.

HyperPACS in the mobile MRI-unit.

All images acquired with MRI are entered into the teleradiology system by means of a frame grabber. The analogue videosignal from the MRI which is sent to the multiformat camera, is intercepted and digitized. Information on the patient (name and birthdate) can be typed in, and can be stored together with the image file in the HyperPACS database. By means of HyperPACS software it is possible :

- to store MRI images and additional patient data on magnetic disc,
- to present images on the colour screen, and manipulate them (e.g. level and window, zoom/minify, flip, rotate)
- to transfer images by means of a modem connection to another HyperPACS station, and
- to remove images.

For most of these functions the system has the possibility to make selections out of the existing database: firstly from one or more patients, secondly from one or more MRI examinations of the patient(s) and finally from one or more images of the selected examinations.
The magnetic disc can contain all images acquired by the MRI during approximately two working days. The input module shows how much discspace is still available and a signal is given when less discspace than required for a complete MRI procedure is available.

HyperPACS in the Academic Expert Center.

The AZVU system has a high resolution screen (Mega Scan) which can present 20 MRI images at the same time in full resolution. With the HyperPACS module, which is specially made for this hardware, one can navigate through all 60 - 100 images of a MRI examination in a rather user-friendly way. With a single command one can look at the next or the previous 20 images. Like in the mobile unit, one can also select a series of images. The selected images can be removed or presented on the screen.

Technical aspects

It takes about 90 minutes to transfer a MRI examination of 80 images. The images are made of 300 kBytes without compression and of approximately 30 kBytes after JPEG compression. To send images by means of an 18 Kbit/sec modem (nett approximately 1.5 kByte/sec) takes about 20 seconds. The sending of images is interrupted as a result of compressing each image, which also takes about 20 seconds. Before sending the images, the local database of the HyperPACS station at the AZVU is prepared, which all together takes about 20 to 30 minutes. Overall it takes about 80 to 90 minutes to tranfer a complete MRI examination by means of teleradiology. It appeared that the standard telephone lines are sufficiently stable to allow such long transfer times.

It was not possible to carry out two HyperPACS functions at the same time. For example: When an examination was sent by modem (80 to 90 minutes), the images of a new examination could not be entered into the system. This means that the sending of examinations was put off till the end of the working day. When a new series of images was printed on film, this could not be done in parallel with image entry into the HyperPACS system.

To set-up a modemconnection between the mobile unit and the AZVU-system, the AZVU-system had to be in a neutral position, in other words: the main menu had to be presented. When the AZVU system was in use, was not put back in the neutral position, or was still carrying out a decompression, a connection between the mobile unit and the AZVU-system could not be realised.

ORGANISATIONAL ASPECTS

Because of the long transfertime of a MRI examination by means of teleradiology, hospitals were recommended to install two telephonelines in the mobile MRI unit. In only two hospitals this was realised. The consequence was that in the other hospitals it was not possible to send images to the expert center during working hours, because the only available telephoneline was needed for the daily work. In such a case the best solution was to send the images immediately after the last MRI procedure of that day. However, sometimes there was no time left, because the MRI unit was already moved to another location. Then it often happened that one had to wait untill again a hospital was visited with two telephonelines available for the mobile MRI-unit. Two problems occurred in this situations:
1) The technician in charge the day after was not well informed about the case, or
2) The images were removed accidentally.

Daily project management was necessary:
- In the morning the technicians had to be reminded to use the system.
- In the afternoon the radiologist had to be reminded to select the examinations that ought to be transfered (because when such a decision was made the next day it was be possible that the images were already automatically removed from disc).

When an expert consultation was requested, the project manager had to check if:
- the technician transfered the selected images,
- the form with patient data was filled in properly and sent by fax together next to the images,
- the expert radiologist recieved the images and the fax.

DIAGNOSTIC EFFICACY

An expert consultation for mobile MRI by means of teleradiology was realised during nine months. Although the number of cases for which expert consultation was requested was less than expected (n=43), the clinical impact was high.
To answer the clinical question concerning the diagnostic efficacy of working with the teleradiology consulting system the answers of the MRI-expert radiologist were divided into five groups namely:
1. The expert opinion differs significantly,
2. The expert opinion gave adjacent information,
3. The expert opinion was confirmative,
4. The expert had too few clinical data to respond adequately,
5. The expert recommended to extend the MRI procedure.

The expert opinion differed significantly in eight cases (see Table 1), gave adjacent information in six cases and was confirmative in 23 cases. Two times there were too few clinical data to respond adequately and four times an extension of the MRI procedure was recommended.

In four of the eight cases in which the expert opinion differed from the original report, this had direct impact on patient management.

Table 1.
The eight cases where the expert opinion differed significantly.
The four cases leading to a change in management are marked as *.

Opinion of the radiologist on duty at the mobile MRI	Expert opinion at the AZVU
- white matter disease without aspect of Borrelia infection	- white matter disease with aspect of Borrelia infection
- no recurrent tumor	* oligodendroglioma
- arachnoid cyst	* epidermoidoma
- infarction or ischemia	- normal aspect; artefacts
- subcortical lesions seen in elderly	- subcortical lesions seen in toxic encephalopathy
- no compression of the spinal cord	* compression of the spinal cord
- ependymoma, astrocytoma or multiple sclerosis	- extramedullary tumor (meningeal tumor)
- infarction	* infection

DISCUSSION

The management input concerning the technical and organisational aspects to realise a continuous teleradiology service appeared to be very high. This was caused e.g. by the number of people and locations involved. Maybe this was one of the reasons that the number of requests for expert consultation was less than expected. From March 1992 to December 1992 the mobile MRI unit functioned 118 days, in which 43 requests for expert consultation were made and completed. It appeared that only for neuroradiological problems an expert consultation was requested. The reason therefore was probably that the expert radiologist is a well known expert MRI neuroradiologist.
Furthermore the question arises if the patient population in mobile MRI was representative. It seemed that a lot of cases were rather "simple", because
(1) the waiting time for a mobile MRI procedure is rather long, and
(2) the limitation of using a mobile MRI unit (no respiration- or ECG triggering).

CONCLUSIONS

This Dutch project of expert consultation for mobile MRI teleradiology showed that:
- the clinical impact in terms of diagnostic efficacy is rather high (in eight of 43 cases the MRI expert opinion differed significantly).
- the number of MRI procedures in which an expert opinion was requested, was less than expected.
- teleradiology in the Netherlands is possible by using regular telephone lines (despite the long transfer times of 80 - 90 minutes)

REFERENCES

1. Golgberg, M.A.; Rosenthal, D.I.; Chew, F.S. et al. New High-Resolution Teleradiology System: Prospective Study of Diagnostic Accuracy in 685 Transmitted Clinical Cases. Radiology 1993; 186: 429-434.
2. Bidgood, W.D. and Staab, E.V. Understanding and Using Teleradiology. Seminars in Ultrasound, CT and MRI, Vol 13, No 2, April, 1992: 102-112.

The Irrelevance of Geography and Time in Assuring High Quality Patient Care Resulting from Proliferation of Teleradiology in the Pacific Rim

A. GELISH[1], & DR S. K. MUN[2]

[1] HQ PACAF/SGAL 25 "E" Street, Suite F318, Hickam AFB, HI

[2] Georgetown University Medical Center, Department of Imaging Physics, Washington, DC

Summary

Factors involved in selecting remote teleradiology sites and establishing their proliferation priority validates the concept of a virtual radiology department. Factors include mission, access to alternative care, workload and types of care, patient population, environmental concerns, transportation and isolation. Research topics have been proposed for using a three way digital satellite communication link of the massive, parallel-processing computer at the University of Hawaii computer laboratory, Maui, Hawaii, Georgetown University Medical Center's imaging physics laboratory, Washington DC, and Tripler Army Medical Center, Honolulu, Hawaii, as a source for digital image data for analysis.

Military Scenario

In a hushed, dimly lit room on a United States military installation somewhere near Seoul, Republic of South Korea (ROK), military officers hover in front of their computer monitors their faces bathed in an eerie glow. They peer intently at moving images . Each officer speaks quietly into a microphone oblivious of the activities of others in the room.

Overextended Radiology Services

On the South Korean peninsula, an area the size of Indiana (roughly 38,230 square miles), nearly 316,000 active duty military men and women in the US Air Force, Army and Navy are totally dependent on three radiologists for their diagnostic needs. Similar situations exist in Alaska, Guam, Honshu and Okinawa islands, Japan, Singapore, Madagascar, Navy ships, and for military personnel deployed on humanitarian missions in Thailand, Somalia and other remote areas.

High Technology

The room near Seoul isn't some secret bunker in the heart of a military intelligence headquarters nor a mission control center for a classified weapons launch facility. This scenario describes the hub of a virtual radiology department soon to encompass all DoD medical imaging sites in the ROK in an invisible telecommunications web. This communications web will ultimately extend from Korea to Japan, Hawaii and mainland US. This "...dimly lit room..." is presently being configured at the 121st US Army Medical

Evacuation Hospital, YongSan, Seoul, ROK in the office of the Chief, Radiology Services. Strands of the virtual radiology department's communication web will first extend to medical sites on Kunsan and Osan (SongTan) Air Bases ; Camps Casey (Tongduchon) and Walker (Taegu) ROK, Tripler Army Medical Center (AMC) and the University of Hawaii (UH) Honolulu, Hawaii, massive parallel-processor computer center and on to Georgetown University Medical Center (GUM), Washington, District of Columbia (DC). Subsequent phases will extend spokes throughout the South Korean peninsula creating a network of 16 US armed forces medical treatment facilities (MTFs) and links to Japan, Hawaii, and the mainland US. US military forces have a significant presence in the Pacific region. They represent a massive healthcare workload on widely dispersed military sites across an area covering roughly 51 percent of the earth's surface or approximately 103 million square miles.

Standard Evaluation Factors

Linking all these far flung military sites by a teleradiology (T/R) network is clearly not financially feasible, nor does it make logistical sense. As a result, needs must be assessed at each location and determine those for which T/R is the best solution. The following factors, listed in order of descending priority, determine whether T/R is appropriate:

- War fighting requirements and level of the threat. — The greater the threat, the greater the need to provide comprehensive peacetime medical care in order to maintain a fully ready warfighting military force and prepare adequately for contingency operations.
- Access to healthcare in a DoD facility. — The greater the availability of full specialty DoD healthcare, the less need for electronic access to remote specialty consults.
- Access to reasonably priced, quality healthcare in the private sector. — The greater the availability of reasonably priced healthcare, comparable to that available in US MTFs, the less the need for electronic access to remote specialty consults.
- Historical workload. — The busier the small MTF, the greater the need for electronic access to remote specialty consults.
- Categories of care. — The more kinds of specialty care a small MTF performs, the greater the need for electronic access to remote specialty consults.
- Type and length of DoD military mission — Small forces on temporary duty only need assistance with minor ailments and emergency care to stabilize and transport the more severely ill. Permanent installations staffed by personnel with families have needs for all ranges of healthcare.
- Size of total eligible patient population. — The greater the size of the beneficiary population and the smaller the facility, the greater the need for electronic access to remote specialty consults.

- Environmental conditions. — The more difficult the terrain, the more demanding and unpredictable the weather, and the more fragile the infrastructure of roads and public transportation, the greater the need for reliable electronic access to remote specialty consults.
- Transportation feasibility. — Variety of available transportation, the actual cost to transport each patient, the health risk of transporting them, and length of time in transit influence a networking decision.
- Isolation, both psychological and cultural. — Language barriers, cultural traditions, unavailability of acceptable healthcare, unfamiliar diseases and risk, and limited opportunities for healthcare providers in isolated locations to maintain skill currency affect quality patient care. With increased isolation from US mainland healthcare services comes increased need for electronic access to remote specialty consults.

Using standard evaluation factors, optimizing the communication network, creating a single digital hub archive, and exploiting two-way communication for consults and research permits you to organize a virtual radiology department. Such a department optimizes patient care, acts as a force multiplier for limited radiologist availability, and establishes an enhanced referral network making remote locations more acceptable assignments.

The Medical Diagnostic Imaging Support (MDIS) System

At the November 1988 Radiological Society of North America annual meeting in Chicago, Illinois, a group of visionaries within the DoD healthcare system met and agreed to pursue award of a joint service contract for acquisition of a digital imaging technology system; the MDIS System including T/R. In September 1991 a four-year contract for MDIS was awarded to Loral Western Development Laboratories, San Jose, California, and Siemens Gammasonics, Inc., (SGI) Hoffman Estates, Illinois. DoD MTFs which now have MDIS include Madigan AMC, Fort Lewis, Tacoma, Washington; US Air Force Medical Center Wright-Patterson Air Force Base (AFB), Dayton, Ohio; and Brooke AMC, Fort Sam Houston, San Antonio, Texas. There are T/R systems linking K.I. Sawyer AFB, Michigan to Wright-Patterson AFB. A T/R system whose hub is the 58th Medical Group (MG), Luke AFB, Phoenix, Arizona, is linked to the 836th MG Davis-Monthan AFB, Tucson, Arizona, and awaits linkage to six more spokes. The Veteran's Administration Hospital, Baltimore, Maryland, has an MDIS system in place. Loral and SGI have also put this technology in place at Hammersmith Hospital London, United Kingdom, and are actively pursuing proliferation of MDIS at other private sector healthcare institutions in the United States, Europe, and Asia.

Virtual Radiology Department

The foremost location calling for continual readiness to fulfill the war fighting mission is the ROK. Because of this strategic importance, the first US operational overseas deployment of

T/R will occur there. In conventional T/R, spokes usually consist of small, remote MTFs networked to a full specialty tertiary care MTF hub acting as a source for consultations, which may or may not be the referral center, forming an inter-MTF connection. Usually, these efforts involve relatively short distances and small geographic areas and digital image transfer is one way (i.e., spoke to hub). Spokes usually maintain their own digital archives, or x-ray film files, simply relying on the hub for image interpretation and reporting by facsimile or phone. The hub is usually a PACS site and maintains its own digital archive. A virtual radiology department is an order of magnitude more sophisticated than simple T/R. It ignores "real world" geographic and time boundaries. It makes the hub of a T/R network the single referral center for network images . Each spoke is treated as though it is merely part of one large medical facility. Images are captured using direct digital output capture (e.g., CT scanner) computed radiography or film, subsequently digitized, or frame grab, in the case of some older and less sophisticated ultrasound equipment.

The hub appears to a satellite site to be merely part of the small facility's Department of Radiology. The hub maintains the only digital archive. A single digital archive saves the high cost (e.g., acquisition and maintenance) of multiple archives , greatly simplifies management of the image and demographics data base and reduces the required maintenance staff. The support staff can "circuit ride" from this central location to spokes to provide other services as required. This is a two way communication network so images and consultations can migrate from spoke to hub or hub to spoke an important feature for patient triage, healthcare provider CME, and research. Some of the cost savings realized are used to acquire the exceptionally robust network communication links. While all of these features may at first appear to be an obvious improvement it is clear from the literature that few installed systems have recognized the economies to be realized from the differentiation between T/R and virtual radiology.

Communication Links

In the ROK there are US government furnished equipment (GFE) T-1 grade communication links between the hub and each spoke in the first implementation phase. For example, Kunsan to Osan is a T1 GFE microwave link. In subsequent phases the links are a mix of T1, 512 and 256 KBPS. In a limited number of cases x-ray film will be transported from very remote, low volume sites to a digital access spoke for digitization, transmission and interpretation. The network connectivity to Japan, Hawaii and US mainland are a mix of GFE T1 satellite communication links, leased T1 commercial submarine cable (using on demand, dial-up, fractal bandwidth), and commercial satellite bandwidth.

Ongoing Efforts

The DoD in conjunction with GUM and UH will use collected digital images to conduct research in a number of areas including creation of sophisticated pattern recognition algorithms. These algorithms may contribute to a search for advances in artificial intelligence applications to computer aided diagnosis. These research efforts would use the three way digital satellite communication link between the UH computer, GUM's imaging physics laboratory and Tripler AMC as a source for digital image data from ROK for analysis. Additional research would focus on provider referral patterns and the incidence of aeromedical evacuation both tactical (local) and strategic (long distance) before and after the virtual radiology department is put in place. This research could be especially significant when assessing referral patterns of physician extenders (e.g., independent duty medical technicians, physician assistants, etc.) who are most dependent on the diagnostic disciplines.

Conclusion

The virtual radiology department has applicability for Health Maintenance Organizations (HMO), large hospitals with numerous local remote primary care clinics, and large hospital management organizations which concentrate specialties (e.g., thoracic surgery, burn management, etc.) at geographically separated individual facilities within the system. An example is Humana HMO's heart transplant facility in Louisville, Kentucky, where all transplants in their system are referred. In the DoD, the virtual radiology department may eventually include every location where US servicemen and women serve ... any image, any time, any where, to any authorized user.

References

1. Baerson, Kevin M.; Federal Computer Week, "Army Brings New Life to Medical Imaging", Jun 92, pps S8-S9.

2. Baxter, Kirkman G.;Wetzel, Louis H.; Murphey, Mark D.; Rosenthal, Stanton J.; Haines, John E.; Batnitzky, Solomon; Caresio, Joseph F.; Templeton, Arch W.; Dwyer, Samuel J.; Journal of Digital Imaging, "Wide Area Networks for Teleradiology", Feb 91, Vol 4, No 1, pps 51-59.

3. Brice, James; Diagnostic Radiology, "Options Multiply for Teleradiology Systems", Jun 91, pps 139-148.

4. Buelva, Alma; IDG International News Service, "Hospital Relies on International Net to Feed Images to Doctors", Feb 92.

6. Cerva, John R.; "Technical Specification for a Teleradiology System", Nov 85, a report produced under contract for the Uniformed Services University of Health Sciences (USUHS) by MITRE Corporation.

7. Crawford, Ken; Wilson, Judy; Army Times, "Computerized Diagnostic System", Nov 90.

8. D'Agostino, Janet; Air Force Times, "AF Buys Advanced X-Ray System", Feb 92, pp 11.

9. Franken, Jr., E.A.; Driscoll, Charles E.; Berbaum, Kevin S.; Smith, Wilbur L.; Sato, Yutaka; Steinkraus, Lawrence W.; Kao, Simon C.S., Journal of the American Medical Association, "Teleradiology for a Family Practice Center", May 89, Vol 261, No. 20, pps 3014-3015.

10. Frazer, Hilary A.; Diagnostic Imaging, "Teleradiology Teamwork a Boost to Rural Care", Mar 91, Vol 13, No 3, pps 67-68.

11. Goeringer, Frederick, Proceedings, Medical Imaging V: Image Capture, Formatting, and Display, "Medical Diagnostic Imaging Support System for Military Medicine and Other Federal Agencies", Feb 91, pps 340-350.

12. Honeyman, Janice C.; Staab, Edward V.; Syllabus: A Special Course in Computers for Clinical Practice and Education in Radiology, "Teleradiology", Nov 92, pps 135-146.

13. Lear, James L.; Manco-Johnson, Michael; Feyerabend, Angela; Anderson, Gene; Robinson, David; Radiology, "Ultra-High-Speed Teleradiology with ISDN Technology", Jun 89, Vol 171, No 3, pps 862-863.

14. Mun, Seong K.; Gelish, Anthony; DeTreville, Robert; Sheehy, Monet; Hansen, Mark; Hill, Mac; Zacharia, Elizabeth; Sullivan, Mike; Sebera, Wayne; Proceedings: Society of Photo-optical Instrumentation Engineers (SPIE) Symposium, "Health Care Using High-Bandwidth Communication to Overcome Distance and Time Barriers for the Department of Defense", Sep 92, Boston, MA.

15. Purvis, Andrew; TIME, "Healing by Wire", May 92, Vol 140, No 20, pp 68.

16. Roth, Margaret; Army Times, "New X-Ray May Link Hospitals Worldwide", Apr 90, pp 16.

17. Seshadri, Sridhar; Arenson, Ronald L.; Sprague, Douglas F.; Gitlin, Joseph N.; Proceedings, Medical Imaging II: Image Capture, Formatting, and Display, "Satellite Transmission of Medical Images", Feb 88, Vol 914, pps 1416-1423.

18. Siegel, Eliot L.; Diagnostic Imaging, "Hospital Takes Plunge Into Full-fledged PACS", Feb 93, Vol 15 No 2, pps 69-71.

19. Skjei, Eric; Diagnostic Imaging, "Military Sends CTs to Saudi Field Hospitals", Mar 91, Vol 13, No 3, pps 113-122.

20. Taylor, Ted; Chu, Wei-Kom; Morien, Marsha; Administrative Radiology, "Teleradiology at the University of Nebraska", Jul 89, Vol VIII, No VII, pp 12.

21. Webb, Mark; Walgren, Harold; Peterson, Dale; Wayda, Jack; Van Dzura, Mike; Patch, Dan; Saxton, Matt; Tactical Air Command, 1912th Computer Systems Group, Advanced Systems Division, "Teleradiology", an unpublished background paper, Jun/Jul 89.

Hospital Information System

Radiological Information System

The Totally Digitalized Radiological Department, the Viborg Project

Finn E. Lindhardt

Department of Radiology
Viborg Sygehus
Denmark

I n t r o d u c t i o n:
This paper describes from a clinical radiological point of view the change over a period of 3 years of an ordinary radiological department into a totally digitalized medical imaging department complet with RIS/HIS and PACS in the daily working routine.
In contrast to the well described large installations at university hospitals, it is possible to see the Viborg project as a very interesting prototype for many smaller departments, moving towards a digital future.

C o n d i t i o n s:
Viborg Sygehus is a general hospital of 400 beds. The X-ray department performes 48.000 examinations per year, CT, US and isotope studies included. 60 % are performed on outpatients.
A very thorough analysis of the department, done in 1989, revealed bottlenecks in the production, that had to be solved before digitalization. The analysis was most important for purpose of making the staff realize the demands of the coming changes in the organization.
In 1990 we got the first filmless fluoroscopic room. We began the translation and implementation of a RIS-(Radiological Information System)-system, Simedos. A surprisingly heavy task, which will hopefully be finished in the spring of 1993. Work with PACS began late 1992.
We have recieved much help and advise from the supplier, Siemens. No one in the department had any special experience of computer work beforehand. We formed serveral specialized groups and "super-users", who have the responsibility of daily maintenance and education of staff in the different modalities, RIS and PACS as well. A very rewarding and effective structure. The digitalization has been made alongside an increasing radiological workload.
Siemens was chosen as supplier for the whole project. No other coherent solution appeared possible in the fase of planning. It might seeem a problematic and limitating disposition, but the inevitable complications, that arise between several differnet suppliers, are definitely avoided. And system and standards can be equally open as well.
Financing is done by leasing. Funds came from the annual amount put aside for renewal of X-ray equipment by the County, from the expected reduction in costs of film, chemicals and such. But also from a reduction in personnel, i.e. secretaries. This can not be recommended. The very long transitional period from analog to digital department with RIS and PACS is extremely demanding for the whole staff.

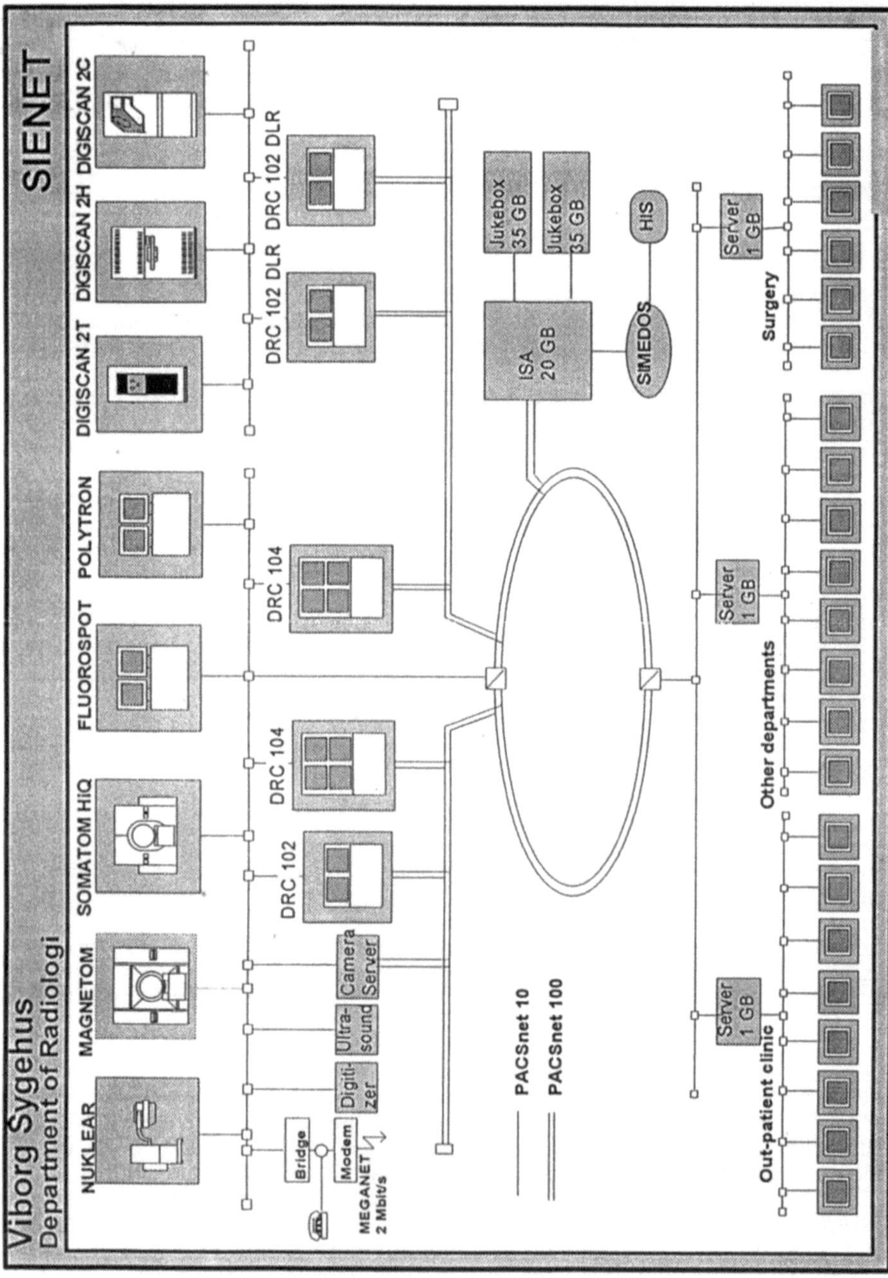
Viborg Sygehus
Department of Radiologi
SIENET
NUKLEAR
MAGNETOM
SOMATOM HiQ
FLUOROSPOT
POLYTRON
DIGISCAN 2T
DIGISCAN 2H
DIGISCAN 2C
Bridge
Modem
MEGANET
2 Mbit/s
Digiti-
zer
Ultra-
sound
Camera
Server
DRC 102
DRC 104
DRC 104
DRC 102 DLR
DRC 102 DLR
ISA
20 GB
Jukebox
35 GB
Jukebox
35 GB
SIMEDOS
HIS
PACSnet 10
PACSnet 100
Server
1 GB
Server
1 GB
Server
1 GB
Out-patient clinic
Other departments
Surgery

fig. 1., comments:
The two jukeboxes will be upgraded to 180 GB in all. Since we estimate to produce 1,8 GB/day and 170 GB (compressed) per year, we shall have one years production on-line. ISA, Image Storage Archive, is the fast intermediate storage. The next generation of the three servers will give our users at the clinical wards the possibility of requesting their own X-rays from our digital archive. 2H, 2C and 2T (Thorax) are all DLR-systems, Digital Luminiscens Radiography (or CR, Computed Rdiography). 2T is working without cassettes, feeding directly into PACS. We intend to digitize as few old films as possible. Research has shown very little demand for analog pictures. When reporting on PACS, we always have previous reports, all by trained specialists, on the Simedos monitor within close reach.

The DLR system, Digiscan.
Our expectations for the phosphor plate system have been fullfilled over the last two years. The light weight cassettes are easy to work with. Much better examinations of ICU and pediatric patients, of the urinary and G-I tract as well. The overall diagnostic quality is good, especially now that we have got the Simomed monitors af 1K to 2K in PACS. Digiscan 2H has its limitations in speed: 60 plates/hour. We had to move the smaller Digiscan 2C (40 pl/h) from the emergency X-ray room to the main department, in order to get the necessary capacity during peak hours. A substantial reduction in film can be achieved with clever picture composition on the plates. A nearby department with a PCR-system (Philips) has reduced the number of hardcopies to 0.9 pr. exmination in average. We did not expect to produce many hardcopies on our only laserprinter, when we started softcopy reporting, but the demand for documented examination results at the university hospitals with cardiac and neurosurgery is surprising. We pay a high prize for being digital and filmless, when our neighbours and referring hospitals are not. Nobody thought of that !
We try to do termoprints on photographic paper instead, reducing the hardcopy expence to one third, but the quality is not yet good enuogh for diagnostic purposes.

The clinical wards:
Relations are still good. The information and education of nurses and younger doctors has stimulated the interest for monitor display. But when it comes to the changing of habits, of conferences and reporting, there has been little understanding or forgiveness. It has been extremely difficult to move doctors and conferences five minutes in any direction! Great impatience is the result, when the inevitable software problems occur. The wards have got the great advantage of PACS: fast and secure display of X-rays. It is very important to establish the PACS net throughout the hospital as quickly as possible. The acceptance of digitalization will get higher. We do not hesitate to send

the softcopy to the wards without an accompanying report. We still have the daily conferences in the X-ray department with all the clinical wards, a great nordic tradition. Every examination is reported within 24 hours and the result transferred to HIS, the IBM based Hospital Information System, where the patients general record will be kept.

Working with PACS:

We are constantly expanding softcopy reading. Our aim is to have every type af exmination on PACS by the end of the year. The diagnostic quality is very good. 2K monitors and the many possibilities of postprocessing have been convincing. Lots of ROC analyses elsewhere have proven the same.
A clinical conference takes just as long as previously,but the meticulous preparation of a conference "package" by the secretaries tends to last longer than before. The rewards are that when the conference is over, there is no cleaning up. We have made shorter and more frequent conferences; the bigger ones tend to be chaotic. In the future we shall do conferences on requested examinations only. It is difficult to perform a fluent conference with a mixture of analog and digital pictures. Another reason for omitting old examinations. Analyses in the department have described the demand of previous X-rays in conferences. It is more time consuming to have every old picture displayed, than to produce a few from the digital archive, especially when decompression will be done in millesec.s
Former complaints, that working with PACS is too slow, may not be true, and if so, only in the sense that the many possibilities of postprocessing tend to engage the doctors. But this delay is justified, if the patient is now having fewer and better examinations. We have an excellent overview of each examination with 4 monitors of up to 64 token views on the DRC 104 conference consols. That makes comparison with previous results fairly easy. We always read prepared "packages", so nothing is lost unreported in the archives. Until now we have never lost a picture in PACS.
Pilot studies in the department have verified that mammography on DLR with High Resolution plates displayed on 2K monitor is equivalent to film/screen in quality, also regarding microcalcifications. But clinical mammography could be quite difficult to perform, since there are no monitors in the examination room. Reading screening mammography would be easy enough.

Working in a digital department:

The workload for the secretaries has changed, but not disminished. Much of the heavy and dusty archiving has disappered. Paperwork will disminish, instead there will be the handling with files and opticals discs. Preparing the conferences and reporting packages lasts longer. During the long transitional period it will, as mentioned before, be very risky to reduce this kind of staff. This would seem a matter for consideration, when the whole project is finished.

In our department we have always had a very well functioning archive and very few examinations temporarely missing, so we have not gained much with PACS there. Copying and posting examinations on hardcopy has risen dramatically for reasons mentioned earlier.
The radiographers' skill is not to be forgotten.In our department studies of film waste have revealed that 40 % of discharged examinations are rejected because of faulty positioning, no less with the DLR system. So there is still much to be improved. The flow of patients through the filmless X-ray room is definitely higher. More examinations can be performed during the day. But we are very carefull, that patient care does not deteriorate on that account.
Apart from working in the X-ray room (and reading the softcopies) almost every function could be done by anybody in the filmless digitalised department. This has evoked many discussions about changing areas of responsibility for the different types af staff.
The doctors workload has increased with PACS, no doubt. Former reports on a 30 % increase in time seems to be exaggerated, though. There has been a tendency of isolation, when we do the softcopy reading. The changing of working places during the day and maintaining the clinical conferences will hopefully prevent that.

For the future there are plans af connecting the 5 X-ray departments in the County of Viborg in a common PACS and Simedos net. Extension of a well functioning 2 MB net to more than the one university hospital is in progress.Smaller X-ray clinics and the radiological specialist on duty may be equipped with teleradiological devices.

Prototyping of HIS/PACS Integration - MIDAM Approach

I. R. YOUNG, M. H. SADDREDINI, R. DALE-JONES, D.A.BELL

Dept. of Information Systems,
University of Ulster, Newtownabbey, Northern Ireland

Abstract

The use of an extension of the distributed database approach to handling of HIS/RIS data with images is outlined along with details of the implementation of such a system as a hospital- based prototype. Details of the way in which the architecture of the system facilitates the decomposition of a query into subqueries with the subresults are merged to produce a final result are given. Details of further laboratory-based prototypes are also reported.

Introduction

Recent advances in database technology and image processing, coupled with the advent of high bandwidth networks and powerful workstations have brought about the possibility for the integration of heterogeneous collections multimedia data. In the hospital environment Hospital Information Systems and Radiological Information Systems are well developed and much effort has gone into Picture Archiving and Communication Systems (PACS).

There are a number of approaches to interoperability bewteen these media, and one is an extension of the distributed database managment system (DDBMS) [1] approach in the form of a multimedia DDBMS [2]. In the context of HIS/RIS data and PACS, this involves integration of HIS/RIS information and image location information extracted from the image indexes.

MIDAM and the EURIPACS Prototype

The EURIPACS project is an EC-funded collaboration, under the AIM initiative, of partners from a number of EC and non-EC countries. One of its objectives is to produce a prototype to demonstrate the integration of PACS systems with conventional HIS and RIS, which are text-based systems. This would facilitate the retrieval of a patients' medical records along with medical images generated for diagnostic purposes.

The initial scenario for the hospital environment is that of the request for both HIS/RIS information and images by a doctor from a workstation. The prototype mentioned above will demonstrate the feasibility of such an approach in the hospital environment.

The prototype will be installed at the AZ-VUB hospital, the site of the prime contractor of the EC project, using information from the HIS/RIS combined with CT images for the initial working version (October 1993). There are a number of functionalities provided for the prototype including the NOSS network management system, built by PRIMIS, which will control the image movement on a dedicated high-speed image network and provide a facility for the doctor to make a request for patient information with corresponding images. In addition the BAZIS group are providing the HIPIN interface [3] between the HIS/RIS and PACS environments.

Our own contribution is the MIDAM (Medical Image Data Access Module). This has the capacity to receive a doctor's request from a workstation via NOSS and decompose it into subqueries, one for text (HIS/RIS) and the other for the image index. The HIS/RIS subquery will be passed to the HIPIN box and the image index subquery to the Image Management System (IMS) which accesses the Image Index (INX). Format conversion is provided by the Local Data Module (LDM) which allows access to HIPIN or IMS. The prototype will relate the use of particular modules of our global architecture [2] with the environment provided by our partners.The relationship between MIDAM, NOSS and HIPIN is shown in Figure 1.

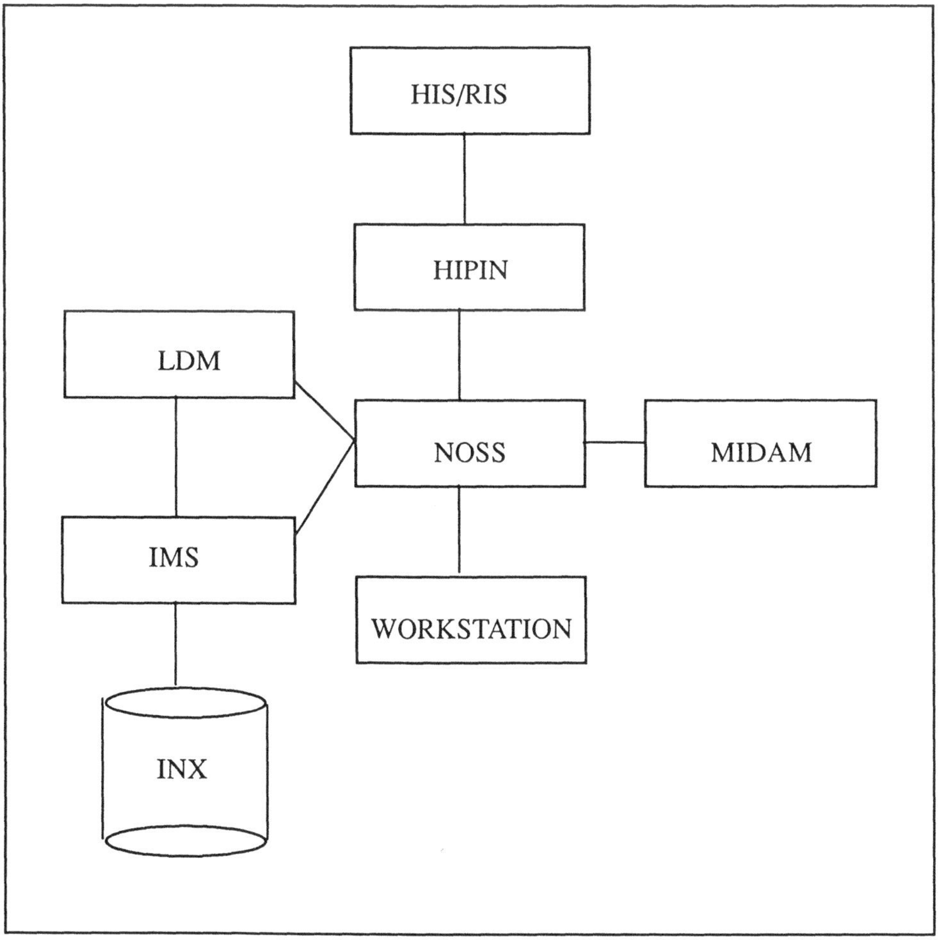

FIGURE 1 - Interaction of MIDAM with NOSS and HIPIN

MIDAM itself is an extension of the distributed database concept and has a global query language which is a form of SQL. This provides a global representation of a query which can be transformed into relational algebra expressions in a query tree and decomposed into one or more subqueries according to the location of the data as shown in the global data dictionary. The system is thus split into a global component and a number of local components according to the number of local sites which need to be accessed. The functions provided by these components are:

GLOBAL COMPONENT:

- Query Interface: providing an SQL interface or interface with the NOSS system

- Query Interpreter: validation of query correctness according to SQL syntax and data availability according to the global data dictionary

- Query Decomposer : formation of a query tree to indicate how the query may be broken down into subqueries by location and providing information on how to join the subresults when they return to the global site

- Communications Manager: oversees the packaging and sending of subqueries to the local components and the receiving and unpackaging of subresults

- Concatenator: provides the global joining facility to merge subresults into a single table

- Presentation Manager: provides an interface for the subresult with the NOSS system

LOCAL COMPONENT

- Local Communications Manager: provides unpackaging of subqueries and packaging and return of subresults to the Global Module

- Local Translation Facility: provides translation from local SQL facility into the approriate query language and receives subresult (from HIPIN via NOSS or from Image Index system via NOSS)

MIDAM is built in a UNIX environment (SUN SPARC or Microvax) using C. Communication between the global MIDAM and the LDM is on ethernet using TCP/IP.

Thus a typical query such as:

'Get the patient name, image name and image location from HIS and the Image Index '

might be translated into the SQL query in Figure 2 by MIDAM. This demonstrates how a single global query may be decomposed into twosubqueries S1 & S2. These subqueries are sent to the HIS/RIS sysyem and CT index via HIPIN and NOSS. The corresponding subresults R1 & R2 are received by MIDAM and merged to produce a single result table using the global join facility. This permits both HIS/RIS and image location data to be viewed as a single table.

This approach will be particularly useful if multi-modality images are required for a particular patient where the image addresses are held in a number of image indexes at different PACS islands.

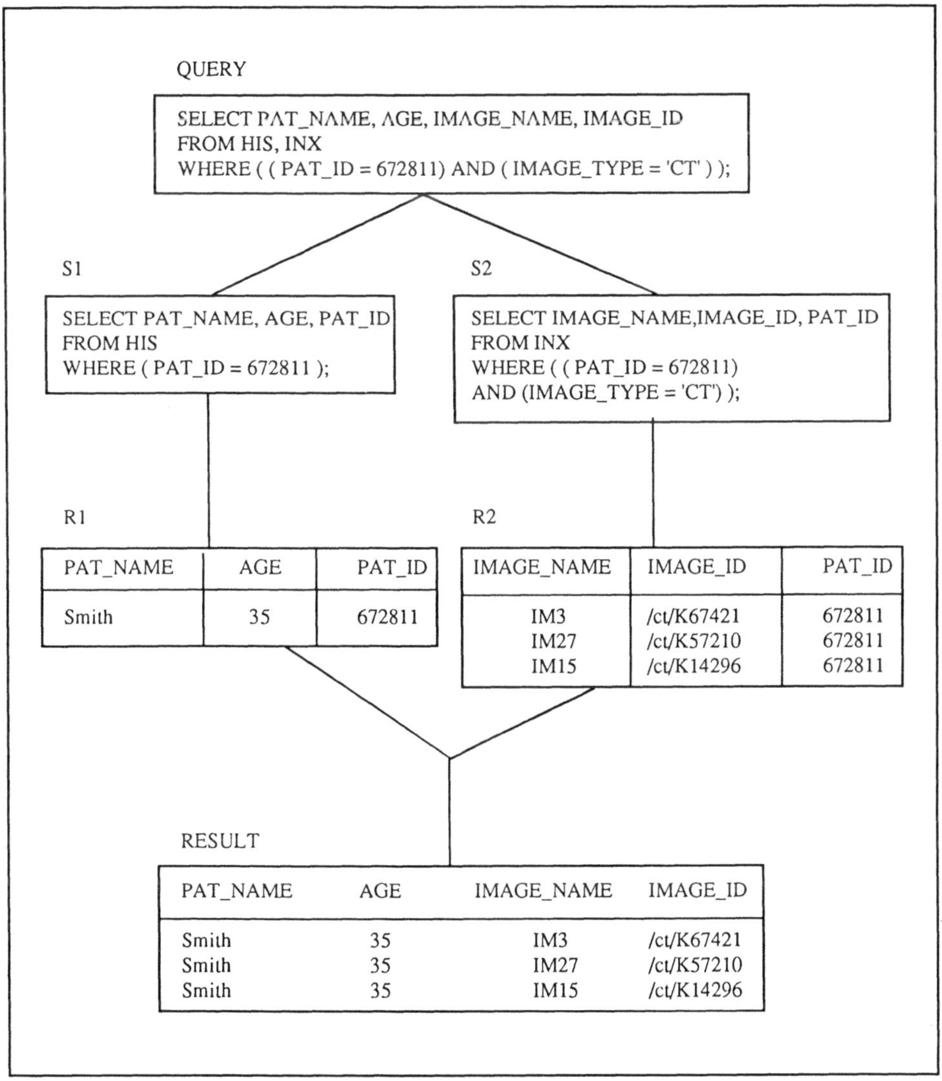

Figure 2 - Processing of a Distributed Query by MIDAM

Further Development of MIDAM

Currently our MIDAM laboratory prototype is able to access HIS data and images for individual patients. However for use in the hospital environment a number of extensions have been designed to facilitate queries involving a list of patients where there are a range of image modalities and PACS islands. Four important ares for extension [4] are:

LIST PROCESSING: An entension of MIDAM to allow a list of patients for a clinic session to be used in a prefetching query

IMAGE INDEX HANDLING: In order to handle multiple modalities at a variety of network locations it is necessary to handle a number of distributed image indexes containing data about local image addresses. There sre two possible approaches to this problem:

1. Global Image Index: All indexes could be copied to the global MIDAM site to facilitate fast image address retrieval. This would improve query performance greatly but would require the maintenance and update of a potentially very large index.

2. Local Image Handling: All queries to images could be broadcast to all local indexes and a temporary global index containing the union of image address data for only those patients required for one query is generated. From this temporary index the required images can be selected.

TRANSACTION MANAGEMENT: Since recovery is a very important issue in any distributed system, the need to provide suitable mechanism for ensuring theat a query has been successfully completed is required. In the EURIPACS prototype, this is the responsibility of NOSS. However, from a distributed database standpoint some transaction management will be required (e.g. ensuring successful responses to all subqueries, resending subqueries if no results materialise).

QUERY OPTIMISATION: Distributed query performance can be optimised by a combination of data placement and query execution strategies. In particular, the presence of multiple copies of images lends itself to access strategies based on location and network load.

Summary

A MIDAM prototype integrating requests for HIS/RIS data and CT images is to be installed at AZ-VUB Hospital, linking with other EURIPACS project modules. A laboratory-based prototype is being extended to incorporate further extensions.

References

1. D.A.Bell, D.H.O.Ling, I.R.Young, Databases' in ' HIPACS, A Second Generation PACS Concept' pp211-238 (Ed. M.Osteaux) Springer Verlag (1992)

2. I.R.Young, D.H.O.Ling, D.A.Bell, 'MIDAM - A Generalised System for Integrating Images and Patient Records, Lecture Notes in Medical Informatics pp 174-178 (Eds. K.P.Adlassnig et al.) Springer Verlag (1991)

3. F.P.Ottes et al., Functional Specification of the HIPIN HIS/RIS-PACS Interface, Proceedings of SPIE Medical Imaging Conference, Newport Beach,CA,1993 (in press)

4. M.H. Saddredini, D.A.Bell, I.R.Young, Architectural Considerations for Integrating HIS/RIS PACS in a distributed and heterogeneous environment, Proceedings of SPIE Medical Imaging Conference , Newport Beach, CA,1993 (in press)

Design of a Generic HIS-RIS/PACS Interface based on the Radiodiagnostic Working Methods

FP OTTES[1], R VAN DEN BROECK[2], P DICKE[3], E LIST-HELLWIG[3], FJ MARTENS[1], R RECHID[3], CA SCHULZ[1], M STOCKMANN[3], R VAN DER VELDE[2], E VERLINDEN[2], F VERHELLE[2].

[1]BAZIS, Central Development and Support Group HIS, Leiden, NL,
[2]Brussels Free University Hospital, Brussels, B,
[3]Phillips University Hospital, Marburg, D

Summary

The objective of the EurIPACS/HIPIN topic is to realize a *generic* HIS/RIS-PACS interface. The advantage of the generic interface concept is that it minimizes the efforts to realize a new instantiation of the HIPIN interface for a specific hospital. For a new instantiation only minimal adaptations in the software of the HIPIN interface itself and of the local HIS/RIS and PACS systems are required.

Our strategy to design the HIPIN interface is based among other things on an analysis of the radiodiagnostic working methods in european hospitals. The types of messages that can be sent or received by the HIPIN interface are defined as a generic set of message types, the so-called HIPIN message catalog. For each specific hospital a limited number of message types, selected from the HIPIN catalog, will be sent and received by the HIPIN interface.

Will each new hospital require extra site-specific messages or will a situation be reached where the HIPIN catalog contains all messages required in any hospital? Based on experiences in the HIPIN topic this question is discussed here.

Introduction

The AIM/EurIPACS project, which runs from January 1992 until December 1994, is subdivided in a dozen topics. One of the key topics, namely HIPIN, is concerned with the integration of HIS/RIS and PACS. The objective of the HIPIN topic is to develop and realize a generic HIS/RIS-PACS interface, i.e. an interface that can be widely used in european hospitals to integrate any kind of HIS/RIS with any kind of PACS. To demonstrate and evaluate its applicability the HIPIN interface will be implemented in clinical daily routine in two european hospitals, namely the Phillips University Hospital in Marburg and the Free University Hospital in Brussels [1].

The user requirements for a HIS/RIS-PACS interface comprise [2]:

a. it should be possible to search PACS images using features of the patient, the case and the examination to which the images belong, as search keys. For that purpose the images in the PACS database should be coupled with the patient, the case and the exam data in the HIS/RIS. Retyping that data, e.g. patient number, into the PACS should be avoided, because it is time consuming and will lead to errors.

b. medical information that is stored in the HIS/RIS, e.g. diagnosis, lab test results, reports and discharge letters, should be accessible for the PACS workstation user together with the images of that patient. And vice versa information in the PACS on the actually performed examinations, e.g. status, number of images actually acquired, should be available for the HIS/RIS user.

c. when images have to be retrieved from the PACS archive the response times will be unacceptable long using currently available information technology; optical disk manipulation in a jukebox requires 10-20 seconds. These response times can be significantly reduced by the application of image migration strategies in the PACS, that are fed by HIS/RIS data [3,4]. The prefetching strategy, for example, automatically retrieves historical images from the archive and stores them in the local storage of the workstation, where it is anticipated that these images will be needed for display. The prefetching actions will be triggered by events in the hospital that are registered in the HIS/RIS, e.g. an admittance of a patient or an appointment for an outpatient visit or for a radiological examination. Not bluntly all images of the patient will be prefetched, but only a limited set which is selected based on HIS/RIS data, such as exam execution date, referring specialist, or body part.

These requirements are covered by the functions of the HIPIN interface [5,6].

Design of the generic HIPIN interface

Following the generic interface concept (also in [7]) each HIPIN interface consists of a common part, which is identical in all incarnations, and specific parts which are different among the incarnations. Furthermore, the use of the HIPIN interface requires some adaptations in the local HIS/RIS and in the local PACS.

All incarnations contain the following (identical) ingredients:

- the hardware/software environment (currently a PC with a 486-processor running SCO UNIX),
- the module that provides message scheduling, including error handling,

- the message processing module (to split queries, to join responses, to split and/or join data elements, to translate data elements),
- the module that logs all messages,
- the module that creates the image migration requests.

The following features are site-specific:
- the handling of the communication (depending on used standards, e.g. TCP/IP) and the conversion of the message syntax (standards, e.g. ACR/NEMA, HL7),
- the parameter setting of the image migration algorithms,
- the actually used message types, selected from the HIPIN catalog,
- the actually used data elements in these messages, defined in the HIPIN data dictionary.

Radiodiagnostic scenario

The future radiodiagnostic working methods in hospitals, *where a HIS/RIS-PACS is used*, are described below. This description focuses on those aspects which effect the HIPIN messages, i.e. on the user actions that implie the retrieval, creation, or alteration of data that is shared between HIS/RIS and PACS. It is based on the film-based working methods as performed at the HIPIN sites and descriptions in the literature. Our description of radiodiagnostic working methods is structured as a scenario for a movie. The scenario is divided in scenes, and each scene is subdivided in takes. A scene occurs at a single location, and covers a contiguous time interval. In a take the interaction of a single medical professional with a single system is described.

SCENE: BOOKING A HOSPITAL STAY/VISIT*

An appointment is made prior to an inpatient hospitalization.

An appointment is made prior to an outpatient visit.

SCENE: PATIENT ENTRY*

When an inpatient arrives at the hospital for a hospitalization the admission data are registered in the HIS/RIS at the admission desk. Similarly, when the responsibility for an inpatient shifts from one inpatient department to another the transfer data are registered in the HIS/RIS.

When an outpatient arrives at the hospital for consultation this in general will be registered in the HIS/RIS at the reception desk of the outpatient department. An outpatient may be referred for a radiodiagnostic examination by a physician outside the hospital, in which case the patient skips this scene and the CLINICAL CONSULTATION scene.

For an emergency patient a temporary (often incomplete) registration in the HIS/RIS will be performed. When the patient data can not be obtained dummy information will be entered.

SCENE: CLINICAL CONSULTATION

During a clinical consultation of a patient at the outpatient or inpatient department, the hospital physician could decide to request a one or more radiodiagnostic examination(s). This request is entered into the HIS/RIS.

An emergency patient is immediately investigated at the emergency department in a fast way, and the physician orally requests the radiographer to perform the required exam. The request for the radiodiagnostic examination of an emergency patient will be entered into the HIS/RIS in a very abbreviated way.

SCENE: EXAMINATION SCHEDULING*

When the request is accepted, the exam(s) will be scheduled by an authorized professional at the desk of the radiology department. In some cases exams can be scheduled automatically by the HIS/RIS (in this case this scene merges with the previous scene).

For an emergency patient the examinations are automatically scheduled to be performed immediately.

SCENE: PATIENT ARRIVAL AT RADIOLOGY DEPARTMENT*

Upon arrival the patient is registered in the HIS/RIS at the radiology reception desk.

For an emergency patient this is not done, it is assumed that this patient will come to the radiology department immediately after registration of the request.

SCENE: PERFORM EXAMINATION*

The radiographer prepares the examination using a PACS viewing station, and performs the examination using the acquisition modality in an examination room.

For the digital modalities the images can be entered into the PACS automatically and for the conventional X-ray technique the films must be digitized. The images entered in the PACS will be coupled to HIS/RIS data of the patient, of the case and of the exam in question. After the acquisition the details of its execution (e.g. actual mobility, used contrast media), will be entered into the PACS or the HIS/RIS. After approval of the examination and the acquired images the patient may leave the radiology department.

The emergency patient will either be hospitalized (and then handled as an inpatient) or not (and then handled as an outpatient).

SCENE: REPORTING

The radiologist reports the new examination in the reporting room using a PACS viewing station based on the images of the examination, images and reports of previous examinations, and other medical data of that patient. The report is authorized after being typed into the RIS.

SCENE: REVIEWING

During conferencing, patient consultation, or treatment planning the requests, the acquisition parameters, the images and the reports of the radiodiagnostic examination, together with other patient data, can be viewed using a PACS workstation. Using a HIS/RIS terminal the status of the requested examinations can be accessed.

SCENE: PATIENT DEPARTURE*

After a hospitalization period the inpatient is discharged, and the discharge data are entered into the HIS/RIS.

When the treatment and/or the control consultation of an outpatient is completed, no new appointments for visits will be made.

* The events in these scenes can be used as a trigger for image migration.

The HIPIN message types are defined based on the user manipulations of data during each take of the radiodiagnostic scenario. The data relevant here is that data which is shared between the HIS/RIS and the PACS, namely some patient, request, examination and report data. This strategy is demonstrated here for the PERFORM EXAMINATION scene (see Table I).

Table I: For each take of the Perform Examination scene, the data interaction of the user and the HIPIN messages to support those interactions are listed.

TAKES	INTERACTIONS	HIPIN MESSAGES as seen from the PACS
Take 1: The radiographer displays a list of exams that are scheduled on the console of the digital modality. To couple the acquired images to the data of the actual exam, it is selected from that list. (see note below). If additional information on the scheduled examination and/or patient, is of interest for the radiographer or must be stored in the PACS database, it can be retrieved from the HIS/RIS.	View Acquisition Worklist	Request ACQUISITION WORKLIST Receive ACQUISITION WORKLIST
	View Patient or Update Patient,	Request PATIENT Receive PATIENT
	View Exam or Update Exam	Request EXAM, Receive EXAM
Take 2: To prepare the examination the radiographer will: display the request, display the information of previous examinations, display images of previous examinations, display the reports of previous examinations.	View Request	Request REQUEST, Receive REQUEST
	View Exam	Request EXAM, Receive EXAM
	Display Images	
	View Report	Request REPORT, Receive REPORT
Take 3: Radiographer acquires images. As soon as all images have been acquired and entered into the PACS a message could be sent to the RIS that the exam has been completed.	Complete Exam	Send CONFIRMATION EXAM
Take 4: Radiographer enters details of the actually performed exam into the PACS (message required), or into the HIS/RIS (no message required).	Enter Exam Parameters	Send EXAM PARAMETERS
Take 5: A radiologist approves the execution of the examination. A message will be sent to the HIPIN to trigger the preloading of the images to the reporting workstation and to the HIS/RIS to update the status of the examination.	Approve Exam	Send APPROVAL EXAM

Note on take 1: To identify films from conventional modalities, when they are digitized, machine-readable labels are printed by the PACS based on HIS/RIS data sent to the PACS during a previous scene.

Discussion

We found that descriptions of the film-based radiodiagnostic working methods concern the same activities and are similarly structured ([8], personal communication with EurIPACS partners and with members of the WG4 "Medical imaging and multimedia" of CEN/TC251 and internal BAZIS documentation), although the wording may be different (see definition of terms in [6]). This similarity appeared especially true for the requirements for HIS/RIS information when viewing images, and vice versa for alphanumerical data from the PACS when using a HIS/RIS terminal. Also the hospital events registered in the HIS/RIS, that can be used to trigger the image migration, appear to be comparable in the descriptions. Based on this we expect that also in the PACS-based situation the working methods will be similar with respect to the requirements for HIPIN messages, and thus that the HIPIN catalog will soon cover the requirements for any hospital.

To define the HIPIN messages an assumption is made about the storage location of the relevant data elements across the HIS/RIS and the PACS databases: the patient, request, exam, report data are stored in the HIS/RIS, some patient and exam data are copied in the PACS. It has not been investigated yet to what extent other existing distributions of data across the HIS/RIS and the PACS may require extra messages in the HIPIN catalog.

Acknowledgements
We gratefully acknowledge Otto Rienhoff (Marburg), Albert Bakker and Kees Bijl and others (Leiden) for their support of the HIPIN topic. This research has received financial support from the EC through its AIM program.

References

1. FP Ottes, AR Bakker, CA Schulz, FJ Martens, Prototyping a PACS-RIS/HIS interface in Europe, in: Medical Imaging VI Conf, SPIE 1654, 551-558, 1992.
2. K Bijl, ML Koens, AR Bakker, JPJ de Valk, Medical PACS and HIS : Integration needed! In: Proc. Medical Imaging Conf, SPIE 767, pp. 765-770, 1987.
3. Bakker AR, Didden H, de Valk JPJ, Massar ADA, Considerations on an algorithm for activation of images in a multi-layered storage system within PACS. In: Proc. PACS IV Conf, SPIE 626, 1986, 637-640.
4. BM van Poppel, H Lodder, PH Elsakkers, ML Koens, AR Bakker, Prefetching offers prospects for a quick responding PACS. In: Proc. 5th International Symposium CAR '91, Springer Verlag Berlin, pp. 464-470, 1991.
5. FP Ottes, P Dicke, FJ Martens, R Rechid, CA Schulz, M Stockmann, R van de Velde, R van den Broeck, E Verlinden, Functional specification of the HIPIN HIS/RIS-PACS interface, In: Proc. Medical Imaging VII, 1993, in print.
6. FP Ottes, CA Schulz, FJ Martens, R van den Broeck, R van de Velde, E Verlinden, M Stockmann, R Rechid, P Dicke, Report on the Functional Specification of the HIPIN interface, EurIPACS-HIPIN deliverable, 1992 (available through BAZIS).
7. K Soehlke, P Fisher, Evaluation of a generic RIS-PACS interface, In: Proc. Medical Imaging VI: PACS design and evaluation, SPIE 1654, 455-466, 1992.
8. O Rienhoff, CFG Greinacher (eds.); A General PACS-RIS Interface, an analytical approach to information use in radiology, Springer Verlag Berlin Heidelberg, Lecture Notes in Medical Informatics Nr. 37, 1988.

The Integration of Radiographic Images with other Hospital Information System Patient Data

J. W. LONDON PH.D., D. E. MORTON M.S., and H. KESSLER M.D.

Fox Chase Cancer Center, 7701 Burholme Avenue, Philadelphia, PA 19111 USA

Summary

At the Fox Chase Cancer Center a Hospital Information System (HIS) has been developed which presents images from CT and MRI along with other patient information at more than 75 locations in the hospital. X-terminals are used to display these images along with text and plots of other clinical data, such as radiology and pathology reports, progress notes, clinical laboratory results, and patient orders. The X-terminals are located in the clinic, at nursing stations, physician offices, and in radiology. The HIS is implemented on a distributed client-server local area network (LAN) of Reduced Instruction Set Computer (RISC) workstations. The CT and MRI scanners, as well as the film printers, are also configured as network nodes. Whenever possible, non-proprietary software and hardware standards are used (e.g., Unix operating system, "C" programming language, OSF Motif X-Windows toolkit, ACR-NEMA image protocol). In the Picture Archiving and Communication System (PACS) developed at Fox Chase, image data is transferred from the scanners to image management RISC workstations where radiology technicians perform image modifications (e.g., window and level setting, zoom). The workstation image data is then queued over the network to the film printers. Images are also archived (using digital audio tape) at the image management workstations. The images can be displayed on the many X-terminals on the LAN, located throughout the institution. Physicians can review the CT and MRI images at the X-terminals while simultaneously accessing additional relevant clinical information. Additionally, radiologists in the reading area can view images at remote X-terminals while a patient is still being scanned, and thereby communicate any needed study modifications to the technicians. We will present the details of our system's design and implementation, and the experiences of the past two years during which it has been in daily use by our medical staff.

Introduction

The Fox Chase Cancer Center is one of twenty-eight comprehensive cancer centers in the United States. These institutions have been designated as "comprehensive" by the U.S. National Cancer Institute because of their combination of basic research with clinical treatment of cancer patients. Over the past decade a hospital information system has been developed at Fox Chase which supports the medical staff by providing computer access to much of the patient's medical record [1]. Being a cancer center, oncology-specific software has also been developed and incorporated in the hospital information system [2,3]. Recognizing that radiographic images are an essential element of patient data, work was begun in 1989 to present such images along with the text and graphs that comprise the rest of clinical information [4].

The goal of our system design is to present all the information needed by the clinician at the point-of-care. Furthermore, information to support the clinician in their treatment of patients should be presented in a rational and efficient manner [5,6]. In this paper we will describe the methods we used to accomplish this goal. Also, since we now have two years of day-to-day operational experience, we will discuss our experiences and the lessons we have learned.

System Design

The computer systems at Fox Chase Cancer Center have a client-server local area network architecture, which is depicted in Figure 1. A committment was made in 1989 to use non-proprietary, open standards whenever possible. Committments were also made to providing a "point-and-click" window-oriented graphical user interface, and to the utilization of the optimum "price-performance" hardware and software. These committments have resulted in a network of Reduced Instruction Set Computers (RISC) [7] acting as compute and database servers for X-terminal display stations throughout the institution. The X-Windows system [8] is the graphical user interface used.

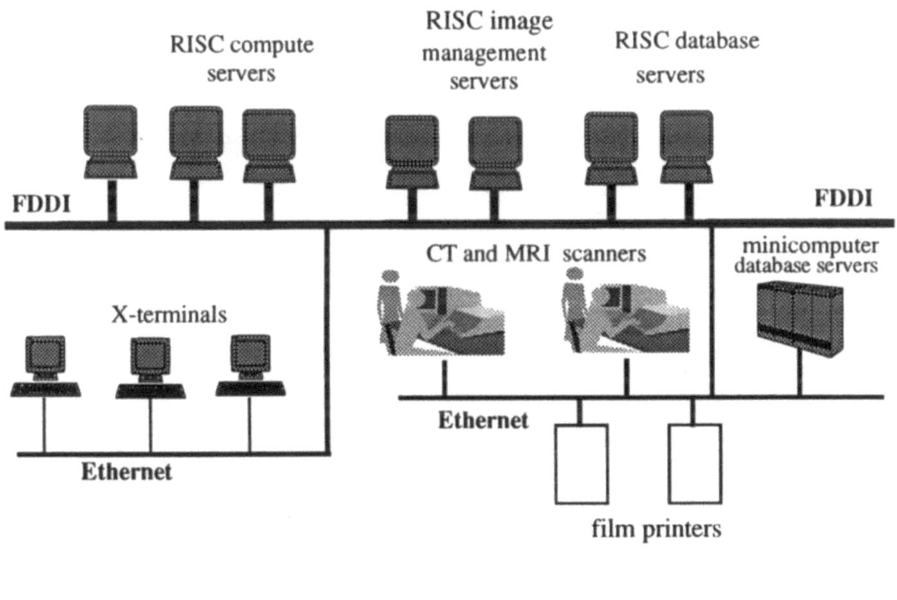

Figure 1. Schematic describing client-server network at the Fox Chase Cancer Center.

Physicians, nurses, and administrators all have access to X-terminals, either in their office, at a nurse station, in the clinic, or in their laboratory. These X-terminals can connect to over a dozen RISC workstations over Ethernet. The RISC machines can communicate via the 100 million bits per second Fiber Distributed Data Interface (FDDI) protocol. From an X-terminal various applications can be initiated on one or more of the network RISC platforms. Each RISC machine is dedicated to providing one or more compute services (*i.e.*, applications) or database information. At the time we embarked on our migration to a network of RISC processors and X-terminals our hospital information system utilized a minicomputer cluster and character-cell terminals. We developed a two-phase plan which would enable us to introduce the newer technology with a minimum of disturbance to the many users of our system [9]. The minicomputer cluster is still part of our network today (although there are less cluster members), serving database information that has yet to be ported to RISC database servers.

A distributed client-server network architecture, based on open standards, was also used in our systems for acquiring and managing radiographic images [10]. It should be noted that we have limited our scope to modalities that are inherently digital, specifically CT and MRI. The CT and MRI scanners and the film printers are independent network nodes. The scanners transmit image data files upon request from RISC image management workstations. Software on the workstation converts the scanner vendor proprietary header data structure to a format consistent with the ACR-NEMA protocol standard [11]. The technicians have software available on their workstation to perform image manipulations (*e.g.*, window and level settings). Images can be sent from the image management workstations (or any network node or X-terminal) to any film printer on the network. Radiologists can display images while a study is being performed from any X-terminal (such as those in the reading areas).

System Description

There are currently nine UNIX RISC processors on the Fox Chase Cancer Center network which serve the medical applications (other nodes are present for serving other applications and users). Four are Digital Equipment Corporation DECstation 5000/200's, three are model 5000/25's, one is a 5000/240 and the last is a model 3000/400 "Alpha" processor. The minicomputer cluster consists of four Digital Equipment VAX's: one model 8800 and three 785's. This cluster functions as a database server, and is being phased out as databases are being ported to RISC platforms. There are over 75 X-terminals placed throughout the institution at nurse stations, in the Outpatient Clinic, in MRI, CT, and Radiology reading areas, and in offices. The X-terminals are manufactured by Network Computing Devices, Inc., and are either color, greyscale, or monochrome, with 15, 17, or 19-inch monitors.

All software is written in "C." The X-Windows Motif toolkit is used for X-terminal displays. Fiber Distributed Data Interface (FDDI) is used for 100 megabit per second communication between the RISC processors, while 10 megabit Ethernet communication is used to communicate with X-terminals, minicomputers, and scanner computers. The film printers are connected to the image management workstations via DR11-W interfaces.

Experiences

Protocol conversion --

As described above, our system design stipulates the use of "standards" whenever feasible. Accordingly our workstation software for image manipulation, printing, and archiving was written for the ACR-NEMA image protocol. While manufacturers of CT and MRI scanners have a growing commitment to using this standard, it has not yet been universally implemented. Initially we decided to achieve the conversion of vendor-specific image protocol to the ACR-NEMA standard by purchasing a protocol conversion system. The system we used consisted of 386-DOS PC's and a software library. PC's received images from a CT or MRI scanner, converted the proprietary header to the ACR-NEMA format, and passed the data along to the UNIX workstation. Similarly, a second set of PC's received ACR-NEMA image data from the workstation, managed the printer communication protocol, and sent the data to a film printer. Our plan was that these systems would enable us to *buy* a solution to our protocol conversion problem, rather than having to develop a solution ourselves.

After using these PC-based protocol converters for over a year, it was clear to us that they were incapable of meeting our needs. Those PC's dedicated to converting proprietary scanner protocol to ACR-NEMA were subject to communication breakdowns somewhere along the scanner-PC-workstation path. The PC's serving the film printers had signification problems meeting the multiple tasks of receiving image data while printing previous image data. Overall, we feel that DOS is inappropriate for these applications, and that the intermediary PC's only provide an unnecassary single point of failure.

We now have removed the PC systems and directly communicate from our workstations to the CT and MRI scanners and the film printers. To accomplish this we had to develop software for the scanner computers and the workstations for image transfer, as well as workstation software for image protocol conversion (vendor proprietary --> ACR-NEMA). This software development can be characterized more as *tedious* than *difficult*. It should be noted that the scanner vendor was very cooperative, providing us with the documentation needed to perform the protocol conversion. Likewise, our film printer vendor provided excellent technical support.

Image Archiving --

We had decided to use a patient-based archiving strategy, that would use a single piece of media -- 4mm digital audio tape (DAT) -- to archive image data for each patient. Additionally, each day's studies for all patients would be backed-up to DAT's. We therefore had redundacy in that a patient's DAT, if lost or otherwise unreadable, could be reconstructed from the daily back-up DAT's. DAT's were chosen as the media because of their compact size (easy to store in the patients' master folder jackets), and their low-cost (about $12 US). While we did not have on-line access to previous images, we did have fast access to all of a single patient's images because only a single piece of media had to be retrieved.

Our two years' experience have uncovered flaws in our archiving plan. First, the use of a DAT per patient is very costly. The DAT's have at least 1 gigabyte capacities, and so no patient had more than one DAT. On the contrary, we had many patients with only a very small fraction of their DAT being used (i.e., with only a single study). However, the cost of supplying each patient with a DAT became a very large expenditure. Furthermore, the advantage of being able to retrieve multiple past studies from a single piece of media was not often utilized as only the most recent previous study was commonly requested.

Currently we are investigating the use of CD-ROM's for image archiving. Because of the permanency of optical versus magnetic media we will rely on the daily backups of all studies, and not have the redundacy of an additional patient-centered archive. When past images are requested, we will in most instances still only have to retrieve a single piece of media, since usually only the most recent prior study is desired. The media and system costs of CD's compares favorably with optical disks.

Image Manipulation Algorithms --

In keeping with our design goal of developing practical, cost-effective system, we rely on the RISC workstations to perform image functions, such as "zoom" (magnification). No image array processors are utilized. The RISC processors have performance ratings in the 25 spec-mark neighborhood. We learned that with these computing resources it is important that the most efficient algorithms be used for image manipulation, if satisfactory computation times are to be realized. We are now exploring a new generation of RISC workstations with 125 spec-mark ratings, but similar cost to our current machines. These faster machines may greatly reduce the sensitivity to algorithmic efficiency when developing useable image manipulation software.

Image Extensions to X (XIE) --

We have experimented with early implementations of XIE [12]. XIE supports many features useful to medical image presentation that are not available in X alone. XIE provides a facility for mapping 12-bit image data into 8 bits within the X server, reducing network traffic when adjusting image window and level. This same feature provides for independent image manipulation, which X does not support. Additionally, XIE provides a mechanism for "zoom" entirely within the server, reducing the computational load on the host for image presentation. Finally, XIE supports JPEG image compression, with image decompression occuring again within the server. Lossless JPEG modes will meet the needs of the medical imaging community. While the specification awaits final adoption as a standard, it promises to be a very useful tool.

Image Presentation Toolkits --

We have developed our own presentation software for non-radiologist physicians. This software allows them to view CT and MRI images along with the radiologist's interpretations in their offices or at nurse stations. Our presentation software is somewhat rudimentary, so we have an interest in utilizing more extensive toolkits developed by others. To accomplish this we arranged with the University of Geneva to become a site for their "Osiris" image presentation toolkit. We are working on porting their package to our RISC workstations and extending the useability of Osiris to X-terminals. Unlike workstations, X-terminals do not have disk storage, and must utilize available memory solely for images.

In a similar fashion we are working with the Helios Image Toolkit group at the German Cancer Research Center, to utilize their "Basic Image Processing Service" (BIPS) for image presentation. They have successfully ported their package to our system, since their system was designed to work with X-terminals (and furthermore, we both use the same vendor's RISC platforms). We are evaluating BIPS in our environment.

Network film printers --

A significant benefit has been realized by having the film printers disassociated from the scanners, and instead connected directly to our local area network. This architecture allows film printing to be initiated from any network workstation (or X-terminal). If the film printer in the CT department is not operable, CT films can be printed in MRI, and *vice versa.* We recently also experienced another benefit from this design when a CT unit was obtained for Radiation Therapy treatment planning. We were able to save the institution the cost of purchasing another film printer for this application, as well as the expense and space allocation needed for a film printer installation, by simply connecting this scanner to the network and having the films printed on existing printers.

Conclusion

Our experience of the past two and one-half years with a distributed network of client-server RISC processors with X-terminal point-of-care displays has proven that this architecture is suitable for hospitable information systems. We have also found this approach to be cost-effective. Our CT/MRI image acquisition and management subsystem has also proven to be efficient and practical. While the total system is in daily, routine use, work continues on improving and expanding it.

Acknowledgements

The aspects of our project which involve medical image presentation under X-Windows are supported by an External Research Grant from Digital Equipment Corporation.

References

[1] London JW, A Cluster Solution to Clinical and Research Computing Needs. In: *Proceedings of The Ninth Annual Symposium on Computer Applications in Medical Care.* Ackerman MJ(ed.). Baltimore: IEEE Computer Society Press, 1985: 722-726.

[2] London JW, Speer JF, Werbitzky D, Munic R, Computer Software for Cancer Staging. In: *Proceedings of the American Association for Medical Systems and Informatics Seventh Annual Joint National Congress.* Hammond WE (ed.). San Francisco: AAMSI, 1988: 31-35.

[3] Martin G, Padavic K, Toomey C, Yeslow G, Needleman R, London J. A Computer Program for Side Effects of Chemotherapy. Oncology Nursing Society National Congress 1990. San Antonio.

[4] London JW and Morton DE: The Integration of Text, Graphics, and Radiographic Images on X-Terminal Clinical Workstations. In: *Proceedings of the Seventh World Congress on Medical Informatics.* Lun KC (ed.). Geneva: Elsevier Science Publishers, 1992: 41-46.

[5] Silva, JS, Zawilski, AJ, O'Brian, J, Gunby, N, Siegel, J, Lauteren, M. and Halley, R. The Physician Workstation: An Intelligent "FrontEnd" to a Hospital Information System. In: *Proceedings of The Fourteenth Annual Symposium on Computer Applications in Medical Care.* Miller RA (ed.). Washington, D.C.: IEEE Computer Society Press, 1990: 764-767.

[6] Shortliffe EH, Wulfman CE, Rindfleisch TC, Carlson RW: Final Report: An Integrated Oncology Workstation. National Cancer Institute, contract NIH 263-89-C-0080, 1990.

[7]Patterson D. Reduced Instruction Set Computers. *Communications of the ACM* 1985, 28:8-21.

[8] Scheifler RW, and Gettys J. The X-Window System. *ACM Transactions on Graphics* 1986, 5:79-109.

[9] London JW, Morton DE, and Polekoff D, A Migration Path to Distributed Clinical Computing. In: *Proceedings of the The Fifth Annual Information Technology in the Health Sciences Conference.* Clark FC (ed.). Memphis, 1990: 44.

[10] London JW and Morton DE: A Network Scanner Image Management Station. *J. Digital Imaging* 1992, 5:7-13.

[11] Spilker C: The ACR-NEMA Digital Imaging and Communications Standard: A Non-technical Description. *J. Digital Imaging* 1989,2:127-131.

[12] Morton DE, London JW, Huang L: Early Experiences with X-Window Imaging Extensions (XIE), Symposium for Computer Assisted Radiology, Baltimore 1992.

Blueprint for Success: Designing and Implementing the Self-Paced Instructional System for PACS-RIS Training

L. C. WRIGHT, Ph. D.
15th Medical Group/SGHMM
Hickam Air Force Base, Hawaii

Summary

Use of a self-paced instructional program designed to train medical staff to use Picture Archiving and Communication Systems - Radiology Information Systems (PACS-RIS) equipment is an economic, time saving, and convenient method for a medical facility to quickly train its staff with minimum disruption to on-going health care delivery. A well-designed self-paced program requires only brief assistance from outside personnel until the first group of students graduate from the program. After the first group has graduated, each group of graduates can help train the next group. With the use of computer-assisted formal testing and computer-based proficiency tests, accurate academic test results are rapidly obtained and re-certification testing is easy to conduct. A computer-assisted self-paced instructional program is easily monitored by credentialing agencies.

Characteristics of Self-paced Instruction

The self-paced instructional system is designed for use by individuals at a time and place of their own choosing and with minimal help from a teaching assistant. Each lesson is a self-contained package in the form of a booklet, a video tape, or a computer file. Each lesson contains two alternate forms of a self-administered quiz so each student can privately assess his or her mastery of a lesson before either returning to the lesson for additional study or proceeding to the next lesson. When the self-paced program also teaches an actual skill and or use of specific equipment, the program includes practice exercises in a learning laboratory.

Self-paced instruction is convenient for students with limited free time and an inflexible work schedule. Students progress at their own pace, studying at a time and in a place convenient for them, and mastering each step before progressing to the next. Students can temporarily set aside their self-paced lessons without risk of missing crucial lessons. Bright or highly motivated students can complete a self-paced course much sooner than other students so do not experience the boredom or frustration of slow-paced classroom lectures designed for the average student. Students have the opportunity of privately assessing their mastery of lessons before taking formal tests, so experience less anxiety before formal tests.

The classic medical school practical training model of "see one, do one, teach one" is applicable to self-paced instruction in PACS-RIS. Recent graduates of the self-paced instructional course can serve their turn as teaching assistants to the next group of students. An organization that uses some of its own staff to train all potential students within the organization need not hire additional staff as teaching assistants or enter into a long-term contract with a professional teaching team.

Formal tests in a self-paced program are designed in a multiple-choice answer format that allows for rapid and accurate machine scoring. In PACS-RIS training, formal tests can be administered through the computer terminal with the student typing in an answer choice as each multiple choice test item appears on the computer screen. Tests can be scored by the computer and instantaneous results provided to the student and the course supervisor. Proficiency tests of actual mastery of skills, such as PACS-RIS computer terminal use, can be monitored by teaching assistants who observe the student performing a sample of the learned skill.

The self-paced program is not meant for the unmotivated student as there are too many opportunities to procrastinate. Procrastination is unlikely to be a problem if the student is a responsible adult, is motivated to master the course material, and if the study materials are intrinsically interesting. The self-paced program requires thoughtful planning to ensure each lesson contains adequate information and is presented in a format comprehensible, or "user-friendly," to all likely students. The program cannot rely upon a teacher to compensate for any deficiencies within each self-paced lesson. The self-paced PACS-RIS training program must have a learning laboratory setting where students can practice with the equipment and, when ready, perform their proficiency testing.

The sites for PACS-RIS training range from small clinics in remote locations to large medical centers in heavily populated areas. At a given site, the number of staff to be trained in PACS-RIS use may be just a few or may be many dozens. Most PACS-RIS users will be general practice medical professionals (family practice physicians, physicians assistants, nurse practitioners, etc.) who will only occasionally and briefly use the PACS-RIS system. Some PACS-RIS users will be medical specialists such as radiologists and orthopedic specialists who will spend considerable time each day using the system.

<u>Components of PACS-RIS Self-Paced Training Package</u>

Self-paced programs typically rely upon reading materials and written quizzes to provide most of the study material and self-assessment. A self-paced program for PACS-RIS can provide a portion of its instructional material in a computerized format, if students have convenient access to PACS-RIS computer terminals. Supplemental materials for advanced students may take the form of additional and more complex study materials, quizzes, and practical training exercises. The study materials may be text books, journal articles, additional booklets, computer programs, or video tapes.

For PACS-RIS training, the learning laboratory can be the actual PACS-RIS work stations at each test site. At any given time, the learning laboratory may be in use by students at various stages in the self-paced program. Ideally, the room should be able to accommodate at least one-fourth of the students currently enrolled in the program. The learning laboratory should be accessible before work, during the lunch hour, and after work hours. Students who want help from teaching assistants in using PACS-RIS equipment, should be able to schedule a tutorial time in the learning laboratory.

The self-administered quizzes should cover the important information in the lesson. Correct answers to quizzes should be available. By completing a quiz and checking the answers for accuracy, the student should be able to assess how well he or she recalls key information from the lesson and know whether more study is needed before proceeding to the next lesson. The formal tests should be composed only of items taken from the quizzes. A student who has learned enough from the lessons to pass all self-administered quizzes should have no difficulty passing the formal test or tests. The items for each computer-administered formal test can be randomly selected by the computer from the pool of quiz items so that each student has an unique test. Use of unique tests for each student prevents early testers from giving late testers inside information about test answers so effectively prevents cheating and helps maintain the validity of test scores.

Proficiency tests should be as close as possible to the actual skilled behavior the graduate will be expected to perform after training. A proficiency test should not require the student to perform a task he or she has not already mastered in the learning laboratory. In PACS-RIS training, some of the many computer-assisted training exercises previously mastered in the learning laboratory can be randomly selected and used as a proficiency test.

Design Team for the PACS-RIS Self-Paced Training Program

The quality and ultimate success of a self-paced program are greatly influenced by the expertise of the design team and the effort they expend during the developmental stage. Careful selection of the design team is crucial. The team leader should be a generalist who knows how to coordinate group efforts and can adequately represent the design team to other agencies. The team leader must be committed to producing a quality product but must also be practical in deciding when to sacrifice some quality in order to maintain timely completion of design goals. Other design team members should be knowledgeable in areas crucial to the self-paced program: computer programming, clear communication of information, effective study, academic course design, academic performance testing, research design, organizational management, and PACS-RIS use.

No team member is expected to be highly skilled in all areas of training program design, but each is expected to be expert in at least one area. Each must be able to perform his or her specific tasks in an efficient manner and must be able to work cooperatively with other team members to solve design problems that affect the areas outside an individual's expertise.

Expert opinion is not an adequate substitute for a trial run of the instructional program. Team members who are expert in human research, academic instruction, and academic performance testing should conduct pilot-testing of the self-paced program. This group within the design team must assess the adequacy of instructional materials, quizzes, the test item pool, and proficiency testing so any needed design changes can be made before the final version of the instructional program is released.

<u>Testing and Fine-tuning the Training Package</u>

The pilot test is simply a trial run of the self-paced program on a small sample of persons selected for their similarity to the type of students who will be using the finished product. The pilot test should closely approximate the actual conditions under which the training program will operate. Persons involved in pilot testing can give design team members valuable information about the adequacy of instructional materials, practice quizzes, and formal tests in addition to detecting simple typographical or sequencing errors in training materials. Test item selection for quizzes can be based upon whether the majority of pilot test subjects are able to select the correct answer. Additional pilot test information can be gathered through questionnaires, individual interviews, from written editorial comments made on the pages of the instructional materials, etc.

The design team determines what constitutes a satisfactory score on formal tests as well as the number of proficiency tasks and level of task difficulty required for the student to complete in order to graduate at a basic or advanced level and will provide grade criteria to training sites. After the instructional program is released for general use, the design team should periodically assess the effectiveness of the program and make any necessary adjustments to incorporate technological advances or improved technique. The team should also develop self-paced refresher courses to help program graduates keep abreast of any significant changes in PACS-RIS use.

<u>Implementing the PACS-RIS Self-Paced Training Package</u>

The program must be properly introduced to the medical staff of a test site so that potential students (staff members) see the training program as credible and of benefit to them. Senior staff members should be the first to graduate from the test site's self-paced program. Their initial completion of the program will signal their support of the program, will provide them with the knowledge to make informed policy decisions about use of the instructional program on their training site, and will send a clear message to junior staff that the instructional program is valuable.

Assessing Effectiveness of the PACS-RIS Training Program

A training program's effectiveness can be assessed indirectly or directly. Indirect measures include the ratio of the number of those who successfully complete the program to those who enroll in the program (a ratio near "1" is ideal, such as 83/85), percentage of students who obtain formal test scores well above the level needed for graduation, satisfaction surveys of graduates, a comparison of malpractice/critical incident rates of program graduates vs. non-graduates, etc. A direct measure of the training program's effectiveness would involve comparing graduates' performance of a PACS-RIS task with the performance, on the same task, of those trained by other methods.

Periodic inspections of the various PACS-RIS training sites should be undertaken by representatives of the program's design team to ensure the program is being properly administered at a test site and produces graduates who are adequately trained. A third party, such as a health care credentialing agency or professional organization, that conducts site visits of residency training programs would likely also assess the adequacy of the PACS-RIS self-paced instructional program.

The design team should survey its students immediately after each has completed or left the self-paced program and should again survey all program graduates after they have had time to do professional work using PACS-RIS. Survey data help the design team assess whether further refinements should be made to the instructional program.

Proficiency Testing of PACS-RIS Skills for Credentialing

Training directors and chief administrators of training sites are responsible for ensuring only students who master PACS-RIS skills at their sites are allowed to graduate from the self-paced program. In addition to demonstrating sufficient knowledge about PACS-RIS through formal testing, students should demonstrate proficiency with the PACS-RIS equipment as a requirement for graduation. Requiring periodic proof of proficiency ensures graduates maintain at least basic proficiency in order to retain certification.

Reference

1. “Human Variables Affecting Assimilation of a New Technology such as the Medical Diagnostic Imaging Support (MDIS) System”, Major Linda C. Wright, Ph.D.; paper presented at Euro PACS' 92 Conference, London, September 1992.

A Computer System for Requesting and Reporting Radiologic Examinations, Verifying Receipt of Reports by Referring Physicians, and Tracking Patients with Significant Abnormal Findings

Mahfuz Ahmed, M.D., Ph.D., Curt Anderson, R.T. and Charles Stuewe, B.S.

Kaiser Permanente Medical Center - Anaheim,
441 Lakeview Avenue, Anaheim, California 92807, U.S.A.

Summary

Verification of the receipt of a report by the referring physician when significant abnormalities are seen on an imaging examination, and the timely follow-up of these patients are not only good medical practices, but have recently been deemed a duty of the Radiologist by several courts in the United States[1]. This paper describes a computer system developed several years ago at Kaiser Permanente Medical Center, Anaheim, which fulfills these needs. In addition to many other functions, the system can track the 160,000 imaging studies per year performed at the hospital and at eight (8) outlying clinics (the furthest one being 26 miles), print reports at all these locations in the offices of the referring physicians, track timely interpretation of all examinations, attach acknowledgement letters to reports containing significant abnormal findings which the referring physician is expected to sign and return, track these acknowledgement letters, and track patients with these abnormal findings so that they are not "lost" to follow-up. To increase patient satisfaction a "normal" letter is mailed to patients with normal examinations. It also tracks the ordering practices of the referring physicians, and provides them with feedback so that resources are utilized optimally.

System Design

The system is designed as an automated electronic patient record, for inpatients and outpatients, with the ability to track any type of data. All requests for an imaging examination are made through the system and the results are transcribed on it. The results are then matched to the requests. Tracking can be instituted at multiple levels. For example, when an abnormality is seen on a screening mammogram, the system tracks the patient until a definitive treatment has been provided. This would include any or all of the following: additional views in the Radiology Department, six-month follow-up, verification of receipt of report by the referring physician, verification of appointment with a surgeon, needle localization, biopsy result, treatment for cancer or follow-up mammography.

Hardware and Software

The hardware consists of Data General Minicomputer Distributed Processing System consisting of sixteen (16) DG S-280 minicomputers, 400 terminals, 100 printers and four (4) Gigabytes of main storage. The system operates on the RDOS/ CHAMPS operating system, with applications written in Hybrid "Basic/Assembly" language with random access file structure.

Operation of the system

The Radiology department consists of the main department at the hospital, and several outlying clinics. All patients referred for imaging studies are first registered on the system with demographic data such as name, age, sex, address, telephone number etc., and assigned a unique eight(8) digit number, called the CHAMPS number, under which the patient's entire medical record is filed. Requests for imaging studies are then generated electronically using this number, with each request being assigned another unique number in a sequential fashion, so that a history of all the imaging studies on a particular patient is maintained on the system. In addition, each type of examination performed in the department is assigned a unique number. All physicians and non-physician providers (nurse practitioners, physician assistants, midwives etc.) who are involved in the care of a patient are assigned a unique five(5) digit number under which requests for imaging studies can be generated and reports sent to the appropriate person.

Requesting an examination

To request an examination a physician (or his/her designee) logs into the system and pulls up a menu screen on the CRT which lists the different types of examination performed by the department. Each of these types of examination has a unique number assigned to it. The physician then selects the appropriate number for the examination he/she wishes to request and input it into the patient file with an appropriate history. The system then creates a predefined electronic request form, which remains in the system and is printed out at the appropriate radiology clinic when the patient arrives for the examination. The examination is performed and a report is generated using the patient's CHAMPS number, the requesting physician number, and the unique examination sequence number.

Reconciliation of examinations

Reconciliation is the matching of examinations with reports and ensures that every examination has a report. Reconciliation is accomplished by using the CHAMPS number to identify the patient, and the unique examination sequence number. At the time of transcription the assigned sequence number (which is printed on the requisition) is input into the system which enables the system to "know" that a report has been generated. A list of all patients and examinations, which have not been transcribed, is generated weekly for a supervisor to research and act on the missing reports.

All reports are printed in the physician's office. When a significant abnormality is identified on an examination, the acknowledgement letter shown in Figure 1 is automatically printed with the report.

Dear Dr.________________

The information contained in the above report is of great importance to the patient. In order to verify receipt, please sign this pre-addressed slip and place it in the inter-office mail.

Signature of Physician____________ Date:_________

Figure 1

The physician is expected to sign and return the letter to the Radiology department. A file is automatically created in the system to track all acknowledgement letters. If a response is not received within ten(10) working days, a second computer generated letter is sent. If no response is received within another ten(10) working days, a letter is sent to the chief of the service for follow-up action. Once the chief of service is notified, the patient is automatically taken off the primary tracking list and placed on a secondary chief's list which is maintained for record purposes only.

Mammography tracking

Every mammogram performed in the department is reconciled as described above. Some screening mammograms require the patient to return for additional examination. To ensure that these patients have the required workup the radiologist dictates a preliminary report to create a tracking system. When the words "preliminary report", are typed in the first field of the mammogram report by the transcribers, the system creates a tracking list. This list is generated every five(5) days for review and action by a supervisor.

If an abnormality on a mammogram requires six month/one year follow-up, or a biopsy, an acknowledgement letter (Figure 2) is sent to the referring physician to ensure appropriate follow-up and therapy, and this is tracked in the manner described above. When the box: [ ] surgery consult requested is checked, a second tracking system is initiated until the surgical consultation and appropriate therapy is completed.

Dear Dr. ___________

The information contained in the above report is of great importance to your patient. Please check your response below, take appropriate action and sign this pre-addressed slip and place it in the inter-office mail.

[ ] Surgery consult requested. (Physician to fill out and send surgery referral.)
[ ] Repeat study in six months.
[ ] Repeat study in one year.
[ ] Return surgery appointment requested.
[ ] Report received and results noted, no action required.

Physician signature__________________ Date___________

Figure 2

If the mammogram is normal, a computer generated letter is attached to the report which is sent to the patient by the referring physician.

Management Reports

The information required to generate management reports is present in the system database. This includes the physician identification number, the patient demographics, and the type of examination. It is then a simple matter to generate different types of reports, such as total volume of procedures, the volume of any particular type of examination, the requesting practice of individual physicians e.g. the age distribution of patients referred for mammogram etc. This feedback is helpful to the referring physicians in the optimal utilization of resources.

Conclusion

As mentioned previously, the basic design and configuration of the system allows tracking of any type of data present in the patient's electronic record. The utilization of the system is at its infancy and the Radiology department is working on additional applications which will be reported at future meetings.

References

1. **Berlin, L.: AJR 159: 1335-1339**

Technology Assessment and Implications

Clinical Implications of PACS

RICHARD M. FRIEDENBERG

Department of Radiological Sciences
University of California, Irvine

Accurate diagnosis, rapid throughput, immediate reporting, and efficient retrieval of both images and reports are the goals of all radiology departments. In most large institutions reports reach the charts of referring physicians 24 to 72 hours after the examination. In many institutions the telephone report has replaced the written report as the instrument affecting decision making. Therefore, for reporting, accessing and storing images, the implementation of an electronic digital picture archival and communications system (PACS) should be efficient, and most people now consider it inevitable in the near future.

The slow development of integrated PACS systems relates primarily to the considerable cost involved and the hesitation by physicians to change their mode of practice, and their concerns about making significant investments in computers and computer programs which are changing rapidly. Many have first instituted teleradiography without film digitization as an inexpensive interim procedure. The beneficial effects of computer archiving and communication systems are numerous and well-known. They include the ability to produce readable films without significant waste, archiving of films with easy recall, prompt dispatch of films to the clinical services, the ability to report from distant stations including home, the ability to consult on films at a distance, and the ability to recall multiple examinations for review. The shift in emphasis from complete departmental or hospital PACS systems to the concept of networking individual areas gradually into a system has made PACS more acceptable both from the point of gradually introducing the system into the department as well as spreading the expense over a longer period of time and diminishing the concern about purchasing total systems which shortly become outmoded.

The sequence of installing network systems is based upon which areas produce most problems for the department and which areas provide necessary immediate service to other physicians, where prompt display and easier recall of films would speed patient treatments. Various sequences have been instituted but a popular one is: 1) intensive care, 2) satellites, 3) operating room, 4) emergency room, 5) clinics, 6) hospital floors phased in over time, 7) consulting with neighboring hospitals or groups. The gradual organization of a digital file room should be completed before all of the inpatient hospital services are brought on line.

As the department gradually increases the networking of digital information, they must also reorganize those functions related to all areas of imaging within the department; this is frequently left to happen by chance rather than planned in advance. If ignored, they can cause considerable confusion, particularly in hospitals where teaching of residents is a primary concern. Some of the more important items which must be considered as a teaching department changes to digital radiology include:

1. The immediate reporting of images as soon as they are obtained. If other departments are aggressive and if you are furnishing images without reports, these departments may report the examinations and bill for it. In the United States, any physician may bill for radiologic interpretation.
2. Inaccurate interpretation by non-radiologists. If the images are furnished to the physicians without interpretation, errors of comission or omission may be made which will affect patient treatment. This may be compounded if substandard monitors are furnished to the floors.
3. Problems associated with education of residents. In most residency programs films are reviewed with residents at specified times. The resident may preview the films and then go over them with the radiologist or the resident and the radiologist may review them together. They are usually reported by organ system, so that a resident assigned to the chest board reviews those chest films obtained the previous day. If images are furnished

to the referring physician as they are obtained and not immediately reported, then the report will reach the physician frequently after he has decided upon his therapy. If reporting is done on-line so that the image and report reach the referring physician at the same time, then the educational process must be altered and either the resident will review films after they have already been reported or will have to report them with the radiologist on-line. This diminishes the level of teaching since image reporting will not be by organ system but by continuous on-line reading. Logically, it would seem that it will be necessary to report films as soon as the image is displayed to avoid misinterpretation by the referring physician and to help guide him in patient treatment. In either case, resident education will be significantly affected.

4. When digital radiology is in effect, there are certain clinical departments, particularly in the United States, where these departments will desire to report films on their own. This is particularly true with the emergency room physicians and orthopods. This may be partially mitigated by prompt reporting and by having reading units in the radiologist's home for night and weekend reporting.
5. As digital radiology develops, groups of radiologists will tend to expand their area of influence over several hospitals without increasing significantly the number of radiologists within the group. Central reading stations can be established which can read the basic films from multiple hospitals and clinics with a minimum number of radiologists within the institution. Obviously, interventionalists and consulting radiologists need to be present but certainly staffing can be significantly decreased by bulk reading in a central site. This may lead to increased competition between radiology groups and to a decrease in the number of radiologists required.

It is important that the department of radiology consider how film digitization and digital radiology will affect their function while they are considering the implementation of the systems. There is a hard core of radiologists who will not willingly switch from films to monitors. It will be necessary to work with this group *before* installing systems to obtain their cooperation. It would be wise to form an interdepartmental committee to share the functioning responsibilities with the clinical services. The radiology department should have control of the system, its modification and upgrading. In most instances it would be wiser to start with limited areas of networking such as the intensive care units, satellite units and the operating rooms, and initially to maintain film files until the system functions smoothly. Many individuals will find it easier to start with digitization of films and later switch to digital radiology. This will tend to keep costs moderate and allow for gradual enlargement and expansion of function.

Technology Assessment of PACS: Case Study of the New Osaka University Hospital

K.INAMURA*, K.SATOH#, H.KONDOH#, J.IKEZOE#, and T.KOZUKA#

*Osaka University, Colledge of Biomedical Technology, Technology, Toyonaka, Osaka, Japan
#Osaka University, School of Medicine

Summary

PACS, RIS (radiological information system) and HIS (hospital information system) will be rebuilt and operated in the new Osaka University Hospital of 1100 beds which is going to be moved from the present building at down town Osaka to the new building at the another campus. In order to complete technology assessment of our PACS, more than 24 items of measurement to grasp present situation of PACS related performance in the present hospital were carried out. Those data will be used to compare with the result which will be measured after the new PACS is operated in the new hospital. Definitions of variables, parameters, formulae and methods of measurement are common or same as possible between before and after new PACS operation. These definitions are carefully selected and prepared for the purpose of generalization of methodology in technology assessment of PACS in hospitals at large here in Japan. This paper describes the philosophy, methods of our technology assessment and some result of the survey of present situation.

Schedule of Technology Assessment

Table.1 shows our history and schedule of technology assessment of PACS in Osaka University Hospital. From 1986 to March of 1993, the evaluations of experimental PACS and the survey of present situation of image diagnosis were carried out. Evacuation of the present hospital begins in April of 1993 and the new hospital will open in September 1993. RIS and HIS will be gradually in operation, and the end of March of 1994 these two systems will be in full sized operation. PACS installation and operation are now behind schedule as shown in Table 1, since system design was revised rather drastically recently in order to adopt the latest versions of technologies developed. Variables will be measured according to the parameters and conditions defined at the time of measurement in the old hospital, in the process of operation progress of new PACS. Our target of completion of technology assessment is in March of 1997, which will be followed by comparison and adjustment of measured data between before and after PACS operation.

Philosophy and Methodology

Our philosophy and methodology of PACS technology assessment are itemized as follows.

1. Efficacy and effectiveness are completely different each other.[1] Efficacy is composed of speed, capacity and image quality such as resolution and MTF. We must measure these for the purpose of acceptance check of hardware and software of purchased systems. But this is only the process of technology assessment. Our final measurements are the effectiveness evaluations. These are the time study for a patient who spends their times in the central clinic of radiology, cost evaluation such as expenditure for image diagnosis, and the effectiveness of imaging diagnosis which influences probabilities of diagnosis by referring physicians.[2]

Table 1. The history and schedule of technology assessment of PACS in Osaka Univ. Hosp.

Year	Evalustion and Development	Survey	Experimetal PACS
1986	Evalustion of Stand-alone film digitation, storage and display	Review of the current medical image information	MediFile-100 of NEC made
1988	Clinical evaluation of high resolution storage phospher system.	Start of measurement and analysis of present medical diagnosis system	Fileace 4100 of Mitsubishi made Kodak CR 4Kx 5Kx 12bits
1989	Evaluation and Consensus of medical image quality. Test run & evaluation of small PACS TDIS	↑	TDIS of Toshiba made
1990	Demonstration and consensus in image quality and speed in the Hospital	Measurement and analysis of present medical diagnosis system	
1991	Test run of a small PACS FCR-Medifile to identify problems	↓	FCR (CR by Fuji) MediFile-500
1992	System design in detail and software development .		
1993	Move to the new hospital Start of RIS operation Start of new PACS installation	End of the survey. Preparation for Technology Assessment	
1994	Start of new PACS operation	The survey of the new PACS	
1995	Link to RIS and HIS		
1996	Full size PACS operation	Comparison of data between before and after PACS	
1977		Adjustment and completion of technology assessment	

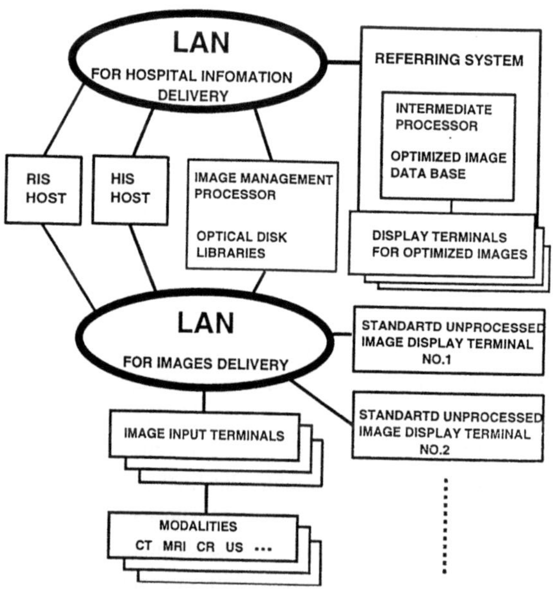

Fig.1. Outline configuration of our new PACS in Osaka University Hospital

2. Especially, the change of accuracy of predetermined diagnosis by referring physicians is the most significant parameters of PACS effectiveness. We are trying to express quantitative relation between PACS effectiveness with every layer of the 6 tired model of the image diagnosis efficacy. [2]
3. Compassion between before and after PACS operation is the most understandable method to express effectiveness of PACS. Consequently, the measurement of every detail of variables with the parameter and the conditions in the present situation of image diagnosis in the old hospital is indispensable. The definitions of parameters and variables must be also applied to the measurement of future situation after PACS is operated.

Outline of PACS in Osaka University Hospital

Fig.1. shows outline of configuration diagram of our PACS. The PACS has two categories of terminals. The first one is the "standard unprocessed image display terminal" which is dedicated to diagnosis by radiologists. These terminals have high image quality with image processing capability, and are rather expensive and comparatively of large size, so that number of installation is limited to within ten units in the radiology department. The second one is the "display terminals for optimized images" which are used for reference display of images together with the radiological report of main concern of referring physicians. These simpler image workstations have plain image quantity and simple function, so that less expensive enough to install them at most clinics and wards together with HIS or RIS terminals.

The Survey of the Situation before the New PACS Operation

Following items of survey of present status related to PACS have been done in order to compare them with that after the new PACS is operated. [3], [4]

1. Number of images per examination and per modality. (Film taken and film loans)
2. Number of light box screens and their distributions in clinics and wards of every department in the whole hospital.
3. Number of urgent cases of image examination at every clinic and ward in all departments, and ratio of them to non-urgent cases for each.
4. Number of outside films at clinics and wards and ratio of them to inside films.
5. Number of referred previous images at diagnosis, and ratio of them to that of today's images.
6. Number of referred previous images of patient of long term follow-up.

7. Flow studies of examination orders, images, reports and patients.
8. Flow study and time study of reservation and scheduling of image examinations.
9. Flow study and time study of patients in the central clinic of radiology.
10. Patients waiting time in front of examination rooms in the central clinic of radiology.

11. Time study of radiological diagnosis and reporting.
12. Turn around time from the time of order of image examination placed to the time of radiological reports acquired together with films. These were measured at the 9 clinics of out patient and 4 wards sampled.
13. Time study and its analysis in the process of image examination of every modality and from 9 clinics and 4 wards sampled.

14. Time study of previous image retrieval.
15. Time study of film delivery.

16. Number of words, characters and schematics contained in a text of image examination order and in a text of radiological report.
17. Process analysis of image observation by radiologists at the light box screen.

18. Time dependent number of image generation in terms of the hours in a day, and detection of the peaks in the time dependent numbers.
19. Time dependent number of images interpreted and reported in terms of the hours in a day, and detection of the peaks in the time dependent numbers.
20. Time delay of the peaks between above items 19. and 20.

21. Cost of image examinations: time dependent change of the amount of image examination cost and its ratio to the whole cost. (from 1982 to 1991)
22. Cost of image examinations: time dependent change of the purchased amount spent for image diagnosis in terms of purchase of films, isotopes and contrast agents.

23. Image quality evaluation: comparison between screen-film system and computed radiography in terms of detecion of subtle interstitial legion in chest images. ROC anylysis was applied.
24. Image quality evaluation: optimization of data compression parameters and compression ratio of medical images. (ROC analysis) [ ]

Fig.2 illustrates a result of the survey item 1 of above mentioned. Number of films taken per every department and Number of film loans per every department are shown in the upper figure and the lower figure respectively. An adequate correlation between these two figures is clearly shown. The upper side of Fig.3 is the result of the survey item 2. One screen is defined as an area for displaying a adult's chest film of 14"x17". These data must be compared with those of number of films taken in each department as illustrated in the lower side of Fig.3. Correlation between these two data tells availability and throughput of light boxes in their installed position. Our comparison of Fig.3. obviously indicates very loose correlation between them, which means a department having many light box screens does not necessarily have many films taken. We expect the availability and the troughput of image workstations will be closely correlated with the medical images taken by each department after new PACS is operated. This will be realized by proper distribution of image workstation installation throughout hospital.

The items 5 and 18 of above list are illustrated in Fig.4. The vertical axis is the number of read-out images per every 15 minutes in the reading room in the radiology department. We expect to increase the number of referred images of previously taken when our new PACS is operated.

Fig.5 shows an example of the survey results of item 12. Also we anticipate our new PACS shorten the time interval between order entry of image examination and image delivery to wards or clinics.

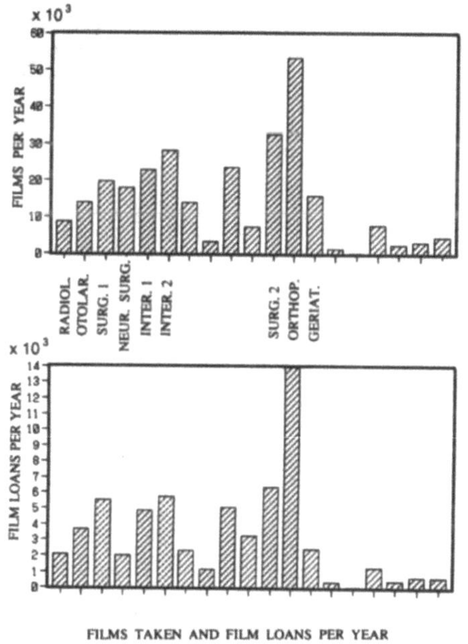

Fig.2. Number of films taken per year in each department (the upper) and film loans per year by the same department (the lower)

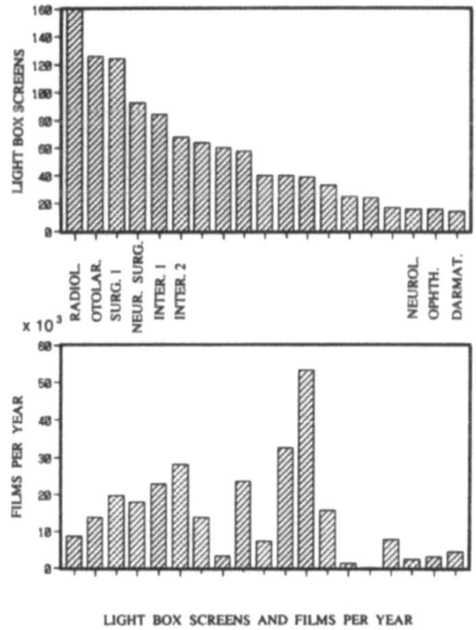

Fig.3 Number of light box screens installed in each department (the upper) and film loans per year by the same department (the lower)

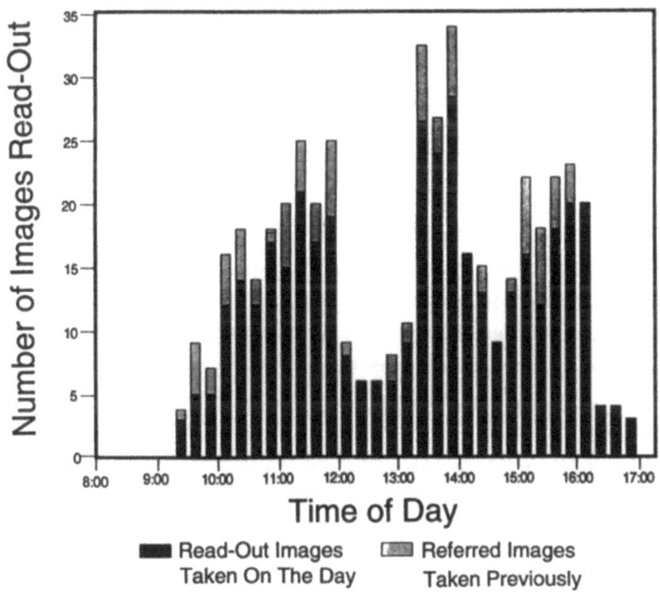

Fig.4. Number of images read out in terms of time of day as the avarage of 5 days measurement.

The item 21 of above list is illustrated in Fig.6. These change does not always reflects the effectiveness of the system operated, but is closely affected by the political factors such as revised health insurance system. However, the improvement of the PACS effectiveness will affirmatively

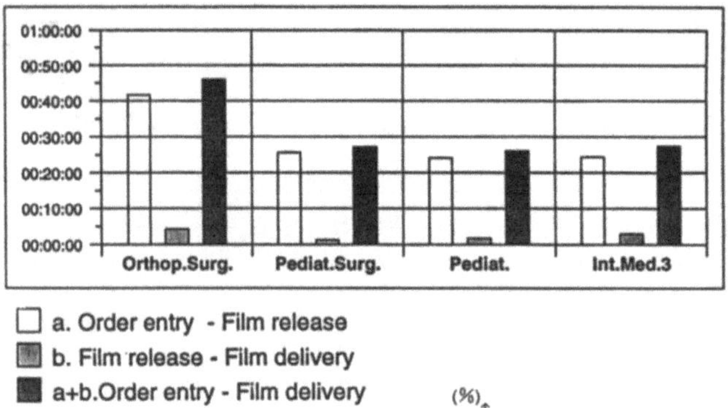

Fig.5. Turn around time from the time of order entry of image examination at wards to the time of radiological reports acquired together with films at the ward.

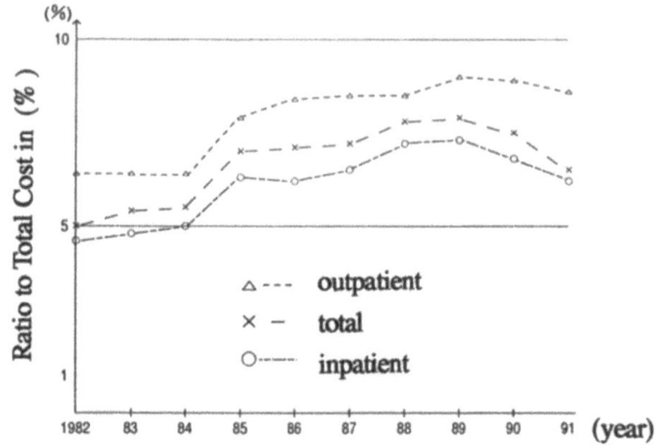

Fig. 6. Change in the composition of costs of image diagnosis in Osaka Univ. Hosp.

contributes to the decrease of cost of medical examinations. We would enjoy to observe the curve of next five years as the extension of Fig.6 .

Conclusion

We have finished the first half of the technology assessment of our PACS, namely all necessary preparations for the quantitative comparrison were completed. Our next action of the latter half of technology assessment will be to measure variables under the same units and parameters and methods listed up in this paper. Furhter investigation on the quantitative relation between the image diagnosis efficacy and PACS effectiveness will be carried out.

References

[1] M.F.Drummond, et al, Methods for the economic evaluation of health care programs, Oxford University Press, 1987.

[2] D.E.Fryback and J.R. Thrnbury, The efficacy of diagnostic imaging, Med. Dec. Making, 11, 88-94 , 1991

[3] K.Inamura, System design and technology assessment in IMAC/PAC, Proc. IMAC 91, IEEE Comp. Soc. Press 1991. Washington.

[4] K.Inamura, et al, Research and development activity of medical image information systems in College of Biomedical Technology, Osaka University, J.J. Med. Imag. Inf. Sci., 9, 101-109, 1992.

Imaging Department Design: Rikshospital Oslo, Norway

RAY L. BROVOLD, AIA
Senior Medical Planner, Ellerbe Becket, Inc.
800 LaSalle Avenue
Minneapolis, Minnesota, USA, 55402-2014

ARVID OTTAR, MNAL
Medplan AS, Architects
Stensberggt 29
N-0170 Oslo, Norway

Overview

The new Norwegian State Teaching hospital "Nytt Rikshospital" will replace four existing hospitals in Oslo. It will have approximately 110,000 square meters (1,180,000 square feet) floor area and will contain approximately 600 beds.

The hospital is intended to serve as a state-of-the-art referral center for all of Norway. It will provide all the diagnostic and treatment services ordinarily found in a major medical center and teaching hospital, with the exception of Radiation Oncology.

The new hospital has been given a beautiful, undeveloped site which is centrally located and easily accessible. It is located only 2 km from downtown Oslo and has Oslo University as its next-door-neighbor. It is placed on a hillside overlooking both the city and the Oslofjord. The forests that surround Oslo come down to the site along the north and west border.

For architectural reasons a low rise structure with 3-4 floors above ground has been chosen giving an optimal integration into the local environment as well as giving patients and staff the best possible contact with the surrounding landscape.

The hospital complex has been broken up into several identifiable buildings, reflecting the many different functions within the project.

These buildings are interconnected by glass covered bridges, pedestrian streets and squares to create short and convenient interdepartmental connections. It is designed to create a stimulating environment with voluminous daylight and visual contact with the exterior landscape.

The slope and slight bowl shape of the site is utilized to ensure both traffic segregation between services, outpatients and inpatients. This makes it possible to give the hospital a low profile to the exterior by placing massive building volumes demanding technical interstitial floors centrally and lower buildings on the perimeter.

Diagnostic Imaging

The imaging department is located on the 2nd Floor of the Diagnostic and Treatment Building which is central to the Rikshospital complex. Surgery and clinical laboratories are located on the 3rd Floor. The emergency department and the outpatient clinics are located on 1st Floor. The research and teaching facilities are in an adjacent building with direct connection to the hospital.

This is a large and complex project. The design and construction phases will span approximately six years. The imaging department at the new Rikshospital is designed to anticipate change and adapt to new technology. Much of the imaging equipment that will be installed in this facility is not available today.

The totally digital concept has had a significant influence on the design process although the optimism of its advocates is not shared by everyone. This is an open issue which will continue to be studied throughout the design and construction process and may not be resolved until the imaging equipment is purchased in 1996. If the equipment was purchased today, the department would probably not be totally digital. However, if digital technology continues to advance at the same rate as it has in the past ten years, the department will probably be totally digital when it opens in 1997. Since the decision to go totally digital is deferred to a later date, the design of the department must be responsive to the requirements of both film and digital technology.

The imaging department is subdivided into 5 units or subspecialties:

1. General imaging.
2. Pediatric imaging.
3. CT.
4. MRI.
5. Angiography.

Ultrasound is located at the interface of pediatrics and general imaging.

Nuclear medicine and diagnostic imaging are to be a part of the clinical chemistry department in accordance with Norwegian tradition. Clinical chemistry is therefore placed close to and on the same floor as radiology, with nuclear medicine next to radiology. Discussion of the nuclear medicine department is not included in this paper.

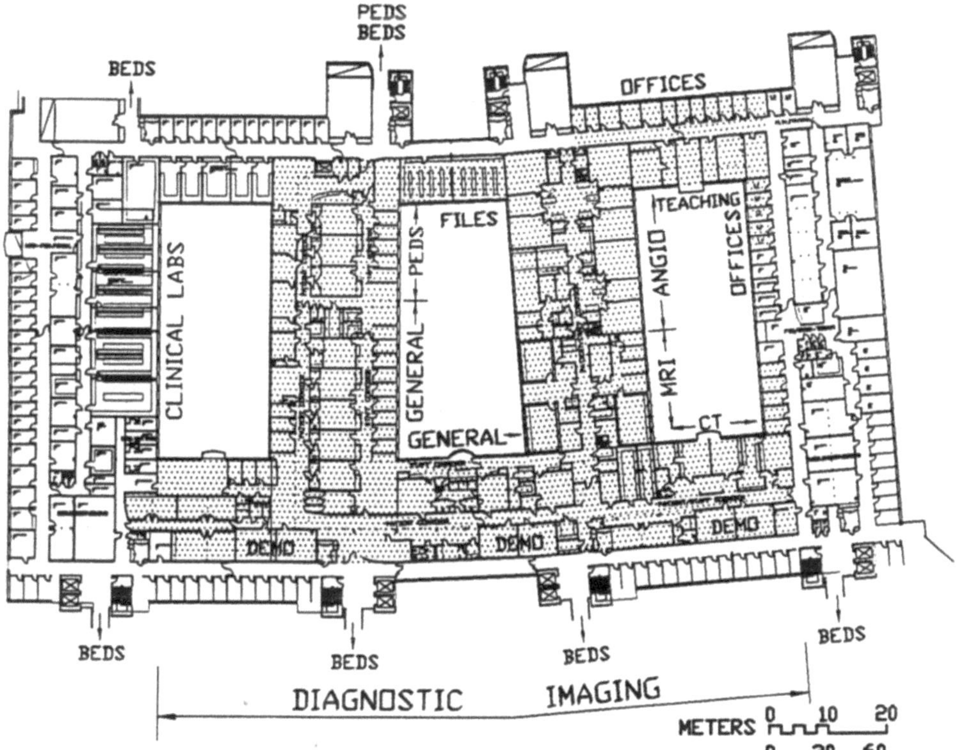

FIG. 1.

The final schematic plan (Figure 1) is the result of testing and revising the design for each subspecialty separately with respect to patient and staff traffic, imaging technology and image management and general work flow.

A modified concentric design philosophy was implemented in the high volume general imaging and pediatrics units to provide separation of patient and non-patient traffic. Some offices have been located in the patient area to satisfy the daylight requirements.

Daylight is highly valued in all the Scandinavian countries and the right to daylight in the work place is required by law. To provide this daylight the department is organized around two large open 20 x 40 meter courts with relatively narrow building volumes.

The separation of patient and non-patient traffic in the high tech special imaging areas was not given high priority. Some of the factors that led to this decision were the relatively low volume of patients seen in these areas and the fact that these areas will be the first to go totally digital. Staff traffic in these areas will be significantly reduced when all images and related reports are transmitted by cable and there is no longer a need to hand carry film and reports from point to point.

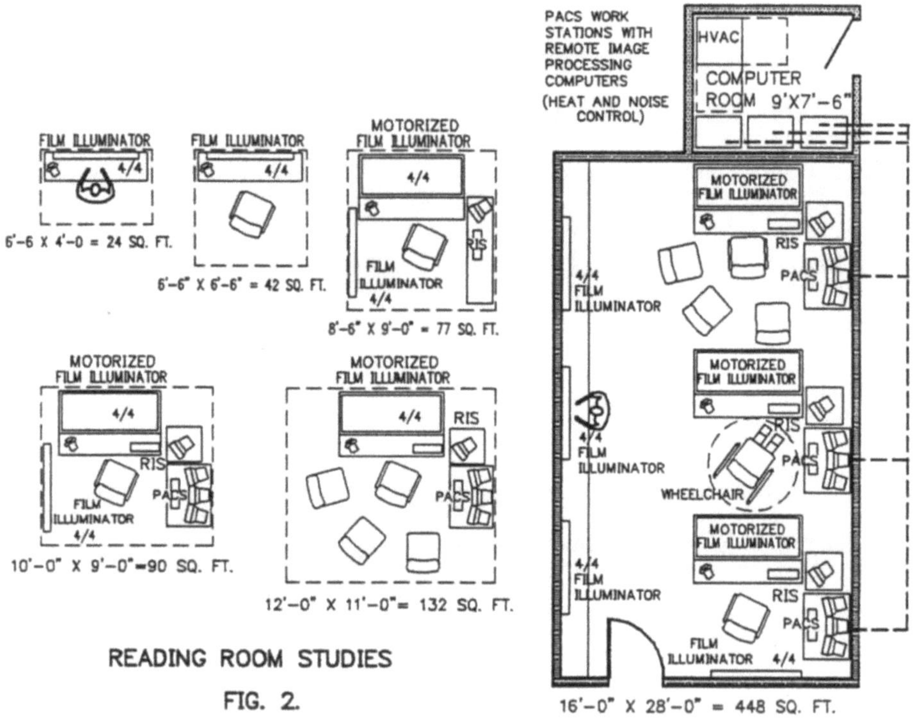

READING ROOM STUDIES

FIG. 2.

Image processing rooms are located throughout the department serving groups of imaging rooms or specific imaging modalities. Here again, the selection of equipment is deferred to a much later date keeping the options open so that new technology can continue to be evaluated. The locations were selected to support film based imaging and a decision to go totally digital would allow some of these spaces to be reassigned to other functions.
Reading rooms are sized to provide space for film reading stations alongside digital reading stations so that the radiologist can work with both technologies during the transition period (Figure 2). The image processing computers that support each PACS reading station generate both noise and heat. In most cases, these computers do not need to be located alongside the reading station and should be remotely located in a computer room to improve the reading room environment.

During the transition there will be a need to store both film and digital images. The plan provides both a film file room and a picture archiving and communication room (PACS). Theoretically, there will be less need for film as individual modalities complete the transition to digital imaging. As the transition to digital imaging progresses, the volume of film will decrease and eventually the film file room will become available for reassignment.

The Imaging Rooms are designed to allow the installation of similar equipment by any of the major manufacturers by using worst case scenario. If a fluoroscopy room will meet the requirements of a fluoroscopy system that requires the most space, the most power, generates the most heat and weights the most, it will meet the needs of the others also. The flaw in this concept is that the system that generates the most heat may not be the largest in terms of space requirement. Application of the worst case scenario implies that we need to know all the requirements of all the manufacturers and then select the most demanding criteria to set up generic requirements for each room type. This is fairly easy to do for general radiographic rooms but nearly impossible for angiographic rooms.

It is very helpful if the imaging department staff can narrow the scope of this task by selecting two or three manufacturers to use as a guide to develop a generic data base. It must be understood that this set of generic criteria is only a guide that may change at any time prior to the day the equipment is purchased.

We know that we cannot completely avoid changes but the intent is to minimize costly changes to the building and the building systems when the equipment is finally selected.

The direction that new technology will take is unpredictable. Therefore, we must base our design on what we know about today's technology and provide as much flexibility as we can justify and at the same time contain cost. A design based on the worst case scenario automatically has some reserve capacity built into it. Providing additional flexibility by adding reserve capacity to the building systems is purely a judgement call that will vary from one imaging department to the next.

Legal Aspects of Digital Image Generation, Communication and Archiving

O.Riehhoff

Institute of Medical Informatics
Bunsenstr. 3
Philipps University Marburg
Germany

Abstract

Based on a historical perspective and own experiences in operating a PACS (since 1989) some suggestions for legal approaches are formulated. Three 'legal compartments' are differentiated and a modified concept for the definition of the medical record is recommended.

1. Historical Perspective

A few years after the initial appearance of the term PACS, first papers were published pointing out clearly that there are legal obstacles for medical imaging and PACS (e.g. [1]). Since then, only few articles have described more detailed legal aspects of the matter (e.g. [2,3,4,7]). Nevertheless, these questions have neither been followed up nor has any country solved them. Most of the published material deals with technical or medical aspects; there are only few papers that try to structure the technical phenomenon to allow support from the legal side. Further, as 'PACS' is defined differently in the underlying literature, the use of the term will be avoided in this paper, if possible.

Within the EURIPACS-AIM project one topic (LEGIS), for which research has only recently been started will be dealing with the problem. Within the AIM project SEISMED a summary of general legal aspects of data protection regarding medical files is compiled.

The introduction of data protection laws in several European countries has made life more complicated for computer suppliers and computer users. There is a strong objection in most of these countries against new regulations in the field of medical imaging. Therefore, it has become increasingly difficult to approach the imaging matter in a balanced way.

Historically, another problem has evolved: the legal constructs of the sixties were based on a mainframe-oriented computer environment with clear point-to-point connections between machines. The environment of the outgoing century with PERSONAL computers everywhere, networks everywhere and a growing culture of information systems could not be foreseen. Thus, many regulations are technically outdated. Meanwhile, the number of legal regulations in the computer business has grown considerably so that we are aiming at a new codex of information laws - comparable to the social laws following the industrial revolution of the 19th century.

This development leaves the individual physician, patient and systems supplier in a frightening status: he/she is more and more working with and dependent on instruments which might cause major and personally unforseeable legal problems. The fast spread of information technology resulted in a widely distributed feeling of legal insecurity. We are far from an accepted legal culture regarding the handling of personal information within medical information systems[1].

This general picture also reflects legal aspects of medical imaging - only contrasts are even sharper: moving from film to digital imaging parallels opening the Pandorra Box because the legal cornerstone - the analog film - is disappearing. Our own group has published some ideas [4] how to redifine images in the electronic world.

Within this context it is of key relevance that some of the legal problems have been tackled 20 years ago when microfilm technology was utilized to minimize storage space for patient x-rays in hospitals. At that time an important concept was introduced: the relevant document (image) is the one a clinical decision was based on[2].

This paper will try to develop an overall framework of thinking which might be helpful for the ongoing discussion.

2. Relevant Legal Constructs in Medical Imaging and PACS

At a first glance one underestimates the number of different legal issues which interfere with medical imaging. Further, this situation differs from country to country:

1) Similar problems exist in dealing with specimens of patients. A philosophical, well debated approach to this matter has been published by Kluge in "Methods of Information in Medicine" recently.
2) In the case of the German Military Hospital at Koblenz the microfilm became the original, because it was used for diagnosing while the original film could be destroyed as it was not used for decision making but only as an intermediate technical step to produce the microfilm x-ray image.

1. Privacy related regulations
2. Medical Documentation related regulations
3. Liability related regulations
4. Technical safety of operations related regulations
5. Professional ordinances.

Regarding a comparative international evaluation of this matter it is surprising how many different specific laws are 'hiding' regulations - not to mention court decisions. To give an example:

Privacy related regulations (Datenschutzgesetze) in Germany are differently defined on the Länder (State) level and for institutions on a federal level. Details about when to disclose personal information are specified in the penalty law and social law In addition, there are different regulations in most hospital laws of the Länder. Further several professional regulations by various official physician chambers and institutions have been issued. Finally, there are a few fundamental court decisions which formulated general requirements for handling of personal data. All this is dealt with by professional data security officers who have developed a privacy culture which sometimes bears fruit and delivers practical solutions for everyday problems.

3. A Framework for a General Approach

Steadily, our own PACS environment has become an important part of the everyday routine operations in the hospital. Linking several distant functions of image production, analysis and demonstration it has grown to an important communication tool. Summarizing our experiences one can distinguish three 'legal compartments'(see also [8] for similar concepts):

1) **Image generation** (modalities)

A physical phenomenon is applied to a patient, the results measured and an image generated through software technology (raw data set).

2) **Image transformation** (modalities, PACS, image workstations)

The raw data set is transformed by various manipulation processes into reversably or irreversably changed copies of the raw data to perform reporting, diagnosing, therapeutic planning, therapeutic action and quality control of these processes (manipulation data set).

3) **Image distribution** (HIS, Teleconferencing, viewing stations)

The images representative for clinical decisions are distributed to those care units which base their further work primarily on the report and need images for illustration (medical record data set).

These 'legal compartments' are already known to us from other areas of clinical medicine: chemical path labs are following this order as well intensive care units. Only in Radiology we have been hesitating to apply the scheme.

It is obvious that the various 'legal compartments' need different legal regulations to balance the interests of patients, staff, industry and care institutions. Looking at the raw data compartment, technical safety and liability play key roles. Within the manipulation compartment it is absolutely essential to keep the data consistent and document the image manipulation process (e.g.in the image header). During the distribution all privacy aspects are dominating.

Thinking in these 'legal compartments' eases system design considerably because legal requirements become closer linked to system functionality. But, this also means that simple LANs would not be sufficient for linking modalities and workstations in the manipulation compartment as they cannot guarantee the required image (data-base) consistency in the interest of both physicians and patients.

4. The Definition of the Medical Record

Since World War II the definition of what a medical record is, what has to be included, by whom, for whose sake and who is allowed to have a look at all the information has constantly changed. In some countries the medical record is still a memory aid for physicians, in others it is a legal obligation for physicians to document data for the patient in his/her record. The most important contributions to a definition have been made by WEED and WESTIN:

1. Weed defined the problem oriented medical record and thus introduced the concept of later ratifying medical decisions.

2. Westin formulated the idea that objective medical data belongs to the patient and subjective data to the physician.

Various authors requested a change from a paternalistic physician role to a partner role by giving the decision to look into the medical record to the patient.

In the international scientific discussion of the matter these ideas have widely been accepted (at least in Northern Europe). Together with the 'legal compartment model' the combination of the three does not require to always document all details - but rather to make sure that the decision process within the compartments can be reconstructed [2]. Instead of storing all images - an impossible task - the focus lies on those which were basis to the clinical decisions at the time being, including their generation process [5,6].

Consequently, the medical record has to be defined in new terms. So far, the paper and pencil version has physically reminded us of a book, a folder or a dossier. Within medical information systems the medical record is a set of pointers linked to a personal treatment episode which links various kinds of data to the episode. The linkage is defined by a legal algorithm (or standards [1]) defining software which may be different for each legal culture (see also [2]).

Regarding the value of the various components of this electronic medical record as legal document two steps could be envisaged:

1) Combination of WORM-technology with legal certifications as for microfilming for decision relevant data,

2) Combination of data with electronic signature systems that guarantee the correctness and completeness of the data.

5. Images

A concept for the legal definition of images has already been published [4]. Quality assurance measures have to be documented within an imaging department (as in laboratories) which guarantee the highest feasible level of performance within the image manipulation and interpretation process.

Departments have to document the used software versions as well as the persons responsible for the manipulation and interpretation process which led to a clincally relevant decision.

6. Conclusions

For several years legal aspects of image processing have rarely been discussed. Today, further progress on the technical side has to be accompanied by a clarification of legal requirements for the various system components. It is expected that some input may be generated by the LEGIS-EURIPACS-AIM project. Nevertheless, any result of theoretical considerations has to be translated into the legal culture of the various nations. Medical Informatics can only provide a framework of thinking to ease this transition process.

It is necessary that lawyers reflect the impact of their suggestions on the practicability of medical action. It might even be valuable to reconsider some privacy considerations from the sixties and transform them into appropriate versions for a technical society in which information technology has become part of the everyday live.

REFERENCES

1. Kilian, W.: Rechtliche Aspekte der digitalen medizinischen Archivierung von Rötgenunterlagen. In: Riemann, H.E., Kollath, J., Rienhoff, O.: Digitale Radiographie, Bad Nauheim 1986, Konstanz: Byk Gulden 1986, 183-190; also published in: Neue Juristische Wochenschrift 1987, 695 ff.

2. Lutterbeck, B., Wilhelm, R., Hadjerouit, S.: PACS and the need for a new legal structure. In: Lemke, H.U., Rhodes, M.L., Jaffe, C.C., Felix, R.: Computer Assisted Radiology CAR '89, Proceedings, Heidelberg: Springer 1989, 810-814

3. Piccolo, U., Rienhoff, O.: Legal considerations regarding PACS - International Perspective. In: IMAC'91 2nd International Conference on Image Management and Communication in Patient Care, Los Alamitos, California: IEEE Computer Society Press, 1991, 335-339

4. Retter, K., Rienhoff, O., Karsten, C., Prince, H.E.: Requirements for a documentation of the image processes within PACS. In: Dwyer, J.: Medical imaging IV - PACS system design and evaluation, SPIE vol. 1234, 795-805

5. Reichertz, P.L.: Kontextabhängigkeit medizinischer Informationen. In: Reichertz, P.L. und Kilian, W. (Hrsg.): Arztgeheimnis - Datenbanken - Datenschutz. Berlin: Springer 1982, 204-219

6. Rienhoff, O.: Zur Beurteilung der 'Richtigkeit' patientenbezogener Daten in der Medizinischen Dokumentation. In: Reichertz, P.L. und Kilian, W. (Hrsg.): Arztgeheimnis - Datenbanken - Datenschutz. Berlin: Springer 1982, 196-203.

7. Rienhoff, O., Lutterbeck, B., Retter, K., Reinecke, E.: Rechtsunsicherheiten bei der digitalen Bildverarbeitung und Archivierung. In: Riemann, H.E., Kollath, J., Rienhoff, O. (Hrsg.): Digitale Radiographie, Konstanz: Schnetztor 1991, 159-164.

8. Velde van der, R.: Hospital Information Systems, Springer Verlag, Berlin (1992)

Ethical and Legal Issues about IMAC in Different European Countries

B. BEOMONTE ZOBEL *, B. WEIN ^, M. OSTEAUX °, R. PASSARIELLO §

* Department of Radiology, University of L'Aquila, Italy

^ Department of Radiology, University of Aachen, Germany

° Department of Radiology, Vrije Universiteit Brussel, Belgium

§ Department of Radiology, University of Rome, Italy

Summary

The electronic technology available today is capable of providing significant improvements in the management of Diagnostic Imaging departments. Unfortunately, the adoption of such technology is inhibited by legislation, regulations and legal principles in many of European Community (EC) countries. With the introduction of the deregulated flow of goods and persons in Europe and with the future realization of the political European Union, it is mandatory to ensure the enacting of specific regulations at European level. The objective of this paper is to review the legal concerns related to the use of IMAC systems and to evaluate the work already done by international organizations to produce a legal harmonization in Europe. The assessment of the legal status in the European countries today shows that the lack of homogeneity is mainly due to different national laws but in addition there is a complete absence of rules about the use of modern information technology in the field of Diagnostic Imaging. In this context, an European convention, covering all the different topics related to the use of electronic medical records, seems to be the best proposal to enable a correct future development of health care informatics within EC countries.

Introduction

The use of computers in the medical field raised the question of how to match the important benefits deriving from the immediate availability of clinical data, both inside and outside the hospital departments, with the disadvantage to offer an easy access to medical information to non authorized people (1).

The several innovations brought about by the diffusion of informatics in health care, especially in the Diagnostic Imaging departments, and the organization of international electronic networks, with the possibility of a free exchange of patient data among the EC (European Community) countries, certainly make the achievement of an adequate European legal regulation a very important issue.

In fact, in most of member states of EC, the adoption of electronic technology for medical records is not allowed by legislation and legal principles (2) while in other

countries there is the possibility to use automatic processing of medical data, although with some limitations, so it seems appropriate the suggestion to reach an homogeneity in legislation at European level.
Some proposals were made by international organizations, such as the Council of Europe (3-4-5-6), but these actions do not have a real legal power because specific recommendations have first to be ratified by member states and then included in each national body of laws by enacting a specific regulation.
In addition, we have to consider a group of ethical issues, recognized by most of penal codes, that although derived by the old Hippocratic rules, usually show significant differences from a country to another (7).
Legal and ethical matters, related to the use of information technology in the management of diagnostic images, may be conveniently divided into four categories:

a) Personal data protection;
b) Medical databases;
c) Medical secrecy;
d) Image manipulation, storage and archiving.

<u>Personal data protection</u>

The present situation of the European legal environment about data protection is constituted by both international and national documents.
Among the different international organizations, the most active in the field of data protection in health informatics is the Council of Europe that emanated several recommendations (n. 108/1981 Protection of individuals with regard to automatic processing of personal data; n. 1/1981 Regulations for automated medical data banks; n. 10/1983 Protection of personal data used for scientific research and statistics; n. 23/1987 Hospital Information Systems). The EC action in this field has consisted mainly of the Commission Recommendation (July 1981), in which all the EC countries are advised to proceed with the rapid implementation of the principles contained in the recommendation n. 108/1981. But the EC recommendation was signed by only seven countries, that is Denmark, France, Germany, Ireland, Luxembourg, United Kingdom and Spain, among which, till now, Spain does not have any law. Of the southern countries only Portugal has recently enacted a data protection law, while Netherlands has its own law although it did not sign the EC recommendation. If we consider the different laws of these eight countries, we can see that different solutions are found for the different problems raised, and different procedures are considered to open a file, to protect the individuals, to notify to authorities and to acquire, control and transfer personal data.

Medical databases

As far as health data banks are concerned, the recommendations n. 1/1981 and n. 23/1987 express only general principles that are often non specific and sometimes in contrast with national regulations. Till now no detailed national laws have been enacted in any of the EC countries. However, specific rules on health data may be found in the prevailing data protection law. Usually these specific rules reinforce safeguards on data protection, as, for istance, we have in the DANISH PUBLIC AUTHORITIES ACT, which contemplates that health data must not be registered except where specifically required for the purposes of the registration. The DUTCH DATA PROTECTION ACT includes a special chapter on personal data files in the areas of government, education, health care, and social services. The LUXEMBOURG LAW limits in one of its article the processing of health data. The PORTUGUESE LAW and the DUCH ACT prohibit the automated treatment of health care data in principle. The ITALIAN DEONTOLOGY CODE prohibits the organization of health data banks that are not able to guarrantee the rights to privacy and confidentiality of the patients.
A certain uniformity can be found in the various legal text of the EC countries as they tried to follow the general parameters of data protection included in the Council of Europe Recommendation, but this kind of uniformity is not sufficient to obtain a common legal environment in the field of electronic medical records. In fact, this would require the existence of the same specific principles and the same derogations. For istance in Denmark the passing on of data to private individuals or public authorities is allowed when required for carrying out scientific or statistical investigations of social relevance, but this is forbidden in France where such a request must be first approved by the "Commission Nationale de L'Informatique et des Libertes". This commission was constituted after Franch Parliament enacted a specific regulation on "Automatic data processing and individual liberties" (January 6, 1978). Some articles of this law have specific influence on health care data such as the one that prohibits the use of national health number to access to medical data files or the other that states the need, for a patient, to ask the doctor to obtain the access to his own medical data file. The Franch Commission is an important example of a flexible approach to the problems that very often arose for the safeguard of citizen's privacy but unfortunately do not exist similar Commissions outside France.

Medical secrecy

Other difficulties originate also on the topic of professional secrecy. It seems to be a consensual issue in all the EC countries as all of them record it in penal or, as in Ireland and UK, in deontology codes, but the legal consideration of this individual right is different . Actually, the EC countries see the professional secrecy in a different way. For istance, in Belgium and France, breach of secrecy is considered a crime against the community and not against the individual and so is punished most severely. In case of detection of a crime, the right of the patient is recognized only in Ireland, Italy and Luxembourg. In Belgium, Denmark, France, Germany and Netherlands the judge can oblige a physician to reveal a professional secret also against the patient's and doctor's will, while in Ireland and Luxembourg only the patient can authorize the physician to reveal a secret. In addition, in Portugal, France and Germany doctors are allowed, by professional codes or by other kinds of regulations, not to reveal the real clinical situation to their patients about a bad prognosis or a fatal illness, even if there is a specific request made by the patient himself. The same policy was confirmed by a sentence of the German High Federal Court in 1983. In that case a patient with mental illness, who underwent a surgical intervention in a hospital, made a formal request to obtain a copy of his clinical folder. The hospital refused to give the copy of the folder because the clinical condition of the patient did not allow a complete understanding of his health data. The legal decision of the Court is interesting to evaluate the legal position in Germany. The Court affirmed two important principles. First of all the right of the patient to access to his own clinical data must be always recognized even if there is a lack of capability to understand them. The second one is that a doctor can reject the access of the patient if there are good reasons to avoid the knowledge of his clinical data such as in presence of a bad prognosis or a fatal illness. This kind of interpretation of the personal relationship between the patient and the physician is different from the one that exists in other countries, such as Italy, where in presence of a specific request made by the patient himself, the physician is obliged to respect his/her right to know the truth.

The balancing of the public interest in obtaining access to health management information system and individual privacy concerns may involve a reassessment of the appropriateness of traditional legal principles of confidentiality. This is the case of a new law, enacted by German parliament in December 1992, that determined a wide spectrum of possibilities for insurance companies and public organizations to obtain a free access to personal clinical data from the government, who is considered the administerer of citizen's health data. But this role of custodian of health information have been severely criticized by German medical association who expressed the opinion that the existence of a public interest has to be clearly demonstrated every time before authorizing the access to information.

Action programmes in Health Care outside EC

In some countries, such as the Nordic countries, the use of safe computing in health care has been considered for many years. The Swedish Medical Association, in 1986, was the first organization to estabilish an action programme for computers in health services. In 1991 the Norwegian Medical Associations followed with a similar comprehensive action programme. Both the associations consider ethics, integrity and confidentiality to be the key issues related to the use computerised health information systems (8). Their basic position concerning data files which contain individual patient information is that this files should remain at the base unit level. A base unit refers to a clinical department, a community health centre or similar unit. So the harmful effects of possible misuse of information can be limited when less information is stored in different units and when communication protocols work to control the accesses.

Conclusions

To achieve a homogeneous legal environment on an EC basis we need to stimulate some countries to enact a national regulation about data protection, following the recommendations of the Council of Europe, and the Community to develop a directive that, covering all the various topics related to the use of electronic medical records, will be accepted and ratified by the members. Recently a new draft of the recommendation n. 108/1981 has been produced but it seems to be not sufficient to answer to all the questions arose about the use of IMAC systems.
As the utilization of automated images management in health services is able to provide a positive impact on the delivery of health care, it is essential that radiologists are actively involved in the development process including the problems of ethical and legal issues.

This research activity is part of the EurIPACS project , financed by EC in the framework of AIM program.

References

1) Griesser G., Bakker A., Danielsson J., Hirel J. C., Schneider W., Wassermann A. I. Data protection in health informatics systems: considerations and guidelines Amsterdam, Elsevier Science Publisher B. V. (North Holland) 1980
2) Commission of the European Communities DG XIII/F AIM Data protection and confidentiality in health informatics: a survey of legal issues in the EC Community Amsterdam, IOS Press 1992
3) Council of Europe. Convention for the protection of individuals with regard to automatic processing of personal data. Strasbourg 1981
4) Council of Europe. Recommendation on regulations for automated medical data banks. Strasbourg 1981
5) Council of Europe. Recommendation on protection of personal data used for scientific research and statistics. Strasbourg 1983

6) Council of Europe. Recommendation on Hospital Information Systems.
Strasbourg 1987
7) Brazier M.
Medicine, patient and the law
London, Penguin Book Edt. 1987
8) Bengtsson S., Solheim B. G.
Enforcement of data protection, privacy and security in medical informatics
IMIA Proceedings, Elsevier Science Publisher B. V. (North Holland) 1992

Evaluation of PACS at Hammersmith Hospital: Baseline Assessment of Costs and other Resource-use Parameters within the Radiology Department

S. BRYAN, G. WEATHERBURN, J. KEEN, M. BUXTON

Health Economics Research Group

Brunel University
Uxbridge
Middlesex UB8 3PH

Introduction

The first hospital-wide picture archiving and communication system (PACS) in the United Kingdom is to be implemented at Hammersmith Hospital, London and is expected to be fully operational in late 1993. An evaluation of the Hammersmith PACS has been commissioned by the UK Government Department of Health, who are also funding the PACS implementation. The evaluation is being undertaken by the Health Economics Research Group (HERG), Brunel University. The scope and nature of the overall evaluation exercise has been detailed elsewhere [1,2]. This paper concentrates on the methods being employed to assess the resource requirements of a hospital radiology service operating with film and those of a similar service operating with PACS. Early results relating to one aspect of the resource use measurements are provided.

Methods

A comparison is being made of the radiology costs before the PACS implementation with the costs once the PACS is operational and running in a 'steady-state' setting. For the two alternative radiology department systems all costs relating to both management and operation are being measured. The analysis is not restricting its attention only to expenditures falling on the radiology budget but will include all resource inputs that go towards the provision of the radiology service. In addition it will look at the cost of the PACS *implementation* process which is a highly resource-intensive activity. There are three principal categories of resources required for implementation: the purchase and installation of the technology and any associated building works; the time of staff formally involved in the implementation process and of those involved on an informal basis; and the training and development of staff in the use of the new equipment.

The *capital* costs (or strictly speaking prices) associated with all the required imaging equipment, both conventional and digital, are being recorded. For the before PACS period this includes all equipment in operation in the radiology department at Hammersmith. For the post-implementation period this will include those items that are purchased as part of the PACS project capital expenditure and those that were previously at the Hammersmith but which will form part of the PACS system. Expenditure on capital items represents an investment in assets which, it is assumed, are to be used over a number of years. From an economist's perspective, capital costs have two components: the 'opportunity costs' of the resources tied up in the capital assets (ie. the value of the next best alternative use of the resources) and the depreciation over time in the value of the assets. Whilst several methods exist for measuring and valuing capital costs, the approach being taken in this evaluation is to annuitize the initial capital outlay over the 'useful life' of the assets such that an 'equivalent annual cost' can be calculated. An obvious difficulty here relates to specifying, with any degree of certainty, what the useful life of the new technology is likely to be, this being a function of both 'robustness of capital' and rate of technical change.

The costs of *running* the radiology service before and after implementation are also being measured to allow direct comparison. This analysis considers such resource components as the health care professionals and other department staff, supplies of film and associated consumables, the number and size of the department's rooms, and equipment maintenance. The impact of the PACS developments on the workload of radiologists, radiographers and clerical staff is also being assessed.

One of the most significant factors in determining the running cost of a radiology service is the number of radiologists required. This requirement is obviously a function of a number of variables, possibly one of the most important of which is the time taken for the radiologist to view a series of images relating to an examination and then to produce a report. There is an expectation that radiologists will, in the short-term at least, require a longer period of time to report from PACS workstations than they do currently from hard copy conventional films. This empirical issue is being addressed in the evaluation exercise. As part of the baseline, an observational study of reporting by senior radiologists (consultants and senior registrars) has been undertaken for a period of 25 days over 3 months such that each day of the week was observed on five separate occasions. Hammersmith has a reporting arrangement whereby a senior radiologist is always available between 9am and 5pm Monday to Friday to report on films being generated within the department. A rota system operates with change-over of radiologist and secretary, to whom the report is dictated, at 11am, 1pm and 3pm. Study observers also rotated at these times. Reporting in this setting was observed since it provided the opportunity to study both 'hot' reporting (where the report is required immediately) and 'cold' reporting (where the report is not required with such urgency) by senior radiologists. An observer recorded, for each report of every examination, details of the number and type of new images viewed, the number and type of previous images viewed, the time taken for the report to be produced (measured from when the film packet was picked up by the radiologist to when the film packet was passed to the secretary), and the number, type, and length of all disturbances which occurred. This exercise will be repeated at regular intervals to include the implementation and transition period, and the period when PACS is operational and running smoothly.

Other resource-related measurements being made within the radiology department include the staff time devoted to clinico-radiological sessions, where both radiologists and other clinicians meet to discuss selected cases. The junior radiologists responsible for preparing for these sessions are completing a log. In addition to their time input to these sessions they also record the number of images required for each session, the number that could not be found and the number that are required on consecutive weeks. Again, these measurements will be repeated on several occasions during and following the completion of the PACS implementation process.

Results

Results are available for the baseline measurements of the study of reporting by senior radiologists. Data were collected on a total of 2432 observations, each observation referring to one or a series of x-ray examinations of an individual patient that were subsequently viewed and reported on together by the radiologist. The frequency distribution of report times (REPTIME) is shown in Figure 1. This data is clearly not normally distributed and when the log of the report time is taken, a more normal distribution results (Figure 2). The median report time for all observations is 117 seconds: 113 seconds for 'cold' (non-urgent) reports and 119 seconds for 'hot' (urgent) reports. This difference is not statistically significant ($p=0.12$). The most frequently viewed image was the chest and the median report time for observations where the only images viewed were of the chest was 112 seconds. A total of 1249 (51%) report time observations related to images of the chest only. A total of 16 different radiologists were observed over the 25 days, and very different reporting styles were identified. In terms of speed of report production the range was from a median of 52 seconds to a median of 216 seconds per report produced. Report times for each individual radiologist when compared to those for all other radiologists were statistically significantly different ($p<0.05$) for eleven of the fifteen radiologists studied. Ten of the eleven radiologists for whom the difference was statistically significant tended towards a longer report time.

The baseline observational data collected for this aspect of the research have been used to explore the way report time is influenced by several other factors. Ordinary least squares (OLS) multi-variate regression analysis techniques have been applied in specifying the model. As the PACS technology comes into operation, the data collection exercise will be repeated on several occasions and the model will be re-run with additional dummy variables to represent the stages of introduction and operation of the technology. Early analysis suggests that the model will have as its dependent variable the log of report time which will be regressed on:

- the number of new conventional films viewed (NONEWF) and that variable squared;
- the number of new non-conventional modality films viewed, eg. CT, (NONEW1) and that variable squared;
- the number of previous conventional films viewed (NOPREVF) and that variable squared;
- the number of previous non-conventional films viewed (NOPREV1) and that variable squared;
- the number of disturbances (NODIST) and that variable squared;
- the time occupied in disturbances (TIMDIST) and that variable squared;
- a series of dummy variables for the body areas viewed;

- a series of dummy variables for the radiologist observed;
- two dummy variables for the observer;
- two dummy variables for the time of day.

This model specification results from a backwards stepwise OLS regression procedure which identified those variables that contributed significantly towards the explanatory power of the model. Variables that were dropped from the model as non-significant related to the day of the week, the secretary to whom the report was being dictated and whether the report was considered to be 'hot' or 'cold'. The adjusted R^2 statistic is 44% which reflects the proportion of the variation in report time that is explained by the model.

The signs on the coefficients of the continuous variables (NONEWF, NONEW1, NOPREVF, NOPREV1, NODIST and TIMDIST) are of interest. The coefficient signs on all of these variables are positive and the coefficient signs on the same variables squared are all negative. This implies that report time initially increases as the continuous variable, for example number of new films viewed, increases but then reaches a maximum before decreasing with increases in the continuous variable. The signs on the coefficients of the dummy variables are also of interest. If a film viewed in the reporting process was one of the following: the abdomen, the chest, a hand, the pelvis, the cervical spine, the lumbar spine, a knee, a mammogram, or a shoulder then the report time was significantly longer than reports where one of these images was not viewed. If, however, a film viewed in the reporting process was an OPG then the report time was significantly shorter. A significantly shorter report time was also found for reports completed between 9.00 and 11.00 and those completed between 13.00 and 15.00, when compared to reports completed at other times of the day.

Results from the baseline assessment of capital and running costs for the conventional radiology department at Hammersmith, and of the implementation costs, are not currently available.

Discussion

The installation of the PACS technology at Hammersmith is not complete and thus, at this stage, only evaluation methods and baseline results can be reported. The initial baseline study of reporting detailed in this paper has attempted to identify all the important factors that should be controlled for in an analysis of the relationship between report times and the change in technology from conventional radiology to PACS. A backwards stepwise regression procedure was employed since data were collected on all variables thought potentially to be important explanatory factors. From the data available, it appears that there is no requirement to control for which secretary is being dictated to, the day of the week the report is being provided or whether the report is considered as 'hot' or 'cold'. However, our results do suggest that there are a number of other factors that must be taken into consideration when analysing report times. These include the number of films viewed, the number of and time devoted to disturbances, and the body areas viewed. In addition, it is clear that there are very different styles of reporting between radiologists and this must be controlled for; the time of day the report is produced is also an important factor and, inevitably, the individual observing appears to have an effect on length of report times.

Finally, a comment on the overall 'goodness of fit' of the model is appropriate. An adjusted R^2 figure of less than 50% implies that more than half the variation in report times are currently not explained by the independent variables in the model. Our response to this will be to consider and discuss with others whether there are other explanatory variables that should have data collected upon them. When additional potential explanatory variables have been identified then the baseline data collection will be completed with a second round of observational study prior to the introduction of the first software release of the Hammersmith PACS.

References

1. Bryan, S.; Keen, J.; Buxton, M.J.; Weatherburn, G.C.: Evaluation of a hospital-wide PACS: costs and benefits of the Hammersmith PACS installation. SPIE (Medical Imaging VI: PACS Design and Evaluation). 1654 (1992) 573-576.

2. Keen, J.; Bryan, S.; Buxton, M.J.: Evaluation of PACS, In Todd-Pokropek A (ed) Computer Aided Management of Medical Images, Spinger-Verlag, Heidelberg (forthcoming).

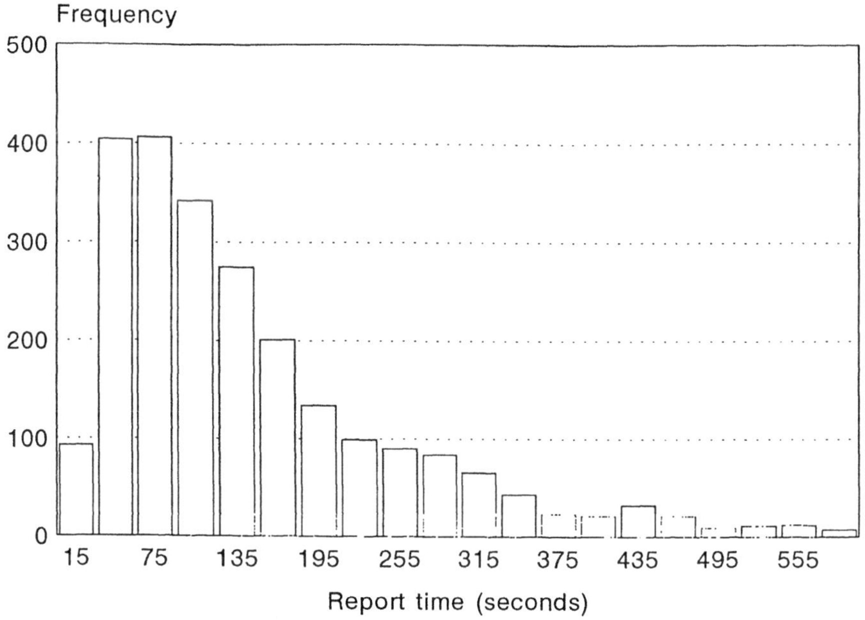

Figure 1: Frequency distribution of report times

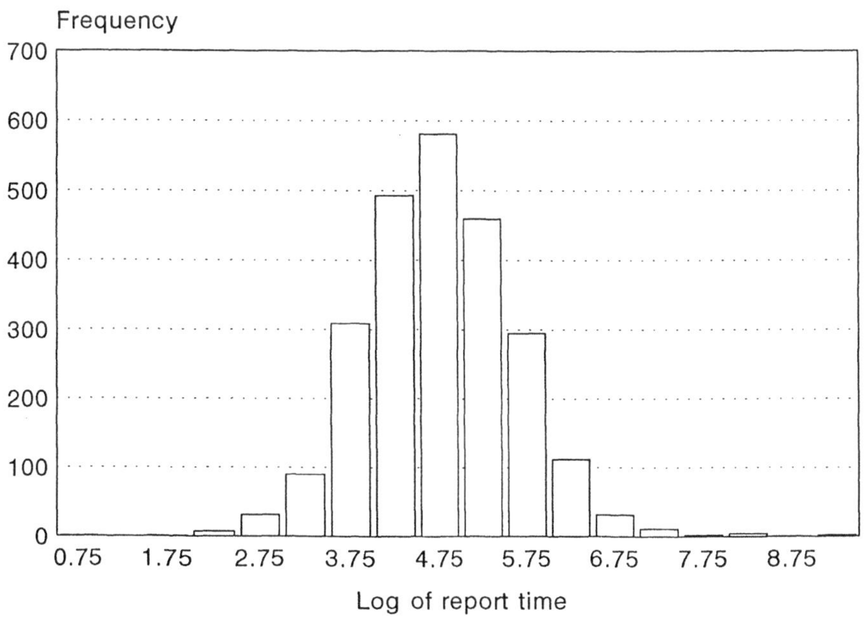

Figure 2: Frequency distribution of log report times

Shared Medical Imaging Digital Archiving

JOHN VANDEN BRINK

Technology Marketing Group
Des Plaines, Illinois 60016

Perceptions of the primary benefits of PACS technology have remained unchanged over the last 4 years, based on the TMG IMACS/PACS tracking study. The primary perceived attributes of the technology are to provide accessibility to images, save space, improve communications between health care providers, improve efficiency of the diagnostic imaging process and reduce film cost.

1988	%	1990	%	1992	%
Access	49	Access	50	Access	52
Space	49	Space	49	Space	34
Efficiency	21	Communications	26	Communications	23
Communications	12	Film Cost	17	Efficiency	16

When asked what are the primary obstacles to acquiring PACS technology, the response is consistently cost justification, funding, improved archiving technology, etc.

What issues, if any, need to be addressed before this technology can be implemented in your environment?

Radiology Professionals

(Respondent Base: 1992=237; 1990=243)

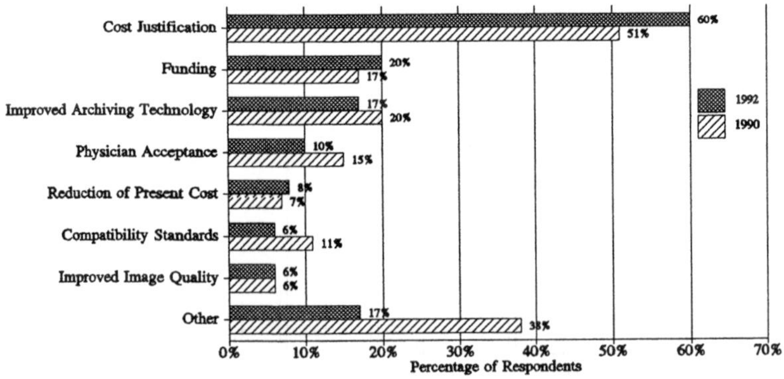

An alternative way to provide the benefits of PACS in whole or in part and to reduce costs is through a shared medical image archive where users pay on a transaction basis rather than on a capital investment basis.

Shared archiving, organized on an urban or regional basis, can provide the following benefits:

- Shares costs of hardware, software, personnel, capital equipment.
- Provides low cost, off-site storage charged on a transaction basis.
- Provides accessibility, communications and space saving benefits which may, in whole or in part, be equivalent to the benefits derived from an on-site PAC system.
- Advances the development of an archiving and communication infrastructure necessary to achieve the benefits of image and medical record telecommunication between health care providers and buyers.
- Provides a low cost way to introduce PACS technology within the hospital and to the medical community.
- Enables third party investment to provide capitalization for the project if desired.
- Can easily accommodate the other components of the medical record as required.

To provide an acceptable alternative to present methods, shared electronic archiving has to be cost competitive, utilizing methods which can provide superior reliability, accessibility and high speedy communications.

Several basic assumptions accompanied the decision by TMG to propose a feasibility study of Shared Medical Image Archiving for the City of Chicago, a program which began in 1987. They are as follows:

- The most competitive bit storage costs will be achieved by capital intensive technology and shared costs of expensive scarce professional labor required to develop, manage and maintain the archive.
- The high volume of imaging data required to achieve the lowest cost/bit must be addressed in an urban or regional area with sufficient activity to support a capital intensive investment.
- The time to evaluate the feasibility of a shared archive (two to four years) would be consistent with the pace of technology development.
- A feature of a shared archiving strategy that permits migration over time to optimal performance/cost media and architecture can only be implemented in a large scale archive.
- The cost justification issue for capital investment in PACS will not be sufficiently addressed to compete with transaction based strategies for a good time to come.
- The Chicago Area provides a unique environment in which to conduct such a prototype development because of the diversity and depth of its medical and scientific community, the commitment to developing the telecommunications infrastructure, and the national medical organizations headquartered in Chicago that would benefit from a successful demonstration of this capability.

To explore financial support for the project an initial contact was made with the Information Industry Council of Metropolitan Chicago formed by the Civic Committee of the Commercial Club of Chicago to attract information technology companies and individuals to the Chicago area. Through this relationship, the University of Illinois at Chicago (UIC) became involved, and a grant request was successfully submitted to the Technology Challenge Grant Program administered by the Department of Commerce and Community Affairs (DECCA), an agency of the State of Illinois for funding of a feasibility study. Concurrently, industry participants were solicited for additional funding which resulted in support from Digital Equipment, Kodak, Siemens, and TelePort Communications Corporations.

Separate industry and user advisory boards were empaneled to provide guidance for the feasibility study.

The Metropolitan Chicago Health Care Council, which provides shared services for 91 hospitals in the Greater Chicago Area, participated in the project and solicited support for the feasibility study among hospitals in the area. Initial participation was solicited through a mail questionnaire survey with telephone follow-up.

Thirty-nine hospitals provided the information requested. Individual hospital cost modeling, follow-up telephone interviews and a general meeting were used to complete the feasibility study.

<u>Results</u>

Respondents identified the following benefits they perceive to improve the quality of health care through improved archiving methods mentioned in order of frequency:

- Ease of access to past record for comparison, consultation and improved diagnosis.
- Improved radiology report turnaround time
- Improved patient care
- Improved medical history, satisfying legal requirements,
- Elimination of lost films
- Better use of Personnel
- Reduced Cost

Following are key operating variables and estimated costs abstracted from the study based on averages of aggregate data:

	MEAN
Estimated annual cost of archive, retrieve, and re-archive images	$293,013
Space dedicated to film archiving	3,160 ft^2
Perceived archiving costs per procedure/year for 5 year archive	$ 0.67
Annual Procedure volumes	
Digital (CT, MRI, Nuclear Medicine, Ultrasound, DSA)	21,183
Radio graphic and/or flouro	65,969
TOTAL PROCEDURE VOLUME	87,152
Image retrieval-procedures/ day	191

Based on the computer cost model and using the average procedure volume of all hospitals responding, the annual cost of a five-year archive based on the cost model is $572,855 for an average cost of $1.31 per procedure per year or twice the respondents perceived cost of archiving.

ARCHIVING AS A PORTION OF MANUAL IMAGE MANAGEMENT COSTS
TMG Cost Model
Average Procedure Volume 87,152/year – 5 Year Archive – 435,760 Procedures
(Mean Procedure Volume of Respondents)
Overhead = 150% of Labor
Space Costs = $60/ft^2

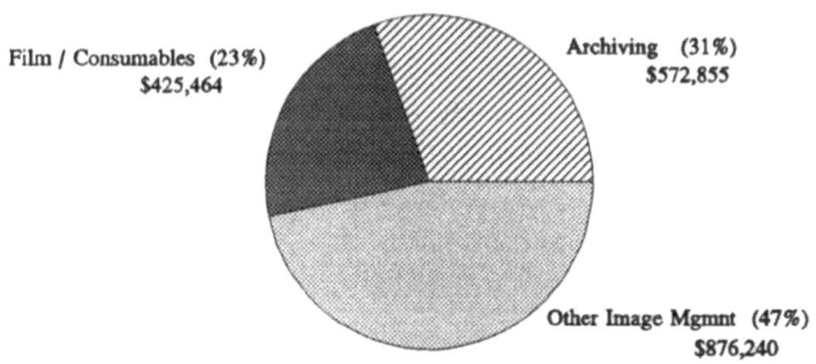

With respect to the overall priority of archiving relative to other issues, about one-third of the hospitals thought it was a high priority.

Very Critical	5	2
	4	5
	3	8
	2	12
Not a Critical Issue	1	12
	Responses	39

Two-thirds of the respondents indicated modalities to be chosen for the first phase of the archiving would be the digital modalities, i.e., CT MRI, Ultrasound and Nuclear Medicine.

Questions and concerns about the service were:

- Cost
- Confidentiality/Security
- System downtime and backup
- Special training and protocols required
- Reliability and company standing
- Legality
- Image quality and turn around time

Project Scoping

For shared archiving to be successful, it has to be cost competitive with other alternatives. This includes present methods, off-site manual storage and on-site electronic archiving. Following is a summary of costs for these alternative methods.

	TMG Cost Model	Perceived Costs Feasibility Study	*On-Site Optical Disc Disc Archiving	**Off-Site Manual Storage	***Shared Archive
Annual storage Cost/procedure	$1.31	$.67	$1.44	$.30-$.74	$.44
Retrieval (2-8 hrs)	Included	Included	Included	$11-$29	$3-$12
Re-archive/ Destroy	Included	Included	Included	$1.75-$2.75	N/A

* 4 T-bytes of uncompressed data, $500K capital investment, $600K/year in labor, overhead, service, software support. Media cost $.08/procedure

** Average of 15 service companies

*** Includes telecommunications. Requires workstation/gateway

Based on the data provided by the 37 participating hospitals, total imaging volume is equivalent to 37 terabytes/year or a total of 185 terabytes for five years, which represents the maximum archiving requirement for this group of hospitals.

The industry advisory board projected a cost of $1.5M for prototype pilot site development. If the prototype is successful, the preliminary proforma suggests conservatively an additional investment of $2.5-$3.0M would be required to cover captial investment and operating capital. Given the assumptions in the previous table on costs charged/procedure, the operation would become profitable in the 4th year.

	Year 1	Year 2	Year 3	Year 4	Year 5
Profits Cumulative	($843,102) ($843,102)	($845,221) ($1,688,323)	($11,294) ($1,699,616)	$560,188 ($1,139,428)	$1,046,266 ($93,162)
Terabytes Stored	1.44	2.88	6.30	9.72	13.14

From the 37 initial participants, 11 expressed a continuing interest and 4 have agreed to participate in a planning phase for pilot site installations. The planning commitments from these four major hospitals in the Chicago area will provide adequate information to determine the need and value in proceeding to installation of pilot sites and eventual full development of the shared archive concept.

At this writing, the funding is being solicited to support a planning activity for prototype development. A progress report will be provided at time of presentation of the paper.

REFERENCES

1. Vanden Brink, J.A.: "Cheshire Cat of Medical Imaging", Administrative Radiology, Vol. 8, no. 5, p. 32-35, May 1989

2. Vanden Brink, J.A., Cywinski, J.: "Experience with Cost Analysis Computer Model for PACS Evaluation". In CAR International Symposium and Exhibit, Berlin, 1989

3. Saarinen, A.O., Haynor, D.R., Loop, J.W., et al: "Modeling the Economic of PACS: What is Important?", Society of Photo Inst. Engrs., 1093: 62-73, 1989

4. Straub, W.H., Gur, D.: The Hidden Costs of Delayed Access to Diagnostic Imaging Information: Impact on PACS Implementation. American Journal of Radiology, 155: 163-616, 1991

5. Hindel, R.: Impact of Image Storage Organization on the Effectiveness of PACS. Journal Digital Imaging, 3: 205-120, 1990

6. Vanden Brink, J.A.: "Image Management and Communications". Diagnostic Imaging Special Editorial Supplement, September 1991

7. Hilsenrath, P.E., Smith, W.L., Berbaum, K.S., et al: Analysis of the Cost-Effectiveness of PACS. American Journal of Radiology, 156: 177-180, 1991

8. Vanden Brink, J.A.: "The Market for PACS". SCAR News, Vol. 3, No. 2, November 1992

Facility Design for Medical Imaging Computer Workstations

BILL ROSTENBERG, AIA

Stone Marraccini Patterson Architects and Planners
One Market Plaza - Spear Street Tower
San Francisco, California 94105 USA

Abstract
The purpose of this paper is to describe design considerations for medical imaging computer workstations and their environment. Although many computer workstations are in use today, they are often placed within a physical space which is not designed appropriately for their use. A well designed workstation environment well help radiologists and technologists work both safely and efficiently.

1. Overview of Medical Imaging Computer Workstations

• General Considerations

While considerable effort and expense accompany design and planning of most examination and treatment spaces in medical imaging facilities, technologist's and radiologist's workstations seldom enjoy the same level of concern and detail. For example, the configuration, arrangement of equipment and quality of light found in interventional radiographic exam rooms is usually the result of comprehensive analysis of imaging techniques; ergonomics; and the flow of people, supplies and instruments. In contrast, computer workstations in the same department are frequently placed in whatever space is available, with little consideration for lighting comfort and work habits.

Compared to examination and treatment rooms, however, the workstation is occupied for longer durations and its design is just as critical to the successful delivery of quality medical care. In addition, properly designed workstations can result in reduced incidence of occupational injuries and improved efficiency and work performance.

• The Workstation Environment

As electronic image acquisition, storage, transfer and manipulation are embraced and accepted as fundamental components of modern medical imaging services, the computer workstation replaces more traditional equipment, such as film illuminators and alternators, as a tool of choice for viewing—and more important—for manipulating medical images. The workstation differs from film illuminators and alternators in that it is more interactive and it represents the point of access between radiologists and an entire automated system of radiographic and medical information. To date, most computer workstations also differ from film illuminators in their lower spatial resolution and luminance, and therefore, less desirable image quality[1].

To compensate for the workstation's lower luminance, ambient lighting must be dimmable to increase contrast and reduce glare. Visual display terminals (VDTs) and data input devices (keyboard, mouse, trackball, etc.) must also be arranged for the user's comfort and in such a way that the user does not need to frequently and repeatedly shift his or her focus.

Because radiographic image interpretation often involves multiple radiologists reviewing and conferring with each other, the room in which diagnostic workstations are located should be designed for acoustic control. It is just as important to contain conversational sounds within the space, in an effort to maintain privacy, as it is to prevent external noises from entering the room and disturbing ongoing work.

Finally, both seating and work surfaces must be designed for comfortable use over long durations. This suggests adjustability, so a variety of users can work comfortably, and durability, so the equipment and furnishings will continue to provide adequate support for each user over time.

• Workstation Uses and Locations

Radiology workstations have many uses and, therefore, may be found in various configurations and locations. Two different types include diagnostic workstations used for primary or secondary diagnosis based on image interpretation, and clinical workstations where diagnostic images are reviewed along with other pertinent non-radiographic medical data.

Image interpretation is usually based on comparison of recent images with previous ones. Therefore, most radiographic workstations will contain multiple monitors. Often the monitors are arranged in a configuration similar to film illuminators (i.e., two–over–two or four–over–four). The U.S. Army's Medical Diagnostic Imaging Support (MDIS) system, for example, includes diagnostic workstations with two to eight monitors and clinical workstations with one to four monitors[2].

Multiple computer workstations may be found in reading rooms within the radiology department. The problems associated with glare and noise are often compounded in such large installations. Individual workstations may be found in radiologist's offices and the offices of other physicians. Here the challenge is to design a physical environment which accommodates both activities performed at the workstation and other activities such as reading, writing and consultation. The ergonomic, lighting and acoustic requirements for each activity are different. An office with an exterior window, for example, might be ideal for reading but presents problems of glare when using the workstation.

As another example, a partial picture archiving and communications system (PACS) located in a critical care suite should provide adequate acoustic control to prevent consultation conversations from being overheard by patients or their family. The workstation should be located close to the activity core so users do not need to leave the unit, but far enough from the patient rooms to provide privacy and to prevent disruption of ongoing critical activities.

Other workstation locations might include the emergency, surgery, oncology, orthopedic, urology and obstetric departments or other locations where medical providers rely heavily on viewing or manipulating medical images. Management of non-radiologic images—such as endoscopy and pathology images—is growing in popularity and will have additional impact on the use of computerized medical workstations.

2. Design of Workstations and Their Environments

• Ergonomics

Improper design of workstations or improper use of well-designed workstations can lead to myriad health problems ranging from occasional headaches to chronic orthopedic and neurological disorders. "Most experts agree that three points of contact are the chief concerns in maintaining a healthful workplace: where your eye meets the monitor screen, where your fingers press against the keyboard, and where your body rests in its chair.[3]"

Eyes

Many eye problems can be avoided by proper work habits and proper ergonomic design. One study reported in the *New York State Journal of Medicine* suggested that eye irritation of workers using video display terminals was due to poor ergonomic conditions, such as improper lighting, glare, work materials which were poorly arranged on the work surface and improperly corrected vision[4].

Hands

The incidence of hand and wrist–related health problems has escalated since computer workstations have become increasingly common. One such type of malady, repetitive strain injuries (RSI), are caused by repeating the same physical motion over and over. Multiple adjustments of seating, work surfaces, key boards and monitors can help reduce the likelihood of such injuries. The proper height of these workstation components can further reduce injury.

In 1988 the American National Standards Institute and the Human Factors Society published a voluntary national standard (ANSI/HFS Standard 100–1988) with recommended visual display terminal (VDT) workstation dimensions (Figure 1). Some of the ANSI/HFS Standard 100–1988 requirements are:[5]:

COMPONENT	ANSI/HFS 100–1988 REQUIREMENT
Keyboard Support (see figure 1, dimension A):	• Keyboard support shall range in height from 23" to 28"
Seat Height (see figure 1, dimension B):	• Minimum range of adjustment shall be 16" to 20.5"
Seat Depth:	• Seat pan shall provide relief to the back of the knee if seat depth is greater than 16"
Seat Width:	• Seat cushions shall be at least 18.2" wide
Seat Pan Angle:	• Seat pan angle shall allow thighs to be at an angle somewhere between 10^0 above to 30^0 below horizontal
Angle Between Seat Back and Seat Pan:	• Fixed back: Angle shall be 90^0 or more • Adjustable back: Range shall include some part of range 90^0 to 105^0
Backrest:	• A backrest shall be provided • Support in the lumbar region shall be provided
Armrests:	• Inside distance between armrests shall be at least 18.2"

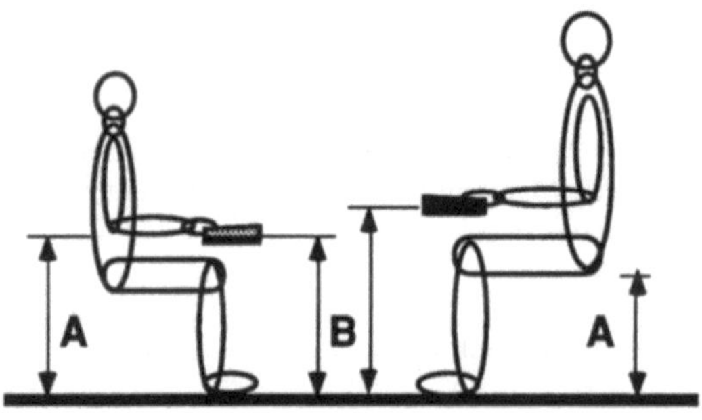

FIGURE 1: Range of height for seat and keyboard support.

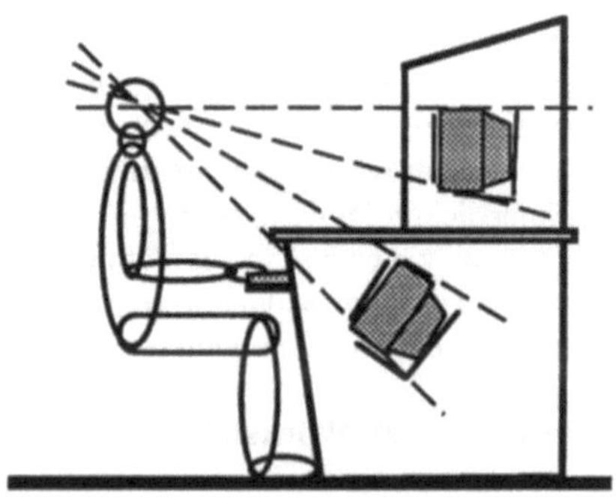

FIGURE 2: Primary viewing angle between 0 and 60 degrees below the horizontal angle of sight.

Seating

Proper seating should conform to the ANSI/HFS required dimensions and ranges of adjustability. The ability of workers to occasionally modify their posture and seating position can reduce the likelihood of workplace–related discomfort. "Except for specially designed ergonomic computer chairs, standard office equipment is typically not designed to adjust to the full range of settings required to accommodate the diversity of human beings.[6]"

- **Function**

There are two key differences between medical image–management computer workstations used by radiologists and technologists, and text–management computer workstations used by administrative and clerical staff. First, functions performed on image–management workstations usually involve less direct data entry, especially extended amounts of keyboard entry. Second, image–management workstations are generally used in rooms with lower light levels than their clerical counterparts. To accommodate these differences, medical workstations require both a work surface large enough to accommodate special tasks such as graphic data selection and editing, and a special approach to lighting design.

- **Lighting**

The greatest challenge facing designers of radiographic workstation environments is the low luminance of most visual display terminals. An over-lighted environment—or one in which the lighting is designed for paper work instead of computer work—creates a lighting imbalance between the monitor and the work surface. Ambient lighting for computer workstations must be provided in such a way that the contrast between the screen and surrounding surfaces is not so great that it causes the viewer eye strain, yet the room must be dark enough for the screen image to be clearly visible.

Room lighting which is three times brighter than the display terminal itself has been suggested for text–entry workstations; for image–management workstations ambient luminance equal to that of the display terminal has been recommended[7]. In addition, separate task lighting is advised for such low light conditions. Task lighting, however, must be arranged so it does not result in additional screen glare (excessive brightness reflected onto the screen).

Screen glare not only hinders image interpretation, it also causes unnecessary eyestrain. The amount of glare on the screen results from the physical properties of both the computer monitor and the surrounding environment. Although anti–glare filters, screens and coatings are available for most monitors their use implies placing an additional element between the screen image and the viewer, further compromising the image.

A better approach is to design the surrounding environment to minimize incidents of reflected light. Neutral colored surfaces with low reflectivity will help. "Selecting surface finish materials with low reflecting values for work surfaces, keyboards, armatures, shelves, and the like also cuts down on glare[8]" If an exterior window is present, it should be capable of being completely blocked and should have adjustable shades or louvers. Monitors should be placed away from the window or at an angle which will reduce the amount of light reflected onto them—directly or indirectly—from the window.

The single most effective way to control glare, however, is with the proper selection and placement of artificial lighting within the room. Dimmable indirect ambient lighting fixtures should be used in conjunction with carefully placed dimmable narrow beam task lights. Indirect light sources project light which is bounced off of wall or ceiling surfaces to soften its intensity. The actual light source itself should be completely screened from the workstation. Where indirect lighting is not practical, recessed lamps with a narrow beam can be used. Controls should be dimmable to limit the amount of ambient light.

Task lights can be strategically placed to illuminate individual work areas without causing excessive screen glare. Where multiple workstations are provided in one room, low partitions—just slightly higher than the viewer's eye level— on two or three sides of each workstation can help to control both glare and acoustics.

Task lighting fixtures mounted on articulating arms allow the fixture to be positioned in such a way that ..."light is projected onto the task from the side instead of from in front of the user. This reduces veiling reflections by bouncing light across a task and not back into the user's eyes.[9]" Veiling glare or veiling reflection results from a source of illumination—such as a light fixture or a window—which is reflected over a glossy surface[10].

Fixed task lighting can be attached directly to the wall or to modular workstations. Special lenses are available to minimize glare and reduce fixture brightness. Some fixtures include controls which allow the light intensity to be set at different levels for various activities or for the special lighting needs of different users.

Other Considerations

A properly designed workstation environment should consider multiple physical aspects. The heat generated from computer equipment must be removed for proper functioning of the equipment and for user comfort. Ventilation systems should not cause a draft over the work area.

Another consideration is clean power. Electronic circuitry can be easily damaged by slight variations in voltage. Dedicated power circuits and a filtered power supply are recommended.

• A Different Approach

One unique approach to workstation design has been developed for executive offices and security command centers[11]. Its application, however, could be adapted to radiographic workstations. Rather than placing the video display terminal on top of the work surface, it is placed beneath at an angle of 20^0 to 40^0 below the user's horizontal eye level (Figure 2). Multiple monitors can be placed both vertically and horizontally. The design was developed, in part, to allow the viewer to see multiple monitors without excessive head movement.

3. Conclusions

As computer workstations become more common—in both clerical and radiology workplaces—awareness of what constitutes proper environmental design should also increase. Key aspects to consider are ergonomics, lighting, acoustics, ventilation and the selection of proper materials and finishes for both furnishings and wall surfaces. The workstation and the workstation environment should be designed with an understanding of the unique tasks which will be performed on the workstation. The design criteria for image–management radiology workstations can be significantly different from those of text–management clerical workstations.

---

[1]Horii, SC, Electronic Imaging Workstations: Ergonomic Issues and the User Interface – Syllabus: A Special Course in Computers for Clinical Practice and Education in Radiology; Chicago; RSNA; 1992; p. 125.

[2]Smith, DV; Smith, S; Sauls, F; Cawthon,MA; Telepak, RJ; Design strategy and Implementation of the Medical Diagnostic Image Support system at Two large military medical centers; Frederick, MD; US Army Medical Materiel Agency; 1992

[3]Rosch, WN; Does your PC–or how you use it–cause health problems?; PC Magazine, 1991; p. 491.

[4]Rosch, WN; Does your PC–or how you use it–cause health problems?; PC Magazine, 1991; p. 492.

[5]Herman Miller, Inc.; Ergonomic and Technology Support Products; Zeeland, MI; Herman Miller, Inc.;March 1992; p. 51.

[6]Rosch, WN; Does your PC–or how you use it–cause health problems?; PC Magazine, 1991; p. 494.

[7]Horii, SC, Electronic Imaging Workststions: Ergonomic Issues and the User Interface – Syllabus: A Special Course in Computers for Clinical Practice and Education in Radiology; Chicago; RSNA; 1992; p. 131.
[8]Herman Miller, Inc.; A few simple facts on the risky business of office ergonomics; Zeeland, MI; Herman Miller, Inc.;1992; p. 13.
[9]Herman Miller, Inc.; Ergonomic and Technology Support Products; Zeeland, MI; Herman Miller, Inc.;March 1992; p. 51.
[10]Pile, JF; Interior Design; New York; Harry N. Abrams, Inc.; 1988; p. 214.
[11]Command Centers Brochure; Engineered Data Products, Incorporated; Broomfield, CO; 1992.

Formal Rules for Funding of PACS in German Teaching Hospitals

S. J. PÖPPL, PH.D; O.RIENHOFF MD; W. BRÖCKER, PH.D

Inst. f. Medical Informatics, Univ. Lübeck; Inst. f. Medical Informatics, Philipps-Univ. Marburg; DFG Bonn (FRG)

The Deutsche Forschungsgemeinschaft (DFG) serves science in all fields by financing research projects. In many scientific areas, the Deutsche Forschungsgemeinschaft has taken over the task of strengthening cooperation among researchers and coordinating basic research with state research promotion. The Deutsche Forschungsgemeinschaft advises parliaments and public authorities on scientific matters, cultivates relations between science and economy and promotes cooperation between German scientists and foreign research. Special attention is paid to promoting young scientists.

For these purposes the Deutsche Forschungsgemeinschaft has established several councils and commissions.

According to the University Building Promotion Act (HBFG) the Computer Commission is evaluating all proposals regarding computers in medicine.

In September 92 the DFG had invited to a round table meeting about Picture Archiving and Computer Systems (PACS). Besides representatives of the Computer Commission two members of the Committee Board from the DFG, further medical informaticians, as well as twelve radiologists of the entire Federal Republic took part at it.

In reference to the recommendations of the Computer Commission a 10 point programm was discussed and then passed.

The draft submitted in October 92 was presented at the 5th. Annual Meeting of Digital Radiography (October 7th to 10th, 92, Bad Nauheim) and discussed in detail with users, as well as with industrial representatives.

All suggestions were worked in, so that the present paper of recommendations was approved by all members of the round table meeting. The Computer Commission of the DFG passed and enacted the submitted recommendations by the end of November 92.

My special personal thanks go to Dr. Werner Bröcker from the office of the DFG. Without his enforced help neither the above mentioned round table meeting nor the present draft of recommendations would have been able to be managed in this period of time.

I also would like to express my special thanks to my colleague, MD Prof. Dr. Dr. O. Rienhoff, who made it possible to present the draft of recommendations at the 5th Annual Meeting of Digital Radiography and gave significant suggestions for the final form after critical examination.

The original text of the recommendations is stated as follows:

Recommendations of the Computer Commission of the Deutsche Forschungsgemeinschaft for making proposals in Accordance with the University Building Promotion Act (HBFG)

Picture Archiving and Communication Systems (PACS)

PACS means picture archiving and communication systems which are divided into organized functional aspects of medical care, research and science, and is binded in between visual picture tools (modalities), radiological information systems (RIS) and hospital information systems (HIS).

1. **Overall Clinical Concept**

   Each proposal should describe its embedding into the whole clinical concept. The introduction of PACS into the hospital communication and information systems, as well as the radiological information system is to be described in the acquisition. A PAC-system should not be installed without RIS. Due to the profitability it is desirable that few hospitals decide on a common proceeding (i.e. a common invitation for tender and ordering). Regional or country concepts which are basic for the procedure, are to be described in the proposal.

2. **Migration Path to open PAC-Systems**

The country and clinical concept should demonstrate the migration path to distributed and open systems in hospitals with common data basis.

3. **Steps to Plan the Concept of Supply**

The concept of supply should derive from the relevant organizational requirements, the application software; the application software should determine the system software and system hardware. Longterm technical requirements in the areas network technology and digital picture archiving, are to be taken into consideration.

4. **Scientific Systems**

Scientific application systems should be handled seperately from the clinical support system, whereas the cooperation with the support system has to be ensured.

5. **Data models**

In relation to the data storage of the whole system (HIS, RIS, PACS) the data models to be implemented have to be derived. With regard to SQL relational and/or object oriented models are to be preferred.

6. **System Characteristics**

The following characteristics are expected from the systems:

a) Standard interfaces between the imaging devices and PACS, RIS and/or HIS and PACS. If such standard interfaces are not yet available, the system concept must permit future standardizations.

b) Orientation to the "client-server" concept

c) Intelligent client devices with graphical user interfaces

d) Modular architecture

e) Independence of proprietary system software.

7. **Standard Software**

For the basic supply standard application programms which are available on the market are to be used, provided that no evidently scientific, functional of technological reasons are contradictory.

8. **Improvements and New Developments**

Compared to new developments improvements and supplementations of systems available in the market will be principally preferred. New developments are important, if - considering also the reduction of competitive distortion - modern organization and system concepts are developed, pilot users are available and the system developer brings in significant advances and contributions to costs.

9. **Calls for Tenders**

Acquisitions should be handled by calls for tenders. The cost for hard- and software should be stated seperately.

10. **Requirements for the Proposal**

The basis for the proposal should be the concept for the whole clinical supply, which is to be described in a shortened form. The proposal should at least state the following topics:

a) Global description of data storage and data models for radiological relevant subsystems, including the connectivity of HIS and RIS.

b) Calculation of the capacity requirements for mass data storage including the portions of conventional storage techniques (i.e.film).

c) Performance- and main storage requirements in relation to the planned transaction rates.

d) Concept of organization of infra-structure including the demand of personnel, as well as the current state and planned development of the network.

e) Project management and project plan

f) Training concept

g) Integration of information services (i.e. linkage to reference image data basis, literature information systems).

Standards

Using ISO/IEC's Image Interchange Facility (IIF) for Medical Image Data Communication

Christof Blum
Fraunhofer Institute for Computer Graphics
Darmstadt, Germany

Eva–Maria Stephan
Technical University of Darmstadt
Germany

Abstract

Most of today's imaging fields support their own application–specific image data format, which prevents data exchange and network communication among different domains.

The Image Interchange Facility (IIF) is being developed by an ISO/IEC joint technical committee as part of the Image Processing and Interchange (IPI) standard. The IIF has been designed to define an overall image format and communication basis for enabling data transfer between many different image application domains, including the medical domain. A scenario could be a medical image data exchange for teaching or publishing.

This work concentrates on analyzing the use of the IIF data format for representing medical image data stored in the common ACR–NEMA standard format. It develops a representation for medical image data using abstract data types of the IIF.

The IIF not only overcomes some of the ACR–NEMA limitations, as it allows the structuring of multidimensional data in complex data types including image data and image related data (e.g. text and sound). The IIF is also intended to be used as a content architecture of future OSI services. The syntax of the IIF is specified using ISO/IEC's Abstract Syntax Notation One.

This work includes the definition of the *Medical Image Profile* of the IIF tailored to the interchange of medical imaging data, and capable of representing all the information contained in ACR–NEMA files. The motivation for this work is to provide a translation facility between ACR–NEMA and the IIF in order to create a link from medical imaging networks to general purpose imaging facilities which use the IIF.

1 Introduction and Motivation

Due to the increasing performance of computers during the last years, imaging is becoming more and more promising in medical science. Improved technologies in image management: capture, networking, storage, and retrieval enable important facilities for medical imaging and make it possible to distribute captured images over wide or local area networks. Image management requires a fundamental *image format*, which is flexible enough to hold all information and which supports all the basic functions essential to image management. The ARC–NEMA Digital Imaging and Communications Standard for Medical Imaging – the Publication No.300–1988 [1] defines such an image format, called ACR–NEMA, which is used by medical imaging devices for compound documents, e.g. including text and image data.

The ACR–NEMA format is used solely in the field of medical imaging. However, there is an increasing number of application areas where different kinds of digital images are generated and processed; furthermore, there is an increasing demand for the exchange of digital images between different areas. Until now, images have been exchanged mainly among applications within the same area and usually each application area provides and uses its own image format. Often there are even multiple formats existing within a specific field. A major reason for this is the varying requirements that are imposed on images from the distinct areas.

Different demands conclude in different capabilities of image formats. This complicates or even prevents the communication and exchange of image data and ancilary application data.

1.1 The Image Processing and Interchange Standard

In order to fulfill the demands from various application areas. a new *Image Processing and Interchange (IPI)* standard [9], [10], [11] is being developed, which includes the the definition of mainstream imaging functionalities and the definition of an overall, universal image data format. This standard will be a generic imaging standard, to be used in many different domains of imaging.

This future standard is being defined by the ISO/IEC JTC1/SC24/WG7 (a joint technical committee of the International Standardization Organization and the International Electrotechnical Commission). Part of the standard is the specification of a unique image format that can be used in many different imaging application areas[1] to represent and interchange images. Its scope is to enable *lossless data communication* between different application areas as well as *shared access of common image algorithms.* The new IPI standard is divided into three parts.

Part 1: *The Common Architecture for Imaging (CAI)* This document provides the top level description of the IPI and defines the Common Architecture for Imaging. It defines a set of data types and a common image representation for use with all other parts of the standard; it defines Profile Conformance Principles and general implementation rules.

Part 2: *The Programmer's Imaging Kernel System (PIKS)* This document contains the functional specification for operators, tools, utilities, management and control structures, profiles, error handling and functional element specification within PIKS, and the application programmer's interface.

Part 3: *The Image Interchange Facility (IIF).* This document describes all aspects of the Image Interchange Facility; it defines both, a data format (called the IIF–DF), syntax and semantics, profiles, data compression methods and functional description of the application programmer's interface (called the IIF Gateway).

2 Technical Overview of the IIF

2.1 Scope and Field of Application

The IIF–DF applies equally to storage and communication applications (LAN and WAN) and is independent of physical network or storage media. It may be used for image data interchange across application boundaries as well as within a certain imaging application.

Corncerning image structures and iconic attributes, the IIF–DF may be regarded as a superset of existing domain–specific image formats. Due to its flexibility regarding the encoded representation of pixel–fields, the conversion of an application–specific image format into the IIF–DF requires a minimum of computation.

The IIF effort does not intend to prevent applications from using arbitrary internal image data representations. Instead, the IIF–DF may be regarded as the "common language for the interchange of iconic information among different imaging applications" [7].

[1]Possible imaging application areas are: Medical imaging, Publishing, Industrial vision, Computer graphics/animation, Scientific visualization, Remote sensing and Insurance of property.

2.2 IIF Objects

The concept of the IIF–DF is realized by supporting a set of data objects. Images described within the IIF are represented by a hierarchy of objects. The syntax description and the format of those objects are defined in an abstract syntax, which will be explained in section 2.3 . The encoding of the abstract data types will be outlined in section 2.4 .

Basically, there are four different kinds of objects within the IIF–DF:

Image data objects : Image data objects may be of three different types:

- *compound structured objects*
- *basic structured objects*
- *elementary objects* describing one pixel value

Image data objects may be composed of an aggregation of objects of the similar type or a simple object of the next lower type. Aggregation of types may be composed as records, lists or arrays. By defining a *recursive image data structure* based on specific data types the IIF–DF offers a flexible concept to store any kind of compound rastered images, such as multidimensional and sequences of images. For example, an object could represent a sequence of related images, a single image, or even a single pixel value. In the medical case one IIF compound structured object may represent, for example all images belonging to a hospital station, a patient's study, or multiple series of sliced scans; basic structured objects may represent images of arbitrary dimensionality.

Image attributes: Image attributes for medical images contain for instance the metric description of the image.

Image–related data objects: Image related data objects include the specification of match points, look–up–tables or regions of interests.

Image annotation: Image Annotation objects describe any textual, graphical or application specific information for explanation/commentary on the encoded image.

All three types of objects representing non–image data are optional.

The IIF–DF splits the transmitted image into the description of the image structure and the real image data[2]. This is of special importance, as some image processing algorithms, data compression for example, then can be applied only on the real image data.

2.3 IIF Syntax

The IIF–DF syntax is described using the international standard *Abstract Syntax Notation One (ASN.1)* [3] [4]. ASN.1 was originally designed for the specification of OSI layer 6 protocols [13]. It provides elementary and constructed data types.

Using ASN.1, the definition of encoding rules is separated from the syntax specification: the syntax defines what kind of data entity may occur at what place of the format, whereas the encoding determines the digital representation of every entity. Basically, the IIF–DF syntax was created by defining a sequential representation for every IPI image data object.

2.4 IIF Encoding

For encoding IIF–DF syntax entities, the *Basic Encoding Rules (BER)* [5], [6] of ASN.1 are employed. Because there is a direct mapping between ASN.1 and its BER, specifying the IIF encoding using ASN.1 results in an unambiguous, machine independent coding for raster images and associated information. For efficient encoding of pixel fields, the IIF–DF supports the following standardized compression schemes: JBIG, JPEG, MPEG.

[2] *Real image data* refers to the actual pixel values of the images.

2.5 IIF Profiles

In order to support the various requirements of different application areas, the IIF provides a *complex syntax*, which is able to encode diverse images from multiple sources. If certain structures are not necessary within some application areas a sub-set of IIF data structures can be defined in a *profile.*

Profiles specific to a certain domain restrict the IIF syntax description. They simplify the data format, since only data types which are actually used are kept in the profile. This allows simpler parsers, generators, etc. to operate on the data. They also enable a lossless data conversion from the IIF into the application specific data format. Restricting profiles to a subset of the full IIF syntax guarantees that image files conforming to the profile also conform to the full syntax. With this concept a specific imaging domain is able to support the IIF format even by using only a sub-set of the complete syntax.

The IIF-DF supports two different kinds of profiles: application profiles and conformance profiles. Application profiles restrict the IIF syntax for specific kinds of images, like color-image-profile, or the binary-image-profile.

Although the complete IIF syntax supports many different data types, it is outside of the scope of the IIF to define image-related data elements that are used by only one distinct domain. Instead of specifying these application-specific data structures, the IIF provides generic data fields which may be customized in conformance profiles.

On the one hand conformance profiles reduce the complex IIF-DF to types which are actually needed for a specific domain, to make the IIF more distinct and better manageable. On the other hand they specify an exact meaning for IIF generic structures, which are not explicitly defined in the IIF-DF. This kind of specification is an extension of the IIF from the point of view of semantics, however the definition of an explicit syntax for the generic ASN.1 types is a syntactical restriction.

3 The Medical Profile of the IIF

The definition of the medical profile is achieved within three steps :

- The analysis and comparison of data structuring capabilities supported by both formats.
- The mapping of data elements that are present in both data formats.
- The definition of a representation for domain-specific medical image data in the IIF.

3.1 Comparison IIF / ACR-NEMA

Scope of the work was to define a representation of medical image data in the IIF-DF, and also to enable a conversion from IIF data structures into ACR-NEMA. Therefor, it has to be made sure that every possible data element in ACR-NEMA has an equivalent representation in the IIF-DF. By restricting the full IIF syntax to only this kind of data, that also can be described in ACR-NEMA, we assure a lossless data conversion between ACR-NEMA and the medical profile in both directions. The equivalent representation in the two formats may be syntactically different from the original data structuring, but will have the same semantic meaning. The next sections will describe the paradigm of mapping between the two data formats.

The main goal, when converting data between different formats, is to assure that there is *no loss of data.* This can only be guaranteed, if all information represented in one format can be represented in the other. Both, syntactical and semantical meaning of data have to be considered and transferred when data is converted into a different representation.

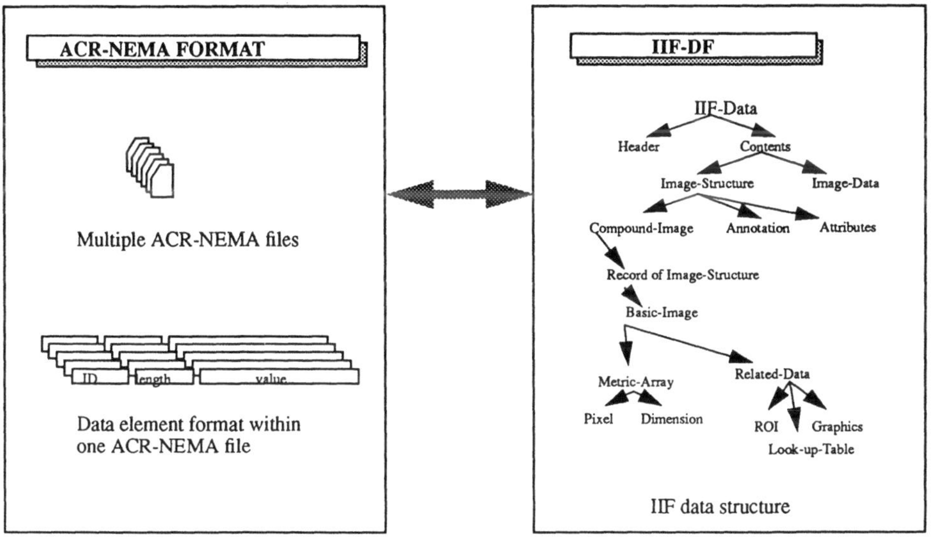

The figure shows the different structuring of both image data formats.

3.2 Data Mapping

Three cases were encountered in the definition of the medical imaging profile:

1. There were elements in both formats with equivalent semantics. This permitted a direct mapping of data between the two formats with only a syntactical difference.

2. The IIF provided functionality that was not needed in the profile, or offered multiple methods for encoding particular types of information. In this case, the IIF syntax was restricted to eliminate structures not needed in the profile and redundant encoding methods.

3. Finally, there were elements which had no exact semantic equivalent in the other format but which could be represented in general purpose place-holder elements. In the case of conversion ACR-NEMA data into the IIF-DF, this application specific medical imaging data was represented using the IIF generic structures. In order to guarantee a one-to-one mapping, the current profile did not allow IIF constructs without equivalent ACR-NEMA elements. However, future "upwardly-compatible" profiles may define elements not present in ACR-NEMA; in this case, shadow elements could be defined to hold such new information.

The next section discusses methods used to deal with the first two cases, i.e. the mapping of image data and image-related data in which the IIF-DF already contained constructs to encode the ACR-NEMA information. The following section addresses the third case, the representation of application-specific medical imaging information in the IIF-DF.

3.2.1 Mapping image and image-related data

The IIF-DF's support for generic imaging data types is used to encode the majority of information contained in ACR-NEMA's Image Presentation, Relationship, Overlay and Pixel Data groups. All information needed to represent the image physically can be mapped into generic IIF types. The few image-related data elements not covered by the IIF (such as window values and ROI statistics) are encoded using the techniques to be described in the next section.

But what is of greater importance, is that the IIF does not offer any way to "pack" multiple, distinct images into a single pixel array the way ACR–NEMA does with its images,overlays and ROIs[3]. Although the IIF–DF supports structured pixels, each of the structure's components must be composed of elementary IIF types, which are limited to multiples of 8–bits long. In the current profile, the overlay and ROI information is separated into individual data structures and encoded separately. Currently the IPI standardization committee is elaborating on new methods for efficient representation of heterogenious pixel–fileds (band–interleaved, pixel–interleaved, f.ex.) [12] in order to meet compression needs discussed in the medical domain [14], [15].

IIF objects can contain attributes that describe the object's semantics and carry information about the object's interpretation. Two classes of attributes are used in the medical profile. Metric–descriptions are used to define the image's position in space and time (including its coordinate system, dimension mappings and metric transformations). Freeform–descriptions are provided to allow future extensions.

In ACR–NEMA, volumetric studies are represented as multiple slices through the body; each slice is stored in a separate file. In contrast, the IIF–DF provides constructs which allow a hierarchical structure of multiple images to be stored in one compound structure. The medical profile uses these constructs to store related ACR–NEMA images in one IIF image file based on the ACR–NEMA hierarchy (station, date, study, series, acquisition and image level). Grouping multiple files together into one compound structure makes it possible to eliminate redundant information, such as information about the image presentation, pixel representation and image annotations, that are duplicated in each slice file, thus avoiding sources of error.

3.2.2 Mapping domain–specific non–image data

There is a great deal of application–specific information in ACR–NEMA files that have no corresponding structure in the generic IIF syntax. The majority of this information is contained in the Identifying Information, Patient Information, Acquisition Information, and Text groups. As was mentioned previously, these elements are handled by customizing the IIF's generic "image annotation"syntax elements to incorporate the necessary semantics.

When defining a concrete syntax specific to medical imaging, we tried to maintain the global structure of the ACR–NEMA data format. Thus, for each ACR–NEMA standard group, except the Overlay and the Pixel Data groups, we defined a corresponding IIF structure. The IIF–DF structures have the same names as the corresponding ACR–NEMA standard groups and contain syntax elements for representing standard ACR–NEMA elements. Of course the profile does not include a new definition for all the elements defined by ACR–NEMA. Many of them are mapped directly into corresponding IIF objects.

The profile also provides additional IIF structures to represent non–standard ACR–NEMA groups and elements. Their syntax specifications allow two types of private information.

- *undefined–elements:* These structures are used if an ACR–NEMA element's group ID or element ID are not known. They contain an attached data field which can be used to encapsulate the ACR–NEMA data without any attached semantics.
- *defined–elements:* These structures are used as place–holders to allow private ACR–NEMA extensions to be defined, which is in many ways similar to the way in which the general IIF syntax allowed profiles to be defined.

4 Implementation

The work was developed and a set of converter programs were implemented and tested on a UNIX workstation at SIEMENS Corporate Research in Princeton, NJ. The implementation consists of two conversion programs, a converter from ACR–NEMA to the medical profile and one to do the inverse transformation. All programs are written in the C programming langue and run on a Sun Sparc Station. In order to be prepared to syntax changes the converter programs were developed with the automatic syntax analysis tool *ISO Developement Environment* [8].

[3]To our knowledge, ACR–NEMA is the only file format to use this technique.

References

[1] American College of Radiology (ACR) and National Electrical Manufacturers Association (NEMA): *Digital Imaging and Communications.* Standard Publication Nr. 300–1988, Version 2.0.

[2] K. Andress, E.–M. Stephan; *Representation of ACR–NEMA Data using the ISO/IEC Image Interchange Facility.* Spie's Conference on Medical Imaging VII, Newport Beach, 1993.

[3] ISO/IEC: *Information processing systems – Open Systems Interconnection (OSI) – Specification of Abstract Syntax Notation One (ASN.1).* International Standard 8824, 1987.

[4] ISO/IEC: *Information processing systems – Open Systems Interconnection (OSI) – Abstract Syntax Notation One (ASN.1) – Draft Addendum 1: Extensions to ASN.1.* International Standard 8824, 1987.

[5] ISO/IEC: *Information processing systems – Open Systems Interconnection (OSI) – Specification of Basic Encoding Rules for Abstract Syntax Notation One (ASN.1).* International Standard 8825, 1987.

[6] ISO/IEC: *Information processing systems – Open Systems Interconnection (OSI) – Abstract Syntax Notation One (ASN.1) – Draft Addendum 1: Extensions to ASN.1 Basic Encoding Rules.* International Standard 8825, 1987.

[7] C. Blum ; *Design Principles and Applications of ISO/IEC's Image Interchange Facility (IIF).* ImageCom Conference, held in Bordeaux, March 1993.

[8] Marshall T. Rose; Performance Systems International, Inc.: *The ISO Development Environment.* User's Manual, Version 7.0.

[9] ISO/IEC: *Image Processing and Interchange (IPI), Part 1: Common Architecture for Imaging (CAI).* Draft International Standard (DIS) 12087-1, November 1992.

[10] ISO/IEC: *Image Processing and Interchange (IPI), Part 2: Programmer's Imaging Kernel System (PIKS).* Draft International Standard (DIS) 12087-2, November 1992.

[11] ISO/IEC: *Image Processing and Interchange (IPI), Part 3: Image Interchange Facility (IIF).* Draft International Standard (DIS) 12087-3, November 1992.

[12] ISO/IEC: *Information technology – Computer Graphics and Image Processing – Encoding for the Image Interchange Facility (IIF).* Committee Draft (CD) 12089, March 1993.

[13] Marshall T. Rose: *The OPEN Book – A Practical Perspective on OSI.* Prentice Hall 1990, Englewood Cliffs, NJ 07632.

[14] SPIE Proceedings: Science and Engineering of Medical Imaging Vol.1137, April 1989; Robert Di Paola: High compression coding to dynamic medical image series.

[15] SPIE Proceedings: Medical Imaging V, Vol.1137, Feb. 1990; Compression.

[16] A.F.Clark, *Image processing and interchange - the image model*, Proceedings of the SPIE/IS T's Conference on Electronic Imaging, Image Processing and Interchange, San Jose, CA., February 1992.

[17] W.K.Pratt, *Overview of the ISO/IEC Programmer's Imaging Kernel System: applications programmers interface*, Proceedings of the SPIE/IS T's Conference on Electronic Imaging, Image Processing and Interchange, San Jose, CA., February 1992.

[18] E.–M. Stephan: *Representation of Medical Image Data in IIF.* Diploma Thesis, Technical University of Darmstadt, 1992.

An ACR-Nema 2.0 Gateway for IMAC Systems

S. L. Fritz[a], W. T. DeJarnette[b], S. R. Roys[a]

[a]University of Maryland, Baltimore, MD, USA

[b]DeJarnette Research Systems, Towson, MD, USA

Summary

The authors have implemented a gateway based on the ACR-NEMA standard for connecting devices and local area networks which do not share a common protocol. The gateway supports several protocols: ACR-NEMA 2.0 (50-pin), PACSNet, TCP/IP and a proprietary modem protocol. The TCP/IP protocol is based on the draft ACR-NEMA version 3.0 (DICOM) standard. This gateway has been in operation since June of 1992 at the University of Maryland School of Medicine connecting a General Electric ID/Net 2 network, a Siemens PACSNet network and a third party imaging workstation which supports the ACR-NEMA 50-pin parallel interface. The software design is discussed in detail and our experience to date presented.

Introduction

In 1989 NEMA published version 2.0 of the ACR-NEMA standard. Originally designed to allow connecting modalities to a network which did not support the modality data formats nor a network protocol, the standard specified a 50-pin parallel interface and a protocol stack modelled after the ISO Open System Interconnect (OSI) design. In 1990 at Georgetown University six manufacturers participated in an interconnectivity workshop which demonstrated that the standard had been reduced to practice on at least a[1] minimal level by several groups. In the time since development of practical interconnectivity solutions for IMAC systems has continued.

In 1991 the University of Maryland at Baltimore (UMAB) decided to build a network of imaging systems based on the ACR-NEMA 2.0 standard. DeJarnette Research Systems(DRS) of Towson, Maryland agreed to provide a network hub capable of connecting two local area networks together - one a General Electric IDNet 2 network and one a Siemens PACSNet network. The software to provide this interconnectivity was developed jointly by UMAB and DRS. General Electric made available an IDLink 2 ACR-NEMA gateway linking their network via a 50-pin adapter. Siemens concluded an agreement with DRS to allow the DRS gateway to connect directly to a PACSNet system.

In June 1992 the ImageShare was used to integrate the two LANs and a Columbia Scientific, Inc. (CSI) ImageMaster 101 workstation. The Siemens LAN was connected via PACSNet and the GE LAN and the ImageMaster workstation were connected via 50-pin connections. Software for the ACR-NEMA 50-pin protocol on the ImageMaster workstation was written jointly by University of Maryland research staff members and CSI staff members. Since then the University of Maryland researchers have been using the network to access images from both GE and Siemens networks for a variety of research projects. One such project is the stereotactic positioning evaluation presented elsewhere in this volume.

System Design

The gateway system, the ImageShare 910, is a PC-based platform using standard PC-compatible hardware for all functions except the ACR-NEMA 50-pin interface. This interface uses the DRS ANSIF, a custom ACR-NEMA interface board designed by DRS and demonstrated at the Georgetown workshop. The ImageShare supports multiple network protocols in addition to the ACR-NEMA 2.0 protocol, in anticipation of the version 3.0 ACR-NEMA (DICOM) standard, which allows multiple lower layers under a common Presentation layer interface.

The initial design of the gateway determined that multiple tasks would be required. Support of this complex software architecture requires more than DOS can provide. Multitasking support is provided by a DRS proprietary multitasking manager, DRX. DRX allows sharing of hardware resources by multiple tasks within a single 640K memory map provided by DOS. DRX provides intertask communication, memory management and timing functions to whatever tasks are running.

The overall software design incorporates several tasks: one for each protocol stack in operation and a routing task which also manages local disk storage. These tasks communicate via DRX primitives. Incoming messages from network peers are accepted over a particular protocol and sent to the router task. The address of each ACR-NEMA message is then used to direct the message out to a destination task over another protocol stack. The system design also provides the capability to direct any message to an internal destination via an intertask communication message. Thus the ImageShare may support internal and external network destinations.

Implementation

The desired network configuration is shown in figure 1. The GE LAN consisted of two modalities, a High Speed Advantage CT scanner, a Signa 1.5 T MRI system and an independent viewing console. This network was interfaced using a GE IDLink 2 gateway. The IDLink gateway supported both the Ethernet-based IDNet 2 and the ACR-NEMA 50-pin interface. The ImageShare system were connnected to the GE LAN via such a 50-pin interface. The Siemens PACSNet network incorporated a Somatom

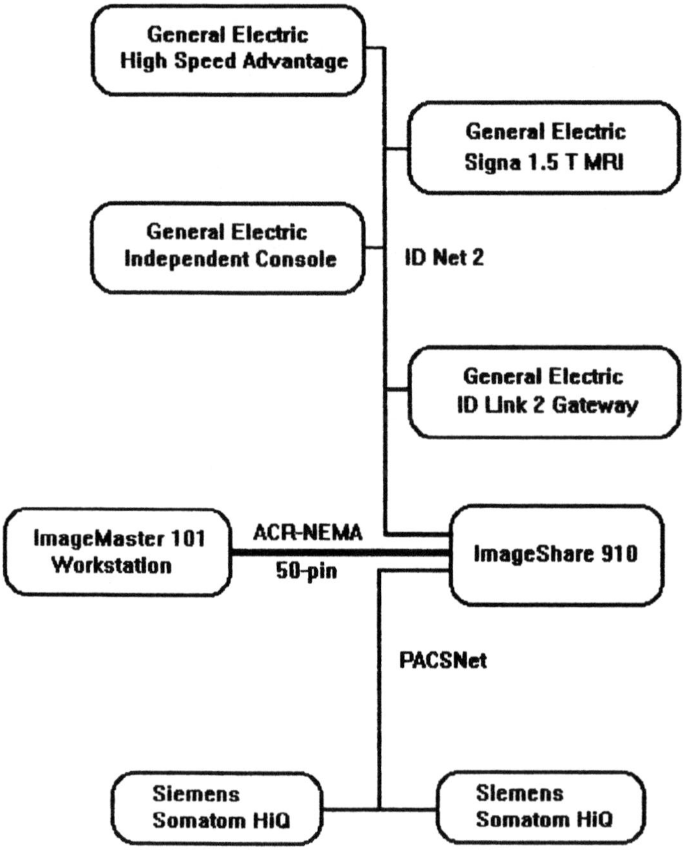

Figure 1. UMAB Network Configuration

HiQ CT scanner and a LiteBox workstation. The Siemens LAN allowed connection of the ImageShare 910 as a PACSNet node, eliminating the need for an additional gateway. The ImageMaster workstation was connected to the ImageShare via a second 50-pin link. Thus the ImageShare was running two protocols simultaneously: ACR-NEMA 50-pin and PACSNet.

Connection of the two networks and the workstation to the ImageShare required very little effort. The primary difficulties were in configuring each of the LANs with the appropriate network addresses of the ImageShare and the ImageMaster workstation. The entire connection was accomplished in one day.

Transfer of images from either network to the ImageMaster workstation were straightforward. Both GE and Siemens protocols had been tested in advance at the respective manufacturer's facilities and few problems were anticipated.

Since the GE gateway was a send-only gateway it was not possible to send images into the IDNet LAN. However, the first transfer of GE images to the Siemens LiteBox workstation worked without difficulty.

There were a few problems with implementation profiles. These profiles are subsets of the information defined in the ACR-NEMA data dictionary which the manufacturers supported. The GE pixel data retained informtion from the overlay planes in the high order bits of the image which were not interpreted in the same way on the Siemens LiteBox. Consequently the image brightness in the image outside the field of view of CT and MR images was not displayed as completely black and the overlay text written on the images by the LiteBox did not display the same way for GE as for Siemens images. However these did not obscure the images in any meaningful way. The GE orientation data was included in a shadow group and not displayed on the Siemens display system. Both of these problems were minor.

Transfer of individual images to the ImageMaster workstation seldom proved problematic. However, certain images were not able to be transferred. This problem turned out to be the presence of large shadow groups in certain images. The ACR-NEMA protocol software had been written under the assumption that headers would be relatively small compared to the image pixel data. This problem is currently being resolved.

In summary, implementation problems were virtually nonexistent. This was ensured by close cooperation by DRS staff members with their counterparts at General Electric and Siemens.

Discussion

Performance of the gateway was essentially limited by the Ethernet links. The overall transfer rate of images was approximately one 512 x 512 x 16 bit image every 6 - 8 seconds. This is more than adequate to keep up with the rate of acquisition from the modalities surveyed when averaged over the day. It does represent a limitation in the ability of radiologists at a remote workstation to access the images in real time. As the impact of IMAC systems is felt on the practice of radiology and real-time reading becomes the norm, this limitation can be addressed.

In addition to the protocols in use for the UMAB network, the ImageShare also supports three additional protocols: TCP/IP, modem and disk. The TCP/IP implementation is a prototype DICOM implementation which provides the session capabilities of DICOM spelled out in chapter 8 of the DICOM draft standard. The modem protocol is designed to support a teleradiology display system via a 14.4 Kbps V4.2bis modem. The teleradiology display system is simply a variant of the ImageShare design with a high quality display system and a local destination process. Evaluation of this system using the modem protocol in ongoing at UMAB in 1993. The disk protocol is a method of accessing local disk using exactly the same primitives as are used for accessing remote network nodes. This may seem to levy an unwarranted performance penalty on local disk accesses but has proven to be a valuable means of testing application protocol software without the complications of a network.

Future Efforts

The University of Maryland is currently devoting a substantial share of its research in radiology to IMAC-related issues. The intention is to add additional ImageShare systems to interface digital subtraction angiography, ultrasound imaging, nuclear medicine and computed radiography to the core network for research and clinical purposes. An example of such a project is the use of DSA and MR images for Gamma Knife radiotherapy treatment planning which is currently underway.

1. "Digital Imaging and Communications", ACR/NEMA Standards Publication No. 300, National Electrical Manufacturers' Association, Washington, D.C., 1989.

Automating Image Format Conversion - Twelve Years and Twenty-Five Formats Later

David P. Reddy[†], Gerald Q. Maguire Jr.[‡], Marilyn E. Noz* and Robert Kenny[◦]

[†]Radio Logic Inc., P.O. Box 9665, New Haven, CT 06536

[‡]Department of Computer Science, Columbia University, New York, NY 10027

*Department of Radiology, New York University, New York, NY 10016

[◦]Mirage Ltd., Folly Gate, Okehampton, Devon EX20 3AF. England

Summary

We describe an integrated approach to extracting image data from diverse manufacturer's formats into a common file type. The interface to the actual extraction process has been standardized through use of the MIT X11 Window System™ with the Open Software Foundation's Motif™ graphical user environment.

Introduction

The purpose of this project is to address the needs of clinicians, researchers, and computer systems personnel involved in the handling of medical images from heterogeneous source machines.

When, over a decade ago, tomographic brain studies from different image sources (positron emission tomography (PET) and computer assisted tomography (CT) scanners) were combined by researchers both at NYU and at other institutions,[1, 2] it was found that one of the most important obstacles to be overcome was the conversion of the images, both the transfer of the images onto the same computer system and the coersion of the images into a common format. Even images from the same manufacturer do not have the same image format (for example, single photon emission computed tomopgraphy (SPECT) images obtained from a GE Starcam nuclear medicine system are in a completely different image format from CT images obtained from a GE 9800 or HighLight Advantage system). Even though the respective manufacturers are generally cooperative in providing written documentation on specific image formats, the actual reformatting is a formidable task. Additionally, the media on which the images are available is often physically incompatible. Fortunately, most images are available on 9-track magnetic tape which has been well standardized by computer vendors and national standards. Alternatives to this consist of floppy disks, which have highly incompatible formats, and serial line transmision. In the last one to two years, some image source manufacturers have begun using standard computer hardware which frequently has a network interface, and the manufacturers have begun to allow images to be transported over the network, often using the standard Internet File Transfer Protocol (FTP) or Sun Microsystems' Network File System (NFS)™. However, most manufacturers of image scanners and viewing stations still provide limited facilities for exporting and importing images.

Also over the past decade, conversion of medical image formats has become increasingly desirable as multiple image sets from the same or different image sources are being ported to non-manufacturer supplied computers (UNIX™ or VMS™ workstations, IBM-compatibles or Apple Macintoshes™) for image viewing, processing and analysis, or for correlation between studies taken at different times and by different image source machines. Indeed, analysis techniques which use tomographic images from the same or different image sources in concert with each other not only enhance the information content of each image, but are rapidly becoming clinically significant.[3, 4]

Furthermore, as picture archiving and communcation systems (PACS) become more commonplace, there is an increasing need to distribute images from various image source machines across the network to a wide variety of clinician viewing stations. In any practical PACS system it is essential that digital and digitized (from film) images be archived as quickly and easily as possible. It is also useful if during this process the native image format can be processed to conform to the local common image format. This is especially important to provide imaging departments with the freedom to choose the best image generating machine for a particular type of image (modality - such PET, SPECT, CT) irrespective of specific manufacturers.

We first implemented a set of programs for medical image format conversion from multiple vendor-specific formats to the AAPM (American Association of Physicists in Medicine) Standard Format[5]. This format was chosen because of the "lossless" nature of the conversion. Recently, the Canadian, European COST B2 group, and the Australian and New Zealand Societies of Nuclear Medicine have adopted the AAPM standard and are continuing to define keys. This standard, now called Interfile,[6] is rapidly being adopted for all radiology (several manufacturers are making their images available outside their machines, recently over the network, in Interfile format) and other (such as electron microscopy) images. A competing stardard file format for radiology images is that proposed by the Amercian College of Radiology and the National Electrical Manufacturer Association (ACR-NEMA).[6, 7]

MaterialsandMethods

Since 1980, we have been involved in the creation and use of software modules, known collectively as *qsh*.[8] This software necessarily includes a large number (twenty-five) of conversion routines from maufacturer (or other) specific image formats to the *qsh* specific format.

To accomplish the goal of automating the conversion process, the problem of obtaining the images either via tape format (e.g., 1/2 inch, 1/4 inch), a serial line, or over the network, must be solved. Then the particular original image format at hand must be understood or "decoded" automatically and the proper conversion routine invoked.

We sought to solve this problem by using a single user interface and automatically detecting the image format to be converted. This software, developed by one of the present authors (DPR) runs under UNIX or VMS and is written in the C programming language. The interface to this interactive program has been standardized through the use of the MIT X11 Window System and the Motif graphical user interface. A character-only interface is also available for automated conversion, inclusion in scripts, and terminal or serial-line communications.

The only parameter necessary to use this program is to indicate where the images to be converted are located. This can be done on the command line or by defining an "environment" variable QSH_TAPE which specifies the path to the image location (e.g., "/usr/local/images/patient_1"). If no location is specified for the images, a nine track tape transport (for example on a UNIX system, "/dev/nrmt0") is assumed. The QSH_TAPE variable or command line argument may point to the header or directory file of one of the more rationally organized manufacturer formats which might have been obtained via the network or standard tape archiving programs and would be currently located on disk, or may be a set of numbered files which are an image of the files which originally were located on a tape (e.g. patient_1.000, patient_1.001, etc.) such as one might obtain with the UNIX program "dd", this option being especially useful for systems which do not have a local tape drive of the correct format or to speed conversion program development by replacing sequential tape seek and read delays with disk seeks and reads, or may be a physical nine-track or cassettte tape.

The program reads the first block or blocks of the first file it finds on the device specified to determine the identity of the image source machine, and returns a format identifier, which may indicate that the format is unknown or unsupported. If the format is supported, it invokes the

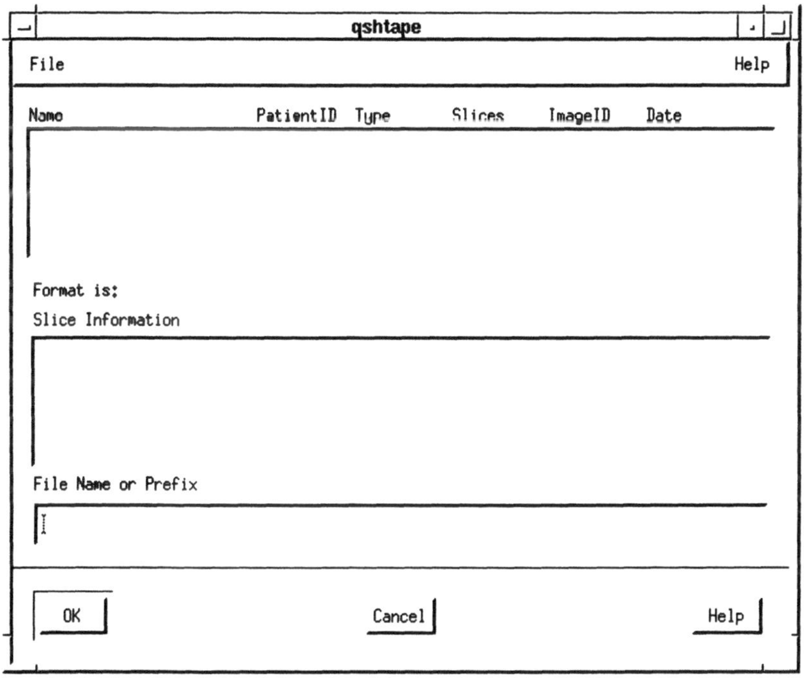

Figure 1: Initial menu for image conversion. When the choice "File" is chosen from the main menu bar, the user is able to choose "New Tape or File", "Rewind Tape or File" "Unload Tape or File" and "Quit"

routines required to reformat the manufacturer image directory into an internal directory list, which is then presented to the user in a consistent format, with each three-dimensional image presented as an entry in the list.

Figure 1 shows the initial menu that is invoked by the program specification. When the user clicks on "File", the choices "New Tape or File", "Rewind Tape or File", "Unload Tape or File" and "Quit" are displayed. When "New File or Tape" is invoked, the program follows the path to the specified image files, and inserts a list of available patient images in the list area as shown in Figure 2. The user chooses an image or images from the list, and for each three-dimensional image selected, a list of the two-dimensional images which comprise it is presented. The user selects "OK" when satisfied with the selections, and the specific routines are invoked to convert to our standard *qsh* file format[6] or into another local format.

There is one header file and three routines associated with the image format section, one to invoke the format identification routine, the directory conversion routine, and the image conversion routine for each individual format. The individual conversion routines are contained each in a separate module with appropiate header files. The conversion routines are designed to be completely portable with repsect to the byte alignment, byte order and floating-point formats, and are designed to automatically convert from those of the image source machine into those of the local machine. To accomplish this, all manufacturer image directory and image header information is accessed through C structures defined in the header files for each specific format.

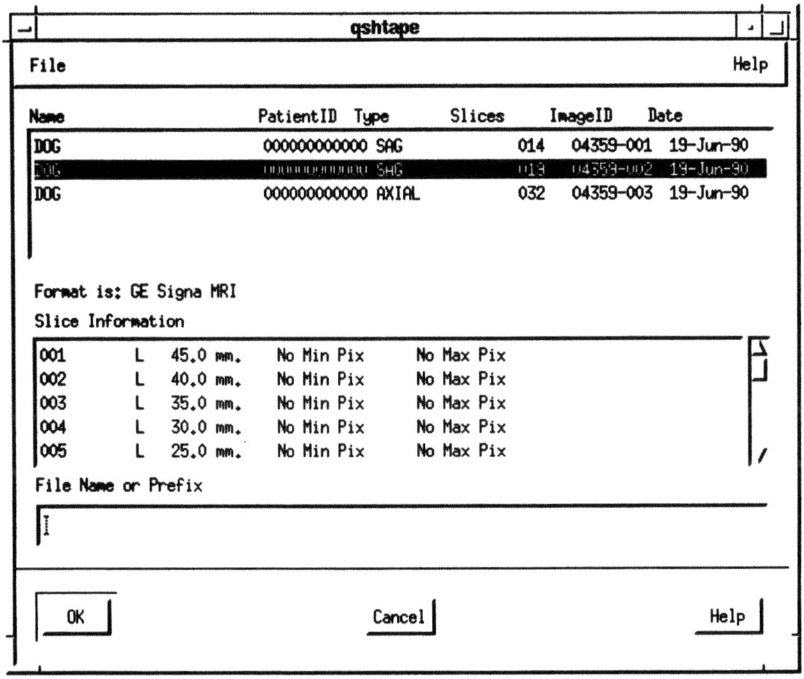

Figure 2: The available image sets and the available images for a particular set are displayed in the list windows after the user chooses "New File or Tape". When the user has chosen the patient and images to be converted, he/she clicks "OK" and the conversion process commences.

In these structures, any objects other than characters or character strings are defined as a structure consisting of an appropriate number of characters, and are accessed only through routines which automatically byte align, byte swap, and floating-point convert as needed. The program asserts that the sizes of each of the individual and aggregate structures is correct at execution time before any conversions are attempted, to ensure that the program was compiled without any inappropriate byte packing or byte alignment of the structures.

Results

This set of automated conversion programs now supports seven different file formats (GE Signa, GE 9800_AS, GE9800_AB, GE HighLight Advantage, Technicare2000, Positron, Trionix Triad) and others are being incorporated (e.g., Siemens and GE HighLightAdvantage MR) as the need arises. It is in use at four different locations and is in the process of being adopted by other institutions.

Discussion

A previous attempt to automate this image conversion process was the development of the Aura "Babel box" by one of the authors (RK) and Barry Winston. This software converts nuclear medicine images from Scintronix, MDS, ADAC and Nodecrest to Interfile and back. It runs on a PC, generally uses a serial line to obtain images, has a driver for the CompatiCard under Interactive UNIX and an eight inch floppy disc drive to read Data General floppies as used by Scintronix.

In conjunction with the PACS system devloped and used at the University of Geneva, an image parser was developed for the "OSIRIS" image processing system by Osman Ratib and co-workers.[9]

Conclusion

The most requested aspect of *qsh* has been for the image conversion routines, and while on the surface, image conversion appears to have the lowest intellectual effort required, i.e., it is not very fancy or flashy, it is essential if one wants to actually process images associated with real patients in order both to provide clinical care and to develop clinical research protocols.

Acknowledgements

Macintosh is a trademark of Apple Computer Company, Inc. UNIX is a registered trademark of UNIX System Laboratories. SUN Microsystems and NFS are registered trademark of SUN Microsystems, Inc. licensed to SunSoft, Inc. VMS is a trademark of Digital Equipment Corporation. MSDOS is a trademark of Microsoft, Inc. MIT X11 Window System is a trademark of the Massachusetts Institute of Technology. Open Software Foundation, OSF, OSF/1, OSF/Motif and Motif are trademarks of the Open Software Foundation, Inc. This research was supported in part by an IBM Faculty Development award (GQM), and equipment grants from American Telegraph and Telephone Company, Hewlett-Packard Company, Philips Medical Systems and SUN Microsystems Inc.

References

[1] T. Farkas, A.P. Wolf, J. Jaeger, J.D. Brodie, D.R. Christman, and J.S. Fowler. Regional Brain Glucose Metabolism in the Study of Chronic Schizophrenia. *Arch. of General Psychiatry*, 41(1):293–300, March 1984.

[2] A. Wolkin, J. Jaeger, J.D. Brodie, F. Gomez-Mont, J. Rotrosen, A.P. Wolf, and *et al.* Persistent Cerebral Metabolic Abnormalities in Schizophrenia as Determined by PET. *Am. Journ. of Psychiatry*, 142(5):564–571, May 1985.

[3] E.L. Kramer, M.E. Noz, J.J. Sanger, A.J. Megibow, and G.Q. Maguire Jr. CT-SPECT Fusion to Correlate Radiolabeled Monoclonal Antibody Uptake with Abdominal CT Findings. *Radiology*, 172(3):861–865, September 1989.

[4] Kramer.E.L. and Noz. M.E. CT/SPECT Fusion of Analysis of Radiolabeled Antibodies: Applications in Gastrointestinal and Lung Carconoma. *International Journal of Radiation Applications and Instrumentation, Part B Nuclear Medicine and Biology*, 18(1):27–42, January 1991.

[5] B.S. Baxter, L.E. Hitchner, and G.Q. Maguire Jr. *A Standard Format for Digital Image Exchange.* American Association of Physicists in Medicine (AAPM) Report No. 10. Am. Inst. of Phys., New York, NY, October 1982. 10pp.

[6] G.Q. Maguire Jr. and M.E. Noz. Image Formats: Five Years after the AAPM Standard Format for Digital Image Interchange. *Medical Physics*, 16(5):818–823, September/October 1989.

[7] ACR-NEMA Standards Publication No. 300-1985. Digital Imaging and Communication. National Electrical Manufacturers Association (NEMA), Washington, DC, February 1986. 112 pp.

[8] M.E. Noz and G.Q. Maguire Jr. QSH: A Minimal but Highly Portable Image Display and Handling Toolkit. *Computer Methods and Programs in Biomedicine*, 27(6):229–240, November 1988.

[9] Y. Ligier, O.M. Ratib, M. Funk, R. Perrier, C. Girard, and M. Logean. Portable Image-Manipulation Software: What is the Extra Development Cost? *Journal of Digital Imaging*, 5(3):176–184, August 1992.

Medical Imaging and Multi-Media: The View of CEC on R&D and Standardization

R. Mattheus CEN TC251/4, L. Beolchi CEC/AIM

The technological developments in informatics and communication technologies, when properly adjusted to the specific demands of the health care sector, offer the opportunity to process and extract the relevant information from the rapidly increasing amount of patient data and to communicate this information effectively where ever needed. The CEC is stimulating these activities in two concrete ways: the AIM programme and the standardization efforts.

Advanced Informatics in Medicine (AIM)

In promoting international cooperation in research and development the Commission of the European Communities (CEC) has a number of objectives. Of primary importance is the realisation of a common and uniform market, and the creation of opportunities for the own industries on that market.

Advance Informatics in Medicine (AIM) is the medical programme inside the CEC, dealing with health care informatics and has a leading role concerning precompetitive research in the imaging and multi-media domain.

The AIM Programme, then, aims at the promotion of the applications of these technologies in health care and medicine. The general objectives are: the improvement of the efficiency of health care, the reinforcement of the position of the European Community in the field of medical, biological and health care informatics, and the realisation of a favourable climate for a fast implementation and a proper application of informatics in health care. Furthermore, it was considered that as the costs of medical care are high and still rising, the applications of informatics and telecommunications form an ideal opportunity to improve the quality, the accessibility, the efficiency, and the cost-effectiveness of health care.

Five projects are dealing directly with medical image processing, visualisation, handling, or communication. AIM has pointed towards different homogeneous activities reflected into project lines, including medical imaging and multi-media.

COVIRA - Computer Vision in Radiology
EurIPACS - European Integrated PACS
Helios Hospital Object Tools
MILORD - Multi-media Interaction with Radiology Database
SAMMIE - Software for Multi-Modal Images and Education

We consider it a strategic activity, not only of technical and scientific interest, but also as regards its impact on health care expenses, the big expectations of health care providers and purchasers, policy makers, but mainly of the citizen, and finally as regards the real competence and traditional awareness of the radiological community. For this reason we consider the clinical acceptability as one of the main issues for AIM. Solutions for the medical imaging domain should be integrated into the medical informatics architecture of the hospital.

Aim is trying to contribute to avoid the isolation of a radiological domain, and to integrate the radiological world into the departmental systems and the health care informatics systems. Telematic services will be set up soon allowing access inside and outside the hospital of different classes of services.

AIM is concentrating on usability, clinical acceptance and cost-effectiveness, in order to contribute to an effective improvement of the quality of health care.

AIM stimulation towards standardization activities

AIM has undertaken two types of activities to advance standardization. On a long-term objective, contributions to the definition and use of standardization procedures in health care informatics were to be made. It was, therefore, essential to follow the European Community policy on standards and to align with as well as promote European and international standards. Addressing the standardization activities in coordination with the international standardization bodies would ensure the coherence of the results, facilitate the widest possible acceptance, and assure the preservation of all parties interested. A short-term objective was to obtain concrete results during the AIM actions. Common technical specification produced during AIM projects were to be made available to the standardization bodies.

- The urgent need, expressed by all groups that participated in the preparatory activities of AIM is to identify the needs and make progress in this domain. It became apparent too, that the effective involvement of actors from all the relevant sections in this field was an essential requirement for success.
- The importance of the definition of minimum standards for the development of health care information systems. This will add to their interoperatability, and the effectiveness in their functioning will reduce the development and implementation costs and provide the industry with the common European-wide standards for the coming internal market.
- The need to collaborate with other international organizations and to build upon progress already achieved in related domains are a basic element for any effective action on standards in health care informatics.

CEN TC 251: The scope and objectives

With respect to standardization in the medical imaging field this cannot be looked at separately from the overall prospect of standardization in medical informatics. It has become clear that, for instance, the successful implementation of PACS (or IMACS) depends on the integration of these systems in an overall Hospital Information System and other departmental systems. Medical Informatics should take into account all relevant work outside the field. The necessity for integration of systems and for communication of information in the health care sector is evident when studying the variety of interested parties, the multitude of applications and their importance. Prototyping should be done before a standard is produced, ensuring that the standard work is implementable.

Health care communication standards should build on existing open systems standards including OSI, taking full advantage of the investments that have already been made by the computing industry. It is necessary for the support of the standards by the manufacturers of systems platforms, that these standards for health care communications be international. Standardization should not be limited to communication issues but should pertain to terminology, coding, privacy, and safety too.

Working group 4 „Medical Imaging and Multi-Media" of CEN TC 251 and EWOS EG-MED is working hard to come up with standards fulfilling the users' and European needs in this field. Of course, we should take into account the work that is currently done. Project teams were created and started working in April and May 1992. Under the coordination of CEN a project team is set up to deal with „Medical Image and Related Data Interchange Format Standards", another, under the coordinationof EWOS, deals with „Profiles for Medical Image Interchange".

The framework model for medical image data interchange describes and recommends a partitioning consisting of an information model, a minimal set of interchange formats, and a minimal set of communications profiles that span the needs of health care, which will be based on ISO/IPI.

The creation of CEN 251 is a major step into the direction of the coordination of these diverse activities in Europe with a strong relation to the international standardization work groups in the U.S. and Japan. We do not have to re-invent the wheel. We need one standard now, but we must also work on a next step preparing the adaption to the driving stream in IT&T sectors.

Conclusions

Since the European Health System has to quickly become more cost-efficient and offer better quality of care, more effort should be put into information exchange and telematic services.

The work done through the AIM projects of DG XIII of the Commission of the European Communities and other actions has provided an enormous platform for the coordination of research activities and even the development of draft standard proposals. We have the possibility to build up efficiently and need to coordinate all activities to come up with standards. At the end of 1993, drafts will be available for image formats and profiling, which are two major needs in this fast growing imaging world. Joint efforts between the international standardization institutes must ensure that we end with one international standard.

Data Bases

Medical Image Management Using a Semantic Approach: Image Description

F. AUBRY[1], V. CHAMEROY[1], F. LAVAIRE[2], J.P. RAMOND[3], I.E. SAIDANE[1], A. GIRON[1], Y. BIZAIS[2], A. TODD-POKROPEK[4],R. DI PAOLA[1]

[1] INSERM U66, IGR Rue C. Desmoulins, 94805 Villejuif, Cedex, France;
[2] CNRS URA 823, Nantes, France; [3] GE-CGR, Buc, France; [4] UCL, London, UK.

SUMMARY

PACS would allow sharing of knowledge about medical images among a large number of biomedical actors (acquisition, consultation, administrative and archiving systems), but each one has its own viewpoint on this knowledge.

As a coherent sharing of the knowledge must be based on its semantics, it necessitates a multi-layer structure.

The resulting repository allows coherent specification and implementation of the archiving system (image database) and of the interfaces with acquisition or consultation/processing systems.

The data model and a part of the functional and of organisational models can be used to specify and implement image database and interfaces with image handling systems. Consequently, the users of the image database are also the users of the kernel, as they access the image database through the kernel.

INTRODUCTION

Medical imaging not only include 2D or 3D spatial pictures, but also curves (e.g., ECG), histograms, graphics, etc... Once digitized, these objects, by nature numerical, can be represented as multidimensional pixel arrays. and they can be usually processed using mathematical algorithms which can be deduced, for instance, from some image algebra [1].

These arrays can not be interpreted as medical tokens without reference to a collection of associated parameters describing their environment. Parameters are assertions, and the collection structure is declarative and relational.

Furthermore, according to the context, an individual image taken off a set of images, may be meaningless by itself, and only the whole data set can be clinically interpreted. It is particularly true for dynamic acquisitions for which only the knowledge of the position of the image inside the image set gives a meaning to the image contents.

Thus, an efficient management process of medical images must be based on the relational structure of associated parameters, and must allow the consideration of individual objects as well as of data sets [2,3]. Such a model, able to be used as interface between storage and image handling equipments, can aid in the specification and the diffusion of PACS [4]. In order to avoid confusion between these complex entities, which define the kernel of our semantic model of the medical image, and the displayed data arrays, these entities have been named 'image object'.

This work, which belongs to the MIMOSA [5, 6] topic part of the EC project EurIPACS (AIM A2009), aims at proposing a generic model of medical image objects, which can be used to implement a medical image database or to specify interfaces between the image server and image processing or display equipments The final model is expressed in the NIAM formalism [7] thanks to the CASE tool ISW® (ITS, Brussels).

I. STRUCTURAL DESCRIPTION OF MEDICAL IMAGES

An image object is composed of three main layers (Fig. 1) which are related to :

- the semantics of the image object, that is its description in terms of clinical or medical object. It is related to the clinical context of the image object. Patient description, clinical context are described here. It leads to define the type of examination to be performed in terms of goals, modalities, etc...
- the structure of the image object in technical terms. It is independent of the modality, and of the implementation. This work is focused on this second level.
- the data instance, which is formed by the array and its set of associated parameters. This layer represents the implementation. It describes the data structure on computers (e.g., file formats) and, the computer resources required to handle the image object.

Image object structure can be divided into three parts :

- the logical description of the image object which is directly related to its semantics;
- the version description which is based on presentation characteristics of the image object. It is useful for display and processing specifications.
- the physical storage and retrieval process oriented layer which describes operations and mechanisms put into action during access procedures.

I.1. Logical description

The logical description includes three modules :

- the acquisition condition module which includes the modality description (physical process at the origin of the image, e.g., nuclear medicine, nuclear magnetic resonance), and the description of the acquisition or generation equipment (e.g., location, identity, characteristics, specific setting).
- the position of reference which indicates the relative position between the image source, or image device, and the patient. It allows image registration. It is close to the DICOM frame of reference module.

- the mathematical description module which describes the kind of image object, shown as a function, the grouping of image objects into data sets and the relationships between image objects within the data set.

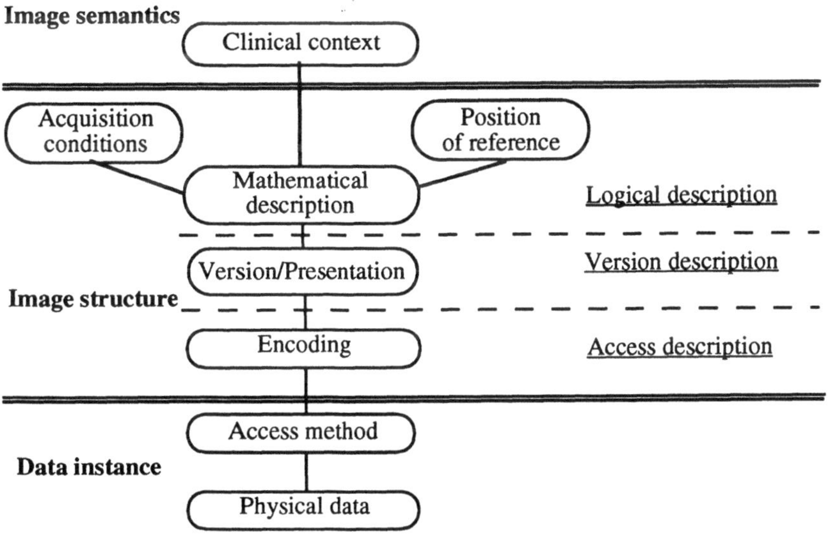

Figure 2 : Structural model of medical images

I.2. Version description

The version/presentation module is related to the characteristics of the image object which are useful in display and in image processing procedures. It allows the description of both analogical images and of digitized image objects.

I.3. Access description

It aims at providing information about the physical way to access data, and the computer resources required in such operations.

The encoding module describes the way data are coded, for instance using conventional multidimensional arrays, or using compression techniques.

The access method module indicates the location and the way to access physical data. It also describes the kind of storage units (e.g., UNIX files) and how image objects are structured and gathered within files.

II. FORMAL DESCRIPTION OF THE IMAGE OBJECT

Not all features of the model can be described here. Main features related to its formal basis, i.e. its mathematical description (Fig. 2) and the sequence definition, to its representation and the subsequent concept of version (Fig. 3), and to access are discussed.

I.1. Mathematical description

An image object is physically a set of measurements portraying chemical, physiological or physical processes, which are usually associated with the patient state, as a function of variables defined on physical domains. The measurements are the image object components.

Such a definition is very comprehensive as it includes all kinds of data obtained from acquisition equipments, physiological signals (e.g., ECG) included, and data resulting from image processing. Moreover, it is generic because it is independent of the acquisition conditions, especially of the modality.

Because domain of variables and of components are physical domains, they can be considered as two specialisations of the same concepts, called elements of observation.

Formally, an image object is a function F defined on a definition domain of dimension N, and its range is of dimension M :

$$F : X_1 \times ... \times X_N \text{------>} Y_1 \times ... \times Y_M$$

$$F : (x_1, ..., x_N) \text{--->} (y_1, ..., y_M) = F(x_1, ..., x_N)$$

where $F(x_1, ..., x_N) = (F_1(x_1, ..., x_N), ..., F_M(x_1, ..., x_N))$.

The F_i, i =1 to M, are the components of the function. The X_i are the physical domains on which the image object is defined, and the Y_i are the image object components.

$$F_i : X_1 \times ... \times X_N \text{------>} Y_i$$

Examples :

- Tomographic cardiac equilibrium acquisition is defined as :

$$TC : X \times Y \times Z \times T_f \text{------>} S$$

where X x Y x Z stands for the 3D space, T_f is the phase in the R-R interval, and S is the acquired signal (e.g., NMR, SPECT signal according to the modality).

Because, the order of physical domains, both of variables and of components have no signification, the object TK defined by :

$$TK : T_f \times X \times Y \times Z \text{------>} S$$

represents the same object as TC.

- ECG is defined as :

$$ECG : T \text{------>} S$$

where T is the time, and S is the ECG value in mV.

II.2. Image sequences

Usually, 2D pictures which are ordered in a 3-indexed data set according to the third variable, for instance time in dynamic studies, define a sequence.

If the number of variables is greater than three, it is possible to recursively define sequences of sequences. They define trees in which the depth is the number of variables able to define sequences [8]. For instance, a dynamic tomographic acquisition is a 2-depth tree and it can be represented as a spatial stack of dynamics or as a temporal sequence of 3-D images.

However, the internal structure of the tree is defined by the use. Only the ability to be represented as a tree is a property of the image object. Therefore, the concept of sequence is an user's view, and the sequence constructor belongs to the interface between the image model and the users.

II.3. Qualitative vs Quantitative variables

By nature, all elements can be measured, that is they can be represented by a value defined according to an unit.

However, this value can be measured if we are interested in their numerical value. Also, it can be identified by a label, because the label is either more informative, or the exact knowledge of the value is not attainable. For instance, defining view by the label anterior, posterior or lateral is more understandable than giving an angle in a coordinate system associated with the patient. Similarly, in a multichannel color images, the red, green and blue labels represent the predominating color of the acquisition filter, because we are not able to indicate the exact response of the filter.

Qualitative variables are defined by a list of labels.

Quantitative variables can be irregularly sampled, and described by a list of values, or piecewise uniformly sampled.

II.4. Version

The above descriptions are valid both for analogical and for digitized data. In fact, for digitized data, all the variables are sampled and image object components are quantified. For analogical data, variables are not all sampled, and components are not sampled.

For digitized data, various sampling schemes can be used. Resampling or interpolation can occur. But, these sampling schemes refered to the same image object. Then, the version defines the sampling and the quantization scheme. It results that the same image object can have many versions.

II.5. Access

The access modules describe the physical access of data, or copies. The manner data are stored and the computer resources to be used in order to perform the I/O operations are described in the access description. The encoding module describes whether a compression method and its description, or the conventional representation.

The access method module is implementation dependent. It describes pathways to data.

CONCLUSION

We have defined a semantic model of the image, able to be implemented in a relational way. The layered structuration of image environment has allowed us to build a generic model which is independent of the modalities, contrary to other current works, such as DICOM, which distinguishes the various origins of the images.

However, this model is compatible with DICOM work as our concept of image object is close to the multi-frame image of DICOM, and our whole model considering image semantics, can be considered as a generalisation of the various image modules of DICOM associated with every modality [9].

ACKNOWLEDGEMENTS

We acknowledge Charles Parisot, from GE-Medical System, for their helpful discussions and criticisms.

REFERENCES

1. G. X. Ritter, J. N. Wilson and J. L. Davidson, "Image algebra : An overview", CVGIP, 49, 3, pp 297-331, 1990.
2. Y. Bizais, A. M. Forte, B. Gibaud and F. Aubry, "Storage and retrieval of medical image data, relationships to image processing and analysis", Picture Archiving and Communication System (PACS) in Medicine, H. K. Huang, O. Ratib, A. R. Bakker and G. Witte (Eds), 74, pp 357-359, Springer Verlag, Berlin, 1990.
3. F. Aubry, V. Chameroy, H. Benali and R. Di Paola, "Modèle pour la manipulation de l'image médicale", 3èmes Journées 'Symboliques-Numériques', E. Diday, Y. Kodratoff and S. Pinson (Eds), pp 359-371, 1992.
4. V. Chameroy, F. Aubry, B. Gibaud, Y. Bizais, O. Ratib, A. Todd-Pokropek and R. Di Paola, "Picture Archiving and Communication Systems : from myth to reality ?", IEEE-EMBS, Tutorials - Frontiers of Computers in Biomedical Engineering, S. Laxminarayan and J.-L. Coatrieux (Eds), IEEE, 1992.
5. F. Aubry, V. Chameroy, A. Giron, R. Di Paola, B. Gibaud, Y. Bizais, D. Vital, R. Liénard, A. Todd-Pokropek, R. Kanz, F. Deconinck and O. Ratib, "MIMOSA : Conceptual modelling of Medical Image Management in an Open System Architecture", IEEE-EMBS, S. Laxminarayan and J.-L. Coatrieux (Eds), pp 1199-1201, IEEE, 1992.
6. B. Gibaud, Y. Bizais, N. Morcet, Y. Guandon, T. Buhé, A.-M. Forte, F. Aubry, J. Chabriais, V. Chameroy, R. Di Paola, O. Ratib, A. Todd-Pokropek, R. Kanz, M. Guitel, D. Vital and J.-P. Ramond, "Toward modelling image management in PACS : lessons from a preliminary review of end-users requirements", IEEE-EMBS, S. Laxminarayan and J.-L. Coatrieux (Eds), pp 1202-1203, IEEE, 1992.

7. G. M. Nijssen and T. A. Halpin, Conceptual schema and relational database design. A fact oriented approach, Prentice-Hall, New York, 1989.
8. F. Aubry, H. Kaplan and R. Di Paola, "An image handling system for medical images", Science and Engineering of Medical Imaging, M. Viergever (Ed), 1137, pp 131-138, SPIE, 1989.
9. C. Parisot, "DICOM V3.0 : Overview and progress report", Personnal communication, 1993.

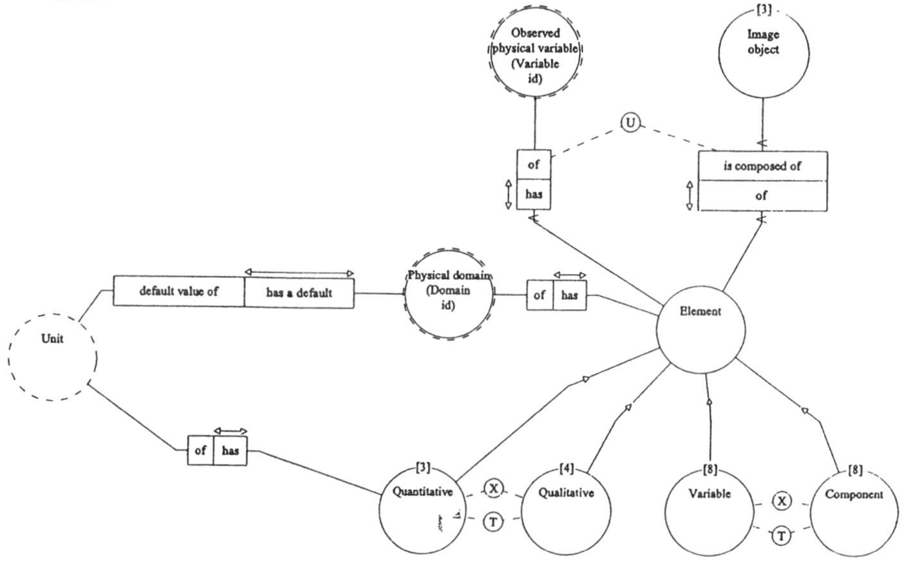

Figure 2 : Mathematical description of medical images

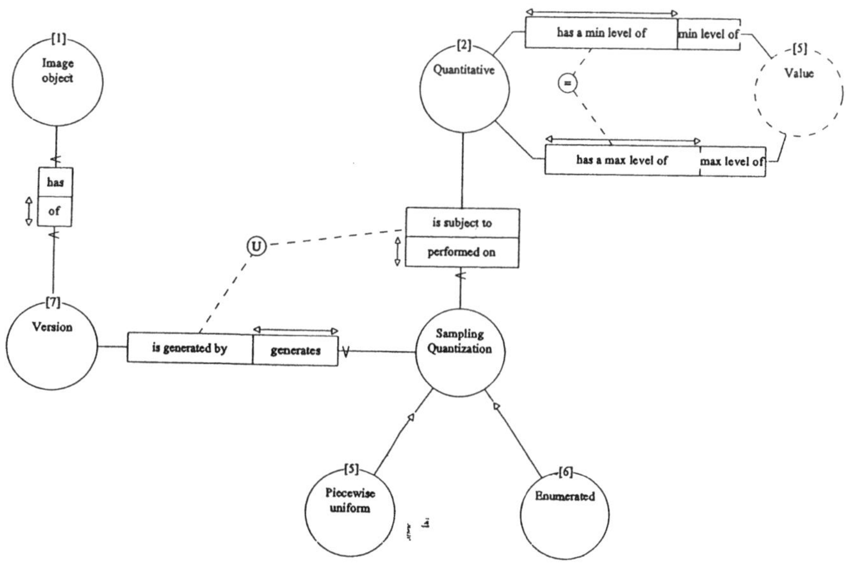

Figure 3 : Version and sampling relationships

A Novel Approach to the Management of Multidimensional Discrete Data in PACS

PETER BAUMANN*
Softlab GmbH
Munich / Germany

PAULA FURTADO
Graphical Systems and Methods Group
University of Coimbra
Coimbra / Portugal

1 Introduction

Current PACS technology relies on separate storage for images on the one hand and image descriptions and other hospital data on the other hand [MwLW-89, Oste-92]. Besides the well-known problems arising from such a loose data coupling, this separation usually leads to poor operational support for image structuring and accessing.

This paper presents a novel approach to the management of the variety of multidimensional discrete data (MDD) - in particular two-dimensional images obtained from CT, MRI, NMI, and other scanners - coming up in PACS database systems [AlFr-92, BeMB-92]. The concept has been designed to achieve the following goals:

- Efficient access to subimages, regardless of the cutout size and position specified,
- management of huge data objects of arbitrary dimensionality, and
- integration of MDD with non-pictorial information into a single coherent database.

In the next section, we first briefly sketch the APRIL object model [BaKö-89], which serves as our development basis. Our approach on MDD management is introduced in Section 3, followed by a conclusion in Section 4.

2 The APRIL Object Model

Objects are identified by an automatically generated, externally visible *object key*. An object type determines the following additional object constituents:

---

* This work has been performed at the Fraunhofer Computer Graphics Research Group in Darmstadt, sponsored by DFN Assoc. (German Research Net Association) under grant no. TK558-VA014

- A set of *attributes* for which enumeration, records and arrays with arbitrary nesting are available.
- One optional long field per instance, the so-called *object contents*. Representing state of the art, the contents up to now was viewed as a byte string with no further semantics imposed.
- A *successors clause* [ZhBa-92] describing the set of admissible object references wrt. referenced object types, cardinality, and possible variants. Both object hierarchies and general object graphs are modelled this way.

As a very simple example, let a physician conduct examinations where X-ray, CT, MRI, or sonogram images are produced. Each image is described by patient id, the radiation dose (s)he was exposed to, and the image's geometric resolution and pixel type; the pixel matrix itself is stored in the object contents; Fig. 1 shows a graphical view into a sample database. An (abbreviated) type definition might look as follows:

```
typedef object
{   successors 0 .. * Examination;
} Physician;

typedef object
{   char patientId[20];
    float radiationDose;
    enum { X_RAY, CT, MRI, SONOGRAM } technique;
    unsigned int xResolution, yResolution;
    contents;
} Examination;
```

The APRIL query language leans itself towards common SQL. The `where` clause is enhanced with path expressions to state conditions on the object graph. There is no `from` clause, as the types involved can be deduced from the other query parts. For example, the query *"Retrieve the keys of those physicians working with MRI"* is expressed as

```
select Physician
where  Physician downto Examination.technique = MRI
```

3 MDD Definition, Manipulation, and Implementation

Instead of mapping n-dimensional arrays to one-dimensional byte streams, we aim at preserving structure information on both the external and the internal system layer. Essentially, this means providing the array abstraction for arbitrary sizes - limited only by disk space - and arbitrary base types.

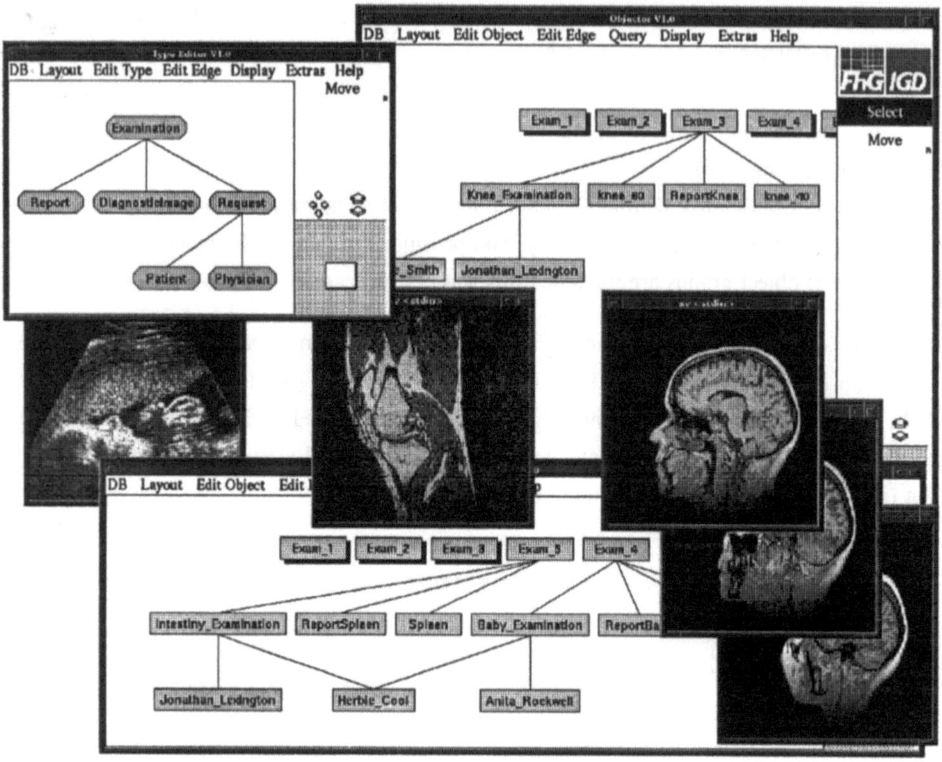

Fig. 1: APRIL object editor view on a clinical database

To define such huge arrays, the contents is overlayed with a structure definition; syntactically, the single keyword `contents` is extended to become an array definition for the special attribute named `contents`.

For instance, a volume tomogram (VT) can be defined as

```
typedef object
{    unsigned int contents[256][256][256];
} VolumeTomogram;
```

If an array boundary is specified as '#', then variable limits are assumed in that dimension.

Three kinds of operations are especially important [Baum-93] and, therefore, enjoy special support:

- The current array limits of an MDD attribute can be queried through a family of builtin functions where each one returns the index range of one array dimension.
  Example: `range1(VolumeTomogram.contents)` returns the number of pixels per line (which here is 256).

- *Trimming*, i.e. extraction of a cutout of an array with axis-parallel boundaries.
  Example: `Examination.contents[#,80..130]` extracts a full-width band of 51 lines starting with line 80.
- *Induced operations*. Operations available on the base type are lifted to the array structure by applying the operation to each array element.
  Example: `Examination.contents+5` yields an examination image with value 5 added to each pixel.

An important service of DBMSs is the ability to query a database in an unanticipated way. Let us demonstrate this with our VT example. Insertion of the `i`th of all 256 slices into a VT object `v` is done by statements like

```
update VolumeTomogram
set    contents[0..255][0..255][<i>] = <scan data>
where  VolumeTomogram = v
```

Now suppose we want a cut through the volume along the x/z plane at a certain position `y0`. The query "*extract all pixels in the x/z plane with y position* `y0` *of VolumeTomogram* `v`" is written as

```
select VolumeTomogram.contents[#][y0][#]
where  VolumeTomogram = v
```

`y0` is short for `y0..y0`. Note that data ordering is orthogonal to the way the VT has been stored.

On the internal level, such queries are supported by a combination of two different techniques. First, MDD objects are decomposed into rectangular tiles of selectable size. Second, in order to provide efficient access to image parts, a spatial index is constructed on these tile data speeding up access of the tiles affected by a particular query. The spatial index adopted for our implementation is the R+-Tree [FaSR-87].

In our current implementation of the MDD manager, both the spatial index and the raster tiles are stored in tables of the relational database system ORACLE; clustering and (conventional) indexing facilities offered by ORACLE are exploited to minimize disk access and disk head movement.

Preliminary evaluations yielded a performance degradation of roughly a factor 3 compared to the conventional way of reading images from a sequential file. In our opinion, this is encouraging, as (1) the potential of optimizations has not been fully exploited yet, and (2) performance is not the only criterion - there are substantial advantages of our approach over pure file system implementations, e.g., enhanced modeling capabilities, flexible query facilities, multi-user synchronization, and recovery.

4 Conclusion

In summary, we feel that MDD database capabilities will lead to an improvement of usability, consistency, and hence cost effectiveness of PACS. Due to its generality, the MDD manager can handle a wide variety of application data types. It not only supports (two-dimensional) images, but also (one-dimensional) time series like electrocardiogram data, (three-dimensional) voxel fields such as volume CTs, and the five-dimensional `BasicImage` type of the forthcoming ISO *Image Interchange Facility* Standard [ISO-92]. Besides, advanced registration data like colormaps can be associated appropriately.

References

[AlFr-92]

Allen, L.; Frieder, O.: *Exploiting Database Technology in the Medical Area.* IEEE Eng. in Medicine and Biology, March 1992, pp. 42 - 49

[BaKö-89]

Baumann, P.; Köhler, D.: *APRIL - Another PRODAT ImpLementation.* FhG-AGD Darmstadt, FhG Report FAGD-89i007, June 1989

[Baum-93]

Baumann, P.: *Database Support for Multidimensional Discrete Data.* Proc. 3rd Symposium on Large Spatial Databases, Singapore, June 1993 (accepted for publication)

[BeMB-92]

Beltrame, F.; Marcenaro, G.; Bonadonna, F.: *Integrated Imaging for Neurosurgery.* IEEE Eng. in Medicine and Biology, March 1992, pp. 51 - 66

[FaSR-87]

Faloutsos, C.; Sellis, T.; Roussopoulos, N.: *Analysis of Object Oriented Spatial Access Methods.* Proc. ACM SIGMOD 1987 Annual Conference, San Francisco, June 1987, pp. 426-439

[ISO-92]

International Organization for Standardization (ISO): *Information technology - Computer Graphics and Image Processing - Image Processing and Interchange - Functional Specification - Part 3: Image Interchange Facility (IIF).* ISO/IEC JTC1 SC24 Document IM-157, Draft International Standard (DIS), October 1992

[MwLW-89]

Meyer-Wegener, K.; Lum, V.; Wu, C.: *Image Management in a Multimedia Database.* Proc. Working Conference on Visual Database Systems, Tokyo/Japan, April 1989, Springer 1989, pp. 497 - 523

[Oste-92]

Osteaux, M. (ed.): *Hospital Integrated Picture Archiving and Communication Systems: A Second Generation PACS Concept.* Springer 1992

[ZhBa-92]

Zhou, J.; Baumann, P.: *Evaluation of Complex Cardinality Constraints.* Proc. 11th Int. Conf. on Entity-Relationship Approach, Karlsruhe/ Germany 1992, LNCS 645, Springer 1992, pp. 24 - 40

The Architecture of a System for the Indexing of Images by Content

S. Kostomanolakis, M. Lourakis, C. Chronaki, Y. Kavaklis, and S. C. Orphanoudakis
Computer Vision and Robotics Laboratory
Institute of Computer Science, Foundation for Research and Technology, HELLAS [1]
and Department of Computer Science, University of Crete
P.O. Box 1385, GR-71110, Heraklion, Crete, Greece

Summary

This paper presents the architecture of an image database system which provides a platform for the accommodation of various algorithms for interactive and automatic indexing, storage, and retrieval of medical images by content. The system maintains a dynamic hierarchy of image classes. The class hierarchy is used to narrow down the search to images of the same modality, anatomical characteristics, etc. Each image is classified into an image class based on information supplied by the user or obtained from the image itself. An important feature of the system is its ability to support multiple image indexing by content methods, in the form of *description types*. During system installation, one or more description types are selected for each image class based on the inherent characteristics of the class. Then, for each description type, a content description of each image in the class is generated and inserted in the description database. In the case of a query, the system generates one or more descriptions of the query image, automatically or interactively, and searches the logical database for similar descriptions. The images whose descriptions match those of the query image are retrieved for browsing by the user. This system may also be used as a platform for evaluating image content description methods, since new methods can be added to it easily.

Introduction

An Image Database (IDB) is an integrated system for the storage and retrieval of large amounts of image data and related information[3]. Image indexing and retrieval by content addresses the problem of retrieving from an IDB images similar to a query image.

We believe that image retrieval by content will prove valuable to the medical imaging field by providing considerable clinical decision support and increased efficiency and effectiveness in diagnostic image interpretations. In this paper, we present the architecture of a system that serves as a platform for the implementation, testing, and evaluation of class-based indexing by content algorithms, since such algorithms of general applicability are not currently available.

Our system is built around the concept of image classes to achieve effectiveness and efficiency. We achieve effectiveness by employing algorithms tuned to perform well on

[1]This work was funded in part under project EURIPACS of the CEC AIM programme.

a particular image class and efficiency in retrievals by directing queries to appropriate image classes.

The remainder of the paper is organized as follows: First, we present a brief overview of the proposed architecture. Then, we focus on the indexing by content module of our system, and conclude with plans for future work.

The Architecture

Since the main goal of this work is the development of a platform for the implementation, testing, and evaluation of indexing by content methods, this paper does not address image content description, or search and retrieval strategies, directly. In our system, distinct approaches to content description generation, storage, and matching are encapsulated in description type objects. The emphasis is on the development of an object-oriented environment (*core*) capable of easily integrating different description types and image processing algorithms. This core (see Figure 1), acts as a host to various state-of-the-art algorithms which are incorporated to the system, at regular system maintainance. The main modules of the system core are the *image visualization module* that hosts

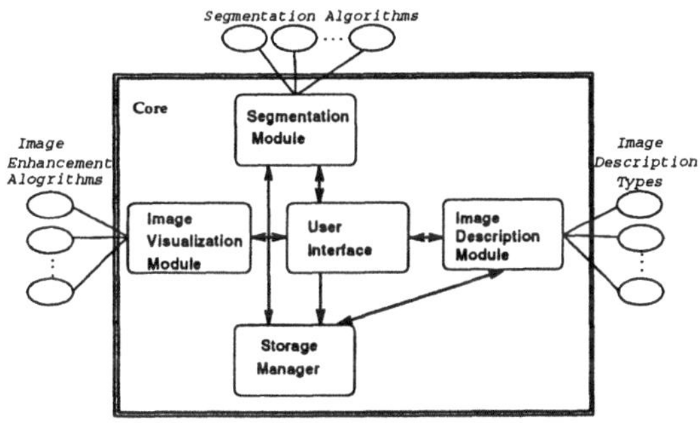

Figure 1: The system architecture

image enhancement algorithms, the *segmentation module* that hosts segmentation algorithms, the *image description module* that hosts description types, and the *storage manager* that maintains the image classes known to the system and their attributes in the form of persistent objects (persistent is an object whose lifetime exceeds that of the program that created it.). In the current version of the system, the storage manager is based on EXODUS[1]. Appropriate User Interface tools[2], such as image browser, the contour editor, etc., support user friendly input and access to the results of algorithms. This design approach resulted in a loosely coupled architecture, consisting of a set of autonomous communicating modules. Various modules request services from each other using specialized communication objects.

Image Classes

Images are assigned to classes using criteria such as imaging modality, anatomical characteristics, orientation, plain of cut, etc. For each image class, the system maintains a class object that encapsulates attributes relevant to the images in it. Class attributes are segmentation algorithm types, description types and general information on the class. A segmentation algorithm type identifies a segmentation algorithm object (see section on segmentation) and the class default values for the input parameters of the segmentation algorithm.

We expect that images in the same class have the same source and represent similar objects, hence call for similar treatment. Specialized algorithms can be developed and tuned for the images in a specific class. Such algorithms are likely to yield far better performance than general purpose algorithms, since they can exploit knowledge of the content of images in the class.

For efficient retrieval by content, we impose a class based partitioning of the image space. Each description type supported by a given class maintains its own database of content descriptions of the images in the class. In this way, a CT image of the head is never compared for similarity to a CT image of the abdomen.

The association of segmentation algorithm types and description types to image classes is dynamic. The user may bind (or unbind) an algorithm or description type to a class.

Segmentation

Usually, the first step towards description generation is image segmentation. Segmentation derives a set of disjoint regions or objects from an input image. The *segmentation module* provides the support for class-based segmentation of images. The introduction of a new segmentation algorithm to the system is an off-line operation that registers the new algorithm with the segmentation module, usually with minimal impact to the rest of the system. For each segmentation algorithm known to the system, the segmentation module maintains a persistent object, namely, a *segmentation algorithm object.* This object contains the name of the algorithm, the name, type, and default value for each of its input parameters, and a means of identifying the code which implements the algorithm.

A segmentation algorithm takes an image object as input and produces a contour object as output. The role of the derived contour object is twofold. First, it is used as an alternative input to the description generation algorithms and, second, it is stored in the database and used in browsing through the contours of images retrieved by a query.

Segmentation may proceed either as an interactive or as an automatic process. In interactive segmentation, the user first enhances the appearance of the image in the image visualization module (see Figure 2). Then, the user experiments with one or more available segmentation algorithms, changing their default values, comparing results, etc. When a contour image that is relatively satisfactory is produced, it can also be edited interactively using the *contour editor* (see Figure 2).

Since interactive segmentation is a time consuming process, class-based automatic segmentation offers a viable alternative. We view our system as evolving: users may change the default parameters of a segmentation algorithm within an image class, change bindings of algorithms to classes, and introduce new class-specialized segmentation algorithms.

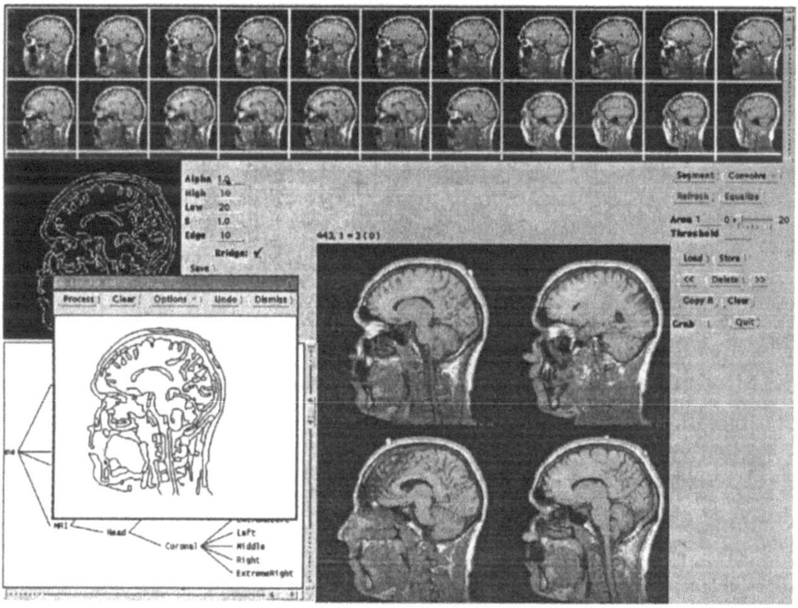

Figure 2: The user interface of the prototype.

Image Description

The description generation method, the description manager, and the description matching method are the basic components of a description type (see Figure 3). The description generation method produces the content description consisting of a set of persistent objects. The description manager stores, retrieves, and modifies content descriptions in the logical database of the description type. Finally, the description matching method compares the content description of a query image to content descriptions of images in the logical database and identifies images similar to the query image.

The object-oriented approach we followed imposes minimal constraints on the description methods that can be supported by the system. Each description type is viewed as an object capable of generating image content descriptions, storing them and retrieving images with similar descriptions from the database. Thus, the system does not concern itself with the schema of the logical database or the search and retrieval strategies associated with a specific indexing and retrieval by content algorithm. Such knowledge is embedded in the description type object of each individual indexing by content algorithm. What the system provides is a set of primitive operations that enable indexing by content

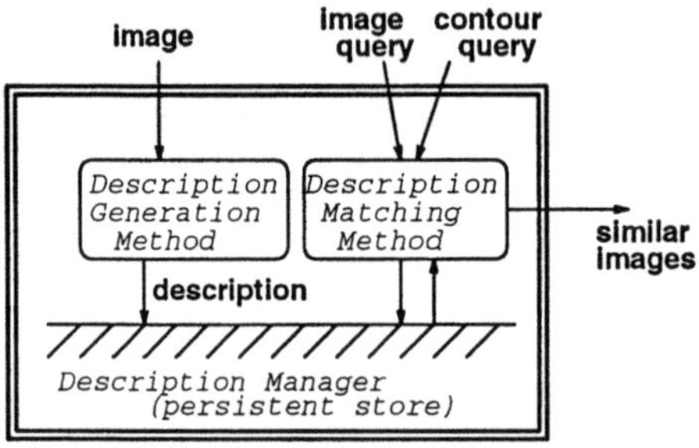

Figure 3: Description type structure

methods to create persistent clusters of objects, maintain indices on them, retrieve/insert objects from/to a cluster, etc. Indexing by content methods make use of these primitives to define their storage and retrieval strategies.

The description generation process associated with a description type can be either interactive or automatic. The description generation and description matching processes, are generally automatic. However, when the content description is contour-based, the user may interact with the system for the generation of the contour object.

Two basic forms of query by content are currently supported. First, the user displays an image in the image visualization module, and selects one of the description types available to the relevant class to serve the query. The retrieved images are displayed in the image browser as miniatures. Second, the user draws a rough sketch on the contour editor, and again selects an image class and a description type to serve the query.

Conclusions

This paper presented the architecture of an image database system for the indexing of images by content. It forms the basis for a prototype that is currently under development in our laboratory. Furthermore it is expected to give added value to integrated Picture Archiving and Communication Systems and to support computer aided diagnosis, medical training, and research.

References

[1] Using the EXODUS Storage Manager V2.2. Technical report, Computer Science Department, University of Wisconsin Madison, November 92.

[2] Petros Kofakis and Stelios C. Orphanoudakis. Graphical Tools and Retrieval Strategies for Medical Image Databases. In *Proceedings of the International Symposium on Computer Assisted Radiology 91*, pages 519–524. Springer-Verlang, 1991.

[3] Hideyuki Tamura and Naokazu Yokaya. Image Database Systems: A survey. *Pattern Recognition*, 17(1):29–49, 1984.

iLAN* - A New Path to a Filmless Radiology Department

K.D.Foord and N.Tomlinson

Radiology Department,
Conquest Hospital,
Hastings, East Sussex,
United Kingdom

Summary

iLAN is a semi-intelligent image network based on the Sybase relational database. It has Sun viewing and work stations for radiologist reporting and complex image analysis. It can acquire, store, display, archive and hard copy images under 'rules book' intelligent control. Archiving is on optical disks, mounted in change units.

iLAN has been developed to meet the 'PACS' needs of the smaller hospital. Via a FDDI bidirectional triangular network backbone it links image acquisition equipment and image hard copy devices including digital screening and angiography units, CT, gamma cameras, a computed radiography unit, film digitisers and laser imagers from a number of manufacturers. It is also connected to an image send/receive unit to communicate images over the ISDN telephone network.

This paper outlines its design philosophy and structure.

Design Philosophy

Initially we assumed that our new hospital, which opened in July 1992 would have conventional film/cassette daylight and laser film processing. However we decided to explore the possibility of installation of a partial PACS system, which of necessity had to be at least cost neutral in capital and revenue terms to be accepted by our hospital management board, as well as fitting a number of technical and clinical requirements.

These requirements were:

1. The system should mimic a conventional radiology department, but have clinical features improving clinical radiological input to patient care.

2. It must be easy and intuitive to use by persons with no specialist computer skills.

3. The view stations should allow images to be viewed at the highest resolution possible, up to that of the image acquiring modality.

4. Images must remain available to radiologists at all times.

5. There should be no loss of perceptible image data on decompression of data - it must be possible to make the same diagnosis again.

6. The image archive must be secure and lossless.

7. It, for us, is clinically acceptable to have slower recall of images more than a few months old, but they must be easily accessible.

8. Old hard copy images must be convertible to the new format.

9. Conversely, it must be possible to produce edited hard copy , including of multimodality studies, from the network.

10. A facility to transmit full image data between centres is required.

11. A facility for complex image studies, for example superimpositions and 3-D studies must be incorporated.

After exploring a number of options it was finally decided that the Regional Hospital Authority, who were at that time responsible for the hospital building project, would enter into a collaborative design and development project to develop a system with the above features, suitable for a small to medium sized hospital with economic constraints. iLAN was the result of this project, which has been carried out in partnership with Simis Medical Imaging Ltd., with the full cooperation of Siemens, Philips Medical Systems and 3M.

Design Features

The most important feature of iLAN is that it uses computer industry standard products in an 'open' environment, so that we are not tied to specific suppliers.

iLAN uses Unix as its operating system and is based on the Sybase relational database, which was chosen as it is the fastest of four high performance open system databases currently available using SQL with standard interfaces and a network library.

It handles image sizes ranging from 128x128 to 4000x4000 pixels - 0.25 to 10 MBytes uncompressed, 8 to 16 bits deep and will be developed to handle 3x8 bits colour.

Initially images are stored on multi-array Winchester 'Raid' disks, the image being partitioned between disks for security against image loss. Raid disk capacity is 5GBytes.

Once the Raid disks are full, images automatically transfer to WORM 12 inch 5.6 GBytes optical disks in 5 disk change units. Our system currently has two such change units, but more can be added if required. Once the 10 on-line ODs are full the oldest is taken off-line and stored in a disk safe, being replaced with a new disk.

iLAN makes considerable use of intelligence in lieu of on-line storage, it knows where images are even if off-line. The worst estimate of numbers of disk changes per day in our hospital is 18.

There is a bidirectional, triangular FDDI 100 Mbits/sec backbone, with state of the art bridges and concentrators at each apex.

Data compression is in development, the first stage being lossless 2:1 compression with 5:1 compression being considered at WORM transfer stage.

Departmental practices are matched with image ordering and search systems based on patient family name, forename(s), date of birth, imaging number, acquisition modality, physiological/anatomical groups, age of image and image audit value. There is an image folder approach with structured reporting and at reporting an audit of perceived clinical value of the image can be entered for later clinical review.

There is a wide area ISDN network capability.

Each radiologist has a Sun viewstation with an 800 MByte local hard disk which acts as a local multiple image store buffer to facilitate rapid review, but with the facility to directly interrogate the database for specific images if required. Each also has the ability to directly hard copy images to the laser imagers after image manipulation or image collaging.

Dual viewstations for clinician image reviews and a workstation for complex image analysis are also available on the network. The network is extended to the adjacent orthopaedic and trauma department, rendering the hospital partially filmless. Hard copy will still be produced for the rest of the hospital but in smaller formats. Both these features mean that film consumption will be reduced by approximately 50%, a necessary saving to help pay for the system!

Old films will be digitised as patients attend for new consultations and gradually the old film archives will reduce in size.

The final image network is shown as Figure 1. Items of equipment networked include Siemens AR.T CT scanner and Diacam 120/Icon nuclear medicine system, Philips DSI, Integris V3000, and ACe computed radiography system and 3M laser imagers. It is also intended to link ultrasound and mobile image intensifiers into the network via 3M disk reader systems. Additionally the iLAN network will be fully integrated with the departmental management system.

Clinical Benefits

A number of clinical benefits ensue from such a development.

Firstly images are available at all times to radiologists who can directly input into clinical management with greater ease than before. The reporting process is speeded as once an image is transferred to the network it is available for review and the time consuming clerical collation process is by-passed. It has been estimated that a hospital can make strategic efficiency savings with such a system, increasing bed occupancy rates by 1 - 3%.

There are also potential value added facilities, listed below:

1. Multimodality studies e.g Thyroid ultrasound and isotope scans and chest films as one image collage.
2. Multitemporal copy e.g. a collage of a series of Chest radiographs.
3. Mixed multimodality and multitemporal copy.
4. Teaching/ research/ interesting case collages.
5. Colour studies
6. Static superimposition studies e.g Tc bone scan on a radiograph for occult fracture study.
7. Dynamic superimposition studies e.g. Tc HMPOA SPECT study on a brain CT or MR scan.
8. 3D studies from any appropriate image set from nuclear medicine, CT or MR.

Conclusion

We have in partnership developed a semi-intelligent, flexible, relatively cheap image network, particularly suited to the small or medium sized hospital which has already or is about to acquire a significant amount of inherently digital equipment. As part of the development an image send-receive system has been constructed which can use standard ISDN line transmission to send and receive full image data sets, not just video images. The key to economic operation of iLAN is the incorporation of computed radiography, the linkage of at least one high level image user in the network, the ability to produce smaller hard copy for unlinked areas and a realistic use of on-line image storage capacity based on image recall data analysis.

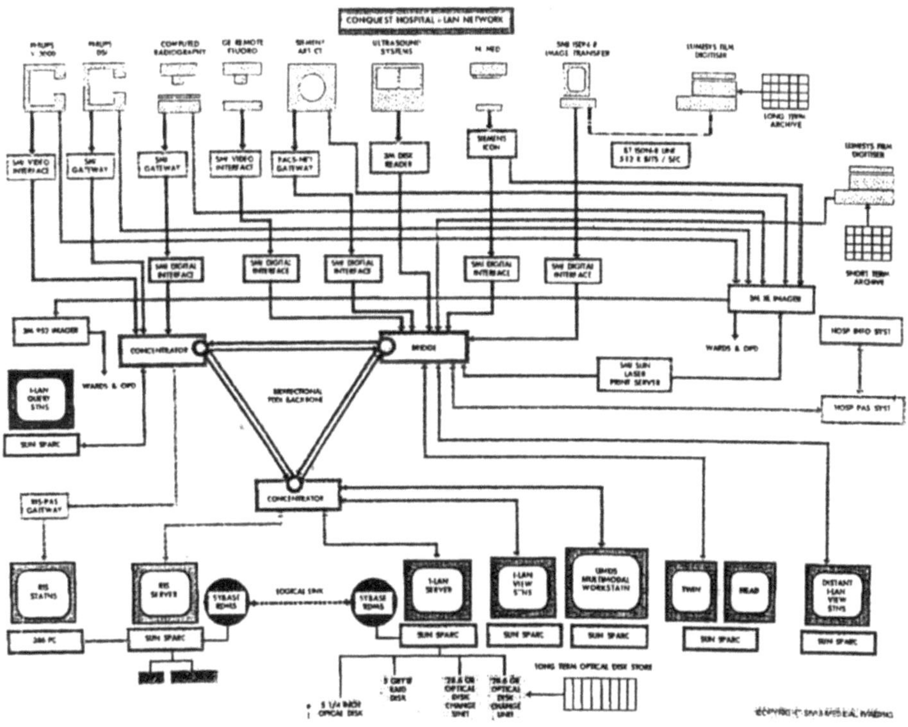

Figure 1

* iLAN is a Trademark registered by Simis Medical Imaging Ltd.

Medical Workstations

Computer Vision

Digital TomoSynthesis (DTS) with a Conventional X-Ray System

R. KOPPE, E. KLOTZ, G.J. LAURENSSEN*, J. LOODEN*

Philips GmbH Forschungslaboratorien, Forschungsabteilung Technische Systeme
Hamburg, Vogt-Koelln-Str. 30, D-2000 Hamburg 54, FRG
*Philips Medical Systems, P.O. Box 10.000, 5680 DA Best, the Netherlands

Summary

Today, film tomography is still useful in many settings because of its easy availability, high spatial resolution, and short examination time. A disadvantage is that only one image can be acquired from a selected layer and that for every other layer the tomographic procedure has to be repeated. This can be overcome by so-called DTS, which uses a finite number of common digital radiographs taken from different angles during a single tomographic sweep by pulsing the X-ray generator. From these radiographs the reconstruction of arbitrary layers in a specific volume is possible retrospectively, which reduces the X-ray dose, the time and cost of the examination. We performed both phantom and clinical DTS experiments on a Philips Diagnost 96. The results have been compared with film tomography for some many occuring examinations. For DTS it is necessary to know the geometric parameters of the different projections. Additionally the geometric pincushion distortion of the image intensifier II/ TV system and a change in magnification, caused by the arc-plane movement, have to be corrected for each projection.

Introduction

The main disadvantage in film tomography is that for each layer the tomographic procedure has to be repeated. In order to register all medically relevant structures, positioned in different layers, several exposures are needed. This results in a high skin dose for the patient. This problem can be overcome by so-called tomosynthesis [1-7]. This technique uses a finite number of common radiographs of an object taken from different angles during a single tomographic sweep by pulsing the voltage supply of the X-ray generator. From these radiographs the reconstruction of arbitrary layers in the whole volume is possible retrospectively.

A study on the possibilities of DTS on a Diagnost 96 (D-96) Remote Control system has been carried out. The D-96 is designed for universal routine medical examinations (Fig.1). It allows for tomographic images in an arc-plane movement of 40° (-20° to +20°). The system is equipped with a 38 cm (II) and a Plumbicon TV-camera. The TV signals are digitized and stored in a Philips Digital Spot Imaging (DSI) system. The X-ray tube moves along an arc while the film and the (II) move linearly along a straight line (Fig. 2). In this case the magnification is not constant, therefore the imaged layer is blurred.

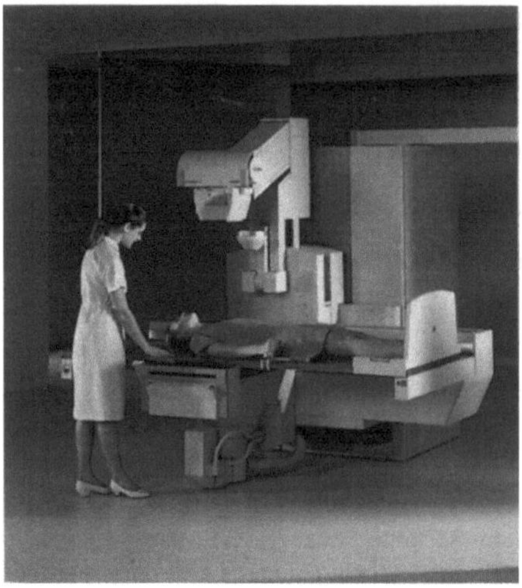

Fig. 1: Diagnost 96 (D-96) at 0°-position

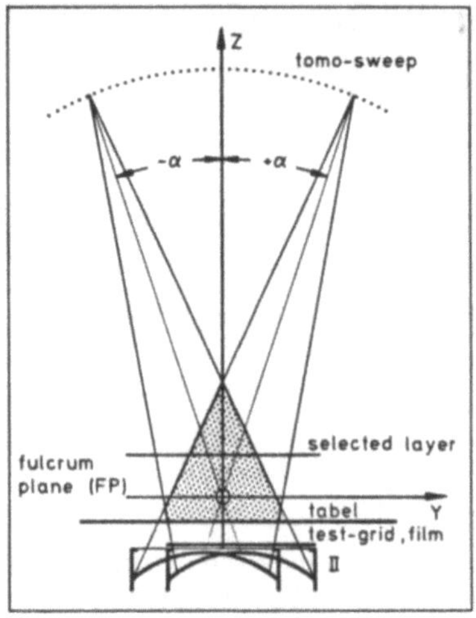

Fig. 2: D-96 model simulation

Methods and Material

Verification of the geometry parameters

For reconstructing a selected layer in DTS the geometry of the recording system has to be known for all projections. We have designed a 3-dimensional geometric phantom for verifying the accuracy of the reconstruction. It consists of a number of point markers and cross-bar patterns at different layer heights. This phantom is imaged with the D-96 and additionally simulated by a software model. The measured data can be analyzed in order to estimate the geometry parameters. The measurement of angles has shown that in a first approximation the increment in angulation between projections can be regarded as constant and the reproductivity of the angular positions was splendid. With these conditions the reconstruction can be performed with the parameters estimated previously.

Correction of imperfections of the D-96

Geometric distortion correction: For DTS the distortion of the image intensifier has to be corrected for all 49 projection angles in the range of ±20° in order to guarantee correlation of corresponding pixels in the different projections. The geometric distortion is due to the curved input screen (pincushion distortion) and due to the oblique projection. The later is the most significant one. The distortion is measured with a test-grid, which is mounted in front of the (II). The test-grid has to be adjusted perpendicular to the central axis of the X-ray system. By this adjustment the coordinate system is matched with the coordinate system of the simulation model, which is a necessary condition for an accurate reconstruction. In the correction step first a pattern recognition procedure detects the imaged steel balls of the test-grid of all 49 projections automatically. The functional relation between the imaged grid points and the known test-grid is determined by modelling the distortions along the x- and y-axis by bivariate polynoms [8,9]. The distortion correction is performed with subpixel interpolation methods, which significantly improve the image quality, compared with well-known point sampling (nearest neighbour) methods [10].

Gray-value correction: Besides the geometric distortion, grey scale imperfections occur in the II/TV-system. The different types of errors are described in detail in reference [9]. Vignetting and shading are the most significant effects. Both reduce the brightness of the image on the border. For DTS the shading effect is much higher than in commonly used X-ray systems resulting from the oblique projection geometry. However, shading will be widely compensated in DTS, if the tomo sweep is performed symmetrically to the 0°-angulation. The residual shading is comparable to that of the central projection. We have corrected the nonuniformities of shading by subtracting from each projection of the tomo-sweep a corresponding *empty image* mask acquired without any object. Thereby noise is increased in the projections but will be reduced by DTS, where an averaging effect is obtained by adding the different projections.

Magnification correction: Due to the arc-plane movement of the D-96 the magnification changes with varying tomo-angles. The magnification factor depends on the tomo-angle, the position of the fulcrum plane (FP) and the selected layer height for reconstruction. The error is about a few pixels of a 512^2-image. In the case of film tomography this error remains uncorrected. In DTS this error can be corrected.

Digital TomoSynthesis (DTS) reconstruction

Principle of DTS: Fig. 3a-d illustrates the principle of DTS. Fig. 3a shows the imaging geometry with three projection directions 1-3. The object consists of a structure in the fulcrum plane (in-FP) and above the fulcrum plane (off-FP). Fig.3b shows the resulting projection images 1'-3'. In Fig. 3c the reconstruction R of the in-FP is illustrated based on a simple adding of the three projections 1'-3'. Fig. 3d shows the reconstruction of the layer lying in off-FP, which is performed by shifting projection 1' and 3' with respect to the projection 2', which is the projection at 0°-angulation. Afterwards the projections 1'-3' are added to the layer R. With shift and addition arbitrary reconstruction planes above and below the FP can be obtained.

Calculation of DTS layers: Within the reconstruction algorithm a series of transformations for distortion correction, magnification correction and shifts for the selected layer has to be applied to each pixel of each input projection image before adding the images to obtain the tomosynthetic image. The image transformations are performed in a single step in order to avoid unnecessary interpolation errors. The interpolation of the projection images in the input-matrix (8-bit) is performed in a loop over all projection images in the range from -20° to +20°. The interpolation is performed with 32-bit accuracy. The reconstructed layer images are stored in a memory at 16-bit level. Afterwards these images can be enhanced using image processing techniques.

Enhancement of DTS layers: With digital image enhancement [11] the image quality can be improved. The method is based on a suppression of low frequencies. The enhancement is a separate part of our tomographic reconstruction and will be done with 16-bit accuracy. The displayed output is an 8-bit image. The enhancement can be done in different ways, e.g. in the following way:

$$\mathbf{out = in + C^* (in - low\text{-}pass (in))}$$

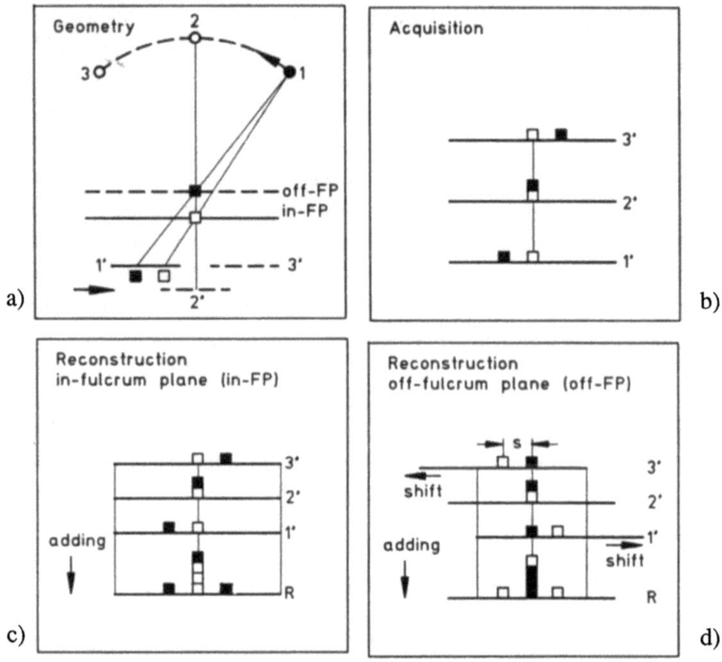

Fig. 3: Principle of DTS

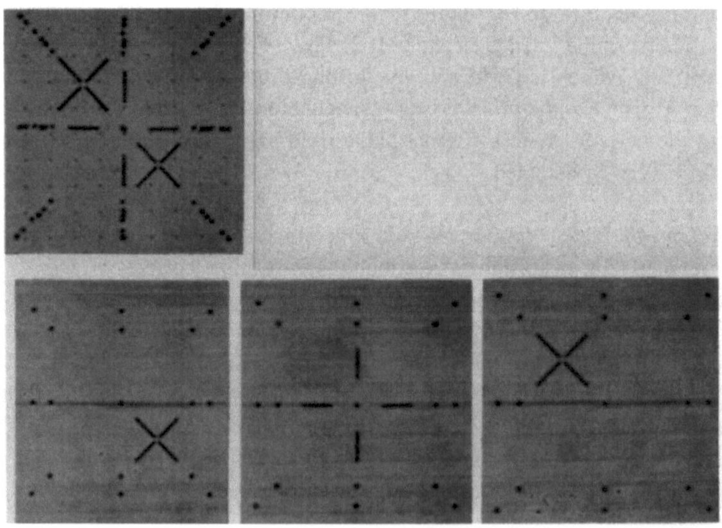

b)

$L_1 = 0$ mm $L_2 = 97$ mm $L_3 = 150$ mm

Fig. 4: DTS of the 3-plane phantom: Original projection at 0° (a) and reconstructions (b) at different layer heights L_n.

Reconstruction volume: The effective reconstruction volume in DTS (shadowed area in Fig.2) will be defined by the max. projection angles of the tomo-sweep. With an increase of the table-(FP) distance the shape of the volume becomes smaller and higher. The setting of the FP has to be adapted to the region of medical interest.

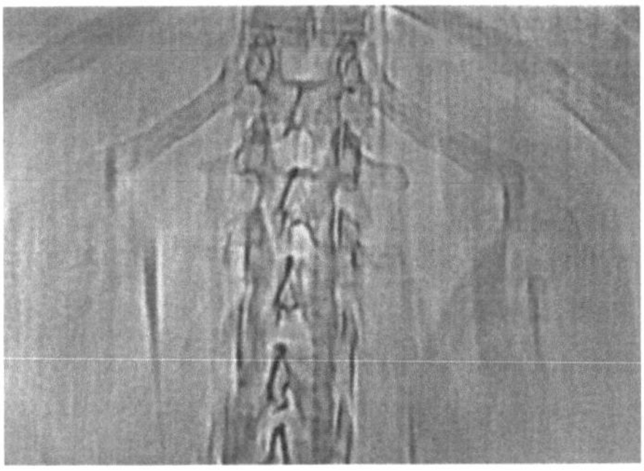

$L_1 = 25$ mm

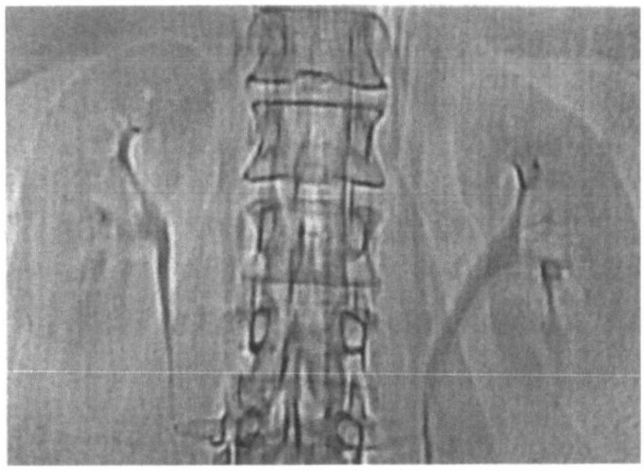

$L_2 = 70$ mm

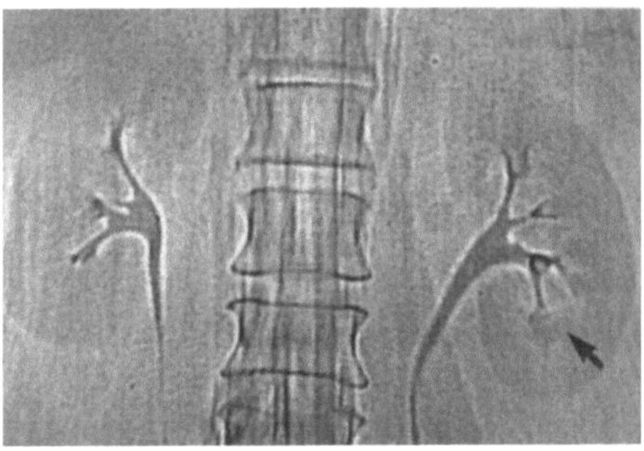

$L_3 = 72$ mm

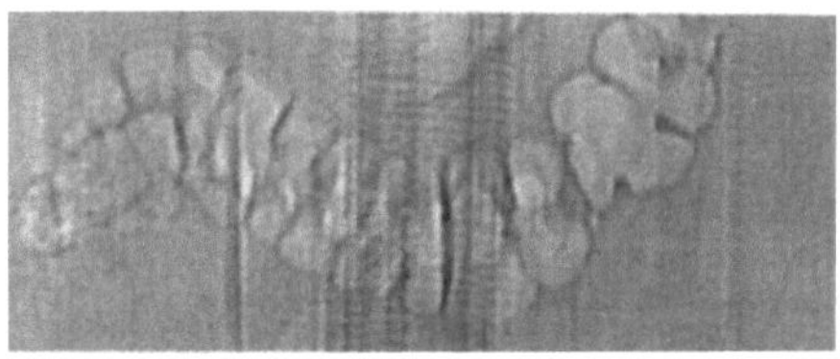

$L_4 = 220$ mm

Fig. 5a: DTS of a *kidney*: The clinically relevant information is the contour of the *kidney* and the ends of the *calices renales* (L_3). In (L_4) the *transversal colon* is reconstructed.

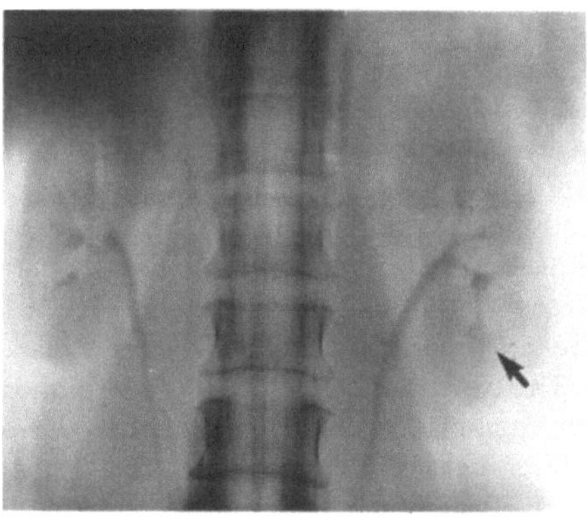

Fig. 5b: Film tomogram of a *kidney* at L = 80 mm

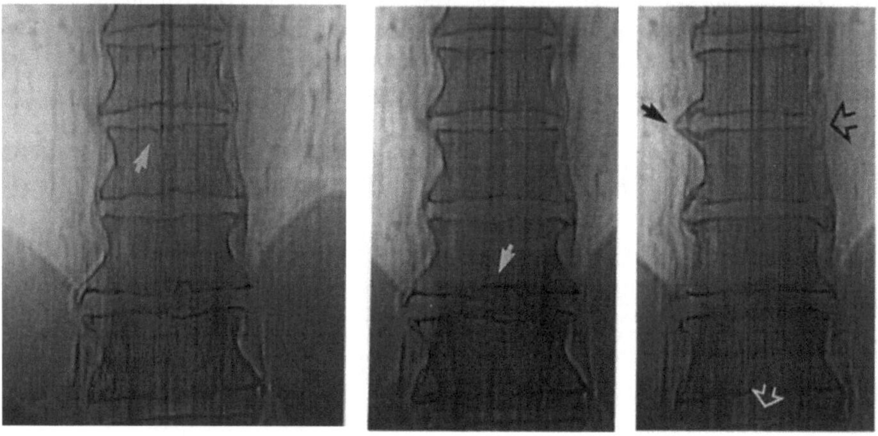

L_1 = 68 mm L_2 = 70 mm L_3 = 80 mm

Fig. 6a: DTS of a *vertebral column* at different layer heights L_n.

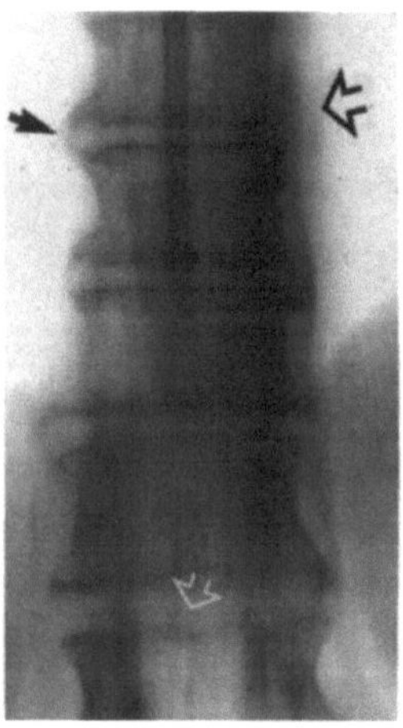

Fig. 6b: Film tomogram of a *vertebral column* at L = 80 mm.

Results and Discussion

3-plane phantom: The accuracy of DTS was controlled by reconstructing the geometric 3-plane phantom (Fig. 4). All 3 planes L_1-L_3 have been reconstructed with a sufficient image sharpness.

For a clinical application study, in a first step film tomography was performed. Afterwards the DTS examinations were acquired by 8 frames per second.

Kidney investigation (intravenous pyelogram): Fig. 5a shows 4 selected DTS images of a patient scan (L_1-L_4 = 25-220 mm). The clinically relevant information in this investigation is the contour of the *kidney* and the ends of the *calices renales (L_3 = 72mm)*. Both are equally good or even better visible in DTS images than in the film tomogram (Fig. 5b). Fig. 5d(L_4) shows that even structures (*transversal colon*) far above the fulcrum plane (220 mm = FP+140 mm) can be reconstructed. The benefit of DTS is the high resolution in depth in steps of 1 mm which allows better examination of the ends of the *calices renales*. Especially the high contrast of the DTS images is appreciated. The *stripes* (blurring artefacts) in the DTS images caused by the linear tomographic movement appear to be acceptable in these images, but they are stronger than on film tomograms, where the artefacts are blurred continuously by an infinite number of projections.

Vertebral column investigation: The contours of the *corpus vertebrae* were equally good or slightly better visible in the DTS images (Fig. 6a) as compared to the film tomogram (Fig. 6b). DTS shows defects on the *corpus vertebra* in steps of 1 mm. Due to this high resolvability in depth the chance in missing small defect is reduced in comparison to film tomography. However, the fine bone structures which can be seen on the film tomogram are lost in the DTS images which is caused mainly by the limited resolution of 512^2 pixels. A further loss of information caused partially from the blurring artefacts. The blur can be reduced by 1-dimensional spatial filtering [12] in a direction parallel to the tomographic movement. First experiments have demonstrated the reduction but it has to be noted, that the visibility of anatomic details in the direction of the movement will be reduced.

Advantages and applications: A main advantage of DTS is that the radiologist can reconstruct clinical relevant layers, over a large image depth, from a single tomo-sweep. This reduces the X-ray dose, the time and cost of the examination. While one single tomo-sweep is used for DTS, the relative position of the layers in depth is assured. In case of film tomography where several tomo-sweeps are made in time, layer positions also depend on the breathing depth of the patient. The possibility of DTS to reconstruct contiguous layers can be of great medical value for special applications. It also opens new perspectives in the field of trauma diagnostics, because patient handling can be performed with more care and much faster than with other examinations requiring positioning of the patient, e.g. in a Computer Tomography (CT) system. The tomo-sweep may be speeded up in order to suppress artefacts due to possible patient movement.

References

1. D.G. Grant: Tomosynthesis: A three dimensional radiographic imaging technique. IEEE Trans Biomed Eng. 19, 20-28 (1972).
2. K.R. Maravilla, et al.: Digital Tomosynthesis: Technique for electronic reconstructive tomography. AJNR 4, 883-888 (1983).
3. D. Rimkues, et al.: Digital tomosynthesis: Phantom and patient studies with a prototype unit. Computerized Med. Imaging and Graphics, Vol. 13, No. 4, 307-318 (1989).
4. M. Takahashi, et al.: Digital TV tomography: Description and physical assessment. Med. Phy. 17 (4), 681-685 (1990).
5. C. Düber, et al.: Die Digitale Mehrschicht-Tomographie: Erste klinische Ergebnisse. Electromedica 57, Heft 1, 36-39 (1989).
6. S. Yoshioka, M.D., et al.: Clinical evaluation of digital TV tomography. Radiation Medicine, Vol. 8, No. 2, 40-45 (1990).
7. S. Sone, M.D., et al.: Development of a hight-resolution digital tomosynthesis system and its clinical application. Radio Graphics, Vol. 11, No. 5, 807-822 (September 1991).
8. P. Haaker, et al. : Real-time distortion correction of digital X-ray II/TV-systems: An application example for Digital Flashing Tomosynthesis DFTS. International Journal of Cardiac Imaging 6: 39-45 (1990/91).
9. J.M. Boone, et al.: Analysis and correction of imperfections in the image intensifier - TV - digitizer imaging chain. Med. Phys. 18 (2), 236-242 (Mar/Apr 1991).
10. A. Rosenfeld; A. Kak: Digital picture processing. 2. Edition, Vol. 2, 31-33 (1982).
11. A. Rosenfeld; A. Kak: Digital Picture processing. 2. Edition, Vol. 1, 209-210 (1982).
12. P.R. Edholm; L. Quiding: Reduction of linear blurring in tomography. Radiology 92, 1115-1118 (1969).

An Interactive System for High-Speed X-Ray Image Filtering

B. PETERS*, B. WEIN**

* Lehrstuhl für Meßtechnik, RWTH Aachen, D-W-5100 Aachen

** Klinik für Radiologische Diagnostik, RWTH Aachen, D-W-5100 Aachen

Summary

Due to technical restrictions, Digital Image Filtering as a restoration and enhancement technique for medical x-ray images has not been investigated systematically yet. Striving to close the technical gap, the system presented in this paper consists of high performance hardware modules and adequate software components, allowing the design and application of any digital filter the user chooses. Combining FIR- (Finite Impulse Response) and IIR- (Infinite Impulse Response) filter techniques permits to use small filter kernels and therefore leads to a very efficient hardware implementation of the filter algorithm, complemented by suitable filter approximation strategies. The system´s user interface will be adapted to the DIBA-concept (Digitaler Bildarbeitsplatz) [1], at the same time aiming to expand DIBA´s outline. As a matter of consequence, the central aspect of this approach is the interactive realization of the filter variation, i.e. the rapid visualization of image modifications caused by altered filter characterictics.

Introduction

The enormous time expense of software-implemented filter algorithms seems to be the main obstacle to an application as an image restoration and enhancement tool for medical practice. Well-known primitives, e.g. unsharp-mask techniques [2, 7], imply the advantage of high processing speeds, but, on the other hand, offer suitable solutions only for a few diagnostic problems. Variations of unsharp-mask filter kernel sizes and weight factors as well as the superposition of images being filtered with different kernel sizes represented first steps to expand the scope of this technique [8]. The same applies to combinations of greyscale transformations and unsharp-mask filter algorithms [4]. Nevertheless, these techniques only apply to some cases and do not suit a general and systematical evaluation of spatial frequency filters as an image processing tool in medical practice.

Provided that no time limits existed, the most straight-forward step in digital x-ray image optimization would be to restore digital images, i.e. to compensate sub-optimum detector characteristics by Wiener filtering. As shown by Rupp [6], this can be done very efficiently with a combination of 2-D FIR and IIR filters. Comparing this approach with pure 2-D FIR filters, the approximation of a required OTF (Optical Transfer Function) can work more precisely and with much smaller filter kernels.

Following this way, after having analyzed the OTF and noise spectrum of a digital radiography system by means of phantom images, we designed suitable filters for the restoration of x-ray images coming from this modality. Furthermore we are developing more complex, application-specific image enhancement methods for difficult differential diagnostics. As well as Wiener filtering, these optimization methods usually suffer from huge time requirements, not acceptable in clinical practice. Our approach promises to overcome these restrictions. A massively parallel hardware implementation of the filter algorithm and a sophisticated filter design subsystem represent the suitable components of an adequate image processing system, allowing fast automatic image restoration operations as well as interactive image enhancement procedures.

Scope of the system

In order to gain first experience with the system, chest a.-p. images coming from a digital radiography system shall be restored and enhanced. Provided that the OTF of the display medium is known, it will be taken into account, too. In addition to the inverse OTFs of the detector and the display, restoration filters depend on their noise spectra and on an estimated spectrum of the image details to be restored. For reason of simplification these relationships are not shown in Fig. 1.

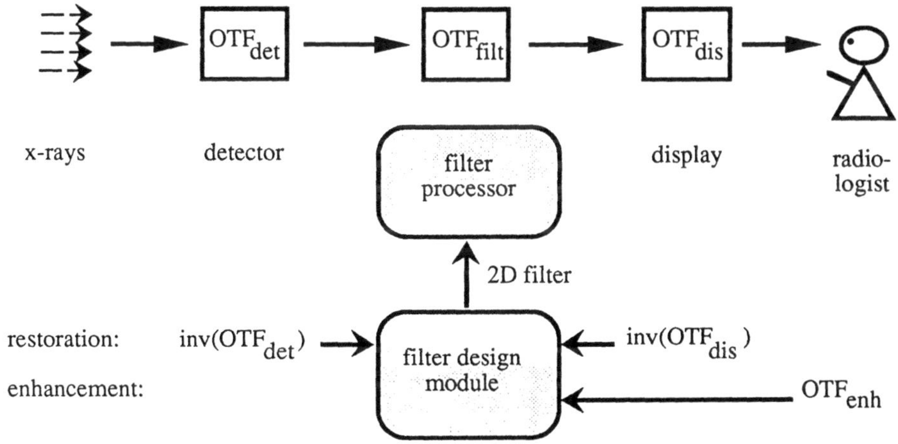

Fig. 1: Simplified model of image restoration and enhancement processes

Images containing objects of interest with known geometry can automatically be enhanced. For example, the correct position of central venous catheters could be controlled with an increased certainty, as it is often useful in intensive care.
Other interesting details in chest images, e.g. pulmonary nodules, vary in size and shape. In order to avoid an impasse, a first step will emphasize those objects interactively, because the

visibility enhancement of a single nodule would probably suppress others. The filters with OTFs found to be well suited for an application-specific enhancement will be stored and can be recalled. Following sessions on similar cases may profit from the recalled filters and lead to a finer filter specification. Later on, rules may be derived from the interactively found optima to correlate specific tasks with suitable filters. The investigation of adaptive filtering techniques and how to realize those techniques by means of the presented system, will be the next step towards an optimum processing method for medical x-ray images. A first concept for suitable adaptive filter methods has been proposed earlier by one of the authors [5].

System Architecture

According to the requirements, we have developed a special 2-D IIR filter processor board with a fast asynchronous interface, compatible to the TAX (Transparent Asynchronous Xmitter/Receiver) interface specification of IMNET/2 (Image Network 2) [3]. A VME/VSbus-DMA-interface for TAX is commercially available, other interfaces, e.g. SCSI, are under development. The filter board´s maximum processing performance of 2,9 GOPS is achieved by means of 12 integrated filter processors inmos IMS A100, performing 384 multiply/accumulate operations in parallel.
The second basic component of the system (Fig. 2) is the filter design module, which designs 2-D IIR filters by way of genetic algorithms and iterative gradient methods. In most cases, our genetic algorithm was superior to gradient methods as well in accuracy as in design speed.

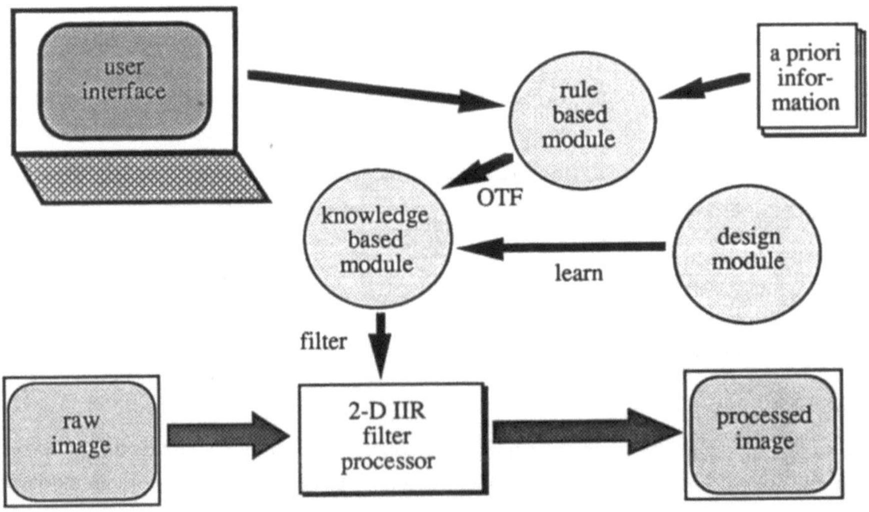

Fig. 2: Architecture of the high-speed filter system

Since the design process is currently the most time-consuming task, we are implementing an advanced genetic algorithm, that works in the spatial domain and therefore profits from the high efficiency of the filter processor board. This implementation may result in acceptable design speeds for some dedicated applications or for single high quality designs of restoration filters specific for a new imaging modality. Due to the fact that interactive enhancement methods require much faster solutions, a knowledge based module is necessary. The input of this module is the required OTF for the image processing task, the output is a 2-D IIR filter that approximates this OTF. At the moment, a data base and a MSE- (mean square error) criterion for the filter selection is used here. Because of their generalization properties, we are now investigating neural networks with regard to their suitability for this approximation task.
A-priori information, e.g. detector and display related parameters and application-specific information, both determining the OTF of the optimum filter to be applied, will be converted and combined by the rule based module. While information conversion is a basic task of the rule based module, the combination of a-priori information and application-specific information will be the subject of our future research work.
The actual output of this module highly depends on the wishes of the operating physician. In this context, DIBA´s user interface [1] is the global environment for image handling and display organization. Additionally, the user may select between restoration or enhancement filtering and recall previously found enhancement filters for the current application. Image restoration requires no further interaction, while image enhancement is performed via interactive forming of the filter´s OTF. Two concurrent methods for function modelling may be chosen, the first one operating with forming tools of different sizes, the second one using symbolic handles with variable distances between them.

Conclusion and Future Work

The quantitative validation of an interactive system like the one being presented here comprises problems. Nevertheless, comparisons of observers´ performance between original, restored and interactively enhanced images should be possible provided that the system permits the physician to recall application-specific filter characteristics he had found during earlier sessions. Starting from this point, he should try to optimize the parameters further and to adapt them to the actual image. The result of this optimization session again flows into the next starting point and so on. At the same time, the physician becomes more familiar with enhanced images. Thus, even if it is quite unlikely for the physician to reach the same experience with enhanced images as with the unprocessed ones, subsequent ROC (Receiver Operating Characteristic) tests will be fairer and more reliable.
As a consequence of the iterative procedure described above, non-measurable influences, e.g. the observer´s individual visual perception characteristics, are automatically included. This should result in better operating performance and, perhaps, lead us to a better understanding of what really happens with an image during its long way from an x-ray pattern to a mental representation of the reporting physician.

Acknowledgements

This work is supported in part by the Graduiertenkolleg "Informatik und Technik - Methoden und Werkzeuge der Informatik für technische Anwendungen", RWTH Aachen.

References

[1] M. Dahm, K. Glaser, H. Jansen-Dittmer, A. Keizers, D. Meyer-Ebrecht, K. Münker-Kaupp, H. Rudolf, C. Schilling, A. Selbmann, W. Winkler, "DIBA: Der digitale Bildarbeitsplatz für die Medizin", Software-Ergonomie '91, Teubner, S. 342-351, 1991

[2] W.D. Foley, C.R. Wilson and M. San Dretto, "The effect of varying high pass filter parameters on the detectability of isolated pulmonary nodules in digitized chest images", Investigative Radiology, Vol. 21, S. 305-310, 1986

[3] A. Keizers, D. Meyer-Ebrecht, F. Vossebürger, "A Fiber-Optic Line-Switching Network with a 140 Mb/s User Data Rate", IEEE Journal on Selected Areas in Communications, Vol. 10, Nr. 7, S. 1197-1202, 1992

[4] I. Maack, U. Neitzel, "Optimized Image Processing for Routine Digital Radiography", Proceedings of the International Symposium CAR '91, Computer Assisted Radiology, Springer, S. 109-114, 1991

[5] B. Peters, "Entwurfsverfahren und Methoden zur Aufbereitung medizinisch-diagnostischen Bildmaterials durch digitale Bildverarbeitung", Aachener Informatik-Berichte, Nr. 92-16, 1992

[6] S. Rupp, "Digitale Radiographie - Optimierung der Bildqualität durch Bildverarbeitung", Fortschrittberichte VDI, Reihe 10: Informatik/Kommunikationstechnik, Nr. 161, 1991

[7] R.H. Sherrier, C. Chiles, W.E. Wilkinson, G.A. Johnson and C.E. Ravin, "Effects of image processing on nodule detection rates in digitized chest radiographs: ROC study of observer performance", Radiology, Vol. 166, S. 447-450, 1988

[8] P.G. Tahoces, J. Correa, M. Souto, C. Gonzalez, L. Gomez and J.J. Vidal, "Enhancement of chest and breast radiographs by automatic spatial filtering", IEEE Transactions on Medical Imaging, Vol. 10, Nr. 3, S. 330-335, 1991

Optimum Edge Detection in Quantitative Coronary Arteriography

W. WUNDERLICH, T. LINDERER, B. BACKS, F. FISCHER, R.SCHRÖDER

Div. of Cardiology, Med. Klinik, Klinikum Steglitz, Freie Universität Berlin, Berlin

Summary

Optimum geometric edge detection in quantitative coronary arteriography is determined by the accuracy of measurements of known phantom diameters. We optimized the edge detection for different calibration methods, diameter ranges, accuracy measures and performance strategies. We found, that the optimization process itself is ill-defined, depending on the calibration method used, the included range of small diameters and the accuracy measure. For detectability of diameter changes accuracy-optimized edge detection is inferior to precision-optimum approach. The optimization of both, accuracy and precision can only be achieved with a precision-optimum approach and subsequent diameter correction.

Introduction

Computer based evaluation of coronary angiograms (quantitative coronary arteriography, QCA) is critically affected by accuracy and precision of vessel edge detection (ED). Since the true vessel edge in digitized angiograms is not detectable, geometric ED scans vessel density profiles perpendicularly to the vessel midline (Fig. 1. left.) and computes the 1st and 2nd derivative functions along each profile. Their extrema are then located as the detectable "landmarks" inside and outside the vessel (Fig. 1. middle.), providing an inner- and outermost edge of the vessel (Fig. 1. right.). The true vessel edge is located somewhere between both extrema. Thus, the need for optimization of geometric ED is obvious: one has to define a ratio of both derivatives, which returns the most likely true edge. This ratio is derived from the measurement accuracy of known diameters of phantoms, for calibration purposes filmed together with a cm-grid or in the center of the X-ray system. The deviation of the diameter curve "measured-true" from the line of identity is estimated by an accuracy measure and then used to optimize the ratio of derivatives and its attendant edge positioning.

However, due to limitations of the imaging process, the theoretical assumptions on the inside- and outside-the-vessel location of the derivatives are violated increasingly, when the vessel diameters decrease. In the small diameter range ($d \leq 1.5$ mm) the pointspread function, X-ray scatter and contrast lowering cause an increasing outward shift of the maxima of both derivatives and therefore an increasing overestimation of diameters. Since the location

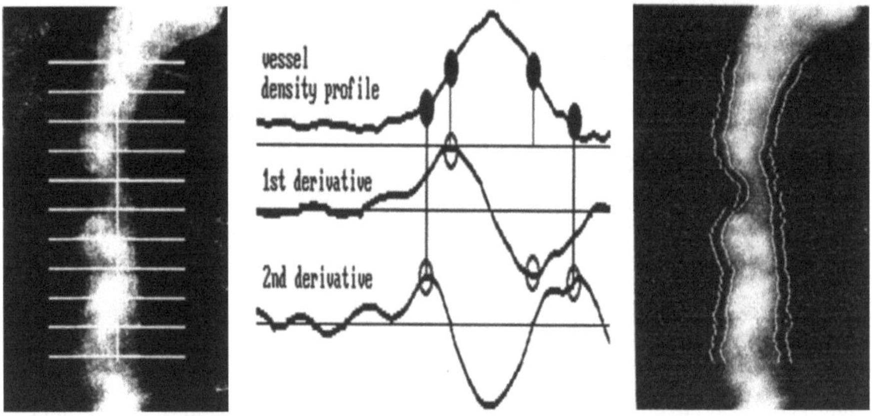

Fig. 1. Vessel edge detection: Scannig of vessel density profiles (left). Computing the extrema of derivatives (middle). Locating the inner- and outermost vessel edge (right).

of the derivatives is diameter dependend, the problem of ED-optimization cannot be defined as "positioning on the probably t r u e edge of a vessel" but as "positioning on an edge being an a p p r o x i m a t i o n of the probably true over the whole diameter range". Thus, the optimization problem becomes rather complex, which in turn may explain different results of the several groups [2]. We analyzed the main factors, inferring with the optimization process: the calibration method, the diameter range, the accuracy measure and at least the performance strategy.

Material and Methods

A perspex bloc of drilled vessel phantoms (0.3 to 5.0 millimeter diameter), filled with 100 percent contrast dye was filmed biplane with and without a scattering phantom (10 cm saline). The edges of each phantom were detected in nonmagnified digitized angiograms on a Kontron Cardio 500. All individual diameters, contributing to a given phantom contour were measured and reported as mean phantom diameter and standard deviation (SD) of one phantom measurement. Furthermore, all phantoms were measured 10 times in 24 consecutive frames. For each phantom these 240 means were averaged, providing an uncalibrated diameter curve "measured average diameter in pixels - true diameter in millimeters" over all phantoms. The whole procedure was performed stepwise varying the ratio of the 1st and 2nd derivatives in ED from 100:0 to 0:100, thus providing to a set of diameter curves.

Results

Inference of the calibration method with ED

All measurements were transformed from pixels into millimeters by isocenter [1] and catheter calibration [2] (catheter 2.5 mm diameter), providing a set of calibrated diameter curves "measured average diameter - true diameter [mm]" (Fig. 2). Isocenter calibration (ISO; Fig

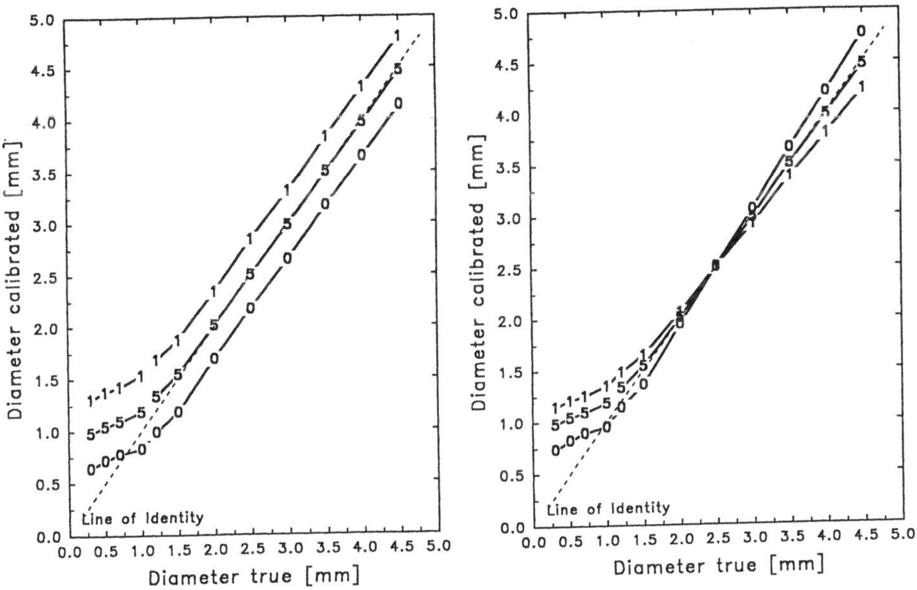

Fig.2. Impact of calibration method on diameter curves for varying ED: Isocenter (left) and catheter calibration (right). Percentage 1st:2nd derivative 100:0 (0), 50:50 (5), 0:100 (1).

2. left.) computes the exact radiological magnification from the data of the X-ray system and transforms the pixel measurements constantly and independend of the ED used. With increasing percentage of the 2nd derivative in ED, the diameter curves are shifted towards the line of identity and above it. In catheter calibration (CAT) the calibration ratio true/apparent catheter size becomes a variable itself, since the apparent catheter size itself is detected by the ED. With increasing contribution of the 2nd derivative in ED, CAT rotates the diameter curves around the line of identity instead of shifting towards them.

Impact of diameter range and accuracy measure on optimum ED

ED-optimization quantifies the deviations between measured and true diameters by an accuracy measure and minimizes them over all ratios of the derivatives. The main problem in ED-optimization is the overestimation of the small diameters, shifting the diameter curves away from the line of identity (Fig. 2, lower left). Since these deviations have to be compensated by corrections in the normal diameter range (d >1.5 mm), the ED-optimization becomes dependend on the range of small diameters included. Furthermore, the accuracy measure, controlling the compensation becomes important too. To compare the impact of the diameter range and the accuracy measure on optimum ED, we optimized the ED for different diameter ranges (≥0.3, ≥1.0, ≥1.5 mm), based on two accuracy measures, the usual "signed" accuracy (mean of signed deviations from the line of identity) and the "absolute" accuracy (mean of absolute deviations).

ED-optimization of isocenter calibrated measurements performs steady:

ISO:	optimum 1st:2nd [%]			accuracy [mm]		
ϕ range	≥0.3	≥1.0	≥1.5	≥0.3	≥1.0	≥1.5
"signed"	75:25	55:45	50:50	-0.012	-0.003	-0.005
"absolute"	55:45	50:50	50:50	0.169	0.050	0.020

With a decreasing range of small diameters, the percentage of the 2nd derivative increases, leading to an improved accuracy of normal diameters by a more outward edge positioning. ED, optimized on signed accuracy includes a higher percentage of the 1st derivative, leading to a more inward edge positioning. This virtual error compensation means a compromise between the overestimation of small diameters and the underestimation of normal diameters. In ED, optimized on absolute accuracy this error compensation is based on the remaining real deviations. Its higher percentage of the 2nd derivative yields to a more outward edge positioning and improved accuracy of normal diameters.

ED-optimization of catheter calibrated measurements performs unsteady, when signed accuracy is used as the quality criterion (Fig. 3.) :

CAT:	optimum 1st:2nd [%]			accuracy [mm]		
ϕ range	≥0.3	≥1.0	≥1.5	≥0.3	≥1.0	≥1.5
"signed"	100:0	10:90!	60:40	0.108	0.009	0.001
"absolute"	70:30	60:40	60:40	0.143	0.031	0.008

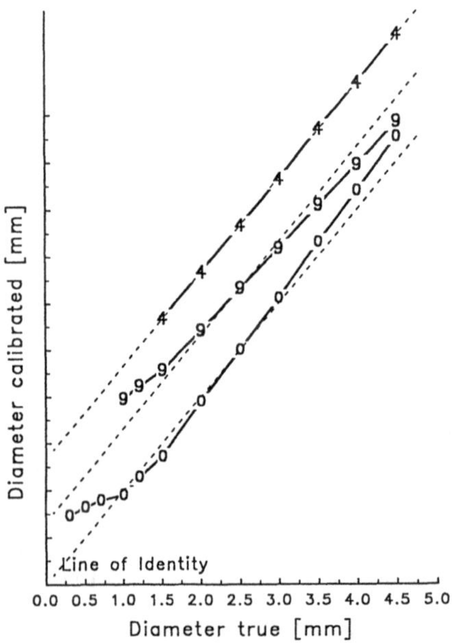

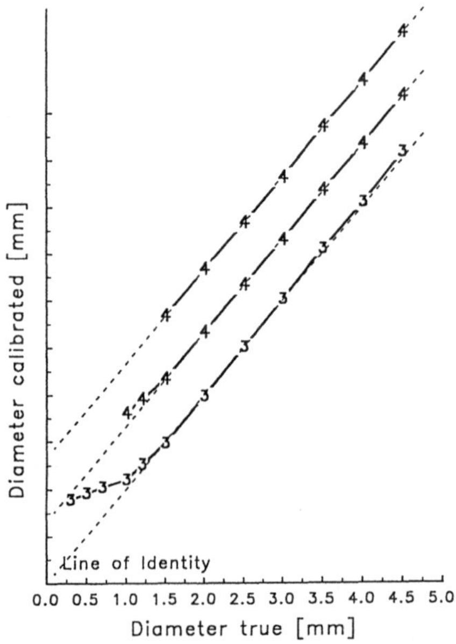

Fig.3. CAT diameter curves of optimum ED for the diameter ranges ≥0.3, ≥1.0, ≥1.5 mm (superimposed), based on signed (left) and absolute accuracy (right)

With a decreasing range of small diameters, the optimum ED switches from the innermost to the outermost edge positioning and finally converges to a middle positioning. This surprising finding for the most widely used calibration method and accuracy measure can be explained by the calibration caused pixel transformation (rotation of diameter curves around the line of identity) and the accuracy measure used (compromise between over- and underestimation). However, optimum ED based on absolute accuracy, performs steady indicating its preference.

Impact of performance strategy on optimum ED

The aim of ED-optimization described is an overall minimization of deviation of measured from true diameters, i.e. this performance approach is accuracy-guided. Although this approach may be helpful for singular ad-hoc measurements, the power of long-term studies on progression/regression of atherosclerosis is dependend also on the detectability of diameter c h a n g e s . Under these auspices, accuracy-guided optimum ED may not be optimally with regard to measurement variance. We analyzed the measurement variance of optimum EDs, defined by different ratios of 1st:2nd derivatives, as it occurs for different zooming conditions [3] or catheter sizes. Using vessel phantoms (1.0 - 5.0 mm diameter, filmed with a scattering backround), we compared the standard deviation (SD) of all individual diameters contributing to a given phantom contour with the SD of repeated mean phantom diameter measurements.

ratio 1st:2nd derivatives:	100:0	50:50	20:80	0:100
SD mean diameters	0.034	0.039	0.044	0.050
SD single diameters	0.092	0.100	0.117	0.134

As the data show and the 2nd derivative located edge in Fig. 1. left. indicates: an increasing percentage of the 2nd derivative in optimum ED deteriorates the precision of all measurements. This can be confirmed by the high pass characteristics of the gradient filter, applied twice for computation of the 2nd derivative. Compared with mean diameter measurements, precision decreases nearly 3 fold, when taken from single measurements.

Discussion and Conclusions

Since the c a l i b r a t i o n m e t h o d leads to basic differencies in diameter calculation, the use of the s a m e calibration method for ED optimization a n d daily practice later on is mandatory. Results of ED optimization, based on isocenter or grid calibration, are not transferable to later on used catheter calibration.

ED-Optimization is ill-defined and depends on the subjectively choosen d i a m e t e r r a n g e and a c c u r a c y m e a s u r e . Generally, the more of overestimated small diameters are included in the optimimization, the more of normal diameters are underestimated. Despite the real deviations, signed accuracy vanishes when the ratio of derivatives is varied sufficently small. Therefore, it is misleading to report its values as the real QCA performance. Since ED, optimized on signed accuracy, decreases the overestimation of small diameters, it might be an adequate quality criterion for quantitative assessment of stenosis. However, when the clinical endpoint is percent stenosis, relating then overestimated small diameters on underestimated normal diameters, signed accuracy fits the problem poor. Furthermore, considering the unsteady optimization process for catheter calibrated data, we suggest absolute accuracy instead of this quality criterion.

All the different demands on QCA cannot be satisfied by one performance strategy, i.e. accuracy of ED. For QCA in clinical studies the performance strategy of precision is important, since it provides a variance-minimum ED. Precision decreases markedly, when taken from single measurements. However, only the SD of individual diameters is equivalent to the real, one time performed vessel measurement procedure. Thus, the conventionally reported SD of mean diameters feigns a false optimistic detectability of diameter changes [4] which in turn affects the planning/evaluating of clinical trials.

Since accuracy-optimum ED is ill-defined, leading to a poor compromise between over- and underestimation of diameters, we suggest the avoiding of optimization as a whole. This can be achieved with 1st-derivative-based ED and subsequent diameter correction, as proposed by Kirkeeide [1]. As implemented in our system (Kontron Cardio 500), the vessel edge is detected precision optimally and the isocenter-calibrated diameters are corrected for their deviations from the line of identity [3]. These deviations are calculated from a diameter curve "measured - true diameters" of foregoing extensive phantom measurements. This approach considers QCA as a stochastic process: accuracy reflects the systemic and precision the stochastic component of the measuring process. Since the systemic properties of the measuring process are easier to control, than the sea of variability sources, it seems reasonable to optimize on precision and correct for accuracy.

<u>References</u>

1. Büchi, M.; Hess, O.M.; Kirkeeide, R.L.; Suter, T.; Muser, M.; Osenberg, H.P.; Niederer, P.; Anliker, M.; Gould, K.L.; Krayenbühl, H.P.: Validation of a new automatic system for biplane quantitative coronary arteriography. Int. J. Card. Im. 5 (1990) 93-103.

2. Reiber, J.H.C.; Serruys, P.W. (eds.) Quantitative coronary arteriography. Dordrecht, Boston, London: Kluwer Academic Publishers 1991.

3. Linderer, T.; Wunderlich, W.; Backs, B.; Nöring, J.; Schröder, R.: Edge detection in quantitative coronary arteriography (QCA): The impact of image zoom on edge positioning, accuracy, precision and discrimination. Comput. Cardiol. (1992) 99-102.

4. Koning, G.; van der Zwet, P.M.J.; von Land, C.D..; Reiber, J.H.C.: Angiographic assessment of dimensions of 6F and 7F Mallinckrodt softouch coronary contrast catheters from digital and cine arteriograms. Int. J. Card. Im. 8 (1992) 153-161.

Multicriteria Regularizing Neural Network Approach to Implicit Image Information Extraction from two Projections

Lu Weixue and Wang YuanMei

Institute of Biomedical Engineering
Zhejiang University, Hangzhou, P.R.China

Abstract
This paper describes attempts to model the implicit image information extraction from two projections in terms of minimizing energy function of multicriteria regularizing neural network. It examines the success of using multicriteria-style analog network for solving such problem. Finally it discusses the computer simulation results of the energy function approach.

1. Introduction

The extraction of the implicit image information from two orthogonal projections is the problem of image reconstruction from line integrals when the data available are very few and sparse. This is, for instance, the case when reconstructing cross section from a pair of biplane angiograms. Although, total cross section information is implicit in its two orthogonal projections, in general, it is not possible to reconstruct an arbitrary two-dimensional cross section image from two projections. Therefore, certain assumptions must be made on the cross section.

The reconstruction of images is an inverse problem and is usually severely ill-posed. The basic strategy for regularizing ill-posed problem is to restrict the class of feasible solutions by imposing constraints that exploit any additional information concerning the properties of the solution and/or noise. The most popular approach to regularization consists in reformulating the problem as a well-posed, well-conditioned, single objective constrained optimization problem. These methods have been considered, both theoretically and experimentally, and widely reported in computer vision [1], image restoration [2] and reconstruction [3].

In biplane angiography a two-dimensional cross section image is reconstructed from its two one-dimensional projections. At same time, we state our problem as the reconstruction of a binary matrix from its row and column sums. In general, there is no unique solution to this problem. A priori knowledge has to be used to reduce the ambiguity. A priori knowledge can be exploited, for example, under the form of multicriteria optimization principles that impose constraints on the possible solutions or as statistical properties of the solution space. We will use the general term multicriteria regularization for any method used to make an ill-posed problem well-posed under multiobjective decision making framework [3].

2. Model

This problem is the reconstruction of a binary image matrix from its row and column sums. Let Z denotes an mxn binary matrix with entries $z_{ij} \in \{0,1\}$. The elements of Z are to be determined from the following set of equations:

$$\begin{cases} \sum_{j=1}^{n} z_{ij} = p_i, \ i = 1,2,\dots,m, \\ \sum_{i=1}^{m} z_{ij} = q_j, \ j = 1,2,\dots,n \end{cases} \tag{1}$$

where p_i, i=1,2,...,m are the elements of the vector p containing the row sums of Z and q_j, j=1,2,...,n the elements of the vector q containing the column sums of Z. It is clear that our problem is underdetermined, ill-posed. The structure of the problem (1) finally can be written more clearly by writing the constraint equation in standard form:

$$Az=y \tag{2}$$

where $z=[z^1,z^2,\dots,z^m]^T$, $z^i=[z_{i1},z_{i2},\dots,z_{in}]^T$, and $y=(p_1,p_2,\dots,p_m,q_1,q_2,\dots,q_n)^T$, superscript T denotes transpose.

The standard regularization of the ill-posed problem (2) of finding z from 'data' y requires the choice of norm ||.|| and of a stabilizing functional || Cz ||. In conventional regularization theory, A is a linear operator, the norms are quadratic and C is linear. Two methods that can be applied are [1]: (1) among z that satisfy || Az−y ||<e find z that minimizes (e depends on the measurement errors and is zero if the data are noiseless)

$$\|Cz\|^2 \tag{3}$$

(2) find z that minimizes

$$\|Az-y\|^2+\lambda\|Cz\|^2 \tag{4}$$

where λ is a so called regularization parameter.

The first method computes the function z that is sufficiently close to the data and is most 'regular', that is minimizes the 'criterion' $\|Cz\|^2$. In second method, λ controls the ompromise between the degree of regularization and its closeness to the data. Therefore, the standard regularization theory is based on single objective optimization theory.

The multicriteria regularization methods impose the constraints on the reconstruction problem by multiple conflicting performance criteria, such as the cross entropy measure between the reconstructed image and a prior estimated image, the squared norm between the original projection data and reprojection data due to the reconstructed image, and the sum of the peakedness and nonuniformity function of the reconstructed image, et al. [3]. These criteria functions that are minimized simultaneously reflect physical constraints about what represents a best-compromized solution: it has to be close to the data and regular by making the multiple criteria functions small. The choice of the multiple criteria involved is dictated both by mathematical properties and by physical plausibility.
Multiple criteria regularization model for image reconstruction from two projections is of the form:

$$\begin{aligned} \min\ & f(z) = (f_1(z),\dots,f_l(z)) \\ s.t.\ & Az = y,\ z \geq 0 \end{aligned} \tag{5}$$

where $z \in R^{m\times n}$ is an mn-dimensional vector of decision variables; f_i, i=1,...,l, is a real-valued function that represents the ith criterion.
We used the weighting methods to handle multiobjective optimization problem (5). Mathematically, the weighting method can be stated as follows:

$$\min_{z\in Z} \phi(z) = \sum_{k=1}^{l} w_k f_k(z) \tag{6}$$

where $w \in W \underline{\underline{\Delta}} \left\{ w \middle| w \in R^l, w_k > 0 \text{ and } \sum_{k=1}^{l} 1 \right\}$. w_k is the weight of kth criterion function, $f_k(z)$ is kth criterion function. It is customary to assume that in the weighting problem all $f_k(z)$, k=1,...,l, are convex functions and constraint set z is convex. See Chankong et.al [4], Yuanmei Wang and Weixue Lu [3] for a detailed review.
In our multicriteria regularization method, the first criterion function is the cross entropy between the reconstructed image and an prior image density, the second criterion is the sum of the nonuniformity and peakedness functions of an image, the third criterion function is defined as the sum of the squared error between the original projection data set and the reprojection data due to the reconstructed image.
Here, we use an neural network to solve optimization problem (6). Let the output variable v_j for neuron j have the $0 \leq v_j \leq 1$ and be a continuous and monotonic increasing function of the internal state variable u_j of the neuron j: $v_j=h(u_j)$. A typical choice is $h(u_j) = \left(1+e^{-2u_j/\alpha}\right)^{-1}$, where α determines the steepness of gain. The energy function, E(v), can be expressed as

$$E(v)=\left\{w_1\gamma_1\sum_{j=1}^{n}v_j\ln\left(v_j/z_j^0\right)+\frac{1}{2}w_2\gamma_2\left(v^TSv+v^Tv\right)\right.$$
$$\left.+\frac{1}{2}w_3\gamma_3\sum_{i=1}^{q}\sum_{j=1}^{p}\left(\alpha_{ij}v_j-y_i\right)^2\right\}+\frac{1}{2}\eta\sum_{i=1}^{q}\sum_{j=1}^{p}\left(\alpha_{ij}v_j-y_i\right)^2 \quad (7)$$
$$+\sum_{j=1}^{p}\frac{1}{R_j}\int_0^{v_j}h^{-1}(v)dv$$

where each neuron has an associated capacitance Cj and resistance Rj (with $\tau=R_jC_j$). The binary image variables is mapped into continuous variables v_j. p=mxn, q=m+n. The umber of neurons in the network is p.

The first term {.} in (7) is interactive energy among neurons. The second term in (7) encourages the net to operate in the interior of the p-dimensional unit cube $\{0\le v_j\le 1\}$ that forms the state space of the system.

Like Hopfield's model, the "strength" of the connection between neuron i and neuron j is given by the matrix element $T_{ij}=-\frac{\partial^2 E(v)}{\partial v_i \partial v_j}$. The neuron activities v_j, as well as the input signals, u_j, depend on time t. The evolution of the neuron j of the net is determined by the n coupled ordinary differential equations

$$C_j\frac{du_j}{dt}=-\frac{\partial E(v)}{\partial v_j} \quad (8)$$

For simplicity, in this paper we have always chosen Rj=1.0 and Cj=1.0, independent of j, although this is not necessary.

A brief summary of the net's behavior is (1) E(v) is bounded from below and (2) E(v) is nonincreasing (i.e., $dE/dt\le 0$), with dE/dt=0 only when $dv_j/dt=0$ for all j. Therefore, E must reach a minimum.

Use Euler's method to the systems of equations, the system converges to a stationary state. The only stopping criterion we use is when the changes in the firing rates become insignificant, i.e. when all $\left|v_j(t+\Delta t)-v_j(t)\right|<\varepsilon$, where $\varepsilon<<1$. After the network converges to a solution, we must check if it is a valid solution that satisfies the condition, i.e. for every point j

we must have $v_j=1$ or $v_j=0$. In analog network the activity of a neuron can never become exactly 0 or 1 and can only be close to these limits. Therefore, for $v_j<\beta_0$ we take $v_j=0$, and if $v_j>1-\beta_1$ we take $v_j=1$, where β_0 and β_1 are small positive numbers. In simulations we have chosen the following parameter values: time step $\Delta t=10^{-2}$, convergence parameter $\varepsilon=10^{-3}$, conditional parameters $\beta_0 = \beta_1 = 0.25$, and gain steepness parameter $\alpha = 0.5$.

3. Simulation

To study the computational performance of a multicriteria regularizing neural net, we simulated state transitions of neurons by using a PC/AT serial microcomputer. We used two intersect ellipses in a given plane. The image of the mathematical phantom is shown in Fig.1. Using this model, we analytically determine the line integral data y for this phantom for 127 rays in each of two orthogonal projections, and the image is constructed on a 127x127 sampling lattice.

We have systematic method for finding the best combination of the weighting coefficients w_1, w_2 and w_3 under multiobjective optimization framework, see Y.Wang and W.Lu [3]. Here, it should be emphasized that the ability to obtain a best-compromise solution depends strongly on making choices for w_k's. After about two hundreds iterations, we reached feasible solutions satisfied all the constraints. Fig.2 shows the binary model reconstruction performed by a multicriteria regularizing net with p=127x127 neurons.

Fig.1. Phantom.

Fig.2. Reconstructed Image

References

[1].T.A.Poggio, V.Torre and C.Koch, "Computational Vision and Regularization Theory", Nature, Vol.317, pp.314-319, 1985.

[2].A.K.Katsaggelos, J.Biemond, R.W.Schafer, and R.M.Mersereau, "A Regularized Iterative Image Reconstruction Algorithm", IEEE Trans. On Signal Processing, Vol.39, No.4, pp.914-929, 1991.

[3].Yuanmei Wang and Weixue Lu, "Multiobjective Decision Making Approach to Image Reconstruction from Projections," JOSA A, Optics and Image Science, Vol.8, No.10, pp.1649-1656, 1991.

[4].V.Chankong and Y.Y.Haimes, Multiobjective Decision Making: Theory and Methodology, North Holland, N.Y. 1983.

[5].H.J.Ryser, Combinatorial Mathematics, Wiley, New York, 1963.

[6].S.K.Chang, "The reconstruction of binary patterns from their projections," Comm. ACM Vol.14, pp.21-25,1971.

[7].C.H.Slump and J.J.Gerbrand, "A Network Approach to Reconstruction of the Left or Right Ventricular from Two Projections," Computer Vision, Graphics and Image Processing, Vol.18, pp.18-36, 1982.

[8].Yongjian Bao, On Three-Dimensional Reconstruction and Modeling of Coronary Arteries from Biplane X-Ray Angiograms, Ph.D. Thesis, 1991.

[9].E.T.Jaynes, "Information Theory and Statistical Mechanics I," Physical Review, Vol.106, pp.620-630, 1957.

[10].R.L.Kashayp and M.C.Mittal, "Picture Reconstruction from Projections," IEEE Trans. on Computer, Vol.C-24, No.9, pp.915-923, 1975.

[11].J.J.Hopfield and D.W.Tank, "Computing with Neural Circuits: A Model," Science, Vol.223, pp.625-633, 1986.

„Neurovision" - a Multimodality Image Fusion Package for Neuroradiological Diagnosis and Neurosurgical Planning

R. GRAUMANN*, C.BERTRAM*, T. HILDEBRAND*,D. HENTSCHEL*,P.PLETS*,C. SINDEL*,V. ZOURLIDES*,D.PETERSEN**,H.RUDER***,J.GYBELS+,P.SUETENS++

* Siemens AG, Medical Engineering Group, Erlangen
** University Tübingen, University Hospitals, Neuroradiology
*** University Tübingen, Inst. of Theoretical Astrophysics
\+ University Leuven, University Hospitals, Neurosurgery
++ University Leuven, ESAT

Introduction

Today patients are frequently examined with different imaging modalities for diagnostic and therapeutic purposes in order to provide complementary information about deseases. The imaging devices have to be connected to a common computer platform for further image processing like image fusion or trajectory planning. Image fusion allows the combination of complementary image information from different modalities to improve the quality of diagnosis and surgery planning.

Being a member of the European AIM (Advanced Informatics in Medicine) project COVIRA (COmputer VIsion in RAdiology), we are developing a multimodality application package for use in neuroradiology and neurosurgery. In stereotactic surgery planning computer assistance is required for biopsy planning, the implementation of depths electrodes, and pain surgery. The evaluation of aneurysms, arterio-venous malformations (AVM) and tumor vascularizations are of potential benefit for neuroradiological diagnosis.

This presentation gives an overview over our activities with a focus on Multimodality Image Registration. Image fusion as well as stereotactic surgery planning require a high spatial accuracy which is in contradiction to the inherent geometrical distortions in MR and DSA images. Therefore a procedure for evaluating and correcting geometrical distortions has been developed and applied to MR and DSA images. Beside this several other problems have to be addressed for the fusion of multiple data sets from different modalities [2,3,4].

Distortion Evaluation and Correction

Clinical application like stereotactic neurosurgery planning and image data fusion require a high geometric accuracy of the order of a milimeter. In contrast to that requirement imaging modalities like MR and DSA are not free of image distortions. Scanner related geometric distortions in MRI images are caused by main magnetic field inhomogeneities and gradient field nonlinearities while patient related distortions are caused by chemical shift artefacts, flow artefacts, and tissue susceptibility artefacts [1]. Geometric distortions in DSA images are mainly due to the image intensifier. We have investigated and quantitated the geometrical distortions inherent in MR and DSA image data and designed an approach to minimize distortions or to correct them. In MR images patient induced effects can in principle be minimized by suitable measurements using dedicated MR pulse sequences but most of such approaches require at least a twofold increase of measurement time for the patient. Field mapping methods have been designed in order to perform an in vivo evaluation and quantitation of patient related distortions. It is under investigation what kind of spatial resolution can be achieved for such a method.

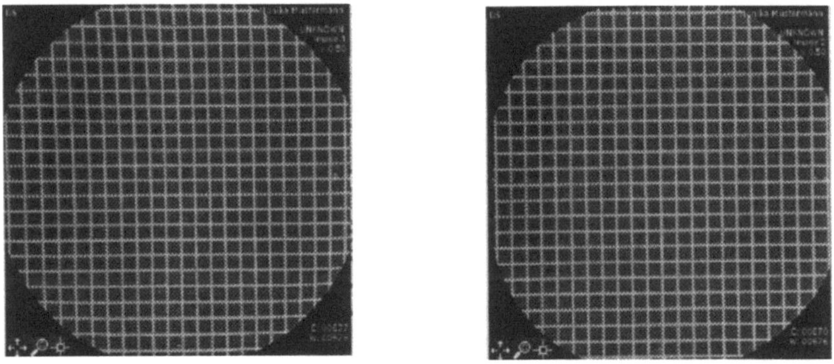

Fig. 1 DSA images of the grid phantom before and after applying the correction procec

Scanner related distortions have been quantified by phantom studies [5]. In MR imaging a rectangular grid phantom of tubes filled with contrast medium is used for evaluation of the spatial distortions. Measurements using this phantom are made at different positions within the magnet to cover the 3D volume of interest. The phantom rods are identified automatically over the series of images. Using the known and the measured positions of the tubes the correction polynomials are calculated. The patient data can than be corrected by applying the polynomials. Since the scanner related image distortions depend on the pulse sequence used both phantom and patient measurements have been carried out using identical pulse sequences and parameters

As for MR, phantom measurements are useful to evaluate and correct DSA image distortions [6]. The main source for the spatial nonlinearity is the pincushion distortion of the image intensifier. The evaluation and correction can be done by using an equidistant rectangular lead grid fixed to the image intensifier. In analogous to MR the deviation between the measured and the actual positions of the crosspoints are used to compute the correction function. The cross points of the lead grid are identified automatically and than used for the correction procedure in the same way as for MR. Fig. 1 shows the DSA images of the grid phantom before and after applying the correction procedure. Additionally, the image distortions caused by the image intensifier may be depend on the position of the intensifier due to changes in the earth magnetic field or other external magnetic fields.

Image Registration

In many clinical situations the fusion of image data from multiple modalities is a prerequisite for an optimum diagnostic and therapeutic procedure. Image data sets acquired from different modalities differ only in size and orientation after applying algorithms for correcting for inherent distortions as mentioned in the previous section. Differences in size and orientations can be treated using only rigid transformations like scaling, translation, and rotation. We consider therefore only rigid registration via a transformation matrix which we derived from common points of reference in the different data sets. Two kinds of reference points have been taken into account:

1. Prospective registration using fiducial markers attached to a stereotactic frame which is rigidly fixed to the patient during data acquisistion.

2. Retrospective registration using internal anatomical landmarks which are interactively detected in the images.

1. Prospective Registration

For prospective registration a stereotactic frame is fixed to the patient's skull. It defines a frame coordinate system for the measurements and serves as a support for surgical instruments during subsequent surgical treatment. During image acquistion localicer markers are mounted onto the base frame. In MR tubes filled with contrast medium are used. The known frame geometry and the marker positions identified in the image data sets are used to compute a global transformation between image and frame coordinate system. To date, only the geometrical data of the BRW stereotactic frame is incorporated in the algorithm. The fiducial

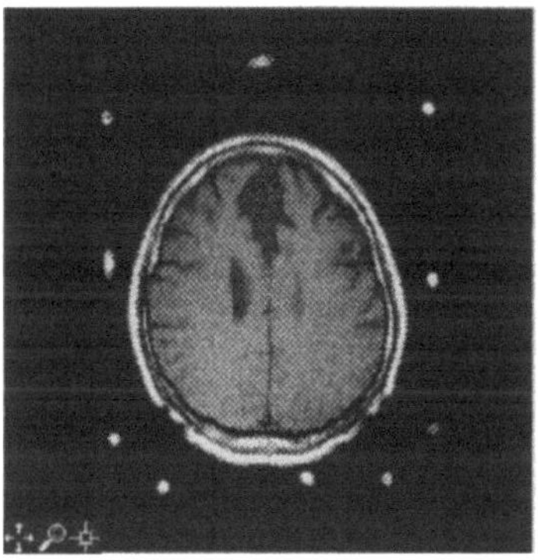

Fig. 2 Axial MR slice of a 3D data set showing the fiducial markers of the BRW frame

markers (Fig. 2) have to be identified in the images with a sufficient accuracy to get precise information about the geometry of the stereotactic frame. The positions of the markers have to identified only approximately. The exact positions of the markers are determined by calculating the center of gravity of each marker. To assist the user in the process of identifying markers in different slices, tools have been developed. The physician identifies the localizer positions only within the first slice, for all adjacent slices the marker positions are traced automatically. The addidtional information from multiple slices improves the accuracy of the calculated transformation. Generally an increased number of markers leads to an improvement of registration accuracy but decrease the clinical acceptance. The registration procedure has been repeated for all images of the same patient using different imaging modalities. The procedure described above results in a common reference coordinate system defined by the stereotactic frame. Merging is eventually done by applying the calculated transformation matrix to the imaging data. Although matching methods based on stereotactic frames may be used with alomost all imaging modalities and yield a high level of accuracy, their application is limited to cases where stereotactic surgery is indicated. This is due to reasons of patient discomfort due to the invasive nature of this procedure.

2. Retrospective Registration

Registration using anatomical structures has the advantage of being non invasively and patient friendly but suffers from the fact that the accuracy is significantly reduced. The definition of internal anatomical landmarks crucially depends on the detectability of the respective structures in the images. Good contrast and sufficient spatial resolution is required in all data sets obtained with different modalities. Internal markers are suited very well if the data sets are registered not successive in time, e.g. for comparison of images acquired pre- and postoperative. The accuracy of the registration process using anatomical landmarks is limited to about 2-3 mm depending on the number and size of suitable landmarks used.

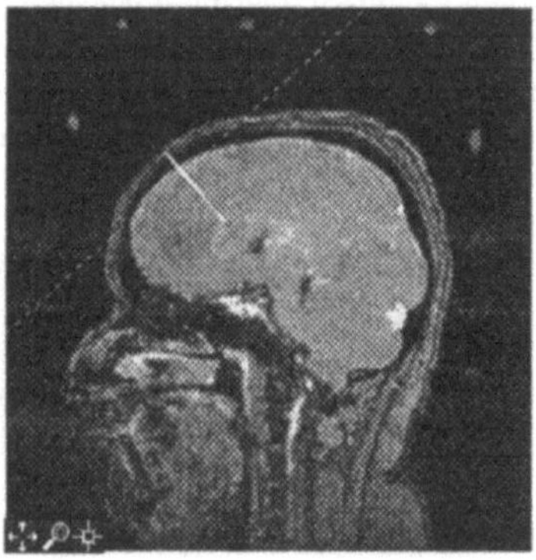

Fig.3 Sagittal MRI image showing an arbitrary trajectory. The slice orthogonal to the trajectory line is indicated as a broken line. The fiducial markers of the BRW frame are also shown

Neuroradiology

In neuroradiological diagnosis the fusion of morphological and vascular information from MRI, MRA, and DSA is of great interest for improving the evaluation of the anatomy in cases of aneurysms, arteriovenous malformations or vascular structure of different tumorous lesions. The registration of MRA and DSA can be done by calculating an MRA image data set into the DSA imaging geometry. This is done in two steps. First, the three dimensional imaging geometry of the DSA device has to be determined by attaching a frame loosely to the patient's head. The frame carries markers which are visible in the DSA image. The known geometry of the marker within the frame and their position in the DSA image allows to calculate the DSA imageing geometry. This includes the distance between the x-ray source and the image intensifier and the position and orientation of the frame relative to the x-ray source. The second step is the identification of anatomical landmarks visible both in MRA and DSA. The transformation of the MRA data into the DSA imaging geometry can than be done by using the calculated imaging geometry and the internal landmarks. .Fig.4 shows an MRA image of the intracranial vessels using the maximum intensity projection (MIP).

In DSA or MRA, a 3D display of the cerebral vascular tree will greatly improve evaluation of the anatomy. The reconstruction of the 3D vessel tree from several 2D DSA projections is currently under development. Generally the 3D representation improves diagnosis. Images acquired or reconstructed under stereoscopic conditions and displaying them on a stereo monitor is a step in this direction.

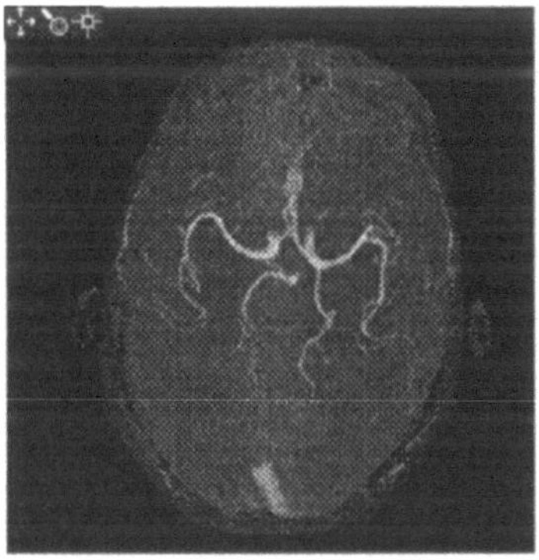

Fig.4 MRA image of the intracranial vessels using maximum intensity projection

Neurosurgery Planning

In stereotactic neurosurgery an intracerebral target is approached with a probe by way of a small burr hole in the skull. The technique is applied for the implantation of depth electrodes to record the electrical activity of different cerebral structures or to carry out tumor biopsies using a stereotactic frame as described before. Target structures and possible trajectories are determined using MRI, MRA, and DSA image data. For trajectory planning the display of 2D slices orthogonal and parallel to the trajectory is implemented to avoid damaging of important cerebral structures or vessels (Fig. 3). Needle trajectories can be integrated in arteriographic data via image fusion. Functional neurosurgery can be improved by matching the anatomical data to a digital stereotactic atlas.

Results and Discussion

Two prototypes of the multimodality application package have been in clinical evaluation since January 1993. The first clinical experiences indicate that a reduction and a simplification of the user interaction are crucial for clinical acceptance of the image registration. A trained user needs about five minutes to identify eight frame markers in a DSA image. Assigning common landmarks in DSA and MRA takes even longer. For clinical routine computer assistance to identify landmarks is necessary to speed up registration by reducing user interaction. Registration using stereotactic frames shows more accurate results due to the automatic identification of frame localizers.

References

1. E. Bellon,E.Haacke, P.Coleman, et.al.: „MR artifacts: A review“ *American Journal of Radiology*, 147, 1271-1281, 1986

2. G.Chen, M.Kessler, S.Pitluck: „Structure Transfer between sets of three-dimensional imaging data“, *Proc. Computer Graphics, National Computer Graphics Association, Dallas, USA* 171-175, 198

3. D.Ney, E.Fishman, L.Dickens: „Interactive multidimensional display of MRI data“, *Journal of Digital Imaging*,3,254-260,1990

4. C.Pelizarri, G.Chen, D.Spelbring, et.al.: Accurate three-dimensional registration of CT,PET and/or MR images of the brain, *Journal Computer Assisted Radiology, 13, 20-26, 1989*

5. F.Schmitt, Correction of geometrical distortions in MR images, H.Lemke (ed), *Computer Assisted Radiology, Berlin, Springer Verlag,* 15-23,1985

6. S.Tehrani, M.LeFree, J.Sitomer, et.al.: High-speed digital radiography pincushion distortion correction using an array processor, *Computers in Cariology, IEEE Computer Society Press,* 615-618, 1986

Innovative Interactive Methods for Image Segmentation

F. FONTANA, A. BONOMI and G. VERNAZZA

Department of Biophysical & Electronic Engineering, University of Genoa
Via all'Opera Pia 11a, I-16145, GENOVA - ITALY

Abstract

Two novel approaches to interactive image segmentation are presented. Both methods aim to extract the boundary of a connected object. Accuracy and real-time processing are the main objectives to be accomplished. The speed requirement is met by a) drastically reducing the user interaction during processing and, b) performing pseudo 3D processing (propagation of information between slices). Accuracy is improved through multilevel segmentation Results are reported, using the COVIRA [1] reference data set.

1. Introduction

In the medical field, the attention of physicians is focused on few details in a scene. In common practice, they delineate objects of interest manually, slice by slice, for a whole spatial sequence. Usually, this procedure requires too long a time and does not ensure accuracy, nor objective decisions. As a consequence, and merely for practical reasons, user-assisted segmentation constitutes a good compromise to fulfil physicians'requirements. At present, there is a growing interest in all techniques that allow the user to perform more reliable and real-time processes.

In this paper, two approaches are presented, which aim to extract object boundaries in a short time and to a good degree of precision. Our methods lie midway between automatic and interactive segmentations. The general idea is to limit as much as possible the use of a-priori knowledge and to devote considerable efforts to increasing the robustness of algorithms by reducing the number of parameters that may influence their performances. In addition, such interactive methods do not require any specific knowledge of technical details.

In the first method, the user is provided with a mouse-driven interface to depict an approximate shape (i.e. a polygon) of the object searched for. By accomplishing an integration of various kinds of information, the algorithm iteratively modifies the initial object shape, trying to adapt it to the actual boundary. When data are available as a set of 2D slices, the contour extracted from one slice (e.g., the middle one) is propagated to the next slice and to the previous one and the process is iterated until a complete segmentation of the whole volume is obtained. A key task is to correlate the information derived from various slices. To this end, we plan to infer 3D information by performing orthogonal cuts, once the contours have been extracted from all slices . The constraint of surface continuity is imposed to detect possible errors incurred during contour extraction from a single slice. Since the processing is performed on orthogonal planes, the speed requirement can also be met; at the same time, 3D information is exploited. The basic idea of this future work is also introduced in this paper.

In the second method, the user is asked to point (by clicking a mouse button) to a part of the object he is interested in. This provides an initial point, namely, a seed, from which the algorithm starts. The subsequent growing process aims to produce a multilevel map, by which a number of possible segmentation solutions can be obtained. In this context, "isoregions" and "isocontours" are defined and ordered according to a given degree of reliability. In this way, the problem of dependence on thresholds and parameters can be overcome. Indeed, as a postprocessing step, the user can quickly move through the map, searching for the most appropriate solution.

The next Section provides brief descriptions of the two methods. In Section 3, preliminary results are reported. Finally, some concluding remarks are made in Section 4.

2. Interactive Methods for Image Segmentation

2.1 Contour extraction from a rough approximation

The method of contour extraction from an approximate shape integrates edges with elastic deformation. An iterative processing is performed that allows an alternative exploitation of edge and shape information.

The initial state is given by the user in the form of an approximate contour (reference contour) of the object shape. Segmentation is performed in some iteration cycles, consisting in local searches for good contour portions and in a global elastic deformation. The first step is edge extraction from the input image, typically by using the Canny's algorithm [2]. Then, a first edge selection is made, using the Chamfer distance [3]: all the edges whose spatial distances from the reference contour are greater than a fixed threshold are not considered in the next processing. Subsequently, we use a radial methodology to analyze all the edges and to select those with the best features. To this end, a new parameter, i.e. "persistency" (previously introduced in [4]), is used, which takes into account edges primitives, like length and strength, and the morphological evolution of an edge w.r.t. the reference contour. The selected edges so obtained constitute a non-competing set because only one edge is chosen for each direction. To provide a closed contour and to eliminate possible errors on the selection based on persistency, a closing operation is then performed, driven by morphological information derived from the starting reference contour. Finally, the method of snake [5] is applied to impose the constraint of contour continuity.

If data are available as a set of 2D data, the contour previously extracted can be propagated to neighbouring slices (Fig. 1) and regarded as the new reference contour. In some way, it predicts the shape of the object contour in the next slice. The greater the distance between slices, the larger the number of cycles required. The process is iterated on all the slices until a complete segmentation is performed. It is worth noting that an iterative processing on 2D planes is much faster than the processing of all 3D data. However, accuracy may be limited by lack of knowledge about 3D spatial information. As future activity, we plan to perform a processing on 2D orthogonal planes (Fig. 2), which allow us to impose continuity constraints and , at the same time, to ensure real-time processing. One can assume that an error has been made, when one detects a discontinuity in the surface of a reconstructed volume (e.g., by using the algorithm suggested by Pavlidis [6]). To correct possible errors, different strategies can be adopted: for example, the snake algorithm can be utilized to force the boundary to pass through the gradient points. Alternatively, the contour responsible for the discontinuity can be recomputed by a backtracking operation, utilizing the edges discarded by previous processes.

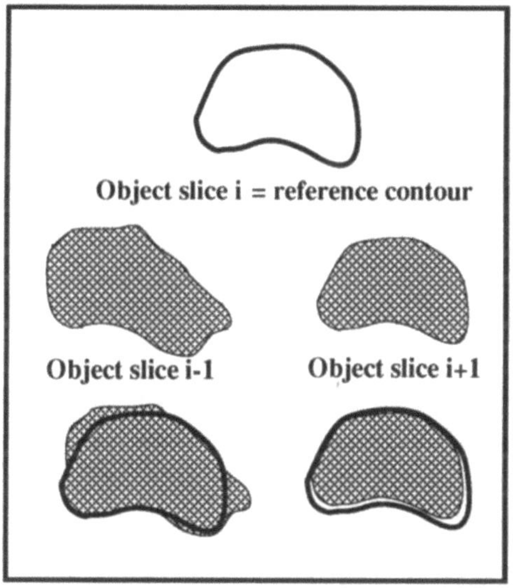

Fig.1-Propagation of 2D contours between slices

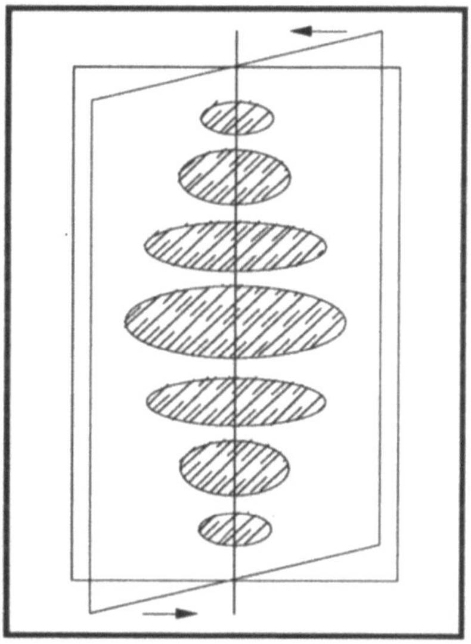
Fig.2- Pseudo-3D processing

2.2 Isocontour Segmentation

Isocontour segmentation utilizes contextual information, both local and global, to grow a region, starting from an internal seed point. The seed point is suggested by the user, at the beginning of the processing session. This point constitutes an "initial state", and is regarded as surely belonging to the object to be segmented.

At each pixel site, a label is extracted that indicates the degree of membership in the region searched for. Since labels take on values between 0 and 1, a region is actually a fuzzy set, R, made up of a kernel (i.e., a seed area around the seed point, characterized by membership values equal to 1) and of surrounding pixels with decreasing membership values.

The result of isocontour segmentation does not establish a univocal correspondence between a region boundary and the contour of the object searched for. It can rather be interpreted as a topological map. Then the highest membership values correspond to points surely belonging to the searched object.

In this context, we originally define *isoregions* and *isocontours* at various levels for the region Reg_i, which will be detected at the end of the processing. Every point characterized by a membership value higher than a fixed value α is said to belong to the "isoregion at level α of Reg_i". The related boundary is then named "isocontour at level α of Reg_i".

The detection of multiple isocontours by $\Delta\alpha$ steps results in a map of possible contours of the searched object. The user can select from the set of isocontours the most appropriate one, according to his own knowledge and purposes, or an automatic mechanism can be applied to make the best choice on the basis of some training or specific information about the object class.

For a detailed description of the method, the reader is referred to [7].

3. Experimental Results

We applied our algorithms to a number of real images. The results reported in this section were achieved by using the reference data set of the COVIRA project. Fig.3 shows the results obtained by applying the contour-extraction algorithm to a CT Image (Fig. 3a). Figs. 3c-3h present the results of two experiments aimed at segmenting the left and right temporal lobes. By exploiting the information derived from the initial polygon depicted by the user, a set of non-competing edges were first extracted (Figs. 3d, 3g). Final results (Fig. 3e and 3h) represent the evolution of the contour after some iteration cycles.

In Fig. 4, isocontour segmentation is applied to a Magnetic Resonance (MR) Image, aimed at segmenting a radiotherapy target. Three isocontours are displayed at different levels of reliability. After an initial evolution, the isocontours relaxes on an equilibrium position (bottom right). Notice that, the greater the membership value, the smaller the isoregion.

4. Conclusion

In this paper, two interactive methods for segmenting single connected objects have been presented. An initial state, provided by the user, is always required, in the form of an approximate shape of the target object (for the first approach) and of a starting point (a seed, in

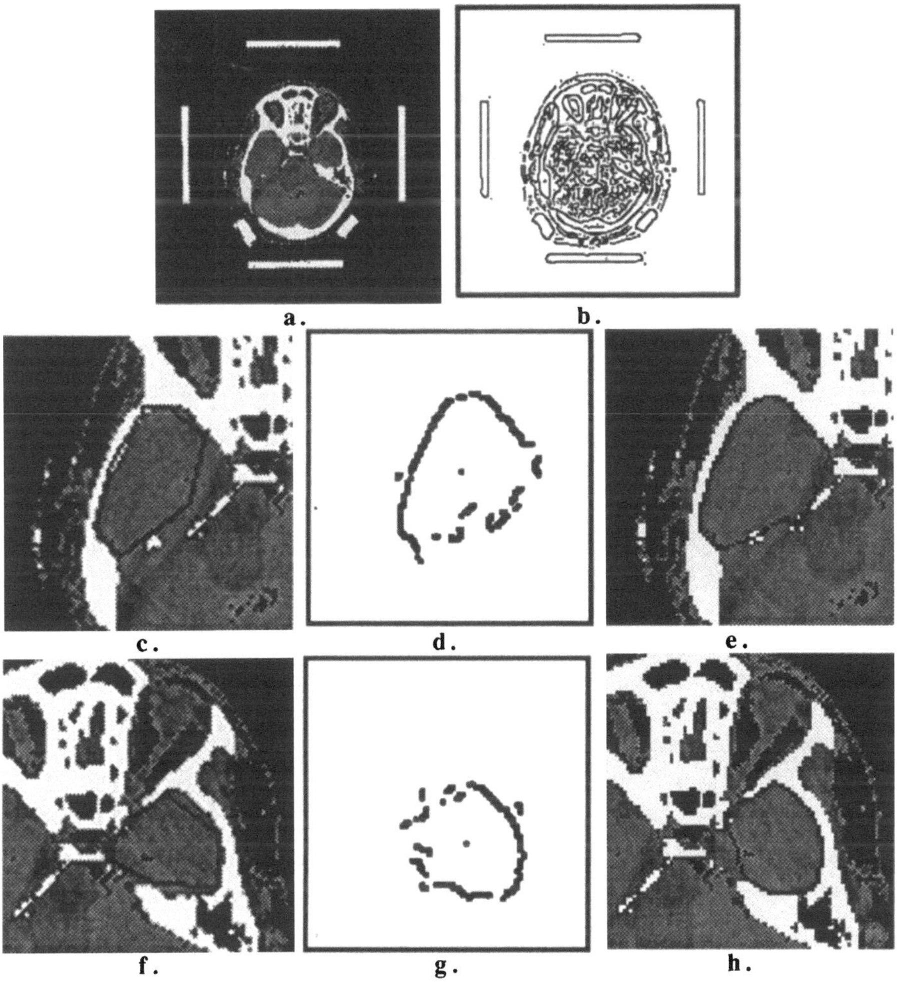

Fig. 3 - Contour extraction from a rough approximation - a) Original CT image - b) Edges extracted by the Canny's algorithm at sigma=0.2 - c) User provided contour approximation of the left lobe - d) Set of non-competing edges, after edge selection step at first iteration- e) Final result, after 5 iterations - f) User provided contour approximation of the right lobe - g) Same as d) - h)Final result, after 7 iterations

fact, for the second approach). Concerning the first method, a preliminary idea (to be developed in the future) has been introduced, which consists in obtaining 3D information by processing on 2D orthogonal planes. Moreover, it has been shown that isocontour segmentation provides a multilevel result, so the user can select the best contour from those presented. One of the most

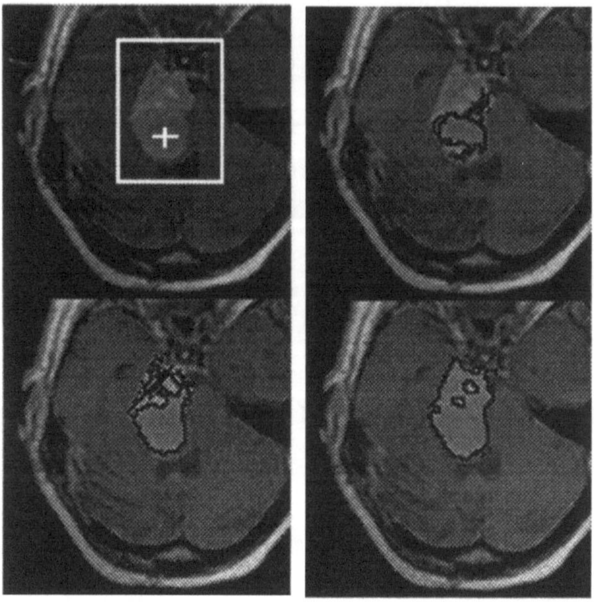

Fig.4 - Isocontour segmentation of a Magnetic Resonance Image. Top left: the area of interest selected by the user for the segmentation of a radiotherapy target; the white cross indicates the seed point. From top right to bottom right: three isocontours at levels 0.97, 0.95, 0.85, respectively.

important aspects of this algorithm is that multilevel segmentation results represent all possible results achievable by changing parameter and threshold values.

Results demonstrating the effectiveness of the proposed methods have been reported, using the reference data set of the COVIRA project.

5. Acknowledgement

This work was supported by the COVIRA project within the framework of W.P. 1.3.4 (Interactive 2D/3D Segmentation).

References

1. COVIRA - Project A2003 of the AIM programme of the European Community, Consortium Partners:
- Philips Medical Systems, Best (NL) and Madrid (E); Corporate Research, Hamburg (D) (prime contractor) - Siemens AG, Erlangen (D) and Munich (D) - IBM UK Scientific Centre, Winchester (UK) - Gregorio Maranon General Hospital, Madrid (E) - University of Tuebingen, Neuroradiology and Theoretical Astrophysics (D) - German Cancer Research Centre, Heidelberg (D) - University of Leuven, Neurosurgery, Radiology and Electrical Engineering (B) - University of Utrecht, Neurosurgery and Computer Vision (NL) - Royal Marsden Hospital/Institute of Cancer Research, Sutton (UK) - National Hospital for Neurology and

Neurosurgery, London (UK) - Foundation of Research and Technology, Crete (GR) - University of Sheffield (UK) - University of Genoa (I) - University of Aachen (D) - University of Hamburg (D) - Federal Institute of Technology, Zurich (CH)
2. J. Canny, "A computational approach to edge detection", IEEE Trans. on PAMI, vol. PAMI-8, n° 6, Nov. 1986.
3. G. Borgefors, "Hierarchical Chamfer Matching: a Parametric Edge Matching Algortihm", IEEE Trans. on PAMI, volPAMI-10, no. 6, Nov. 1988
4.S. Battista, C. Rosa, G. Venturi, A. Righini, "A method for fast segmentation of 3D connected objects", Proceedings of the Symposium for Computer Assisted Radiology, Baltimore, Maryland, pp.702-708, June 1992.
5.M.Kass, A.Witkin, D.Terzopoulos, "Snakes: Active Contour Models", Proc. First Int. Conf. Computer Vision, pp.259-268, 1987
6.T. Pavlidis, S.L.Horowitz, "Pictures segmentation by tree trasversal algorithm", Journal of A.C.M., pp.368-388, April 1976.
7. F. Fontana, S. Dellepiane, "Context-Dependent Region Growing by a Label-Driven Method", Proceedings of 8th Scandinavian Conference on Image Analysis, Tromso, Norway, May 25-28, 1993 (in press).

Interactive Image Segmentation Applied to CT and MR Images

Gordon J. Sivewright, John M. Knapman, Will Dickson and Peter J. Elliott.

IBM UK Scientific Centre, Winchester, UK.

Summary

In radiation treatment planning, it is necessary to mark out selected parts of CT or MR images which are clinically relevant, before designing the treatment plan. This process is usually done manually with a mouse, and for a typical data set of some 40 slices it can be a time consuming and tedious process. Automatic image segmentation techniques have been investigated, but are of little use in this field since it is difficult to incorporate the relevant clinical knowledge. We describe here two methods : Interactive Volume Growing and Hierarchical Probabilistic Segmentation, which assist the clinical user in outlining relevant volumes whilst allowing him full control of the process. Both methods have been integrated into one of the pilot systems developed as part of the COVIRA project.

Interactive Region and Volume Growing

In this section we describe an interactive region growing method which can be used in two modes: 2d region growth applied slice by slice or direct 3d growth applied to a stack of images. Region growth starts from a seed pixel selected using a pointing device such as a mouse. The seed pixel is marked in a work image and the 4 neighbours (or 6 for 3d growth) are added to a neighbour list. Each pixel in the neighbour list is then tested according to the similarity criteria given below. If a neighbour passes the test, that pixel is marked in the work image and any new neighbours are determined and added to the end of the neighbour list. This process is repeated until none of the neighbours in the list pass the similarity criteria or until the region has grown by a predetermined size. Region growth can be constrained so that it does not encroach on existing regions or edges. This is done by marking the results of previous segmentations or edge detection processes [1] into the work image.

A pixel $P_{x,y,z}$ of intensity $I_{x,y,z}$ is included in a region if the following condition is satisfied,

$$| L - I_{x,y,z} | \leq W \qquad (1)$$

where L is a level control parameter which will act as a central value of the growing region, and the window control parameter W is a threshold distance in pixel intensity units. The level, L, and window, W, control parameters for the region growing algorithm can be computed automatically by using the statistics of a small sample region, S, of about 20 pixels in size which is located centrally over the seed point of the region. The level control parameter is set equal to the mean of the sample region. The window threshold parameter, W, is computed by multiplying the standard deviation of the sample region, $S_{s.d.}$, with a given scaling factor, C, which is dependent on the contrast to noise ratio in the image. A value for C of 2.0 has been found to give reasonable results for CT and MR images.

The values of the parameters W and L are displayed using two slider bars forming part of a GUI control panel for the algorithm. These sliders can be operated independently in manual or automatic mode according to the setting of two toggle buttons. In automatic mode, the values of L and W are computed from the sample region as explained above and displayed using the slider bars. In manual mode, adjustment by the user of either slider bar causes the algorithm to be restarted using the new parameter values and the previous seed location. In this way, the extra step of pointing at the image to specify the seed location is avoided and a range of results can be quickly observed simply by adjusting the sliders. This feature is possible because region growth is very fast, (typically less than a second) and there is little overhead in restarting growth from scratch many times.

2d region growth can be extended into neighbouring slices by pushing an up or down arrow icon button. By successively repeating this procedure, a connected 3d volume can be computed. The user may identify 2d disconnected regions in one slice that will merge in further slices by clicking with the mouse. To transfer a set of regions from slice 1 to slice 2, all pixels lying within the interior of the marked regions in slice 1 are formed into a new list which is then assigned as a list of neighbouring pixels of a set of potential regions in slice 2. Region growth then proceeds in slice 2 exactly as before by testing each member of the new neighbour list. This approach allows the shape of the volume to be checked slice by slice but involves more user interaction than the direct 3d approach.

Upon completion of region growth, the boundaries are computed by tracking around regions marked in a work image. Some preprocessing is done to reduce the number of small spurious boundaries that are detected. This consists of searching out and filling in holes in the region that are 2 pixels thick or less. In 3d growth mode an option is provided to allow volume growth to be restarted from the existing boundaries. This feature is useful because it avoids the extra time needed to regrow a large volume.

Performance of the Algorithm

The speed of the algorithm is dependent on the size in pixels/voxels of the region/volume to be grown. Region growth usually takes less than 1 second and so can be used truly interactively. A brain volume, occupying a bounding box of approximately 120x100x35 voxels can be grown and all inner and outer boundaries computed and displayed in about 25 seconds on an IBM RISC System/6000 Model 320H, using unoptimised code.

The algorithm has been tested on unsmoothed 3d CT image data and was able to segment eyes, sinuses, brain, ventricles and the skull. These results have been visualized in 3d and appear satisfactory. Region growing works well in segmenting 2d regions in MR data which exhibit high contrast. The brainstem can be segmented in some slices, in particular where there is contrast provided by the surrounding ventricular CSF. However segmentation generally fails in slices where there are direct connections to structures with similar intensities, such as the connections to the cerebellum via the pons. In these slices there is often no visible border and anatomical knowledge must be applied to define a suitable boundary location.

Hierarchical and Probabilistic Segmentation (HPS)

The HPS system consists of two image processing stages: an offline preprocessing stage, which constructs a hierarchy of possible segmentations from an image plus the relevant calibration data; and an interactive stage where the user moves through the hierarchy of segmentations and identifies the regions and contours which are required for the current task. A more complete description of the method can be found in [2].

Preprocessing Stage

The preprocessing stage consists of the following operations applied to the raw image data. Edge points are detected as points of high contrast and a calibrated probability is assigned to each edge point. The edge detection algorithm used in HPS is the well-established algorithm described by Canny in [1]. This algorithm can be executed on a

256 × 256 image in one or two seconds on a modern workstation, and was shown in [3] to recover 85 per cent of the edges marked manually by a team of clinical experts. The usual hysteresis stage of the algorithm has been modified to assign probabilities to the resulting edge points.

While the Canny edge detector is very successful in finding extended edges, its performance is poor at corners or other points of high curvature; thus some form of gap closing is required to provide a set of closed contours. A relatively simple but effective straight line interpolation algorithm is used to join edge terminators to each other or to other edge points.

Once we have the set of closed contours we can label the regions using a simple flooding procedure, as well as labelling all the points which separate a pair of regions as belonging to the same edge. For each region we determine the mean and variance; then for each pair of regions separated by an edge a variance statistic is calculated. The results of the edge detection and gap closing are combined with this variance statistic using Baysean methods to assign a combined probability to each edge.

A hierarchy of segmentations is built up by successively merging the regions either side of the weakest remaining edge and recalculating the probabilities of the neighbouring edges. The resulting hierarchy is encoded in an image where the edges have values between 1 and 100: the percentile edge image. The percentile values are arranged so that each value is assigned to an equal number of edges. The percentile image is the only output from the preprocessing stage.

Interactive Stage

Much of the power and usability of the HPS system derives from the effective use of the segmentation hierarchy in an interactive system. In the interactive stage, the edges from the percentile image are presented to the user, overlaid on the original image. The user can dynamically move up or down the segmentation hierarchy by changing a threshold value, below which edges from the percentile image are not displayed. Figure 1 shows the percentile image as displayed at different thresholds.

Once the best level in the hierarchy has been reached, the user can select the required contours with only moderate recourse to hand/eye coordination and with predictable results. Regions can be selected and merged at a particular level in the hierarchy; they

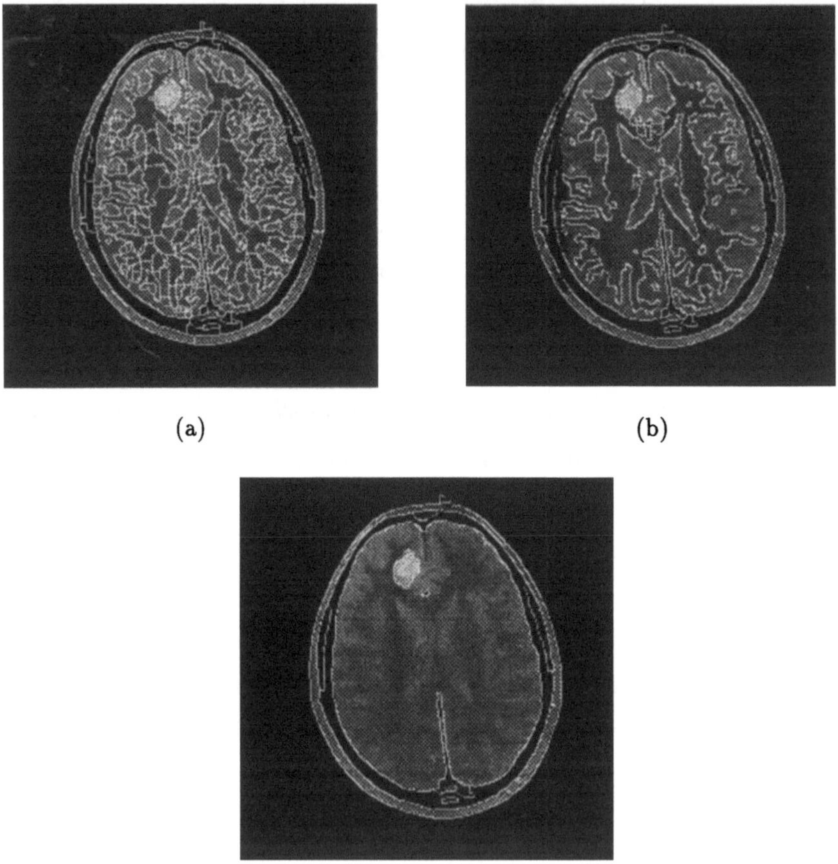

(a) (b) (c)

Figure 1: Percentile edges at the lowest threshold (a), at an intermediate threshold (b) and at a high threshold (c).

then remain frozen and are not affected by the choice of percentile threshold. The user also has the option of manually adding edges where they have not been detected by the preprocessing.

Conclusions

We have described two alternative methods for outlining features of interest in image data. In general the region growing methods will work for those subclasses of problems where there is a sufficient contrast to noise ratio to discriminate between adjacent anatomical structures. The method is fast, requires little user interaction and can operate directly in

3d. The HPS method allows the user to rapidly and easily traverse an entire segmentation hierarchy. In the accurate determination of complex contours this system offers useful results with considerably lower levels of visual concentration than required by manual outlining systems.

Acknowledgements

This work is part of COVIRA (COmputer VIsion in RAdiology), project A2003 of the AIM (Advanced Informatics in Medicine) programme of the European Commission.

Participants in the COVIRA consortium are: Philips Medical Systems, Best (NL) (prime contractor) and Madrid (E); Philips Corporate Research, Hamburg (D); Siemens AG, Erlangen (D) and Munich (D); IBM UK Scientific Centre, Winchester (UK); Gregorio Maranon General Hospital, Madrid (E); University of Tuebingen, Neuroradiology and Theoretical Astrophysics (D); German Cancer Research Centre, Heidelberg (D); University of Leuven, Neurosurgery, Radiology and Electrical Engineering (B); University of Utrecht, Neurosurgery and Computer Vision (NL); Royal Marsden Hospital/Institute of Cancer Research, Sutton (UK); National Hospital for Neurology and Neurosurgery, London (UK); Foundation of Research and Technology, Crete (GR); University of Sheffield (UK); University of Genoa (I); University of Aachen (D); University of Hamburg (D); Federal Institute of Technology, Zurich (CH).

References

[1] J.F. Canny. A Computational Approach to Edge Detection. *IEEE Trans. Pattern Analysis Machine Intell.*, 6:679–698, 1986.

[2] J. Knapman and W. Dickson. Hierarchical Probabilistic Image Segmentation. *Image and Vision Computing (submitted)*, 1993.

[3] G.J. Brelstaff, M.C. Ibison, and P.J. Elliott. Edge-Region Integration for Segmentation of MR Images. In *Proc. British Machine Vision Conference*, pages 139–144, Oxford, England, September 1990.

Extraction of Closed Boundaries from Fragmented Edge Maps Using Shape-oriented Grouping Procedures

R. Ogniewicz, T. Vehkomäki and G. Székely

IKT/Image Science Laboratory
ETH–Zurich, CH–8092 Zurich, Switzerland

Summary

Segmentation techniques which employ edge detection suffer from the fact that the generated edge maps usually do not represent connected boundaries. However, in order to perform a segmentation of the image one needs image regions defined by closed contours. We present an approach to segment 2D radiological images (MR and CT slices) based on the Canny edge detector followed by a gap closing operation which exploits the geometrical characteristics of the Voronoi tessellation of the edge points. Thereafter, closed contours are extracted by global path optimization based on a dynamic programming technique.

1 Introduction

Unfortunately, common edge detectors do not always produce closed contours. Since the extraction of image regions requires closed boundaries, a mechanism to deal with fragmented edge maps is mandatory. Our contribution, a segmentation tool for gray-valued 2D images, presently provides two methods to handle disjoint edge fragments:

- *perceptual grouping of edge fragments*: disjoint edge fragments are grouped according to criteria such as collinearity or local density anisotropy of the edge points. Additional attributes such as local edge orientation assist the grouping process. The behavior of the grouping can be interactively controlled by means of a threshold parameter.
- *user intervention*: links between edge fragments can be inserted or deleted interactively.

After the necessary modifications of the edge map have been achieved by a (possibly iterative) combination of the above two methods, the user has to provide a starting location for a subsequent boundary tracking algorithm. This algorithm traverses the edge map and extracts a single optimal closed boundary of the region indicated.

Section 2 sketches the theoretical background of our perceptual grouping concept. Section 3 presents a brief description of the segmentation tool. The segmentation procedure is illustrated by an example in Section 4. In Section 5 plans for future development of the present algorithms are outlined.

2 Perceptual grouping of dot patterns

Many perceptual grouping methods published so far ([3], [8], [5]) share a common philosophy in determining which points of a dot pattern to connect. First, a pair of points is chosen according to the underlying neighborhood criterion. Second, a clustering criterion is applied involving additional neighboring points. Finally, the decision to link or not to link is made upon examination of the intrinsic characteristics of the neighborhood, some global minimization considerations, or the result of applying a parametric threshold value. In [1], each dot is assigned a label expressing the probability that the dot belongs to a specific category (e.g., border or non-border). A relaxation labeling technique is then employed to obtain locally consistent interpretations.

However, the *resulting* connections are typically regarded as *'hard' links*, i.e., a pair of points is either connected or disconnected, no intermediate degrees of connectivity are allowed. Instead of establishing binary connections between neighboring dots, we postulate the assignment of one or more values of belief to potential links such that the required flexibility is retained. Such 'vectors' of belief or *fuzzy links* naturally rank-order the possible connections.

The usage of the continuous MAT as shape descriptor was proposed by Blum [2]. Due to the semilocal nature of the MAT changes of its originating elements have only a local effect. This type of a shape descriptor introduces the possibility of generating hypotheses about possible continuations and completions of the boundary, e.g., according to the Gestalt laws [10]. A recent implementation of the MAT in [4] relies on the close relationship between the MAT of a (discrete) set of boundary points and its Voronoi diagram (VD) [6]. The implementation in [4] introduced not only Euclidean metrics but also a regularization method to both make the MAT immune to noise or discretization effects induced by affine transformations and to guarantee correct topology.

2.1 Fuzzy links

In order to retain the local characteristics of potential links between (edge) points we devise a *fuzzy link* $\mathcal{F} = \{\mathcal{F}_1, \mathcal{F}_2, \mathcal{F}_3\}$, which consists of three components and which is computed and attached to each pair of neighboring points (the neighborhood of a point is defined by its circumscribing Voronoi polygon). This approach is equivalent to first generating the straight line dual of the VD, the so-called Delaunay graph (DeG) and than attributing each edge of the triangulation with a set of features. The first component of $\mathcal{F}$ embodies the property of *good continuation* and exploits a feature of the MAT. The second reflects the local *density distribution* of the points. The last component refers to the local *convex/concave deviation* from an otherwise smooth course of the boundary.

2.1.1 Good continuation

According to its definition, every branch of the MAT indicates the direction of a significant flow of matter resulting from the propagation of the wavefront. In Fig. 1(a), let P_i be an arbitrary point of the hypothesized discrete boundary. Then, a first approximation of the

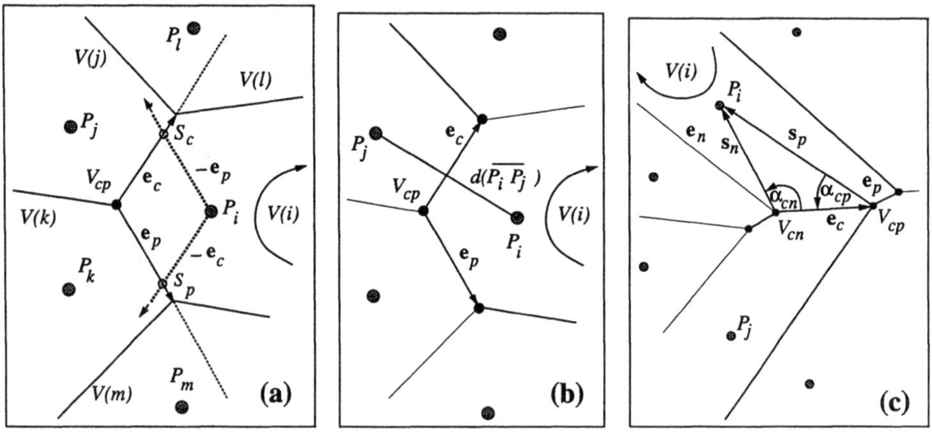

Figure 1: Measures for the fuzzy links (a) $\mathcal{F}_1$: *'Good Continuation'.* (b) $\mathcal{F}_2$: *'Local Density Anisotropy'.* (c) $\mathcal{F}_3$: *'Potential Boundary Deviation'.*

continuation of the boundary in P_i or P_j will be the projection of the MAT (here equivalent to Voronoi edge $\mathbf{e}_c$), namely a segment of the bisector of the straight line connection $\overline{P_i\,P_j}$.

Each point (P_i to P_m) is assigned its proximal (convex) Voronoi polygon ($V(i)$ to $V(m)$). The edges of the polygons are scanned in a clockwise fashion. Let us assume that the scan of $V(i)$ has proceeded up to edge $\mathbf{e}_c$, with $\mathbf{e}_p$ being the edge encountered immediately before. Their respective direction is established such that they point away from the common vertex V_{cp}. Edge $\mathbf{e}_p$ is then used to estimate a possible continuation of a perceptual boundary in P_i, namely the ray emanating from P_i and direction $-\mathbf{e}_p$. Next, the intersection S_c of the ray with the prolongation of $\mathbf{e}_c$ is determined. Finally, $\mathcal{F}_1$ is computed according to the following formula ($d(A\,B)$ denotes the Euclidean distance between points A and B):

$$\mathcal{F}_1(P_i, V_{cp}, \mathbf{e}_c, \mathbf{e}_p) = \frac{|\mathbf{e}_c|}{d(S_c\,V_{cp})}. \tag{1}$$

A high value of $\mathcal{F}_1$ indicates a high likelihood of a connection between P_i and P_j.

2.1.2 Density distribution

The second component $\mathcal{F}_2$ reflects the local density distribution of the points (local density anisotropy) as exemplified in Fig. 1(b). The length $|\mathbf{e}_c|$ is an estimator for the density of points along $\mathbf{e}_c$. Thus P_i and P_j should be rather connected for larger values of

$$\mathcal{F}_2(P_i, P_j, \mathbf{e}_c) = \frac{|\mathbf{e}_c|}{d(P_i\,P_j)}. \tag{2}$$

2.1.3 Local deviation

The third component $\mathcal{F}_3$ introduces a modification to the criterion proposed by O'Callaghan [3] (Fig. 1(c)). In order to establish a connection between P_i and P_j the original criterion requires the straight line segment $\overline{P_iP_j}$ to intersect edge $\mathbf{e}_c$. Hence, two points will be connected if the common edge of their respective zones of influence is close enough to their straight line connection. Our approach defines a value $\mathcal{F}_3$ which is more general but includes the original criterion: For each vertex (V_{cn} and V_{cp}) of the edge $\mathbf{e}_c$ we compute the cosines of angles α_{cn} and α_{cp} between $\mathbf{e}_c$ and $\mathbf{s}_n$ from V_{cn} to P_i or, respectively, between $\mathbf{e}_c$ and $\mathbf{s}_p$ from V_{cp} to P_i. $\mathcal{F}_3$ is then assigned the minimum value of the angles according to

$$\mathcal{F}_3(P_i, V_{cn}, V_{cp}, \mathbf{e}_c) = \min \left(\underbrace{\frac{\overrightarrow{V_{cn}P_i} \cdot \mathbf{e}_c}{|\mathbf{e}_c| \cdot |\overrightarrow{V_{cn}P_i}|}}_{\cos\alpha_{cn}}, \underbrace{\frac{-\overrightarrow{V_{cp}P_i} \cdot \mathbf{e}_c}{|\mathbf{e}_c| \cdot |\overrightarrow{V_{cp}P_i}|}}_{\cos\alpha_{cp}} \right). \tag{3}$$

The equivalent of O'Callaghan's criterion is then: connect P_i and P_j if $\mathcal{F}_3 > 0$. O'Callaghan's approach eventually leads to the so-called Gabriel graph.

3 Segmentation by perceptual grouping

Our segmentation package is composed of the following functional units:

- *edge detection* returning a binary raster image of edge pixels, which are optionally attributed different edge properties (local edge strength, local edge orientation etc.).
- *raster-to-vector transform* of the edge map resulting in a graph-like structure of the edge fragments, the 'edge graph'.
- *computation of the fuzzy link values.* These values are combined with additional edge properties and incorporated into the edge graph. The result is an enhanced edge graph, i.e., an edge graph containing auxiliary link information for each edge point.
- *manual editing* of the enhanced edge graph.
- *extraction* of closed contour outlines from the graph.

3.1 Computation of edge graph and fuzzy links

Edge detection was accomplished by means of the Canny operator, followed by hysteresis thresholding. Our algorithms do not, however, require a particular edge filter, so that the Canny operator could be replaced by any other edge detection scheme.

The output of the Canny operator are thinned strings of edge pixels which are attributed their local contrast and orientation. Since the subsequent processing steps require the edge map to be represented as a list structure rather than as uncorrelated edge pixels, the binary edge map is transformed into a graph-like structure, the edge graph.

Next, the Voronoi diagram and the Delaunay graph of the edge points is generated. In the sequel, the Delaunay graph is merged with the edge graph. Thereafter, the fuzzy links are evaluated according to Section 2.1. These values and further edge point properties such as local contrast, the orientation of the edge elements and the Euclidean distance to the closest endpoint of the current contour fragment are combined into a single link strength $\mathcal{L}$, which attributes the corresponding edge of the edge graph. In the current version, $\mathcal{L}$ turns out to be a weighted sum of fuzzy link components and optional edge attributes. This link strength can be used to extract a subset of the most promising connections just by comparing each $\mathcal{L}$ value to a user-defined threshold.

Since the extraction of salient links just by mere thresholding will hardly result in perfect gap closing, the user must be allowed to correct the results of the previous steps manually. According to the user's interaction, unwanted contour fragments or fuzzy links will be deleted and new connections will be established.

3.2 Contour tracking

The final extraction of closed contours is triggered by the user who selects a seed point within the image. Starting from the closest contour fragment, a contour tracking algorithm traverses the graph structure searching for closed loops. Typically, several closed contours can be found from a single starting point, which calls for optimization methods to select the most salient boundary.

Grid-oriented relaxation techniques for global path optimization have been proposed by Udupa [9] and Sha'ashua [7]. Udupa's dynamic programming method requires that the likelihood of the existence of a path from a pixel to each of its neighbors is evaluated first. Such a local likelihood measure can be derived from the edge strength. In the following optimization step the best route between two given points is searched by maximizing the cumulative likelihood measure of the whole path. In Sha'ashua's method the measure optimized is the length and the smoothness of the potential contour. This optimization does not only return a single path but a saliency map for the whole image. The obvious advantage of both techniques is the incorporation of rather global criteria into the contour extraction process.

We have adapted the dynamic programming scheme to our symbolic edge representation. In our approach, the likelihood of a connection between disjoint edge fragments is defined by the respective fuzzy link weight. Because of the smaller number of possible connections, the optimization phase is remarkably faster compared to the formerly proposed grid-oriented methods.

4 Example

As an example, we segmented the gray-valued image of an axial brain slice. The image is a linear combination of the two echoes of a double spin echo MRI brain slice. The goal was the detection of the brain and skin outline and the segmentation of the large brain tumor. The upper row of Fig. 2 depicts the edge map computed with the Canny filter (left) and the (most salient) additional connections of the attributed Delaunay graph (right).

While the contours of the brain (Fig. 2, middle left image) could be extracted without the computation of fuzzy links (the contours are already connected in the original edge map), the successful extraction of the skin outline (Fig. 2, middle right image) required link information to bridge gaps. The extraction of the tumor is more difficult. Because of the very small distance between the tumor boundary and the ventricular system, the Canny operator does not detect the corresponding line edge. This portion of the edge map had to be edited manually as shown in the bottom left part of Fig. 2. Thereafter, the contours of the tumor could be tracked successfully (Fig. 2, bottom right).

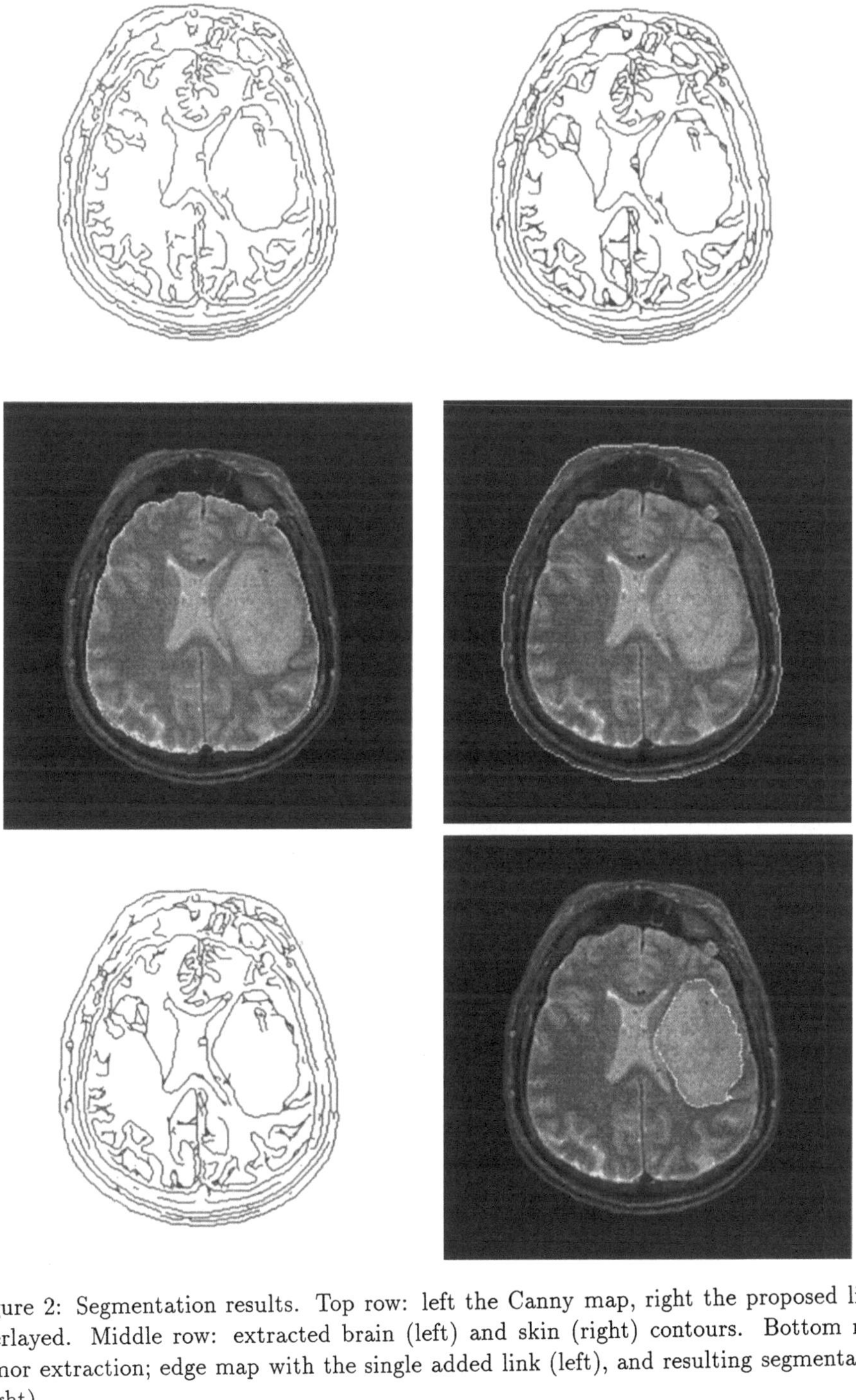

Figure 2: Segmentation results. Top row: left the Canny map, right the proposed links overlayed. Middle row: extracted brain (left) and skin (right) contours. Bottom row: tumor extraction; edge map with the single added link (left), and resulting segmentation (right).

5 Outlook

The current version of our segmentation package provides all the functionality needed to perform the segmentation of 2D gray-valued images. However, some of the algorithms used in this first prototype are very simple, so that their applicability in a clinical environment would be limited. In order to obtain a more robust segmentation, the evaluation of the link strength should exploit further image features, such as the gray-level evidence in the original image (edge strength in the neighborhood of a link, iso-intensity contours, etc.) or (if 3D data sets are to be segmented slice by slice) information about successfully segmented neighboring slices.

Acknowledgements

Part of this research work was funded by COVIRA (Computer Vision in Radiology), project A2003 of the AIM (Advanced Informatics in Medicine) programme of the European Commission. Participants in the COVIRA consortium are: Philips Medical Systems, Best (NL) and Madrid (E), Corporate Research, Hamburg (D) (prime contractor); Siemens AG, Erlangen (D) and Munich (D); IBM UK Scientific Centre, Winchester (UK); Gregorio Maranon General Hospital, Madrid (E); University of Tübingen, Neuroradiology and Theoretical Astrophysics (D); German Cancer Research Centre, Heidelberg (D); University of Leuven, Neurosurgery, Radiology and Electrical Engineering (B); University of Utrecht, Neurosurgery and Computer Vision (NL); Royal Marsden Hospital/Institute of Cancer Research, Sutton (UK); National Hospital for Neurology and Neurosurgery, London (UK); Foundation of Research and Technology, Crete (GR); University of Sheffield (UK); University of Genoa (I); University of Aachen (D); University of Hamburg (D); Swiss Federal Institute of Technology, Zurich (CH)

References

[1] AHUJA, N. and TUCERYAN, M. Extraction of early perceptual structure in dot patterns: Integrating region, boundary, and component gestalt. *Computer Vision, Graphics, and Image Processing*, 48:304–356, 1989.

[2] BLUM, H. A transformation for extracting new descriptors of shape. In WATHEN-DUNN, W., editor, *Models for the Perception of Speech and Visual Form.* MIT Press, Cambridge MA, 1967.

[3] O'CALLAGHAN, F. Computing the perceptual boundaries of dot patterns. *Computer Graphics and Image Processing*, 3:141–162, 1974.

[4] OGNIEWICZ, R. and ILG, M. Voronoi skeletons: Theory and applications. In *Proceedings Conf. on Computer Vision and Pattern Recognition, Champaign, Illinois*, pages 63–69, June 1992.

[5] O'ROURKE, J., BOOTH, H., and WASHINGTON, R. Connect-the-dots: A new heuristics. *Computer Vision, Graphics, and Image Processing*, 39:258–266, 1987.

[6] PREPARATA, F. P. and SHAMOS, M. I. *Computational Geometry.* Texts and Monographs in Computer Science. Springer-Verlag, New York, second edition, 1990.

[7] SHA'ASHUA, A. and ULLMAN, S. The detection of globally salient structures using a locally connected network. In *ICCV*, pages 321–327, 1988.

[8] TOUSSAINT, G. T. Pattern recognition and geometrical complexity. In *Proc. 5th Int. Conf. Pattern Recognition, Miami Beach, FL*, pages 1324–1347, 1980.

[9] UDUPA, J. K., SAMARASEKERA, S., and BARRETT, W. A. Boundary detection via dynamic programming. In *Proceedings 2th Conf. on Visualization in Biomedical Computing, Chapel Hill, NC*, pages 33–39, October 1992.

[10] WERTHEIMER, M. Untersuchungen zur Lehre von der Gestalt II. *Psychol. Forsch.*, 4, 1923. Translated as "Principles of perceptual organization" in *Readings in Perception*, David Beardslee and Michael Wertheimer, Eds., (Princeton, N.J.: 1958), pages 115–135.

Organ Segmentation by Means of Neural Networks

M. Rucci and E. Lorello

ARTS Lab
Scuola Superiore S.Anna Pisa
Italy

Abstract

In Medical Imaging, segmentation of organs and structures of interest is the basis for a large number of operations such as three-dimensional analysis, surgery planning and automated diagnosis. Unfortunately, image noise and the variability of biological structures contribute to make segmentation a very hard task. In order to achieve this goal, a proper fusion of information extracted from the image (bottom-up processing) with *a priori* knowledge of the scene in terms of models of interesting structures (top-down processing) is extremely important.
In this paper we propose a neural network based approach to the segmentation of structures of interest, which integrates bottom-up and top-down processing. The system includes four modules for each examined organ: a Focusing Module (FM) locates the organ of interest in the input image and the edges of the selected area are evaluated by a Low-Level Segmenting Module (LLSM). A High-Level Segmenting Module (HLSM) receives the output of the LLSM and gives a memorized generalized shape. Finally, an Organ Extracting Module (OEM) merges the high-level and the low-level data, producing the final organ segmentation. Both the FM and the OEM include feed-forward nets trained with the back-propagation algorithm, while the HLSM is an associative self-organized memory. The proposed approach has been applied to Computed Tomography and Magnetic Resonance images.

1. Introduction

With the term *organ segmentation* we mean the identification of the pixels of the image which belong to a considered organ. Organ segmentation is a higher level process than image segmentation as it is commonly intended by computer vision researchers [1], due to the fact that it requires the understanding of the considered image, at least as concerns the organ to extract. It is far too evident that organ segmentation can play a major role in medical imaging, being the basis for more general applications like three-dimensional analysis, surgery planning and automated diagnosis.

Unfortunately, the extraction of structures of interest is a very hard task, mainly due to the high level of noise which is typical of medical images, and to the large biological variability. In general, due to these factors and to the complexity of biological structures, organ segmentation cannot be accomplished by means of standard segmentation methods based on similarity criteria of adjacent pixels [2]. On the contrary, we believe that a proper fusion of information extracted from the image (bottom-up processing) with *a priori* knowledge of the scene in terms of models of interesting structures (top-down processing) is extremely important [3].

In this respect Artificial Neural Networks (ANNs) techniques [4] [5] [6] can be suitable implementing tools. In fact, thanks to the capabilities of some ANNs paradigms of learning on the basis of examples, it is possible to include in the networks in a distributed form a high level knowledge which can be used as a guide in the segmentation process.

In this paper we propose a neural network based-approach to the problem of organ segmentation, which integrates bottom-up and top-down processing. The proposed system architecture is described in detail in section 2, and the results obtained by applying it to a specific segmentation case are illustrated in section 3.

2. System Architecture

The system architecture that we propose is characterized by a high modularity degree and it includes several modules for each considered organ. A general scheme of the architecture illustrating the modules dedicated to a specific organ is shown in Figure 1.

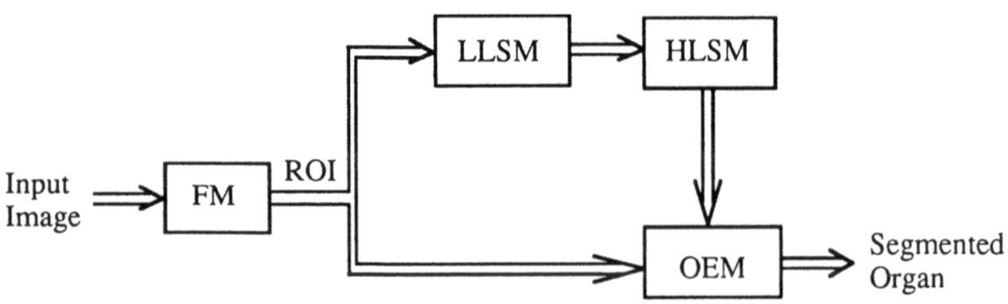

Fig. 1: The proposed system architecture: only the modules dedicated to a single organ are shown.

As depicted in the figure, for each organ to segment the system includes a Focusing Module (FM), a Low Level Segmenting Module (LLSM), a High Level Segmenting Module (HLSM) and an Organ Extracting Module (OEM).

When an image is presented as input to the system, each FM finds its organ of interest by producing an output Region Of Interest (ROI), which is a window of the input image centered on the organ. The ROI is then accepted as input by both the corresponding LLSM and OEM. The LLSM performs a bottom-up segmentation of the ROI by applying standard computer vision segmenting algorithm. In the actual version of the system the LLSM carries out an edge extraction by means of a gradient operator and it produces an output binary image by using a suitable threshold. The output of the

LLSM is a rough organ segmentation and is then refined by the HLSM, which acts as an hetero-associative memory associating to the rough organ segmentation a generalized shape of the organ, which has been included in the module by means of a training session. Finally, the OEM integrates low-level image information present in the ROI with the high-level generalized organ shape, in order to produce the output segmented organ.

All the modules, excepted the LLSM, are based on neural networks paradigms. The architecture of the FM is shown in Fig.2.

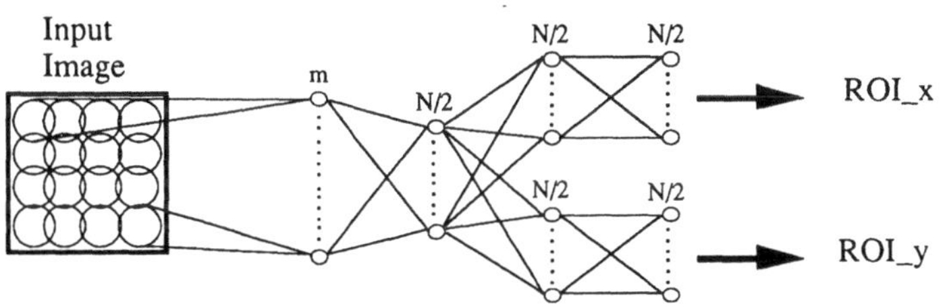

Fig. 2: Scheme of the Focusing Module.

It includes a feed-forward five-layered net which has been trained with the backpropagation algorithm [6]. In the first layer weights are *a priori* set so as to carry out a gaussian decimation of the image. In this way the units of the second layer receive as input a lower resolution image which is processed by the following layers in order to locate the position of the ROI. ROI dimensions are adjusted on the basis of the considered organ.

A general scheme of the HLSM is illustrated in Fig.3.

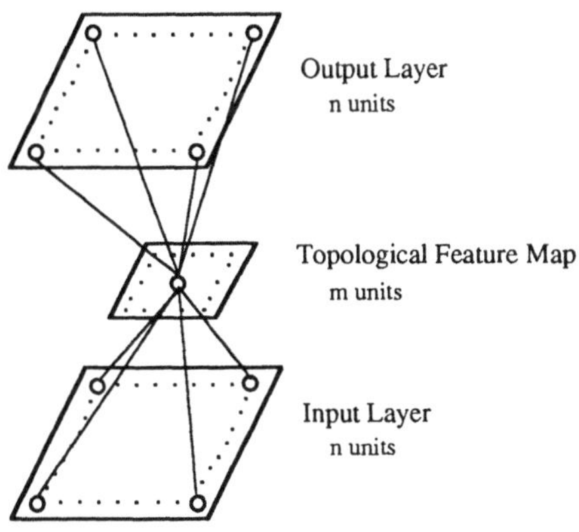

Fig. 3: The High Level Segmenting Module.

The HLSM is basically a self-organized associative memory based on a topological self-organizing feature map [4] [5]. As shown in the figure, it includes a feed-forward net which is composed of three full-connected layers of units. The self-organized map is the central layer and it is trained by means of "ideal" edge segmented images, that is images where the boundaries of the organ of interest were manually extracted by the physician. The third layer has been trained so as to associate to a generalized edge segmented image a corresponding region segmented image, where all the pixels which belong to the organ are set to 255. When a rough organ edge segmentation (such as the one produced by the LLSM) is presented to the HLSM, the most similar generalized edge segmented image stored in the map is recalled and an output generalized organ segmentation is produced. This output can be seen as the protypical generalized shape of the organ and can be used as a guide in the organ segmentation.

The final segmentation is carried out by the OEM, which, as shown in Fig.4, processes at the same time the ROI and the generalized shape produced by the HLSM.

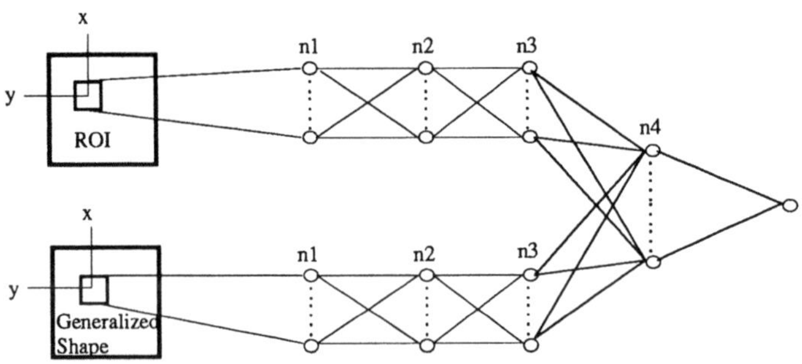

Fig. 4: The Organ Extracting Module.

The OEM includes a feed-forward net composed of four layers of weights. Also this net has been trained with the back-propagation algorithm. As can be observed in Fig.4, the input to the net are the pixels included in two masks having the same positions on the ROI and on the generalized shape image. The network is structured in two branches so as to first process independently the two input images. The results are then merged in the last two layers in order to determine whether the central pixel of the mask belongs to the considered organ. The two masks are shifted on the image as a filtering operation, and each pixel is thus classified. During the training session segmented images produced manually by the physician were used as reference outputs.

3. Results and Conclusions

As an example of application of the system, in this section we describe the results obtained in the segmentation of the spinal cord in Computed Tomography images.

Both the HLSM and the OEM were trained on a training set composed of 25 images of 6 patients. Other 17 images were used for testing the system. As concerns networks parameters we have for the FM, m=256 and N=32, the parameters of HLSM where n=1024, m=25 (see Fig.3), and the OEM was composed of n_1=25 (5x5) input units, n_2=10, n_3=3 and n_4=3 units in the hidden layers and 1 output units. System performance are good: on the images of the training set, 99% of the pixels were correctly classified by the system, that is, they were in agreement with physician classification; on test images, not included in the training set, a score of 96.6% correct classification was achieved. These results confirm that a proper integration of high-level and low-level information is effective for organ segmentation in medical imaging.

At the moment we are extending the approach to other organs, and we are evaluating more accurately system performances, in particular in the presence of different levels of noise.

Acknowledgements:

The authors have been supported by a fellowship from Istituto per la Ricostruzione Industriale.

References

[1] Haralik, Shapiro, "Image Segmentation Techniques" Comp. Vis. Graph. Image Proc. 29 (1985) 100-132.

[2] D.H.Ballard, C.M.Brown *Computer Vision*, Prentice-Hall Inc., New Jersey, 1982.

[3] G.Coppini, R.Poli, M.Rucci and G.Valli "A Neural Network Architecture for Understanding Discrete 3D Scenes in Medical Imaging", Comp. Biom. Res. 25, 569-582, 1992.

[4] T. Kohonen "Self-Organized Formation of Topologically Correct Feature Maps" Biological Cybernetics 43 (1982) 59-69.

[5] T. Kohonen *Self-Organization and Associative Memory* Springer-Verlag Berlin Heidelberg Ney York London Paris Tokyo second edition.

[6] Rumelhart, D., Hinton, G. & Williams, R. Learning internal representation by error propagation. In Rumelart, D. E. & McClelland, J. L.(eds.), *Parallel Distributed Processing: Explorations in the Microstructure of Cognition* (Cambridge, Mass: MIT Press, 1986).

A 3D Segmentation Algorithm on a Massively Parallel Computer

S. Wegner, H. Oswald, E. Fleck, R. Felix*

Deutsches Herzzentrum Berlin
Augustenburger Platz 1, D-W 1000 Berlin 65, Germany
*Universitätsklinikum Rudolf Virchow, Strahlenklinik und Poliklinik

Abstract

In this paper the implementation of a 3D segmentation approach for CT and MR image sequences on a massively parallel computer is described. The 3D segmentation approach is based on a volume growing technique driven by statistical features. Estimation techniques are used to calculate several statistical properties of the neighborhood of a point within the tissue of interest. The volume of interest is specified interactively and used as a seed volume for the growing algorithm. The required homogeneity criterion is obtained with regard to the estimated statistics. For the specific example of a MasPar MP-1102 computer an efficient data assignment mapping for this algorithm is selected and justified.

Introduction

A fundamental operation in computer assisted analysis of medical data is the segmentation of anatomical objects from digital images. For CT images of a head a 3D visualisation of the liquid system provides the opportunity to locate the spatial structure of the system within the head. This kind of 3D visualization can be very helpful in the area of craniofacial surgery. For 3D scenes a segmentation technique based on special object characteristics was developed and tested on CT sequences. The 3D segmentation algorithm is a volume growing algorithm combined with statistical features. Testing this approach for a 512x512x100 CT image sequence of a head leads to an acceptable result. However under practical considerations the implementation on a multipurpose hardware, e.g. SunSparc Station, is very time consuming. Therefore the 3D segmentation approach has been implemented on a massively parallel SIMD (single instruction multiple data) machine. An efficient data mapping scheme leads to a more practicable performance.

A 3D Segmentation Approach

The 3D segmentation approach for MR and CT image sequences is a statistical based volume growing algorithm. Due to the scanning modalities the object intensities vary as a result of noise and artefacts. Obtaining a region description despite these uncertainties can be achieved using the conditional probability distribution for modelling. The conditional probability distribution describes the probability of the intensities based on the available information out of the image data.

The 3D segmentation algorithm is twofolded. Based on an interactively obtained tissue point the statistical features of an 5x5x5 neighborhood are estimated. This marked region is the starting volume for the volume growing algorithm. The homogeneity criterion is acquired based on the estimated statistics.

Feature Description

Using statistical models it is generally assumed that the obtained image intensities $\mathbf{x} = \{x_1, ..., x_n\}$ are in the following relation with the real state vector

$$\mathbf{x} = f(\theta, \mathbf{v}),$$

where f is an n-dimensional function modelling the characteristics of the scanning device and the n-dimensional random vector v represents the error. The goal of this modelling is to obtain an efficient estimation of θ based on the intensities $\mathbf{x} = \{x_1, ..., x_n\}$. To simplify this description an additive error is often assumed

$$\mathbf{x} = f(\theta) + \mathbf{v}.$$

In the case that θ and $\mathbf{v}$ are independently of each other the following representation can be chosen for the conditional probability distribution

$$p(\mathbf{x}|\theta) = p_{\mathbf{v}}(\mathbf{x} - f(\theta)|\theta)$$

i.e. $p(\mathbf{x}|\theta)$ describes the probability of the intensity $\mathbf{x}$ under the assumption that the real value is θ. Since CT scanners are electronic devices the image sequence is disturbed by device influences. The exact shape of $p(\mathbf{x}|\theta)$ is based on these influences. A precise description of the probability distribution of these intensities can only be considered for simple cases. In most cases an exact modelling is not practicable since these models are nearly impossible to calculate. Moreover they are not appropriate to describe errors which are not due to noise (for instance artefacts). As a result of the central limit theorem of probability theory the distribution of the intensities can be assumed to be asymtotically gaussian. For the Maximum-Likelihood (ML) estimation the state vector is regarded as a fixed quantity with unknown value. The best estimate is then defined as the value for which the probability that the sample represents the real state is maximal. Suppose n samples, $\mathbf{x} = \{x_1, ..., x_n\}$, have been drawn independently under the probability law $p(\mathbf{x}|\theta)$. Assuming $p(\mathbf{x}|\theta)$ is a function of θ then $p(\mathbf{x}|\theta)$ is called a Likelihood function of the samples. The ML estimate of θ is defined as the value which maximises $p(\mathbf{x}|\theta)$.

Using the ML estimation for the volume-growing algorithm a gaussian probability distribution with unknown mean μ and variance σ^2 is assumed. Based on the interactively acquired tissue point mean and variance of a 5x5x5 neighborhood are estimated. θ consists therefore of $\theta_1 = \mu$ and $\theta_2 = \sigma^2$. The estimated values can be obtained using the following equations:

$$\hat{\mu} = \frac{1}{n} \sum_{k=1}^{n} x_k$$

$$\hat{\sigma}^2 = \frac{1}{n} \sum_{k=1}^{n} (x_k - \hat{\mu})^2$$

Volume Growing Algorithm

The homogeneity criterion for the volume growing algorithm is determined using a confidence interval. Based on the statistical features, the mean and the variance, this interval is computed. Let $x_1, \ldots, x_n$ be samples of a gaussian random variable X with unknown mean and known variance. For the estimation $\hat{\theta}$ follows that

$$(\hat{\theta} - k_\gamma \hat{\sigma}_{\hat{\theta}}, \hat{\theta} + k_\gamma \hat{\sigma}_{\hat{\theta}})$$

is defined as confidence interval with confidence level γ where k_γ can be acquired using a gaussian distribution tabulation.

From the 5x5x5 neighborhood the 3D volume-growing is started. Defining O(3) [1] as the 3D 8-connectivity the homogeneity criterion consists of the O(3) neighborhood and the mean of the O(3) neighborhood is within the confidence interval. Each point satisfying these two criteria becomes a member of the volume.

```
procedure volume_growing {
  if (current pixel is not marked within mask) {
    mark pixel in mask;
    compute mean of O(3) neighbors of current pixel; (2. criterion)
    if (mean is within the confidence interval) {
      pixel becomes member of volume;
      call volume_growing for all O(3) neighbors; (1. criterion)
} } }
```

Obviously both criteria can be changed and more criteria can be added, e.g. local variance, gradient or model based.

Massively Parallel Processing

The MasPar is a massively parallel SIMD (single instruction multiple data) computer, in our case consisting of 2048 PEs (processor elements). The architecture consists of a workstation front end (Decstation 5000), an array control unit (ACU), and the array of processors. Each PE is a custom load/store processor with its own local registers and memory. Interprocessor communication is possible employing the X-Net, an eight-way connection with toroidal wraparound, or using an arbitrary connection router. The ACU executes scalar instructions, broadcasts instructions to the PE array, and communicates with the front end workstation.

If the number of data points in an image is greater than the number of PEs in the MasPar machine, then some kind of virtualization technique to store the image in the PE array has to be used. The two most common mappings are called cut-and-stack and hierarchical. The first segments the problem by spatial areas onto memory locations, whereas the second takes spatial regions from the problem of interest and maps them to a single processor.

P0 v0	P1 v0	P0 v1	P1 v1
P2 v0	P3 v0	P2 v1	P3 v1
P0 v2	P1 v2	P0 v3	P1 v3
P2 v2	P3 v2	P2 v3	P3 v3

Fig. 1: Cut-and-stack mapping scheme for a 4x4 image by 4PEs

In the case of the volume growing algorithm the cut-and-stack mapping scheme is more appropriate. Figure 1 shows this mapping scheme for a 4x4 image by 4 PEs (P0, P1, P2 and P4). v0, v1, v2 and v3 stands for the virtualization index on the processor. In the 3D case each image is mapped onto the PEs in this way. The more efficient X-Net communication can be used and therefore no index conversion between the image index (x y z coordinates) and the virtualization location has to be done for the neighborhood connectivity in x-y direction. Furthermore the toroidal wraparound provides the opportunity to use the X-Net construct even beyond the processor array boundaries with an additionally computation of the virtualization index. Toroidal wraparound defines the leftmost column of PEs to be east neighbors of the rightmost column of PEs and it defines the top row of PEs to be the south neighbors of the bottom row of PEs. For instance the north neighbor can be found in the following way.

```
north {
 if (current pixel is not mapped on an upper boundary of the PE array) {
   if (current pixel is not marked)
     test pixel on XnetN[1] PE with same virtualization index;
 }
 else
   if (current pixel is not the upper image boundary)
     test pixel on XnetN[1] PE with virtualization index - 2;
}
```

XnetN[1] means next processor in north direction and the minus 2 (number of pixels for each PE in x direction) is due to a PE array of 2x2 using a 4x4 image. In our case of a 64x32 PE array and 512x512 images it is minus 8. For the z direction no processor communication is necessary since the neighbors are on the same processor. In contrast to hierarchical mapping the workload is quickly distributed over more processors.

The algorithm remains basically the same however now all current neighbors are parallel tested. Current neighbors are pixels which are neighbors of region pixels of the last turn and which had not been tested before.

Fig. 2: Visualization of the liquid system

Results and Conclusion

Figure 2 shows a result of the 3D segmentation algorithm in the case of a 512x512x100 CT image sequence of a baby's head. AVS (Advanced Visualization System) was used for the visualization of the liquid system. The result was obtained on the MasPar machine. The SparcStation version leads to a similar result but takes 3 times longer. Similar results have been obtained for several CT image sequences.

The segmetation strategy described in this paper is a successful method for tissue volumes in CT image sequences. Using the MasPar means a big time advantage over multipurpose machines and provides more practicable solutions . Current efforts focuses on the development of a more sophisticated segmentation approach on the MasPar.

References

[1] J.K. Udupa, S.N. Srihari, G.T Hermam, *Boundary Detection in Multdimensions*,
IEEE Transactions on Pattern Analysis and Machine Intelligence, January 1982, pp. 41 - 50

[2] G. Bamberg, F. Baur, *Statistik*
R. Oldenbourg Verlag, 1980

[3] K.R. Koch, Parameterschätzung und Hypothesentests in linearen Modellen,
Dümmler, 1980

[4] C. Morioka, K.K. Chan, H.K. Huang, *Development of an Image Segmentation and Registration Algorithm on a SIMD Parallel Processor*, In SPIE Medical Imaging 4: Image Processing, 1990, pp. 51 - 57

[5] R. O. Duda, P. E. Hart,*Pattern Classification and Scene Analysis*
John Wiley and Sons, Inc., 1973

A New Technique for Three-Dimensional Regression-Based Segmentation in Medical Image Processing

P. H. BLAND, S.-P. LIOU, C. R. MEYER, R. C. JAIN

Department of Radiology, University of Michigan Hospitals (PHB, CRM); Siemens Corporate Research, Inc. (S-PL); Department of Electrical Engineering and Computer Science, University of Michigan (RCJ)

Abstract

This work describes the application of the Liou-Jain segmentation algorithm to medical imaging for automated detection of anatomical surfaces. The algorithm avoids problems inherent in statistical segmentation and 2D techniques. Results are shown for MRI data from the human abdomen and a rat brain glioma. For comparison, the human dataset was processed by a statistical classifier. Quantitative accuracy of the algorithm is assessed by the use of a phantom. This work demonstrates that the Liou-Jain technique is effective in automatically detecting anatomical organ and lesion surfaces in 3D datasets and in obtaining volume estimates.

Introduction

Surface detection in medical image datasets arising from CT (computed tomography) and MRI (magnetic resonance imaging) is important in radiation therapy treatment planning, quantitative assessment of chemotherapeutic approaches to cancer therapy, and planning surgical procedures. These applications have in common the need to identify organ and lesion boundaries in three dimensions for the purposes of visualization and quantification.

To date, the most prevalent techniques for surface detection in medical imaging have relied on statistical classification [1] - [4]. Although relatively straightforward in their application and efficient in their computation, these techniques have two serious flaws: They are adversely affected by spatial non-stationarities in the input image datasets, and spatial relationships are considered separately during postprocessing. In CT images, spatial non-stationarities can arise from cupping artifact [5] and from circulation of the contrast bolus [6]. Spatial inhomogeneity of certain RF field coils used in MRI scanners may produce non-stationarities [7], causing severe errors in classifying MRI datasets by statistical methods. Typically, clusters found by statistical methods do not have a one-to-one correspondence with anatomical structures. Although statistical classification is useful in some situations, its general utility is limited in MRI since significant manual editing of the clusters is often necessary to obtain useful results. Further, these algorithms typically operate at least partially in two dimensions even when the goal is to classify three-dimensional structures [8]. Since spatial relationships are not considered, clusters in adjacent slices may be spatially inconsistent.

The Liou-Jain Segmentation Algorithm

Liou and Jain [9] - [11] have developed a three-dimensional regression-based segmentation algorithm that overcomes these fundamental deficiencies in statistical classifiers. The L-J algorithm is three-dimensional in nature, and thus avoids spatial inconsistencies. By using a regression model to define volumes, non-stationarities are inherently accommodated. A brief description of the algorithm is presented here along with some notes about its use in the medical environment, including pre- and postprocessing issues.

The input data consists of a series of contiguous two-dimensional images (typically from a tomographic imaging modality such as CT or MRI) that define a three-dimensional rectangular volume dataset. Preprocessing includes two-dimensional low pass filtering on each slice of the original data: The elements in the kernel are calculated from slice thickness and in-plane resolution to “blur” the data in the in-plane dimensions to an extent that approximates the partial volume effect that occurs during data collection due to the slice thickness. Thus the remainder of the algorithm operates on voxels with isotropic partial volume contributions; this is particularly important in computing the gradient magnitude and in generating the volume hypotheses. Three-dimensional median filtering may be performed to reduce MRI speckle noise while preserving edges, another important consideration in volume hypothesis generation. A weighted three-dimensional gradient operator is used to compute the isotropic gradient, whose magnitude is passed into the algorithm along with the 3D filtered intensity dataset.

The segmentation process involves two closely related steps: partitioning and identification. Together, they are used to implement a uniformity or smoothness predicate, P(•). Consider a set of correctly identified, non-overlapping volumes that completely fill the 3D dataset. P(•) is defined as follows: *Within* each volume, P(•) is "true", meaning that the voxels within the volume satisfy some uniformity criterion. In addition, when P(•) operates on the union of two adjacent but discontinuous volumes, it is "false"; that is, the intensity values of the voxels comprising the two volumes are such that they do not satisfy the uniformity criterion.

In the partitioning step, a series of volume hypotheses is generated based on discontinuities that typically separate volumes having different uniformity characteristics. In the L-J algorithm discontinuities are detected by thresholding the 3D gradient magnitude data. However, edge strengths can vary; in medical images this may be caused by (1) partial volume effects and (2) orientation and profile characteristics of organ-lesion and organ-organ boundaries. Thus, a *set* of thresholds, α_i, is needed to locate different boundary surfaces optimally. In the partitioning step, the 3D gradient magnitude is thresholded at an α_i, resulting in a new grayscale dataset at this α_i which contains a set of three dimensionally connected regions, the *volume hypotheses*. This process, repeated over a set of α_i, is known as *α-partitioning*. The set of α_i is applied using the largest value first and continuing sequentially to the smallest. At each α_i, an independent set of volume hypotheses is generated.

In the identification or volume filtering step, each volume hypothesis generated by the partitioning step is tested for correctness, where a "correct" partition is one that satisfies P(•). Incorrect hypotheses are discarded; however, the voxels of these incorrect hypotheses are available for consideration at lower thresholds for possible inclusion in smaller volumes. Voxels that are never identified as belonging to a region remain uncommitted.

Uniformity within a volume hypothesis is assessed by a regression technique. Intensity values of the voxels within the volume being tested are fit to an approximating function whose independent variables are spatial coordinates in three-space. At the present time, algorithm uses a polynomial of the form

$$\hat{f}(v, \vec{a}; x,y,z) = \sum_{i+j+k \leq v} a_{ijk} x^i y^j z^k$$

The maximum variation order, v, of the polynomial is chosen to allow a desired degree of non-stationarity or variation in the uniformity of the volume to be accommodated. In the work presented here, v=4 is used. Each volume at each α_i is fit separately and independently and is thus characterized by its own set of coefficients $\vec{a}$.

Regression analysis typically applies a goodness-of-fit test to assess the suitability of the chosen model. In the volume filter of the L-J algorithm, the unbiased estimate S^2 of the true measurement variance is calculated. If S is less than the noise in the input data, the volume hypothesis is accepted as correct.

Results and Discussion

The volumes identified by the L-J algorithm for an MRI exam are presented in Figure 1 (upper left and upper right). The non-stationarity resulting from RF field inhomogeneity causes a difference in intensity between the left and right sides of the liver image. While this causes a statistical classifier to commit errors in detecting the liver, the L-J algorithm identifies the entire liver as a single volume.

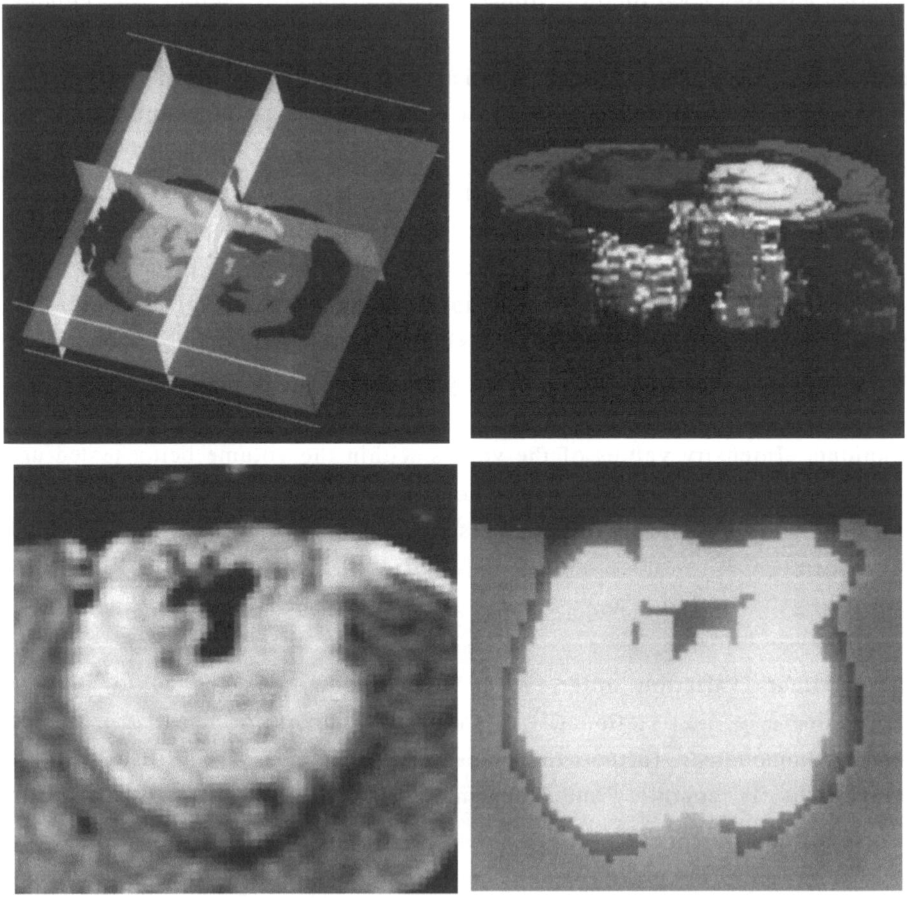

Figure 1: L-J analysis of an abdominal MR scan displayed using orthogonal slicing planes (upper left) and alpha-blend/gradient shading technique (upper right). Intensity data from rat brain glioma (lower left) and the result of L-J analysis postprocessed by intelligent recruitment superimposed over median filtered intensity data (lower right).

(see also in color on page 813)

Orthogonal slicing planes passing through the segmented volume dataset allow viewing of the volumes in three dimensions. Note that although some volumes may appear to consist of a number of disconnected pieces, they are in fact connected. The algorithm discriminates between the vessels in the liver and liver tissue itself. However, some small detail is lost due to volume underestimation, as is apparent in the small vessels of the liver. The volume loss is in the form of a reduction of volume at the surfaces of the volumes as a result of the thickness of the thresholded gradient edges; this is another source of uncommitted voxels.

The algorithm was also applied to a dataset obtained from an MR scan of a rat brain containing a 9L glioma [12]. (See Figure 1, lower left and lower right.) Median filtering reduced the texture noise which appears in both normal and diseased neural tissue. Without texture reduction, regions identified in the brain and glioma are sparsely filled rather than solid because of the difficulty in forming the volume hypotheses.

Volume underestimation is corrected as a postprocessing operation. Although regions could be fattened using a conventional binary morphological operator, this method does not use the input data intensity values and therefore may fatten a region inappropriately. Instead, an intelligent recruitment operation, based in part on morphological fattening, has been developed. An uncommitted voxel is assigned to a neighboring region if (1) it is 26-connected to the region and (2) it does not cause P(•) to become false for that region.

In order to quantify the accuracy of the algorithm, an MR scan of a vinyl glove phantom filled with copper sulfate solution was analyzed. The volume of the fluid within the glove is 749 cm^3. After postprocessing, the volume of the glove's region estimated by the L-J algorithm is 748 cm^3.

Conclusion

We have shown that the Liou-Jain algorithm is immune to problems that are inherent in statistical classifiers: It performs well in the presence of spatial non-stationarities and guarantees connectivity in three-space. The use of the algorithm in identifying structures in the human abdomen and in detecting a tumor in the rat brain are demonstrated. Qualitative assessment shows very good results in correctly identifying anatomical surfaces, and quantitative assessment of volume estimates by the use of a phantom shows a high degree of accuracy. With attention paid to pre- and postprocessing operations, it is well-suited for use in automated surface detection in the medical imaging environment.

This work was supported in part by DHHS PHS NIH grants 1R01-CA52709-01A1 and 1P30-CA46592-01.

References

1. Vannier, M.H.; Butterfield, R.L.; Jordan, D.; Murphy, W.A.; Levitt, R.G.; Mokhtar, G.: Multispectral analysis of magnetic resonance images. Radiology 154 (1985) 221-224.

2. Wharton, S.W.: A generalized histogram clustering scheme for multidimensional image data. Pattern Recognition 16 (1983) 193-199.

3. Farjo, L.F.; Williams, D.M.; Bland, P.H.; Francis, I.R.; Meyer, C.R.: Liver volume determination from CT images using histogram cluster analysis. J. Comp. Assist. Tomography 16 (1992) 674-683.

4. Kohn, M.I.; Tanna, N.K.; Herman, G.T.; Resnick, S.M.; Mozley, P.D.; Gur; R.E.; Alavi, A.; Zimmerman, R.A.; Gur, R.C.: Analysis of brain and cerebrospinal fluid volumes with MR imaging. Radiology 178 (1991) 115-122.

5. Barrett, H.H.; Swindell, W.: Radiological imaging; the theory of image formation, detection, and processing, vol. 2. New York: Academic Press 1981.

6. Nelson, R.C.; Moyers, J.H.; Chezmer, J.L.; Hoel, M.J.; Jones, E.C.; Peterson, J.E.; Cork, R.D.; Bernadino M.E.: Hepatic dynamic sequential CT: section enhancement profiles with a bolus of ionic and nonionic contrast agents. Radiology 178 (1991) 499-502.

7. Lufkin, R.B.; Sharpless, T.; Flannigan, B.; Hanafee W.: Dynamic-range compression in surface-coil MRI. Am. J. Radiology 147 (1986) 379-382.

8. Cline, H.E.; Lorensen, W.E.; Kikinis, R.; Jolesz, F.: Three-dimensional segmentation of MR images of the head using probability and connectivity. J. Comp. Assist. Tomography 14 (1990) 1037-1045.

9. Liou, S.-P.: Qualitative motion analysis using a spatio-temporal approach (dissertation). Ann Arbor: University of Michigan 1990.

10. Liou, S.-P.; Jain, R.C.: An approach to three-dimensional image segmentation (invited paper). Comp. Vision Graphics and Imag. Proc.: Image Understanding 53 (1991) 237-252.

11. Liou, S.-P.; Chiu, A.H.; Jain, R.C.: A parallel technique for signal-level perceptual organization. IEEE Trans. Pattern Anal. and Machine Intell. 13 (1991) 317-325.

12. Kim, B.; Ross, B.D.; Simerville, S.J.; Chenevert T.L.: MRI determination of intracranial tumor doubling time (abstract). Berlin: Society of Magnetic Resonance in Medicine 1992.

Segmentation and Symbolic Description of Cerebral Vessel Structure based on MR Angiography Volume Data

G. Gerig, Th. Koller, G. Székely, Ch. Brechbühler and O. Kübler

Communication Technology Laboratory, Image Science,
ETH-Zentrum, CH-8092 Zurich, Switzerland
e-mail: gerig@vision.ethz.ch

Abstract

Therapy and planning require more information from digital imagery than simply the presence of disease, since the medical task being the measurement and identification of structures. The present paper focuses on the conversion of three-dimensional image structures to an object-centered, abstract description encoding shape features and structure relationships. We describe a prototype system that extracts three-dimensional (3-D) curvilinear structures from volume image data and transforms them into a symbolic description which represents topological and geometrical features of tree-like, filamentous objects. The initial segmentation is performed by 3-D line filtering and/or 3-D hysteresis thresholding. A center-line representation is derived by 3-D binary thinning and by compilation (raster-to-vector transformation) into a vector description. The final graph data-structure encodes the spatial course of line sections, the estimate of the local diameter, and the topology at important key locations like branchings and end-points. The analysis system is applied to the characterization of the cerebral vascular system segmented from magnetic resonance angiography (MRA).

1 Introduction

The value of computer assisted systems for presentation, manipulation and quantitation of objects obtained from multidimensional image data depends critically on the ability to *segment* and also *describe* structures in images. This paper draws the attention to the symbolic description of image structures.

The development of the prototype system is driven by the application of assessing the cerebral vascular tree in magnetic resonance angiography (MRA) volume data. Whereas the MRA acquisition technology has improved considerably during the last years, the evaluation of vascular information is still limited by the lack of appropriate computerized analysis tools regarding visualization and description of structures. Maximum intensity projection (MIP) for visualization obscures features and creates irregular depth cues [1, 2, 3]. Segmentation of the vascular tree followed by 3-D rendering represents an alternative and could already demonstrate its advantage over MIP, but the segmentation of fine vessels still needs improvement [4, 5]. A successful segmentation creates new possibilities for 3-D planning, intervention and for quantitative volumetric studies, but it doesn't provide explicit information characterizing the gross form or some detailed shape features answering particular clinical questions regarding vessel morphology and local flow information.

Offering access to geometrical and topological properties, the binary array representation of the segmented vascular tree has to be transformed into a more appropriate data structure. Volumetric primitives of tubular structures strongly resemble the model of "generalized cylinders" with circular cross-section, suggesting the appropriateness of a skeleton or medial axis description (MAT [6]). Further, the biological variability of the vasculature favors a description based on *topological features* rather than on exact geometry [7].

2 Segmentation and description of 3-D lines

2.1 Segmentation by hysteresis thresholding

The special MRA acquisition technique simplifies the segmentation of large vessels by producing high contrast for vascular structures. A binarization technique suffices to segment the major vessels. However, small vessels with diameters of one voxel ($\approx 1mm$) and smaller are subject to partial voluming and noise fluctuations. Similar problems with binarization of the matched filter output in an edge detection method led to the development of a hysteresis thresholding procedure [8]. Hysteresis thresholding works simultaneously with two threshold values. The upper threshold identifies a certainty level above which voxels most probably belong to the structure of interest, acting as seed regions. The lower threshold specifies the noise level below which voxels will be excluded. Growing from seed regions, voxels with intensities between the two levels will be selected and connected to image structures. The 3-D application of this technique to detect connected edge-surfaces has been already previously investigated [9], demonstrating only limited success due to the high connectivity of surface fragments. The sparse space occupation of the blood vessel structures, however, reduces the number of random connections and proved to be well suited for hysteresis thresholding (Figure 1).

2.2 Segmentation by matched filtering

Limitations of segmentation based on thresholding are approached by a structural segmentation approach. 3-D line features are extracted in a multistep segmentation scheme comprising local feature extraction by multiple convolution with a set of directional Gaussian derivatives, feature synthesis to combine the filter results, and 3D hysteresis thresholding and connected component labeling.

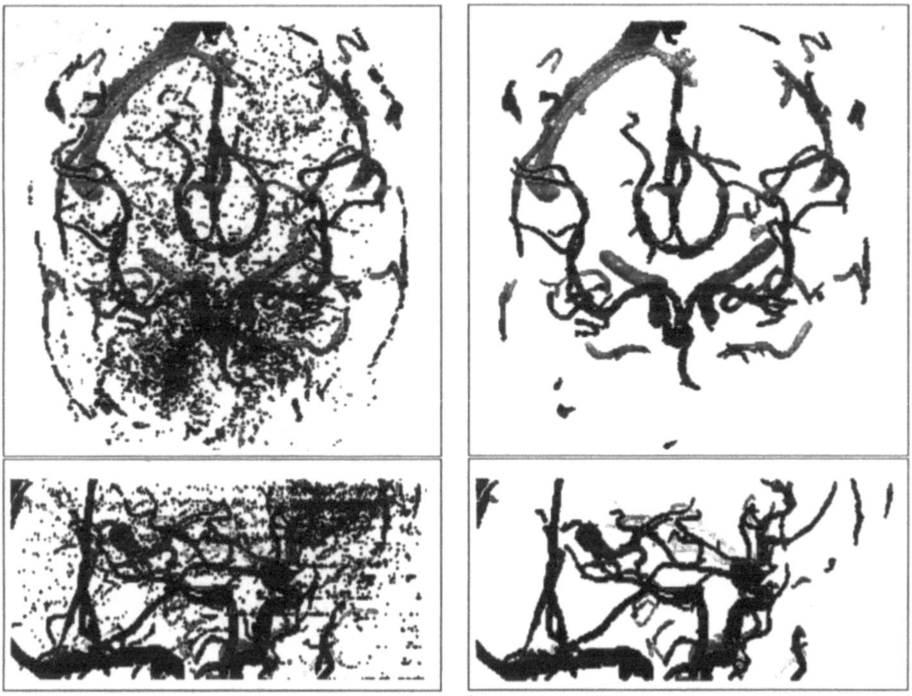

a b

Figure 1: Three-dimensional rendering of the segmented vessel tree (top and side views): a) simple thresholding, b) hysteresis thresholding. The lower threshold in b equals the single threshold used in Figure a, it corresponds to 50% of the higher threshold generating the seed regions.

2.3 Three-dimensional thinning of structures to medial curves

Blum's classical proposal [6] presented the description of shapes in 2-D binary images by the medial axis transformation (MAT). The MAT generates an object-centered, invariant description of shapes in terms of their medial axis. Additionally, the technique decomposes structures and sub-structures by generating a tree-like representation. Herein, binary thinning is chosen for the generation of topologically correct skeletons for filamentous tree-like structures, considering that the preservation of topology offered by these techniques is more important than a perfect medialness.

A 3-D thinning algorithm was implemented based on three requirements: preserving topology, thinning 3-D structures to voxel lists (maximal thinning), and approximating the ideal Euclidean skeleton (center line).

3-D thinning algorithms published so far either do not preserve topology [10, 11] and the major geometrical structure or do not provide a skeleton that is thinned down to a center-line of one voxel width [12], as required for a subsequent raster-to-vector compilation of line structures. A detailed description of different thinning algorithms is found in [13]. We have implemented a pseudo-parallel binary thinning scheme that erodes binary image structures in a sequence of spatially oriented sub-cycles. The resulting skeleton is topologically correct but still contains some segments which are not maximally thinned. A subsequent sequential 3-D thinning achieves a maximal erosion of the skeleton to one voxel width. Line-ends are preserved from shrinking by special criteria. The 3-D thinning scheme has been applied to the binary segmentation result of the cerebral vascular system obtained from MR angiography data (Figure 1b). The thinned structures are displayed using surface rendering (Figure 2a). It is obvious that the vessel structures are replaced by a reasonable approximation to its medial axes.

2.4 Estimation of object width by 3-D distance transform

For a complete geometrical description of the cerebral vessel tree, the midlines resulting from the thinning procedure have to be attributed with the local description of the cross section of the vessel. In the simplest case, if the vessels were approximated with circular cylinders, the cross section can be sufficiently described with the radius of the circle, i.e. the vessel width. A Euclidean distance transform [14] generalized to 3-D is applied to the binary segmentation result. It calculates the minimum distance of each voxel to the background and serves as a reasonable approximation of the local vessel radius.

2.5 Compilation of thinned objects to a graph structure

The abstraction of objects to line structures and the compilation into image graphs would enable a structural analysis by means of graph theory tools. Higher level image analysis, which assigns semantic meaning to image structures using model knowledge, becomes feasible through the selection of relevant image structures by *subgraphs labeling.* The skeleton represented by voxel lists is compiled to a graph structure by connecting segments of pixels to sequences and assigning them to edges, vertices, and faces of a graph data structure.

The output of the 3D thinning and compilation stages is a data structure containing both the topological description of the vessel tree and its geometric properties. It is implemented in an object-oriented design. Each of the different elements is represented by a class, these are : *Vessel Node* representing the branching and end points, *Vessel Link* representing the topological linkage between the *Nodes*, *Vessel Point* holding all the geometric information available from the segmentation scheme such as local width, as well as additional attributes that can be computed (local direction, length, volume), and *Vessel Trajectory* representing a sequence of points and as such the geometric path between *Nodes.* The *Vessel Tree* is implemented as a class containing lists of the above elements. Figure 2b represents a graphical visualization of the compiled vessel-tree. End nodes and branching nodes are displayed as spheres, whereas voxel-lists between nodes are represented as sequences of small tubes.

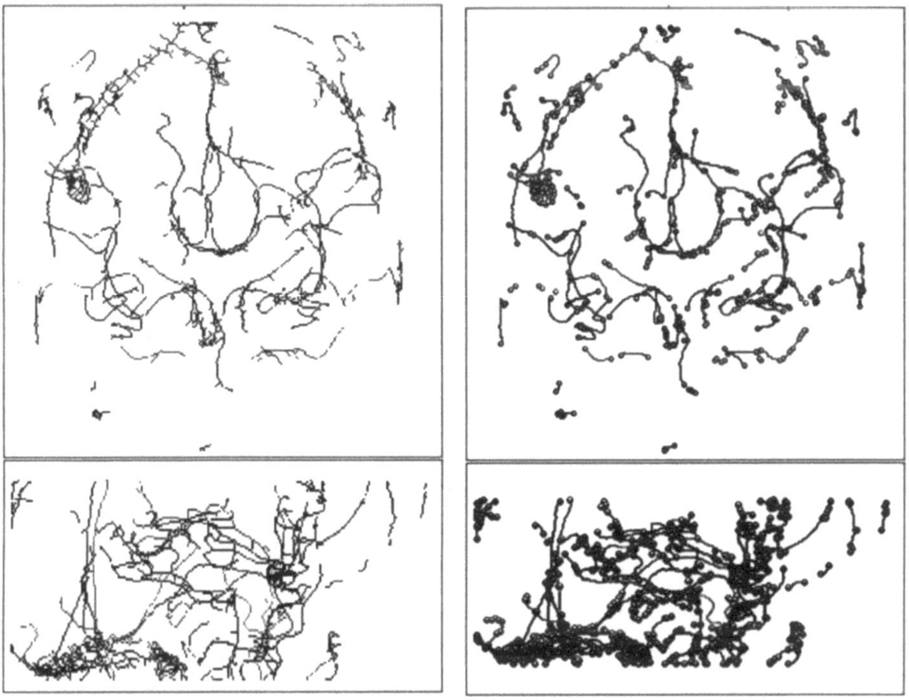

a b

Figure 2: a) Result of 3-D binary thinning. The segmented vessel tree is maximally thinned to one voxel width by preserving topology and approximating center-lines. b) Graphical display of graph data structure describing the cerebral vessel-tree: End nodes and branching nodes are displayed as spheres, interconnecting vessel arcs as flexible pipes.

3 Discussion

The present paper discusses a prototype image analysis system that extracts and describes curvilinear structures from volume data. Tests with several 3-D MRA data sets demonstrated that segmentation by hysteresis thresholding was superior to simple thresholding. We are well aware that this low-level segmentation of vessels from MRA is still insufficient, depicting only the largest vessels and creating gaps. Current developments investigate multi-resolution matched filter techniques based on higher order directional derivatives of the Gaussian. Binary thinning of tree-like structures to maximally thinned skeletons required the development of a new algorithm. The preservation of topology has high priority as one wants to avoid creating gaps. The approximation of the center-line, however, is important too in order to generate a medial axis description. Careful tests on highly complex structures proved that the topology is preserved while object structures are thinned to voxel lists. The latter is a necessary property with regard to a raster-to-vector transformation. The compilation method extends the 2-D λ-graph concept proposed in [15] to 3-D and is well suited to represent tree-like object structures. One of the important local features of a vessel tree is the vessel diameter. Estimation of the local width is obtained by a 3-D Euclidean distance transform applied to the binary segmentation result.

The final output of the 3-D segmentation, thinning and compilation stages is a data structure representing voxel-lists, branchings and approximate local diameter. This symbolic representation codes the geometrical and the topological information of the vessel tree. An advantage of the object-centered description is its invariance (up to an error introduced by the pseudo-parallel thinning) with respect to standard transformations, i.e. translation and rotation, and even to nonlinear distortions.

The methods developed in this project could be useful in application areas where not only a display of blood vessels, but *structural information* about their spatial extension and the relationship among vascular components and to anatomical objects is required. Potential applications could be found in neuroradiological diagnosis, interventional radiology and stereotactic neurosurgery. Two issues will be considered in future developments, an improved segmentation that extracts vessels down to one voxel diameter and the implementation of graph-analysis tools for structural interpretation and model-based analysis.

Acknowledgements

Part of this research work was funded by COVIRA (Computer Vision in Radiology), project A2003 of the AIM (Advanced Informatics in Medicine) programme of the European Commission. Participants in the COVIRA consortium are:

Philips Medical Systems, Best (NL) and Madrid (E), Corporate Research, Hamburg (D) (prime contractor); Siemens AG, Erlangen (D) and Munich (D); IBM UK Scientific Centre, Winchester (UK); Gregorio Maranon General Hospital, Madrid (E); University of Tübingen, Neuroradiology and Theoretical Astrophysics (D); German Cancer Research Centre, Heidelberg (D); University of Leuven, Neurosurgery, Radiology and Electrical Engineering (B); University of Utrecht, Neurosurgery and Computer Vision (NL); Royal Marsden Hospital/Institute of Cancer Research, Sutton (UK); National Hospital for Neurology and Neurosurgery, London (UK); Foundation of Research and Technology, Crete (GR); University of Sheffield (UK); University of Genoa (I); University of Aachen (D); University of Hamburg (D); Swiss Federal Institute of Technology, Zurich (CH)

The MRA data were provided by Siemens AG, Erlangen, Germany and by Dr. F. Jolesz, Brigham and Women's Hospital, Boston, USA.

References

[1] H-H. Ehricke and G. Laub. Combined 3D-display of cerebral vasculature and neuroanatomic structures in mri. In K.H. Höhne, H. Fuchs, and S.M. Pizer, editors, *3D Imaging in Medicine*, pages 229–239, Berlin Heidelberg, June 1990. Springer-Verlag.

[2] K.H. Höhne, M. Bomans, A. Pommert, M. Riemer, et al. Rendering tomographic volume data: Adequacy of methods for different modalities and organs. In K.H. Höhne, H. Fuchs, and S.M. Pizer, editors, *3D Imaging in Medicine*, pages 197–215, Berlin Heidelberg, 1990. Springer-Verlag.

[3] D. Vandermeulen, D. Delaere, P. Suetens, H. Bosmans, and G. Marchal. Local filtering and global optimisation methods for 3d magnetic resonance angiography (mra). In Richard A. Robb, editor, *Visualization in Biomedical Computing*, pages 274–288. SPIE, October 1992.

[4] H.E. Cline et al. Vascular morphology by three-dimensional magnetic resonance imaging. *Mag. Res. Im., Pergamon Press*, 7:45–54, 1989.

[5] D.N. Levin et al. The brain: integrated three-dimensional display of MR and PET images. *Radiology*, 17:783–789, 1989.

[6] H. Blum. A transformation for extracting new descriptors of shape. In E. Dunn, W., editor, *Models for the Perception of Speech and Visual Form*, Cambridge, MA, 1967. M.I.T. Press.

[7] J. Niggemann. Analysis and representation of neuroanatomical knowledge. *Applied Artificial Intelligence*, 4:309–336, 1990.

[8] J.F. Canny. A computational approach to edge detection. *IEEE Transactions on Pattern Analysis and Machine Intelligence*, 8(6):679–698, 1986.

[9] O. Monga et al. Recursive filtering and edge closing: two primary tools for 3D edge detection. In O. Faugeras, editor, *Proc. First European Conference on Computer Vision - ECCV'90*, pages 56–65, Berlin-Heidelberg, May 1990. Springer-Verlag.

[10] S. Lobregt, P.W. Verbeek, and F.C.A. Groen. Three- dimensional skeletonization, principle and algorithm. *IEEE Transactions on Pattern Analysis and Machine Intelligence*, 2(1):75–77, January 1980.

[11] G. Malandain, G. Bertrand, and N. Ayache. Topological classification in digital space. In A.C.F. Colchester and D.J. Hawkes, editors, *Information Processing in Medical Imaging, IPMI'91*, pages 300–313, Berlin-Heidelberg, 1991. Springer-Verlag.

[12] Y.F. Tsao and K.S. Fu. A parallel thinning algorithm for 3-D pictures. *Computer Graphics and Image Processing*, 17:315–331, 1981.

[13] G. Gerig, Th. Koller, G. Székely, Ch. Brechbühler, and O. Kübler. Symbolic description of 3-d structures applied to cerebral vessel tree obtained from mr angiography volume data. accepted by IPMI'93 conference, to appear June 1993.

[14] P.E. Danielson. Euclidean distance mapping. *Computer Graphics and Image Processing*, 14:227–248, 1980.

[15] P.T. Speck. *Übersetzung von Linien und Flächenstrukturen in kombinatorisch-relationale Datenstrukturen zur automatischen Mustererkennung in Digitalbildern.* PhD thesis, ETH Zurich, 1984. Ph.D. thesis No. 7508.

Computer Graphics

Imaging Transforms for 3D Biomedical Imaging: an Open, Transportable System (3DVIEWNIX) Approach

Imaging Transforms for 3D Biomedical Imaging: An Open, Transportable System (3DVIEWNIX) Approach

J.K. UDUPA, R.J. GONCALVES, K. IYER, S. NARENDULA, D. ODHNER, S. SAMARASEKERA AND S. SHARMA

Medical Image Processing Group,
Department of Radiology
University of Pennsylvania, Philadelphia, PA - 19104

Summary

3DVIEWNIX is a data-, machine-, and application-independent software system, developed and maintained on an ongoing basis by the Medical Image Processing Group. It is aimed at serving the needs of biomedical visualization researchers as well as biomedical end users. We have identified and incorporated in 3DVIEWNIX a set of basic imaging transforms that are required in most visualization, manipulation, and analysis methods. The result is a powerful exploratory environment that provides not only the commonly used standard tools but also an immense variety of others. In addition to visualization, it incorporates a variety of multidimensional structure manipulation and analysis methods. We have tried to make its design as much as possible image-dimensionality-independent to make it just as convenient to process 2D and 3D data as it is to process 4D data. It is based on UNIX, C, X-Window and our own multidimensional generalization of the 2D ACR-NEMA standards for image data representation.

Introduction

Three-dimensional imaging in biomedicine encompasses visualization, manipulation, and analysis of structure information captured in three-dimensional digital images. In recent years, this activity has become an established discipline in biomedicine [5].

Although hardware systems specially designed for 3D imaging have been built, it is the software packages providing 3D imaging tools that have made the widespread application of 3D imaging possible. The first software package developed for imaging is a package called DISPLAY [3], released in 1980, developed by the Medical Image Processing Group. This was subsequently improved significantly and released in 1983 as DISPLAY82 [4]. This latter package was released for public use worldwide with source code. In spite of its highly machine-specific features, it played a key role in fostering widespread research in and application of 3D imaging. The first attempt to industrialize 3D imaging was in 1980 when DISPLAY was implemented on the General Electric Physician Display console of the 8800 CT/T scanner which was subsequently made into a product [6] by GE. DISPLAY was also implemented at the Biodynamics Research

Unit in Mayo Clinic in 1980, where subsequently the ANALYZE package [2] was developed. Around 1984, most major scanner manufacturers provided a 3D imaging software option on their display consoles. From then on, a number of independent 3D imaging companies started providing complete workstations with appropriate software for doing 3D imaging. A recent survey of available 3D imaging software (and hardware) products can be found in [1].

A variety of considerations have motivated us in developing 3DVIEWNIX. First, as is clear from the above paragraph, the development of 3DVIEWNIX is a continuation of our past activity of developing 'open' software and distributing it with source code for promoting science. Second, it is an attempt to impose a discipline on the software development activity going on within MIPG in a variety of research and application projects. Third, it is an attempt to provide a unique standard software system to the scientific community on which further development can be based and freely exchanged without restrictions. To this end, a multidimensional generalization of the 2D ACR-NEMA Standards has been developed for the representation of multidimensional image and non-image data and incorporated into 3DVIEWNIX. Also, 3DVIEWNIX is based on a standard software environment provided by UNIX, C and the X Window system. Finally, 3DVIEWNIX attempts to fill a number of gaps that exist in commercially available packages, including the following: (i) provision of numerous alternative techniques for each visualization and analysis task via generic imaging operators, as opposed to implementing a fixed technique for each task, (ii) designed for a wide spectrum of machines supporting the standard software environment as opposed to optimized for a particular brand of machines or machines of a certain capability, (iii) designed for basic research and a wide variety of applications as opposed to a particular application, (iv) open for further development by the user as opposed to being closed with numerous proprietary restrictions, (v) most algorithms and their implementation made image dimensionality independent as opposed to supporting just 3D processing operations only.

Design Considerations

The core of 3DVIEWNIX consists of data interface, graphics interface, and process interface. *Data interface* defines how image data acquired from outside and/or generated internally are handled by 3DVIEWNIX. *Graphics interface* defines how 3DVIEWNIX interacts with the display device. *Process interface* defines how processes related to visualization, manipulation, and analysis interact with 3DVIEWNIX and between themselves. In the next level, we have imaging transforms. Here "imaging" refers to any of the processes related to visualization, manipulation, and analysis and "transform" refers to an operation done on images or on structures derived from images. An imaging transform is *basic* if it cannot be broken up into two or more simpler transforms and *composite* if it is composed of two or more basic transforms. For example, image interpolation is a basic imaging transform, whereas surface detection in a 3D scene at a resolution different from that of the scene is a composite transform which consists of the following basic transforms: interpolation, segmentation, and surface formation. Most imaging methods can be expressed as a sequence of basic imaging transforms. As we see in the next section, once a comprehensive set of basic transforms is available, a powerful environment results in which hundreds of imaging methods can be realized which allow numerous processing pathways for data.

Data Interface

It is designed for a data representation protocol that is a generalization of the ACR-NEMA standards developed earlier for the archival and communication of medical 2D digital image data. The new protocol [7] defines all data to be one of three types: SCENE data, which typify n-dimensional ($n \geq 1$) vector-valued digital image data with a very general sampling scheme; STRUCTURE data, that constitute n-dimensional image-derived non-image structure information; DISPLAY data, that designate a visual representation of any information in the form of pictures ready to be displayed.

Graphics Interface

It is built on the X Window system. Its design is governed by some assumed minimum graphics display capabilities of the class of machines on which 3DVIEWNIX is intended to operate and on the nature of the user interface. In order to provide a consistent, clear, uncluttered appearance to the screen, 3DVIEWNIX considers the screen as being divided into several functionally independent regions. This division is done automatically in a 3DVIEWNIX installation procedure (with interactive verification with the user) based on the size of the screen and the size of the character fonts selected for the various regions. The graphics interface contains a comprehensive set of functions to achieve the functionality associated with each of these regions.

Process Interface

It contains all those functions that are independent of any specific image processing or graphics operations done on data but that are common to processes related to visualization, manipulation, and analysis. All of these functions are implemented in a standard fashion for portability, although often machine-specific, more efficient implementations and even functionalities would have been possible.

Imaging Operations

In this section, we will describe some of the basic transforms that have been incorporated in 3DVIEWNIX and demonstrate how they can be combined to produce a variety of imaging methods. Most of the operations described here are generic and the reader should bear in mind that there are several specific instances for each of these operations.

Basic Transforms

We identify each transform with an operator. The operators are generic in the sense that they do not distinguish between the different variants of the same class of transforms. For example, n-dimensional (nD) image interpolation is a transform that converts one nD image into another with volume elements of possibly a different size. It will be represented by a single operator, although many interpolating functions - linear, bilinear, trilinear, and various cubic forms – can be used with the transform. In what follows, we refer to an nD digital image as a *scene*, to the image intensity value as *scene density*, and to the set of all image elements in the scene as *scene domain*. Most of what is described here applies equally well to scenes and binary scenes (scenes with densities 0 and 1). Therefore, unless otherwise stated, the notion of a scene always subsumes the notion of a binary scene. A *structure system* is a set of related structures (derived from possibly different modalities, subjects, and time instants in a longitudinal or dynamic system) together with a specification of the role of the individual structure in the system. A *structure plan* is a structure system together with a tree indicating the hierarchical relationship of the structures in the system. Each node in the tree has an affine transformation matrix associated with it.

Volume of Interest (A): It maps a scene to another scene, the purpose of the transform being to reduce the size of the data that need to be processed subsequently. It is also possible to specify an image intensity interval of interest on a histogram of the region of interest. This operation often allows reducing drastically the number of bits required to store the resulting scene.

Filtering (F_s) : It also maps a scene to a scene whose purpose is either to smooth or enhance the given scene. It is a powerful operation that forms a bridge between surface and volume rendering. Several types of filters are available within 3DVIEWNIX. The user can verify the result of filtering on sample slices before filtering the whole scene.

Interpolation (I_s, I_b) : I_s is a scene to scene mapping in which the size of the volume elements in the output scene may be different from those in the input scene and output scene density is obtained by interpolating input scene density. I_b maps an nD binary scene to an nD binary scene with possibly a different size for its volume elements via a method called shape-based interpolation [5]. In many situations, the latter methodology gives more accurate results than the former. 3DVIEWNIX allows using different interpolating functions in different dimensions.

Segmentation (S_b) : S_b is a binary segmentation operator that maps a scene to a binary scene. Several segmentation operations are available in 3DVIEWNIX. Thresholding allows identifying a structure or multiple structures in a structure system based on multiple threshold intervals. A gesture-controlled method called "live-wire" allows segmenting optimal boundary segments at interactive rates based on a few points specified by the user. A method based on partitioning a 2D feature space allows segmentation based on vector-valued scene densities or on feature vectors derived from scenes [5].

Masking (M_s) : It converts a scene to another scene whose purpose is to exclude objects of no interest so that when S_b is applied to the resulting scene, the objects of interest may be captured automatically. 3DVIEWNIX allows specifying mask regions quickly using paint brushes of various sizes as well as by outlining. Mask regions can be cut and added using any combination of these modes.

Structure Definition (B_s): Two forms of structure representation are used in 3DVIEWNIX at present. Shell0 uses voxels in the vicinity of the structure boundary to represent the structure while Shell1 uses voxel faces. Both these structures represent structures with connected, closed, oriented surfaces.

Surface Normal (N_g, N_s): N_g converts a structure to a structure by assigning a normal vector to every structure element based on the local geometry of the structure. N_s is a similar operation except that the normal vector is computed using a gradient operator on a specified scene. Thus, N_s operates on a <structure, scene> pair to produce a structure.

Classification (C): This operation converts a scene to a fuzzily defined structure [5] (Shell0). The fuzzy membership of the structure elements is indicated by an opacity value. Classification based on scene density and scene density and density gradient are available.

Structure to Structure (B_o): The purpose of these transforms is to convert one form of structure description to another. They also allow forming a structure system given a set of structures that satisfy certain conditions.

Structure to Scene (G_s): G_s converts a structure description to a scene by assigning densities to volume elements interior and exterior to the structure in different ways. An example of this density is the distance of the element from the surface (positive for interior elements and negative for exterior elements) of the structure.

Slice Display (E): This operation converts any given collection of scene and given display data sets into a "display". A "display" consists of an arbitrary layout of the images in the display data and of certain "slices" of the scene data. The slices may be selected in any arbitrary plane through the scene domain. The images in the display can be magnified arbitrarily. A mode called "montage" allows a disciplined, static layout of the 2D images in the given scene and display data sets while another mode called "cycle" permits any layout chosen by the user and a dynamic cine display within the layout.

Surface Rendering (R_{ss}, R_{sv}): These operations convert a structure system or a structure plan into a rendition (a 2D image depicting the form of and relationship between structures). Optical properties, such as color, reflectivity, matteness, of the structures can be defined independently for each structure for rendition. When a structure system (or plan) constitutes a dynamic organ system such as a moving joint assembly, its rendition will actually depict the dynamics. How close this depiction of dynamics is to truth in terms of the speed of movement depends on the computational power of the workstation on which 3DVIEWNIX is installed. R_{ss} renders structures using the normal vectors given in the structure description. R_{sv} ignores this information and estimates normal vectors in view space [5] from a distance map of the structure elements from the viewing plane. This latter method is not useful for rendering surfaces translucently.

Volume Rendering (R_v): These operations convert a structure system or plan consisting of fuzzily defined structures into a rendition using a method called shell rendering. As in Surface Rendering, some of the optical properties of the structures can be changed independently for each structure.

Structure Manipulation (P): These operations convert a structure system or a plan into a plan by altering selected structures. The alterations include cutting and removing a part of a structure, fragmenting a structure into multiple structures, moving (translation and rotation) of structures independently, and reflecting a structure about a specified plane.

Structure Measurement (H): These operations allow measuring volume of structures and distance and angle between points selected in the structure. We emphasize that these measurements are made on the entire structure system. Therefore, in case of dynamic systems, how these entities change with time can also be determined.

Density Profile (J): These operations allow computing scene density along any curve specified in the scene domain for any given scene. For dynamic (4D) scenes, this means that time density curves can be computed and displayed.

ROI Statistics (K): These operations make it possible to compute scene density statistics including mean, minimum, maximum, and standard deviation of density and its gradient magnitude within user specified regions. The input is a scene, or a structure system and the associated scenes. The ROI may be specified directly on the scene or via a rendition of the structure system.

Registration (L): Given two structure systems, or two structure plans, or a structure system and a plan, or these entities and the associated scenes, this operation produces a structure plan. Registration is based on matching homologous features such as points and line segments in the structures, and entire structures. Selection of the features is made via renditions of the structures or slice display of the associated scenes.

Motion Determination (Q): This operation converts a given structure system into a structure plan by determining the translation and rotation needed to match structures in the system that have successive time coordinates. The structures are assumed to be rigid bodies.

Data Format Conversion (T): These operations allow converting data coming from external sources into one of scene, structure, or display data. Interactive programs exist within 3DVIEWNIX that allow doing this conversion conveniently.

Imaging Methods

We assume that the operations are from right to left. Thus if X and Y are operators on scenes, and if D is a given scene, then $XY(D) = X(Y(D))$. We encourage the reader to carefully study these operator sequences, and even to substitute mentally specific operators for each generic operator. Some examples from a vast variety of possible visualization methods are given below. These sequences assume a scene to be given.

$R_{sv}B_sI_s,\ \ R_{sv}B_sS_bI_s,\ \ R_{sv}B_sS_bF_sI_sA,$

$R_{ss}N_gB_sS_bI_s,\ \ R_{ss}N_s(B_sS_bI_s,\ \ I_s),$

$R_vN_s(CI_s,\ \ I_s),$

$R_{ss}N_gB_sI_bS_bM_s,\ \ R_{ss}N_s(B_sI_bS_bM_s,\ \ M_s),$

$R_{sv}B_sF_sS_bI_s,\ \ R_vN_s(CF_sI_s,\ \ I_s),\ \ R_{ss}N_s(B_sF_sI_bS_b,\ \ I_s),$

$R_{ss}N_s(B_sF_sI_sG_sB_s,\ \ I_s),\ \ R_vN_s(CI_sF_s,\ \ F_sI_s),$

$R_{ss}N_s(B_sF_sI_bS_bF_s,\ \ F_sI_s),\ \ R_{ss}N_s(B_sF_sI_bS_b,\ \ F_sI_bS_b),\ R_vN_s(CF_sI_bS_b,\ \ F_sI_bS_b),$

$R_{ss}N_s(B_sF_sG_sB_sI_bS_bF_s,\ \ F_sI_bS_b),\ \ R_vN_s(CF_sI_bS_bF_s,\ \ F_sI_bS_b).$

Scenes may be divided into two classes based on their characteristics for visualization: *robust scenes* wherein object boundaries are fairly well defined and characterized by high gradient magnitude, and *frail scenes* which do not have these characteristics. CT data and MR data, both of a bony structure, are examples of robust and frail scenes. We have identified classes of new transform sequences that produce significantly better renditions than the existing methods for both these classes of scenes through 3DVIEWNIX operations.

It is clear that a vast variety of imaging methods can also be generated along similar lines. Some examples are given below. In these operations, we use D, D_1, D_2 to denote a scene and W, W_1, W_2 to denote a structure system or plan.

$R_{ss}P(W),\ \ R_{ss}PPP(W),$

$HPP(W),$

$K(P(W), D),$

$R_{ss}PL(W_1, W_2, D_1, D_2),$

$R_{ss}Q(W),$

$^*K(PL(W_1, W_2), D_1, D_2).$

To give an understanding of the application of the various imaging methods that can be generated, we explain the sequence marked "*". Given two structure systems W_1 and W_2 extracted from scenes D_1 and D_2 that were acquired from two modalities, say MR and PET, $L(W_1, W_2)$ creates a single structure plan. We can extract a specific substructure in the underlying organ system via 3D interactive manipulation and the result is $PL(W_1, W_2)$. We can subsequently compute the PET count density within this substructure via the operator K using the associated scenes.

Conclusions

We have been developing a software system called 3DVIEWNIX designed around well established software standards and a data representation protocol that is a multidimensional generalization of the ACR-NEMA standards. In addition to these standards, it is based on some unique design principles to provide the functionalities associated with multidimensional image visualization, manipulation, and analysis. The rich, comprehensive environment created by this portable system, we believe, will promote cooperative research and development in both basic and application research in biomedical imaging in a free and open fashion.

Acknowledgments

The research reported here is supported by NIH Grants CA50851 and CA56071. The authors are grateful to Mary A. Blue for typing the Manuscript.

References

[1] Resources, I. *The 1991 3-D/MPR Market Report.* Lebanon, NJ, 1991.

[2] Robb, R., and Barillot, C. Interactive display and analysis of 3-D medical images. *IEEE Transactions on Medical Imaging MI-8* (1989), 217–226.

[3] Udupa, J. DISPLAY: A system of programs for two- and three-dimensional display of medical objects from CT data. Tech. Rep. MIPG41, Medical Image Processing Group, Department of Computer Science, SUNY at Buffalo, Buffalo, NY, 1980.

[4] Udupa, J. DISPLAY82 - a system of programs for the display of three-dimensional information in CT data. Tech. Rep. MIPG67, Medical Image Processing Group, Department of Radiology, University of Pennsylvania, Philadelphia, PA, 1983.

[5] Udupa, J., and Herman, G., Eds. *3D Imaging in Medicine.* CRC Press, Boca Raton, FL, 1991.

[6] Udupa, J., Herman, G., Margasahayam, P., Chen, L., and Meyer, C. 3D98: A turnkey system for the display and analysis of 3D medical objects. In *SPIE Proceedings* (1986), vol. 671, pp. 154–168.

[7] Udupa, J., Hung, H., Odhner, D., and Goncalves, R. 3DVIEWNIX: A machine-, data-, and application-independent software environment for the visualization and analysis of biomedical structures, SCANNING. *The Journal of Scanning Microscopy 13, Supplement I* (1991), I32–I33.

An Interactive Tool for Manipulation and Presentation of 3D Tomographic Data

I.HOLLÄNDER, M.ŠRÁMEK [1]

Institut für Informationsverarbeitung,
Österreichische Akademie der Wissenschaften
Sonnenfelsgasse 19/2
A–1010 Wien, Austria

Summary

An interactive software tool allowing manipulation and presentation of 3D tomographic data is described. The core idea is to use a 3D cubic visual model of the data set that can be manipulated interactively in an intuitive way. This approach provides possibilities for easy morphometric operations in 3D, comfortable visual object detection and localization (e.g. EEG electrode mapping) and intuitive setting of geometrical parameters for subsequent 3D rendering.

The program uses a sophisticated 3D volumetric rendering algorithm that make fast 3D reconstruction possible: adaptive undersampling, view-to-view coherence, new fast voxel traversal based on cubic macro regions etc. The program is implemented on an X11-based workstation using OSF/Motif.

Introduction

Visualization of 3D tomographic data undoubtedly brings new quality into the branch of medical imaging. Whilst 3D image acquisition techniques develop rapidly and it is not uncommon to meet high-quality and highly attractive 3D-rendered images in medical documentation, the phase of interactive image manipulation often does not meet the requirements of the users. With exception of top-tech products (such as ANALYZE [2]), the user is not given the opportunity to preview the 3D data set in a satisfactorily fast mode and to tune the rendering parameters (the viewing geometry, for example) using intuitive tools.

Our work stems from the cooperation with neurological research ([1]) where a proband has been subject to an EEG examination and the exact positions of the EEG electrodes with respect to the brain structures have been visualized using 3D-MRI. These positions were then 'manually' picked up and used for topological studies (among others, 3D images have been rendered showing the shape of the brain in combination with EEG activity maps on its surface).

The process of finding the electrodes required great deal of visual skills. We therefore tried to write an interactive tool for performing the task. Later on this tool has been enhanced to provide true volumetric rendering. (The rendering algorithms included are themselves product of a research and are also described in the contribution.)

[1] Also affiliated with the Institute of Measurement Science, Slovak Academy of Sciences, Dúbravská cesta 9, 842 19 Bratislava, Slovakia

The interactive paradigm

The 3D image manipulation is based on the concept of a 'data cube'. The stack of image slices forms a rectangular parallelepiped that could be completed to a cube. This cube could then be manipulated using rotation and scaling. To provide direct feedback to the operator, simple wire model (including visibility) of the cube is drawn on the screen and manipulated in real-time using a pointing device (mouse). The sides of the cube are marked (numbered) and the marks are also included in the wire model to avoid confusion during rotation.

The wire model itself would, however, not bring the required level of visual feedback. The ideal feedback would, of course, be reached by real-time 3D-rendering but due to obvious time problems with rendering algorithms, we had to find another solution. This have been solved by using six pre-computed rendered images corresponding to perpendicular views of the cube sides. These images are then 'glued' to the walls of the cube and they are imaged simultaneously (Fig. 1). The imaging procedure avoids rendering and consists of pure image warping. Hence, the complete cube including wall images appears almost immediately after the dragging operation.

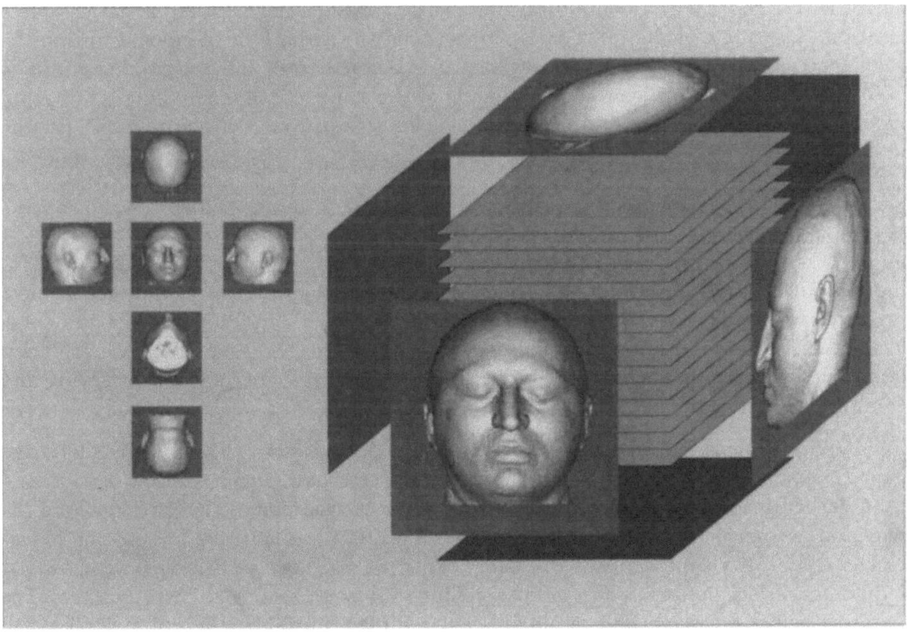

Fig. 1: Covering the data cube with precomputed rendered views.

The other mode of 3D data presentation is the true volumetric mode (called 'sculpt' mode in order to distinguish it from the 'cube' mode). At any stage of manipulation, the user can switch to the 'sculpt' mode and see the object (including pre-selected cuts) in

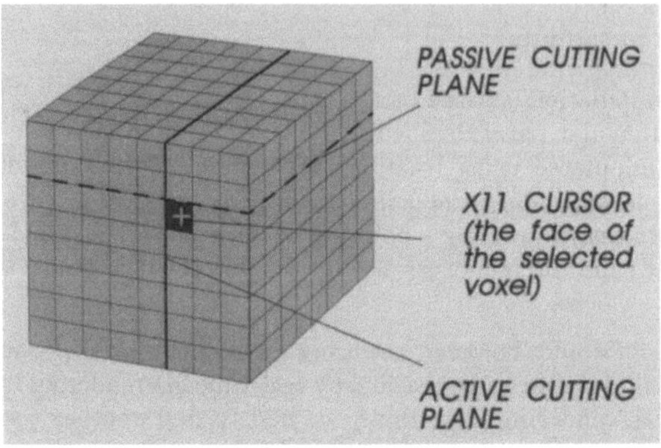

Fig. 2: 3D crosshair cursor.

realistic appearance. The cubic envelope is still present in order to retain the coherence of the manipulation.

The cube has only two rotational degrees of freedom (rotation and elevation) because the need to use all three degrees is very rare in the practice and the two degrees allow intuitive control with a mouse: Horizontal dragging causes rotation and vertical dragging causes elevation. The angle of the rotation is proportional to the path of the mouse during the dragging operation. The transforming factor is derived from the actual characteristic size of the cube. (For a bigger cube, dragging of the same length means smaller rotation angle). Thus, the user has the feeling of touching the surface of the mouse with the cursor.

Cutting and measuring options

The most important feature of the program is the cutting option. In order to inspect the inside of the data volume, the user can either 'stretch' any dimension of the cube, thus actually select the uppermost slice to be displayed, or to use a 3D crosshair cursor sliding along the cube surface and indicating the potential cut plane, position it and then perform cutting.

The 3D crosshair cursor is bound to the standard X11 cursor. Whilst the X11 (cross-shaped) cursor points to a face (wall) of a voxel, the crosshair cursor (consisting of a solid and a dashed trace lines) indicates the two possible cutting planes that are touching the voxel and are perpendicular to the given face (see Fig. 2). The trace lines of the planes are shown on all visible walls of the cube. This makes exact positioning with respect to different object possible. The solid trace line indicates the actual cutting plane. The operator toggles between the two possible cutting planes using the mouse button.

No interpolation is done between slices. Even at large scales, the voxels are depicted as cubes. The aim of this approach is to restrict interpolation artifact generation.

The program provides an option for defining marked points inside the data volume. These landmarks can be subsequently used for measuring operations. Moreover, interesting positions in the data volume can instantly be logged in the special log file (this option has especially been added for the EEG electrode location).

The Rendering Algorithm

For the sake of algorithmic compatibility of the 'cube' and 'sculpt' modes, we have chosen the following concept of 3D rendering in the 'sculpt' mode: An object is projected onto 3 visible bounding cube faces and thus obtained images are warped in the same way as cut planes or 'glued' orthographic projections.

A front-to-back algorithm based on ray casting with object space graylevel gradient shading was chosen.

In order to minimize the rendering time following speed-ups were implemented:

Fast voxel traversal: While for each image pixel a single ray is fired, a concept of voxel traversal based on *ray templates* [4] was chosen. A ray template represented as a sequence of 26-connected line voxels is computed only once for the given viewing direction and than is applied for tracing of rays passing through all image plane pixels. In the preprocessing phase, the scene is binarized (e.g. by thresholding) and to each non-object voxel its *chessboard distance* (CD) [3] from the nearest object voxel is assigned. The CD equal n represents a cubic macro-region of non object voxels with size $2n + 1$ with center in actual voxel, which enables to jump in the ray template voxel sequence to the n-th one, thus saving traversal time.

Ray tracing based on 26-connected rays may cause artifacts due to the fact that they can penetrate too deep into the object. Therefore, to each 26-ray voxel up to two following 6-connected ray voxels are assigned, which are, of course, checked only if voxel with CD equal 0 is found.

Adaptive image undersampling: The rendered image often consists of large areas of constant intensity (e.g. background). Therefore, regions with more object details are rendered with higher density of rays than those with less details. Rendered image is sampled recursively starting from, say, 32x32 pixel regions, which are subdivided until some predefined image homogeneity is found.

Conclusions

The presented program allows simple manipulation and visualization of 3D tomographic data sets. It uses a model of data cube that can arbitrarily be manipulated (rotated, scaled, cut). The novelty in comparison with similar interactive paradigms consists in using covering views on the walls of the cube, thus eliminating the need of time-consuming rendering and still providing realistic feedback for the operator. Moreover, the cutting operation is defined by the 3D crosshair cursor that considerably increases the level of interactivity.

The program itself has been designed for 3D tomographic data in the common sense. However, interesting applications could also be found in other areas where a kind of multivariate data with dimensionality 3 should be manipulated and visualized.

As an example, Fig.3 shows magnetic resonance spectroscopic imaging data set (see research results described in [5]). Here, two dimensions of the data set correspond to common geometry, the third dimension represents spectrum. Thus, in comparison with traditional spectroscopic imaging techniques (see [6], for example), more consistent and uniform visualization paradigm has been obtained.

Acknowledgements

The authors acknowledge the help of Dr.Serge Weis who provided the 3D MRI data sets, and Ján Weis, who provided the spectroscopic data.

The work described has been supported by the grant P8189-MED of the FWF (Austria).

References

[1] Thaller,R.A., Petsche,H., Rappelsberger,P., Pockberger,H., Lindner,K., and Imhof,H.: An approach to a synopsis of EEG parameters, morphology of brain convolutions and mental activities. Brain Topography 4 (1991), 1, 65–73.

[2] Robb,R.A.: 3-D Visualization and analysis of biomedical images using ANALYZE. In: Lemke,H.U., Rhodes,M.L., Jaffe,C.C., and Felix,R. (eds.), Computer Assisted Radiology. Proc.of the Intl.Symposium CAR'91. Berlin: Springer-Verlag 1991, 685–698.

[3] Borgefors,G.: Distance transformations in digital images. Computer Vision, Graphics, and Image Processing, 34 (1986), 3, 344–371.

[4] Yagel,R. and Kaufman,A.: Template-based volume viewing. In: Kilgour,A. and Kjelldahl,I. (eds.), Proceedings of Eurographics '92, 1992, C153–C167.

[5] Weis,J: Magnetic resonance spectroscopic imaging. Internal report. University Hospital, Uppsala, 1992.

[6] Maudsley,A.A., Lin,E., and Weiner,M.W.: Spectroscopic imaging display and analysis. Magnetic Resonance Imaging, 10 (1992), 471–485.

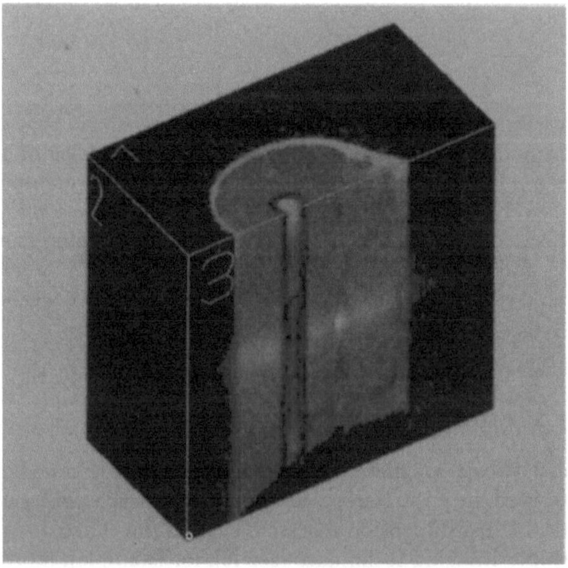

Fig. 3: Spectroscopic imaging – another potential application of the tool.

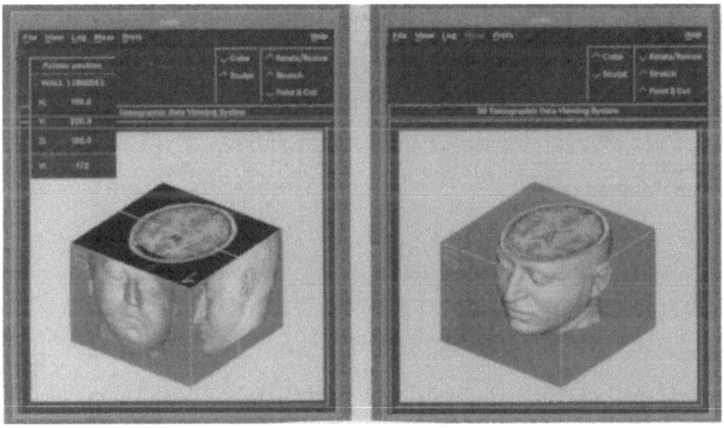

a b

Fig. 4: Two rendering modes: (a) 'cube', (b) 'sculpt'.

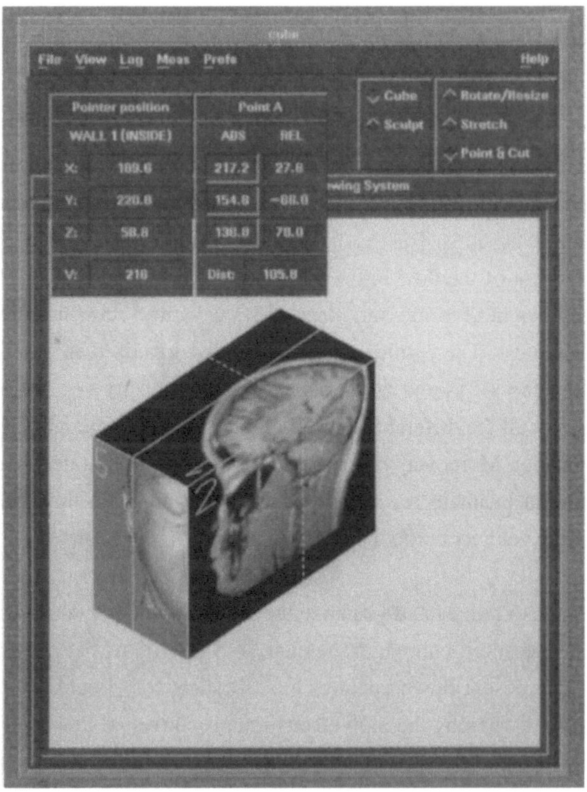

Fig. 5: Measurement using the predefined point.

Visualizing 3D Flow in MR Phase Contrast Angiography

J. F. L. DE BECKER, M. FUDERER and M. KOUWENHOVEN

Philips Medical Systems
Dept. MR-Systems & Methods
P.O. Box 10000
5680 AC Best
The Netherlands

Summary
Phase sensitive MR techniques allow measurement of 3D velocity vector fields inside a 3D volume of interest. Since these data are presented as quantitative maps of 3 velocity components in the chosen velocity encoding directions, visualization and interpretation of the nature of the underlying flow patterns poses a difficult problem. Two techniques to facilitate the comprehension of the velocity data are presented. In the first method we simulate flow by overlaying a time varying pattern to a MRA projection, where the pattern moves in the direction of the local velocity vector. In the second method seed points are tracked through the velocity field and the trajectories of these seeds are displayed overlaid on a MRA projection image. Both methods give additional insight in the 3D flow patterns.

1. Introduction

Phase Contrast Magnetic Resonance Imaging (PC-MRI) is based on the fact that the transverse magnetisation of spins moving in the presence of a magnetic field gradient obtains a phase difference relative to the static spins. The resultant phase shift of the moving spins is proportional to the velocity component of the flow in the direction of the applied gradient [1-3]. Using this property, PC-MRI allows in vivo measurement of 3D velocity vector fields of flowing blood inside a volume of interest. The resulting data is usually organized in 3 sets of quantitative maps, one for each of the x-, y- and z-components of the velocity vectors. It would be very time consuming to view all the different slices of all the velocity components to get an impression of the flow behaviour. Moreover, viewing the 3 components separately makes it very difficult to comprehend and evaluate the nature of the underlying 3D flow patterns, since the correlation between the velocity components is not used.

A convenient technique to reduce and visualize the 3D PCA velocity data is crucial to analyse the blood flow in the measured volume. Yang et al. [4-5] used flow pattern rendering and animation techniques to represent flow structures in a 2D slice. Napel et al. [6] proposed the use of simulated streamlines whereby the 3D velocity data is converted in a set of 3D streamlines that can be viewed with standard 3D display software. We developed two methods to visualize the 3D PCA flow velocity data. In the first method, the flowing pattern method, flow is simulated by overlaying a conventional MR angiogram with a time varying pattern which moves in the direction of the local velocity vector. The second method, the particle tracking method, computes the trajectory of seed particles that are placed interactively or randomly in the velocity field. These trajectories are shown as moving particles, overlaid with a MR angiogram.

2. Methods

Both techniques use animation to visualize the flow characteristics of the blood in the volume of interest. Starting from a "momentary" (non-triggered) acquisition of the blood flow, a set of images (first method) or trajectories of seed particles (second method) is calculated. Displaying such a set in a cine loop gives a good overall impression of motion from which the velocity field can be examined.

In the following, we denote a point in the volume of interest by the vector $\vec{r}$:

$$\vec{r} = (x, y, z), \tag{1}$$

the velocity vector $\vec{v}$ in $\vec{r}$ by:

$$\vec{v}(\vec{r}) = (v_x(x, y, z), v_y(x, y, z), v_z(x, y, z)), \tag{2}$$

the speed $v(\vec{r})$ as the magnitude of velocity vector $\vec{v}(\vec{r})$ in $\vec{r}$,
and the projection $\vec{v}_{pq}$ of a velocity vector $\vec{v}$ into the plane of a MRA projection image by:

$$\vec{v}_{pq}(p, q) = (v_p(p, q), v_q(p, q)) \tag{3}$$

where (p, q) are the coordinates of a pixel of the projection image.

Some assumptions concerning the acquired PCA velocity data were made:

- velocities are identical for all spins in a measured voxel.
- velocities are constant during the whole acquisition.
- higher order effects like acceleration, jerk, etc. are ignored.

2.1. Flowing Pattern Method

This visualization method uses a space-periodic pattern generated in a MRA projection image of the PCA data. To simulate flow, the pattern moves over a time series of images in the direction of the local velocity vector. The pattern discussed below consists of black lines perpendicular to the local flow direction (see fig. 3). Other patterns can be chosen.

The pattern generation and display process consists of the following steps:

1. Projection of the Phase Contrast angio data to a common MR angiogram (e.g. a MIP).
2. Determination of the velocity vector $\vec{v}(x, y, z)$ in the pixels (p, q) of the angiogram.
3. Transformation of the velocity components $v_x(x, y, z)$, $v_y(x, y, z)$ and $v_z(x, y, z)$ to the velocity components $v_p(p, q)$ and $v_q(p, q)$ in the plane of the projection image and determination of the direction of the vector $\vec{v}_{pq}$.
4. Generation of a sequence of N images using:

$$I_n(p, q) = A(p, q) \cdot \left[1 - \delta_M\left(\frac{1}{M}(a_{pq}p + b_{pq}q) + \frac{n}{N}\right)\right] \qquad n = 0, 1, \ldots, N-1 \tag{4}$$

where: • I_n is the intensity of the n-th image;

• $A(p, q)$ is the MRA image;

• $\delta_M(u) = \sum_j \delta(u - jM)$;

• M gives the spatial period of the pattern, N the time period of the series;

• a_{pq} and b_{pq} are functions of the direction of $\vec{v}_{pq}$, as shown in fig. 1.

These N images show a pattern of lines in every vessel perpendicular to the flow direction. If n increments by 1, the pattern in the vessel is shifted one pixel in the direction of the local velocity vector (see fig. 3).

a_{pq} and b_{pq} split up the (v_p, v_q) -plane in 8 discrete octants. This is done to obtain coherent patterns over the images.

5. Display of this sequence of computed images in a cine loop.

Note that the magnitude of the velocity vector is never considered in this algorithm. In fact, the obtained results are qualitative rather than quantitative.

2.2. Particle Tracking Method

In this method, the flow field is examined by tracking the motion of seed particles, injected in a vessel of interest, within the vessel structures. Projecting the positions of these particles for successive simulation time steps on a MR angiogram gives the observer a semi quantitative motion impression according to the acquired velocity field.

The algorithm consists of the following steps:

1. A set of particles is generated within one or more vessels of interest. This can be done either interactively by an observer who chooses the initial positions of the particles, or the initial positions can be spread randomly over the whole 3D measured volume to get a complete overview of the flow characteristics in the volume of interest.

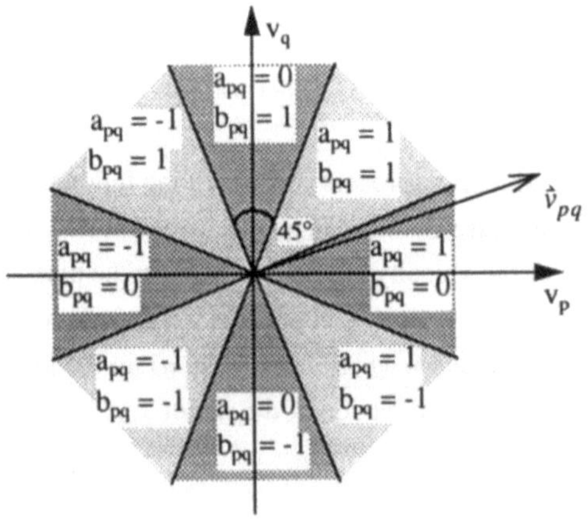

Fig. 1. a_{pq} and b_{pq} as functions of the direction of $\vec{v}_{pq}$.

2. The trajectory of each particle in the volume of interest is computed according to the measured velocity field as a function of the simulation time.
   The used formula is:

$$\vec{r}(nT) = \vec{r}((n-1)\cdot T) + \vec{v}[\vec{r}((n-1)\cdot T)]\cdot T \qquad n = 1, 2, \ldots, N \qquad (5)$$

   where:
   - T is the sampling interval of the simulation time. To ensure that no voxels of the velocity field are skipped in the computation, T is constrained to: $T < d/2v_{max}$ (d is the smallest voxel dimension and v_{max} is the maximum speed in the flow field);
   - $\vec{r}(nT)$ is the new position of the particle ($\vec{r}(0)$ is the initial location);
   - $\vec{r}((n-1)\cdot T)$ is the old position of the particle;
   - $\vec{v}[\vec{r}((n-1)\cdot T)]$ is the velocity vector at the old position of the particle;
   - N is the number of steps for which the trajectory of the particle is calculated.

   The successive 3D-positions $\vec{r}(nT)$ of the particles are stored.
   As mentioned previously higher-order motion terms like acceleration are neglected.
3. Display of a MRA projection image.
4. For each calculated time step separately, the 3D positions of all the generated particles are projected and displayed in the MRA image. The simulated motion of these particles in the vessels represents the measured velocity field. To recognize the particles in the vascular anatomy, they are encoded as black or coloured points.

The particle trajectory is computed using the velocity vector (magnitude and direction). Contrary to the flowing pattern method, blood speed differences can be noticed.
Since the trajectories of the chosen particles will generally not pass through the centre of a measured voxel, interpolation is necessary to compute the local velocities. To reduce computation time, a trilinear interpolation was chosen.

3. Results

The two visualization methods have been applied to clinical data sets obtained on healthy volunteers. All data were acquired with a 3D Phase Contrast MR method on a 1.5 T imaging system (ACS; Philips Medical Systems, Best, The Netherlands). Scan parameters (field of view, resolution, etc.) were set to obtain cubic voxels of 1.0x1.0x1.0 mm. We selected an examination of the neck vessels to illustrate the results of the visualization methods. Acquisition parameters were TR = 29 ms, TE = 10.4 ms, flip angle = 20°, a 275 mm field of view and 25 256x256 slices of 1.1 mm thick. Fig. 2 shows a maximum intensity projection in coronal direction of the angio data. The white box in the angiogram shows the region of interest on which we zoomed in figs. 3 and 4 to clearly display the results of both techniques. Fig. 3 shows two successive frames out of a time series of eight images generated by the flowing pattern method. The line pattern in the second image is shifted one pixel in the direction of the local velocity vector in comparison with the first image. Fig. 4 shows three frames out of a time series of 180 calculated particle locations generated by the particle tracking method. The first image shows the initial locations of a number of selected particles. The two other images show samples of the trajectory of these particles in the vessel of interest.

4. Discussion and Conclusion

The flowing pattern method and the particle tracking method are both convenient techniques for the visualization of 3D flow fields. However, 2D display of the 3D vessel structures and the 3D flow phenomena, like in figs. 3 and 4, do not give a fully, satisfactory insight in the 3D flow patterns. Combining the above described methods with commonly used techniques for displaying 3D data, e.g. object rotation (cine loop of several projections at different viewing angles), provide the information of the flowing patterns and the particle trajectories in a more efficient-way than in a cine loop with all images computed starting from one single projection image. Combined with a 3D data display technique, both methods can give a clear, quick overview of

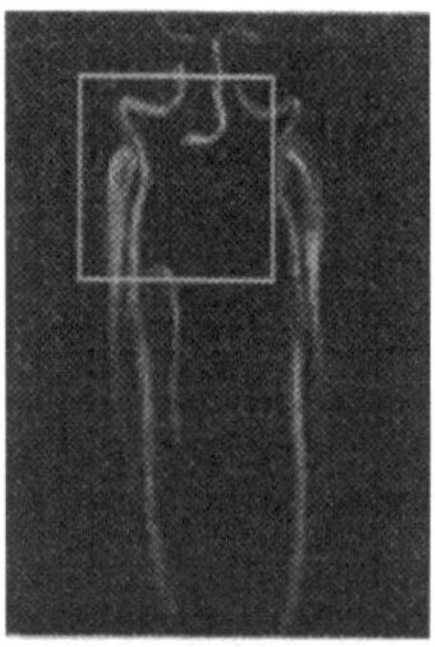

Fig.2. Coronal MIP of a PCA neck examination. The region of interest on which is zoomed in figs. 3 and 4 is surrounded by the white box.

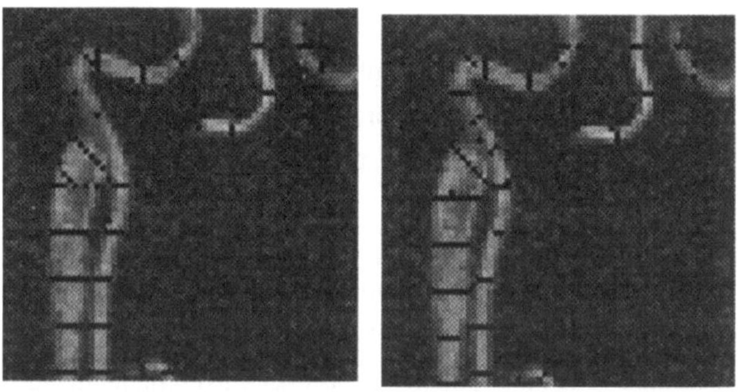

Fig. 3. Two successive frames of a time series generated by the flowing pattern method starting from the PCA neck examination (M=N=8). Compared to the first image the pattern in the second image is shifted one pixel in the flow direction.

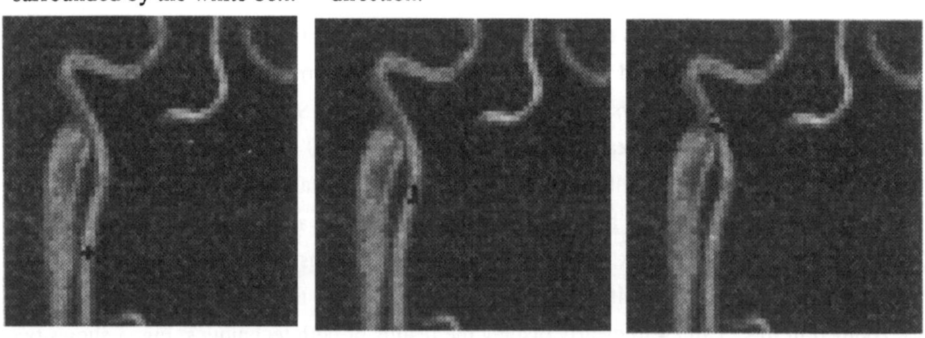

Fig. 4. Three time frames showing samples of the trajectory of a set of particles as calculated by the particle tracking method starting from the PCA neck examination. The first image shows the initial locations of the particles.

the blood flow behaviour in all the vessels of the measured volume of interest, without any significant loss of information about the vessel geometry. In the particle tracking method, this overview can be achieved by generating a large set of seeds randomly spread over all the vessels.

The particle tracking method has some advantages compared to the flowing pattern method:

1. The flowing pattern method in fact represents a 2D flow simulation. Before the actual pattern is computed, the 3D velocity vectors are projected on the plane of the MR angiogram to 2D velocity vectors in that projection plane. Starting from these 2D vectors, the time series of images containing the flowing pattern are computed. The particle tracking method really computes the 3D trajectories of the particles through the vessels, this means a real 3D flow simulation.
2. The velocity magnitude is never considered in the flowing pattern method, just the flow direction is displayed; while a particle trajectory is computed using the 3D velocity vector, i.e. considering magnitude and direction.
3. The 2D velocity vector directions are reduced to 8 discrete directions to obtain coherent patterns over the images. This leads to abrupt changes in the pattern in the neighbourhood of curves of the vessels (examples can be found in fig. 3).

The major problem of the particle tracking method is that particles tend to stick to the boundaries of the vessels or even tend to leave the vessels (they seem to flow through the vessel walls). Contributing to this effect are a.o.: noise in the acquired velocity data; the error caused by the assumption that the 3 velocity components are simultaneously acquired (which is impossible in practice); the error due to the negligence of higher-order motion effects; the error caused by the fact that the interpolated velocities may be the average of velocities of voxels inside and outside the vessel, whereby the contribution of voxels outside the vessel contains more noise.

References

1. van Dijk, P.: Direct cardiac NMR imaging of heart wall and blood flow velocity. Journal of Computer Assisted Tomography 8 (1984) 429-436.
2. Pelc, N.J.; Bernstein, M.A.; Shimakawa, A.; Glover, G.H.: Encoding Strategies for Three-Direction Phase-Contrast MR Imaging of Flow. Journal of Magnetic Resonance Imaging 1 (1991) 405-413.
3. Hausman, R.; Lewin, J.S.; Laub, G.: Phase-Contrast MR Angiography with Reduced Acquisition Time: New Concepts in Sequence Design. Journal of Magnetic Resonance Imaging 1 (1991) 415-422.
4. Yang, G.Z.; Burger, P.; Mohiaddin, R.H.: In Vivo Blood Flow Analysis and Animation for Magnetic Resonance Imaging. Medical Imaging IV, SPIE Vol. 1233 (1990) 176:182.
5. Yang, G.Z.; Burger, P.; Kilner, P.J.; Mohiaddin, R.H.; In Vivo Blood Flow Visualization with Magnetic Resonance Imaging. Proceedings Visualization 1991 202:209.
6. Napel, S.; Lee, D.H.; Frayne, R.; Rutt, B.K.: Visualizing Three-Dimensional Flow with Simulated Streamlines and Three-Dimensional Phase-Contrast MR Imaging. Journal of Magnetic Resonance Imaging 2 (1992) 143:153.

3D Representation of Liver Vein and Portal Vein

TIAN-GE ZHUANG, CHUAN-FEN DONG, HUAN-CHEN ZUO*, YAN-QING HUA*

Shanghai Jiao tong University
Shanghai, People's Republic of China
*Shanghai Medical University

Abstract

A PC-based system and related algorithms for revealing 3D distributions of portal vein and liver vein are discussed. The system is composed of an interface between VAX 11/730 and PC-based image processing system. Data are read from CT tapes then pass through several processing stages such as image segmentation, edge detection, region-filling, data compression, hidden surface removal and shading. Some dynamic data structures are used to implement the FTB algorithm, effectively alleviating memory limitation inherent in personal computers. In order to get a clear vein image certain kind of angiographic agents has to be injected into them and proper window width and window level have to be set.

Introduction

Complete perception of 3D anatomic structure of organs is critical to the understanding of organs per se and is benifit to diagnosis, treatment and medical research. Some structures such as the 3D distributions of portal vein and liver vein are not very clear to people so far. The three dimensional information of these veins can be acquired from CT scans or anatomic sections.

We have developed a 3D surface reconstruction software for that purpose dealt with CT scans, on an personal computer in C language, with the magnetic tape drive in our existing VAX11 /730 as the input peripheral. The CT data on the tape is read out from this device and is transferred to the disk of the personal computer with the help of an interface between them.

The present system is also capable to reconstruct images from anatomic sections.

CT Data Format Conversion

In our example, the CT data was taken from the Model 2060 CT scanner produced by former TECHNICARE CO. There are 257 records in each image file. The record has a length of 2048 bytes. The first record registers the patient data,while the rest of 256 records carry the CT numbers of each individual pixel of the image. The two consecutive bytes are combined to form a single CT number. The first 1024 bytes represent the CT numbers corresponding to the odd rows of the image, while the next 1024 bytes corresponding the even rows. Because the monitor has a limited capability of displaying 256 grey levels only, it is necessary to set a window with suitable window width and window level. Proper selection of these window parameters is very important for noiseless display.

Image Segmentation and Edge Detection

Generally, the organs to be reconstructed contain different type of tissues. To properly reconstruct these tissues, segmentation is necessry. In the light of our ad hoc task, the segmentation can be implemented as follows:

1) Detect the contours of the regions of interest first;

2) Fill the enclosed regions with assigned grey values.

We nave successfully segmented the portal vein, the liver vein and the liver tissue with this algorithm. In order to enhance the veins, some kind of angiographic agents were injected into them. Experiments show that it is hard to distinguish the veins from the liver tissue, it is equally hard to tell the veins from each other by their grey values only , even using different agent for different vein. In order to reveal the edges clearly. We have to carefully select the CT window width and window level.

The edges of veins are detected with Sobel operators.

Region Filling

After edge detection, a set of closed contours and the regions defined by these contours are obtained. By filling these regions with appropriate grey values, an image with finite number of tissues(e.g.3 tissues)is formed.

In implementing the region-filling operation, an algorithm with some trick was proposed, where a stack is used to store coordinates of adjacent pixels which have not been filled with the assigned grey value. The essentials of the algorithm are:

1) Interactively select an initial point (x,y) within certain region, moving this point left till it touches the left border, then moving toward the right border;

2) The new grey value is filled pixel by pixel as the initial point is moving from left to right;

3) As the filling procedure is going on, be sure to check the adjacent line. If this line is not on the border, or it has not been filled up, push the coordinate value of any pixel of that line into the stack;

4) Once the current line has been filled with the assigned grey value, a pair of (x,y) coordinates are pulled out.

5) Repeat the above procedure, starting from this point.

Data Compression

Before exerting the reconstruction,a data compression procedure must be fulfilled to reduce the data involved in the object space. As mentioned above only finite number of regions are concerned, each region with uniform grey value. Thus a run length coding is appropriate to our situation. With this data compression algorithm, a 33.7 compression ratio is achieved.

Hidden Surface Removal and Shading

There are several algorithms in performing hidden surface removal procedure. Among the most popular algorithms are Z-Buffer method, the Back-to -Front method etc. However, the Z-Buffer method needs to encode and compare every voxel in 3D object space. It is rather time consuming. The Back-to- Front method needs to project every voxel to the screen whether they are on the hidden surface or not. Obviously, in this case, the projection of the hidden surface is of no value.

We use a Front-to-Back(FTB) method to overcome the limitations posed by both Z-Buffer and BTF methods. By cleverly making use of the advantages of C language, we established some dynamic data structures to implement the FTB algorithm, effectively alleviating the virtual and physical memory limitations inherent in personal computers. By taking these measures, the overall processing time is reduced by a factor of 2 compared with that realized in FORTRAN 77 on a general image processing system (IIS S600 system on an IIS Model 75 image processor with a VAX11/730 as its host computer).

To achieve the realistic display of 3D object on a 2D screen, some depth cues are required to create the illusions of the 3rd dimension.

Two shading techniques are available in our software package: distant shading and gradient shading.

Distant-only shading produces smooth looking images in relatively short operation time, but the tiny structural features and real edges of the object are blurred.

Gradient shading on the other hand needs to calculate the normals of every point in pre-image obtained via object projections. It is time consuming. The major advantage of this shading mode is that it makes use of the information of the surface, and is suitable to describe the tiny structure of surfaces.

We propose a mixed shading method, distant-only shading for global image description use, gradient shading for describing the surface detail. However in our "liver" reconstruction case the distant-only shading mode is recommended , where only the global distribution of veins are of the main concern.

References

1. G.T. Herman, R.A. Renolds and J.K. Udupa: Computer techniques for the representation of three-dimensional data on a two-dimensional display. Proc. SPIE. Vol.367, 1982,pp.3-14.

2. S.M. Goldwasser, R.A. Renolds, D.A. Talton and E.S.Walsh: Techniques for the rapid display and manipulation of 3-D biomedical data. Comp. Med. Image. and Graphics, Vol.12, No.1, 1988, pp.1-24.

3. D.C. Hemmy, D.J. David, G.T. Herman: Three-dimensional reconstruction of craniofacial deformity using computed tomography. Neurosurgery Vol.13, No.5, 1983, pp.534-551.

3D CT Reconstruction of an Ancient Egyptian Mummy

S.W. Hughes., A. Sofat., D. Whitaker., C. Baldock., R. Davis., W. Wong., K. Tonge., G. Spencer.

Departments of Medical Physics and Radiology, St. Thomas' Hospital, and the Department of Basic Dental Science, University of Wales, Cardiff, and the Department of Egyptian Antiquities, The British Museum, London.

Summary
A mummy of an Egyptian priestess of the 21st dynasty (c1000 BC) was CT scanned and a series 3D reconstructions produced. These demonstrate many features of the embalming technique and funery customs, for example the removal of the brain through the nose. The body is in a very good state of preservation confirming that the art of embalming reached its peak during the 21st dynasty. The remains of organs are seen within the body, including the heart. Several amulets have been placed on the body. The good condition of the teeth, and evidence of uncompleted formation of the third molars indicates that the priestess was only about 20 years of age when she died.

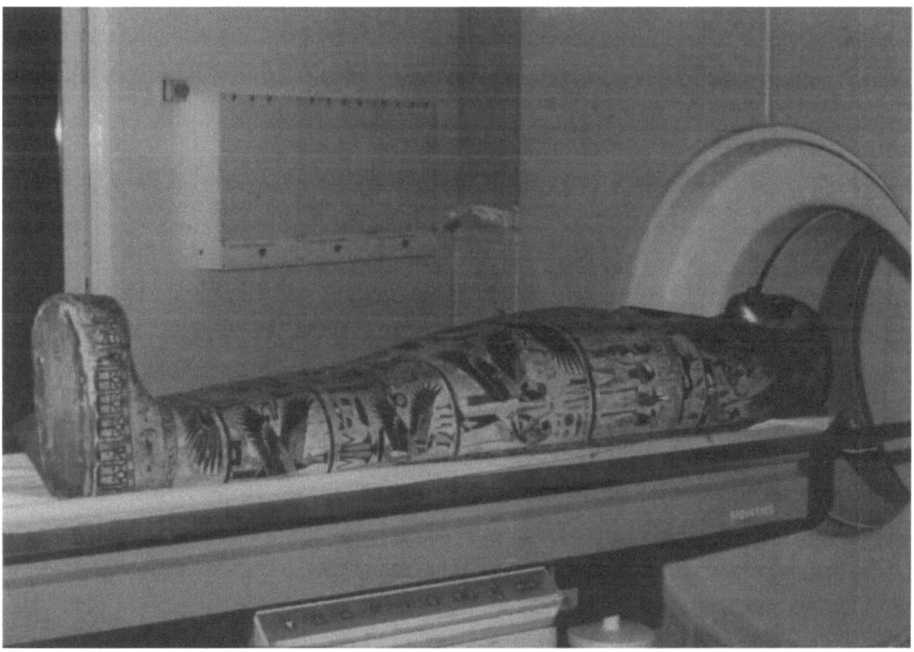

Fig. 1. The mummy on the scanner

(see also in color on page 814)

Introduction

Over the last few years, various research groups around the world have employed X-ray Computed Tomography (CT) imaging in the study of mummies [1-3]. Autopsies have also been carried out [5-6], but these require the destruction of both the outer coffin containing the mummy as well as the mummy itself. Modern archaeologists wish to keep the coffin intact in order to maintain complete preservation of the mummy. Plain X-rays have been used on a number of occassions but these are of limited use due to the problem of the superimposition of structures. CT imaging enables the internal structure of mummies to be examined in detail. The main attraction of CT is that it is non-destructive. The mummy chosen for this project is of a priestess from Karnak (called Tjentmutengebtiu, which will be abbreviated to Jeni. Jeni is sealed in an anthropoid coffin made of cartonnage, a mixture of linen and plaster.

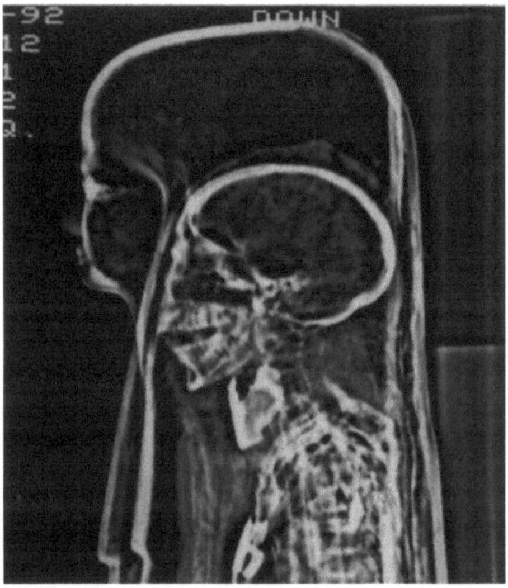

Fig. 2. Lateral X-ray

Method

The mummy was transported from the British Museum to St. Thomas' Hospital on five occasions. The head and neck were scanned with 2mm spacing between slices, using a 512 x 512 matrix, and a magnification factor of 2 pixels per mm. The rest of the body was scanned with 4mm spacing, same matrix size, and a magnification factor of 1.3 pixels per mm. The teeth were rescanned with 1mm between slices, 512 x 512 matrix and a magnification factor of 3.7 pixels per mm. A total of about 700 images were produced. The CT image data was used to produce 3D surface renderings of various features.

Results

The body is in a very good state of presevation, confirming the view that the art of embalming reached its peak during the 21st dynasty. A plate covering the left flank incision is seen. Four main organ packs have been placed in the chest. The abdomen has been packed with linen. Each organ pack encloses a wax figure representing one of the four sons of the god Horus. The remains of an organ which could be the heart has been identified.

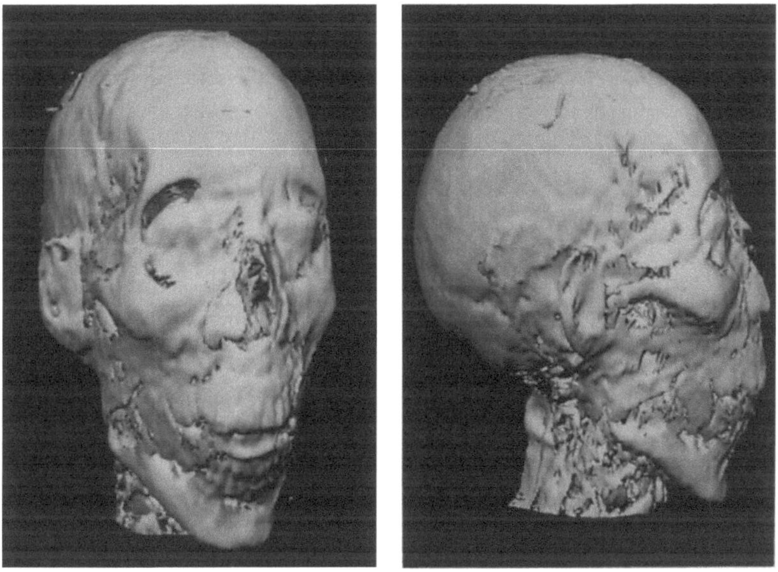

Fig. 3. Surface reconstruction of Jeni's face

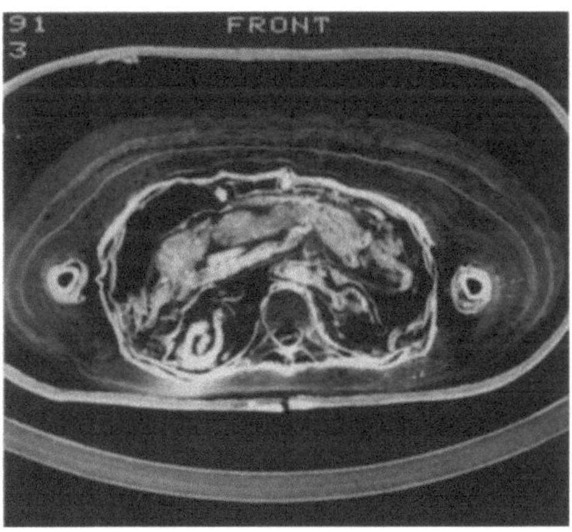

Fig. 4. CT through organ packs

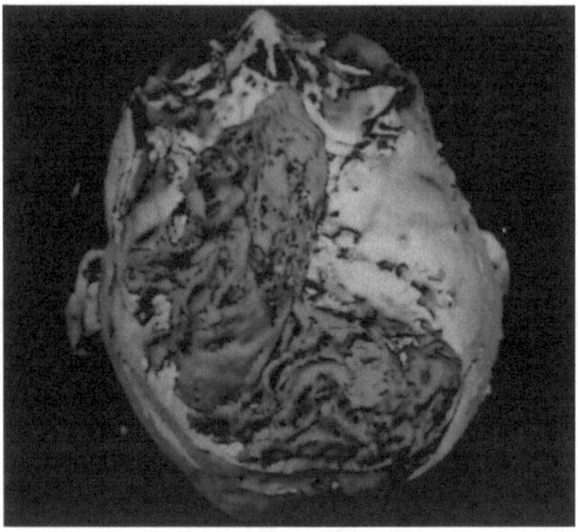

Fig. 5. Linen inside the head converging on the nose

Reconstruction of the skull base shows linen converging onto the nose. This linen must have been pushed through into the cranium after the embalmers had extracted the brain via a fracture in the ethmoid bone. Jeni has artificial eyes made of glass, and amulets across her neck, sternum, naval, and feet.

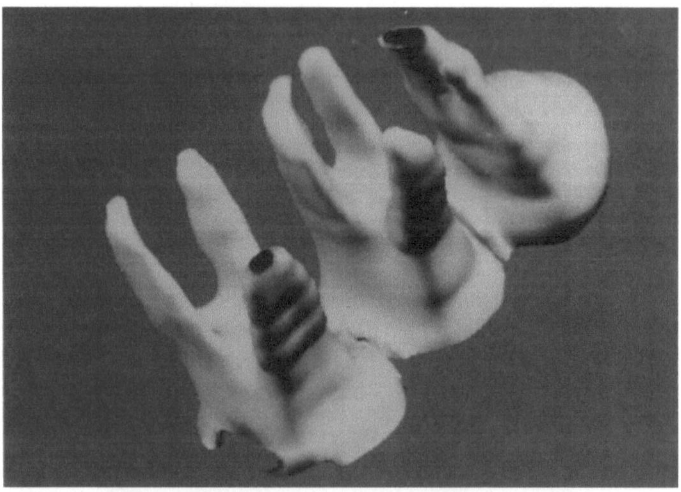

Fig. 6. Apices of the roots of the upper left molars

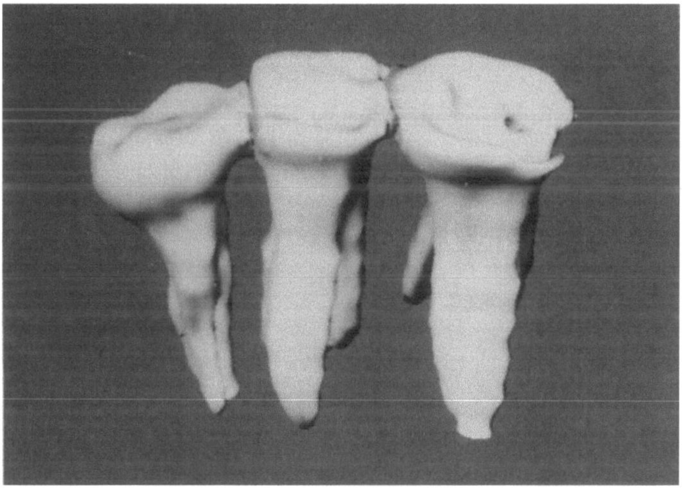

Fig. 7. Occlusal surfaces of the upper left molars

Jeni's teeth are in very good condition, suggesting that Jeni was fairly young when she died. The Egyptian's teeth are known to have worn down extremely quickly, with most of the enamel gone by middle age. This is corroborated by evidence of open apices of the third molars, which do not fully close off until 19-23 years of age. In addition the occlusal surfaces of the third molars are not as worn as the second and third molars suggesting recent eruption.

Conclusion

This study has shown that CT studies produce valuable information about the internal structure of a mummy that would otherwise only be obtainable by opening the case and unwrapping the mummy.

References

1. Marx, M.; D'Auria, H.D.: Three-dimensional CT reconstructions of an ancient Egyptian mummy. American Journal of Radiology 150 (1988) 147-149.

2. Harwood-Nash, D.C.F: Computed tomography of an ancient Egyptian mummies. Journal of Computer Assisted Tomography, 3 (1979) 768-773.

3. David, A.R.; Tapp, E. (eds.) The Mummy's Tale. London: Michael O'Mara Books Ltd 1992.

4. David, A.R. (ed.) The Manchester Mummy Project. Manchester: 1979

5. Cockburn, A,; Barraco, R.A.; Reyman, T.A.; Peck, W.H.: Autopsy of an ancient Egyptian mummy. Science 187 (1975) 1155-1160.

Design and Application

Dynamic Biomechanical Analysis of Deformation Stress during Labour by „Finite Elements Analysis" of MRI Data

K.J. LEHMANN[1], A. WISCHNIK[2], E. NALEPA[3], M. GEORGI[1]

Institut für Klinische Radiologie[1] und Frauenklinik[2] des Klinikum Mannheim, Fakultät für Klinische Medizin Mannheim, Universität Heidelberg
Electronic Data Systems (Deutschland), Rüsselsheim[3]

Summary
Pelvimetry and fetometry give static information about the cephalopelvic dimensions. In order to investigate dynamic aspects of cephalopelvic deformation during labour a method for dynamic biomechanical analysis is presented. MRI pixel matrices of a female pelvis and a foetus (1.5 Tesla) are color-coded and line datas ("Marching Cubes") are created. After sectional attribution of the resulting polygones a three-dimensional mesh of "Finite Elements" is produced, which allows analyses of deformation. The analyses are performed on a CRAY XMP computer (EDS Germany) using I-DEAS, ABAQUS and especially developed software. First analyses of deformation stress show, that under foetal head moulding conditions being considered as normal sensitive structures (cerebellum, brain stem, ventricles) are within the maximum isobars of 480-780 kPa. Maternal soft tissues are within a range of 820 kPa.

Introduction:

New imaging methods (MRI,CT,digital image intensifier radiography) are used for pelvimetry. Additional ultrasound fetometry gives information about foetal diameters. For the detection of cephalopelvic disproportion a dynamic analysis of biomechanical relation during labour is desirable. Therefore a method for computer simulation of vaginal delivery by dynamic postprocessing of static information obtained by MRI has been developed [8,12]. Based on a three-dimensional mesh of Finite Elements deformation and resulting forces can be calculated at any time of the delivery for any point of the anatomical model. Although high requirements towards computer capacity permit no routine use in obstetrics the presented model is suited to outline new parameters for the interpretation of pelvimetric measurements according to the simulation of different cephalopelvic conditions or labour forces.

Material and Methods

Because of high contrast in soft tissue imaging the pelvis of a 30 year old pregnant woman and the head of a new-born were examined with MRI (1.5 Tesla Magnetom SP63/Siemens). Using a specially developed software MRI pixel matrices of the maternal pelvis and the foetal head were colour-coded. The 2^{12} grey-values were translated into 256 colour-values. According to the method "Marching Cubes" [9] line data of equal density were created. After sectional attribution of the resulting polygones a three dimensional mesh of so called "Finite Elements" was developed, which allows deformation analysis [1,2,3,4,11,13]. Hardware and software were provided by EDS Deutschland GmbH, Rüsselsheim. On a CRAY XMP computer the analysis was performed with ABAQUS, the graphic presentation of the results was realized with I-DEAS. Soft tissue spezifications (density, elasticity, etc.) were taken from the literature [10]. On condition that there is a linear-elastic uterus, an incompressible-hyperelastic foetal head and no longitudinal shift of the uterus the movement of the foetal head was defined equivalent to labour conditions. "Slide lines" described the gliding contact between head and uterus. The resulting model is characterized by 277 degrees of freedom, which means the simultaneously solution of 277 equations.

Results

Pelvimetry of the maternal pelvis showed a transverse diameter of the pelvic inlet of 12.05 cm, an interspinous distance of 11.10 cm, an intertuberous distance of 12.00 cm, a sagital pelvic inlet of 11.60 cm and a sagital pelvic outlet of 11.90 cm. The foetal BIP was 9.50 cm, the biparietal diameter 9.50 cm and the fronto-occipital diameter 10.90 cm. At 8 anatomical landmarks of the foetal head time/pressure diagramms were calculated seperating the process of labour into 10 steps (fig.1). Passing the pelvic inlet pressure within the head increased to a maximum of 480 kPa in the region of the cerebellum, tentorium and medulla oblongata. Passing the pelvic outlet pressure within the head increased to a maximum of 780 kPa beside the chiasma opticum. A maximum of 820 kPa was calculated for the maternal soft tissues.

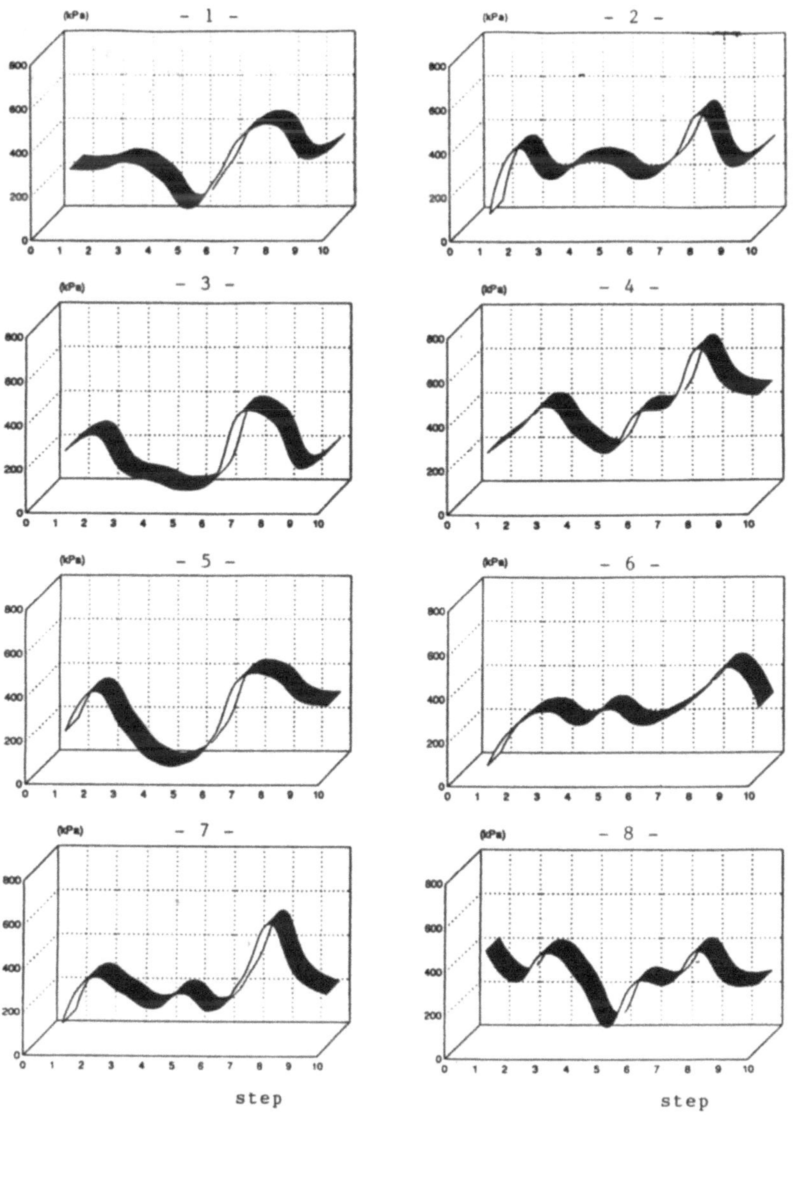

fig. 1: time/pressure diagramm during vaginal delivery in 10 steps at 8 anatomical landmarks: cerebellum (1), plexus choroideus (2), confluens sinuum (3), chiasma opticum (4), calcarina (5), gyrus frontalis sup. (6), tentorium (7), medulla oblongata (8).

Discussion

The static character of the information obtained by pelvimetry is mentioned as an argument against the clinical value of this examination. The development of a computer model based on "Finite Elements analysis" allows a simulation of the dynamic aspects of biomechanical deformation during labour. The cephalopelvic dimensions of the model are defined according to pelvimetry and fetometry data. Changing the cephalopelvic dimensions in the model various conditions can be simulated. Because of high soft-tissue contrast the analysis is based on the post-processing of MRI data. Former biomechanical simulation studies [5,6] were performed using CT data. Moving the foetal head through the birth channel by means of computed simulation implies the consideration of very complex obstetrical rotations and deformations of the foetus. For that reason requirements towards the capacity of the computer are very high. Therefore a routine use in obstetrics is not possible at the moment. Nevertheless, in case of supposed cephalopelvic disproportion an individual simulation of labour should be possible in the future by the input of pelvimetry and fetometry data into the modell.

Even under favourable cephalopelvic conditions (e.g. interspinous distance - BIP = 1.60 cm) and foetal head moulding conditions being considered as normal the Finite Elements analysis proofs high deformation stress of the foetal head and maternal soft tissue during labour. For areas with an increased risk of intracranial hemorrhage [7] time/pressure diagramms are calculated. Anatomical landmarks of high pressure during the passage of the pelvic inlet are the cerebellum, tentorium and medulla oblongata as well as the chiasma opticum and the periventricular region during the passage of the pelvic outlet. To outline borderline conditions of vaginal delivery further studies are necessary. Nevertheless the presented model is suited to develop a correlation between deformation stress during labour and corresponding pelvimetry and fetometry data. Furthermore, in case of supposed cephalopelvic disproportion dynamic computer simulation offers additional information to the obstetrician.

References

1. Argyris,J., H.P. Mlejnek: Die Methode der Finiten Elemente. Bd.I-III, Fiedr.Vieweg,Braunschweig/Wiesbaden 1986-88.

2. Bathe,K.J.: Finite-Elemente-Methoden: Matrizen und lineare Algebra, die Methode der finiten Elemente, Lösung von Gleichgewichtsbedingungen und Bewegungsgleichungen. Springer, Berlin/Heidelberg/New York 1990.

3. Becker,E.B., G.F.Carey,J.T.Oden: Finite elements. Vol.I-VI Prentice-Hall Inc.,Englewood Cliffs, New Jersey 1981-1985.

4. Courant,R., D.Hilbert: Methoden der Mathematischen Physik. Bd.I-II. Springer, Berlin 1967.

5. Endo,B., K.Adachi: Biomechanical simulation study on the forms of the frontal bone and facial bones of the recent human facial skeleton by using a two-dimensional frame modell with stepwise variable cross-section members. Okajimas Folkia Anat. Jpn. 64 (1988) 6.

6. Haskell,B.: Computer-aided modelling in the assessment of the biomechanical determinants of diverse skeletal patterns. Amer. J. Orthodont. 89 (1986) 363-382.

7. Jensen,A.:Das Hirnblutungsrisiko bei Früh- und Reifgeborenen. Geburtsh. Frauenheilk. 52 (1992) 6-20.

8. Lehmann,K.J., A.Wischnik, E.Nalepa, K.U.Wentz, M.Georgi Entwicklung eines Modells zur Simulation der Geburtsmechanik und zur Erarbeitung pelvimetrischer Grenzwerte auf der Basis der "Finite Element-Analyse" kernspintomographischer Daten. Fortschr. Röntgenstr. 157,3 (1992) 317.

9. Lorensen,W.E., H.E.Cline: Marching cubes, a high resolution 3D surface construction algorithm. Computer Graphics 21 (1987) 163-169.

10. Nahum,A.M., J.Melvin (Hrsg.): The Biomechanics of Trauma. Appleton-Century-Crofts, East Norwalk (Conn.) 1985.

11. Timoshenko,S.P., J.N.Goodier: Theory of elasticity. 3rd edition. (Engineering Societies Monographs) McGraw-Hill Kogakusha Ltd., Tokyo 1970.

12. Wischnik,A., K.J.Lehmann, E.Nalepa, K.U.Wentz, M.Georgi, F.Melchert: Die Magnetresonanztomographie als Grundlage biomechanischer Analytik. Fortschr. Röntgenstr. 157 (1992) 333-337.

13. Zienkiewics,O.C.: The finite element method. 3rd.edition McGraw-Hill Book Company, London 1977.

Integrated Modeling of Heterogeneous Information from Medical Images

Klaus D. Toennies

Fachgruppe Computer Graphics
Technische Universität Berlin
Franklinstrasse 28/29, FR 3-3
W-1000 Berlin 10, Germany

Abstract

An increasing number of applications in medicine makes use of multiple information available from radiological images. Representations being employed to describe this information are often designed to serve the specific purpose of a certain diagnosis or treatment. If such information is derived from different sources or if it is used in multiple applications a unified way for its description is necessary. Unification calls for two points to be considered. Firstly, the level of comparison for different representations needs to be investigated and, secondly, tools are needed to treat various kinds of representations in a uniform fashion. Presently, most of the information is used for display and manipulation of patient-specific geometry, thus ensuring a certain level of similarity between possible representations. However, the application of image fusion algorithms and the use of simulation models for treatment planning also requires the representation of analytical and functional entities. A common framework for the two types of representation is presented stressing the importance of an information exchange between analytical and geometrical representations. Comparability without integration, however, impedes the efficient use of represented information. Solutions for integration range from application-dependent approaches to the use of a common base representation. The former was found to be easy to handle while being difficult to change, whereas the latter has the advantage of being application-independent while requiring large overhead for managing the represented data. A solution combining the advantages of the two extreme points of view consists of a heterogeneous representation system based on the paradigms of object-oriented programming. If each representation is treated as an object with its own attributes and methods, two different representations may share just as much as their structures have in common.

Introduction

A 3D model representing knowledge from medical images provides the user with patient-specific information otherwise being unavailable. A number of applications in medicine relies on such information:

- In diagnosis the state and the change of anatomy and pathology needs to be determined. Evaluation through display of a 3D representation permits viewing the information in a true three-dimensional fashion.

- In therapy the interaction of an operator with patient specific anatomy and pathology is simulated. Operator-induced change is evaluated prior to and after treatment. Executing such simulation is not possible without the aid of a computer.

Even such a coarse classification shows that it is not likely that a single type of representation will serve every purpose in 3D computer assisted radiology. Various applications may result in as many representations being not necessarily compatible with each other. However, the need for 3D applications in radiology does not occur frequently enough to justify singular solutions. Integration of different kinds of representations is necessary. For this reason, an object-oriented representation system was developed being capable of integrating various types of geometrical representations from information depicted in medical images.

3D-representations for computer assisted radiology

A physician, evaluating state and perspectives of a certain case, may receive support from a computer offering geometric operations on a three-dimensional representation of a patient's anatomy and physiology. This information can be furnished to the physician in two ways leading to two different kinds of representation. In a *synthetical representation* patient data is displayed leaving its interpretation to the physician. In an *analytical representation* patient data undergoes computer-assisted analysis whose results are presented to the observer.

If display is the purpose of a representation, analytical information is not part of the model (except for simple measurements such as the estimation of volume). The representation should be as exact as possible, because no information is available as to how important elements of the representation are. Evaluation is solely based on the decision of the observing physician.

Several kinds of representation exist for this purpose (see fig.1). They can be differentiated into those representing inner object points (see, e.g., /Dreb88/, /Toen92/) and those describing the object by its boundary (see, e.g., /Herm79/, /Udup91/). All representations have in common, that objects are described through a set of primitives. Neither the primitives nor their arrangement has great semantical value, thus leaving the interpretation almost completely to the user.

An analytical representation, on the other hand, serves a different purpose. Morphological information is incorporated in the structure of the representation and possibly reflects anatomically or physiologically significant features. An analytical representation may either support the user's decision in diagnosis and treatment control or it may serve as a concise description of morphology, e.g., for image fusion. However, it is difficult to establish a relationship between morphology and biological processes because morphological invariants of biological structures are dependent on the kind of structure. Thus, an analytical representation will only approximate such invariants. Existing representations include the

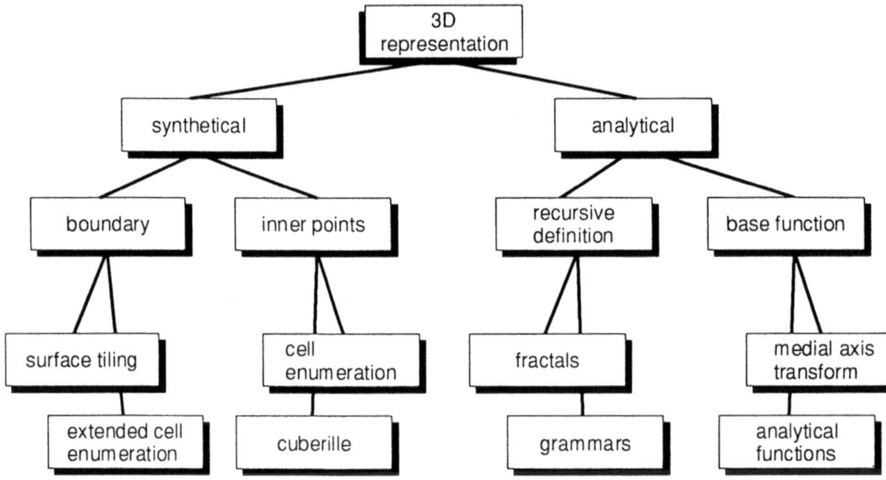

Figure 1: 3D representations of knowledge from medical images

medial axis transform /Nack85/, grammars for the description of morphology /Prus88/, and the description by a set of analytical functions /Toen93/ (see fig.1).

Different but sometimes intertwined problems are solved through synthetical and analytical representations. An example for such a combined need is the image fusion problem where different scenes containing the same object need to be matched. Registration requires identification of morphological features of the object while matching is executed on a synthetical representation. This example shows a need for mechanisms enabling integrated modeling of different representations.

Integration may happen on various levels depending on how much two representations have in common. It ranges from point-wise comparability to the filing of two representations under the same patient name (table 1 shows categories of comparability). If comparison is only needed for a specific purpose then it may suffice to let the methods applied for knowledge

elements of comparability	types of representation
point-wise	cell enumeration representations, analytical and synthetical surface representations
co-ordinate systems	additionally: any representation defined in a co-ordinate system (grammar type representations, ...)
attributes of objects	additionally: any symbolic representation
names	additionally: textual representations

Table 1: Comparability of different types of representation.

retrieval make use of all the information needed /Pize89/. This solution cannot be transferred to another application. Application-independence can be guaranteed, if representations are converted into a common base representation /Toen90/. This, however, requires comparability on a point-wise level as well as conversion of any patient-specific geometric description into the base representation neither of which is always possible.

An object oriented representation system

The high diversity of information types and the fact, that most of these types are exclusively connected to their own manipulation algorithms, has led us to the approach of developing a Heterogeneous Object-Oriented Representation System (HOORS, see /Toen92/). In object-oriented programming an object is an entity consisting of attributes and methods applied to them. Objects sharing the same attributes and methods constitute a class. Classes may have subclasses being a specification of the class and inheriting all its attributes and methods.

In the context of representing knowledge from medical images we are dealing with objects which represent aspects related to anatomy and physiology of the human body. The attributes of these objects consist of parameters to specialise the representation and of methods to create another object for the purpose of rendering the information to the user. E.g., a CT sequence, represented by a cell enumeration representation, would be a 3D data set, based on a volume model, interpolated and represented synthetically by enumerating voxel values, and, for example, used for radiation treatment planning.

Using a single inheritance scheme the object-oriented representation system becomes a hierarchy defining different levels on which information may be represented. On the lowest level the following classes were designed:

- a class *2D-CE* (2D cell enumeration) represents single CT slices or 2D output images.
- a class *3D-CE* describes a three-dimensional isotropic voxel cube.
- a class *3D-vary* contains a voxel cube whose slice thickness may vary from the side length in the slices.

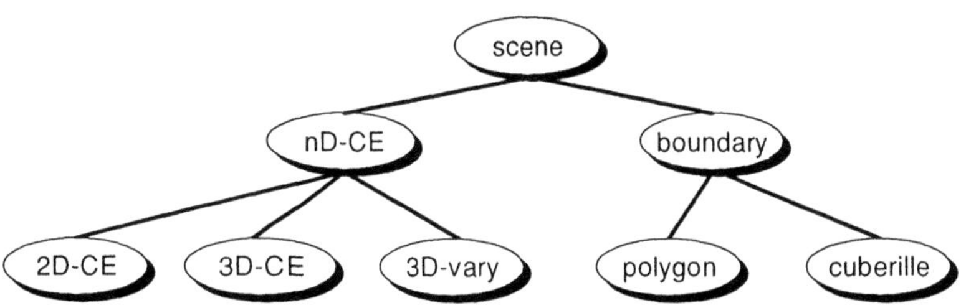

Figure 2: Object classes of the Heterogeneous Object-Oriented Representation System (HOORS)

- a class *cuberille* contains a surface representation by voxel sides
- a class *polygon* describes polygonal surfaces.

These classes have two super-classes *nD-CE* (n-dimensional cell enumeration) and *boundary* which have a super-class *scene* themselves (see fig.2). Besides descriptive information and methods to create, store and destroy objects, several 2D- and 3D-manipulation and -visualisation algorithms are implemented in the object classes (see /Toen92/).

Using the representation system

The representation system described above requires a carefully defined user interface in order to be utilised efficiently. Presently, no data management facilities are included and computation is restricted to data from a single patient, the goal being to develop interaction facilities to access represented information and to investigate manipulation methods of the actual representation structure.

Interaction with the system is realised by an object-oriented window-based interface realised parallel to the object classes /Star93/. The user may through a window manager select a patient, choose a representation object and call methods of this object. Selecting an object results in opening a new interaction window offering methods defined to access attributes in the object. In the simplest case this may be the request for the object name but access also includes calling 3D-visualisation routines, e.g., of a 3D-CE object. Each action opens a new window showing either the requested information or asking for input (e.g., to rotate a 3D-object). During the session the user may open a variety of interaction windows while working with specific patient data. This results in a hierarchy of interaction windows with each window being the child of some other window. Terminating an action associated with an open window is done by closing the window. This, however, causes the closure of all child windows of the terminated window.

In order to avoid confusion, the window manager offers administrative information as to what objects are present for a patient, which objects are presently evaluated and which methods are being executed.

Future developments

The object-oriented representation systems appears to be a powerful tool to handle heterogeneous 3D-representations from patient data. However, the versatility of the system may lead to confusion for users working with it. Therefore, our present plans are twofold:

- Firstly, investigations will be undertaken to facilitate the inclusion of new methods or object classes into the system. Either automatically or through guide lines the developer is enabled to assess the current system state and determine where and how new elements of the system shall be added without being forced to understand the complete structure.

- Secondly, new representation forms (representations for fast display /Udup91/ and analytical representations /Toen93/) will be added to increase versatility of the system.

The final goal is an easily extendible and accessible representations system for users as well as for developers facilitating the exchange of information between various applications and different locations.

Bibliography

/Dreb88/ Drebin, R.A. et al., 'Volume Rendering', *Comp. Graph.*, Vol 22(4), 1988, 65-74.

/Herm79/ Herman, G.T. et al., 'Three-Dimensional Display of Human Organs from Computed Tomograms', *Comp. Graph. and Image Proc.*, Vol.9(1), 1979, 1-21.

/Nack85/ Nackman, L.R. et al., 'Three-Dimensional Shape Description Using the Symmetric Axis Transform I: Theory', *IEEE Trans. PAMI*, Vol.7(2), 1985, 187-201.

/Pize89/ Pizer, S.M. et al., 'Volume Rendering for Display of Multiple Organs, Treatment Objects, and Image Intensities', *Proc. SPIE*, Vol.1137, Paris, 1989, 92-97.

/Prus88/ Prusinkiewicz, P. et al., 'Developmental Models of Herbaceous Plants for Computer Imaging Purposes', *Comp. Graph.*, Vol.22(4), 1988, 141-150.

/Star93/ Starosta, A., *Eine Benutzerschnittstelle für ein objektorientiertes Modelliersystem*, Studienarbeit, Technische Universität Berlin, März 1993.

/Toen90/ Toennies, K.D. et al., '3D Modeling Using an Extended Cell Enumeration Representation', *Comp. Graph.*, Vol.24(5), 1990, 13-20.

/Toen92/ Toennies, K.D. et al., 'Object-Oriented Modeling of Medical Image Data', *Proc. S/CAR'92*, Baltimore, 1992, 239-245.

/Toen93/ Toennies, K.D. et al., 'An Analytical Shape Representation', *Proc. CAIP'93*, Budapest, 1993, to appear.

/Udup91/ Udupa, J.K., 'Fast Visualization, Manipulation and Analysis of Binary Volumetric Objects', *Comp. Graph. and Appl.*, Vol.11(3), 1991, 53-62.

An Hypermedia System to Manage Anatomical Knowledge about Brain

E. MONTABORD[1,2], B. GIBAUD[2], C. BARILLOT[2], S. GARLATTI[1], I. KANELLOS[1], B.S. WU[2], A. BIRABEN[2] , X. MORANDI[2]

[1] LIASC: Laboratoire I.A. et Systèmes Cognitifs, Ecole Nationale Supérieure des Télécommunications de Bretagne, ZI de Kernevent, BP 832, 29285 Brest Cedex, France

[2] INSERM U335, Laboratoire SIM, Faculté de Médecine, 2 avenue du Pr Léon Bernard, 35043 Rennes cedex, France

Summary

This paper reports about first results obtained in a project aiming at developing a computerized system to manage knowledge about brain anatomy. The emphasis is put on the design of a knowledge base including a symbolic model of the cerebral anatomical structures (grey nuclei, cortical structures such as gyri and sulci, ventricles, vessels, etc...) and of browsing facilities allowing to retrieve and display information associated with the objects (texts, drawings, images). Atlas plates digitized from a stereotactic atlas are also used to provide natural and effective communication means between the user and the system.

Introduction

In their daily practice physicians need to refer to some knowledge about the organs and functional systems which may be concerned by the disorders presented by the patients. Most of this knowledge has been assimilated by the doctors during their medical studies and along their personal experience. Nevertheless, the need to use reference data exists and the medical literature provides the information the clinicians need for diagnosis and therapy.

In the particular field of neurosurgery, anatomical information plays a major role and consequently brain atlasses (such as the atlasses published by Talairach and Tournoux [1], Schaltenbrand [2], Ono [3] or Szikla [4]) are very important to provide the information the clinicians need to properly identify anatomical structures like the cortical gyri or to locate a given nucleus within the thalamus, for example. This kind of information is obviously needed to prepare surgical gestures like for instance the introduction of deep electrodes (in order to record the intra-cerebral activity) or to interpret the signals recorded on these deep electrodes. It is also fundamental that surgeons can identify the major functional areas in order to be able to anticipate the functional consequences of a surgical gesture.

In fact, it appears that surgeons have to "navigate" between different structures of information, namely information about the patient (morphological images like MRI, CT or DSA, functional data like PET, or physiological signals like EEG) and reference information available from various atlasses in both *symbolic* (information about the anatomical structures) and *numerical* forms (enabling some kind of modeling of the inter-individual variability). In addition to that, the kind of information to be used and the retrieval modalities may depend on each medical speciality and on the clinical context (interpretation of clinical symptoms, identification of anatomical features within images, surgical planning, intervention, etc).

Computerized brain atlasses have been mostly studied for educational purposes, (with emphasis put either on 3D representation issues [5, 6] or on the organization of symbolic models [7]), or to facilitate image interpretation [8, 9, 10, 11].

In this context, our research aims at improving the conditions of such a medical work. Hypermedia techniques seem appropriate to structure and access multi-typed information in an efficient way, especially if they incorporate dynamical aspects retracing some real cognitive activities of the user. The hypermedia system presented here aims at assisting anatomists and surgeons in the labelling of anatomical structures in MRI and angiography images, by allowing the access to a knowledge base by means of atlas plates.

The Hypermedia system

Hypermedia aim at offering a "natural" access to a wide range of information (texts, symbolic knowledge, images) by allowing the user to find his own path within the whole set of available information. They can be defined as a network made of nodes of information, connected together by links, along which the user can seek the relevant information [12]. Given the growing complexity of the information to be managed, current researches on hypermedia aim at taking into account the user profile to facilitate the access to relevant information [13]. Another general tendency consists in extending the knowledge management capabilities with enhanced man-machine interface integrating information retrieval and manipulation functions. The hypermedia system presented here is based on an artificial intelligence approach. In the first phase of this project, only two aspects have been explored : i) the design and implementation of the knowledge base and ii) the first part of the navigation system.

The hypermedia system currently includes i) the knowledge base itself, ii) a database including illustrations (like texts, schemata and images) and a set of atlas plates digitized from a stereotactic atlas (Talairach and Tournoux), and iii) a navigation system based on the hypermedia intrinsic navigation means (buttons, links, windows). The specific requirements met in surgery planning tasks are not addressed in this paper, since the registration of in vivo image data with atlas plates [14] has not been integrated yet.

The Knowledge base

Many points of view can be used regarding the design of a computerized brain atlas ; in order to consider several of them without scattering to much our efforts we decided to consider the anatomic and functional viewpoints with respect to surgery planning in epilepsy. We have identified the structures concerned in these two approaches and have represented the relationships existing between each other.

From the anatomical viewpoint we have considered structures like grey nuclei, ventricles, blood vessels and cortical structures such as gyri and sulci. These structures are organized in a hierarchy corresponding to the relation of inclusion associating the anatomical features. The following relations have been considered:

- ingredience ("composed of"), applied to gyri, ventricles, sulci and grey nuclei ; these relationships allow to propagate some properties along the ingredience links ;

- spatial relationships between gyri and sulci ("in front of", "behind", "above", "below", etc) ;

- inclusion in a lobe ("in lobe").

These relations aim at representing an anatomical knowledge which helps the identification and the labelling of cortical features (gyri, sulci).

At this moment the functional viewpoint is reduced to the description of the Brodmann's areas and the connections between them and the grey nuclei ; in particular, *"afference"* and *"efference"* relations between grey nuclei and Brodmann's areas have been represented.

The navigation system

The navigation system is based on the toolkit of Y3 (see below) and uses the structures of the knowledge base. Three kinds of information can be reached : brain images (atlas plates), symbolic informations (objects of the knowledge base) and illustration documents (texts, images, etc) ; the system allows to navigate between them. Once a structure has been specified by the user (within a graph, a text or an image) the navigation process identifies the selected structure and relates it with the corresponding object of the knowledge base in order to allow access to the knowledge available on it.

The first graph shows the hierarchy of the structures (Figure 1). The labels of the graph nodes are the names of the brain structures, which allows the user to access very easily to the objects properties by means of the simple selection of the structure of interest. For example the user can display the neighboors of a given gyrus, retrieve to which lobe it belongs to, if it is composed of several parts and from where it is visible. All this information is displayed in windows offering also the possibility to access to the knowledge concerning all the displayed entities (such as illustration data), providing a very powerful navigation means (Figure 2).

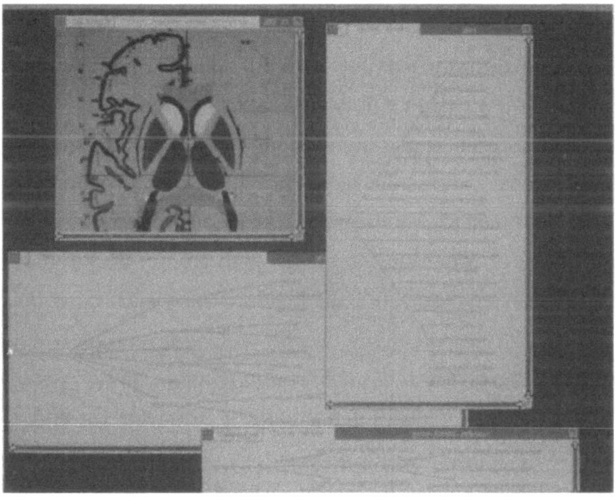

(see also in color on page 814)

Figure 1: Inheritance graph showing the instances of the gyrus object.

A specific navigation facility has been set up to manage the selection of anatomical features by means of atlas plates, in order to take into account the specific way surgeons use such atlasses. The Talairach and Tournoux stereotactic atlas contains three stacks of brain sections (along the axial, sagittal and frontal directions) representing the same brain. The X Y, and Z coordinates are displayed on each atlas plate, allowing to identify a given point in the 3D space and to retrieve it upon axial, sagittal and frontal images. This atlas is routinely used by surgeons to identify anatomical structures on CT or MR images. In order to assist this work, the system allows to display in a window the successive slices along a given direction (e.g. axial) and, upon selection of a point by the user, the system selects the two atlas plates along the two other orientations (e.g. frontal and sagittal) and computes the 2D positions of the 3D point within these two images and display them by means of a cursor. Moreover, once a point has been selected by the user, the system identifies to which anatomical structure it belongs to, allowing then the browsing of the associated symbolic information (Figure 2).

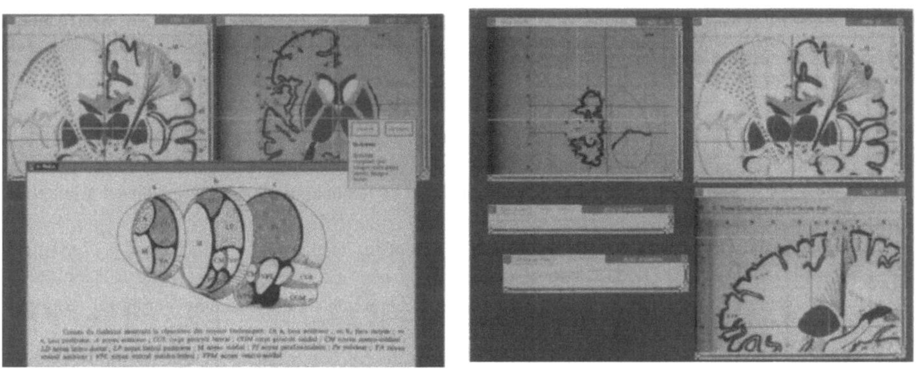

Figure 2 : Examples of utilization of the system:
Left : Selection of the Thalamus and retrieval of related knowledge and illustrations.
Right: Selection of the Pre-Central gyrus and retrieval of part of related knowledge.

(see also in color on page 815)

Implementation issues

The knowledge representation system and the navigation system are built on an inheritance system called Y3 (YAFOOL-YAFEN-YAFLOG) [15]. It is an hybrid object-oriented language based on frames, and involving message passing between objects. Objects are structured in a hierarchy, called inheritance graph, by means of links as "is-a" (between abstract objects) and "member-of" (between abstract objects and individuals). An object is defined by means of a set of properties; some of them are not declared in it but are shared with its ancestors (objects at more abstract levels) ; in a such case they are said to be inherited. Properties are attributes or relations with unique or multiple values. The "is-a" and "member-of" links stand for *inclusion* or *membership* relations and *property sharing* between objects. These characteristics make Y3 particularly relevant for the representation of knowledge about brain : complex relationships can be efficiently represented and the whole set of properties of an object (intrinsic or inherited ones) are accessible from the object itself.

Mechanisms also exist to inhibit the inheritance of properties . This capability is also very interesting because it seems to be very difficult and perhaps impossible to describe the complex anatomical or functional structures of the brain by means of necessary and sufficient conditions (properties).

Conclusion

As one can observe in this presentation, these navigation mechanisms are closely related to the network composed of objects and relations between them. We can consider such mechanisms as fundamental or primary. The navigation system enables users to navigate from atlas plates to symbolic information. Nevertheless, this system does not pretend to address all specific requirements concerning surgery planning. Our objectives for the future consist in offering more convenient and goal-oriented mechanisms of navigation which take into account real cognitive activities of surgeons and in extending the system to cope with different points of view on the brain, depending on the type of users (surgeons, anatomists, physiologists, etc) and different tasks.

References

1. Talairach J., Tournoux P. : Coplanar stereotactic atlas of the human brain. Georg Thieme Verlag Ed., Stuttgart, 1988.
2. Schaltenbrand G., Wahren W. : Atlas for stereotaxy of the human brain. Georg Thieme Verlag Ed., Stuttgart, 1977.
3. Ono M., Kubik S., Abernathey C.D. : Atlas of the cerebral sulci, Georg Thieme Verlag Ed., Stuttgart, 1991.
4. Szikla G., Bouvier G., Hori T. : Angiography of the human cortex, Springer Verlag, Berlin, Heidelberg, New York, 1977.

5. Greitz T., Bohm C., Holte S., Eriksson L. : A computerized brain atlas: contruction, anatomical content, and some applications. Journal of computer assisted tomography, 15 (1), 1991, pp 26-38.

6. Höhne K.H., Bomans M., Riemer M., Schubert R., Tiede U., Lierse W. : A volume based anatomical atlas. IEEE computer graphics and application, 1992, pp 72-78.

7. Bloom F.E. : Databases of brain information., Chapter 13. in Three-Dimensional Neuroimaging edited by A. W. Toga, Raven Press, New York 1990.

8. Cawley M. G., Natarajan K., Newell J.A. : Automatic generation of a dynamic model for use in subsequent radiological image interpretation. CAR'91, Springer Verlag, 1991, pp 651-656.

9. Natarayan K., Cawley M.G., Newell J.A. : A knowledge based system for the automatic interpretation of Xray CT scans, CAR 91', Springer Verlag, 1991, pp 618-623.

10. Evans A. C., Marrett S., Torrescorzo J., Ku S., Collins L. : MRI-PET Correlation in three dimensions using a vomume of interest atlas. Journal of cerebral blood flow and metabolism, Vol 11, 1991, ppA69-A78.

11. Lehmann E., Hawkes D. J., Hill D. L. G., Bird C. F., Robinson G. P., Colchester A. C. F., Maisey M. N. : Computer aided interpretation of SPECT images of the brain using an MRI derived 3D neuro-anatomical atlas. Medical informatics, Vol 16, no 2, 1991, pp 151-166.

12. Balpe J. P.: Hyperdocuments, Hypertextes, Hypermedias, Eyrolles, 1990.

13. Nanard J., Nanard M., Richy H. : Conceptual documents : a mechanism for specifying active views in hypertexts. ACM conf. on document processing systems. Santa Fé, Dec 1988, pp 37-42.

14. Lemoine D., Barillot D., Gibaud B., Pasqualini E. : An anatomical based 3D registration system of multimodality and atlas data in neurosurgery, Lecture notes in computer sciences Vol. 511 , IPMI'1991, A.C.F. Colchester and D.J. Hawkes Eds, Springer Verlag, 1991, pp 154-164.

15. Ducournau R. : Y3: Présentation du système, Sema Group, Montrouge, France, mai, 1988.

A Protocol for Analysis and Visualization of MRI Liver Data

R. STOKKING[1], H.G. VAN MEURS[1], M.S. VAN LEEUWEN[2], H.W. VAN ES[2], M.A. VIERGEVER[1]

3D Computer Vision[1], Room E 02.222/ Radiology Department[2], Room E 01.132
University Hospital Utrecht, Heidelberglaan 100,
3584 CX, Utrecht, The Netherlands.

Summary

We present a protocol for the analysis and visualization of MR image data of the liver. The protocol comprises a validation of anatomical landmarks which are employed in the clinical analysis, and a three-dimensional (3D) visualization strategy to enhance the interaction between the radiologist and the surgeon. The software package Analyze [4] was used as development environment. MRI data of ten healthy volunteers was acquired for evaluation of the protocol. The preliminary conclusion reads that the protocol is useful for landmark validation and beneficial for radiologist-surgeon interaction.

Introduction

In pre-operative planning a 2D MR-image set is generally translated mentally into 3D surgical anatomy by the radiologist. Problems arise in this mental reconstruction owing to large individual variations in interpretation and because complicated spatial relationships are difficult to describe with words.

The exchange of liver anatomy information between radiologist and surgeon is often based on internal and external anatomical landmarks [2]. In order to further improve the interaction a 3D computer-display of the liver and relevant structures can be reconstructed from the 2D images, which gives the surgeon a powerful tool in surgery planning.

The purpose of this study was to develop a protocol for validation of the landmarks and to investigate whether the interaction between the radiologist and the surgeon could be improved by 3D visualization of the MRI liver data.

Liver anatomy and the landmarks

Functionally the liver can be divided into eight hemodynamically independent sections, the boundaries of which are formed by the hepatic veins, while the segmental branches of the portal vein are located in the centre of the sections [2].

The definition of the sections has not yet been standardized. In this presentation we adopt the nomenclature of Couinaud. In this nomenclature the liver is divided by three vertical planes through each of the three hepatic veins with the caval vein as base, and a horizontal plane defined by the left and right portal vein. This results in a total of eight liver sections.[2]

The subdivision of the liver is important in liver surgery, because the section boundaries are relatively avascular (apart from the hepatic veins). Anatomical resections along these boundaries cause little trauma to vital structures such as the portal vein, hepatic artery and bile ducts.

Internal landmarks (hepatic veins, portal vein) are studied on cross-sectional images to judge whether a detected lesion can be resected without inflicting too much damage to vital structures and with reasonable perspective for recovery. The external landmarks located on the surface of the liver (gall-bladder, umbilical fissure, caval vein) are used to guide the surgeon during the operation. There is a presupposed relationship between a number of internal and external landmarks; e.g., the plane through the middle hepatic vein is thought to connect the gall-bladder with the caval vein, and the plane through the left portal vein is thought to encompass the umbilical fissure.

The protocol for analysis and visualization

We developed a protocol for analysis and visualization of MRI liver data with two specific objectives namely

- validation of the anatomical landmarks, and
- improving the interaction between the radiologist and the surgeon using 3D visualizations of the liver with special interest in the vessels.

Ten MRI sets (contiguous 4 mm slices) of healthy volunteers were used for the evaluation of the protocol. The data was first copied from the scanner to our network, and then processed with Analyze for segmentation, object definition, analysis and visualization. The Analyze procedure started by loading a volume of interest (VOI), containing the liver and other important extra-hepatic structures (caval vein, portal vein and gall-bladder) to reduce the total amount of data for faster and easier processing.

Segmentation

Segmentation was performed in a slice-by-slice manner either by manual tracing or by using region growing around a seed-point. It consisted of four steps, namely

(i) segmentation of the outline of the liver and the important extra-hepatic structures,
(ii) segmentation of the two vessels (hepatic and portal veins) and the gall-bladder,
(iii) use of the landmarks in the liver anatomy to define the four planes and thereby the eight sections,
(iv) segmentation of the liver, whereby each section was defined as a separate volume element.

In the case of lesions, an extra classification step was inserted to determine the lesion boundaries. The sections, vessels, gall-bladder, planes, and lesions were subsequently defined as separate objects for easy manipulation.
ad (iii) The definition of the four planes proved to be a difficult problem. Analyze has the option to define three points (and return the coordinates), however there is no possibility to edit the corresponding slice. A workaround would be to transform the volume to force the three points into a transverse, sagittal or coronal slice and subsequently edit this slice. This would mean a transform- and edit-run for each plane, followed by a final transform (optional) to restore the original coordinate setting. Each transform degrades the set because of the required interpolation, while moreover the whole procedure is rather cumbersome and laborious.
The fact that the caval vein tends to follow a vertical path through the abdomen offers another, easier solution. For each of the three vertical planes a straight line between two landmarks (e.g., connecting the middle hepatic vein and the caval vein: the line BC in figure 1) in the appropriate 2D slice was drawn and then automatically transferred to all other slices. The 3D visualization of these lines resulted in the desired vertical plane. The horizontal plane was approximated by editing the 2D slice closest to the corresponding section of the portal vein bifurcation. (see figures 3b and c)

Visualization

For the visualization different shading techniques could be used. Good results were obtained with depth shading (DS), depth gradient shading (DGS) and especially voxel gradient shading (VGS) (see figures 2, 3, 4).
The use of transparency, colour and stereo were other important features for the visualization. Colour is very observer-, display-, and task-dependent. The radiologist used certain colours for reasons of standardization (red and dark blue vessels, white lesions, ochre liverparenchyma, light blue background; see figure 4a), other observers preferred different colours like a green background with contrastive harsh-coloured vessels (see figure 2).
Liver tissue was rendered transparent (in combination with VGS) to give an impression of the vessels, gall-bladder and/or lesion(s) in relation to the liver and the sections (see figure 3a).
Stereo display in combination with a stereo-viewer was used to enhance depth-perception with very nice results (see figure 4c and d).
Finally, movies of rotation-sequences and/or object displacement (e.g., in combination with stereo) were made to further enhance the understanding of the spatial relationships between the different structures.

Results and discussion

Segmentation

Segmentation of the liver was a laborious job; about three hours were needed for the four segmentation steps of a 256*256*56 dataset. The Analyze segmentation procedure is not user-friendly and the liver is not clearly distinguishable from surrounding tissues which have similar grey-values. It proved valuable to have the original MR scans at hand for a quick interrogative look of the consecutive slices. Initially we used region growing around a seed-point, but since this required a lot of editing, simple manual tracing turned out to be preferable.

Validation of the landmarks

Validation of the internal and external landmarks was straightforward. As noted earlier, the caval vein tends to follow a vertical path through the abdomen, which makes it easy to establish the consistency of the various landmarks by investigating the 2D slices. Angles between middle hepatic vein and origo of the gall-bladder with respect to the caval vein could be measured with the caliper option (see figure 1), and the path of the left hepatic vein with respect to the umbilical fissure was easy to trace in the consecutive slices.

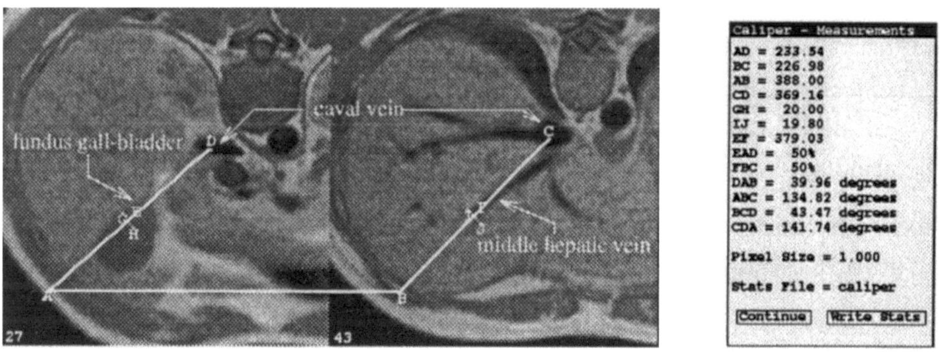

Figure 1: The Analyze caliper measurement option.

Visualization

The pictures had a blocky appearance owing to the thickness of the slices. Smoothing and interpolation were evaluated as strategies to improve the quality of the visualization.

Smoothing provides visually more attractive images but at the same time leads to loss of detail which is unacceptable especially in the visualization of small or thin objects like the vessels. Interpolation was used to calculate intermediate slices with a thickness of 1 mm. The visualization (see figure 2) improved significantly but was still blocky (now 1 mm

blocky instead of 4 mm blocky) while the workload increased significantly owing to the increased number of slices. The interpolation technique available in Analyze is very simple (tri-linear), better techniques would probably improve the visualization, be it at the cost of increased computation time (e.g. shape-based interpolation [3]).
Useful features for the protocol were found to be Oblique Sections (2D display of an oblique slice indicated by a plane in a 3D volume display; see figure 4b), Image Viewer (to quickly scan through 2D slices), and Demonstrator (to make a demo presentation; text and figures can be added); see [4] for more details.

Conclusion

The proposed protocol for analysis and visualization of MRI liver data has proven a valuable tool for validation of anatomical landmarks and for improving the presentation of radiological data to the surgeon [1]. This conclusion, based on the judgement of the involved radiologists and surgeons, should be substantiated by further evaluation studies.

References

[1] M. S. Leeuwen, R. Stokking, H.G. van Meurs, H.W. van Es, E.H. Dillon, and M.A. Feldberg. Liver Anatomy: Protocol for 3D Evaluation. RSNA 1992, exhibit 09-002, 1992. Video presentation, cum laude. Abstract in RSNA Scientific Proceedings 1992, page 379.

[2] M.S. van Leeuwen, H. Obertop, A. Fernandez, E.H. Dillon, M.A.M. Feldberg, and A.H. Hennipman. 3D Display of Liver Anatomy; A Pre-operative Planning Procedure. submitted.

[3] S.P. Raya and J.K. Udupa. Shape-based Interpolation of Multidimensional Objects. IEEE Trans. Med. Imaging, 9(1):32–42, March 1990.

[4] R.A. Robb. 3-D Visualization and Analysis of Biomedical Images Using ANALYZE. Computer Assisted Radiology '91, pages 685–698, 1991. H.U. Lemke, M.L. Rhodes, C.C. Jaffe, and R. Felix (eds.). Springer-Verlag, Berlin.

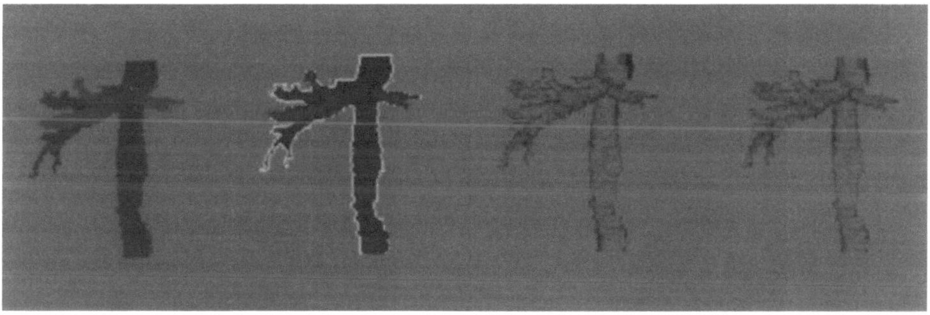

Figure 2: Visualization of an interpolated hepatic vein. (a) DS. (b) DGS. (c) and (d) VGS in stereo-view.

Figure 3: Visualization of the liver with VGS. (a) Use of transparency. (b) Sections and planes. (c) Sections and planes, object-move option.

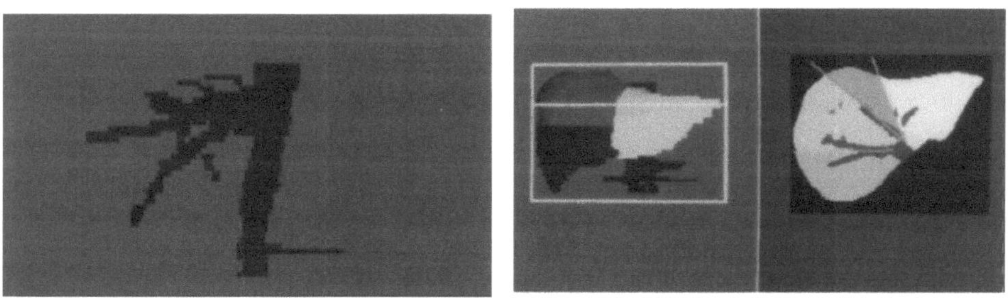

Figure 4: Visualization with VGS. (a) hepatic vein, colours as preferred by the radiologist. (b) Oblique Sections; the horizontal line in the left frame defines the slice shown in the right frame.

(see also in color on page 816)

Man Computer Interaction

Design of High Robust Voice Interface for Speech Activated Neurosurgical Workstation

X. Ying, J. Koivukangas, J. Alakuijala, J. Oikarinen and Y. Louhisalmi

Neurosurgical Research Unit, Department of Neurosurgery, University of Oulu, 90220 Oulu, Finland.

Abstract

Several methods were developed in order to improve reliability and robustness of a voice interface designed for a speech activated neurosurgical workstation that is a command-number-limited system but directly involved in nurosurgical operations. We chose a commercial voice recognizer and synthesizer as basic voice environment of the voice interface. A parallel connection grammar structure in the command recognition and a practical operation procedure oriented logical structure in the command understanding were designed to remove misrecognition and misunderstanding. In the voice recognizing process, by means of setting noise pits and dynamically adjusting tolerance level, the substitution errors caused by uncorrected voice commands and noises in operating room could be greatly reduced. To guarantee the high robust voice control, we employed real-time voice feedback to confirm the accepted commands. Our testing results showed that employing these methods greatly improved voice interface robustness.

Introduction

Applicable voice recognition and synthesis technologies have been introduced in the real world after several decades' research at laboratories [1], but most of these applications are for systems in which a voice recognizing error dose not result in fatal risk, so they don't need very high accuracy rate of voice recognition. However, the neurosurgical workstation is a very demanding system that is directly involved in serious neurosurgical operations. During computer-assisted stereotactic surgery, the system may illustrate the pre-operative CT or MRI scans and its 3D imaging model, display on-line the reformatted CT or MRI slices of intracerebral lesions with guided by the neuronavigator [2], manipulate the real-time intraoperative B-ultrasound image, and align the surgical microscope, even to control a surgical laser beam [3]. It is quite evident that such a vitally important system could not be met by using a simple and low accurate rate voice interface.

However, up to date the accuracy rate of automatic voice recognition is still a problem in most commercial voice devices [4]. We could not directly employ a common voice interface design in such an important system without any improvements. In this study, we searched several ways to improve the reliability and robustness of the voice interface which is specially designed for the neurosurgical workstation [2].

Balancing Choice of Basic Voice Environment

We employed commercial voice products as the basic environment of the voice interface which includes speech input and speech output devices. Since a speaker-dependent recognizer has higher recognition accuracy and is acceptable in our system because we also expect that only the surgeon's or/and an assistant engineer's commands need to be accepted during the surgical operation, we chose speaker-dependent recognizer although a speaker-dependent recognizer needs pre-training and limits general purpose application. Also, while continuous speech is more natural it is error-prone and we can express a short sentence as a long-word for commanding if the recognizer has long-word recognition capability. Thus discrete-words is the better choice here. The neurosurgical workstation is a commands number-limited system. As for voice interface, in principle, the smaller the number of commands in the system, the higher the recognition accuracy that can be reached. However, limiting the number of commands will also limit the performance of the system. In this design we limited the number of active commands by classifying all the commands into several groups, so the minimum requirement for vocabulary size in the interface can be reduced to about 50 words. Since we need plenty of room for the high robustness demand, we defined 200 active words as the minimum requirement for the vocabulary. Considering the fact that the index for vocabulary size in most commercial products presents only the capacity of the product and does not imply its recognition capacity with an adequate accuracy, we sought a voice recognizer with a vocabulary size of 400 - 1000 words. The basic demands for the speech synthesizer in this system are converting ASCII English text to speech in real-time, and working in conjunction with related voice input products. Also, at least 300 characters buffer will be needed for the help message and others.

Based on above designing consideration, we selected the Dragon technique-based *VS 1000 Plus voice recognizer (Cherry Co., USA)* as speech input, and the *Prose 4000* text-to-speech voice module *(SpeechPlus Inc., USA)* as speech output. The system structure of the voice interface adopting the modules is shown in Figure 1.

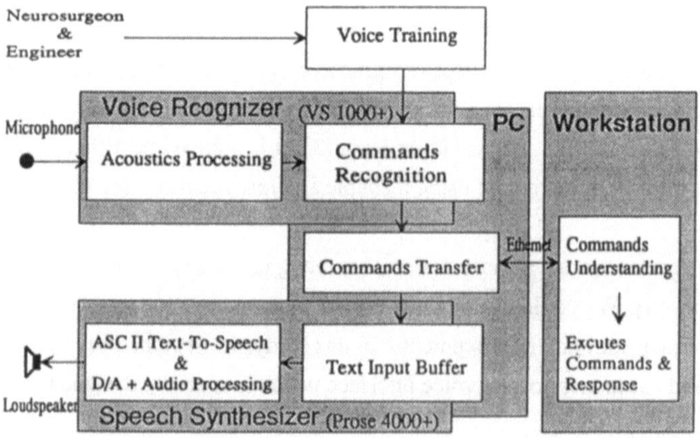

Fig. 1. System structure of the voice interface

Grammar Structure

Adopting structured grammar is an excellent way to obtain the best possible performance from a speech recognizer [5]. Based on analyzing the neurosurgeon-computer-interaction [6], the functions of a neurosurgical workstation can be designed into three states: pre-operative planning, intraoperative aid and post-operative evaluation. The designing emphasis of the voice interface was put on the intraoperative aid in which reliability of the voice interface is most important. In the intraoperative procedure, the possible system applications can be classified into at least following six modules:

Surgical-Poster: Representation of multiple pre-operative CT or MRI scans and/or 3D imaging of the operative target;
Calibration: Matching between the pre-operative imaging and intraoperative coordinates of the patient's head;
Navigator: Navigator-guided representation of the intracerebral imaging;
Ultrasound: Navigator-assisted intraoperative B-ultrasound imaging and corresponding re-formatted CT or MRI slices;
Microscope: Navigator-assisted surgical microscopic operation;
Laser-Operation: Multiple imaging data guided laser removing operation.

Based on this command classification, only dozen commands will be active in one module. Thus, not only can the recognizing accuracy be increased but also the response time can be reduced. In a simulated procedure, we tested two grammar structures for incorporating these function modules. One was series connection, *i.e.* to string the modules together in order of the practical operative procedure. Another was parallel connection, *i.e.* to put all the modules on the same level and set a basic *stand-by* state, enter or exit each module by saying the module name or a *reset* command.

When the interactive procedure was correct, there were no major differences in the running results of the two grammatical structure. The series connection seemed slightly natural than the parallel connection. However, when a recognition error was incurred or the user said a wrong command, the series connection would usually produce grammatical confusion, even resulting in whole voice interaction breakdown. In the parallel connection in this situation, even if grammatical confusion icurred, it was easy to eliminate the confusion by saying the *reset* command to get back to the *stand-by* state, and to re-enter by saying the module name. Also, such design better corresponded to the characteristic of voice interaction because the user wasn't constrained in the grammatical structure and it was easy to find the commands' position [6].

Although it is not necessary to build a grammar for increasing the recognition accuracy in a module which has only a few commands, it is still important in some modules to make a simple rule to incorporate the commands for guaranteeing a correct procedure. In the *calibration* module, for instance, following ***mark-point*** (a voice command to get the computer ready for receiving the coordinate of the pre-marked points on patien's head) should be the point's number. Especially for vital important functions like the *Laser-Operation*, we designed a multi-steps rule, *i.e.*, first say a command was given to active a function then say a command to confirm the command in a given amount of time. Once the command has been confirmed and the function has been performed or time ran out, the system returned to the primary state immediately. There is very little probability that two or three commands will be concurrently misrecognized in a short duration of time. Therefore the safety in voice controlled laser operation can be ensured.

Logic Structure

Based on our testing experiments, the logic structure of the commands in the command-understanding, *i.e.*, the order in which commands can be accepted, is more or less another guarantee of a high robust voice interface. The best structural way is basically in accordance with the practical operation procedures. As an example, a logical structure of the voice commands for brain operation is given in Figure 2. This structure is not wholly equal to the related grammar structure in the voice recognition. The module *Surgical-Poster* and *Laser-Operation* connect separately to other modules in series. This is because in practical computer-assisted brain operation, the multiple images are first needed to be represented. Also the *Laser-Operation*, if it's needed, is after other operative procedures. As only the related voice commands only in the same function module can be understood and translated to system commands, it is beneficial for the robustness of the interface.

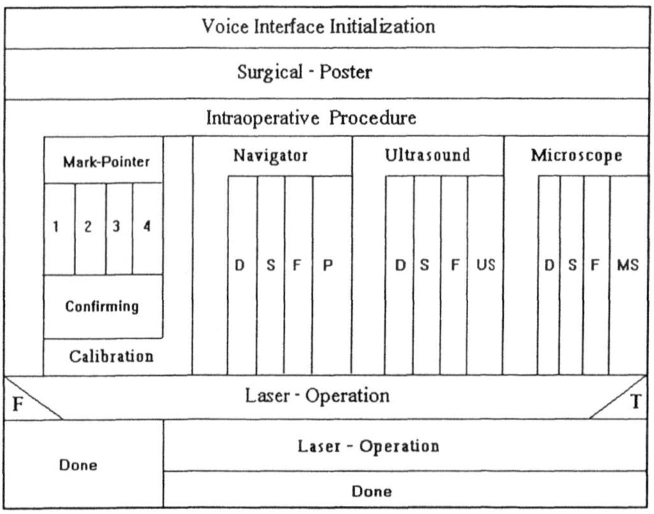

Fig. 2. N-S Charts showing the logic structure of voice commands for brain operation in the commands understanding. D: depth display adjusting; S: slice size changing; F: picture freezing; P: processed image including 3D moduling display; US: realtime ultrasound image manipulations; MS: special display control for microscope.

Setting Noise Pits and Dynamic Tolerance Adjustment

Our experimental results showed that after building the grammar and logical structures, no misrcognition would normally be made in such a small commands group if the interface was trained correctly and a correct command was said. But, inevitably, it would be happen that a uncorrected command was said or a strong noise was made in the practical environment. So two methods were designed to improve the recognition. One was to set several voice 'pits' in each command group, *i.e.*, to insert several words that do nothing in each active command group. The words were trained by using some very short utterances and other voices which may frequently emerge in the operating room. In our testing results, by this method, most noises, even some uncorrected commands which are not in the expected command group would be dropped into the noise pits. Therefore some substitution errors can be avoided by this simple method. Another was the dynamic tolerance adjustment. We found that in a voice interface with a limited number of commands the probability for substitution errors to occur was higher than that false rejections happened. Especially the consequence caused by substitution errors may result in serious confusion, unlike the false rejections that can be easily corrected by only saying the command again. Therefore, it is important to select an optimal tolerance level of speech variability and noises. Based on the voice environment provided by the *DragonKEY* [7], we set several voice commands to interactively adjust the tolerance level. The adjustable tolerance level can be from 1 to 100; we chose 36 as the default according to our testing results, and set 8, 24, 44, 52, 64 and 80 as pre-setting parameters. Normally, the tolerance level was kept at a relatively low level to avoid the substitution errors. Also, this function can be set in the command processing program by using the number of false rejections as the adjusting threshold.

Real-time Voice Feedback and the Microphone-Sleep function

We employed real-time feedback of voice commands as another way to improve the robustness of the voice interface by the speech output device, not only simple real-time repeating of the recognized voice command but also related prompts. Our first experience demonstrated that the real-time speech response is a necessary means for the eyes-busy surgical work, because the user can quickly get the message whether his/her command has been correctly accepted.

In addition, even if the above methods can reduce foreign voice influence, a "microphone-sleep" function is still useful for the user to allow free talk. But the original sleep function provided by the *DragonKEY* [7] was based on the computer stack operation, that is, the function turns the microphone off and saves the current state of the recognizer in the computer stack, then calls it back when waking up the recognizer. The problem is that the application program also needs the stack, and especially if the *wake-up* command is misused without the *sleep* command the voice interface will be fatal. Hence we designed another way to set the sleep function by employing the Recognize routine of the *DragonKEY* to build the sleep function in the grammar. When the *sleep* command was used to close the voice interface listening, only the *wake-up* command module was active, and the previous state was just given up. Once the *wake-up* and the following confirmation commands are spoken, the state can be re-built by the grammar.

Summary

Several practical ways for building a high robust voice interface have been represented in this paper. Our testing results showed that employing these methods with a correct organization can greatly improve voice interface robustness which is a critical factor in the voice control of some vital important systems.

In these improvements, grammatical and logical structure are essential methods. The parallel connection in the grammar not only can increase the recognition accuracy but also allow a clear structure to reduce or eliminate any confusions caused by noise or incorrect commands. However, in the logical structure of voice commands understanding, the most suitable design is according to the practical operative procedure which is usually a series of tasks. Because the higher probability of substitution errors in the voice interface, we designed methods for setting noise pits and dynamically adjusting the tolerance, both of which were shown to be effective for reducing the errors.

The high robustness of the voice interface dose not depend only on the improvements of voice recognition accuracy, because this is a voice command understanding system. Therefore, the accepted voice command feedback was established in this interface. The results proved that this is an excellent way to achieve high robust voice interface.

Although this system was especially designed for the speech activated neurosurgical workstation, it also can be applied to speech control of other computer-based systems based on limited numbers of commands and demanding high robustness.

This work was supported by the Academy of Finland and the Finnish Technological Development Center.

Reference

1. Flanagan, J.L.; Speech processing:a perspective on the science and its applications. AT&T Tech. J., Sept./Oct. pp. 2-13, 1990.
2. Koivukangas, J., *et al*; Ulrtasound-controlled neuronavigator-guided brain surgery, J. Neurosurgery, (in press). 1993.
3. Koivukangas, J. and Louhisalmi, Y.; Computer-guided laser for neurosurgery. Annales Chirurgiae et Gynaecologiae Vol.79, pp. 192-196, 1990.
4. Lee, K., *et al*; The spoken word. BYTE, pp. 225-232, July 1990.
5. Smarte, G. and Penney, W.; Sound and image processing. BYTE, pp. 240-256, Dec., 1989.
6. Ying, X. and Koivukangas, J.; Neurosurgeon-commputer interaction. unpublished report, Neurosurgical Research Unit, the Unversity of Oulu, 1992.
7. Dragon Systems Inc.; Speech Application Designer's Handbook., Dragon Systems Inc., 1986.

Three Dimensional Interaction Techniques of Medical Workstations

Dr. Gabriele Faulkner

Technical University of Berlin,
Department of Computer Graphics, Sekr. FR 3-3,
Franklinstrasse 28/29, D-10587 Berlin.
e-mail: vonvoigt@cs.tu-berlin.de

1. Introduction

Medical workstations are the access medium for the medical staff to the medical data. The patient information consists of various items which can be grouped into the following seven data categories: text, sound, table, curve, video, 2D image and 3D reconstructed image. In order to process these different data classes, multimedia techniques are required. This paper however, concentrates on dealing with the graphical information. In assisting the treating physicians handling the images, the workstation must support this process in an easy and intuitive fashion. The 2D and 3D user interfaces are vital for the acceptance and functionality of the whole system. A first approach in tackling this task has been carried out within the COMED project.

2. COMED Project

The main objective of the COMED project is to assist the radiologist and physician in viewing and manipulating the whole patient folder in a distributed co-operative working situation. Within the framework of this project, methods have been designed for image acquisition, management and processing, hyperthermia treatment planning (which aims at heating the tumour whilst avoiding healthy tissue) as well as data communications and teleconferencing with special regard given to the requirements of the medical environment. The 2D interaction is based on the use of menus and icons in combination with the object oriented "drag and drop" technique. A detailed description of the project can be found in /von Voigt 1993/. The following sections contain descriptions of the more complex forms of interaction.

3. 2 1/2 D Visualization and Interaction Techniques

In order to provide an overview of the CT or MRT patient data which generally is the basis of a planning process, a tool is suggested which displays the original slices (ie. sagittal) in combination with the corresponding orthogonal views (transversal and coronal) /Jackisch, Rudolph 1991/, as well as an arbitrary oblique slice.

The visualization of the orthogonal views also has to include the parameters of a treatment plan. In order to assist for example hyperthermia treatment planning, display of the exact isothermic distribution in each slice from all three orthogonal view directions is required.

The importance of the user interface cannot be too strongly emphasised, when discussing the quality of the whole system. Its layout should be clearly arranged and follow logical rules so that the user is able to easily build his mental model of how to the handle the system. The choice of the position of the orthogonal views can be made by using the familiar and easy to use mouse. The determination of the position of the oblique cutting plane within the 3D cube requires a 3D interaction device (i.e. Geoball) or a technique which simulates the 3D interaction in 2D. In order to achieve a 3D input with a 2D input device like the mouse, the user is typically offered x, y and z sliders which are graphically displayed on the screen to indicate the amount of rotations and translations in each axis independently. The user has to adjust the sliders individually by using the mouse and thus has no feeling for the sum of the transformations. Therefore, for 3D interaction tasks, 3D input devices are recommended.

4. 3D Visualisation Techniques and Interaction

Virtual Reality (VR) can be seen as a totally new human computer interface which enables the user to visualize and interact within a 3D computer generated world in real time /Krauss, von Voigt 1992/. The 3D input and output devices help the user to immerse into the 3D computer generated world and to create the illusion of being part of that world. A variety of 3D input devices and 3D output devices has been developed for VR purposes.

4.1 3D Input Devices

Any discussion of VR input devices leads firstly to a consideration of the variety of gloves available, which can convert hand gestures and positions into a computer-readable form and allow the user to work with the computer-generated objects by reaching, touching and grabbing them. They exist with different accuracy and may or may not be able to give tactile or force feedback - the *VPL DataGlove* is the most popularly used one. They all however, have the same disadvantages in that the user has to wear them on the hand which is cumbersome, they must have an exact fit - otherwise the measurements have some degree of inaccuracy and they need regular re-calibration.

The *Geoball* offers 6 degrees of freedom, 3 for the 3 rotations in pitch, yaw and roll and the other 3 for the 3 translations in x, y and z. The geoball is placed on the desk and is operated by pushing, pulling and twisting the ball in any direction. The forces which the user applies are sensed by six optical position sensitive detectors and so the values for translations and rotations are derived.

The *Logitech 3D Mouse* is a 2D/6D mouse system with maximal 6 degrees of freedom. It can also be used in the ordinary 2D mode as a conventional mouse. In the 6D mode, the device operates using a three dimensional co-ordinate system reading x, y and z axes and pitch, yaw and roll movements in fine increments within a specified area. A triangle with 3 ultrasonic transmitters on the desktop and a triangular piece with 3 corresponding microphones fixed on the front of the mouse are used for sensing its position and orientation in space.

4.2 3D Output Devices

All 3D output devices aim at presenting the user with a three dimensional view which has the illusion of depth.

Head Mounted Displays (HMD) /Brown, Slater 1991/ such as the *VPL EyePhone* are often used within the virtual reality context. They generally have a binocular optical system worn on the user's head presenting one image to each eye. Stereoscopy is used to give the illusion of depth in the overlapping display area between the two eyes. Unfortunately, the resolution of HMDs is rather poor, with for example 720 x 480 pixels of the EyePhone HRX using colour LCD's as the display unit /Aukstakalnis, Blatner 1992/. Head Mounted Displays also have a 3D tracking subsystem which provides for the determination of the location and orientation of the helmet in space. An often used tracking system is the Polhemus 3Space Isotrack which is a magnetic positioning system. It determines the user's viewing direction which is needed for the updating of the presented images. There are two classes of HMDs, firstly the "*total immersion HMD*" where the user wears a helmet like device which presents the computer generated world to the user and secondly, the "*see through HMD*" where the user sees the computer generated world at the same time as the real world.

With HMDs, each image of the stereo pair is presented directly to each eye. If this technique is not available, another possibility as suggested in /Brown, Slater 1991/ is that the multiplexing of the two images on a single display (usually an ordinary video monitor) must be achieved in combination with wearing special glasses to de-multiplex the images for the correct eye. Some systems time-multiplex stereo images on the display, showing each image alternately. The active *shutter glasses* (e.g. "Crystal Eyes") consist of thin lightweight liquid crystal displays for each eye which turn opaque when a voltage is applied. These are synchronised with the display and accompanied by infra-red signals, so that each eye sees the correct image when it appears.

At present, a more suitable 3D output device for medical applications seems to be a head-coupled display like the Fake Space Labs *Binocular Omni Orientation Monitor (BOOM)* /Krauss, von Voigt 1992/ which is a CRT-based stereoscopic viewing device with wide angle optic. It is mounted in a box on a moveable stand and therefore, can be easily placed before or removed from in front of the eyes. Its resolution depends on the resolution of the used CRT and is for the BOOM 1280 x 500 pixels in colour mode /Aukstakalnis, Blatner 1992/.

5. Planned VR Scenario within COMED

In order to assist hyperthermia treatment planning, the 3D treatment plan has to be visualized to the physician. This scene should include the patient as a computerised model, his position, the target volume, all organs involved - especially the critical organs, the temperature distribution, their corresponding antennas and all other variable parameters of the treatment equipment. The aim is to achieve an easy to use feature for interactive antenna positioning and for selecting an arbitrary view point on the patient model in order to simulate the calculated plan. The tool will be developed for the following purposes: simulation of the treatment, interactive changes of the treatment plan, easy understanding of the treatment configuration set up and teaching. A 3D input device should be used in order to achieve a comfortable interaction with the three dimensional data. Within COMED, we are currently exploiting the usefulness of the above mentioned input devices for medical workstations. As a 3D visualization aid for stereoscopic images we used a stereoscopic mirror device. Although this device belongs to the class of "through the window" and not to the group of "total immersion" devices, it has the advantage that its usage is optional and can be easily applied by the user. An evaluation of the usefulness of 3D input and output devices is being carried out at the moment.

6. Final Remark

At present, virtual reality is still at the infancy stage and the described input and output devices have special drawbacks for medical applications. However, in order to create a 3D view of a treatment planning scene which relates as closely as possible to real life, the application of the concept of virtual reality seems useful but the relevant techniques involved are still under development.

Acknowledgement

The work reported in this presentation was done within the COMED project which is one of the BERKOM projects and is funded by the DETECON GmbH (Deutsche Telepost Consulting GmbH, Bonn) and by the Technical University of Berlin.

References

/Aukstakalnis, Blatner 1992/ S.Aukstakalnis, D.Blatner, "*Silicon Mirage: The Art and Science of Virtual Reality*", Peachpit Press, Berkeley, 1992

/Brown, Slater 1991/ M.A.Brown, M.Slater, "*A review of interaction technologies as applied to virtual reality*", Proceedings of Computer Graphics '91, London, 309-330.

/Holloway et al. 1991/ R.Holloway, H.Fuchs, W.Robinett, "*Virtual-Worlds Research at the University of North Carolina at Chapel Hill*", Proceedings of Computer Graphics '91, London, 181-196, 1991.

/Jackisch, Rudolph 1991/ U.Jackisch, M.Rudolph, "*A Highly Interactive 3-D Visualization Tool for Sets of Medical Images in the Environment of Communicating Low and High End Workstations*", Proceedings of ISCAMI'91, London, December 13-15th,1991, in Press.

/Krauss, von Voigt 1992/ M.Krauss, G. von Voigt, "*Virtual Reality in Medicine*", Proceedings of 14th Annual IEEE Conference, Paris, October 30th-November 1st, 1992, 957-958.

/von Voigt 1993/ G. von Voigt, "*Die Entwicklung verteilter medizinischer Arbeitsplätze im Rahmen der COMED-Projektes*", Workshopband "Visualisierung in der Medizin", Freiburg, 10-11th March, 1993.

Digital Image Generation

Digital Radiography

Primary and Secondary Digitizing of X-ray Images in Comparison to Conventional Screen Films

U.Fink, A.Widmann, B.K.Fink, T.Hilbertz, K.H.Englmeier, S.Nissen-Meyer

Dept. of Radiology, Großhadern Clinic, University of Munich and GSF, Medis Institute Neuherberg, Germany

Summary:

Conventional x-ray film examinations still comprise about 80 % of all radiological procedures in spite the well established digital imaging modalities such as MR, CT and DSA. This study concerns the diagnostic value of two digital radiographic procedures in comparison with conventional screen films.
Primary digital procedures were performed with DLR (AC1,Fuji). For secondary digitizing we used the film digitizer system FD 2000 (DuPont). Various phantom studies and clinical studies on 30 intensive care chest radiographs were independently evaluated by 5 radiologists.
In order to test the diagnostic accuracy in the detection of pathologic structures receiver operating characteristic (ROC) analysis was employed.
The results showed that primary and secondary digital imaging were equivalent, to the screen films in those cases where high spatial resolution is not essential. However, the lower spatial resolution of the digital procedures can be compensated by image postprocessing at the PACS-console with little effort.
Our results indicate that the migration to totally digital radiology is feasible in the near future.

1. Introduction

The introduction of modern digital imaging methods as DSA, CT or MRT was the most important advance during the last years in diagnostic imaging. Although these imaging modalities supply digital datas primarily the presentation and archiving will be performed in conventional film technique in most radiological institutes.

Efforts to archive those datas only in digital form failed especially because the conventional x-ray film examinations still comprise the greatest number of all radiological procedures, in our department about 80%, and the loss of acceptance of monitor reporting.

With the introduction of Digital Luminescence Radiography (DLR) an imaging modality is available since some years which allows the presentation of conventional x-ray imaging in digital technique primarily.

Further film digitizers were developed in the last years, which allow secondary digitization of conventional x-ray films in acceptable time and good quality.

Aim of the study was to examine the possibility and diagnostic value of DLR and secondary digitizing of conventional x-ray films in direct comparison to conventional x-ray screen films.

2.Method

2.1. Digital Luminescence Radiographie (DLR)

The examiniations were performed with the Siemens Digiscan system. The documentation consists of two different filter algorithms: One which was similar to the imaging of conventional sreen film (DLR), the other used a special algorithm for edge enhancement (DLRKV).

The documentation was performed on a hardcopy film with a resolution of 5 lp/mm. This hardcopy film was evaluated in the same way as the conventional screen film on a film viewer.

2.2. Film digitizer

We used a Du Pont digitizer FD 2000 with the following technical datas:

- Pixel size 210 μm (film size 35cm x 43cm)
- Resolution 2.35 line pairs/mm
- Image matrix 1682x2084 pixels
- Number of gradations 12 bits (4096 gradations)
- Read density range 0.0 - 3.5
- Read time 12s (film size 35cm x 43cm)

The evaluation of the digitized images was performed on a Siemens PACS-console DRC 40 with four monitors with the following technical datas: 1024 pixels x 1280 lines with 50 cm diagonal, 120 Hz TV frame frequence and 16 MB cache memory. Image post-processing was possible at the monitors like changing of image contrast or film blackening as well as magnification.

2.3. Phantom studies

5 examiners evaluated ten radiographs of a 3M-synthetic resin pelvis phantom with fixed round plastic structures smaller than 5 mm, a 3M hand phantom, a Du Pont breast phantom and a line phantom.

On the pelvis phantom the examiners had to diagnose the number and the localization of the different fixed round structures. At the hand phantom the delineation of the spongiosa and corticalis, the carpal bone and the interphalangeal joints were scored.

At the breast phantom four different structures arranged in four different series had to be evaluated. The study of the line phantom comprised the question up to how many line pairs could be differentiated.

2.4. Chest radiographs

In a clinical study 30 intensive care chest radiographs were independently evaluated by five examiners. The examiners had to delineate special anatomical structures and to diagnose pathological findings like lung infiltration, pulmonary interstitial edema, pneumothorax or pleural effusion each with a special score from 1 - 5.

In order to test the diagnostic accuracy in the detection of pathologic structures receiver operating characteristic (ROC) analysis was applied.

3. Results

3.1. Phantom studies

Fig. 1 summarizes the false negative results in the detection of the fixed round structures on the pelvis phantom. Statistical analysis proved no significant difference (alpha = 0,05) between the conventional screen film and the primary or secondary digitized images.

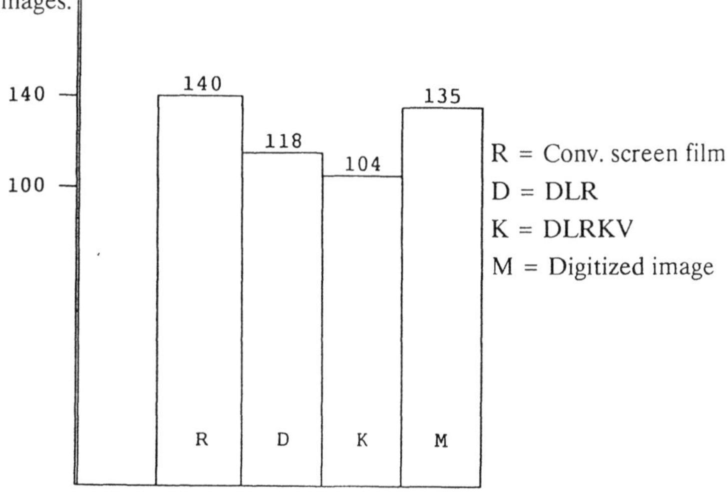

Fig.1. False negative results of the three investigated modalities in the detection of round structures at the pelvis phantom

In the hand phantom study there were also no significant differences using the three imaging methods.

In the evaluation of the breast phantom DLR showed better results than conventional screen films or digitized images. Statistical analysis proved only significant difference (alpha = 0,05) between DLR and the digitized image.
In the evaluation of the line phantom conventional screen film proved to be significant better than the digitized images.

3.2. Chest radiographs

In the detection of six different anatomical structures in chest radiographs DLR presented better results than conventional screen film or the digitized images. Especially in the delineation of the left bronchus, the right and left paravertebral line and the peripheral lung vessels DLR was significantly better than the two other methods.

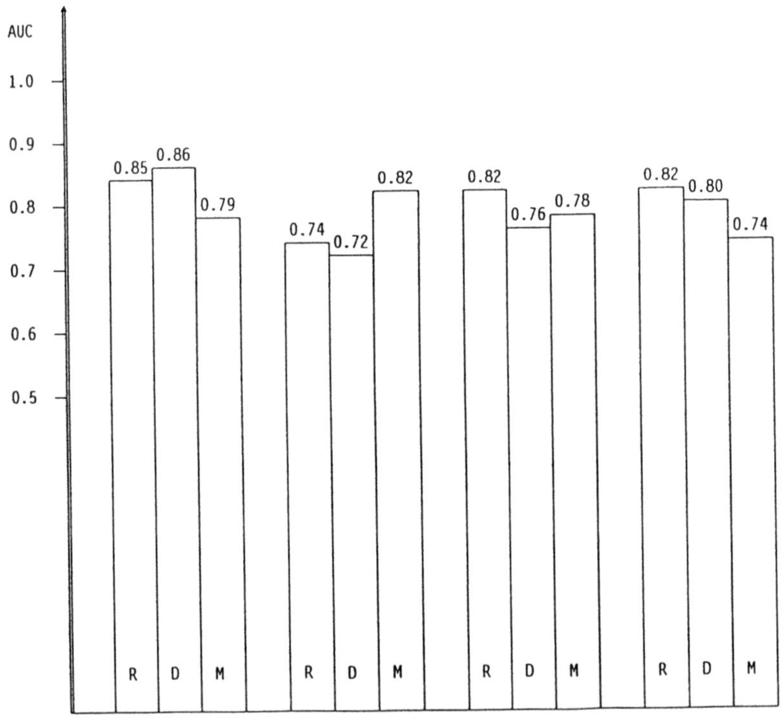

Infiltration Interstitial edema Pneumothorax Pleural effusion
R = Convent. X-ray D = DLR M = Digitized images

Fig.2: Areas below the ROC-curves for the four evaluated pathologies

ROC-analysis showed similar results in the detection of lung infiltration, interstitial edema, pneumothorax and pleural effusion for all three methods.

Fig.2 shows the areas below the ROC-curves for the four pathological findings. There were no statistical significant differences between the three methods.

4. Discussion

Conventional screen films showed the best spatial resolution in comparison to DLR and digitized images. But there were no statistical significant differences between all three methods.

There were also no statistical significant differences in the study of the pelvis phantom. This is in agreement to the literature (2). Especially the detection of round structures is independent to the spatial resolution (1,3). So the digital images are equal to the conventional screen films in the detection of those structures, where one needs a high detail contrast and where high spatial resolution is not essential (4).

The results of the hand phantom showed equal values for the three different methods. There was only a slight superiority of conventional screen film in the delineation of bony structures due to the higher spatial resolution. By using edge enhancement in DLR this disadvantage can be compensated totally.

In nearly all studies the lower spatial resolution of the digitized images can be compensated by using image post-processing algorithms at the monitor. This can also be shown in the detection of pathological findings in intensive care chest radiographs.

So in conclusion conventional screen films showed only better results in the detection of the line pair phantom representing higher spatial resolution of this method in comparison to DLR and digitized images. In the other cases the two digital methods were equal to conventional screen films, or DLR was significantly better than the other two methods. So our results indicate, that the lower spatial resolution of the digital procedures can be compensated by image post-processing for example at the PACS-console with little effort. In our opinion the migration to totally digital radiology is feasible in the near future by using DLR and in special cases digitizing of conventional screen films.

References

1. Hayrapetian A., Aberle D. R., Huang H. W., Fiske R., Morioka C., Valentino D., Boechant M. J.:
Comparison of 2048 - Line Digital Display Formats and Conventional Radiographs: A ROC Study
AJR 152 (1989) 1113 - 1118

2. Krug B., Steinbrich W., Dietlein M., Lorenz R., Altenburg A., Küpper T.:
Abdominelle Röntgendianostik in digitaler Luminineszenzradiographie (DLR) - Vergleich mit konventionellen Filmfoliensystem.
Röntgen-Bl. 43 (1990) 181 - 187

3. Lams P. M., Cocklin M. L.:
Requirements for Digital Chest Radiographs: A ROC Study of Observer Performance in Selected Cases.
Radiology 158 (1986) 11 - 19

4. Schuler M., Rath M., Hahn D., Rienmüller R., Lissner J.:
Digitale Radiographie - Methodik und klinische Anwendung
Digit. Bilddiagn. 5 (1985) 22 - 39

High-Quality Portable Chest Images Using Enhanced Film-Digitization and Computed Radiography

KENNETH R. HOFFMANN, KUNIO DOI, HEBER MACMAHON, MICHAEL CARLIN, XIN-WEI XU, MARYELLEN L. GIGER, ROBERT M. NISHIKAWA, AKIKO KANO
Kurt Rossmann Laboratories for Radiologic Image Research
Department of Radiology, The University of Chicago, Chicago, Illinois

Summary

In order to provide high-quality duplicate portable chest images for the intensive care units, we have implemented two digital radiography systems, an enhanced film-digitization system and a storage phosphor computed radiography system. Films are digitized into 2048x2430x10 bit matrices using a laser film scanner. The densities in the films are corrected to that of properly exposed radiographs based on automated histogram analysis, and non-linear unsharp mask filtering is applied to enhance mediastinal and upper abdominal detail. The storage phosphor computed radiography images are obtained in the standard manner, except that the processing parameters have been modified to produce images similar in terms of density and edge enhancement to those obtained using the film duplication system. Since implementation of digital radiography in our hospital, the repeat rate has decreased by over fifty percent. We have found that the image quality of the digital images is often visually superior to that of the original radiograph as well as consistent from day to day.

Enhanced film digitization and computed radiography

In many hospitals, in order to facilitate patient management, copies of portable radiographs of intensive care unit (ICU) patients are produced using conventional optical duplication techniques and sent to the ICUs. However, the quality of optical copies tends to be inferior to that of the original image unless processing times are changed (ref. 1). In order to provide consistently high-quality duplicate portable chest images for the ICUs, we implemented two digital radiography systems, i.e., an enhanced film-digitization system (ref. 2, 3) and a storage phosphor computed radiography (SPCR) system.

Portable chest radiography is performed in our hospital using Kodak Lanex medium screens and Ortho-C film with a 6:1 grid. After chemical processing, the technologist inserts the film into a Konica laser film scanner (Model KFDR-S). A 14"x17" radiograph is digitized into a 2048x2450 matrix, yielding a pixel size of 0.175 mm, with 1024 grayscale levels. After digitization by the film scanner, fully automated density correction, contrast enhancement, and unsharp mask filtering are performed on a SUN 3/470 computer. Two 8"x10" film copies of the processed images are then printed using a Konica laser film printer (Model KFDR-P); one for interpretation by the radiologist and one for the ICU physicians.

The original radiographs are filed, but have not been used for interpretation. The digitization system is calibrated daily for fluctuations in the chemical processing of the output hardcopy by printing, developing, digitizing, and analyzing a stepwedge pattern.

A Toshiba storage phosphor computed radiography system (Model TCR 3030a) is employed in our department for approximately one-half of our portable exams. A 6:1 grid is also employed with the SPCR acquisitions. In order to provide consistent image quality between images from the two digital modalities, we have modified the processing parameters of the SPCR unit to produce images which closely match visually the enhanced digitized film (ref. 4). In addition, the processing was adjusted so that two identical 8"x10" film copies are produced rather than the standard mildly processed and heavily processed images.

Errors in exposure and day-to-day fluctuations in density levels in film radiographs may make diagnostic comparison difficult. Thus, correction of these density variations may be of clinical value. The mapping between the original pixel values in the digitized image and the pixel values in the density-corrected image can be determined by using the H and D curve. The mapping is determined automatically by estimating a relative exposure factor for the radiograph based on analysis of the histogram of pixel values obtained from the central quarter of the radiograph (ref. 2). The mapping is then applied to obtain the density-corrected image. For the SPCR images, the standard exposure adjustment algorithms are employed.

In low contrast portable film radiographs, which result from large amounts of scatter due to large patient size, the pixel values are adjusted slightly so as to enhance contrast. The slope of the linear curve, i.e., the enhancement factor, to be employed to map the old pixel values into those of the new pixel values is determined by using the width of the pixel-value histogram. The histogram is obtained from a rectangular region in the density-corrected image which overlaps with the upper lobes of the peripheral lung region. Slopes of the mapping curves are 1.0, 1.1, 1.2, and 1.3, and the pivot point of each is at a gray level of 275.

In order to compensate for edge degradation which results from the digitization process and to enhance the visibility of catheters, tubes and mediastinal detail, we employ a non-linear unsharp mask filtering technique (ref. 4). In both systems, the degree of unsharp mask processing is varied throughout the image so that maximum processing is applied in low optical density regions such as the mediastinum, and minimum processing is applied in the high optical density regions such as in the peripheral lung. The relative weighting factors and mask sizes are different for the two systems, however, because of the intrinsic difference in resolution between the two systems (ref. 2, 5).

Results

Our chest radiologists have found that our digital radiography systems produce images of consistently high quality and use the digital images, not the original radiographs, for primary interpretation. The original film radiographs are filed As a result, our repeat rate for portable radiographs has been reduced from 5.3% to 2.0% since implementation of the systems. Since the digital systems virtually obviate the need to retake radiographs for exposure errors, positioning errors are now the dominant reason for rejection.

An additional benefit provided by the systems is that density variations between sequential images has been significantly reduced. The average optical density in the lungs was obtained by averaging the optical densities measured in four intercostal regions of the peripheral lung regions. The distribution of optical densities is centered at 1.87 ± 0.42 before density correction and 1.42 ± 0.14 after correction. The reduction in the variation about the mean indicates the improved uniformity of optical density from image to image. The distribution of optical densities for the SPCR images is comparable to that of the film images after correction.

Acknowledgments

This work was supported by USPHS Grant CA 24806.

References

1. Kim, O.H.; Ahn, M.I.; Bahk, Y.W.: Use of film duplicator to lighten dark radiographs. Radiology 184 (1992) 573.

2. Yoshimura H.; Xu X.W.; Doi K.; MacMahon H.; Hoffmann K.R.; Giger M.L.; Montner S.M.: Development of a high quality film duplication system using a laser digitizer: Comparison with computed radiography. Med Phys 20 (1993) 51-58.

3. MacMahon H.; Xu X.W.; Hoffmann K.R.; Giger M.L.; Yoshimura H.; Doi K.; Carlin M.; Kano A.; Yao L.; Abe K.; Montner S.; Nishikawa RM.; Chen X.: Clinical experience with an advanced laser digitizer for cost effective digital radiography. Radiographics (in press).

4. MacMahon H.; Doi K.; Sanada S.; Carlin M.; Giger M.L.; Montner S.M.: Single new processing algorithm to replace the standard dual image format in computed chest radiography. Radiology 177 (P) (1990) 249.

5. Niklason, L.T.; Chan, H.P.; Cascade, P.N.; Chang C.L.; Chee, P.W.; Mathews, J.F.: Portable chest imaging: comparison of storage phosphor digital, assymetric screen-film, and conventional screen-film systems. Radiology 186 (1993) 387-393.

Comparison Between Conventional Radiography and Storage Phosphor Digital Radiography in the Study of Rheumatoid Arthritis

M.MICELI,P.BURCI,R.STAMATI,S.SARTONI GALLONI,*L.FRIZZIERO

Departments of Radiology I and *Internal Medicine "C"
Maggiore Hospital, Bologna, Italy

Summary

70 patients, mainly females with rheumatoid arthritis, underwent consecutively conventional radiography and storage phosphor digital radiography examination of different skeletal parts. The presence (or absence) of some radiological signs was analyzed on each examination, with both techniques, by three indipendent radiologists, using a 1-4 score as confidence level; ROC analysis of the results was performed. The values resulted always higher with digital modality than with conventional one, but the difference resulted statistically significant only in the detection of soft tissue lesions.

Introduction

Early radiological signs of rheumatoid arthritis (RA) include soft tissue swelling, joint space narrowing (or transient widening), periarticular osteoporosis and marginal erosions (1). Early detection of these signs can be very useful for the diagnosis of RA. Recently digital radiography has been applied to the study of bone and joint disease (2,3,4). Aim of this work is to verify the diagnostic possibilities of a storage phosphor digital radiography (DR) system, compared with conventional radiography (CR), in the study of RA.

Materials and Methods

70 patients, mainly females, 16-84 aged, with RA clinically suspected or proven, previously informed, underwent consecu_ tively a CR and DR examination of different skeletal parts (mostly hand-wrist, foot-ankle, knee). Philips PCR Graphix 1 was used as DR system. CR and DR images were acquired with the following technical parameters (referred to knee examina_ tion):

	CR	DR
Focus	0.6 mm	0.6 mm
Focus-film distance	120 cm	120 cm
Grid	YES	YES
kV	60-70	70
mAs	12-15	8-10
Detector	Kodak TMG	Fuji III St IP
Detector size	18x24-35x43 cm	18x24-35x43 cm
Screens	Kodak Lanex Med.	NO
Spatial resolution	5 lp/mm	5-2.5 lp/mm
Film/hard copy	Kodak TMG	Fuji CR 633
Workstation	NO	Philips 1K DMS 20

For each exposure on DR were obtained different hard copies varying different processing parameters on the workstation (contrast, optical density, spatial frequency, edge enhancem_ ent, electroning zooming etc). Three indipendent radiologists, experienced in DR, analyzed, with no time limits, totally 280 CR and 560 DR films of the same 70 CR and DR examinations, randomly showed, evaluating on each image the presence (or absence) of four pathological findings: a) soft tissue lesi_ ons; b) erosions; c) osteoporosis; d) joint space narrowing. In some patients the presence of these lesions was evaluated also performing a US, CT or MR examination.

The three readers used a 1-4 score as confidence level of the eventual presence of the four signs analyzed. The results un_ derwent a statistical analysis by using ROC curves with Stan_ dard Error calculation and by comparing the areas under the characteristic curves with Student's paired t-test ($p<0.05$).

RESULTS

The statistical analysis showed that the values of the areas under the ROC curves for each of the four pathological findi_ ngs analyzed were higher with DR than with CR (see Fig. 1),

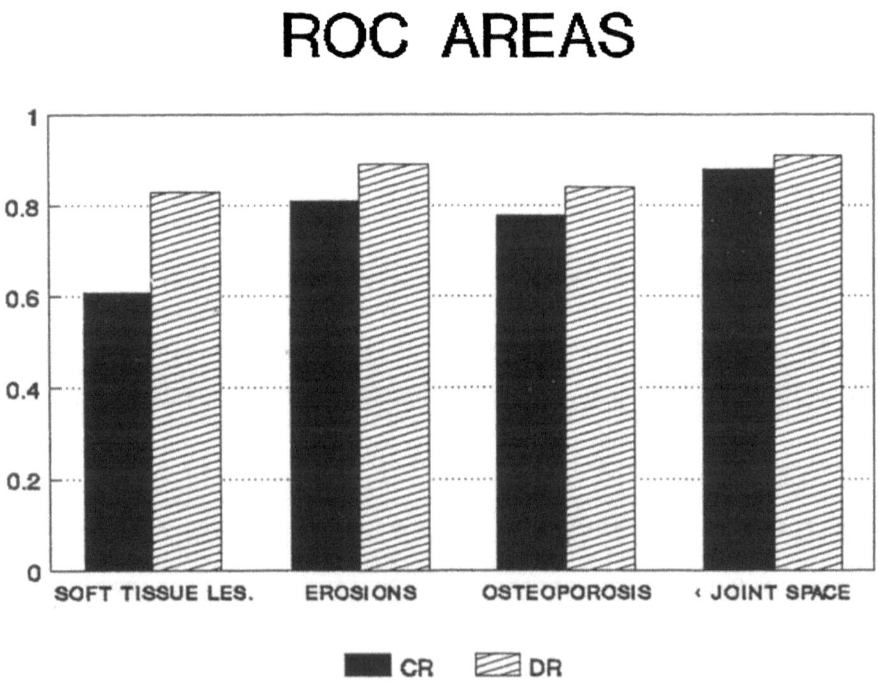

Fig. 1: Values of the areas under the ROC curves for each of the four pathological findings analyzed.

but the difference resulted statistically significant ($p<0.05$) only in the detection of soft tissue lesions (DR ROC area=0.83 +/-0.03; CR ROC area=0.61+/-0.04).

Discussion

Recently different imaging techniques have been applied to the study of arthritis, but plain film radiography is still the mainstay of all examinations in arthritic patients [5] and therefore its optimization is very important.

Storage phosphor DR imaging plates are characterized by a lower spatial resolution and a wider latitude than conventio_ nal film-screen combination. The wider latitude allows to avoid over and under-exposure (and consequently the retakes) and to acquire images with reduction of dose (generally 30% less in our experience in skeletal radiology) with acceptable diagnostic quality and in any case with a good simultaneos de_ piction of both skeletal and soft tissue. This is very impor_ tant in patients with RA who frequently undergo radiological examinations. The lower spatial resolutions of DR detector can be balanced by a correct use of the several postprocess_ ing tools available on the workstation connected to DR syst_ em: in this study we found that high frequency-edge enhanced and electronic magnified images showed very small erosions and trabecular lesions; low frequency-edge enhanced and elec_ tronic magnified images well depicted soft tissue swelling and cystic lesions. Anyway the different postprocessed DR images did not undergo a separate statistical analysis in this study. Furthermore, the density calculation tools available on the workstation were not analyzed in this experience, but we think they can be very useful for a more correct evaluation of the osteoporosis on plain film radiography.

Conclusions

Our study showed conventional radiography and storage phosphor digital radiography to be of equivalent diagnostic value in the detection of bone lesions in rheumatoid arthritis; digital technique was shown to be of higher diagnostic value than conventional one in the detection of soft tissue lesions.

We think that further advantages in the study of rheumatoid arthritis with DR will be reachable by using higher resolution and more sensitive imaging plates and quicker and "friendlier" workstations.

References

1. Renner WR, Weinstein AS: Early changes of rheumatoid arthritis in the hand and wrist. In: Arthritis and other arthropaties. Radiol Clin North Am 26 (1988) 1185-1193.

2. Murphey MD: Digital skeletal radiography: spatial resolution requirements for the detection of subperiosteal resorption. AJR 152 (1989) 541-546.

3. Prokop M, Galanski M, Oestmann JW, Falkenhausen U, Rosent_

hal H, Reimer P, Nischelsky J, Reichelt S: Storage phosph_ or versus screen-film radiography: effect of varying expo_ sure parameters and unsharp mask filtering on the detecta_ bility of cortical bone defects. Radiology 177 (1990) 109-113.

4. Grote R, Rapp U, Rosenthal H, Hendrickx P, Hammer M, Kratz G, Dohring W: Arthritic changes of small and large joints as shown by digital and conventional radiographs. ROFO 156 (1992) 277-281.

5. Kaye JJ: Arthritis: roles of radiography and other imaging techniques in evaluation. Radiology 177 (1990) 601-608.

Quality Control of Storage Phosphor Radiography Machines

M. Freedman M.D., M.B.A, Seong Ki Mun PhD, Einar Pe RTR, John Weiser PhD, John Romlein, S.-C. Benedict Lo PhD, Martha Nelson M.D.
Georgetown University, Washington, D.C., U.S.A.
MDIS Project Office, Ft. Detrick, MD., USA

INTRODUCTION:

Procedures for assuring the quality of image generation by Storage Phosphor Radiography (SR) machines have not been defined by the manufacturer and do not appear to have been described elsewhere. The information given in this article is based on experiences with several Fuji AC-1, several Fuji 7000, and one AGFA ADC. AGFA has designed a test phantom for SR that will be available in the future. The phantom tests geometric consistency and gray scale features of SR.

To assure that all components of the machine are functioning correctly global methods for testing the machine need to be devised, recommended time intervals suggested, and methods for testing individual components need to be developed. Based on our combined experience, the following methods are recommended.

TESTING THE MACHINE FOR PROPER INITIAL CALIBRATION:

The Fuji machines, as delivered and released for clinical use were found not to be intercalibrated. Thus images obtained with the same exposure and same image processing routines had visibly different appearances. The following procedure is recommended for initial checks of calibration:

a. take two imaging plates for each machine you wish to test.

b. erase each plate

c. If available radiograph the AGFA SR phantom, if not place a standard step wedge on each plate. Expose each plate to a standardized, approximately 200 speed exposure. Do not collimate. Be certain that the orientation of the plate and the stepwedge is the same for all exposures. If not using a grid, use the longest possible tube-film distance to minimize the heel effect.

d. Process two plates on each machine. Use the 2 algorithm settings that you expect to use most often (e.g. PA Chest, foot). Use algorithms that are set in the Fuji Auto mode. Use standard processing on the AGFA machine rather than DAM processing

e. Measure the optical density (OD) of each step on the plate. Graph these findings.

f. For each body part selected, the curves of the step wedge should be identical. A variation of up to 5 % may be acceptable.

This test procedure tests the proper function of the image plate reader, the auto mode PRIEF for the body part, the consistency of the Fuji G factors across machines, the settings of the laser camera, and the consistency of function of the film processor.

If the machines pass the tests they are appropriately cross calibrated for the two algorithm settings tested. One should test for cross calibration on all algorithms commonly used.

If the machines fail the test, then tests of individual components will be necessary. These are described below.

GLOBAL TEST OF THE CALIBRATION OF THE FUJI G FACTORS:

These tests can be used to test the calibration of the image processing G factors, testing either the machine calibration or the calibration of a workstation that controls a laser print image.

These tests may or may not be necessary. Because they are time consuming, they are probably not necessary if the first global test demonstrates correct cross calibration, but may be necessary if clinical use suggests that there are problems with cross calibration.

TEST OF THE GS FACTOR: The GS factor is a factor that shifts the gradient curve up or down by the number of OD units represented by the value of the GS factor.

a. set the machine to process the image with a GT of A and a GS of 0.

b. Take a standardized image of a stepwedge or of the AGFA phantom.

c. Process the image

d. Measure and graph the step wedge OD values

e. Reset the machine with all factors remaining the same, except that the GS factor is set at 0.5. Repeat the above process.

f. a graph of both groups of stepwedge values should show that they are 0.5 OD units apart.

TEST OF THE GA FACTOR: The GA factor is the slope of the characteristic curve of the system

a. Set the machine to process the image with a GS of 0, a GT of A and a GA of 1, GC of 1

b. Obtain a standardized image of a stepwedge or of the AGFA phantom.

c. Process the image

d. Repeat with GA factors of 2, and then with GA factor of 3.

e. Measure and graph the OD measurements of the image of the stepwedge.

f. Measure the slope of the mid portion of the characteristic curve.

g. The relative slopes of the three curves should be that the curve with a GA of 2 is twice as steep as that with a GA curve of 1. The slope of the curve with a GA of 3 should with 3 times as steep as that with a GA of 1.

TEST OF THE GC VALUE: The GC value is the rotation point around which the GA factor rotates the gradient curve.

a. Use the same initial data set as use for the determination of GA calibration given above.

b. Set the machine to a GC value of 1.5, a GA of 1, a GT of A and a GS of 0.

c. Expose the stepwedge or AGFA phantom to the standard exposure

d. Reset the machine so that the GC value is 1.5 and a GA of 2, with all other values remaining the same.

e. Take an exposure of the standard stepwedge with the standard exposure.

f. Measure and graph the values.

g. The three different settings of GA set above (with a GC of 1) should cross approximately at an OD of 1. The two settings with the GC of 1.5 and GAs of 1 and 2 should cross at approximately OD 1.5.

TESTS OF CALIBRATION OF THE R FACTORS:

Tests of the calibration of the R factors require the calculations of the modulation transfer function (MTF) or square wave function and a comparison of these values for several different settings of the R factors. Currently, no standardized numbers for the MTF values are available. Tests of R factor intercalibration are not yet devised.

TESTS OF INDIVIDUAL COMPONENTS:

TESTS OF THE IMAGING PLATES:

The imaging plates should be checked for surface defects and for uniformity.

Checks for surface defects. The imaging plates should be examined under white and ultraviolet light for any scratches or other surface unevenness. In some cases the ultraviolet light will demonstrate defects that are not detected in white light.

Checking for uniformity. This is performed by looking at the light emission that occurs from an imaging plate when exposed to radiation. The ST III Fuji imaging plate corresponds to a light output similar to a single 400 speed screen. The uniformity of light output can be tested by placing a single sheet of 400 speed film and a sheet of foam in a cassette against the imaging plate and exposing the plate to a 200 speed exposure. The film is then processed and visual inspection and measurements should show a uniform pattern of optical density apart from any heel effect that may be present.

TESTS OF IMAGING PLATE ERASURE:

Following a standard (50 to 400 speed) exposure of a stepwedge or other phantom, the imaging plate should undergo Fuji secondary erasure (or standard erasure on the AGFA unit). A second 400 speed flat field exposure should be made and the plate processed. Any residual image that appears on a subsequent image would indicate incomplete erasure. If the imaging plate is processed the second time with no additional flat field exposure, then the machine would amplify any small amount of residual information and still create an image whether or not the erasure system is working.

TESTS OF LASER CAMERA:

This is described in the Fuji applications manual, and this description is not repeated here.

TESTS OF THE FILM PROCESSOR:

Testing the film processor is normally done by using a test strip produced by the laser camera. The Fuji recommendation for correcting problems in OD on this computer generated sensitometry strip is to adjust the power of the laser camera. Up to a point, this is acceptable. Beyond some point this procedure becomes unacceptable because, even if you correct processor problems in OD by overexposing or underexposing the film, the noise level on the film will be increased, either from too high a processor temperature or by bad processor chemistry. For this reason, when more than minimal deviations from the OD are found on the computer generated laser printed sensitometry strip, it is suggested that a film sensitometry strip be run through the processor.

The film sensitometry strip: A film sensitometry strip can be made by exposing the laser film to 20 flashes of a blue or green sensitometer. The film is then placed as the top film in the film holder. If a computer generated sensitomtry strip is then run, the standardized light sensitometry strip will appear on its edge. This can then be compared to a standard strip obtained when it is known that the processor chemistry and temperature are correct. If one uses the same box of film for sensitometry, then this procedure would also detect problems in film consistency.

CHECKING THE ALGORITHM SETTINGS:

One of the most common problems encountered in the use of the Fuji machines is the tendency for the technologists to experiment with the image processing settings. It is important to check these settings at fairly frequent intervals to be certain they have not been changed .

A RECOMMENDED INTERVAL FOR CHECKING THE STORAGE PHOSPHOR IMAGING MACHINES:

Initial checks of machine intercalibration should be done when the machines are first delivered and also whenever major repairs are made. Checks of the imaging plates should also be done for each new imaging plate. Once the machines are in operation, the incidence of machine problems has been low. In approximate order of frequency the following types of problems have been detected: 1. Changes in image processing settings., 2. Dirt or scratches of the imaging plate. 3. Problems related to the film processor. 4. Incomplete erasure caused by inadvertent overexposure of the image plate. 5. General system failure.

RECOMMENDED INTERVAL FOR QUALITY CONTROL PROCEDURES:

CONTINUOUS:	Inspection of images as made for obvious machine failure.
DAILY:	Run computer generated strip to check on film processor.
WEEKLY:	Check image processing settings for all commonly used body parts and 1/4 of the less commonly body parts.
	Clean exteriors of cassettes.
MONTHLY:	Clean and check image plates.
	Do global check of stepwedge or AGFA phantom image and compare to baseline image.
NEW FILM LOT:	Run sensitometry strip with old and new film.
MAJOR REPAIRS:	Run global test for machine intercalibration as described above.

SUMMARY:

Quality control problems of storage phosphor imaging systems have been mainly related to lack of intercalibration of the machines at the time of initial delivery. Once intercalibration is accomplished, the major problems have been related to unauthorized changes in the imaging processing settings, scratches and dirt on the imaging plates, and problems with the film processor. Routine quality control checks should be performed at various intervals as listed above. Continuous inspection of images should be made by a supervising technologist for the sporadic failures that can occur.

Automated Detection of Skin Thickening in Mammograms

U. BICK, M. L. GIGER, Z. HUO, R. A. SCHMIDT, K. DOI, R. M. NISHIKAWA, C. J. VYBORNY

Kurt Rossmann Laboratories for Radiologic Image Research,
Department of Radiology, The University of Chicago
Chicago, IL 60637, USA

Summary

As part of a comprehensive computer-aided diagnosis scheme for detection of breast abnormalities in mammograms, we are developing a computer algorithm to detect and measure skin thickening in mammograms. The outside breast contour is located based on local gradient operations and adaptive gray-value thresholding. Following the determination of the internal skin line as a local gradient minimum, the skin thickness is measured perpendicular to the skin surface. The software was written in C and Fortran and is implemented both on a DEC VAXstation 3500 and a IBM Powerstation (RISC 6000 Series 560). As input, digitized mammograms 2k x 2k in size with a pixel size of 0.1 x 0.1 mm^2 and 10 bit grayscale were used.
The developed computer algorithm was shown to be able to automatically measure breast skin thickness in digitized mammograms. Both focal areas of skin thickening as well as generalized skin thickening were detected. Adequate digitization proved to be an important prerequisite for successful operation, since an insufficient dynamic range of digitization may result in loss of the external skin surface.
The described algorithm for detection of skin thickening in digital mammograms shows potential for use in our "intelligent" workstation for the computer-assisted analysis of mammograms, which currently includes the detection of mass lesions and clustered microcalcifications.

We are developing computer-aided diagnosis schemes to assist radiologists in breast cancer diagnosis on mammograms [1,2,7,8]. Although breast skin thickening may occur in a variety of benign disorders like edema, inflammation or scarring, it can also indicate underlying malignant disease and may be the only mammographic sign of an inflammatory carcinoma [4-6]. The purpose of our study is to develop a computer algorithm to automatically detect and measure skin thickening in digital mammograms.

Material and Methods

As input to the skin detection algorithm, conventional screen-film mammograms were digitized (Konica laser film digitizer LD 4500) with 10 bit quantization and a pixel size of 0.1 x 0.1 mm^2 yielding a matrix size of 2048x2580 for a 8" x 10" film. The dynamic range for digitization was set to include 0 - 3.5 optical density. The software was written in C and FORTRAN and can operate both under VMS (DEC VAX 3500 workstation) and UNIX (IBM Powerstation, RISC 6000 Series 560).

The skin detection algorithm consists of the following steps:
(1) Initial coarse segmentation of the image is based on a global histogram analysis establishing mean values for dark (direct exposure) and white (no exposure) image areas.
(2) Determination of the breast region is achieved using adaptive gray-level thresholding based on the local pixel intensity of the direct exposure background.
(3) Potential internal skinline points are identified as a local gradient minimum within a certain distance from the outside breast contour (Fig. 1).
(4) An optimal track through these points is found using an energy function based on connectivity and distance from outside breast contour.
(5) Skin thickness is measured perpendicular to the outside breast contour.

As a preliminary evaluation, two expert mammographers (R.A.S., C.J.V.) were asked to mark the external and internal skin border in five mammograms with skin thickening ranging between 4 mm and 2.2 cm. To assess the accuracy of the computer algorithm, the distance between each point marked by the radiologists and the computer output was calculated.

Results

The program was shown to be able to automatically detect and measure skin thickening in digitized mammograms. Both focal as well as generalized thickening was detected by the program. In the preliminary evaluation of five cases, there was good correlation between the computer results and the points marked by the radiologists (Fig. 2). The mean distance between the markings by the radiologists and the computer output was less than 1 mm in all cases (Table 1). The accuracy was higher in patients with only mild skin thickening of less than 1 cm (case #1, #2, #3). In these patients the maximum difference between radiologists markings and computer output was less than 2 mm. In areas with skin thickening of 1 cm or more (case #4, #5), differences of up to 4 mm for the internal skin line were found.

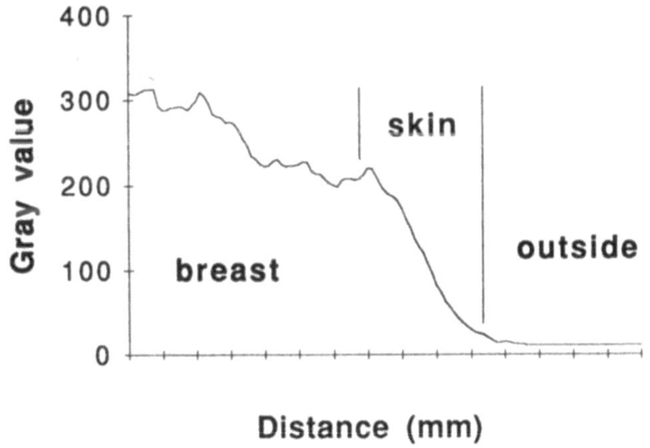

Fig. 1: Gray value profile of the breast perpendicular to the outside breast border. The internal skin contour is identified as a local gradient minimum. Skin thickness in this example (case #1) measures approximately 3 mm.

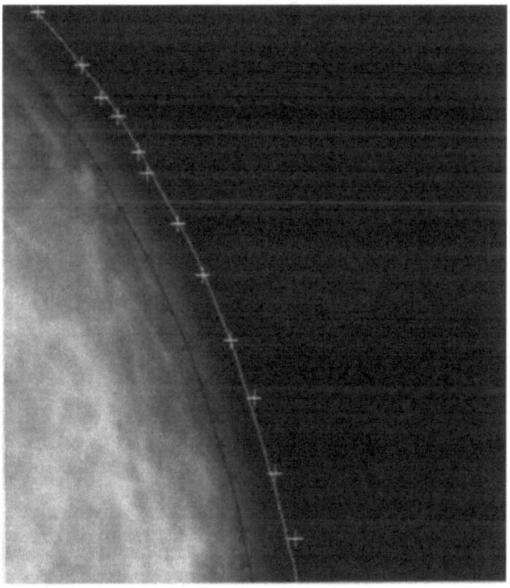

Fig. 2: Magnified view of computer generated internal (black line) and external skin border (white line) in case #2. For comparison, points marked by the radiologists are shown as black (x) along the internal skinline and as white (+) along the external skin contour.

Table 1: Mean and maximum distance between computer-generated border and radiologists markings for both the external and internal skinline. Results from the two radiologists were combined.

Case	Skin Thickness[1]	Distance[2] between Computer Border and Radiologists Markings			
		number of points	mean distance	maximum	skinline
#1	5 mm	n = 146	0.23 ± 0.17 mm	0.7 mm	external
		n = 215	0.21 ± 0.17 mm	1.5 mm	internal
#2	6 mm	n = 105	0.29 ± 0.24 mm	1.5 mm	external
		n = 136	0.29 ± 0.27 mm	1.4 mm	internal
#3	7 mm	n = 161	0.26 ± 0.20 mm	1.0 mm	external
		n = 232	0.22 ± 0.20 mm	1.5 mm	internal
#4	15 mm	n = 144	0.50 ± 0.32 mm	1.6 mm	external
		n = 151	0.81 ± 0.67 mm	2.7 mm	internal
#5	22 mm	n = 145	0.30 ± 0.20 mm	1.5 mm	external
		n = 181	0.74 ± 0.84 mm	3.9 mm	internal

[1] values represent the maximum skin thickness found for each case.

[2] for each point marked by the radiologists, the shortest Euclidean distance to the corresponding computer border was calculated.

Discussion

Breast skin thickening can occur in a variety of benign disorders like edema, inflammation or scarring, however, it may also indicate underlying malignant disease [4-6]. Evaluation of the skin in mammograms is often difficult and may require the use of a hotlight in conventional mammography or special window settings in digital mammography. Thus, skin abnormalities may be easily overlooked and we have seen a case, in which the diagnosis of cancer was significantly delayed by this oversight. As part of a computer-aided diagnosis scheme for detection of breast abnormalities in mamograms, we are developing a computer algorithm to detect and measure skin thickening in mammograms.

Skin thickness in normal mammograms may vary between 0.5 and 3 mm and is usually greater in the inferior and medial portions of the breast [5,6]. Poor radiographic technique may impair visualization of the external skin surface and lead to underestimation of skin thickness [6]. For the same reason adequate digitization with a wide dynamic range is crucial when analyzing the skin in digitized mammograms [3].

Preliminary evaluation of our skin detection algorithm shows good correlation between computer-reported and radiologists-determined skin thickness for five clinical mammograms. The mean difference between radiologists markings and computer results was less than 1 mm in all cases. Only in a few areas with marked skin thickening of more than 1 cm, points with differences of 2 - 4 mm were found.

By outlining areas of skin thickening, the computer results can be used to assist radiologists in detecting skin abnormalities on mammograms. The described segmentation of mammograms with reliable and accurate delineation of the outside breast contour may also serve as an entry point to our other automated analysis schemes in mammography including those for the detection of masses and clustered microcalcifications, which have been developed in our laboratory [1,2,7,8].

References

1 Chan H.-P.; Doi, K.; Galhotra, S.; Vyborny, C.J.; MacMahon, H.; Jokich, P.M.: Image feature analysis and computer-aided diagnosis in digital radiography. I. Automated detection of microcalcifications in mammography. Med. Phys. 14 (1987) 538-548.

2 Doi, K.; Giger, M.L.; MacMahon, H.; Hoffmann, K.R.; Nishikawa R.M.; Schmidt, R.A.; Chua K.-G.; Katsuragawa S.; Nakamori, N.; Sanada, S.; Yoshimura, H.; Metz, C.E.; Montner, S.M.; Matsumoto, T.; Chen, X.: Computer-aided diagnosis: development of automated schemes for quantitative analysis of radiographic images. Sem. Ultrasound CT MRI 13 (1992) 140-152.

3 Fredfeldt, K.E.; Christensen, E.; Conradsen, K.; Ersboll, B.; Stedstrup, S.: Automatic screening of plain film mammography. Sem. Ultrasound CT MRI 13 (1992) 135-139.

4 Leivonen, M.K.: Mammary skin thickening as a prognostic sign in breast cancer. Ann. Chir. Gynaecol. 76 (1987) 209-211.

5 Pope, T.L.; Read, M.E.; Medsker, T.; Buschi, A.J.; Brenbridge, A.N.: Breast skin thickness: normal range and causes of thickening shown on film-screen mammography. J. Can. Assoc. Radiol. 35 (1984) 365-368.

6 Willson, S.A.; Adam, E.J.; Tucker, A.K.: Patterns of breast skin thickness in normal mammograms. Clin. Radiol. 33 (1982) 691-693

7 Wu, Y.; Doi, K.; Giger M.L.; Nishikawa, R.M.: Computerized detection of clustered microcalcifications in digital mammograms: applications of artificial neural networks. Med. Phys. (1992) 555-560.

8 Yin, F.-F.; Giger, M.L.; Doi, K.; Metz, C.E.; Vyborny, C.J.; Schmidt, R.A.: Computerized detection of masses in digital mammograms: analysis of bilateral subtraction images. Med. Phys. 18 (1991) 955-963.

Magnetic Resonance Imaging

Musculoskeletal Neoplasm: 2-D Time of Flight MR Angiography with 2-D and 3-D Display

Philipp Lang, M.D., Harry K. Genant, M.D., James Johnston, M.D., Stephan Grampp, M.D., Michael Jergas, M.D., Charles Gooding, M.D.
Department of Radiology, University of California San Francisco

Summary
We studied 15 patients with primary bone and soft-tissue tumors using MRI and MRA. Seven patients were evaluated prior to and following chemotherapy. The data were transferred to an independent imaging workstation and 3-D reconstructions of the tumor, the surrounding vessels, and the normal bone were obtained using surface and volume rendering techniques. 2-D maximum intensity projections of MR angiograms demonstrate fine vascular detail and help to assess neovascularity. Neovascularity is reduced on follow-up scans of patients who respond favorably to chemotherapy. Neovascularity is seen to persist in non-responders. 2-D MR angiograms may be used as an adjunct imaging modality for evaluating response to chemotherapy. 3-D reconstructions of MR angiograms show the tumor and its relationship to adjacent vascular bundles. 3-D reconstructions appear particularly helpful in defining the need for a vascular graft.

Introduction

Primary bone and soft-tissue tumors pose an immediate threat to the affected body part and eventually to life. The choice between local resection, limb - salvage surgery or amputation in primary musculoskeletal tumors depends on the malignant potential of the tumor, its anatomic extent and relationship to neurovascular bundles, and its histologic grade [1, 2]. Knowledge of the relationship of the neoplasm to vessels is particularly important since it may help in deciding on the need for a vascular graft.

Although MR imaging provides multiplanar imaging capability, it is currently limited by a two-dimensional display of frequently complex three-dimensional anatomy. A three-dimensional image display of the tumor and the surrounding vessels appears therefore desirable when complicated limb-salvage surgery is contemplated.

In the present study, we investigated the use of MR angiography in imaging primary bone and soft-tissue tumors. Two-dimensional maximum intensity projection displays and 3-D reconstructions of the 2-D MR angiographic data were compared. Fifteen patients were evaluated; seven patients were examined prior to and following chemotherapy.

Patients and Methods

We studied 15 patients with various primary neoplasms of the musculoskeletal system. These tumors included 4 Ewing's sarcomas, 6 osteosarcomas, 3 primary lymphomas of bone, 1 fibrosarcomas, and 1 malignant fibrous histiocytoma. Images were obtained on a GE Signa system (1.5T) (General Electric Medical Systems, Milwaukee) using dedicated extremity coils. Precontrast MR sequences included T1-weighted spin-echo images (SE, TR=800 msec, TE=30 msec) and T2*-weighted double-echo gradient-echo images (GRE, TR=600 msec, TE=30 and 60 msec, $\theta=30^0$). Postcontrast images (SE, TR=800 msec, TE=30 msec) were obtained after i.v. administration of 0.1mmol/kg body weight gadopentetate-dimeglumine (Magnevist®, Schering AG, Berlin). MR angiograms (2-D time-of-flight, TR=50 msec, TE=8.7 msec, $\theta=60^0$, 2NEX) were acquired prior to administration of gadopentetate-dimeglumine. Maximum intensity projections of the 2-D MR angiograms were generated using standard scanner software.

The data were then transferred to an independent imaging workstation (MaxiView, Dimensional Medicine Inc, Minnetonka, MN), and 3-D reconstructions of the 2-D MR images and the 2-D MR angiograms were obtained. Tumor, normal bone, and vessels were displayed simultaneously in the 3-D image with different colors and transparencies. 3-D displays were generated using both surface as well as volume rendering techniques.

Results

2-D and 3-D angiographic displays demonstrated neovascularity and feeder vessels (Figs. 1 and 2). 2-D displays were particularly useful for evaluating small vessel neovascularity. 2-D maximum intensity projection angiograms showed only the vascular structures but not the tumor.

In 7 patients, MR angiograms were obtained prior to and following chemotherapy. All 7 patients showed marked neovascularity prior to chemotherapy. Five patients responded to chemotherapy (>95% tumor necrosis at histology); MR angiography showed marked reduction in tumor neovascularity in these patients after chemotherapy. Two patients were non-responders; MR angiography showed unchanged neovascularity in one and increased neovascularity in the other patient (Fig. 1).

3-D displays demonstrated the spatial configuration of tumor, feeder vessels, and normal vascular structures due to the overlay technique (Fig. 2). Tumor encroachment on or encasement of normal vascular structures was well delineated on 3-D displays (Fig. 2).

Discussion and Conclusion

It is difficult to accurately assess the response of a tumor to chemotherapy prior to surgical resection of the tumor and histologic analysis. However, it is clinically important to know whether a patient responds favorably to a given chemotherapeutic agent or whether the drug should be changed. Clinical signs, e.g. decrease in pain, local heat, and soft-tissue mass, are often inadequate in determining the efficacy of preoperative chemotherapy and show poor correlation with the degree of histologic response [3, 4]. Various imaging modalities such as conventional radiography, computed tomography, and scintigraphy have been employed to assess the efficacy of chemotherapy.

Angiography has not only been used to determine the relationship of the neoplasm to adjacent vessels, but also to monitor response to chemotherapy [5]. Reduction in tumor neovascularity following chemotherapy has generally been considered as a sign of response to chemotherapy. Angiography is, however, invasive and is hampered by complications such as bleeding, hematoma, infection, dissection, and arteriovenous fistula.

MR angiography provides information similar to that known from conventional angiography in patients with primary tumors of bone and soft-tissue. The technique is completely noninvasive and adds only ten to fifteen minutes to the overall imaging protocol. In specific, 2-D angiographic displays are helpful for evaluating fine vascular detail and appear useful for assessing tumor necrosis and response to chemotherapy. Decreased neovascularity is seen in patients who respond to chemotherapy. Unchanged or increased neovascularity is present in non-responders.

3-D displays facilitate the understanding of the anatomic relationships between normal vascular structures, feeder vessels, and neoplasm. 2-D and 3-D angiographic displays help in deciding on the need for a vascular graft and, possibly, in defining the length of the graft.

Figure 1:

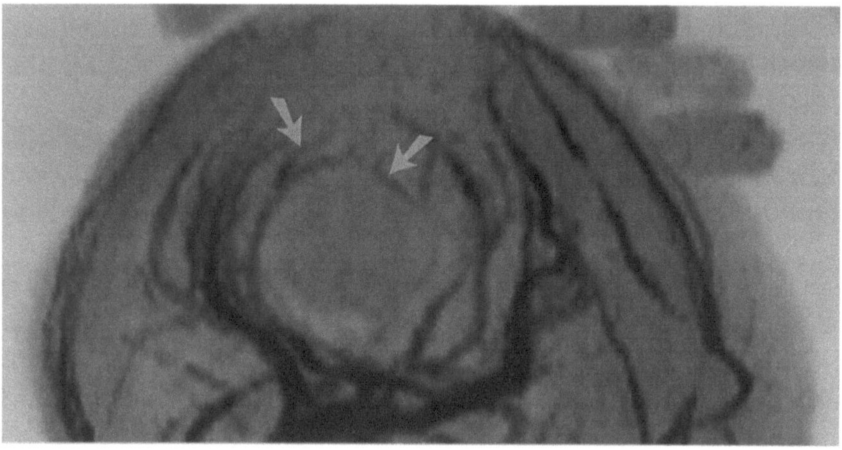

1A. 2-D MRA (maximum intensity projection) (axial orientation) in a patient with osteogenic sarcoma of the distal femur obtained prior to chemotherapy. Marked neovascularity is seen surrounding the tumor (arrows).

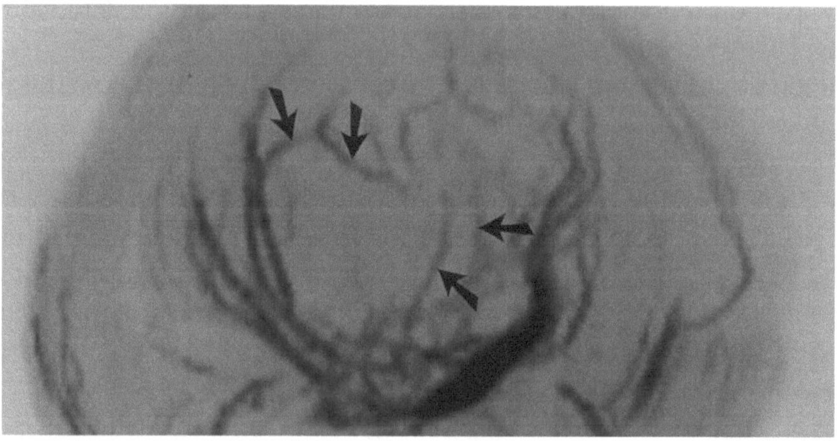

1B. 2-D MRA (maximum intensity projection) (axial orientation) in the same patient obtained following chemotherapy. The tumor has expanded. The vessels are more laterally and medially displaced. Increased neovascularity is seen coursing towards the center of the tumor (arrows). These findings are consistent with non-response to chemotherapy. This was histologically confirmed in the resected specimen.

Figure 2:

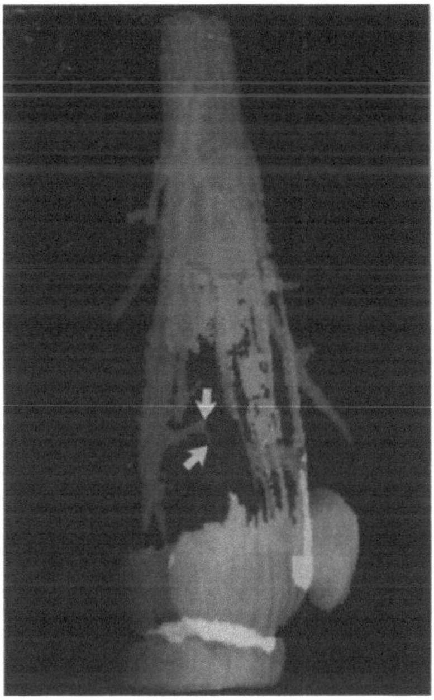

2A.

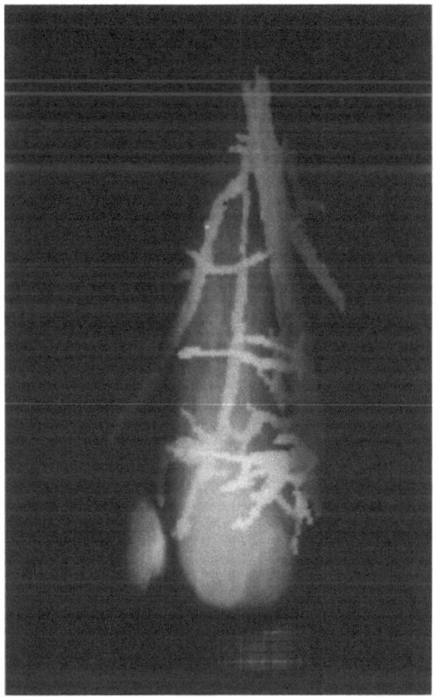

2B.

2A. 3-D surface reconstruction of MR angiograms in another patient with osteogenic sarcoma (oblique posterior projection). Normal bone is displayed in light blue. Tumor extending beyond the cortex is dark blue. Vessels are red. The tumor encroaches immediately on the major vessels and some vascular branches are partially encased (arrows)

2B. In this lateral projection in the same patient as in 2A, the bone is displayed using volume rendering while vessels are shown with surface rendering. The extensive neovascularity becomes apparent.

(see also in color on page 817)

References

1. Baker HW. The surgical treatment of cancer. Cancer 1979; 43: 787-789
2. Gilbert HA, Kagan AR, Winkley J. Management of soft-tissue sarcoma of the extremities. Surg Gyn Obstet 1974; 139: 914-918
3. Jürgens H, Exner U, Gadner H. Multidisciplinary treatment of primary Ewing's sarcoma of bone. Cancer 1988; 61: 23-32
4. Winkler K, Beron G, Delling G. Neoadjuvant chemotherapy of osteosarcoma: result of a randomized cooperative trial (COSS-82) with salvage chemotherapy based on histological tumor response. J Clin Oncol 1988; 6: 329-337
5. Bloem JL, Taminiau AH, Eulderink F, Hermans J, Pauwels EK. Radiologic staging of primary bone sarcoma: MR imaging, scintigraphy, angiography, and CT correlated with pathologic examination. . Radiology 1988; 169: 805-810

Computational Aspects of Blood Flow Measurement by First-Pass Gadolinium-DTPA Enhanced Magnetic Resonance Imaging

Böck JC, Wlodarczyk W, Götze A, Radke KH, Sander B, Vogl, T, and Felix R.

Strahlenklinik und Poliklinik
Universitätsklinikum Rudolf Virchow
Freie Universität Berlin

After an intravenous bolus injection of a paramagnetic contrast agent such as gadolinium-DTPA, the first cerebral passage of the contrast agent can be recorded in the brain tissue by a rapid $T2^*$-weighted pulse sequence. From the recorded data, concentration-time-curves of the contrast medium can be calculated on a pixel-by-pixel basis. The area of the concentration-time-curve is proportional to regional cerebral blood volume. For the computation of regional cerebral blood flow, the mean transit time of the contrast bolus through the brain is required. This mean transit time, however, can not be directly obtained because the tissue concentration-time-curve is affected by the transit of the contrast medium through the brain *and* through veins, the heart and the lungs. To eliminate the latter, a concentration-time-curve must be recorded simultaneously in an artery that supplies the brain (inlet curve) and in the brain tissue. Numerical deconvolution (on a pixel-by-pixel basis) of the inlet curve and the tissue curve yields the transport function of the brain for the contrast agent. On the basis of that transport function the mean transit time through the brain is calculated. Regional cerebral blood flow is obtained as the quotiont of regional cerebral blood volume and regional mean transit time. The methodology was validated in experimental studies with perfused organ phantoms. Initial studies in patients using key-hole subsecond image acquisition demonstrate the clinical applicability of the methodology.

Anatomical Feature Segmentation to Correct Spectral Line Shapes in Magnetic Resonance Spectroscopy

N. SAEED and I.R. YOUNG

Picker International, Hirst Research Centre, East Lane, Wembley, Middlesex, HA9 7PP, U.K.

ABSTRACT

Magnetic field inhomogeneity results in the distortion of spectral line shapes in magnetic resonance spectroscopy (MRS). The field inhomogeneity function in a voxel is computed from the histogram of phase distribution within the voxel. This involves unwrapping the phase information (360^0 discontinuities) using an anatomical feature based approach. The spectrum in the voxel is deconvolved with the point spread function (PSF) representing the phase distribution in the voxel to produce improved spectral line shapes. In the case of MRS of the head, the brain is identified from the magnitude image and the phase unwrapping is confined to this region. The segmentation of the brain is achieved by using texture suppression or thresholding in conjunction with a contour following algorithm. PSF in every voxel in the brain is converted into a unit area. Since deconvolution in the presence of noise can result in an unstable solution, the spectral information in the voxel was enhanced prior to deconvolution. Enhancement has the effect of narrowing the linewidths of the resonances, and the deconvolution process highlights subtle changes in the spectrum.

INTRODUCTION

Magnetic resonance spectroscopy (MRS) exploits the magnetic properties of certain nuclei to give information on the metabolite content in organs [1]. With *in vivo* proton spectroscopy of human organs, such as the brain, some of the metabolites of interest include N-acetylaspartate (NAA), Choline (Cho) and Creatine (Cr). The separation in frequency of resonances (peaks) is a function of the strength of the magnetic field and it can be normalised to the centre frequency (Larmor frequency) so that the resonances have fixed position in the spectrum. This scale is defined as the ppm axis. In the case of proton spectroscopy, NAA, Cho and Cr appear at 2.02, 3.3 and 3.1 ppm, respectively. Under ideal conditions, the peaks approximate to a Lorentzian function, however, due to noise and magnetic field variations, the peak shapes can vary considerably. The effect of noise on the peak shape is less prominent than the field inhomogeneity component. In general, noise produces a slight broadening of the peak, hence, giving it a Gaussian appearance. Since the metabolite peaks are so close to each other (ppm axis), then for proper identification and quantitation of spectra, it is important to get very good field homogeneity in the voxel of interest. Typically, for good spectrum quantitation, the field variation in a voxel ($20*1.5*1.5mm^3$) should be less than 0.1ppm. It has been shown that phase maps can provide information on the magnetic field distribution within a voxel [2]. The field inhomogeneity function in a voxel is computed from the histogram of phase distribution within this voxel. Prior to the production of the phase profile, the phase information has to be unwrapped to remove any 360^0 discontinuities.

In the case of MRS of the head, these discontinuities are most obvious in areas of the scan that lie outside the head region. It was therefore decided to segment the head and the brain, and then limit phase unwrapping to the latter part of the scan. No diagnostic information related to MRS is lost by confining the phase unwrapping process to the brain area. This is especially the case with proton spectroscopy of the head as here the scan area of interest is in all probability confined to the brain. The phase profile is converted into a point spread function (PSF) and then the spectrum in the voxel is deconvolved with the PSF. Prior to deconvolution, the spectrum was enhanced; enhancement has the effect of narrowing the linewidths of the resonances, and the deconvolution process highlights subtle changes in the spectrum.

ANATOMICAL FEATURE SEGMENTATION

The method used to isolate the outer head boundary (OHB) and the brain from the magnitude scan is very much dependent on the MR sequence being employed [3]. In the case of a spin echo sequence which is T1 weighted (TR=600 msec, TE=22msec), the fat region appears as a bright structure and the brain appears grey. In this case, to isolate the OHB, a contour following algorithm (CFA) is initiated at a bright pixel location in the scan. This point is called the initial point (IP). The CFA tracks the OHB and it is terminated on returning to the IP. To isolate the brain in this sequence, a texture suppression procedure is implemented [4] which results in a fairly well defined brain region. A CFA is then implemented to track a range of grey levels which results in the isolation of the brain. T1 weighted images can also be produced by employing a partial saturation sequence (TR=500 msec, TE=17 msec). Here, the brain can be segmented by a thresholding operation [5]. A threshold value is obtained from a grey level histogram computed in a window (6cm*6cm) positioned at the centre of the scan or in areas where there is considerable grey level activity. A threshold value, T, is computed using the following formula:–

$T_L < T < T_H$ (T_L and T_H are lower and upper threshold values, respectively).
$T_L = g_m - 2g_{sd}$ and $T_H = g_m + 2g_{sd}$
g_m and g_{sd} are the mean and standard deviation grey levels in the histogram.

In this sequence, the brain can also be isolated using texture suppression, however, this method is much slower than thresholding. Typical thresholding operation (256*256 pixels) takes 2 secs whereas texture suppression can take up to 15 secs. Fig(1a) shows a T1 weighted sagittal scan of the head. Fig (1b) is the segmented brain after texture suppression. Fig(1c) shows the segmented brain after thresholding. Fig(1d) is the isolated contour from the threshold image, and it defines the region where the phase has to be unwrapped.

PHASE UNWRAPPING AND PSF PRODUCTION

The phase values obtained from the phase map lie between $-\pi$ and $+\pi$ radians. As the magnetic field inhomogeneity gets worse, the phase information could be cyclic with several abrupt crossings from $-\pi$ to $+\pi$ radians. These values (phase discontinuities) are artificial and they distort the computation of the true phase distribution in the voxel. To overcome cyclic phase values, the phase information is unwrapped by identifying this rapid transition in phase based on a threshold setting, t_p. In addition, the phase unwrapping is confined to the brain

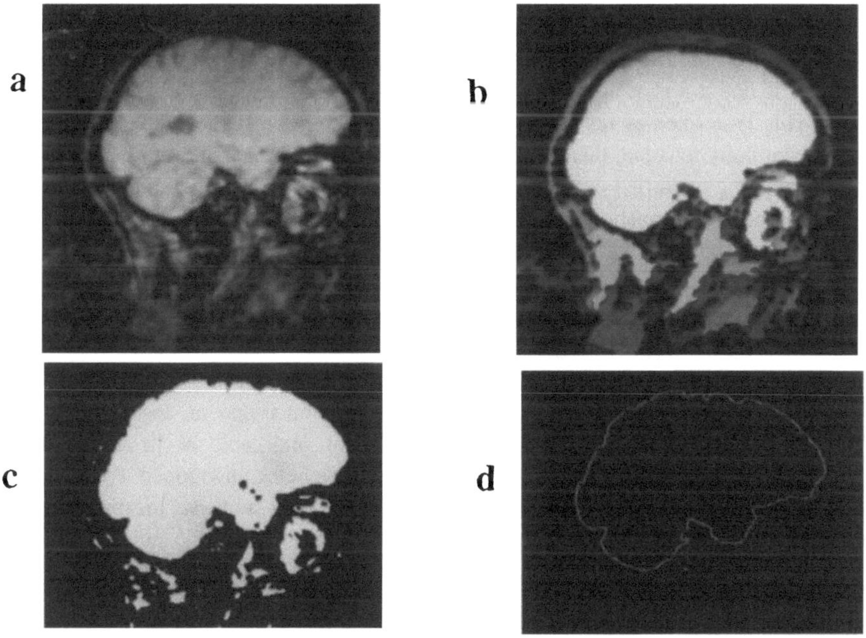

Fig1: (a) T1 weighted sagittal image (b) Texture suppression (c) Thresholding (d) Contour of brain from threshold image.

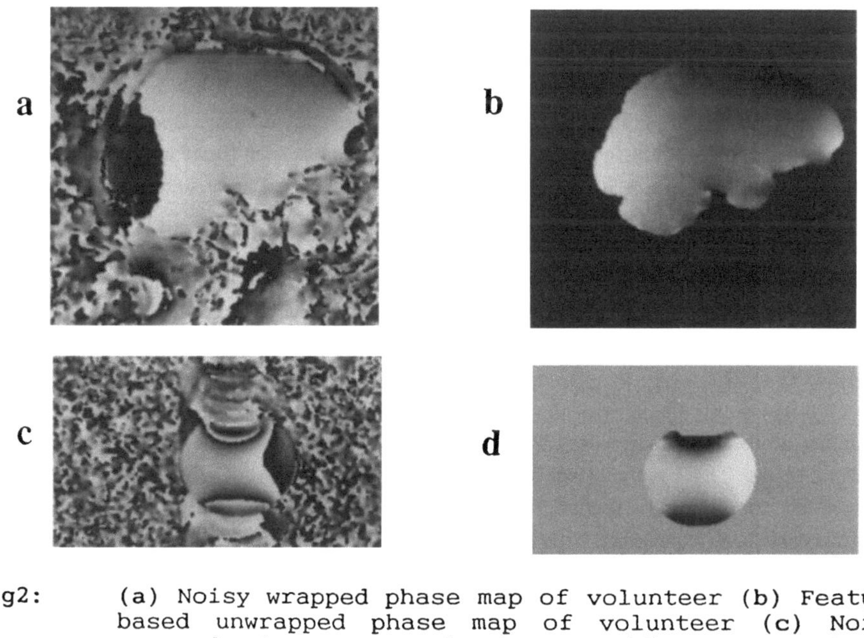

Fig2: (a) Noisy wrapped phase map of volunteer (b) Feature based unwrapped phase map of volunteer (c) Noisy wrapped phase map of phantom (d) Feature based unwrapped phase map of phantom.

region as any unwrapping process that takes into account the background noise region will not be accurate. This is because the background noise is littered with abrupt phase changes. Having identified the brain, the centre of area of this feature is computed. This location, provided it is not at a phase discontinuity location, forms the start of the phase unwrapping process. This is defined as the phase initiation location (PIL). If the centre of area is at a phase discontinuity location, then neighbouring pixels are examined until a PIL is found. The phase unwrapping is carried out in two orthogonal directions (horizontal and vertical axes); the resulting phase information from the two operations is smoothed using a 3*3 pixel mean filter. The phase unwrapping proceeds as follows:–

$$\text{If } [\,|P_w(n) - P_w(m)| \geq t_p] \text{ THEN } P_u(n) = P_u(m)$$
$$\text{ELSE } P_u(n) = P_u(m) + [P_w(n) - P_w(m)]$$

$P_w(n)$ and $P_w(m)$ are the wrapped phase values at location n and m, respectively. n is a neighbouring pixel location to the previous location m, and could be in any of the four directions depending on the direction in which the phase is being unwrapped. $P_u(n)$ and $P_u(m)$ are the corresponding phase values after phase unwrapping at these locations. t_p is the threshold setting and is set to π radians. Fig(2a) and Fig(2b) show the noisy wrapped and noise–free (brain based) unwrapped phase maps, respectively. Fig (2c) and (2d) show the wrapped and unwrapped phase maps for a Lactic acid phantom. An unwrapped phase distribution profile is produced for every voxel. This is then normalized so that it has a unit area and it is called the point spread function (PSF). The width of the PSF which is defined as the frequency axis is proportional to the shift in frequency of the resonances in a particular voxel. The normalized magnitude of the frequency points give the relative contribution of the frequency component at these points to the resonances.

SPECTRA ENHANCEMENT AND DECONVOLUTION

The voxels are isolated and the spectrum in each voxel is deconvolved with the computed PSF in the voxel. This operation is performed in the frequency domain. Deconvolution in the presence of noise can result in an unstable solution [6] and as such the resulting deconvolved spectrum is monitored for instability (intensity overflow). The criteria for identifying instability is when the maximum intensity value in the spectrum after deconvolution exceeds the maximum intensity value in the original spectrum by a factor of 5. Spectrum instability is overcome by compressing the PSF in the frequency direction (width) by a factor of 20% of the previous width until a stable solution is reached after deconvolution. The PSF window width, f_s, for a stable solution is noted. The width of the PSF function prior to compression is defined as f_p. If $f_s << f_p$, then deconvolution will not result in narrowing of linewidths, but may highlight subtle changes in the spectrum. To take full advantage of the dispersion in frequency of the non–compressed PSF, the spectrum, S, is enhanced using the procedure indicated below. This has the effect of narrowing the linewidths, but will not highlight subtle changes. By combining the spectra enhancement and deconvolution operations, the resulting spectra consists of resonances with improved lineshapes. Spectra enhancement proceeds as follows and it employs a compressed PSF of width, f_m (where $f_m=f_p-f_s$) :

$$S(f) = S(f) + g$$

where f= 1 to N (maximum number of frequency points in the spectrum, S).
$S(f+x) = S(f+x) - P(x)S(f)$ where x=1 to f_m
P(x) is the point spread function.

$$g = \sum_{x=1}^{x=f_m} P(x)S(f)$$

The resulting enhanced spectrum is then deconvolved with the PSF, but here the PSF has a compressed width of f_s.

RESULTS

The spectra enhancement and deconvolution procedure was tested on a Lactic acid phantom and proton brain spectra of a volunteer. Fig (3a) shows the doublet corresponding to the Lactate metabolite prior to spectrum lineshape correction. Here the signal to noise (S/N) ratio is 46.2. Under ideal conditions, the two peaks making up the doublet must be identical in height and area. However, due to inhomogeneous magnetic field, peak 1 is less prominent than 2. Fig (3b) shows the resulting spectrum after correcting for the lineshape using the spectra enhancement and deconvolution procedure. It is seen that the peaks are now better resolved, however, the S/N ratio is now 37.4. Curve fitting the peaks in fig(3a) to a Gaussian function showed that the linewidths of peak 1 and 2 were 3.50 Hz and 8.65 Hz, respectively. The area of peak 2 was 5.8 times that of peak 1. With fig (3b), the linewidths of peak 1 and 2 after Gaussian fitting were 4.25 Hz and 4.21 Hz, respectively, and the area ratio of peak 2 to peak 1 was 1.14. Lactate peak was well highlighted in about 50% of the voxels examined after the lineshape correction procedure. Fig (3c) shows the uncorrected *in vivo* spectrum (S/N ratio =18.6). Fig (3d) is the spectrum obtained after spectrum enhancement and deconvolution (S/N ratio= 14.9). It is seen that the Cho and Cr peaks are better resolved in fig(3d) than in fig (3c). Similar results were obtained in about 30% of the voxels examined.

CONCLUSION

Initial studies on the implementation of the spectra enhancement and deconvolution procedure on phantom and proton *in vivo* spectra has provided good results. The peaks are now better resolved and as such quantitation of spectra using line fitting procedures gives fairly accurate integrals and linewidths, however, the method suffers from a decrease in signal to noise ratio when the corrected spectra is compared with the original uncorrected spectra. At the moment, the correction procedure has been implemented on 2 phantoms and 2 different subjects, hence the robustness and accuracy of the method under various signal to noise ratio conditions has yet to be established.

REFERENCES

[1] Gadian D.G., **Nucl. Magn. Res. and its Application to Living Systems**, Clarendon Press, U.K., 1982.
[2] Bailes D.R., et al, **J. Magn. Reson.**, 77, pp460, 1988.
[3] Saeed N. and Young I.R., **Comp. Asst. Radiol.**, pp308, 1991.
[4] Saeed N., **PhD Thesis**, 1990.
[5] Haralick, R.M., **Comp. Vision Graphics Image Proc.**, vol29, pp100, 1985.
[6] Oppenheim, A.V., **Digital Signal Processing**, Prentice-Hall, 1975.

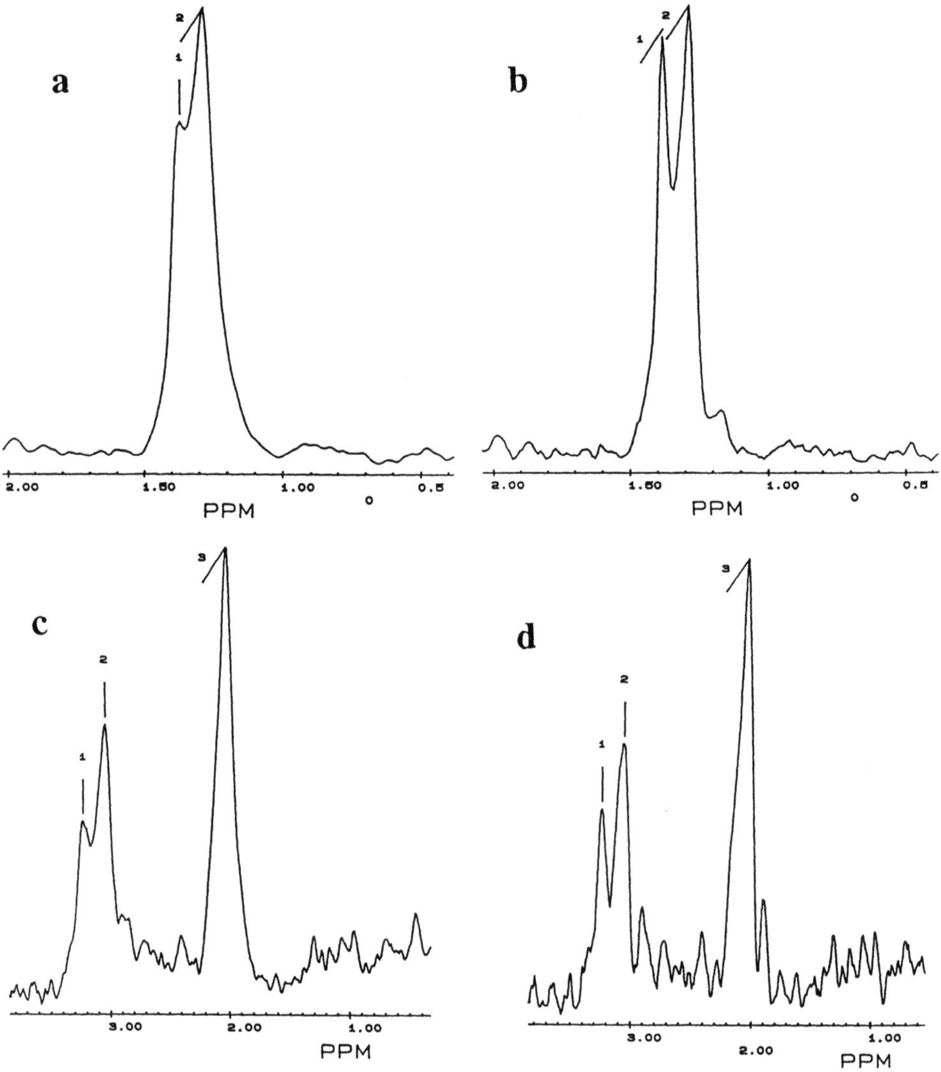

Fig3: (a) Original spectrum of phantom highlighting lactate peak (doublet) (b) Spectrum of phantom after spectrum enhancement and deconvolution (c) Original proton MRS of volunteer (peak 1,2,3 = Cho, Cr, NAA, respectively) (d) Spectrum of volunteer after enhancement and deconvolution.

ACKNOWLEDGEMENTS

The authors would like to thank all the staff at the NMR Unit, Hammersmith Hospital, London, U.K.

Computer Tomography

Contrast Parameters for the Assessment of CT

J. HUESEMANN [1], P. REIMER [2]

[1] Medizinische Hochschule Hannover
Abteilung I : Diagnostische Radiologie I
Konstanty-Gutschow-Straße 8
3000 Hannover 61, Germany

[2] Westfälische-Wilhelms-Universität
Institut für Klinische Radiologie
Zentrale Röntgendiagnostik
Albert-Schweitzer-Straße 33
4400 Münster, Germany

Summary

This study was designed a) to determine which quantitative contrast parameters for CT correlate best with lesion conspicuity obtained by a ROC analysis, b) to establish contrast and size thresholds for lesion detectability, and c) to use these contrast parameters for the assessment of previously introduced novel CT techniques, specifically subsecond/second conventional CT and spiral CT. A plexiglas GS®-phantom was designed to simulate the whole contrast range of clinical studies. A phantom with bores (0.5 - 3 cm) was filled with increasing concentrations of contrast medium thus exhibiting different contrast values. Dynamic CT scans of different scan times and spiral CT (Somatom Plus; Siemens AG, Erlangen) were performed with 8 mm collimations (120 kV) utilizing the same acquisition mode and reconstruction algorithm. Of the six contrast formulas the contrast-to-noise ratio (CNR) yielded the best correlation with identical scoring style as used in a receiver operated characteristic (ROC) analysis (linear correlation coefficient $r = 0.881$). Lesions of 0.5 cm and 1 cm diameter were always detected with a CNR > 4, in 71 % with a CNR between 3 and 4, in 36 % if CNR is between 2 and 3, 11 % if CNR was between 1 and 2, and no lesion was detected if CNR was less than 1. Contrast values increased statistically significant ($p < 0.05$) with higher scan times (doses) in following order : 0.7 s < 1 s < spiral CT (1 s) < 2*1 s < 3*1 s. Spiral CT (1s) showed significant higher contrast values than a dynamic 1 s-scan but no significant higher ($p = 0.115$) lesion detection obtained by the ROC analysis. Only optimal acquisition times (high doses, e.g. 3*1 s) showed significant ($p = 0.034$) better lesion detection than spiral CT. Lesions of 0.5 cm diameter require at least a CNR of 3.42 ± 0.45 and lesions of 1 cm diameter at least a CNR of 2.91 ± 1.01 to be detectable on spiral CT. Contrast parameters allow to assess the clinical performance of novel CT techniques, measure objectively the diagnostic value of CT images, and may obviate clinical studies with the same purpose.

Purpose

Contrast is the major determinant of lesion detectability [1]. Quantitative contrast parameters of lesion contrast can be applied to compare the accuracy of different CT techniques, to find optimal timing parameters, and to establish the efficacy of contrast enhanced CT techniques for the task of tumor detection. Optimally, quantitative contrast parameters derived from CT images should correlate with lesion detection as studied by observer performance tests, which resemble most closely lesion conspicuity. However, these tests are very lengthy, time consuming, and cumbersome for obtaining reliable results.
The goals of this study were to calculate different quantitative contrast parameters which correlate with lesion conspicuity, to determine the best formula, to detect contrast and size thresholds, and to evaluate contrast parameters for the assessment of previously introduced novel CT techniques, specifically subsecond/second conventional CT and spiral CT.

Materials and Methods

We designed a plexiglas GS®-phantom with 0.1 - 3.0 cm bores to simulate lesions of different diameter (Figure 1).

Figure 1 : Plexiglas GS®-phantom

Plexiglas GS® was chosen because of same attenuation of contrast enhanced liver tissue (~120 Houndfield Units; HU). The phantom was filled with water (7000 ml) and increasing concentrations of contrast medium (Ultravist®300; Schering, Berlin; 0.623 g lopromidol/ml) thus exhibiting different contrast values to simulate the whole contrast range of clinical studies between "lesion" (bore) and "liver" (Plexiglas). A total of 12 contrast levels was simulated with hypo-, iso- and hyperintense lesions in comparison with the surrounding tissue (Figure 2).

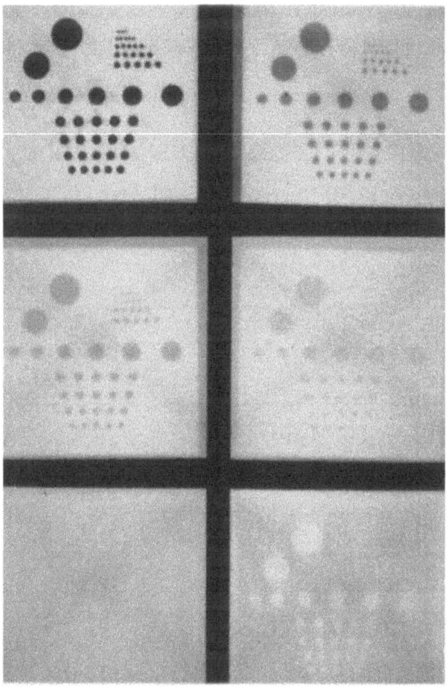

Figure 2 : CT scans of the phantom with six different contrast levels

On every contrast level dynamic and spiral CT scans (Somatom Plus; Siemens AG, Erlangen) were performed with 8 mm collimations (120 kV) utilizing the same acquisition mode and reconstruction algorithm. Dynamic CT was varied using four different scan times (0.7 s/175 mAs, 1 s/165 mAs, 2*1 s/330 mAs, and 3*1 s/495 mAs) in comparison to spiral mode (1 s/8 mm slice/165 mAs). Mean values of Hounsfiels Units (HU) and standard deviation were measured in regions of interest (ROI) in the lesions (0.5 cm, 1 cm, 2 cm and 3 cm), adjacent phantom, and background. A total of 1440 image contrast parameters (6 formulas * 4 lesion diameters * 5 scan times * 12 contrast levels) were calculated using different formulas (Table). Each individual image (168 images without lesion / 240 images within lesion (4 lesion diameters * 5 scan times * 12 contrast levels)) was scored for lesion conspicuity similar to methods employed in receiver operating characteristics

(ROC) analysis [2,3]. Eight radiologists independently reviewed the images and ranked 1 - 5 levels of diagnostic confidence that a lesion existed without time constrains. Correlation coefficients were calculated using standard least squares methology. To assess the concordance of radiologists` subjektive ranking of lesion conspicuity, Kendall's coefficient was applied, accepting significance at $p = 0.001$.

Results

Concordance between the eight radiologists was shown by the absence of statistically significant differences (Kendall's coefficient $W_C = 0.834$, $p < 0.001$) in confidence rankings.
All six image contrast parameters correlated significantly ($p < 0.05$) with the conspicuity scores obtained from ROC analysis (Table).

Formula	correlation coefficient
1. $\lvert HU_{lesion} - HU_{liver}\rvert / SD_{noise}$ (CNR)	0.881
2. $\lvert HU_{lesion} - HU_{liver}\rvert / SD_{lesion}$	0.871
3. $\lvert HU_{lesion} - HU_{liver}\rvert / (HU_{lesion} + HU_{liver})$	0.821
4. $\lvert HU_{lesion} - HU_{liver}\rvert / HU_{liver}$	0.803
5. $\lvert HU_{lesion} - HU_{liver}\rvert$ * diameter / SD_{noise}	0.857
6. $\lvert HU_{lesion} - HU_{liver}\rvert / \sqrt{[(SD_{lesion})^2 + (SD_{liver})^2]/2}$	0.862

HU : mean value of Hounsfield Units; SD : standard deviation

Table : Correlation of quantitative image contrast parameters with lesion detectability judged by 8 radiologists

The image contrast parameters expressed by Formula 1 ("Signal difference (contrast) - to noise ratio"; CNR) showed the best correlation (Figure 3) with lesion conspicuity ($r = 0.881$). High lesion contrast and low image noise resulting in high CNR, which goes along with better lesion detectability.
In a subset of 80 lesions with a diameter of 0.5 cm and 1 cm, detectability was illustrated as a function of CNR (Figure 4). When CNR was greater than 4, 100 % (6/6) of lesions were detected. Lesion detectability was 71 % (10/14) for a CNR between 3 and 4 , 36 % (10/28) for a CNR between 2 and 3, 11 % (2/18) for a CNR between 1 and 2, and no lesion was detected if CNR was less than 1 (0/14).

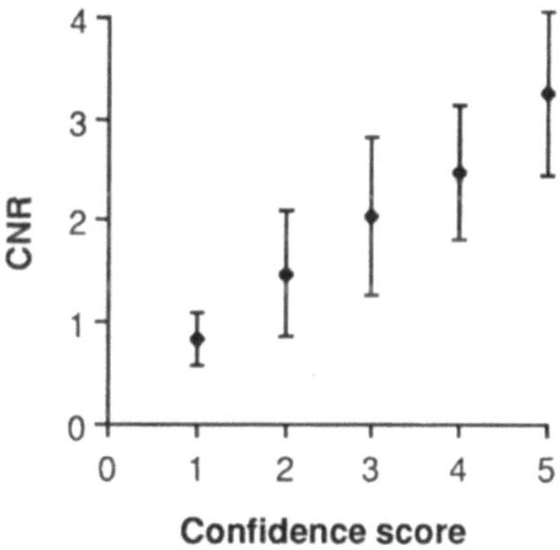

Figure 3: Plot of CNR parameter against radiologists` confidence

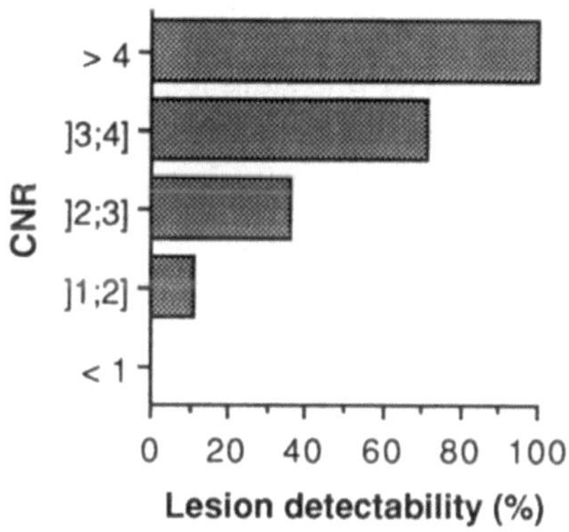

Figure 4 : Plot of CNR parameter against lesion detection

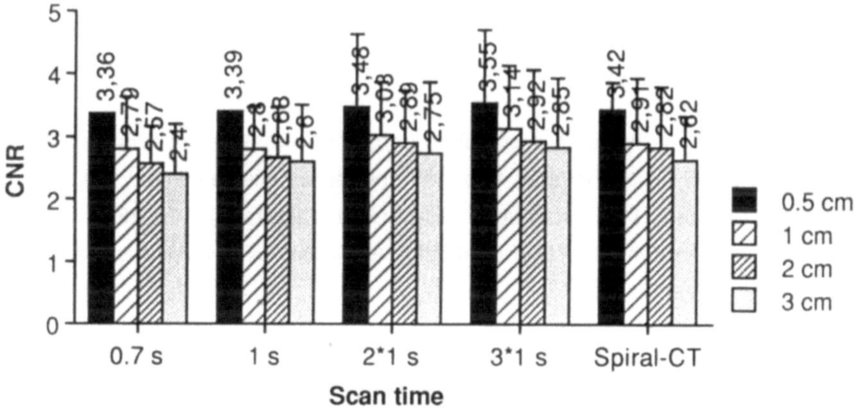

Figure 5 : Correlation of detected lesions as a function of scan time and CNR for each lesion size

Using Formula 1 (CNR) as the best formula of lesion detectability, the threshold of detectability was determined. A lesion was detected if ranked 4 or 5 in level of diagnostic confidence. The correlation of lesion detectability as a function of CNR and size for each scan time is demonstrated in Figure 5. Lesions of 0.5 cm diameter require at least a CNR of 3.42 ± 0.45 and lesions of 1 cm diameter at least a CNR of 2.91 ± 1.01 to be detectable on spiral CT. The larger the lesion, the less contrast is required for detection. Contrast values increased with higher scan times (doses) in following order : 0.7 s < 1 s < spiral CT (1 s) < 2*1 s < 3*1 s. In each case the increase of CNR was statistically significant ($p < 0.05$). Independant of lesion size and contrast to the surrounding tissue spiral CT (1s) showed significant ($p < 0.05$) higher contrast values than a dynamic 1 s-scan (Figure 5) but no significant higher ($p = 0.115$) lesion detection obtained by the ROC analysis. Only optimal acquisition times (high doses, e.g. 3*1 s) showed significant ($p = 0.034$) better lesion detection.

Discussion

Quantitative evaluation of CT images as a method to determine the potentials of novel scan or contrast media administration techniques has not been applied systematically. We investigated six different formulas for the calculation of "contrast" to determine which factors most closely correlate with lesion conspicuity on CT images. The formula with measurements of HU of lesion and organ as well as background noise (Formula 1 / Table) correlated best with lesion conspicuity. Notably formulas with consideration of image noise, an important factor in lesion detection, correlates well. Our results indicate that conspicuity as measured by observer tests correlates well with contrast parameters derived from CT attenuation values. The calculation of suited contrast parameters allows the assessment of novel CT techniques and may obviate extensive clinical studies with the same purpose. The influence of morphologic factors such as lesion size, location, and local anatomy on lesion detection is a more complex issue that requires further investigation.

References

1. Barnes, J. Ed; Characteristics and Control of Contrast in CT; RadioGraphics (1992); 12 : 825 - 837

2. John, V.; Hirche, H.; Callies, R.; Lenz, W.; ROC-Analyse als Testmethode und zur Trainingskontrolle bei Roentgenbildauswertung; Radiologe (1981); 21 : 255 - 259

3. Brandt, G.-A.; Schroeder, H.; Mischke, W.; Tautz, M.; Ueber die Ermittlung von ROC-Kurven und deren Bedeutung für die subjektive Bildguete-bewertung; Radiol. diagn. (4/79); 496 - 505

3D CT Acquisition with Conventional and Volumetric Scan: Comparison of Resolution in Longitudinal Direction

JCM Steenbeek[1], BM ter Haar Romeny[2], MS van Leeuwen[2].

[1] Philips Medical Systems, P.O. Box 10000 5680 DA Best The Netherlands,
[2] University Hospital Utrecht, The Netherlands.

Summary.

With volumetric CT the image quality of the coronal, sagittal and 3D reconstructions is expected to be improved with respect to conventional CT since the acquired volume is more consistent and the sampling in the longitudinal direction is improved. To validate this, phantom experiments has been performed and a simulation program with the help of Mathematica™ has been developed. A spatial frequency phantom and a special hole phantom were used to compare the results. In both cases good agreement was found. The MTF distributions in the longitudinal direction shows that for volumetric scan the best compromise between number of slices and resolution in the z-direction is achieved for slice thicknesses of 3 or 5 mm by a reconstruction index of 2 mm. The simulation program enables the definition of the best protocols to visualize all kinds of patient transients.

Introduction

Conventional Computed Tomography (CT) is able to scan a complete volume by successive, stepwise scans of one slice after the other. To cover a complete volume interscan delays are necessary for table incrementation and, possibly patient breathing. Volumetric scan involves continuous table transportation through the gantry during continuous rotation. This method enables acquisition of a complete volume in one single breath-hold. Additionally slices can be reconstructed at any arbitrary position in the acquired volume.

With volumetric scan the image quality of the coronal, sagittal and 3D reconstructions is expected to be improved with respect to the conventional CT since the acquired volume is more consistent and the sampling in the longitudinal direction is improved.

The image quality in the longitudinal direction is also influenced by the shape of the slice sensitivity profile. The interpolation algorithm used in volumetric CT is based on the reconstruction from 180° opposite views (1,2). This algorithm provides slice sensitivity profiles which are almost similar with the profiles obtained with conventional CT. The effect of the slice sensitivity profiles on the image quality is evaluated.

Goal of this study was a: the validation of the expected improved image quality and b: the optimalisation of the scan parameters for clinical applications.

Phantom Experiments

In order to measure the spatial resolution perpendicular to the scan plane, two phantom experiments were performed on the Philips Tomoscan SR-7000 (Philips Medical Systems, The Netherlands). A special perspex spatial frequency phantom is used with pitch sizes which vary from 20 mm to 1 mm (see fig 1a). Also a special perspex hole phantom is used (see fig 1b) to evaluate the shape of circular transitions. The diameter and the distance in between the holes are equal and vary from 10 to 1 mm. Conventional and volumetric scans were performed with slice thicknesses of 3, 5, and 10 mm. The conventional scans were acquired adjacently. In the volumetric scans the pitch of the helix was selected equal to the slice thickness and the distance between the reconstructed slices (reconstruction index) varied from 1 mm to 3, 5 or 10 mm respectively depending on the used slice thickness.

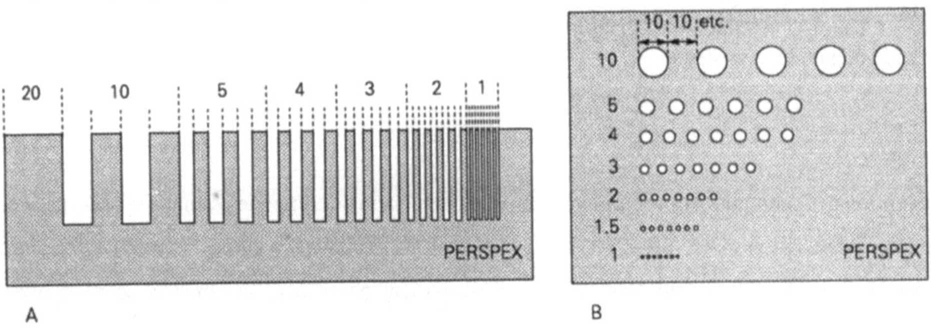

Fig 1: a) Schematic drawing of spatial frequency phantom
b) Schematic drawing of hole phantom
Note: All dimensions are given in mm.

Simulation model

To translate the results of the phantom experiments into the clinical practice a mathematical simulation model has been developed. Such a model has the advantage that the result of the measurement of any input density distribution can be predicted. In this model measured slice sensitivity profiles (1) of the conventional and volumetric scans are mathematically convolved with the 3D phantom data.

The spatial resolution in the longitudinal direction is described by the slice sensitivity profile S(z). The measured density distribution in any object or patient can be computed by a mathematical convolution of the real patient density P(z) with the slice profile S(z). This process mathematically expressed as:

$$\int_{-\infty}^{+\infty} P(x)\ S(z-x)\ dx \quad = \quad P(z) * S(z)$$

where x is an integration variable. In practice the convolution is most conveniently calculated in the Fourier domain, due to the fact that a convolution in the spatial domain is mathematically equivalent with a product in the Fourier domain.

$$F\{P * S\} \ = \ F(P)\, F(S)$$

where F denotes the Fourier transform and F^{-1} the inverse Fourier transform. The convolution can now easily be calculated from the inverse Fourier transform of the product of the Fourier transforms of the slice profile and the patient density data:

$$P(z) * S(z) = F^{-1}\{\ F\{P(\phi)\}\ F\{S(\phi)\}\}$$

where ϕ describes the spatial frequency. The model described is a continuous model. Equivalent formulae can be used to deal with discrete data. A convenient way to model these equations is by the use of a symbolic algebra package Mathematica™ (3).
To validate this simulation program a comparison is made between the measured phantom data and the results of the simulation program obtained with the same phantoms simulated.

Results

In figures 2 and 3 the results of the spatial frequency phantom are shown. In fig 2a, the measured density distribution in the z-direction is given for the volumetric scan with slice thickness 3 mm and reconstruction index of 1 mm. The density distribution is measured over a profile strip with a width of 40 pixels. The pixels were averaged over the width of the strip on order to reduce system noise. In fig 2b, the density distribution computed with the simulation program is given. Comparing these figures, a good agreement between measured data and simulated data can be observed.

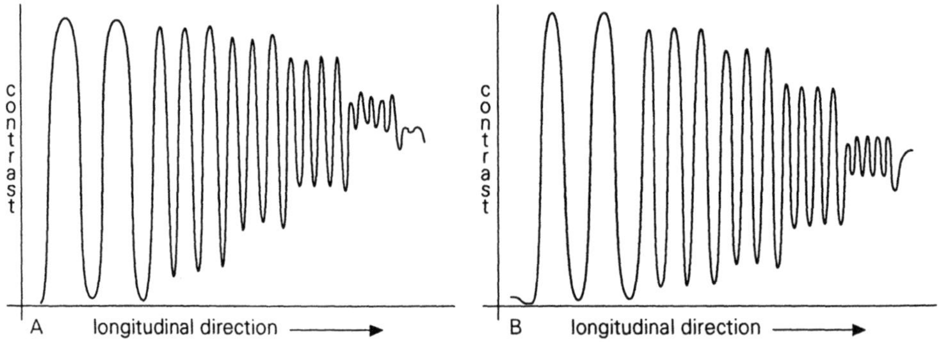

Fig 2: a) Density distribution measured in perspex spatial frequency phantom for a volumetric scan with slice thickness 3 mm, and reconstruction index 1 mm.
b) Density distribution computed for the spatial frequency phantom with the simulation program simulating a volumetric scan with the same parameters as in 2a.

In fig 3, the measured MTF distributions in the z-direction are given for the conventional and the volumetric scans. In fig 3a, the slice thickness is 3 mm and in fig 3b, the slice thickness is 5 mm. The results of the circular hole phantom are shown in figure 4. In fig 4a, the sagittal MPR of a volumetric scan with slice thickness of 3 mm with reconstruction index equal 1 mm is given. The MPR computed with the simulation program using the same scan parameters is shown in figure 4b. In the longitudinal direction, again an good agreement between the measured and the simulated data can be observed.

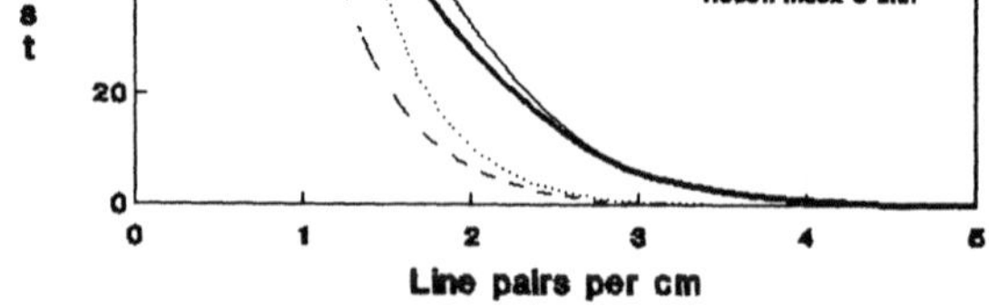

Fig 3: a) MTF distribution in the longitudinal direction measured with the perspex spatial frequency phantom for conventional and volumetric scan with slice thickness of 3 mm. The order in the legend correspond with the right to left order of the curves.

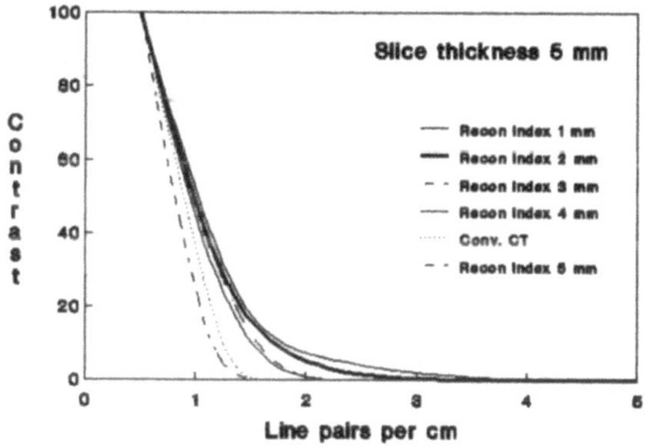

Fig 3: b) MTF distribution in the longitudinal direction measured with the perspex spatial frequency phantom for conventional and volumetric scan with slice thickness of 5 mm. The order in the legend correspond with the right to left order of the curves.

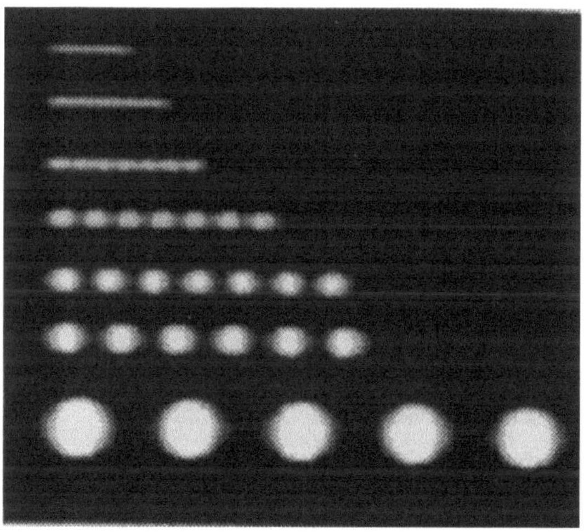

Fig 4: a) Coronal MPR of perspex hole phantom measured with a volumetric scan with slice thickness 3 mm, and reconstruction index 1 mm.

Discussion

The resolution in the longitudinal direction of the image reconstructed from the volumetric scan data using the 180° interpolation algorithm is improved compared with that of the conventional

CT. The analysis of the MTF distributions shows that to get the best compromise between the number of reconstructed slices and the resolution in the z-direction the reconstruction index should be selected at 2 mm, for slice thicknesses of 3 and 5 mm, . The mathematical simulation turned out to be a valid and valuable tool in predicting the effect of a slice thickness profile function on the measured density function. Modern tools are available for these simulations. This enables the prediction of the effect of a wide variety of slice profiles on a similar wide variety of density inputs, e.g. fat-kidney transients, air-skin boundaries etc. The simulation program enables the definition of protocols to have the best compromise between patient indication, required size of examination, scan parameters and image quality of the coronal, sagittal and 3D reconstructions.

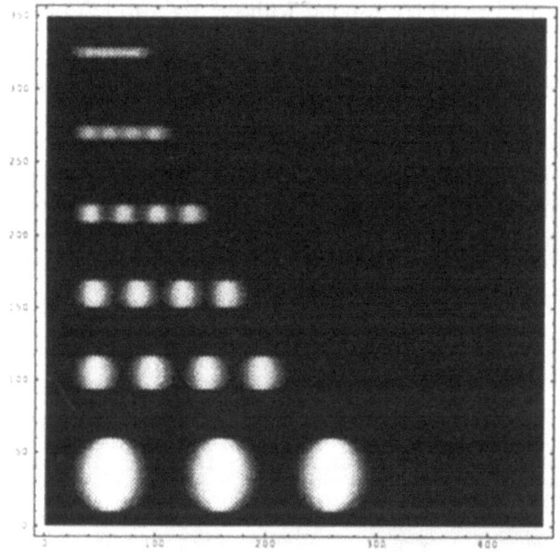

Fig 4: b) Coronal MPR of hole phantom computed with simulation program simulating a volumetric scan with the same parameters as in 4b.

References

1: Steenbeek JCM: Principles and applications of volumetric CT. MedicaMundi 38/1 (1993 in press)

2: Polacin A, Kalender WA, Marchal G: Evaluation of section sensitivity profiles and image noise in Spiral CT. Radiology 185 (1992) 29-35

3: Wolfram S: Mathematica, A system for doing mathematics by computer. Addison-Wesley Version 2.0 1991

3D-Display of Spiral CT Scans - a New Approach to Renal Imaging

Fink B.K., Pentenrieder M., Kohz P., Englmeier K.-H., Haubner M., Fink U., Schmeller N.
Department of Radiology, Klinikum Großhadern, University of Munich
GSF, MEDIS-institute, Neuherberg
Department of Urology, Klinikum Großhadern, University of Munich

Summary

Spiral-CT offers new possibilities of image reconstruction especially 3D-imaging. In this study the value of Spiral-CT and reconstruction methods were examined in 30 patients with renal tumors and 6 patients with staghorn calculi.

The examinations were performed by using Somatom Plus (Siemens Co.), before and after high dose contrast medium application in renal tumors and in staghorn calculi. In all cases 24 cm-volume-imaging was performed.

Spiral-CT with the 3D-reconstruction images proved to be a very valuable method for tumor imaging especially in question of tumor extension and relation to the renal artery and vein. 3D-reconstruction of staghorn calculus gave an excellent information for the surgeons, for the images proved to be a realistic cast of the staghorn calculus. Therefore the total removal of the calculus can be controled intraoperatively.

1. Introduction

The development of computer assisted imaging methods is one of the great advances in the medicine of the last years. Computer assisted imaging methods offer a lot of advantages in comparison to the conventional radiological technology:
By integration in a PACS image data are available in many places and at any time which may be the solution for the storage problem and makes the teamwork of clinicians and radiologists easier. Further computer assisted imaging methods offer a new presentation of anatomical and pathologic structures without superposition and with the possibility of postprocessing. In particular considerable are 3D-presentations which may be a good help in the preoperative planning and which may be an alternative to invasive methods as angiography.
This paper proposes a methology that enables an arbitrary Spiral-CT-scan of the abdomen to be automatically segmented and classified into bony structures, liver, kidneys and vessels. The image sequence is represented at a number of different scales (as in pyramid techniques) where a number of features are associated with each pixel at the particular scale (e.g. density, Hounsfield units, gradient magnitude)

2. Methods

2.1. Image acquisition

We examined 30 patients with renal tumors and 6 patients with staghorn calculi. All patients underwent surgery the day after the examination. The CT-scans were performed as spiral-CT-scans at a Siemens Somatom Plus. Based on slipring technology with continuous data acquisition over a multitude of 360^0 scans and continuous patient transport a volume of a maximum of 24 cm length can be scanned within 24 sec.

Patients with renal tumors were scanned with a table feed of 8 mm/sec, a slice thickness of 8 mm and an increment of 2mm ca. 20 sec after a bolus injection of 150 ml nonionic contrast medium. Bolus injection was done intravenously by an injector with a flow rate of 2 ml/sec.

Staghorn calculi were scanned dependent on their dimension with a table feed and a slice thickness of 4 mm or 2 mm and an increment of 2 mm or 1 mm before and 5 min after bolus injection of 100 ml contrast medium. Bolus injection was done by an injector with a flow rate of 1 ml/sec.

While the 24-sec. spiral-CT-scan the patients were required not to move and to suspend respiration. All scans were taken at 165 mAs and 120 kV.

For the data calculation the algorithm implemented by Siemens was applied.

2.2 Image processing and visualization

The staghorn calculi were reconstructed with the 3D-reconstruction programm, which is implemented in Siemens software. The operating storage for these 3D-reconstructions contains a maximum of 64 slices. True to life reproductions were created in various projections. While the following endoscopic surgery - as a rule combined with ESWL (endoscopic shock wave lithotrypsy) - the 3D-model could be compared with the parts of the surgically removed calculus put together in order to control the completness.

The renal tumors were reconstructed with the reconstruction programm, which is implemented in Siemens software in multiplanar sections. Standardized sections were done in a coronar, a sagittal and an oblique view, the latter in direction of the renal artery and vein. The approximate slice thickness was 10 mm in coronar and sagittal direction and 2 mm in oblique direction. A hardcopy was performed at a laser imager. In order to get a more to true life 3D-reproduction of the renal artery, the renal vein, the tumor and the kidney 3D-reconstructions were reconstructed in a separate hardware-environment. The process of this 3D-visualization as well as morphometry and density measurements are based on the process of segmentation and contour determination. To do this, the acquisitioned images were transferred to a parallel computer (CONVEX C3840) with 4 processors, 1 Gigabyte memory and 36 Gigabyte disc storage. This supercomputing system is connected via Ethernet and Network File System to a workstation cluster consisting of six Silicon Graphics IRIS-INDIGOS R 4000 with true color displays and z-buffers for surface and volume visualization. Using the Distributed Graphics Library (OPEN GL in the near future) both machine types can share the work load, each doing the task it is best suited for, resulting in a more balanced work load and better performance. The implemented software is embedded in a software system with the following specifications (see also 1):

- unique file format for CT, MRI, raw data of MRI, DR, DSA and scintigrams
- easy user interactions using file browsing and pull down menue techniques
- image processing utilities such as filtering, contour detection and segmentation
- image fusion techniques
- 3D-visualization techniques

In order to visualize the kidneys with their vessels, the segmentation of these anatomical structures is necessary. We developed an automatic segmentation method, which reduces the number of user interactions. This method satifies anatomical constraints to achieve structure identification and segmentation. These constraints of anatomical topology are described as follows: Both kidneys are localized side by side with the spinal column. Aorta abdominalis and vena cava inferior are in front of the spinal column. It is assumed, that the crossection of both vessels is nearly elliptical.

In order to localize the region of the both kidneys, firstly the position of the spinal column is detected. The spinal column is a certain locality in the abdominal CT-image, which is noticable brighter than the immediate surroundings. We exploit the advantage of the determination of this bright spot in the transversal CT, because the rectangular regions for the kidney segmentation are correctly localized beside the bright spot of the spinal column.

To detect the spinal column as a bright spot we use the average image pyramids. The algorithm of these pyramids has the characteristic of selecting a point, that is representative of a locally bright area at the image, if we combine the algorithm with a global maximum search. The result of this process step is the position of the maximum density in the lowest resolution image of the image pyramid.

Following that three rectangular regions are created overlaying the localization of the kidneys and vessels. This procedure minimizes the influence of disturbances of the segmentation process caused by structures of similar density (liver). Within the rectangular regions, which overlays the position of the kidneys, a threshold oriented process is applied to the high resolution part of the image pyramid. Based on the a

priori knowledge, that the kidneys are displayed as connected singular region in the CT-image, noise and disturbances are eliminated by recursive object labeling.
Staghorn calculi are characterized by high density values, which are similar to bony structures. Therefore we can use threshold algorithm to localize the staghorn calculi within the region of the kidney.
Based on the a priori knowledge of closed elliptical contours the developed algorithm detects the actual contours of aorta abdominalis and vena cava by means of structure adapted filters via maximizing the cross correlation function (7).

3. Results

In all 6 cases we created a 3D-model of staghorn calculus it could be proven as a true to life model of the surgically obtained concrement or the totality of its composed parts. For the surgeon this model proved to be an excellent possibility to control the complete removal of the staghorn calculus.
In all 30 cases of renal tumor the extension of the tumoral mass could be shown. In contrast to conventional CT-scans the data acquisition is performed with single-breath-hold technique. In this way gaps or double imaged structures - in conventional CT-scans caused by inconstant respiration volumes - can be avoided. Multiplanar sections in each case improved the demonstration of spatial relation especially between kidney or tumor and liver or surrounding tissue. In a great number of cases the possibility of coronal imaging leads to a better delineation of the renal vessels. First results show, that the use of 3D-reconstruction and visualization in frontal, sagittal and oblique view might lead to a better staging of renal tumors.
A difficulty using the implemented software of Siemens often was the fact, that the operating storage for the 3D-reconstruction contains a maximum of 64 images. In many cases this was the reason why the kidneys were only partially imaged or why as a rule only a little part of the surrounding tissue could be delineated. In the 3D-visualization-technique performed at the separate hardware- environment the number of slices which can be utilized for reconstructed images is not limited.
The highest available x-ray power for spiral-CT-scans at the moment is 165 mAs. This leads to a reduced image quality of the transversal images in comparison to the concentional CT-scans obtained at higher energy.

4. Discussion

The 3D-visualization of staghorn calculi is the only imaging method, which shows clearly the true 3D-topography, especially the extension of the calculus into the anterior or posterior calyces. This influences the surgical proceeding.

In the literature 3D-reconstruction is known especially from the visualization of bone structures (8,9). With conventional CT-scanning techniques a sufficient 3D-image of the kidneys mostly was not possible because of the inconstant levels of inspiration and due to this of the kidneys from scan to scan. This often caused omission of some anatomic levels or repeated scanning of others. With Spiral-CT-scanning in single-breathold technique volumes can be scanned continously without any gaps and little influence from patient motion. In this way one of the characterizing advantages of MRI in comparison to other imaging methods now with volume scanning is also available in CT-technique:The multiplanar imaging. In contrast to MRI-imaging this can be performed independently from breathing artefacts.
Another advantage is the short examination time because of the loss of interscan times. This gives the chance of complete using of the bolus of contrast medium and may lead to a much better delineation of vessels.
A disadvantage of Spiral-CT may be the limitation of the available tube current at 165 mAs. This causes a decreased image quality in comparison to conventional CT-scans. With technical proceedings this may be improved in future.
One of the tasks of prime importance for the future seems to be the improvement of the image processing methods. The aim should be an image quality true to life as accurately as possible. Further the user interactions should be reduced and organized as easy as possible. In order to visualize the kidneys with artery and vein, the segmentation of these anatomical structures is necessary. Segmentation is normally described as the partitioning of an image into homogenous regions (characterized by grey values, textures) (3). Previous works on grey value based image segmentation can be considered to have taken two paths:
- edge based segmentation
- region based segmentation (2)

A lot of segmentation procedures (interactive or automatic) are described in the realm of CT and MRI image processing (3,4,5). We developed an automatic segmentation method which may reduce the user interactions and may lead to a better delineation of kidneys, tumor and vessels.

5. References

1. Englmeier K.-H., Fink U., Hilbertz T.: Visualization of Multimodal Image Information in Medicine, Proc. of the Sixteenth Annual Symposium on Computer Applications in Medical Care, 8.-11.11.92, Baltimore, Eds.: M.E.Frisse, 25-29 (1992)

2. Fu K.S., Mui J.K.: A survey on image segmentation. Patt.Recog.3: 3-16

3. Griffin L.D., Colchester A.C.F., Robinson G.P., Hawkes D.J.: Structure sensitive scale and the hierarchical segmentation of grey-level imaging. Proc.of Visualization in Biomedical Computing, 13.-16.10.92, Chapel Hill, Ed. R.A.Robb, 24-32(1992)

4. Schnemann T., Boman T., Tiede U., Höhne K.-H.: Interactive 3D-Segmentation of Tomographic Image Volumes, in: Musterekennung '92, 14.DAGM-Symposium, Dresden.14.-16.9.1992, ed.: S.Fuchs, R.Hofmann, 73-80 (1992)

5. Verazza G.,Serpico S.B., Dellepiane S.G.: A knowledge based system for biomedical image processing and recognition, XEEE transactions on circuits and systems, CAS-34 (11): 1399-1416 (1987)

6. Blanford R.P., Taminoto S.L.: Bright-Spot Detection in Pyramids, Computer Vision, Graphics and Image Processing 43, 133-149 (1988)

7. Englmeier K.-H.,Hecker R., Hötzinger H., Pöppl S.J, Thiel H.: Reconstruction and pseudo-3-dimensioned presentation of the Uterus from Automatic Segmented Transversal Ultrasound Slices - a Fundamental Precondition for Optimizing the individually Adjusted Therapy Planning in Case of Carcinoma of the Body of the Uterus, Poc. CAR '85, eds.: H.U.lemke, M.L.Rhodes, C.C.Jaffe and R.Felix, Springer Verlag, Berlin, 358-370 (1985)

8. Burk D.L.,Jr., Mears D.C., KennedyW.H., CoopersteinL.A., Herbert D.L.: Three-dimensional computed tomography of acetabular fractures. Radiology 155(1985)183

9. Pate D., Resnick D., Andre M., Sartoris D.J., Kursunoglu S., Bielecki D., Dev P., Vassiliadis A.: Perspective: three-dimensional imaging of the musculosceletal system. AJR 147 (1986) 545

In-Vitro Measurement of Stenoses with Spiral CT

G. WITTENBERG, M. JENETT, A. TSCHAMMLER, TH. KRAHE

Institut für Röntgendiagnostik der Universität Würzburg,
Josef-Schneider-Straße 2, D-8700 Würzburg

Summary
After the first promising spiral computed tomography angiographies it is now the aim of this paper to examine the correlation between the real degrees of stenosis and those degrees measured with spiral CT. In the first in-vitro measurements good correlation results (r = 0.89 - 0.99) were obtained depending on the evaluation type. These results, however, need to be confirmed by further test series before spiral CT can be used for vessel diagnostics as a routine method.

Introduction
Angiography is still the method of choice for the evaluation of vascular changes. In rare cases imaging free of superimposed structures of the vessels even when using different projection planes is not possible. Due to the introduction of continuous volume detection within short scanning times by means of spiral CT (S-CT), it is now possible to make three-dimensional (3-D) reconstructions of the vessel system [4,5]. In first clinical applications of the S-CT angiography a good correlation with arterial vessel visualization was demonstrated in addition to a free vessel view [1,2,3].

In our in-vitro measurements we were interested in the correlation between real stenosis degrees and those measured with the S-CT.

Materials and Methods
Measurements were taken with spiral CT on a Somatom Plus Scanner (Siemens). 2 mm were chosen for slice thickness and table speed. Data were acquired with 120 kV and 125 mAs in up to 24-second-long S-CT scans. Image reconstruction was

performed with use of standard algorithm and a reconstruction interval of 1 mm.

Silicone tubes filled with diluted contrast agent (density below 200 HU) with an inner lumen of 4 and 6 mm were used as phantoms. At the 4-mm-tube (Ph A) short-distance, eccentric obstructions with a lumen narrowing of 50 %, 60 %, 75 %, 90 %, and at the 6-mm-tube (Ph B) with a stenosis degree of 50 %, 70 %, 80 %, 95 % (values rounded off) were simulated. After conclusion of the measurement series and evaluation of data the real stenosis degree was determined by sacrificing the phantoms. During scanning the tubes were placed into a water basin. Measurements were performed by aligning the tube models in the x-, y- and z- directions.

Determination of the degree of stenosis was carried out by:

1. electronic lumen width measurement and/or interpolation with the help of single split-images and computation of the degree of stenosis,

2. visual assessment of the degree of stenosis with three-dimensional reconstructions (512 matrix, treshold 180 HU).

For statistical analysis linear correlation coefficients and regression straight line equations were used.

<u>Results</u>

Examination of the correlations for the stenosis degrees determined by measurement techniques - phantom alignment in x-, y- and z- directions viewed together - resulted in a linear correlation coefficient of $r = 0.98$ for Ph A and of $r = 0.99$ for Ph B. The regression straight line equation is $y = 1.0423\ x - 3.54$ for Ph A and $y = 1.06\ x - 5.96$ for Ph B (see chart 1). The individual results are summarized separately in table 1 according to the different phantom alignments.

When estimating the degree of stenosis visually with the help of 3-D reconstructions of the 4- and 6-mm-tube models, it was generally underestimated (see figure 1). Calculation of the linear correlation coefficients - phantom alignment in x-, y- and z- directions viewed together - resulted in r = 0.96 for Ph A and r = 0.89 for Ph B. Calculation of the regression straight line equation resulted in y = 1.0272 x - 21.87 for Ph A and y = 0.9259 x - 13.7 for Ph B (see chart 2). The results of the individual phantom alignments are listed in table 2.

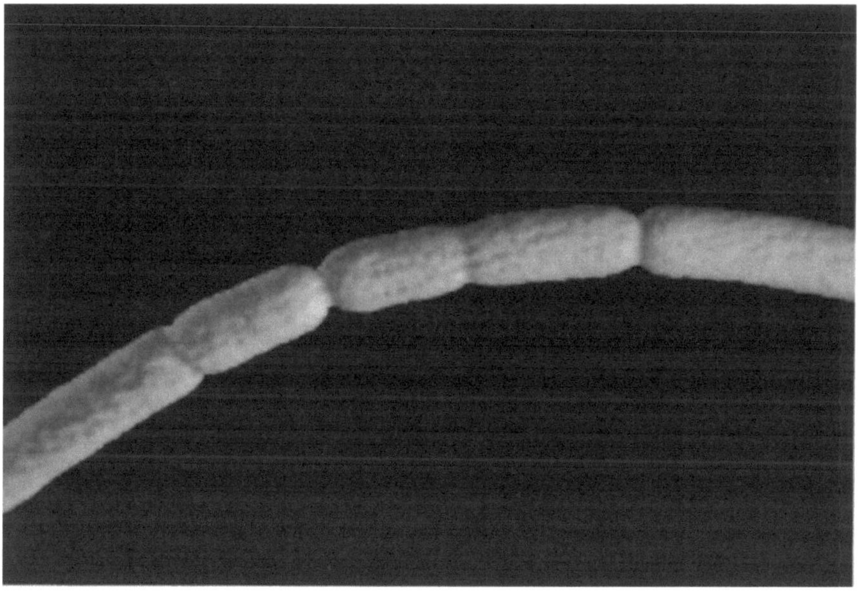

Fig. 1. Three-dimensional reconstruction of the 6-mm-silicone tube with eccentric stenoses (stenosis degree between 50 % and 95 %). (slice thickness/table speed/reconstruction interval 2 mm/2 mm/1 mm)

Discussion

At present, angiography is the standard method when evaluating vessel alterations. However, some patients are afraid of undergoing angiography because of the prevailing risks. Currently, the less invasive spiral CT may be an alternative. This method showed a good morphological correlation between

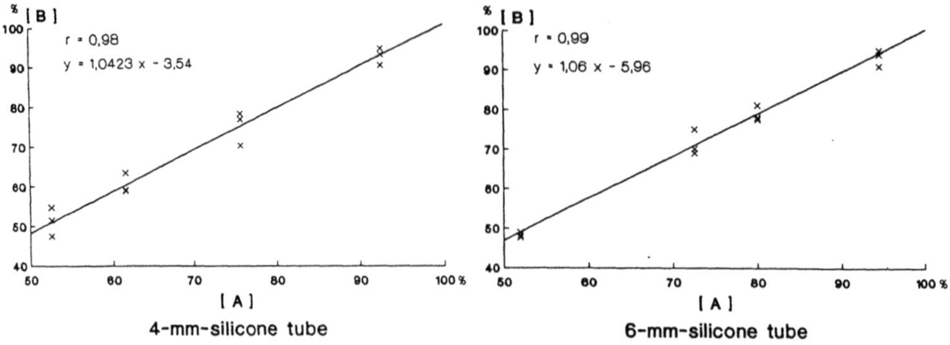

Chart 1a. Stenosis degrees determined by the spiral CT measurement technique (B) versus real stenosis degrees (A)

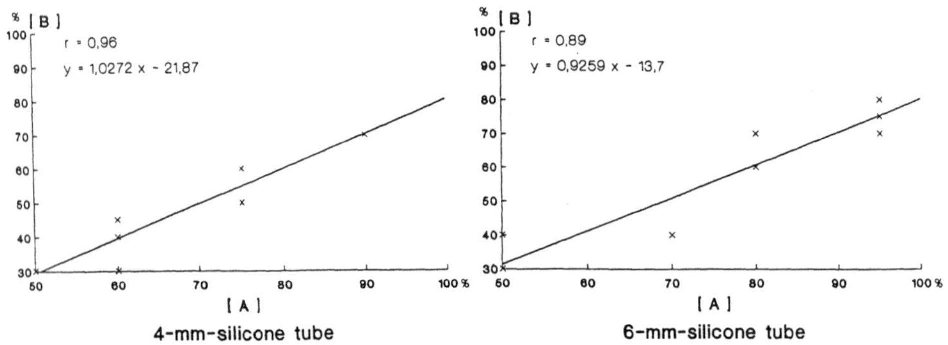

Chart 1b. Assessment of degrees of stenosis with three-dimensional reconstructions of spiral CT data set (B) versus real stenosis degrees (A)

Chart 1. Regression of stenosis degrees determined by spiral CT to real stenosis degrees

angiography and spiral CT in vessel images free of superimposed structures in all required projection planes in its first clinical applications [1,2,3]. Additionally, it also visualizes the surrounding anatomy. The data obtained within the frame of this studie in in-vitro stenosis degree measurements show a high correlation (r = 0.89 - 0.99) between real and measured degree of stenosis. The less positive correlation of the assessment of the degree of stenosis with

Phantom	Phantom alignment	Linear correlation coefficient	Regression straight line equation
A	x-	r = 0.98	y = 1.0011 x - 0.83
B	x-	r = 0.99	y = 1.0914 x - 6.28
A	y-	r = 0.99	y = 1.0405 x - 1.58
B	y-	r = 0.99	y = 1.0276 x - 4.8
A	z-	r = 0.99	y = 1.0853 x - 8.22
B	z-	r = 0.99	y = 1.0914 x - 6.28
A	x-/y-/z-	r = 0.98	y = 1.0423 x - 3.54
B	x-/y-/z-	r = 0.99	y = 1.06 x - 5.96

Phantom A: 4-mm-tube model   Phantom B: 6-mm-tube model

Table 1. Results of determination of stenosis degrees by means of measurement techniques

Phantom	Phantom alignment	Linear correlation coefficient	Regression straight line equation
A	x-	r = 0.99	y = 1.034 x - 21.09
B	x-	r = 0.95	y = 1.193 x - 32.98
A	y-	r = 0.98	y = 0.9864 x - 16.56
B	y-	r = 0.87	y = 0.8713 x - 8.01
A	z-	r = 0.96	y = 1.0612 x - 27.96
B	z-	r = 0.90	y = 0.7135 x - 0.12
A	x-/y-/z-	r = 0.96	y = 1.0272 x - 21.87
B	x-/y-/z-	r = 0.89	y = 0.9259 x - 13.7

Phantom A: 4-mm-tube model   Phantom B: 6-mm-tube model

Table 2. Results of assessment of stenosis degrees with three-dimensional reconstructions

3-D reconstructions compared to the calculation of the degree of stenosis by means of measurement techniques is mainly based on the choice of phantoms. The silicone tubes show higher values (500 HU) in density measurements than the contrast agent used for filling (below 200 HU). Due to this fact, the necessary threshold choice for 3-D reconstructions is more difficult, as an incorrect choice results in a falsified degree of stenosis. At present, new test series with body-adapted models, referenced to the chosen density differences, are performed. If in these measurement series similarly high correlation values between real and measured degree of stenosis are obtained, the way will be paved for the application of spiral CT, e.g. as screening method in arterial hypertension to exclude a renovascular cause.

References

1. Schwartz, R.B.; Jones, K.M.; Chernoff, D.M.; Mukherji, S.K.; Khorasani, R.; Tice, H.M.; Kikinis, R.; Hooton, S.M.; Stieg, P.E.; Polak, J.F.: Common carotid artery bifurcation: evaluation with Spiral CT. Radiology 185 (1992) 513-519.

2. Rubin, G.D.; Dake, M.D.; Napel, S.A.; McDonnell, C.H.; Jeffrey, R.B.: Three-dimensional Spiral CT angiography of the abdomen: Initial clinical experience. Radiology 186 (1993) 147-152.

3. Napel, S.; Marks, M..; Rubin, G.D.; Dake, M.D.; McDonnell, C.H.; Song, S.M.; Enzmann, D.R.; Jeffrey, R.B.: CT angiography with Spiral CT and maximum intensity projection. Radiology 185 (1992) 607-610.

4. Kalender, W.A.; Vock, P.; Polacin, A.; Soucek, M.: Spiral-CT: Eine neue Technik für Volumenaufnahmen. I. Grundlagen und Methodik. Röntgenpraxis 43 (1990) 323-330.

5. Soucek, M.; Vock, P.; Daepp, M.; Kalender, W.: Spiral-CT: Eine neue Technik für Volumenaufnahmen. II. Klinische Anwendungsmöglichkeiten. Röntgenpraxis 43 (1990) 365-375.

Nuclear Medicine

Registration of PET-Images Using Template Matching

P. Plets[1], J. Nuyts, P. Dupont, P. Suetens[2]

Interdisciplinary Research Unit for Radiological Imaging:
ESAT, Kardinaal Mercierlaan 94, B-3001 Heverlee, Belgium and
Department of Nuclear Medicine, University Hospital Gasthuisberg, Herestraat 49, B-3000 Leuven, Belgium and
Labo for neuro- and psychophysiology, Katholieke Universiteit Leuven, Herestraat 49, B-3000 Leuven, Belgium

Abstract

One of the main problems in activation studies with PET (Positron Emission Tomography) is the registration of the different images before one can apply a statistical analysis. The goal of this work is to improve this registration.

A method is presented for aligning the images without the use of external markers or stereotactic frame.

Registration is performed on the original grayvalues, or on the edge enhanced images. As an initial step for the registration the method of the image moments is used, by analogy with the classical mechanics of rigid bodies. The zeroth order image moment is used to calculate an isotropic scale factor, the first order image moment gives a first guess for the translation, and the second order image moment is used to calculate the principal image axes, from which a first guess for the rotation and scale is computed. This result is then optimized in an iterative search: the parameters of a (global) affine transformation function are altered until a point of best match is reached. Some "objective functions", which formulate a criterion for a best match, are described: generalized cross-correlation, the sum of squared differences, SSC (Stochastic Sign Changes) or DSC (Deterministic Sign Changes). At this moment we limit our transformation function to translation, rotation, and scaling according to the main image axes. If only those image pixels belonging to the actual matching target (the human brain) are considered, better and faster converging results can be obtained in cases where the field of view is important.

At this moment, these techniques are successfully tested in the two-dimensional case: PET transaxial and sagittal images, after affine transformation and/or addition of Poisson noise to the sinograms, are registered within 1 pixel accuracy. Further studies will be performed to test the accuracy of match in case of three-dimensional images.

[1]Correspondence address: Peter Plets, Siemens A.G., MRA 1, Henkestraße 127, D-8520 Erlangen, Germany. E-mail address: `plets@ErlH.Siemens.DE`

[2]P. Suetens is also a Senior Research Associate of the National Fund for Scientific Research, Belgium.

1 Introduction

Activation studies with PET are based on differences in brain blood flow between two conditions. The increase in blood flow due to the activation is usually small (about 5 %). Furthermore, the acquisition time is very short (40 - 70 s) which results in noisy images (see figure 1). However, using a statistical technique [1] one can detect these small changes in blood flow. A necessary condition for applying this technique is the registration of the PET images of all subjects.

Registration can be defined as *the process of determining the geometrical relationship between different image data sets* [2]. Mathematically we look for the transformation $T : (u, v, w) \rightarrow (x, y, z) = T(u, v, w)$ so that the two input images $f_1 : F_1 \subset \mathbb{R}^3 \rightarrow \mathbb{R} : (u, v, w) \rightarrow f_1(u, v, w)$ and $f_2 : F_2 \subset \mathbb{R}^3 \rightarrow \mathbb{R} : (x, y, z) \rightarrow f_2(x, y, z)$ with F_1 and F_2 the domains over which the respective images are defined, can be spatially aligned: $f_1(T^{-1}(x, y, z)) \simeq f_2(x, y, z)$, T having a finite set of parameters.

Registration methods can be classified in general according to (see [2]):

- The choice of $f_1(x, y, z)$ and $f_2(x, y, z)$: the grayvalue functions, original or after image processing (e.g. Sobel filtering), or some kind of binary or fuzzy segmentation of the original images;
- The choice of the *objective function*, which measures the correctness of registration;
- The type of mapping T: a rigid (global) or elastic (local) registration.

In this paper we describe a global registration method based on the grayvalues with four different objective functions.

2 Methods

We apply a kind of *template matching*. One of the images is compared pixel by pixel to the other image, which is being transformed under T. The latter image can therefore be considered as a template that is deformed until a point of best match with the first image is reached.

Another approach could be to apply a delineation first – e.g. of the brain surface, the ventricles – and then try to register the crisp segmented structures. For a survey of surface based registration methods, see [3]. This approach introduces the problem of strict segmentation of very noisy, unsharp images prior to the registration step. Since we assume that the grayvalue information is globally similar in the different images, we use the original grayvalues or the Sobel filtered images in the matching process.

For *the mapping* T a global affine transformation is used, permitting translation, rotation and scaling along the orthogonal image axes.

2.1 Objective Function

A few objective functions have been examined:

1. Generalized Cross-Correlation (GCC):

$$\text{maximize: } O_f = \sum_{(x,y,z)\in F_2} \left(f_1(T^{-1}(x,y,z)) \cdot f_2(x,y,z)\right) \tag{1}$$

2. Least of Sum Squares (LSS):

$$\text{minimize: } O_f = \sum_{(x,y,z)\in F_2} \left(f_1(T^{-1}(x,y,z)) - f_2(x,y,z)\right)^2 \tag{2}$$

   One can proof that this yields the same result as the cross-correlation objective function, when no scaling is allowed.

3. Stochastic Sign Changes (SSC) Criterion: This similarity criterion was developed by Venot et al. [4]. Two images are subtracted from each other. The resulting image is scanned in some way (e.g. first the image rows, then columns and planes), and the sign changes according to this scanning direction are counted. This result is then maximized:

   (a) $f_3(x,y,z) = \text{sign}\,(f_1(T^{-1}(x,y,z)) - f_2(x,y,z) \qquad \forall(x,y,z) \in F_2$

   (b) maximize: O_f = number of sign changes in f_3 according to the scanning direction

   This method works best for images containing stochastic uncorrelated noise. Although noise in PET-images is not uncorrelated, this matching criterion also gives good results.

4. Deterministic Sign Changes (DSC) Criterion: (Venot et al. [4]). This method is very similar to the previous one, except that before subtraction, a chessboard pattern is superposed on one image: alternately $+q$, $-q$, $+q$

   (a) $f_3(x,y,z) = \text{sign}\,((f_1(T^{-1}(x,y,z)) \pm q) - f_2(x,y,z)) \qquad \forall(x,y,z) \in F_2 \ , q \in \mathbb{R}$

   (b) maximize: O_f = number of sign changes in f_3 according to the scanning direction

   It is easy to verify that the number of sign changes is maximized when the two original images are exactly identical. (The number of sign changes equals in that case the number of image pixels minus one.) A problem is to find a good value for q. We found that the accuracy of registration was not very sensitive to the choice of q, but best results were obtained with a value of about 1% of the maximum image gray level.

2.2 Limitation in the number of computed pixels

Instead of using all pixels in the two images an additional constraint can be applied to select only a set of pixels in the similarity check. Two constraints were actually tested:

1. *Only Corresponding Points (OCP):* Only those points (x, y, z) are considered for which: $(x, y, z) \in F_2$ and $T^{-1}(x, y, z) \in F_1$ Since the axial "field of view" (FOV) of our PET-scanner is limited to 10 cm, only a part of the brain can be scanned. When a transformed point falls outside F_1, it is not clear what pixel graylevel it should be given. Assigning it for instance the value 0 (for the background) still produces some non-zero terms in the LSS-criterion and in the sign change criteria.

   Without this limitation in computed pixels the algorithm tends to align the different FOV's, instead of the brains themselves.

2. Since almost no information is contained in pixels with low gray values (background + noise), these pixels can be discarded using a threshold: Only those points (x, y, z) are considered for which: $f_2(x, y, z) > t$ and $f_1(T^{-1}(x, y, z))) > t$ t being a real valued threshold.

2.3 Algorithm

The algorithm is an iteration, searching for the mapping T that minimizes (or maximizes) a chosen objective function. The optimalisation algorithm used in this work is the "Powell N-dimensional minimization algorithm" [9]. This method was chosen, because the objective functions depend on multiple variables, and the partial derivates are very difficult to calculate.

2.4 Initial value

For the initial value of the iterative algorithmic search, a good starting value is chosen with the "Method of the Image Moments" (MIM). This method is proposed by some authors ([2], [5] - [8]) as a means for automatic registration of images. Each image is considered as an object mass distribution, and the image grayvalues are thought of being the object mass in a particular grid point. Per image the zeroth, first and second order image moments are calculated. Let $\vec{v} = (v_1, v_2, v_3)$ be the coordinates of the points. The zeroth order image moment $M_0 = \sum_{\vec{v}} f(\vec{v})$ is used for normalization of the grayvalues. The first order image moment $M_1 = \begin{pmatrix} V_1 & V_2 & V_3 \end{pmatrix}$ with $V_i = \frac{\sum_{\vec{v}} v_i \cdot f(\vec{v})}{M_0}$, gives a first guess for the "mass center" or the translation component. The matrix M_2 of the second image moment (moment of inertia) has elements $V_{ij} = \frac{\sum_{\vec{v}} v_i \cdot v_j \cdot f(\vec{v})}{M_0}$, $i, j = 1 \cdots 3$. The initial rotation angles are derived from the relative angles of the eigenvectors of the matrices M_2. This is the same as aligning the principal axes of inertia. The initial non-isotropic scaling factors per axis are the ratios of the respective corresponding eigenvalues of M_2, raised to the 1/4-th power.

3 Evaluation

The algorithm is currently being tested in two dimensions.

3.1 Test images:

Two PET images, A_1 and S_1, have been transformed and distorted in various ways to evaluate the performance and noise sensitivity of the algorithm:

- A_1 is an axial slice (figure 1a).
- S_1 is a sagittal slice.
- In the visual cortex area in image A_1 a small region was drawn. This region was then filled with a constant grayvalue, to which some additional Gaussian noise was added: image A_{1C}.

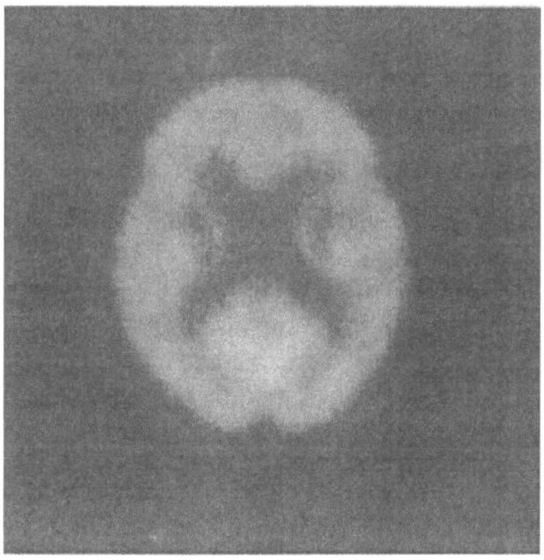

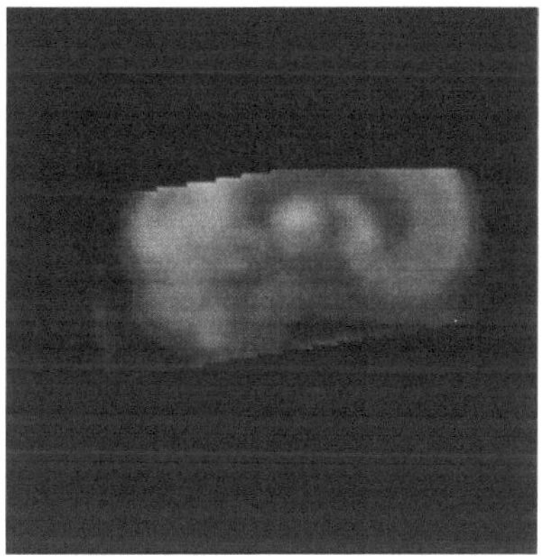

Figure 1: a. Axial cross-section of a PET blood flow image (using $H_2{}^{15}O$ as a tracer), 40 sec acquisition time; b. Image S_{1Ob}.

- Poisson noise was added to the sinogram of image A_1: the sinogram has been rescaled such that the maximum count value equals some real value n. The sinogram pixel values were then replaced by a sample from a Poisson distribution, using the rescaled count value as the expectation value for that distribution. We tested following values for n: 1000 (reconstructing image A_{n1000}) and 1 (A_{n1}). In A_{n1} the noise amplitude equals the signal amplitude.
- From S_1 an oblique portion under 10 degrees (to the original FOV) was cut: image S_{1Ob} (figure 1b).
- S_1 and S_{1Ob} were Sobel filtered: S_{1S} and S_{1ObS}.

All test images have 128 by 128 pixels. Pixel intensities were rescaled to 8-bit integers.

3.2 Test Results:

Tests were performed on an IBM RS/6000 computer.

In figure 2 an overview is given of the accuracy and the execution time in 5 test cases, after MIM and per objective function, each with/without OCP:

Case (1): A_1/A_{1C}

Case (2): A_1/A_{n1000}

Case (3): A_1/A_{n1}

Case (4): S_1/S_{1Ob}

Case (5): S_{1S}/S_{1ObS}

In all cases, a value of 1 % of the maximum grayvalue was used for q in the DSC-criterion, except for case 3: A_1/A_{n1}. In this case only much higher values for q yielded good results: listed accuracies are for $q = 7\%$ of the maximum grayvalue. Different parameters were optimized together in the Powell algorithm: accuracy for translation in x- (= row-) direction and translation in y- (= column-) direction, is given in pixel misalignment distance; rotation error is given in degrees misalignment; the error for the scale misalignment in x- (= row-) direction and the scale in y- (= column-) direction is in %, indicating the scale alignment error, relative to the exact scale. Execution time for MIM and each objective function, without and with OCP, is given in CPU-seconds.

4 Discussion

MIM provided in all cases already a very good initial value in a very short computation time. Only the initial scale values yielded poorer results. The Powell algorithm converges relatively slowly to a slightly better result. Where the FOV was not important (case 1 - 3), the OCP-method did not improve the already good results without OCP, but OCP was generally faster.

In all cases, using a (low) threshold ($t = 0.5\%$ of the maximum grayvalue) did almost not improve the accuracy of the results nor the speed of convergence, compared to the OCP-method.

The GCC almost always converged to a completely wrong result, when we let the scale being changed in the mapping function T. This can easily be understood: blowing up the (template) image until in the limit only one high intensity pixel remains, maximizes indeed the GCC objective function.

The sign change criteria, introduced by Venot et al., are as accurate as the LSS method in our test cases, but sometimes faster. The DSC-method yet requires to optimize the choice of an additional parameter (q). In the test cases where the FOV is important, the OCP option saves the algorithm from converging to an undesired local optimum (LSS, DSC). This is promising for the 3-D case, where we expect the "FOV"-problem to be more important.

The speed of the algorithm increases with the number of parameters being optimized in the mapping T. This could be a real problem when extending to the third dimension:

we would have to optimize in the worst case 9 parameters, instead of 5 in the 2-D case. Adapting the Powell algorithm, so that only in a restricted area is searched, could speed up the algorithm.

We may conclude that the first results are very promising. Further work will be done, extending the methods to the three-dimensional case.

References

[1] K.J. Friston, C.D. Frith, P.F. Liddle and R.S.J. Frackowiak. "Comparing functional (PET) images : the assessment of significant change," *J Cereb Blood Flow Metab* (1991) 11: 690-699

[2] D. Vandermeulen. "Methods for registration, interpolation and interpretation of three-dimensional medical image data for use in 3-D display, 3-D modelling and therapy planning," *PhD Thesis,* Katholieke Universiteit Leuven, Faculteit Toegepaste Wetenschappen, ESAT/MI2, April 1991, chapter 2: "3-D Registration of 3-D images".

[3] A. Collignon, D. Vandermeulen, P. Suetens, G. Marchal. "Surface based registration of 3D medical images," SPIE: Medical Imaging VII (1993), Image Processing: vol. 1898, Newport Beach, California, 14-19 feb. 1993.

[4] A. Venot, J.F. Lebruchec, J.C. Roucayrol. "A New Class of Similarity Measures for Robust Image Registration," *Computer Vision, Graphics, and Image Processing,* 28, 176-184 (1984).

[5] O. Ratib, L. Bidaut, H.R. Schelbert, M.E. Phelps. "A New Technique for Elastic Registration of Tomographic Images," *SPIE Vol. 914 Medical Imaging II* (1988), pp. 452-455.

[6] D. Cyganski, J.A. Orr. "Applications of Tensor Theory to Object Recognition and Orientation Determination," *IEEE Transactions on Pattern Analysis and Machine Intelligence,* vol. PAMI-7, no. 6, November 1985, pp. 662-673.

[7] T.L. Faber, E.M. Stokely. "Affine Transforms Determination for 3-D Objects: A Medical Imaging Application," *Proceedings IEEE Comp. Vis. Patt. Recog.,* (1986) pp. 440-445.

[8] T.L. Faber, E.M. Stokely. "Orientation of 3-D Structures in Medical Images," *IEEE Transactions on Pattern Analysis and Machine Intelligence,* vol. 10, no. 5, September 1988, pp. 626-633.

[9] W.H. Press, B.P. Flannery, S.A. Teukolsky, W.T. Vetterling. "Numerical Recipes in C, The Art of Scientific Computing," *Cambridge University Press,* 1988, pp. 290 ss.

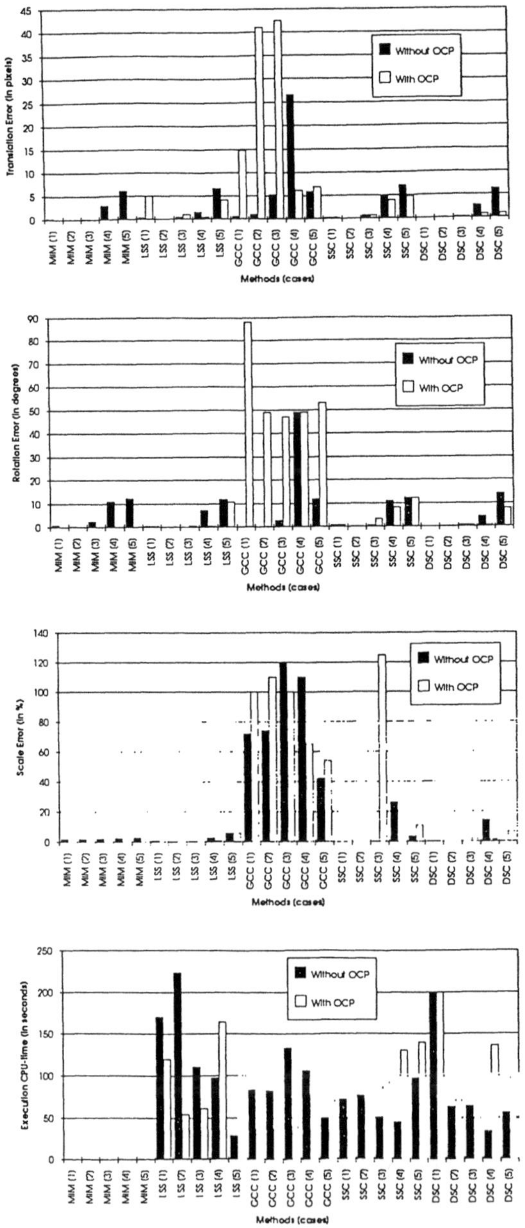

Figure 2: Translation, rotation, scale error and execution time, for MIM and the 4 objective functions, without and with OCP.

High Resolution PET Images through REDISTRIBUTION

U. KNORR, Y. HUANG, G. SCHLAUG, R. J. SEITZ, H. STEINMETZ

Department of Neurology
Heinrich-Heine-University Duesseldorf
Moorenstr. 5
D-4000 Duesseldorf 1

Summary

A method for deconvoluting low-resolution PET images to high-resolution images is presented. By increasing the resolution, the method also corrects partial-volume-effects. This was achieved by estimating a hypothetical high-resolution activity distribution the simulated PET image of which was equal to the measured PET scan. The estimated activity distribution should then reflect an activity distribution in the examined object that could have led to the observed PET image. Starting with an estimated distribution where all pixels have zero values, the REDISTRIBUTION algorithm iteratively adjusts the estimated distribution to minimize the difference between real and simulated PET image. To improve convergence of the iterations, structural information from an individually fitted and segmented high-resolution MR image was used. Simulations on 2D objects showed full recovery in the redistributed images even for small and irregular geometries.

Introduction

Positron-Emission-Tomography (PET) is a powerful tool for imaging human brain function. Due to detector geometry PET images have a low in-plane resolution of several mm (Full Width at Half Maximum, FWHM) and low inter-plane resolution in the range of 6-7 mm (FWHM). Thus, PET is unable to account for anatomical details as visible in high-resolution magnetic resonance (MR) data. Furthermore, the quantitative activity measured with PET underestimates real activity due to the partial-volume-effect (PVE). To give an example, if the field of view of a PET pixel contained equally distributed amounts of gray matter and cerebro-spinal fluid and the activity of gray matter was 100, the PET pixels value would only be 50. For the amelioration of these shortcomings the anatomical information derived from high-resolution MR images can be used. To obtain this information one has to segment the original MR data into gray matter (GM), white matter (WM) and cerebro-spinal fluid (CSF). Thereafter, PET and MR information can be 'fused' in the following two ways.

First, one can estimate the PVE for each PET pixel and then multiply the pixel value by a certain factor for correction [1]. We will call this procedure PVE-Correction (PVEC).

In principle, the PVEC smoothes high-resolution MR information to the resolution of a PET image (by simulated PET scanning). Then this structural PET analogue is taken to correct each individual PET pixel for PVE. However, PVEC leaves the problem of low spatial resolution unresolved.

The second approach (REDISTRIBUTION) that we have developed leaves high-resolution MR information unaltered creating a new image which reflects PET image activity on an MR image matrix. We will call these new images PMR (PET-MR images). They are obtained as follows: After accurate matching of MR and PET data [2] and segmentation of MR data into GM, WM and CSF, the segmented image is used as a template to determine, to what degree each MR pixel contributes to a PET pixel according to the line spread function of PET image resolution (FWHM). Thus, MR pixels close to a given PET pixel contribute more activity than MR pixels farther apart, and a single MR pixel contributes activity to more than one PET pixel. In other words, it is possible to estimate how much activity registered in a PET pixel originates from a single MR pixel. The REDISTRIBUTION algorithm was implemented as 1D version in MATHEMATICA and as 2D version in BorlandC++ under Microsoft Windows, both using an 80486 PC. A 3D version is currently being developed.

Method

The REDISTRIBUTION algorithm consists of two steps: 1) Simulation of a PET image from a hypothetical activity distribution, and 2) Using the pixel-by-pixel difference between simulated image and real PET scan for improving the hypothetical distribution. This is illustrated by the following.

The quantitative values of PET pixels are an expression of the underlying continuous activity distribution of the examined object. Each pixel in a PET image collects activity from the examined object (this is done by a PET-Pixel-Scanner PPS). How much of the activity at a location m in the examined object contributes to the value of the PPS at location p is given by the weight function W(p,m) of the PPS, and the relative position of m and p (see Fig. 1). Here, the Gaussian line-spread function of a PET scanner, characterized by its FWHM with unity area below the curve, is used [3]. So for each PPS the sum off all weights is 1. Since the structural MR data are not continuous but discrete, a discrete model for PET scanning will be used. This discreteness of the structural information implies that MR data are subject to PVE's themselves, but here we will regard MR data as the best available approach to structural information. Thus, it is assumed that the value of a PPS at position p is given by the discrete activity ACT(m) over the discrete structural object MR(m) (see Fig. 1) as follows

$$\mathbf{PET(p)} = \sum_{\mathbf{m}} \left[\mathbf{ACT(m)} \cdot \mathbf{W(p,m)} \right], \forall \mathbf{p}: \sum_{\mathbf{m}} \mathbf{W(p,m)} = \mathbf{1}$$

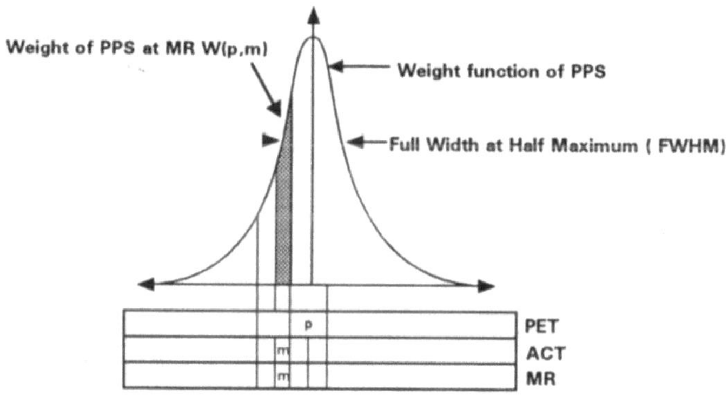

Fig. 1: Gaussian function characterizing the field of view of a PPS. If PET, ACT and MR are aligned, the contribution of the activity of a pixel ACT(m) to a PPS value at p is given by the product of ACT(m) and W(p,m).

To estimate the amount of activity from a PPS that could have originated from the tissue of a certain pixel, the contribution of this MR pixel to the PPS and the share of activity of the MR pixel of the whole activity of the PPS, originating from all involved MR pixels, must be calculated.

For the sake of simplicity, the case of just one PPS in the whole PET image (a PET image containing just one pixel), will be considered first. In this case, the contribution is given by W(p,m). To estimate the share of activity, theoretical contributions were assigned to different tissue compartments of the brain. After segmentation of the MR data into GM,WM and CSF, we can assign numbers to each compartment, that reflect its theoretical contribution to the activity. For example, CSF would receive a value of 0 since it does not contain tissue and cannot contribute to the measured activity. In our simulations, we arbitrarily assigned values of 3 to GM and 1 to WM. This means that GM is assumed to contribute 3 times as much activity to a PPS than does WM. If we now simulate a PET image of this MR image (SPET), we receive a value for each PPS that reflects the theoretical amount of activity that it sees in its neighborhood.

$$\mathbf{SPET(p)} = \sum_{\mathbf{m}} \left[\mathbf{MR(m)} \cdot \mathbf{W(p,m)} \right]$$

Now we can estimate the share of activity of an MR pixel m to a PPS p with MR(m)/SPET(p), and redistribute the activity of the PPS according to this share. So PMR will receive an amount

$$\mathbf{PMR(m)} = \mathbf{PET(p)} \cdot \frac{\mathbf{MR(m)}}{\mathbf{SPET(p)}}$$

of activity from the PPS p. The critical case of SPET(p) being very small occurs if most of the MR pixels in the field of view of PPS p belong to CSF or background. In this case, PMR(m) will be set to 0. In general, the factor MR(m)/SPET(p) in the equation above is a correction of the PVE. If MR(m)=0 (i.e. the pixel belongs to CSF or extracerebral background), PMR(m) won't receive any activity from the PPS. If all pixels in the field of view of the PPS would have the same value MR(m)=const, the value of SPET(p) will be this constant too and the factor is 1. In this case, there is no PVE. Anyway, SPET(p) will be the average theoretical activity as seen by PPS p. If the MR pixel m lies above this average value, PET(p) will be an underestimation and the pixel PMR(m) will receive more activity than given by PET(p). If the MR pixel m lies below this average, PET(p) will be overestimated and the pixel PMR(m) will receive less activity than PET(p). The PET image of the PMR image (PCA) is equal to the real PET scan as shown below:

$$\begin{aligned} \mathrm{PCA}(p) &= \sum_m \left[\mathrm{PMR}(m)\cdot W(p,m)\right] = \sum_m \left[\mathrm{PET}(p)\cdot\frac{\mathrm{MR}(m)}{\mathrm{SPET}(p)}\cdot W(p,m)\right] \\ &= \mathrm{PET}(p)\cdot\frac{\sum_m\left[\mathrm{MR}(m)\cdot W(p,m)\right]}{\mathrm{SPET}(p)} = \mathrm{PET}(p)\cdot\frac{\mathrm{SPET}(p)}{\mathrm{SPET}(p)} = \mathrm{PET}(p) \end{aligned}$$

We succeeded in recovering the PET image completely with respect to both activity and resolution, without the need for an iteration (or after the first step of the iteration). While this PET image had only one pixel, the real situation is only little more complicated. As shown in Fig. 2 there is a great overlap of the fields of view of neighboring PPS's. This means that MR pixels will receive activity from many different PPS's.

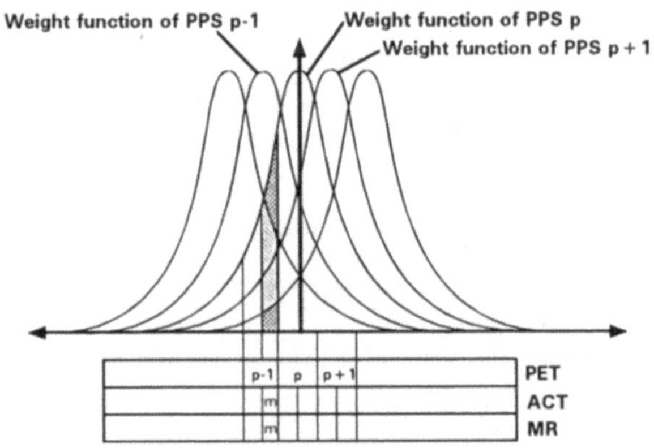

Fig. 2: Neighboring PPS's overlap strongly. The share of activity of PPS p that might originate from MR(m) can be estimated from the ratio of W(p,m) to the sum of all W(q,m) where q runs over all PPS's that may receive activity from ACT(m).

If activity from a PPS was redistributed as in the case of one pixel (that is, without considering neighboring PPS's) the activity in each PMR pixel having more than one PPS contributing to it, would be overstimated considerably. Thus, it has to be calculated how much of the activity of each involved PPS j may originate from the PMR pixel m under consideration. To do this, the weights of all PPS's j that might contribute to the PMR m are added and the ratio of the actual PPS's weight W(q,m) to the sum of all weights is taken as the share of activity of the actual PPS that is redistributed to the PMR. Since the PMR given by these redistributions may not be the best possible approximation, the process will be iterated, and the i.th iteration gives a new image PMR_i from which a simulated PET scan (PCA_i) is calculated. When the iterations start, the images PMR and PCA are empty ($PMR_0(m)=0, PCA_0(p)=0$). During each step of the iteration, the PMR images are incremented as follows:

$$PMR_{i+1}(m) = PMR_i(m) + \Delta PMR_i(m)$$

$$\Delta PMR_i(m) = \sum_q^{PET} \left[\left[PET(q) - PCA_i(q)\right] \cdot \frac{MR(m)}{SPET(p)} \cdot \frac{W(q,m)}{\sum_j^{PET} W(j,m)} \right]$$

$$PCA_i(p) = \sum_m \left[PMR_i(m) \cdot W(p,m)\right]$$

The outer sum (q) is extended over all PPS's that may contribute to the PMR pixel m. The first factor of this sum is the residual error between the real measured PET scan and the actual simulated PET image of the last estimated activity distribution PMR_i. The second factor is the above mentioned estimation of the distribution of PPS q into the PMR m (PVE correction). Finally, the third factor describes the influence of neighboring PPS's. Using the procedure outlined, the deviation from PCA to PET is decreasing in each step of the iteration, but the first step is already sufficient in most of the cases we have examined.

Results

To test the recovery of activity and spatial resolution, we have simulated 2D PET images (6.5 mm FWHM) of several geometrical objects and of human brain slices. A simulated PET image of 10 concentric circles (1 pixel thickness) with distances of 1 to 10 pixels and activities of 100 resulted in an image where the inner circles could not be separated while the activity was 63.9 and the activity of the outer ones where only 27.4 . After REDISTRIBUTION, the original activity distribution was 100% recovered after the first step of the iteration. Of course fine spatial structures in the activity distribution that were below the PET image resolution could not be recovered completely. The utility of REDISTRIBUTION for more complicated geometries was assessed by simulated PET images of human brain slices (Fig. 3) that had been segmented into GM, WM and CSF compartments. The CSF compartment was attributed zero activity, the WM compartment '50', and the GM compartment '150'. To simulate focal activations of GM we assigned arbitrary activities of '230' to different regions in the GM compartment.

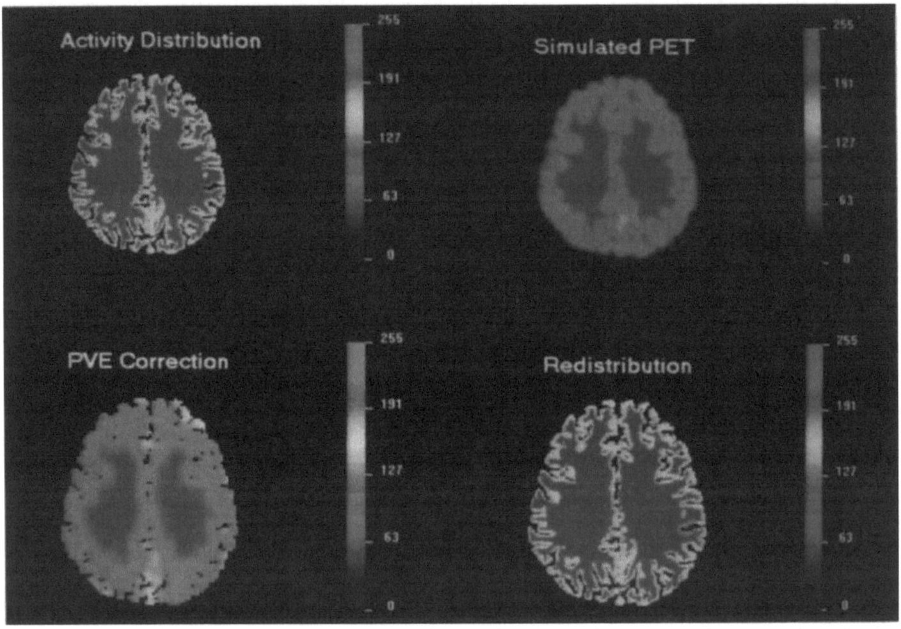

Fig. 3: Image slice of a hypothetical activity distribution, its simulated PET image, and the results of PVE Correction and REDISTRIBUTION. Note, that scaling is identical for all images.

(see also in color on page 817)

A nearly 100% recovery was achieved after 10 iterations. Since the difference in activation between focus and surrounding GM compartment was very steep, there is some over- and undershoot of the redistributed activity near the border.

Discussion

In two model calculations, REDISTRIBUTION significantly improved the in-plane resolution and recovery of activity of a PET image. Of course, this requires accurate matching of MR and PET datasets [2]. It should be stressed, that the increase in out-of-plane resolution will be even greater because of the great slice distances unavoidable with current PET scanners. Recovery of activity could be increased from 10-70% in the simulated PET images to 80-100% in the redistributed images even for small objects. For geometrical objects the recovery could be increased even to 100% in all cases. In PET image slices of the human brain, recovery was 80 - 90% for the redistributed cortex as compared to 10-65% without redistribution. Further improvement is to be expected from operations in 3 dimensions. We are currently working on a 3D implementation of REDISTRIBUTION that will hopefully end up with high-resolution 3D functional images with the resolution of modern MR imaging.

Acknowledgements

This work was supported by grants from the "Deutsche Forschungsgemeinschaft" (SFB 194 / A7,A8). It is part of the doctoral dissertation done by U.K. .

References

[1] Mueller-Gaertner,H.W.; Links,J.M.; Price,J.L.; et al.: Measurement of Radiotracer Concentration in Brain Gray Matter Using Positron Emission Tomography: MRI-Based Correction for Partial Volume Effects; J. Cereb. Blood Flow & Metab. 12 : 571-583 (1992)

[2] Steinmetz,H.; Huang,Y.;Seitz,R.J.;Knorr,U.; et al.: Individual Integration of Positron Emission Tomography and High-Resolution Magnetic Resonance Imaging; J. Cereb. Blood Flow & Metab. 12 : 919-926 (1992)

[3] Rota Kops,E.;Herzog,H.;Schmid,A.; et al.: Performance Characteristics of an Eight-Ring Whole Body PET Scanner; J Comp Assist Tomogr 14 : 437-445 (1990)

Ultrasound

Software for Archiving Ultrasound Images of Thyroid Glad in Children Irradiated after Chernobyl Accident

Alexan K. Sahakyan, Victor K. Ivanov, Sergey A. Ajrapetov, Vitaliy B. Rivkin and Anatoly F. Tsyb

Medical Radiological Research Center RAMS of Russia

Summary

The sotware of the system for archiving and analyzing ultrasound images of thyroid gland in children irradiated as a result of the Chernobyl accident is discussed. In this report we consider the special methods of the image processing (different kind of filters are proposed : high frequency filters, oriented filters, adaptive filters) as well as methods for image compressing and archiving. The comparison with known methods is performed. Some results of image processing are presented.

I.INTRODUCTION

The Chernobyl Nuclear Power Plant accident in 1986 is the most serious ecological disaster in mankind history. More then one million peoples are living in territories of the Russian Federation contaminated with radiation. One of the most important problem of medical monitoring is the estimation of the thyroid gland status in children irradiated with I 131. So the regular ultrasound investigation of the affected population is the basic problem of the physician. It is common knowledge, that ultrasound image of the thyroid gland is projection of 3-dimension objects cross-sections on PC display or another hard carrier. There are some limitations and drawbacks in the ultrasound investigations:

- limitations in registration accuracy and data representation;
- geometric misrepresentations due to various degree pressure scanning headler in an organism;
- presence of noises and artifacts.

Often arise situation, when physician after organ scanning by ultrasound to do another examination. Moreover, considerable quantity of patients with pathology is remained unnoticed. And the diagnosis quality, in fact, determines the fate of a patient.

In present paper the software of the system for archiving and analyzing ultrasound images of thyroid gland in children irradiated as a result of the Chernobyl accident is discussed. The aim of this study is to improve the diagnostic quality ultrasound diagnostics for local hospital physician.

Ultrasound diagnostics hardware for local hospital physician is shown in Fig. 1.

Physician select by ultrasound images scanner or video tape recorder necessary slice and then by video card transform into IBM-compatible personal computer. It make possible for physician to achieve:

- ultrasound images quality improvement;
- archiving and long time storing of information;
- fast scrolling and compare with of past examination;
- recognition of some disease.

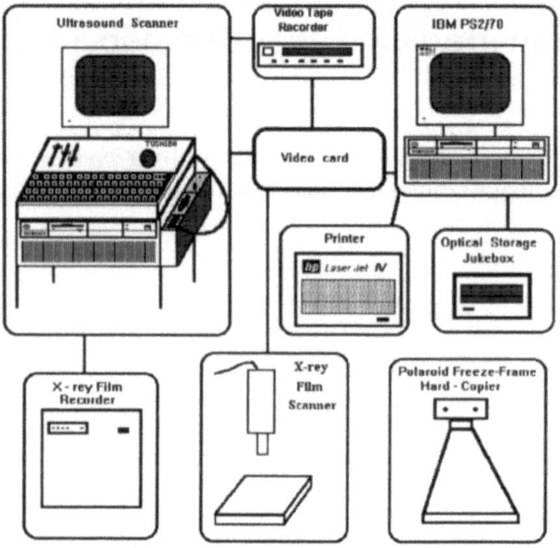

Fig. 1. Ultrasound diagnostics hardware for local hospital physician.

Because of lack of appropriate equipment the most local hospitals have not possibility to transfer data from ultrasound scanner to PC. In such local hospitals the X-ray film or Polaroid paper is used for storing images. Then, in regional hospitals, the ultrasound images are input into PC by means of TV camera.

During analyzing the ultrasound images a doctor usually uses 2 types of processing: visual and quantitative. Therefore the methods for visual and machine processing are included in study. This study, in turn, fall into the following problems:

- the image correction method by optical density taking into account the information distortion on the X-ray film or on the polaroid paper;
- high frequency filter algorithm making possible isolation and contrasting the pathological round form formations;
- the method of directional filtering to isolate the oblong objects;
- the adaptive filtering method to decrease impulse noise and to make an image outline map when having certain parameters values;
- irradiated thyroid gland diseases recognition;
- compressing and archiving of ultrasound images.

II. SOFTWARE FOR PROCESSING AND ARCHIVING OF ULTRASOUND IMAGES THYROID (in children)

The distortion of grey scale is occurred during images storing (X-ray film, Polaroid paper) and further processing. Nowadays there are many ways of image standardization. Most of them are based on table conversion of the image optical density values. But at the section of transfer function where the derivative is nearing zero, information is lost. At the same time the images of pathological formations often belong to these sections.

We have developed the standardization method with recovery according to the density scale which makes it possible to standardize and recover the image parts relating with the parts of uncertainties by using a priori information. On Fig.2 we present the application of the suggested method for recovery and standardization of the real ultrasound images.

The high frequency filters are often used in practice for improving the image quality. It is due to the fact that the spectrum of small formation shadows which are to be isolated mainly belong to the high frequency characteristics. There are many ways of high frequency filtering nowadays. They are mostly based on transforming images into frequency area, reducing the low frequency components and reverse transformation. Hence, there is another approach which was developed by Ar.[1] in 1971 and it realizes the processing in space area. We have summarized the method of high frequency filtering for two - dimensional signals. On the Fig.3 we present the ultrasound image processing scheme method straightening through smoothing (STS). The frequency component passage cocfficicnt m_1,m_2 of two-dimcnsional Furricr spcctrum

$$\Phi_{m_1,m_2} = \sum_{j=0}^{N-1} \sum_{k=0}^{M-1} F_{j,k} \exp(-2\pi i(m_1 j/N + m_2 k/M))$$

after S smoothing and C straightening is show

$$T = \{1-[\cos(\pi m_1/N)\cos(\pi m_2/M)]^{2s}\}^{c}, \quad (1)$$

where T- transfer coefficient of m_1,m_2- frequency component of two-dimensional Furrier spectrum.

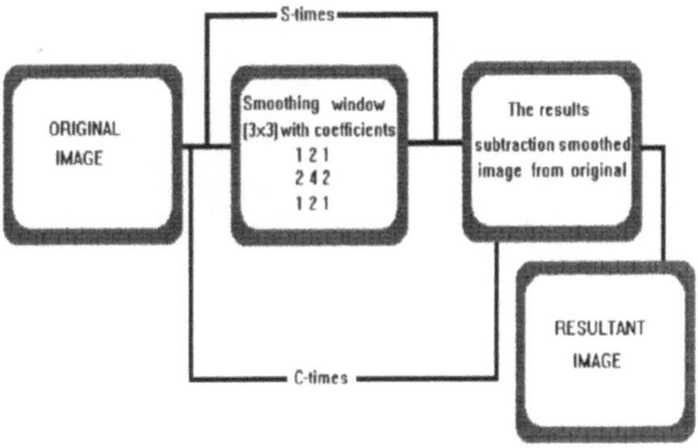

Fig.3 Scheme of processing ultrasound images by STS method.

For visual demonstration of filter properties the diagram on dependence of (passage coefficient) transform from frequency components according to the equation (1) are shown on the Fig.4.

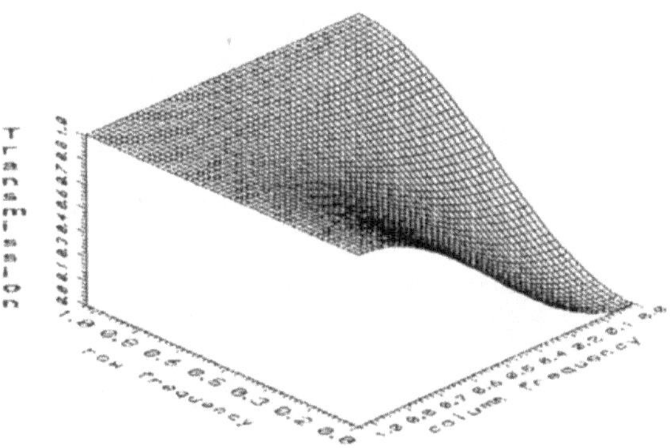

Fig.4 Transfer function (1) for frequency component coefficient m_1,m_2

By varying parameters of smoothing and straightening one can regulate the passband and image contrast. When one increases the number of smoothing in lines the passband of filter widens but by rising the number of straightening one may increase the steepness of the directional function. The Fig.5 demonstrates the results of the real ultrasound image processing by STS filters.

In situations when it is necessary to isolate the space attended objects on the ultrasound slice of organ , for example, isthmus or papillary cancer, the use of high frequency filters results in gaps of contour. In such cases it is more useful to apply

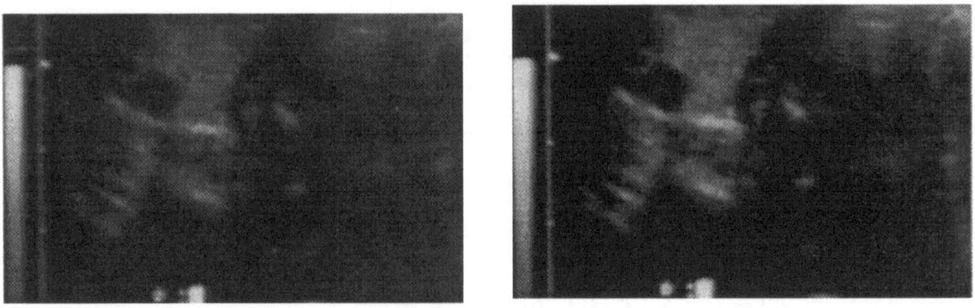

a b

Fig.2 Result of using standardization and gray scale recovering method for real ultrasound images (a - before, b - after processing)

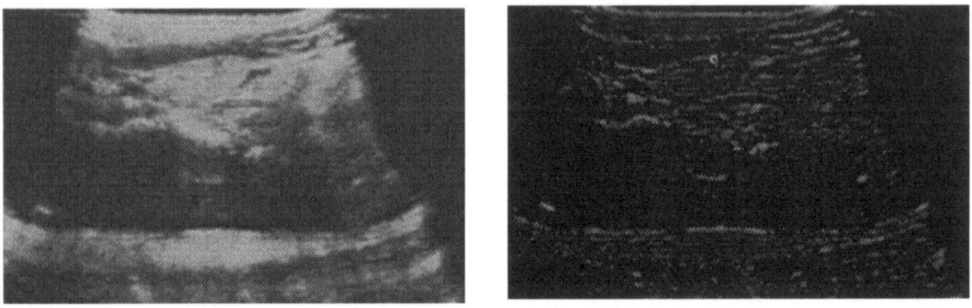

a b

Fig.5 The results of the real ultrasound image processing by STS filters (a - before, b - after processing).

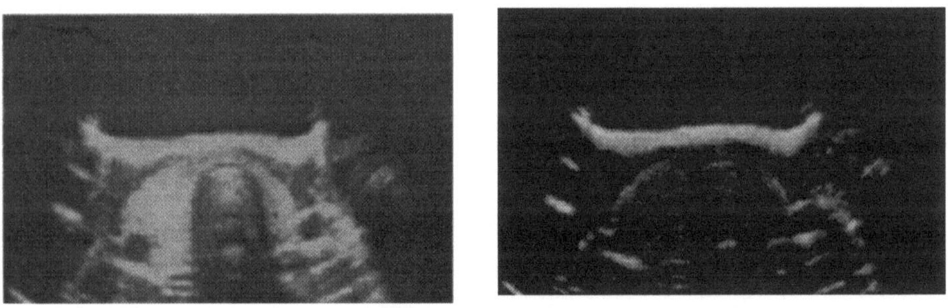

a b

Fig.7 The results of the thyroid glad real ultrasound image processing by DF filters (a - before, b - after processing).

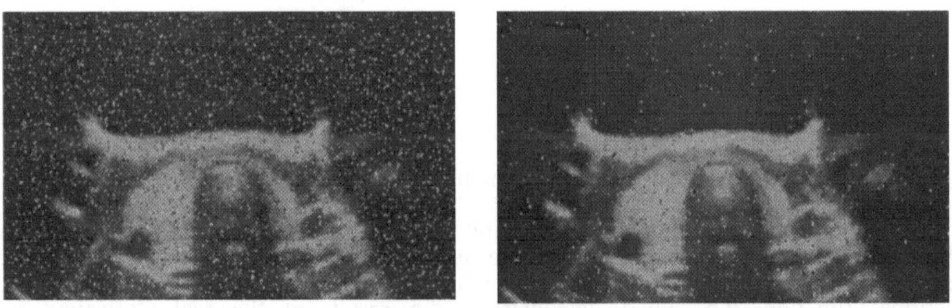

a b

Fig.9 The results of the real ultrasound image processing with added 10%

directional filters (DF). The considerable disadvantage of these filters is their inflexibility. The soft image in one direction and hard image in the orthogonal one are impossible with such filters. We have designed a new filter without the above disadvantages. The algorithm differs from the STS as the smoothing operation is made with various parameters separately in lines and columns. The frequency characteristics of the filter for two-dimensional Furrier spectrum is shown

$$T = \{1-[\cos^{2s_1}(\pi m_1/N)]\ [\cos^{2s_2}(\pi m_2/M)]\ \}^c , \qquad (2)$$

where T- transfer coefficient of m_1,m_2- frequency component of two-dimensional Furrier spectrum; S_1,S_2- numbers of smoothing in lines and columns accordingly. For demonstration the new filter properties the Fig.6 shows diagrams of (passband coefficient) transmission dependence on space frequencies with certain parameters.

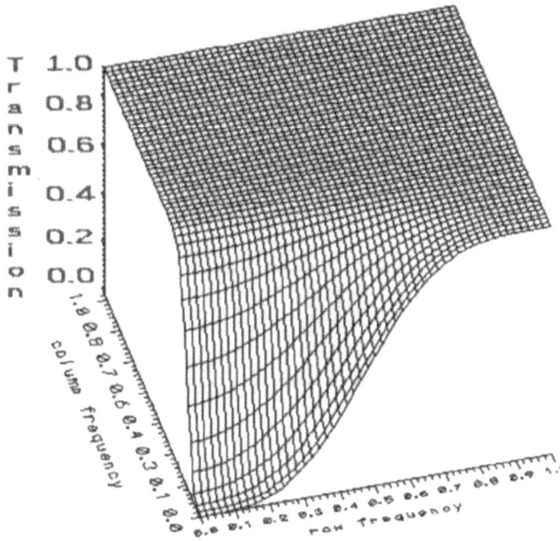

Fig.6 Transfer function (2) for space frequencies with certain value of parameters.

That high frequency components in vertical will be better isolated if be true $S_1>S_2$.

And vice versa if be true $S_1<S_2$, then the horizontal components of the image will be isolated better. In particular when the numbers of smoothing in lines and columns are equal the result of the processing is identical to that by STS. The Fig.7 demonstrates the results of the real ultrasound image processing by STS filters.

In situations when it is necessary storage of ultrasound images on X-ray film or on polaroid paper, not rarely one can come across them defect which results in small diffusively distributed white and black spots. And it often makes a doctor repeat the examination that results in increasing time for examination and, of course, waste a lot of photo material. Such mathematical methods of pulse noise decreasing as low frequency and adaptive filters are used most often. The application of low frequency filters results in destroying the boundaries of organs and pathologies. Most adaptive filters are based on median filters. But their use leads to geometric misrepresentations of objects on the image. There is another adaptive filter which we have designed and named the method of maximum modes(MM).

Definition

If for any odd aperture there exists one element the number of repetitions of which in the aperture is m and there is no other element in this aperture the number of repetitions of which is l and it does not satisfy unequally

$$l < m/2, \qquad (3)$$

then this element is the maximum mode (MM) of this aperture. In any case then the resulting value is unchanged and such points in the image will be called the stable points of the filter MM.

Note that one may conclude from the property for MM that monotonous function is invariant to the MM filtering with arbitrary width of the aperture. These functions are the simplest stable points of the MM filter. It follows from the property that MM method may be useful for isolating the objects contours in the image as the transition boundary from one area into the other is the locally monotonous discrete function. As the possibility of nonlinear filters analysis is limited let us study the filtering property by the MM method that are analogous to the studies made by Pratt [2] for median filtering. The Fig.8 shows several one-dimensional signals the results of their processing by median filter and by MM method.

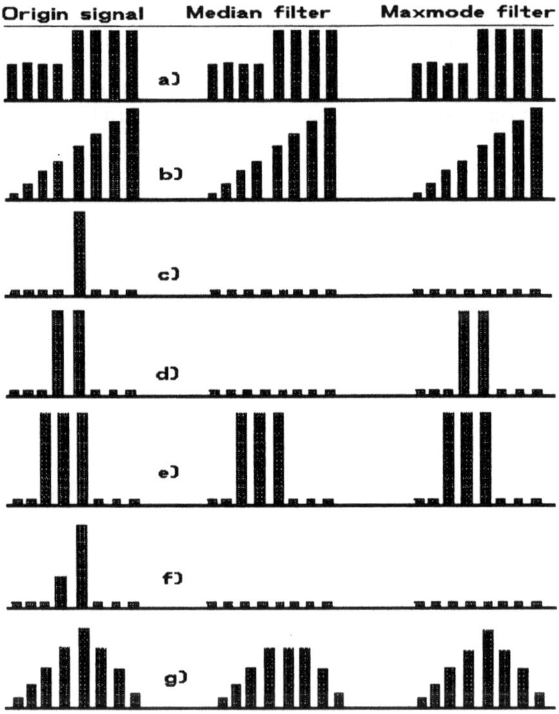

Fig.8 The results of several one-dimensional signals processing by median filter and by MM method.

It is demonstrated here that the MM method decreases the pulse noise and in some cases it distorts the signal less than filtering by the median method.

The Fig.9 demonstrates the results of the real ultrasound image processing. White 10% noise was added to the image beforehand. The application of the filter properties makes it possible to use the MM method both for decreasing of pulse noises and for contours isolation.

It is common knowledge that irradiation of thyroid gland leads to different pathological changes. Then the algorithms for pathological formations recognition was developed by authors. This algorithms are based on modified modes quantum methods [7] and second-generation image coding algorithm [8]. The cartesian distance between vectors of given area features was identification criteria.

The software includes the following subsystems:

- data base of ultrasound images
- archiving and storing of ultrasound images of thyroid
- quick retrieval of the sonogrammes of a definite patient
- quick scrolling of thyroid images

To summarize me must say that the designed algorithms require the acceptable working time in the iterative conditions for a doctor. All means of information exchanges between a user and a computer are organized so that a specialist not having profound knowledge in mathematics and programming can work with the system.

Nowadays the developed software are used in local and regional hospitals for ultrasound images storing, processing and analyzing.

III. ACKNOWLEDGMENT

The authors acknowledge the financial support of QUEST Automation Ltd. company.

REFERENCES

1. E.Bolduc, J.J.Quemener and P.Marmet // Electron-impact excitation of 3s3p6nl states of Ar,- Canadian Journal of Physics, N49, 1971, p. 3095.
2. W.K.Pratt // Digital image processing, 1982,p.790.
3. H.R.Maxon // Med. Clin. N. Amer., vol.69, N5,pp. 1049-1061, 1978.
4. L.J. De Groot, M. Reilly, K. Pinnameneni, S. Refetoff, //Amer. J. Med.,Vol. 74, N 5,pp.852-862,1983.
5. Induction of Thyroid Cancer by Ionizing Radiation (NCRP report N80),Bethesda,1985.
6. P.R. Larsen, R.A. Conard, K.D. Knudsen, et al,// J.A.M.A., vol.247,N 11, pp.1571-1575,1982.
7. L.P.Yaroslavsky//Digital Image processing Introduction, -Moscow: Sow.radio,1979,p.312.
8. M.Kunt, A.Ikonomopoulos, M.Kocher // Second-Generation Image-Coding Techniques, Proceedings of the IEEE,v.73,N4, 1985, pp. 549-574.

3D Reconstruction of Fetal Images Based on Ultrasound

HANS-PETER KLEIN, DR. MARKUS MARQUART, DR. HANS POLZ,
GABI SCHULLER

TomTec Tomographic Technologies GmbH
Breslauer Str. 1-3
8057 Eching

Introduction

3d reconstructions are widely used in the field of radiology by means of CT and MRI. Whether the 3d reconstruction is done mentally in the brain of the physician or a dedicated 3d workstation is used, the information of parallel slices on a multiformat display and / or the 3d surface rendered display is important for the diagnosis to be made based upon these modalities.

Ultrasound so far has not offered 3d information or the important feature of systematic scanning. Usually the probe is handheld and the scanning is made interactively from the operator. This way of acquiring images does not provide the basis for 3d reconstruction and consequently the ultrasound system has to be modified to provide a way of systematic scanning.

Data acquisition

In order to do a 3d data acquisition we have developed a number of different mechanical frames, which can be used in conjunction with normal 2d ultrasound systems, to add the required systematic scanning to such systems. Depending on the organ acquired, the frames range from a parallel shift frame to a rotating frame and, for examinations of the heart even a TEE probe which can be stiffened after the introduction in the esophageous. There is no "ideal frame", as each part of the human body is a another challenge to get good 3d results.

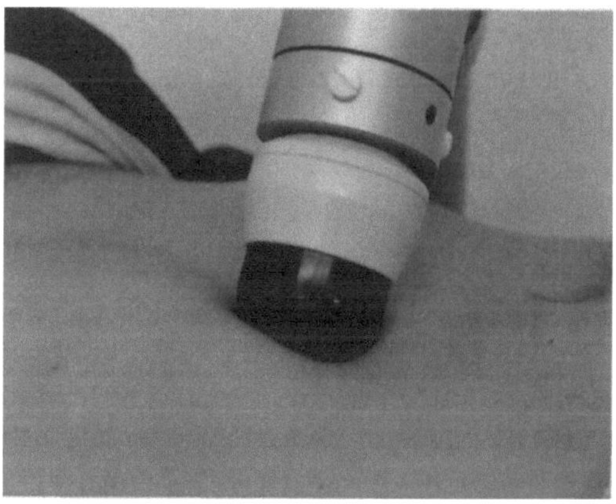

Figure 1: Rotating frame applied to the abdomen

The rotating frame (see figure 1) is used together with sector scanners (mechanical, phased array). It can be used in applications where a small acoustic window is available. This approach is not ideal for objects directly under the surface (e.g. vessels), due to the nearfield problems, and the limited field of view.

In order to address the nearfield problems we also have developed a device which is translating the probe over the skin surface called the parallel shift mover (see figure 2).

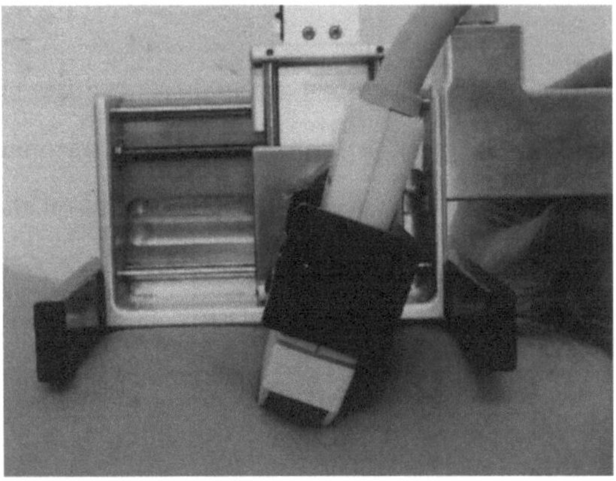

Figure 2: Parallel shift mover applied to the abdomen

This device is equipped with a stepper motor to allow the systematic scanning. As the surface of the skin is usually not flat, the mover has the capability to follow the skin surface. An altitude measurement is done during the scanning of the organ in order to correct for the different altitudes.

Both mechanically frames are driven from a stepper motor which is controlled by the computer (486, 50 Mhz). During the acquisition a frame grabber is digitizing the image information in a 256*256 matrix size. The number of images can be selected, usually 200-350 images are acquired.

As grabbing is performed in real-time the acquisition takes typically about 10 seconds. After the acquisition is completed, a postprocessing step is initiated in order to correct for the altitude, fill in the gaps (rotational frame only), and remove motion artifacts (Reduction Of Spatial Artifacts). This step is performed in about 1 minute.

The following example shows initial clinical results in a gynecology application. The acquisition of this dataset was obtained using a Kretztechnik Combison 300 ultrasound system, which performs the acquisition by means of a sweep. The data is transferred using a SYQUEST disk to a TomTec Echo-View for postprocessing.

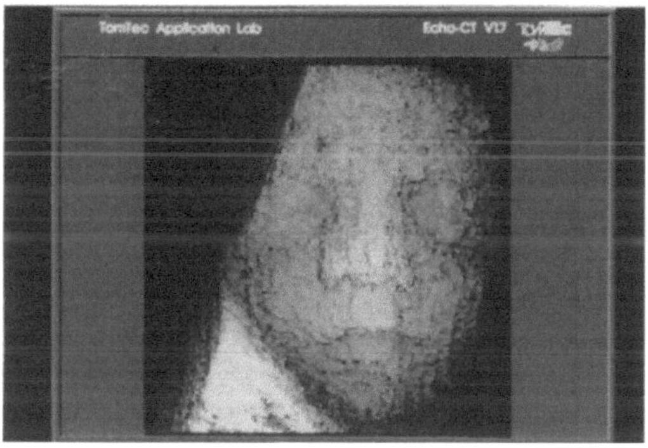

Figure 3: Face of a 23 week old fetus, rendered with CT-algorithms

Applying 3d surface rendering to this dataset, using algorithms like in CT and MRI, is shown in figure 3. The image shows that the different image characteristic of ultrasound data produces also very noisy surface rendered images. A substantial improvement can be achieved, if the whole dataset is filtered by means of a 3 dimensional convolution.

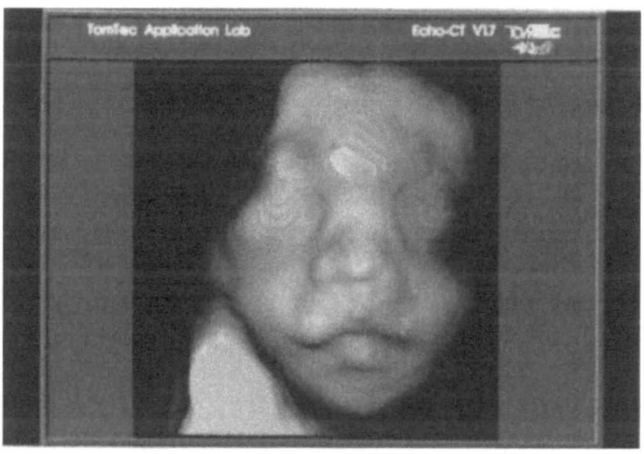

Figure 4: Same case as above (figure 3) filtered with a 3d-convolution

Figure 4 shows the image after that process. A much better definition of the skin surface can be seen. However, the details which have orginally been in the dataset have been destroyed, using this kind of approach.

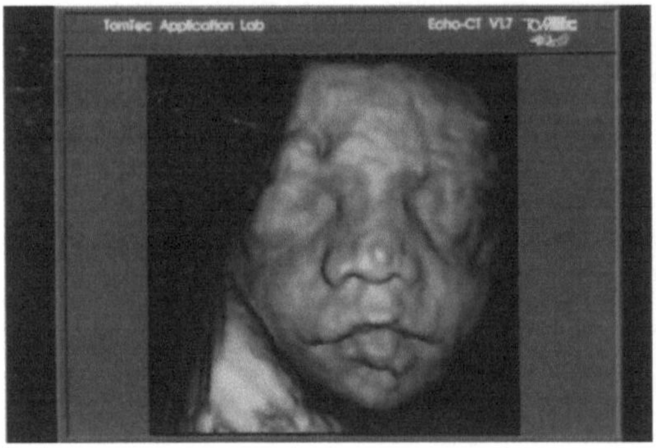

Figure 5: Same case as figure 3, ultrasound specific algorithms were applied

In order to address the specific needs of ultrasound images, we have developed a dedicated 3d surface rendering, which does not require a pre-processing. Also this algorithm does not supress the fine details through smoothing in the original dataset. The results in figure 5 show the final 3d image based on this new algorithm.

Conclusion

The demonstrated data have clearly shown, that 3d ultrasound is feasible, if the important issues are addressed:

- systematic data acquisition using parallel shift, rotation or sweep
- dedicated image processing for ultrasound images

Also, it should be mentioned, that the use of a 486 based system, gives an acceptable price performance ratio, compared to the original ultrasound system.
The number of potential clinical applications still has to be determined, however the results so far indicate a big potential for 3d ultrasound.

Digital Angiography

A Technique for 3D Reconstruction of Arteries from Angiographic Projections

D. Scott, A. G. Davies, A. R. Cowen and A. Workman.

FAXIL, Academic Unit of Medical Physics, Deptartment of Clinical Medicine
The Wellcome Wing, The General Infirmary at Leeds
Great George Street, Leeds LS1 3EX, UK.

Abstract

A practical technique for the 3D reconstruction of arteries from 2D conic X-ray projections is described. This could provide valuable information to clinicians concerning the extent of atherosclerotic disease in both coronary and peripheral vessels. The new generation of angiographic X-ray equipment such as the Philips Integris C2000 allows acquisition of a complete set of rotational data, enabling the use of an analytical reconstruction algorithm. Approximately 25 to 35 projections taken over 180° are required to enable direct conic reconstruction using filtered backprojection. The projections used in this study were taken in a clinical angiography suite which adds particular problems not associated with an experimental jig. The technique was tested on arterial samples and the results provide supportive evidence that future clinical research may be productive in this area; not only has the gross vascular structure been reproduced, but fine surface detail of atherosclerotic lesions has been resolved. This is an application which could be used on existing X-ray equipment in an imaging modality where scanning techniques such as CT are not suitable.

Introduction

Atherosclerosis is the most common disease of arteries in Western Society; being a major predisposing factor for both myocardial infarction and stroke (1). Peripheral vessels are also the subject of lesions of clinical consequence, particularly in the supply to the leg; the iliac, femoral and popliteal arteries are the main sites of disease. Clinical manifestations range from intermittent claudication to gangrene, requiring amputation (2). The main radiological technique for assessing the condition of these vessels is angiographic contrast studies. The information available to the clinician from such X-ray studies could be improved if a 3D representation of the arterial lumen was also available. This would assist both in the planning of balloon angioplasty treatment for the stenosis and in the assessment of its efficacy.

With contrast studies performed on the new generation of angiographic equipment, such as the Philips Integris C2000 in use at the General Infirmary at Leeds, the possibility exists of obtaining suitable projection data to perform an analytical 3D reconstruction algorithm. Here a means of direct vascular 3D reconstruction from angiographic projections is presented. The technique is based on the reconstruction principle employed by the CT scanner except that two dimensional conic projections are used as opposed to a scan.

Reconstruction by Filtered Backprojection

The image data required for reconstruction algorithms was obtained by taking a number of projections around the structure of interest at different angles of rotation (figure 1). Practically there is a choice of two groups of algorithms to use for reconstructing from these source projections (3): Iterative techniques such as algebraic reconstruction, or analytical methods such as filtered backprojection.

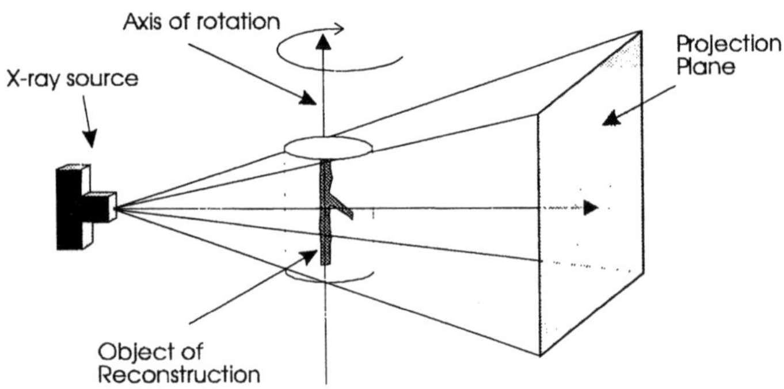

Figure 1 Cone beam projection data used for reconstruction.

The availability of a complete data set enabled vessel reconstruction using an analytical algorithm without need for introduction of any a priori knowledge. The technique used was filtered backprojection with cone beam geometry, based on the algorithm derived by Feldkamp et al (4) with its practical implementation described by Scott (5).

Methods

(i) Construction of a Femoral Artery Test Phantom

Two femoral artery samples were obtained at post mortem for testing of the reconstruction technique. The lumens were filled with solid barium contrast medium as commonly used by pathologists. This enabled radiological examination of the arteries. In order to mimic the problems caused by a leg to a reconstruction algorithm a test phantom was constructed; a bone was introduced and the samples were suspended in a water bath. The sharp edges of the bone produced artefact which could potentially 'pollute' the arterial reconstruction, whilst the water simulated the X-ray scatter effect of soft tissue.

(ii) Image Acquisition

The acquisition system used was a Philips Integris C2000 which has a rotating X-ray gantry, capable of rotation in any plane. The system currently in use at the General Infirmary at Leeds is shown in figure 2. Digital images of resolution 512 x 512 pixels or 1024 x 1024 pixels are produced by the Philips DSI system and the X-ray images stored onto magnetic streamer tape.

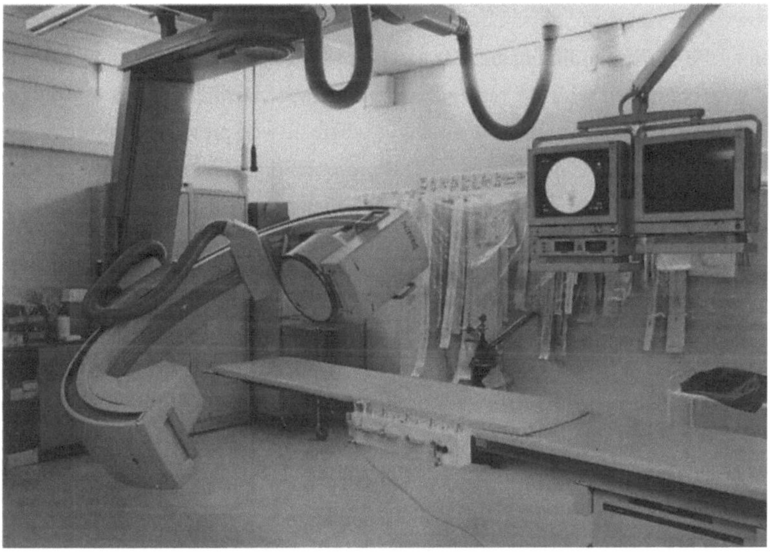

Figure 2 Angiography Suite at Leeds General Infirmary with Philips Integris C2000.

Accurate determination of the geometry of image acquisition and set-up of the apparatus was required as a preliminary to the production of source data suitable for the reconstruction algorithm. The test phantom was placed in the X-ray field and X-ray images were taken by sweeping the arm around through 180°. It was found that a set of source images numbering from 25 to 35 was satisfactory for reconstruction. For the reconstructions presented here 35 projections were taken at 5° increments

(iii) Reconstruction and Display

The images were transferred to a Sun SPARCstation for application of the reconstruction algorithm. The images were reconstructed to a resolution of 255 x 255 x 255 pixels. The resultant volume data was rendered to gain a 3D perspective using the SunVision software. The surface contours were displayed because these correspond to the internal surfaces of the arterial lumen, which is filled by the contrast medium. Any unwanted part of the data could simply be stripped away prior to rendering, and in this way the bone was removed from the images.

Results

The results obtained for the reconstruction of the arterial samples are shown in figures 3-4. Figure 3 shows two views of the same surface rendered vessel alongside the source X-ray images taken at the corresponding angle. It can be seen how a 3D display shows the nature of the vessel wall compared with the rather flat effect gained via simple X-rays. Additionally the surface of the vessel can be seen at an angle where it was previously obscured by the presence of the bone in the X-ray image. The presentations shown here are sub-optimal because the reconstructed arteries are best visualised dynamically using a cine-display of the reconstruction rotating through 360°. With such a format, it is possible to view the artery from all aspects, maximising the chances of spotting abnormalities.

In figure 4 a reconstruction is shown alongside the dissected artery from which it was obtained. It can be seen that the surface shape of the arterial wall is represented faithfully by the reconstructed image. In particular the narrowing caused by two large atherosclerotic plaques have been clearly reproduced in the femoral artery and also a significant stenosis can be seen in the profunda femoris artery, distal to the bifurcation .

Discussion

The proposed technique has been used to successfully reconstruct femoral artery samples. The reconstruction is indicative of the 'column' of blood flowing through the vessel lumen, into which atherosclerotic plaques encroach. The reconstructed image data can be viewed from any aspect via a 3D rendering system, without obstruction from the bone. This would be an invaluable aid to the clinician in assessing the nature of such vessels in order to plan treatment. Measurement of the quantitative size of the vessel could also be made if a contour tracking algorithm such as the one developed by Sun (6) were used to trace around the lumen. Patients, with or without treatment, could be followed up to compare the vessel size at equivalent sites in order to monitor the disease progression.

The research was carried out in a manner which truly reflects the clinical situation of digital angiography; the X-rays were not taken on an experimental jig, but on a working system with the practical limitations associated with this. These include movement of the cone beam centre during rotation and field disturbances produced by either magnetic or electrical fields, or distortion caused by the electro-optical components in the imaging chain. The research thus carries practical implications which are transferable to a clinical study.

The technique may not only be applicable for angiographic studies on peripheral arteries, but also on the coronary and cerebral networks which are of great clinical importance. The main problem faced would be the pulsatile nature of the arteries. This would be a particular problem with coronary angiograms in which the arteries also move significantly during contraction of the heart. This is a problem which could possibly be surmounted by the use of ECG gaiting in the angiographic equipment. If the X-ray timing was linked to the ECG then equivalent shots could be taken at the same point in the heart cycle for each angular projection.

The proposed technique would be an additional application for the existing angiographic hardware in use clinically and hence could be regarded as an evolutionary rather than revolutionary development. It would enable reconstruction in an imaging modality for which CT is unsuitable due to the fast flow of contrast out of the vessels.

References

1. L.M. Buja, The vascular system. **in:** Basic Pathology (4th ed) (Ed. S.L. Robbins & V. Kumar), 285-311, (1987).
2. A.J. Camm, Peripheral vascular disease, **in:** A Textbook of Clinical Medicine (4th Ed.) (Eds. J.P. Kumar & M.L. Clark), 622-626, (1990).
3. M. Davison, X-ray Computed Tomography, **in:** Scientific Basis of Medical Imaging (Ed. P.A. Wells), 54-73, (1981).
4. L.A. Feldkamp, L.C. Davis & J.W. Kress, Practical cone beam algorithm, *J. Opt. Soc. Am.*, **1, No. 6**, 612-619, (1984).
5. D. Scott, A Feasibility Study into the Reconstruction of Arteries from X-ray Projections, MSc Dissertation, University of Bradford (U.K.), (1992).
6. Y. Sun, Automated identification of vessel contours in coronary angiograms by an adaptive tracking algorithm, *IEEE Trans Med. Imag.*, **vol. 8 no. 1,** (1989).

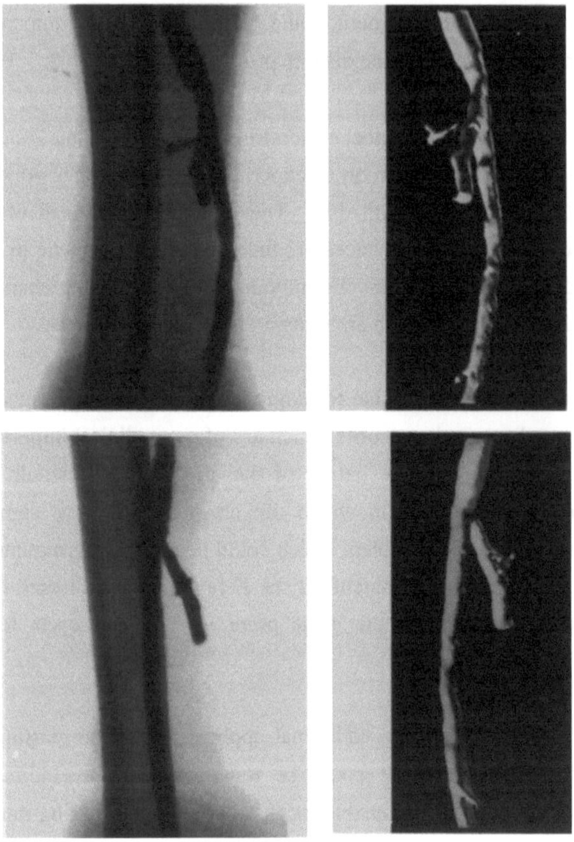

Figure 3 Two views of a reconstructed femoral artery alongside the corresponding source X-ray images.

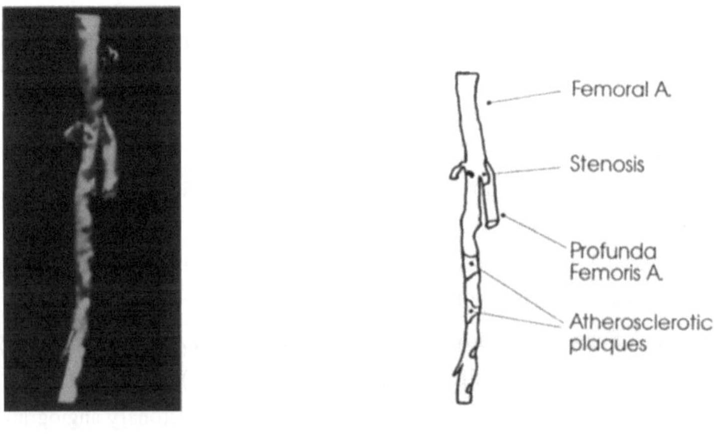

Figure 4 Single view of a reconstructed femoral artery with a diagrammatic representation.

Biplane Angiocardiography: General Solution for Pairing Images Taken from Oblique Views

G. P. M. PRAUSE and D. G. W. ONNASCH

Clinic for Pediatric Cardiology
Section Biomedical Engineering
Christian-Albrechts-University
Schwanenweg 20
D-2300 Kiel 1, FRG

Summary
The identification and matching of details in biplane angiocardiograms taken from an oblique view is complicated by rotations, translations and different magnifications of the images. A procedure for pairing images is proposed that is applicable to arbitrary biplane gantry systems and projections.

Introduction

For visual and computerized evaluations of biplane angiocardiograms it is necessary to find corresponding points and density profiles in the projection images. This task is heavily complicated by the following properties of biplane isocentric X-ray systems:

1. For caudal or cranial views the images are rotated differently depending on the characteristical movement of the gantries.
2. Due to the cone beam divergency, the images are magnified differently, and biplane reconstruction planes are represented by fan-like arranged density profiles.
3. Geometrical image distortions and miscalibration of the gantry system must be determined and corrected when biplane projections are paired.

With knowledge of the projection angles and the imaging distances biplane projections of an arbitrary gantry system can be displayed in a standardized way simplifying the detection of corresponding points and cross sections.

Image Rotation

The global coordinate system used to describe a biplane gantry setting is a right handed three-dimensional Cartesian system with its origin O in the isocenter (Fig. 1). A projection $A = (\alpha, \beta)$ is defined by the rotation angle $\alpha \in (-180°, 180°)$ and the angulation angle $\beta \in (-90°, 90°)$ specifying the center of the image intensifier input screen. Projection $B = (\gamma, \delta)$ is given by the rotation angle γ and the angulation angle δ.

The gantries can be classified with regard to their characteristics of movement (Fig. 2). Wollschläger et al. [1] introduced and analyzed the equipment types I and II with

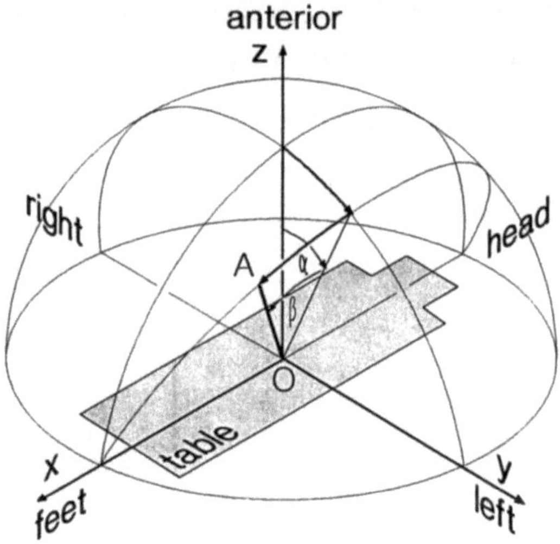

Figure 1: Global coordinate system with a projection $A = (\alpha, \beta)$. Right rotations and cranial angulations are indicated by negative angles, left rotations and caudal angulations by positive angles. The projection $(0°, 0°)$ describes the posterior-anterior view.

the poles located on the y-axis and x-axis, respectively. Type III with poles on the z-axis was defined by Dumay [2]. Widespread systems serving as a paradigm for other manufacturers are the Siemens BICOR with gantry A of type II and B of type III, and the Philips Poly DIAGNOST C / Lateral ARC system with both gantries of type II.

To follow the trace of the image intensifier for a given projection, it is necessary to transform the projection angles α and β into the physical angles of rotation around the gantry's rotational axes. For all types of equipment the left-right rotation is done around the patient's long axis (x-axis), whereas the axes for the subsequently caudal-cranial angulation differ between the three gantry types. A gantry of type I angulates around the y-axis, type II around the already rotated y-axis (the y'-axis) and type III around the z-axis.

Consequently the image intensifiers of different gantry types reach a certain caudal or cranial angulated projection with a different rotation around the central beam, as shown in [3] for the Siemens BICOR system. To adjust a certain biplane projection the central scan lines of both images must be rotated into the reconstruction plane spanned by the two central beams (isocenter plane).

Table 1 contains the formulas to calculate the image rotation angles ρ_A and ρ_B for the three gantry types, both imaging planes and arbitrary biplane projections. These formulas can be directly applied to the projection angles $\alpha, \beta, \gamma, \delta$ without prior transformation into the gantry's physical rotation angles.

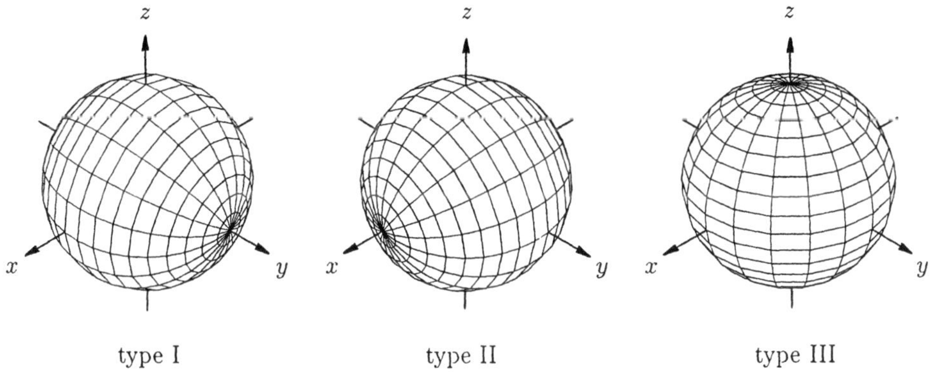

Figure 2: The three gantry types and their characteristics of movement.

X-Ray Divergency

The magnification factors of the images are equalized by iso-center calibration utilizing the distances between focal spot, isocenter and image intensifier input screen.

Due to the divergent X-ray geometry, all reconstruction planes of a biplane projection are running through both focal spots (Fig. 3). Hence they are represented by a fan of density profiles in the projection images. With reference to the isocenter plane the intersection of a reconstruction plane with the image intensifier input plane is described by the projection lines

$$l_A(k_A; d) = \frac{d}{c} \cdot \left(a + k_A \cdot \tan\frac{\epsilon}{2}\right), \qquad l_B(k_B; d) = \frac{d}{c} \cdot \left(b - k_B \cdot \tan\frac{\epsilon}{2}\right), \tag{1}$$

where a and b are the distances between the focal spots F_A, F_B and the central points C_A, C_B of the image intensifier input planes, c the distance between the focal spots and the isocenter O and $\epsilon = \arccos(\sin\beta \cdot \sin\delta + \cos\beta \cdot \cos\delta \cdot \cos(\alpha - \gamma))$ the angle between the projections. The reconstruction plane is specified by its distance d to the isocenter measured perpendicular to the isocenter plane.

The projection lines of the reconstruction planes are in close connection to the epipolar lines which define for a certain point in one view the location of all possible corresponding points in the other view. Projection lines and epipolar lines can be superimposed in the biplane projections to support the identification and matching of image details.

Image Alignment

For biplane image pairing it is not sufficient to correct the individual geometric distortions of the two projections. In addition the images have to be aligned exactly in both views.

Influences of magnetic fields, especially the geomagnetic field, on the moving electrons inside the image intensifier cause translations and rotations in the projection images

Image Rotation of System A

Type	$\sin \rho_A$		*Condition*
I	$\frac{\cos\alpha \cdot \cos\beta \cdot \sin\delta - \sin\beta \cdot \cos\gamma \cdot \cos\delta}{\sin\epsilon \cdot \sqrt{1-\sin^2\alpha \cdot \cos^2\beta}} \cdot \mathrm{sign}(90^\circ - \lvert\alpha\rvert)$		$\lvert\alpha\rvert \neq 90^\circ$
	$-\frac{\cos\gamma \cdot \cos\delta}{\sin\epsilon} \cdot \mathrm{sign}(\beta) \quad (\beta \neq 0^\circ)$	$\frac{\sin\delta}{\sin\epsilon} \quad (\beta = 0^\circ)$	$\lvert\alpha\rvert = 90^\circ$
II	$\frac{\cos\beta \cdot \sin\delta - \sin\beta \cdot \cos\delta \cdot \cos(\alpha-\gamma)}{\sin\epsilon}$		—
III	$\frac{\sin\alpha \cdot \cos\beta \cdot \sin\delta - \sin\beta \cdot \sin\gamma \cdot \cos\delta}{\sin\epsilon \cdot \sqrt{1-\cos^2\alpha \cdot \cos^2\beta}} \cdot \mathrm{sign}(\alpha)$		$\alpha \neq 0^\circ$
	$-\frac{\sin\gamma \cdot \cos\delta}{\sin\epsilon} \cdot \mathrm{sign}(\beta) \quad (\beta \neq 0^\circ)$	$\frac{\sin\delta}{\sin\epsilon} \quad (\beta = 0^\circ)$	$\alpha = 0^\circ$

Image Rotation of System B

Type	$\sin \rho_B$		*Condition*
I	$\frac{\cos\alpha \cdot \cos\beta \cdot \sin\delta - \sin\beta \cdot \cos\gamma \cdot \cos\delta}{\sin\epsilon \cdot \sqrt{1-\sin^2\gamma \cdot \cos^2\delta}} \cdot \mathrm{sign}(90^\circ - \lvert\gamma\rvert)$		$\lvert\gamma\rvert \neq 90^\circ$
	$\frac{\cos\alpha \cdot \cos\beta}{\sin\epsilon} \cdot \mathrm{sign}(\delta) \quad (\delta \neq 0^\circ)$	$-\frac{\sin\beta}{\sin\epsilon} \quad (\delta = 0^\circ)$	$\lvert\gamma\rvert = 90^\circ$
II	$\frac{\cos\beta \cdot \sin\delta \cdot \cos(\alpha-\gamma) - \sin\beta \cdot \cos\delta}{\sin\epsilon}$		—
III	$\frac{\sin\alpha \cdot \cos\beta \cdot \sin\delta - \sin\beta \cdot \sin\gamma \cdot \cos\delta}{\sin\epsilon \cdot \sqrt{1-\cos^2\gamma \cdot \cos^2\delta}} \cdot \mathrm{sign}(\gamma)$		$\gamma \neq 0^\circ$
	$\frac{\sin\alpha \cdot \cos\beta}{\sin\epsilon} \cdot \mathrm{sign}(\delta) \quad (\delta \neq 0^\circ)$	$-\frac{\sin\beta}{\sin\epsilon} \quad (\delta = 0^\circ)$	$\gamma = 0^\circ$

Table 1: Formulas to calculate the image rotation for the gantries A and B using directly the projection angles $\alpha, \beta, \gamma, \delta$ and the angle $\epsilon = \arccos(\sin\beta \cdot \sin\delta + \cos\beta \cdot \cos\delta \cdot \cos(\alpha-\gamma))$ between the two projections. After rotation of image A by ρ_A and B by ρ_B the central scan lines are turned into the isocenter plane.

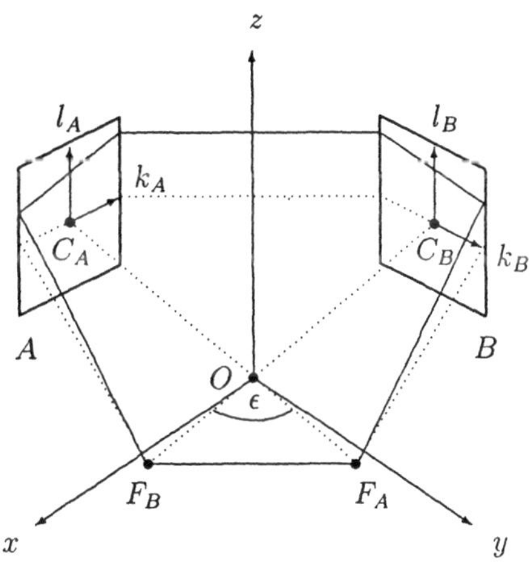

Figure 3: For biplane cone beams a reconstruction plane (solid lines) runs through both focal spots F_A and F_B. The isocenter plane is drawn in dotted lines. The projection lines of different reconstruction planes have different slopes in the images.

depending on the orientation of the gantries. For biplane image alignment we fixed a square of four semitransparent markers on the center of each image intensifier input screen. The markers are located fully automatically in the images and used to derive two non-linear centering polynomials [3].

Remaining deviations are due to geometrical miscalibration of the biplane equipment. They can be compensated by an image shift using previously measured calibration data of the gantry system or at least two interactively defined reference points in both projection images [4].

Results

As an example an enddiastolic ventriculogram of a three days old girl is presented in Fig. 4. The images were taken with the BICOR system under $A = (-16°$ RAO, $-5°$ cranial) and $B = (70°$ LAO, $-20°$ cranial) with an intersectional angle of $\epsilon = 84.5°$. The distance from the focal spot to the image intensifier was 85cm (A) and 95cm (B), respectively, the distance to the isocenter 70cm. The subtracted and edge enhanced originals are shown in the upper panel of Fig. 4.

The paired images in the lower panel are centered, calibrated and rotated by $\rho_A = -19.7°$ and $\rho_B = 10.4°$, respectively (applying formulas II for system A and III for system B of Tab. 1). Here the correction of miscalibration was carried out automatically by shifting the centers of gravity in both images to the nearest common reconstruction plane. The superimposed projection lines mark the reconstruction planes through the apex and a prominent bifurcation of the right pulmonary artery.

Conclusions

The described operations for biplane image pairing simplify the identification of corresponding points in both projections. The proposed standardized image display is a basis for further geometric or densitometric evaluations. The mathematical tools can be easily implemented in the cardiac workstation for use in the clinical routine.

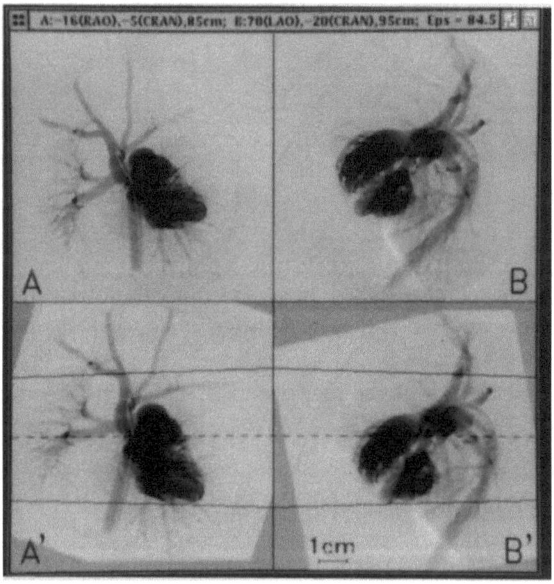

Figure 4: Enddiastolic biplane ventriculogram of a three days old girl with a ventricular septal defect. The upper panel shows the subtracted original images, the lower panel the geometrically corrected, equally scaled and aligned images with superimposed projection lines of two reconstruction planes. The dashed line marks the isocenter plane.

References

1. Wollschläger, H.; Lee, P.; Zeiher, A.; Solzbach, U.; Bonzel, T.; Just, H.: Mathematical tools for spatial computations with biplane isocentric x-ray equipment. Biomedizinische Technik 31(5) (1986) 101-106.

2. Dumay, A.C.M.: Image reconstruction from projections. Ph.D. thesis, Technical University Delft, Den Haag: CIP-Data Koninklijke Bibliotheek 1992.

3. Onnasch, D.G.W.; Prause, G.P.M.: Geometric image correction and iso-center calibration at oblique biplane angiographic views. Computers in Cardiology 1992, IEEE Comp. Soc. Press, Los Alamitos (1992) 647-650.

4. Wahle, A.; Oswald, H.; Wellnhofer, E.; Beier, J.; Schulze, G.A.; Fleck, E.: Accuracy in three-dimensional reconstruction of coronary vessel systems from biplane angiographic views. Computer Applications to Assist Radiology S/CAR'92, Symposia Foundation, Baltimore (1992) 695-699.

Reconstruction of the 3D Structure of Cerebral Vessel Trees from few DSA Projections

M.BAHNER, J.DICK, B.KARDATZKI, H.RUDER, M.SCHMIDT, A.STEITZ

Lehr- und Forschungsbereich Theoretische Astrophysik Universität Tübingen, Arbeitsgruppe Medizinische Bildverarbeitung

Summary

A new model based approach is presented which permits the 3D reconstruction of blood vessels with multiple branches from few DSA projections by iteratively projecting single truncated cones, local optimization of an appropriate cost function and with the usage of anatomical constraints. In contrast to voxel based methods [5] our algorithm produces and works on a symbolic object representation. Simulations on synthetic data sets and real projections of a MRA/DSA flux phantom confirmed the feasibility of our approach. As a simulation result it has turned out that usually not more than six accurate projections are necessary to reconstruct complex synthetic vessel structures comparable to real cerebral vessel subtrees. Due to inadequate DSA imaging devices difficulties arise from inaccuracies of the camera parameters and asynchronously taken projections. With the use of a fiducial marker system for the reconstruction of the camera parameters and image preprocessing, i.e. distortion correction and segmentation of the DSA images, these problems can be overcome.

Introduction

The global three dimensional reconstruction of the brain from just a few projections is - expressed in a mathematical term - a highly underdetermined problem and not feasible in a reasonable manner. If, however, the objective of the reconstruction is limited to sparse structures in the 3D-volume as represented by the segmented cerebral vessel structures in DSA projection images or subregions thereof, the task is considerably simplified.

A prior condition is the assumption that the cerebral vessel tree is a structure in the 3D volume which can be sufficiently approximated with a sequence of generalized cones.

The basic idea for the reconstruction method is to iteratively construct a three dimensional object, whose shades exactly match the original projections as obtained from the imaging devices. In addition the algorithm uses the anatomical and topological properties of blood vessels as essential constraints, e.g. the vessel diameter and curvature radius just changes slightly and the vessel diameter is small in comparison to the average vessel distance.

Prerequisites

Imaging geometry

The algorithm's basic operation is the projection of a test vessel element on each DSA image and its valuation by comparing this projected shade with the underlying image. Therefore the accurate knowledge of the camera geometry of each DSA projection is substantial. To apply the method on real data an external marker system is used to permit the reconstruction of the imaging geometry because a commercial DSA equipment does not record this information in a practicable manner. The marker system consists of 20 aluminium pellets and is fixed to the patient's head. This technique has two major advantages: any conventional X-ray angiographic equipment can be used and the immobilization of the patient is only neccessary during the data aquisition time to avoid motion artefacts. The applied algorithm for computing the imaging geometry requires at least six different image coordinates of marker points and their corresponding 3D coordinates as input to produce practicable results.

Image preprocessing

Another important condition for a successful reconstruction procedure is the preparation of the distortion corrected 2D DSA images by several segmentation routines. In a preceding step the extraction of blood vessels and blood vessel properties must be accomplished [7].

Algorithm

The algorithm consists of the minimization of a local cost function and a global three dimensional tracking procedure to manage geometrical ambiguities.

The local cost function is a combination of the weighted valuation results of the projection of a test vessel element into each reference image.

For a truncated cone as the test vessel element the free parameters can be depicted as:

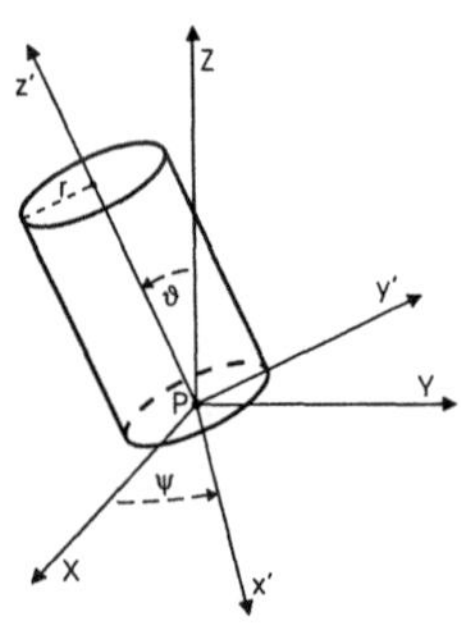

Figure 1: Truncated cone as a test vessel element where r is the upper radius and ψ and ϑ are Euler angles ($\varphi = 0$).

The minimization of the local cost function can be written as

$$C(\psi,\vartheta,r) = 1 - G(\psi,\vartheta,r)\cdot\prod_{i=1}^{N}(1 - g_i\cdot f_i(\psi,\vartheta,r)) \stackrel{!}{=} \min \qquad \text{(N=number of images)}$$

where g_i provides the possibility to weight individual projections and $G(\psi,\vartheta,r)$ represents global constraints on the model parameters.

For binarised reference images, a pixel overlap value can be used for valuation:

$$f_i(\psi,\vartheta,r) = 1 - \frac{n_s + n_r - n_x}{2\cdot n_s}$$

where n_r is the number of black pixels in the working area (a), n_s is the number of pixels in the shade of the projected test vessel element (b) and n_x is the number of black pixels in the XORed image (c).

(a) (b) (c)

Figure 2: Pixel overlap

At a bifurcation point or in case of topological ambiguities there may appear several significant relative minima for different parameter sets beside the absolute minimum of the cost function.

The flow graph shows the central part to reconstruct a single vessel element:

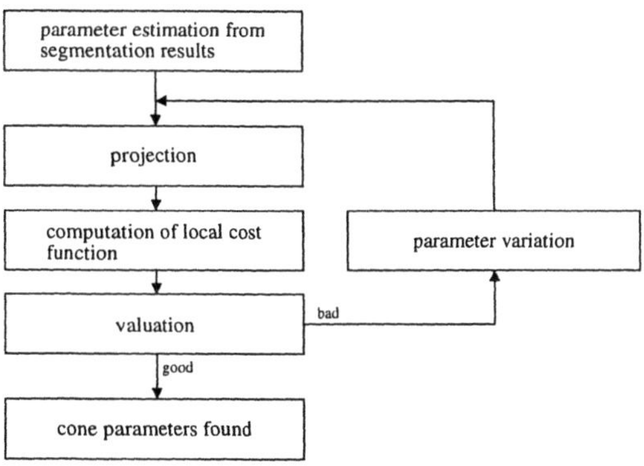

Figure 3: Reconstruction of a single vessel element

The reconstruction process proceeds on two iteration levels. In the inner iteration the minimization step returns the parameters of a single truncated cone which defines the starting point for the next vessel element. The outer iteration sequentially connects the single elements to a 3D vessel branch, and globally resolves geometrical ambiguities.

Bifurcations are directly given as special keypoints from the segmentation methods[7] and are validated as ambiguities in the base reconstruction step. At each detected bifurcation point a reconstruction procedure starts for all identified branches.

Preliminary results and conclusion

The following examples show the feasibility and functionality of two major parts of our algorithm:

- **Calculation of imaging geometry** Figure 4 shows the projection of our flow phantom with integrated X-ray marker system. The marker system consisted of eight aluminium pellets with 2.5 mm diameter. The pellets were arranged at two different z-planes of the phantom (z-distance $= 67.2 \pm 0.2mm$) with distances in x and y direction of (60.0 ± 0.3) mm. The resulting average calculation error for the camera model of this image was 2.7 pixel.

  Different projection geometries were achieved by rotating the phantom.

- **Reconstruction of a 3D object** A helix has been chosen as an easy but sufficiently complex 3D object to show that the reconstruction process resolves local ambiguities in the projections.

  The reconstruction result as presented in figure 6 has been obtained with $N = 6$ synthetic images. A sample of them is shown in figure 5.

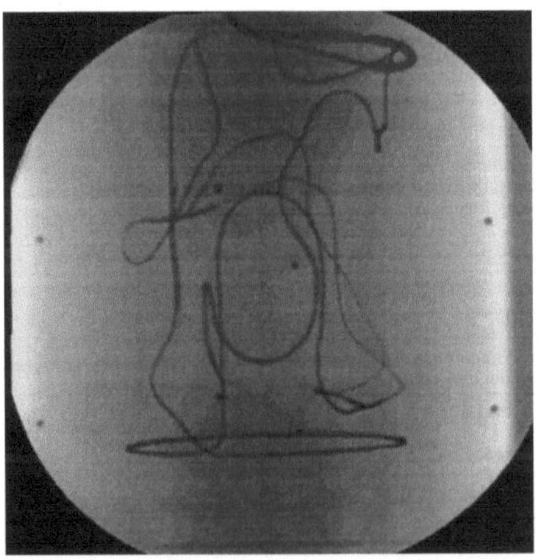

Figure 4: DSA mask projection image of an acquisition set of our combined MRA-DSA flow phantom (recording system: Siemens POLYTRON).

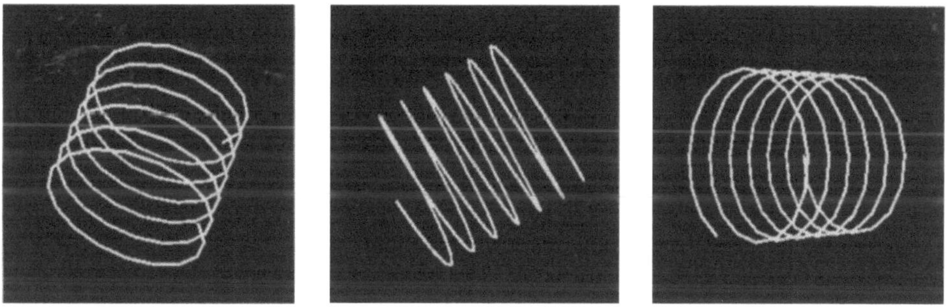

Figure 5: Synthetic projection images for the helix reconstruction.

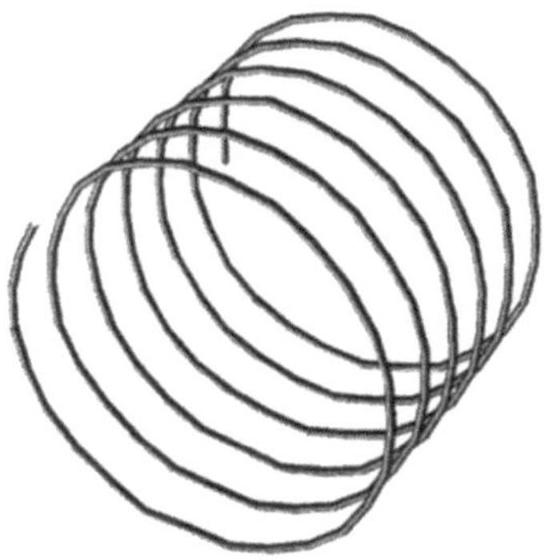

Figure 6: Reconstructed helix.

We pointed out that 3D reconstruction for topologically complex synthetic vessel trees from few projections is possible and yields correct results. The projection geometry for DSA images from a patient fixed marker system can be computed with sufficient accuracy. Our results indicate that the described approach is feasible for real DSA images of the cerebral vessel tree as well.

The produced symbolic description of the vessel tree can serve multiple purposes:

- presentation of the topology of the vessel tree or single selected branches.
- fusion of the 3D reconstructed DSA images with images received by other medical imaging modalities. Therefore a common system of fiducial markers or corresponding anatomical landmarks are prior conditions.
- correction of MRA image distortions caused by flow phenomena and magnetic field inhomogenities.
- geometric information for positioning systems in stereotaxis

References

[1] Barrillot, C.; Glibaud, B.; Scarabin, J.-M.; Coatrieux, J.-L.: 3D reconstruction of cerebral blood vessels. *IEEE Computer Graphics & Applications*, 1985.

[2] Brent, R.P.: Algorithms for minimization without derivatives. Englewood Cliffs, New Jersey: Practice Hall, 1973.

[3] Bresler, Y.; Macovski, A.; 3D reconstruction from projection with incomplete and noisy data by object estimation. *IEEE Transactions ASSP*, (35), 1987.

[4] Kitamura, K.; Tobis J.M.; Sklansky J.: Estimating the 3D skeletons and transverse areas of coronary arteries from biplane angiograms. *IEEE Transactions on Medical Imaging*, 7(3), 1988.

[5] Klotz, E.; Haaker, P.; Koppe, R.; Linde, R.: 3D reconstruction of vascular structures from few X-ray projections. *Computer Assisted Radiology CAR 91*, Berlin, 1991, 55-61.

[6] Krestel, E. (Hrsg.): *Bildgebende Systeme für die Medizinische Diagnostik*, 2. Auflage. Berlin; München: Siemens-Aktiengesellschaft, 1988.

[7] Kutka, R.; Stier, S.: Fast extraction of line properties. *Proc. Computer Assisted Radiology CAR 93*, Berlin, 1993.

[8] Rake, S.; A model-based approach to the automatic reconstruction of three-dimensional arteries from biplane angiograms. Tech. Rept. 239, UKSC Report, IBM UK Scientific Center, Winchester, August, 1990.

[9] Smets, C.: A knowledge-based system for the automatic interpretation of blood vessels on angiograms. Universität Leuven, 1990.

[10] Smets, C.; Vandermeulen D.; Suetens, P.; Oosterlinck, A.: A knowledge-based system for the 3D reconstruction and representation of the cerebral blood vessels from a pair of stereoscopic angiograms.

[11] Suetens, P.: A three-dimensional image of the cerebral blood vessels for use in stereotactic neurosurgery, Universität Leuven: ACCO, 1983.

[12] Trousset, Y.; Picard, C.; Saint-Felix, D.; Rougee, A.: Three-dimensional computerized angiography, *Computer Assisted Radiology CAR 91*, Berlin, 1991, 68-72.

[13] Venaille, C.; Mischler, D.; Collorec, R.; Catros, J.Y.; Coatrieux, J.-L.: Automatic 3D reconstruction of vascular networks from three projections: A simulated annealing approach. *IEEE Eng. in Medicine & Biology*, 11th Annual Int. Conf., 1989.

[14] Wu, J.; Parker, D.L.: Three-dimensional reconstruction of coronary arteries using more than two projections. SPIE Vol. 1233, 1990, 77-84.

Digital Radiographic Analysis of Single-Leg Separations (SLS) in the Outlet Struts of Björk-Shiley Convexo-Concave (BSCC) Heart Valves

K.A. Powell, J.F. Cornhill, S.N. Nissen, J.G. Chandler, A.H. Abolfathi, E.L. LaPresto, E. Herderick, K. Mueller, E. Cesmeli, S. Chandra
Cleveland Clinic Foundation, Shiley Heart Valve Research Center, and the Ohio State University, USA

Summary

An automated radiographic image processing technique has been developed for the identification of SLS in the outlet struts of BSCC heart valves. Clinical studies of explants and SLS valves implanted in sheep suggest that one leg of the outlet strut separates from the valve flange some time prior to the second leg breaking that results in sudden valve failure. Preliminary analysis of digitized (approximately 36.5 microns/pixel spatial resolution) 35 mm cineangiogrpahic data of BSCC mitral valves in the first 60 patients of a 300-patient study indicates that the analysis of differential pixel intensity along the long axis of the leg of the outlet strut is a sensitive measure of outlet strut integrity of these valves. A rapid automated evaluation technique is necessary in order to process the multi-frame cine data. The following criteria were considered in the development of the image processing software:

1) high-speed processing of multi-frame data,
2) minimal user interaction,
3) reproducibility,
4) ease of use, and
5) workstation platform independence.

An image classification technique based on Bayesian statistical theory is used for the rapid accurate identification of outlet strut outlines in the images. A method for automated temporal registration of sequential frames of data is implemented by aligning individual outlet struts along the long axis midline of the strut and the intersection of this midline with the curvature of the valve in the region of the outlet strut flange. Preliminary analysis of the patient data also indicates that a possible increase in specificity of SLS identification in BSCC mitral valves may be achived by digitizing the radiographic images at an 8-fold increased magnification (approximately 4.5 microns/pixel). Quantitative measures of temporal cues such as independent leg motion are also being evaluated as possible methods for identification of SLS in these radiographic images.

Multimodality Imaging

Visualization and Analysis of Brain Function Using Multi-Modality Image Integration

C.-T. CHEN, M. COOPER, J. METZ, J.-S. CHOU, C. E. ORDONEZ, M. JIANG, S.-Y. CHEN, D. DOOLEY, M. WERNICK, X. OUYANG, X. YU, T. BROWN, and J. YAP

PET Center/Frank Center for Image Analysis
Franklin McLean Memorial Research Institute
Department of Radiology, The University of Chicago
MC-1037, 5841 South Maryland Avenue, Chicago, Illinois 60532, U.S.A.

Summary

Multi-modality image integration, in which functional positron emission tomography (PET) and single-photon emission computed tomography (SPECT) images are co-registered spatially with structural x-ray computed tomography (CT) and magnetic resonance (MR) images, can provide new knowledge of correlates of brain function and brain structure that was difficult to access previously. For visualization and analysis of brain function, we have used this concept in the following areas: (1) Image reconstruction: We have developed a Bayesian data augmentation method that uses Gibbs prior for incorporation of high-resolution structural information extracted from the correlated CT and MR images in reconstruction of functional PET images. (2) Image segmentation and visualization: We have developed multi-spectral approaches for segmenting brain images into entities of functional/structural importance based on the integrated PET, CT, and multiple MR images. These segmented regions can be displayed in both 2-D and 3-D by using various visualization techniques. (3) Brain function mapping: Brain activation using PET has become a powerful tool for mapping brain functions in health and disease. Image subtraction techniques applied to a pair of images acquired respectively under baseline and activated conditions form the basis for detecting subtle and localized brain functions. Image integration is useful in correction for possible shifts in subject position between two separate scans, in quantitative interpolation between slices for localizing small and subtle brain functional signals, and in overlaying of the detected brain function signals with structural MR or CT images for providing accurate localization of the physiology/anatomy correlates. These image integration approaches provide improvements in reconstruction accuracy and useful new tools for visualization, as well as powerful new techniques for brain mapping. When used in combination, understanding of certain brain functions that were difficult to assess previously becomes accessible.

Multi-modal Image Reconstruction

We have developed new Bayesian algorithms for PET image reconstruction using Gibbs models and the concept of line process [1,2]. The Gibbs models are designed to incorporate *a priori* information that can be employed for improving the image quality. One potentially fruitful source of *a priori* information in PET image reconstruction is provided by CT and MR images that have high resolution, high image-contrast, and superior noise properties. These structural images can be brought into precise spatial registration with PET images by means of appropriate correlative computations, for example, the surface-matching method that was developed previously in our laboratories

[3]. Once multiple sets of images are in registration in space, a detailed anatomical map corresponding to each PET slice can be extracted from these spatially-correlated CT or MR images. This anatomical map provides a rich source of *a priori* information that can be used with the Bayesian methods as an aid in providing more accurate PET image reconstruction. The high-resolution anatomical maps extracted from the spatially-correlated structural images can be employed in the Bayesian methods for enhancement of boundaries that separate regions with different radioactivity concentrations in the corresponding PET images [4]. This new approach offers novel strategies for taking advantage of the high quality of CT and MRI images and integrating them into the reconstructed PET images. This is particularly useful in improving the detection and quantitation related to small or low-contrast regions. New information, unattainable previously, will become accessible because of this new multi-modal image reconstruction approach.

One can also use the spatially-correlated anatomical maps for the development of new methods that correct for the effect of photon attenuation. The effect of photon attenuation is one of the most detrimental factors affecting the quality of reconstructed PET images. In a typical PET study, a transmission scan is performed which produces data that one can use to derive attenuation correction factors for individual coincidence detector pairs. Because the number of photons collected in these transmission measurements is relatively limited, the noise associated with these attenuation correction factors tends to degrade the quality of the reconstructed PET image significantly. We have developed new techniques that use the anatomical maps extracted from the spatially-correlated CT and MR images to derive attenuation coefficent maps for the corresponding PET images [5]. These attenuation coefficient maps are more accurate and less noisy than those from other techniques. Therefor, this unique approach is expected to produce more accurate reconstructed PET images with improved noise properties. It has been demonstrated that noise level can be reduced as much as 30 - 50% by using this type of new approaches for attenuation correction.

Multi-modal Image Segmentation and Understanding

We have developed an adaptive slit-and-merge image segmentation algorithm for the initial segmentation step in the expert vision system [6-8]. This is a region-based segmentation algorithm which combines the strength of characteristic feature analysis and hypothesis model to produce an initial segmentation. All of the model parameters used in the algorithm are computed automatically based on the characteristic features extracted from the window areas and they only depend on the context of the image under analysis. The computed parameters provide the hypothesis model with appropriate constraints to test the homogeneity in the region. In the hypothesis model, either a heuristic criterion or a statistical test is applied according to the feature distribution of the region in analysis. Two processes, the initial region creation and the final region formation, are implemented in the split-and-merge segmentation scheme based on the hypothesis model. The spilting process is performed independently in each window, while the merging process for adjacent regions is carried out globally.

We have employed the Dempster-Shafer (D-S) theory for integration of uncertain information about the brain anatomy, brain function, imaging devises, and expert's knowledge of interpreting brain function and images [9-11]. The D-S theory employed in this system incorporates the concept of compatible frames and multivariate belief functions. These two features form the core of the reasoning scheme in the expert vision system. This system is capable of using the reaoning process of a human expert to divide a set of CT, MR, and PET images into semantically meaningful entities. The blackboard architecture is selected as the framework to implement this system. Different kinds of evidence provided by various knowledge sources form a hierarchy of evidential space to which the D-S theory is applied. Two algorithms that deal with the computation of hierarchical evidence have been investigated. Preliminary studies, in which a set of spatially correlated CT image, T1-weighted and T2-weighted images are used as the basis for delineating bone structure, subcortical areas, gray matter, and white matter, has demonstrated the feasibility of this approach.

Brain Mapping

A technique pioneered by Fox *et al.* [12-14] have demonstrated that cortical localization of neuronal activation evidenced by changes in regional cerebral blood flow (rCBF) can be achieved with a precision which is not limited by the spatial resolution characteristics of the imaging system. This method employs repeated injections of boluses of O-15 labeled water and PET for measuring rCBF under various physiologic/cognitive conditions. This method provides for multiple repeat studies within reasonable time periods, which allows for averaging techniques to be applied for enhancing signal detection from improved signal-to-noise ratio. Repeat studies are necessary since the basis of localization of signal is image subtraction and local-maximum sampling. The precision of localization is achieved from application of a center of mass algorithm to the regional change signal. An advantage for data analysis is the reduction achieved in the data set since only changed areas are sampled. The data is then further reduced by statistical analysis for discrimination of signal from noise in the change distribution image. The change images with the reduced data set of ROI's described by three location coordinates and a magnitude are converted into stereotactic coordinate space for anatomical localization. For individuals, change images can be created and searched before conversion; but for group data averaging, anatomical normalization must be achieved first. This approach provides a powerful tool for brain mapping research.

The approach proposed by Fox *et al.* is dependent upon the use of a lateral x-ray image and the stereotactic atlas for structural localization and image transformation [15]. However, if functional PET images could be brought into registration spatially with structural CT or MR images from the same subject, the information of structure/function correlation can be given on the individual basis, thus providing more accurate localization and detection. In addition, Fox and his colleagues did not address the issue of possible shift in subject position between repeat scans. In order to meet this need, we used the surface-matching technique developed in our laboratories [3] for integrating CT, MR and PET images, as well as for registering two sets of emission images. It has been repeatedly demonstrated that, although the subject in a PET study is usually immobilized by a face mask attached to the head holder during repeat scans, there is still a significant transaxial shift of the order of 4-6 mm, a lateral move of approximately 1-3 mm, and a slight tilt of a 1 - 2 degree. This position shift can be corrected for *a posteriori* by using the surface-matching method for image registration.

We have performed a series of preliminary studies using FDG/PET method to localize changes in cerebral glucose metabolism as a response to visual and somatosensory stimulations [16]. Using a pair of sets of images obtained under two different conditions, (i.e., visual and somatosensory stimulations, respectively), merely performing direct subtraction without any processing would produce a set of difference images which are very noisy and no obvious signals can be identified. We tested a normalization procedure in which only gray-matter regions were considered. After applying a statistical z-transformation for each pixel identified as in the gray-matter regions, and a registration step to bring two sets of emission PET images into congruence, a set of difference images shows clear signals in the occipital cortex area which corresponds to the response to stimulation given to the visual field, and in the tempero-parietal regions which are indications of responses to somatosensory stimulations. By integrating these difference images with spatially-correlated CT and MR images, the function/anatomy correlates can be easily localized and identified. These techniques used in combination form the basis of a powerful tool for brain mapping.

References

1. Johnson, V.E., Wong, W.H., Hu, X., and Chen, C.-T.: Image restoration using Gibbs priors: boundary modelling, treatment of blurring, and selection of hyperparameters. IEEE Trans. Pattern Analysis and Machine Intelligence PAMI-13 (1991) 413-425.

2. Chen, C.-T., Johnson, V.E., Wong, W.H., Hu, X., Metz, C.E.: Bayesian image reconstruction in positron emission tomography. IEEE Trans. Nucl. Sci. NS-37 (1990) 636-641.

3. Pelizzari, C.A., Chen, G.T.Y., Spelbring, D.R., Weichselbaum, R.R. and Chen, C.-T.: Accurate three-dimensional registration of CT, PET and MR images of the brain. J. Comput. Assist. Tomogr. 13 (1989) 20-26.

4. Chen, C.-T., Ouyang, X., Wong, W.H., Hu, X., Johnson, V.E., Ordonez, C.E., and Metz, C.E.: Sensor fusion in image reconstruction. IEEE Trans. Nucl. Sci. NS-38 (1991) 687-692.

5. Yu, X., Chen, C.-T., Bartlett, R., Pelizzari, C.A., and Ordonez, C.E.: Using correlated CT images in compensation for attenuation in PET image reconstruction. In: SPIE Proceedings Vol. 1396 (1990) 56-58.

6. Chen, S.-Y., Lin, W.-C., and Chen, C.-T.: Split-and-merge image segmentation based on localized feature analysis and statistical test. CVGIP: Graphical Models and Image Processing 53 (1991) 457-475.

7. Chou, J.-S., Chen, C.-T., Chen, S.-Y., and Lin, W.-C.: Three-dimensional adaptive split-and-merge method for medical image segmentation. SPIE Proceedings Vol.1660 (1992) 439-445.

8. Chou, J.-S., Chen, C.-T., Chen, S.-Y., and Lin, W.-C.: Parallel implementation of an adaptive split-and-merge method for medical image segmentation. SPIE Proceedings Vol.1652 (1992) 62-67.

9. Chen, S.-Y., Lin, W.-C., and Chen, C.-T.: An expert image analysis system for recognition of major brain structures. (Submitted to IEEE Trans. Med. Imaging).

10. Chen, S.-Y., Lin, W.-C., and Chen, C.-T.: Medical image understanding system based on Dempster-Shafer reasoning. In: SPIE Proceedings Vol. 1445 (1991) 75-77.

11. Chen, S.-Y., Lin, W.-C., and Chen, C.-T.: Spatial reasoning based on multivariate belief functions. 1992 Proceedings of Computer Vision and Pattern Recognition (CVPR '92), pp.624-626, 1992.

12. Fox, P.T., Mintun, M.A., Raichle, M.E., Miezin, F.M., Allman, J.M. and Van Essen, D.C.: Mapping human visual cortex with positron emission tomography. Nature 323 (1986) 806-809.

13. Fox, P.T., Burton, H. and Raichle, M.E.: Mapping human somatosensory cortex with positron emission tomography. J. Neurosurg. 67 (1987) 34-43.

14. Fox, P.T., Miezin, F.M., Allman, J.M., Van Essen, D.C. and Raichle, M.E.: Retinotopic organization of human visual cortex mapped with positron emission tomography. J. Neurosci. 7 (1987) 913-922.

15. Fox, P.T., Perlmutter, J.S. and Raichle, M.E.: A stereotactic method of anatomical localization for positron emission tomography. J. Comput. Assist. Tomogr. 9 (1985) 141-153.

16. Metz, J.T., Chen, C.-T., Jiang, M., Chou, J.-S., and Cooper, M.: Detection of physiological (PET) signals in individual subjects. Scoiety for Neuroscience Abstracts 18:865, 1992.

An Object Oriented Tool for 3D Multimodality Surface-based Image Registration

André COLLIGNON, Dirk VANDERMEULEN, Paul SUETENS, Guy MARCHAL

Interdisciplinary research unit for radiological imaging[1].
Department of Electrical Engineering, ESAT,
Kardinaal Mercierlaan 94, B-3001 Heverlee, and
Department of Radiologie, University Hospital Gasthuisberg
Herestraat 49, B-3000 Leuven
Katholieke Universiteit Leuven, Belgium
tel: +32-16-34 37 46, fax: +32-16-22 18 55
e-mail: andre.collignon@kuleuven.ac.be

Summary

Three surface-based registration algorithms using different point projection-based matching criteria and using different distance minimization strategies are analyzed: Pelizzari's head-hat algorithm, a variant of Borgefors's hierarchical chamfer matching, and Besl and McKay's iterative closest point algorithm. This work is part of COVIRA (Computer Vision in Radiology), project A2003 of the AIM (Advanced Informatics in Medicine) programme of the European Commission. COVIRA-specific object oriented aspects of the algorithms' implementations are briefly discussed. Finally they are compared with respect to speed, accuracy and flexibility. This comparison is based on the registration results of the COVIRA reference data sets (MRI/MRA and MR/CT), and a database of MR/PET data sets.

1. Introduction

The development of medical work stations for medical diagnosis and therapy planning based on multimodality image analysis requires registration algorithms for defining the geometrical relationship between the different image data sets. Existing registration algorithms can be subdivided in patient-intrinsic (using anatomical markers) and patient-extrinsic (using external artificial markers). Patient-intrinsic methods are advantageous since they are patient-friendly, do not require special provisions during imaging, and can thus be used retrospectively. We have developed a patient-intrinsic, surface-based registration software tool using the Iconic Core, an object oriented implementation of the emerging Image Processing and Interchange (IPI) standard. Currently the tool supports crisp surfaces only, i.e. surfaces of binary objects. Three algorithms have been selected for integration in this tool: Pelizzari's head-hat algorithm[10], a variant of Borgefors's hierarchical chamfer matching[2,3], and Besl and McKay's iterative closest point quaternion-based algorithm[1]. Answers to the following questions are presented in the next sections: 1) Which one of the selected surface-based registration algorithms is the fastest, the most accurate, and the most flexible? 2) What advantages can object oriented technology offer to facilitate the implementation of a registration tool? 3) How accurate can surface-based registration be for different combinations of MR, CT and PET images?

[1]Directors: A.L. Baert, A. Oosterlinck

2. Surface-based Registration Algorithms

This section gives a short overview of the important components of the three selected surface-based registration algorithms. Some of the components are discussed in detail in order to highlight the algorithmic differences and implementation aspects.

2.1. Components of Surface-based Registration Algorithms

Registration algorithms have been classified by Vandermeulen et al[12], and this classification has been extended with references to recent literature by Van den Elsen et al[11]. Surface-based registration algorithms have been analyzed in detail by Collignon et al[4,5]. This analysis reveals that surface-based registration algorithms can be characterized by the following features: surface representation, geometric transformation, matching criterion, optimisation algorithm, outlier treatment, and accuracy. Since surface representations are always chosen in function of implementation considerations and since they are essentially equivalent they will not be discussed here. The accuracy of the different algorithms will be the subject of section 4.

2.2. Geometric Transformation

Pelizzari allows for 3D translation, rotation, and non-isotropic scaling, i.e. 9 degrees of freedom, each of which can be preset by the user. Our extension of the chamfer matching algorithm to 3-D (see also Mangin et al[9], Jiang et al[8]) also allows for 9 degrees of freedom, that can be preset by the user. Besl and McKay's algorithm allows for 6 degrees of freedom only, i.e. 3D translations and rotations. It can be argued that this is sufficient for registration of brain scans because the brain size is fixed of course and because voxel sizes of all related images are available. This might not be entirely true due to the unavailability of identical surfaces. It may be concluded that Besl and McKay's algorithm is less flexible with respect to the geometric transformations allowed during registration.

2.3. Matching Criterion

Two families of matching criteria can be distinguished: moment based and point projection based. Moment based matching criteria measure the registration quality by means of the difference between multiple moments of surfaces or volumes, or derived geometric features (e.g. skeletons). Point projection based criteria use a representative selection of points on the surface of the object to be matched in the low resolution image and a specific projection rule to project them on the same surface in the other image. In this way a list of corresponding points is built up and the corresponding distances are averaged to represent a single component global measure of registration quality. Pelizzari specifically suggests that a minimum of about 250 points should be used. His criterion specifies that the projection should be along straight lines through the points and the centroid of the projection surface. Both Borgefors, and Besl and McKay prescribe the closest point projection rule. The difference between the latter two matching criteria is that Borgefors uses the chamfer distance transform to calculate an approximation of the point distances, while Besl and McKay, as does Pelizzari, calculate the exact Euclidean distances.

2.4. Minimization Strategy

In this study, surface-based registration amounts to the minimization of the point projection based average distance in function of the registration parameters.
Both Pelizzari and Borgefors reason as follows: "Because the distances defined above do not have analytical expressions, general n-dimensional optimization algorithms that do not use derivative information are required to solve the problem. The registration parameters are the optimization variables." In practice the distance function has multiple local minima. Therefore a complete minimization method has two components, i.e. an accurate locally converging algorithm (e.g. Powell's or steepest descent) is used under control of a global optimization algorithm (e.g. simulated annealing, hierarchical optimization).Currently we use Powell's optimization algorithm with both Pelizzari's head-hat distance and Borgefors's chamfer distance, and we do not use a global optimization algorithm because sufficiently accurate estimates of the registration parameters are available.
Besl and McKay on the other hand reason as follows: "If both surfaces are in matching locations then the selected points and their projections will coincide. So, in each iteration, let us consider the projections to be the true corresponding points. In that case in each iteration a point based registration problem with point correspondence information has to be solved instead of a surface based registration problem. So, a least squares estimation of the registration parameters can be calculated." Besl and McKay use quaternions for this calculation. But they also mention the calculation based on a singular value decomposition discussed by Faugeras and Hebert[6]. We used the quaternion-based solution.

2.5. Outlier Treatment and Partially Overlapping Surfaces

Application of surface-based registration algorithms is meaningful only if the following two fundamental problems are taken into account: 1) incorrect segmentation of surface parts has a non-negligible influence on the registration accuracy. 2) the same surfaces may not be completely overlapping in both scans. Points selected for distance calculation may be lying on non-overlapping parts of the surfaces and thus distort the global surface distance calculation. A single solution for both problems was proposed by Collignon and Géraud[7] which is far more superior to the simple threshold proposed by Jiang et al[8]. This solution is an automatic threshold selection algorithm using an optimal approximation of the curve representing the set of increasing point projection distances by a piecewise-linear curve.

3. Object Oriented Implementation

The distributed nature of the entire COVIRA project team and the size of the project make the application of object oriented technology (OOT) very attractive. Indeed it already turned out that integration of software modules that did not use OOT was considerably more difficult. Because different C++ compilers have been used by different project partners some incompatibility problems needed to be overcome, the most important one of which was the difference in template handling. Using the OO approach also offers the possibility to make the user interface more flexible. E.g. intelligent interface objects that control automatic registration or suggest the invocation of specific segmentation methods in order to facilitate the registration process with a minimal amount of work. Such a suggestion could be function of the specific data sets to be registered and the requirements of the clinical application.

4. Evaluation of the Tool

Evaluation of the tool is performed on two levels: on a technical level in order to determine the accuracy of the different algorithms, and on a clinical level in order to determine its value for specific clinical applications, e.g. registration of multimodal data sets in support of radiotherapy planning. The surface based registration tool is intended to be used for matching 3D MRI, MRA, CT and PET images. The tool can be used only if common surfaces can be segmented in all these modalities. Whether this condition can be met is considered part of the evaluation task. The evaluation is based on the registration of real world images. The COVIRA reference data set consists of an MRI/MRA image pair of a patient's head and of an MRI/CT pair of another patient. These data sets were acquired with stereotactic reference frames. For evaluation of matching with PET we use MRI/PET image pairs marked with fiducial markers.

4.1. Technical Evaluation

Technical evaluation of the tool is currently going on. Skin or skull surfaces have been segmented by two different plane by plane thresholding programs. In the near future the same surfaces will be segmented using COVIRA segmentation objects. The need for interactive editing of contours has clearly been established for surface segmentation, especially during segmentation of complete 3D images (128 planes of 256 x 256 pixels). The accuracy of the different algorithms can be evaluated by comparison of the corresponding registration results with those obtained using the available external markers. For example, the registration matrices obtained for the COVIRA MRI/MRA reference data set based on the frame and on our version of the chamfer matching algorithm respectively were both applied to the mass center of the head surface in the MRA image. The resulting positions in the MRI image coordinate system were [129.6 132.2 70.40] and [131.1 132.5 78.21] respectively. This corresponds to a difference of [1.65 0.33 8.59] mm in patient space. The rather large component of the error along the longitudinal axis is probably due to the bad quality of the surfaces (skin in MRI versus fat in MRA). Figure 1 gives a visual impression of the surface based registration quality using chamfer matching. Similar results were obtained with the MRI skin and the PET skull surfaces. In summary, experience teaches us that the accuracy of registration algorithms is limited by the lowest resolution of the images to be registered and by the quality of the surface segmentation algorithms used.

4.2. Clinical Evaluation

Clinical evaluation of the tool is currently in the planning stage. Detailed evaluation procedures are being developed for specific diagnostic, neurosurgery and radiotherapy applications. The actual evaluation will start mid 1993.

5. Conclusion

All surface based registration algorithms have comparable accuracy, determined by the resolution of the images involved, and more importantly by the quality of the segmentation of the surfaces. Differences in efficiency of the algorithms are negligible if segmentation times are taken into account and if good initial estimates for the registration parameter values are available, which is always the case in clinical practice. The extra flexibility offered by scaling parameters can be used to increase registration accuracy in specific cases. Future work will be concerned mainly with further evaluation of the algorithms both technically and clinically. More research will also be done in the domain of application specific segmentation algorithms for surface based registration, in the development of registration algorithms that can input a combination of corresponding points, crisp surfaces, and probabilistically defined surfaces, and in the development of visualization algorithms especially suited to the interactive evaluation of registration accuracy.

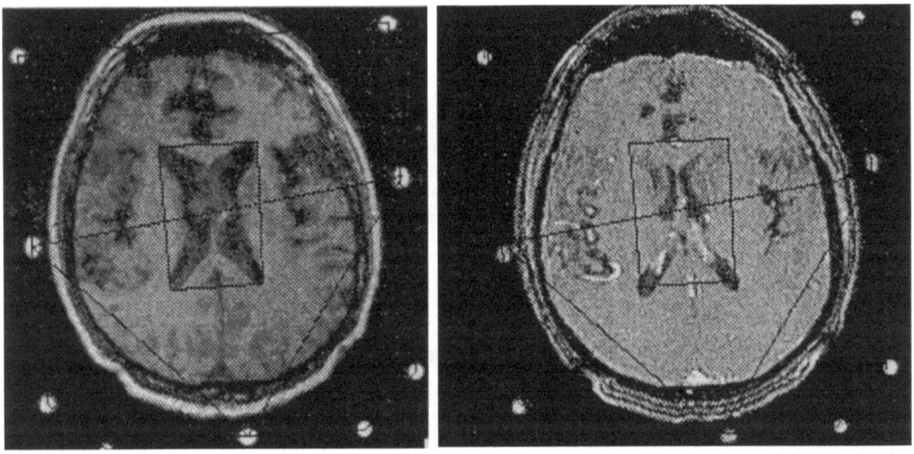

Figure 1: Left: resliced MRI t1-weighted plane with lines indicating the frame marker points and the ventricles added interactively in overlay. Right: corresponding original MRA plane with the same lines overlaid using a linked cursor. The large registration error perpendicular to the image is reflected in the rotation of the triangle connecting the variable frame marker points and in the differences in position and outline of the ventricles.

Acknowledgment

Special thanks go to Jan Baetens for matching the MRI/PET data sets and for his aid in the manual segmentation of the MRA images. Special thanks also go to Johan Michiels for help with the segmentation of the surfaces.

Participants in the COVIRA consortium are: Philips; Medical Systems, Best (NL) (prime contractor) and Madrid (E); Corporate Research, Hamburg (D) Siemens AG, Erlangen (D) and Munich (D) IBM UK Scientific Centre, Winchester (UK) Gregorio Maranon General Hospital, Madrid (E) University of Tuebingen, Neuroradiology and Theoretical Astrophysics (D) German Cancer Research Centre, Heidelberg (D) University of Leuven, Neurosurgery, Radiology and Electrical Engineering (B) University of Utrecht, Neurosurgery and Computer Vision (NL) Royal Marsden Hospital/Institute of Cancer Research, Sutton (UK) National Hospital for Neurology and Neurosurgery, London (UK) Foundation of Research and Technology, Crete (GR) University of Sheffield (UK) University of Genoa (I) University of Aachen (D) University of Hamburg (D) Federal Institute of Technology, Zurich (CH)

Bibliography

1. P.J. Besl, N.D. McKay, "A Method for Registration of 3-D Shapes", IEEE Transactions on Pattern Analysis and Machine Intelligence, Vol.14, No.2, pp. 239-256, February 1992.

2. G. Borgefors, "An Improved Version of The Chamfer Matching Algorithm", *Proceedings of the 7th International Conference on Pattern Recognition*, Vol.2, pp. 1175-1177, Montreal, Canada, IEEE Computer Society Press, 1984.

3. G. Borgefors, "Hierarchical Chamfer Matching: A Parametric Edge Matching Algorithm", IEEE Transactions on Pattern Analysis and Machine Intelligence, Vol.10, No.6, pp. 849-865, November 1988.

4. A. Collignon, D. Vandermeulen, P. Suetens, G. Marchal, "Surface based registration of 3D medical images", SPIE Medical Imaging VII, Vol.1898, 10 pages, Newport Beach, California, USA, 14-19 February, 1993.

5. A. Collignon, D. Vandermeulen, P. Suetens, G. Marchal, "New high-performance 3D registration algorithms for 3D medical images", SPIE Medical Imaging VII, Vol.1898, 10 pages, Newport Beach, California, USA, 14-19 February, 1993.

6. O.D. Faugeras, M. Hebert, "A 3-D Recognition and Positioning Algorithm Using Geometrical Matching Between Primitive Surfaces", *Proceedings of the 8th International Joint Conference on Artificial Intelligence*, Karlsruhe, West-Germany, pp.996-1002, August 1983.

7. T. Géraud, I. Bloch, A. Collignon, H. Maitre, P. Suetens, "Matching Criteria for 3-D Multimodality Medical Image Registration", Internal Report, 46 pages, Katholieke Universiteit Leuven and Télécom Paris, December 1992.

8. H. Jiang, R.A. Robb, K.S. Holton, "A New Approach to 3-D Registration of Multimodality Medical Images by Surface Matching", *Visualization in Biomedical Computing*, edited by R.A. Robb, 13-16 October 1992, Chapel Hill, North Carolina, SPIE Vol.1808, pp.196-213, 1992.

9. J-F. Mangin, V. Frouin, B. Bendriem, "Nonsupervised 3D Registration of PET and MRI Data Using Chamfer Matching", *Conference on Medical Imaging*, Nuclear Science Symposium, 3 p., Orlando, Florida, October 27-31, 1992.

10. C.A. Pelizzari, G.T.Y. Chen, D.R. Spelbring, R.R. Weichselbaum, C-T. Chen, "Accurate Three-Dimensional Registration of CT, PET, and/or MR Images of the Brain", Journal of Computer Assisted Tomography, Vol.13, no.1, pp.20-26, Raven Press, Ltd., New York, January/February 1989.

11. P.A. van den Elsen, E-J.D. Pol, M.A. Viergever, "Medical Image Matching - A Review with Classification", IEEE Engineering in Medicine and Biology, in press.

12. D. Vandermeulen, *Methods for registration, interpolation, and interpretation of three-dimensional medical image data for use in 3-D Display, 3-D Modeling and Therapy Planning*, PhD Thesis, chapter 2, Katholieke Universiteit Leuven, April 1991.

Volume Rendering of Multimodal Images for the Planning of Skull Base Surgery

C.F. RUFF[1], D.L.G. HILL[1], G.P. ROBINSON[2], and D. J. HAWKES[1].

Divisions of Radiological Sciences[1] and Neurology[2],
United Medical and Dental Schools, Guy's Hospital,
London, SE1 9RT, UK.

Abstract

In this paper we present an application in which multimodal images are used to generate pseudo-3D scenes to assist in the planning and execution of skull base surgery. Image data from CT, MRI and MR Angiography are registered such that the relationship between their image coordinates is known. The images are then transformed into occupancy maps for the structures of interest. Commercially available rendering algorithms then require that either the data from the multiple imaging modalities be combined into a single volumetric dataset prior to rendering, or that the data from each modality is rendered separately and subsequently overlaid. We present an algorithm that allows us to bypass this last stage and leave the objects in their original datasets. The 3D scenes are rendered by casting a ray simultaneously through the multi-dimensional space represented by the individual datasets.

Introduction

It is important when planning surgery in the skull base to have precise knowledge of both the extent of the lesion and of its relationship to surrounding anatomical structures. Patients therefore routinely undergo imaging investigations from multiple modalities, these frequently include x-ray Computed Tomography (CT), Magnetic Resonance (MR) and either Digital Subtraction Angiography or Magnetic Resonance Angiography (MRA). It has been shown [1] that additional useful information can be made available when the datasets thus acquired are registered, transformed and subsequently displayed in a combined representation of the region under investigation. The combined data can then be displayed either as a series of 2D slices, or as pseudo-3D scenes.

Viewing the combined data as a series of 2D slices has a number of significant disadvantages. Firstly, the course of sparse, linear structures such as blood vessels and nerve fibres are poorly represented, especially when the structure follows a complex 3D path and/or there is significant noise present in the dataset. Secondly, the 3D spatial relationships between the lesion and the surrounding anatomical structures are poorly visualized.

Various pseudo-3D image rendering algorithms have been proposed to improve display of the 3D spatial relationships between adjacent anatomical structures. These algorithms fall into two general categories, surface renderers [2,3] and volume renderers [4,5]. In surface rendering the assumption is made that the data to be visualized contains certain structures to which surfaces can be fitted. In volume rendering no such assumption is made, and the effects on the rendered scene from any structure elements present within the data is calculated on a voxel by voxel basis.

In our application we require that the rendered scenes accurately portray both the spatial relationships between the anatomical structures of interest, and the integrity of the original image data. Uncertainties in tissue classification can arise from a number of sources, firstly the input data itself is digital in nature, giving rise to digitization errors both from the limited spatial sampling frequency and from the "partial volume" effect, and secondly from uncorrelated noise and image artifacts present in the input data. Uncertainties in the spatial relationship between structures derived from different modalities can arise from both image distortion in the individual modalities and from errors introduced by the registration algorithm [6]. Volume rendering techniques are more suited to the visualization of these uncertainties, as they do not require explicit surface representations.

Data Acquisition and Preprocessing

Gadolinium enhanced MR, CT and phase contrast MRA images are acquired using normal clinical protocols, with no special fixation device or markers. Voxel dimensions are typically $0.9 \times 0.9 \times 2.0$ mm for MR, $0.5 \times 0.5 \times 3.0$ mm for CT and $1.0 \times 1.0 \times 1.0$ mm for MRA. The images are then registered using an interactive point landmark based method [7]. The accuracy of this technique has been confirmed on both computer simulated and phantom data[6].

Structures of interest are then identified in the images from each modality, and an occupancy map for each structure is generated (with 0 representing no occupancy and 255 representing full occupancy). In the case of the CT images the structure of interest is the bony detail of the skull base, and the transformation from image data to occupancy map is accomplished by linear mapping of the Hounsfield numbers associated with bone into the occupancy map. The lesion, as demonstrated in the Gadolinium enhanced MR dataset, is then delineated with the assistance of a neuro-radiologist, using the "autotrace" facility in ANALYZE [8], the template thus derived is then used as a mask on the original data prior to transformation into an occupancy map. Images from MRA, are median filtered to suppress noise and then transformed into an occupancy map by linear mapping of the full gray level data within the filtered image.

The Volume Render

Volume rendering was the chosen technique for pseudo-3D image generation because of its inherent ability to visualize the uncertainty or fuzziness of the tissue characterization. However most volume rendering algorithms are optimized to work on single modality, single object datasets. If rendering of multiple objects in a single scene is required then certain compromises have to be made. This usually involves either assuming a set ordering system for the objects, i.e. skin is always encountered before bone, bone before brain etc., or compressing the gray level range available for each object (thus increasing the digitization noise) and assigning the different objects to different sections of the overall range of gray level values available. Rigid ordering of objects suffers from the drawback that objects encountered out of order will be erroneously classified and visualized. Assigning different objects different ranges in the gray levels available will overcome this particular misclassification problem, but still does not allow overlap, for example due to partial volume effects between different objects. The combined image therefore, has occupancy values for only one tissue type at each voxel location. However due to the uncertainties in tissue classification and registration we frequently find it necessary to visualize data where multiple tissue types have finite occupancy values at a given voxel, in particular where the lesion impinges on the bony structures and where the lesion itself contains vascular structures.

Aliasing artefacts in volume rendering are reduced by super sampling the input data or casting rays into the input data stochastically [9], both of which require interpolation of the data. This interpolation can introduce misclassification errors. Consider the case where soft tissue structures are assigned to the gray level range of 1-127, and bony structures to the range 128-255, with 0 representing void voxels with no structures of interest present. Then voxels on the outside of a bone / void interface may be erroneously assigned a gray level value associated with the soft tissue range when interpolation is performed.

To overcome these limitations we have developed a generalization of the volume rendering algorithm by representing different tissue types, or objects, in different datasets, and then casting a ray simultaneously through the multi-dimensional space thus represented. This ensures that, firstly, all the objects have the full range of gray level values available, and secondly, that any interpolation does not introduce misclassification.

The renderer uses a modification to the Phong surface shading model [10] where the reflected light $\Re$ form a surface element at (x, y, z) is given by:

$$\Re(x, y, z) = K_a + (K_d(\overline{\mathrm{L}} \bullet \overline{\mathrm{N}}) + K_s(\overline{\mathrm{R}} \bullet \overline{\mathrm{O}})^n)(1 - K\frac{z}{z_{max}})$$

where K_a, K_d and K_s are the fractions of ambient, diffuse and specular reflection, K is a depth shading factor, n is an exponent used in rendering highlights, and $\overline{\mathrm{L}}$, $\overline{\mathrm{N}}$, $\overline{\mathrm{R}}$ and $\overline{\mathrm{O}}$ are unit vectors in the direction of the light source, the surface normal, the reflected ray and the observer respectively, and z_{max} is the maximum depth the ray can be cast into the scene.

To generate a scene a ray is cast simultaneously through all of the object datasets and a brightness or reflection value is calculated at each voxel location and integrated along the ray. To modify the Phong surface lighting model to fit a volume rendering technique, the local gray level gradients are calculated at each voxel and the surface normal $\overline{N}$ is then estimated. The value of $\Re$ is modulated by the gradient strength and light intensity reaching the voxel, and the intensity of the light is then modulated by the occupancy value for each object type and their respective opacities, before the ray is then advanced by one unit length along the direction of $\overline{O}$ the observer, and new gradients calculated. This process is repeated until either the ray reaches z_{max} the maximum depth in the scene or the intensity of the light ray illuminating the voxels falls below a threshold value.

Results

Figure 1 shows a rendering of a sphere within a box through which a cylindrical object is passing. Figure 1a was generated with our multimodal renderer, figures 1b and 1c are from a commercial renderer Sunvision (from SUN Microsystems). The object boundaries are blurred to give a gradation of occupancy factors over a 3 pixel range from the object boundary, note how this fuzziness is better visualized in the multimodal renderer. Figures 1b and 1c shows the effect of assigning different ordering priorities for the different objects, figure 1b having the rod occupancy factor of higher precedence than the sphere, and figure 1c having these precedencies reversed causing that portion of the cylinder that lies within the sphere to be removed from the rendered scene. Figure 1d shows the tissue misclassification that can result from interpolation of data where different objects have been allocated separate ranges in the same dataset.

Figure 2 shows a patient with an anterior fossa meningioma rendered using both a commercial renderer (Sunvision from SUN Microsystems) on the left, and the multimodal volume renderer described in this paper on the right. The latter image looks slightly blurred which reflects the uncertainty in the occupancy maps, but aliasing effects are much reduced. Also note how the overlap between bone and tumour, due to partial voluming, misregistration and misclassification is faithfully reproduced.

Discussion

The volume renderer presented in this paper allows effective interpolation of multimodal data without introducing misclassification errors. The rendered scenes generated give a visual representation of the inherent uncertainties in the original data. Further work is in progress to determine whether the observer does indeed perceive an appropriate level of uncertainty in the presence and location of boundaries between structures which cannot be easily achieved with surface rendering techniques.

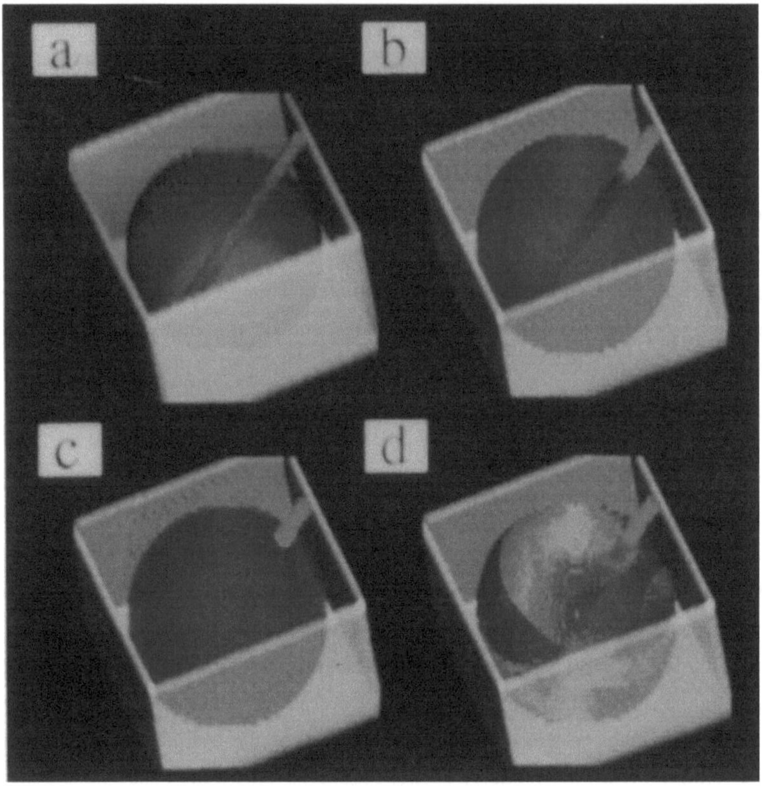

Fig. 1. Rendered scenes of test objects. 1a) generated by our multimodal renderer; 1b) generated by Sunvision (cylinder with higher priority than sphere); 1c) generated by Sunvision (sphere with higher priority than cylinder); 1d) generated by Sunvision (as in 1b, but showing misclassification due to interpolation).

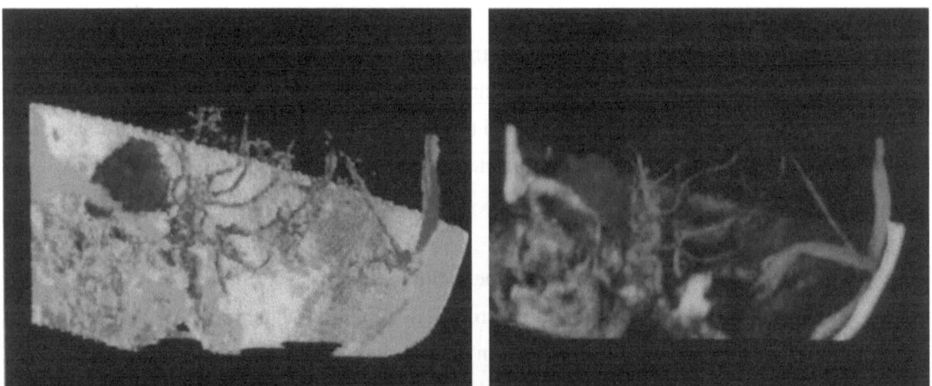

Fig. 2. Rendered scenes of a patient with an anterior fossa meningioma. The image on the left was generated by Sunvision, the one on the right by our multimodal renderer.

(see also in color on page 818)

Further work is also in progress to automate the registration process, improve the derivation of tissue occupancy factors based on fuzzy segmentation techniques, and suppress noise in MRA. Our renderer could be modified to include different rendering and / or lighting models for the different objects rendered, such as maximum intensity projection of angiographic data. The algorithm, with its identical rays being cast through multiple datasets simultaneously is readily transferable to parallel processor architectures.

Acknowledgements

This work was supported by grants from the SERC (grant no. GR/H48170) and the MRC (grant no. G9119735). We would like to thank Tim Cox (neuro-radiologist), Mike Gleeson and Tony Strong (surgeons) from Guy's Hospital. We are also grateful to the scientific and technical staff of Guy's, Maudsley and St. Georges Hospitals for their assistance in image acquisition.

References

1. Hill, D.L.G.; Green S.E.M.; Crossman, J.; et al.: Visualization of multimodal images for the planning of skull based surgery, Proc. of Visualization in Biomedical Computing, Chapel Hill 1992, 564-573.
2. Herman, G.T.; Liu H.K.: Three-dimensional display of human organs from computer tomograms. Comput. Graph. Imag. Proc 9 (1979) 1-29.
3. Cline, H.E.; Lorensen, W.E.; Ludke, S.; et al.: Two algorithms for the three-dimensional reconstruction of tomograms. Med. Phys. 15 (1987) 320-327.
4. Drebin, R.A.; Carpenter, L.; Hanrahan, P.: Volume rendering. Comput. Graph. 22 (1988) 65-74.
5. Levoy, M.: Display of surfaces from volume data. IEEE Comput. graph. Appl. 9 (1988) 29-33.
6. Hill, D.L.G.; Hawkes, D.J.; Hussian, Z.; et al.: Accurate combination of CT and MR data of the head: validation and applications in surgical and therapy planning. Proc. of the IEEE 3D Advanced Image Processing in Medicine, Rennes 1992, 79-83.
7. Hill, D.L.G.; Hawkes, D.J.; Crossman, J.; et al.: Registration of MR and CT images for skull based surgery using point like anatomical features. Br. J. Radiology, vol. 64 (1991) 1030-1035.
8. Robb,R.A.; Stacey, M.C.; McEwan C.N.: A networked workstation approach to multidimensional biomedical image analysis. Information Processing in Medical Imaging, C.N. de Graff and M.A. Viergever (eds.), Plenum Press, New York, (1988) 349-374.
9. Cook, R.L.: Stochastic sampling and distributed ray tracing. An Introduction to Ray Tracing, Glassner, A.S. (ed.), Academic Press Ltd., London, (1989) 161-199.
10. Hohne, K.H.; Bomans, M.; Pommert, A.; et al.: Rendering tomographic volume data: adequacy of methods for different modalities and organs. 3D Imaging in Medicine, Hohne, K.H.; Fuchs, H. and Pizer, S.M. (eds.), Springer-Verlag, Berlin, (1990) 197-215.

Application Systems

Computer Assisted Radiological Diagnosis

Case-Based Tutor for Radiology

R. Macura, K. Macura, V. Toro, E. Binet and J. Trueblood

Department of Radiology
Medical College of Georgia, Augusta, U.S.A.

Summary

Radiologists rely heavily on the visual memory when analyzing new images. Meaningful image patterns that correspond to cases have been developed in memory over years of experience. Our goal is to teach radiology residents to make diagnoses by exposing them to useful cases in ways that will help them integrate cases into memory effectively. We provide the users a case database structured around relevant radiological features that contains a digital image archive with case characteristics. In our approach, we expect the radiologists to use their powers of perception while analyzing radiologic images, and the computers, to store and search for information, and to provide radiologic images relevant to a diagnostic situation. The Case-Based Consultation System is composed of three modules: ***Case Retrieval, Teach Me*** and ***Atlas.*** The program was implemented in 4th Dimension®, version 3.0.1 (ACIUS Inc.) on an Apple Macintosh.

Introduction

An essential part of the diagnostic process in medicine involves referring to accumulated knowledge and experience. This is especially true of radiology in which visual analysis of images relies on a comparison with meaningful image patterns that correspond to cases developed in the radiologist's memory over years of experience (Schmidt et al, 1990). Radiologists rely heavily on the visual memory when analyzing new images. When a familiar pattern is seen, a specific image memory is triggered and the diagnosis (particular cases) comes to mind.

Case-Based Reasoning is a general paradigm for reasoning from experience. It assumes a memory model for representing, indexing, and organizing past cases and a process model for retrieving and modifying old cases and assimilating new ones (Slade, 1991). Conceptually, reasoning from relevant past cases corresponds to the process the experienced radiologist uses to solve new diagnostic problems. When making a diagnosis, the radiologist first extracts from the images representing a new case the features that appear to be significant and then retrieves past cases from memory that include the same features; the past cases contain the prior solution (diagnosis). The next step is to judge the similarity between the images representing a new case and the images retrieved from memory, then to make a decision or appropriate modification to account for changes in various features. The success of the process depends upon the ability of the reasoner to identify those attributes that make a prior case relevant and to draw the appropriate conclusions.

Our goal is to teach residents to make a diagnosis by exposing them to useful images/cases in ways which will help them to integrate images/cases into memory effectively. We provide them a case database structured around relevant features that contains a digital image archive with case characteristics. We want the residents to learn the key features of a subject so they may organize their knowledge around them and construct their own memory structure through the consideration of cases. We also want the user to learn the correct weights for cues, especially for those that have a high discriminatory power and those that are highly specific for certain diseases.

In our project, we expect the radiologists to use their powers of perception while analyzing radiologic images, and the computers, to store and search for information and to provide radiologic images relevant to a diagnostic situation.

Case-Based Consultation System

The Case-Based Consultation System is a tutor for training radiologists and is composed of three modules ***Case Retrieval, Teach Me*** and ***Atlas***. The ***Case Retrieval*** module assists the user in reaching a diagnosis by providing images from cases relevant to a specific case being evaluated. It is designed to simplify and optimize the search for diagnosis. The present library of 120 cases of brain tumors and 640 digitized radiological images (CT and MRI) is integrated with an indexing system that is based on case features (case history and radiological findings). The search direction is controlled by rules incorporating critical diagnostic cues and the judgement of neuroradiology experts regarding the prior probability of diagnoses. In the current version of the system we use the simplest method to identify the most appropriate cases, nearest-neighbor search (Duda and Hart, 1973). This operation compares the new case's description (its features) to the corresponding features of each case in the case library.

Within the ***Case Retrieval*** module, the user works with three types of windows: the *Data Entry Window*, *Diagnostic Window*, and *Show All Window*.

1. *Data Entry Window* (Figure 1.): this window is used to input data to the system. There are three mandatory inputs required to initialize the search: i) image modality (CT or MRI), ii) age of the patient, and iii) location of the tumor. Other inputs (case history and findings/features) are optional, and if entered, narrow the search space.

2. *Diagnostic Window* (Figure 2.): Based on the information input to the system, a list of probable diagnoses is displayed and relevant images/cases with textual description are offered to compare with the case in question. Using this window, the user may browse through cases, change the plane of the brain section, open the Show All Window to view all retrieved cases, go to the Teach Me submodule or open the Atlas submodule for anatomic reference.

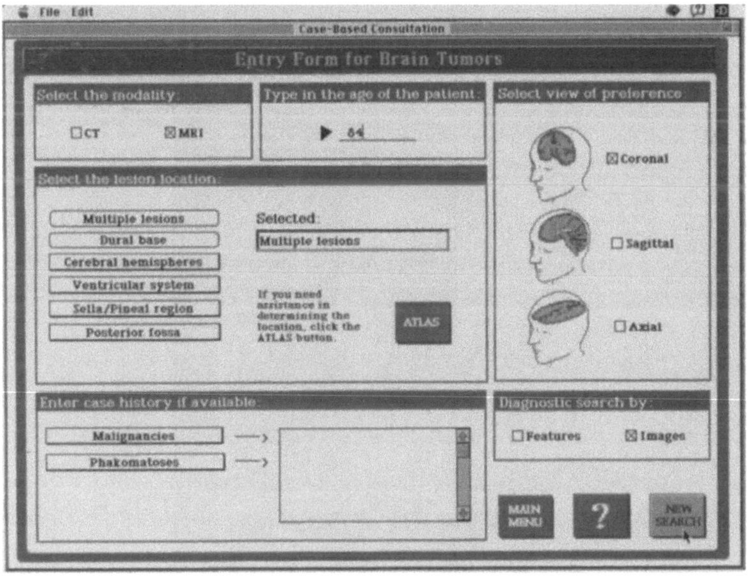

Figure 1. Data Entry Window.

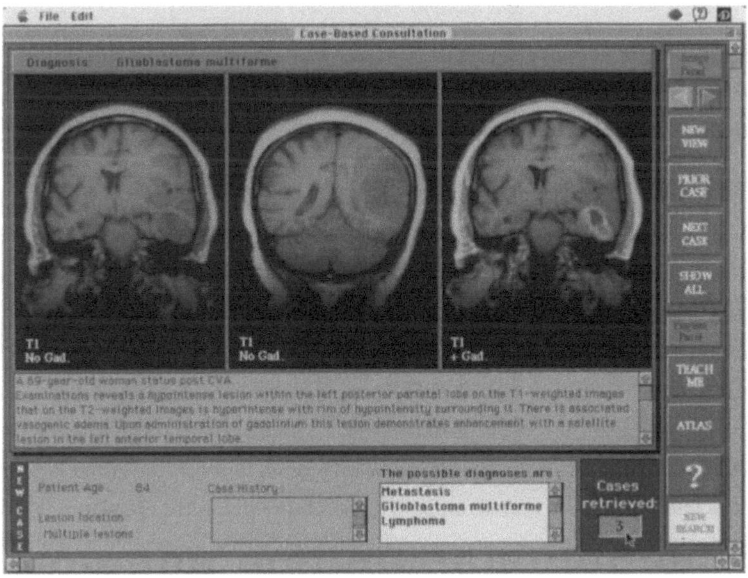

Figure 2. Diagnostic Window.

3. *Show All Window* (Figure 3.): In this window all retrieved images/cases are displayed. The user may select a best match image and go directly to the case of interest.

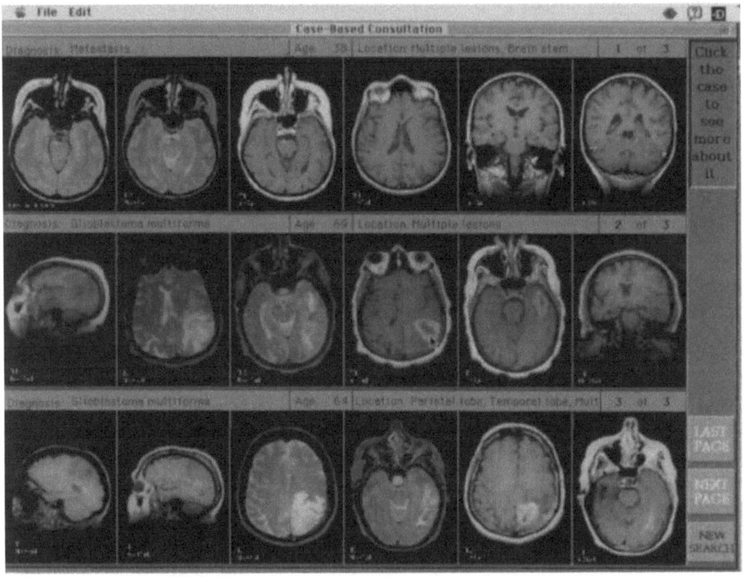

Figure 3. Show All Window.

The ***Teach Me*** module is an educational program that supports visual memorization of a particular tumor pattern. The *Teach Me Window* (Figure 4.) is composed of six parts. In the "Comparative Normal Window" the section of normal brain corresponding to the pathologic image in the "Pathology Window" is displayed. The "Schematic Window" presents, in interactive mode, colored features of tumors superimposed on the pathologic images. The "Features Window" is linked to the "Schematic Window" and, in the interactive mode, displays a textual description of features. The "Text Window" contains general information about particular tumors. The "Color Key Window" contains a colored code of the tumor features.

The ***Atlas*** module is designed to help the user localize a brain lesion. The *Atlas Window* (Figure 5.) displays nine sections of normal brain in one (CT) or three different planes (MRI). The user may change the viewing plane and open, by clicking on the selected cut, the *Atlas Reference Window* (Figure 6.) with detailed explanation of visible anatomical structures.

Hardware and Software Requirements

The program was implemented in 4th Dimension®, version 3.0.1 (ACIUS Inc.) on an Apple

Macintosh with 16MB of RAM, 80MB of hard drive and Futura MX graphics card with E-MACHINES 16'' monitor. Currently the program occupies 60MB. Average search time on a Macintosh IIci is 15 seconds.

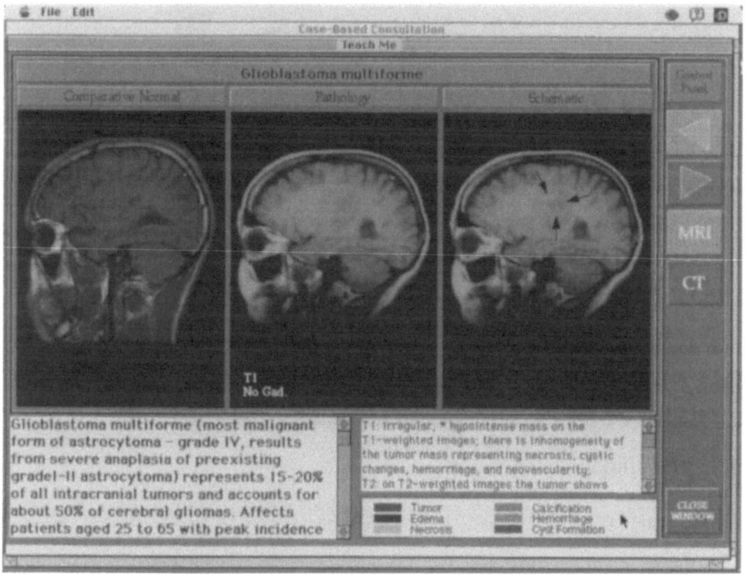

Figure 4. Teach Me Window.

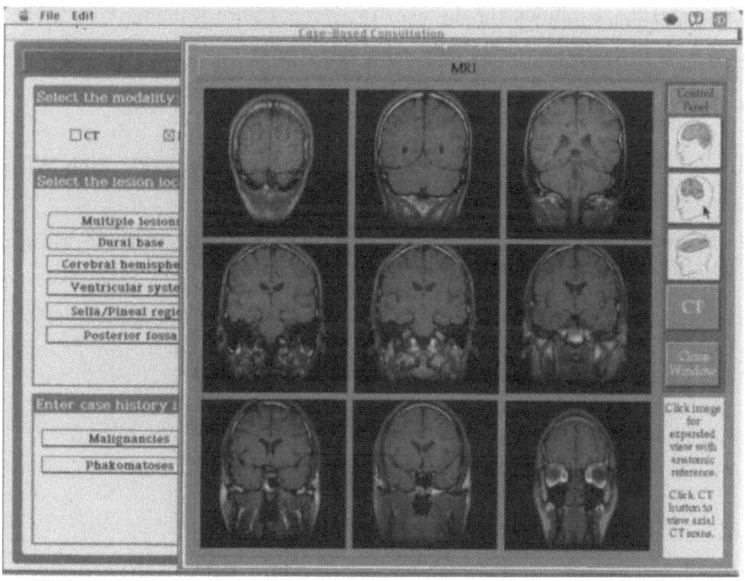

Figure 5. Atlas Window.

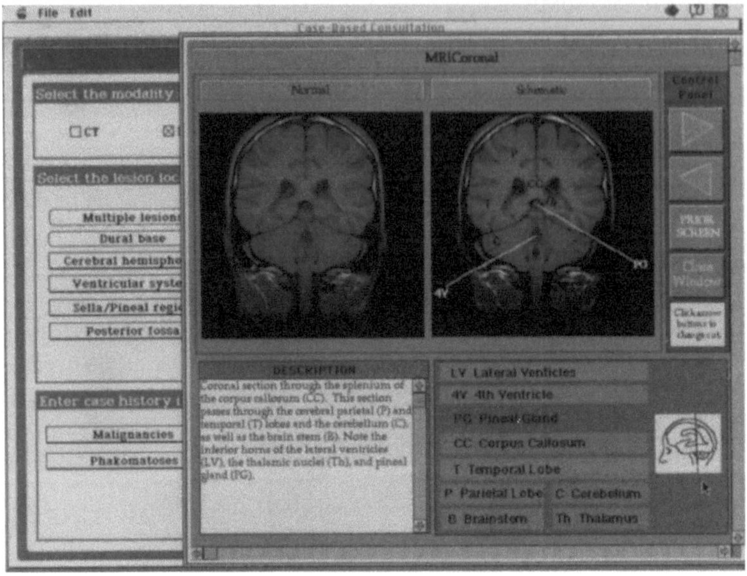

Figure 6. Atlas Reference Window.

Acknowledgments

The authors wish to acknowledge the efforts of Kai Ji who did programming work, Kirah Van Sickle who designed the windows for the user interface, Claire Raeuber who helped develop the atlas, and Cliff Garzzillo who digitized radiological images.

References

1.Duda, R.O.; Hart, P.E.: Pattern Classification and Scene Analysis. New York, John Wiley and Sons 1973.

2. Schmidt, H. G.; Norman, G.R.; Boshuizen, H.P.A.: A Cognitive Perspective on Medical Expertise: Theory and Implications. Academic Medicine, Vol. 65/10 (1990) pp. 611-621.

3. Slade, S.; Case - Based Reasoning: A Research Paradigm. AI Magazine, Spring (1991) pp. 42-55.

Computer Assistance in the Processing of Medical Images to Improve Specific Diagnosis of Peripheral Lung Cancer

T. P. BELIKOVA, V. V. LASHIN and N. I. YASHUNSKAYA

Institute for Information Transmission Problems of Russian Academy of Science, Moscow Medical Academy, Russia

Summary

Computer assisted technology for medical image processing is developed to increase the accuracy and reliability of peripheral lung cancer diagnosis at the early stage of disease. Technology is based on optimal filtering methods that are to emphasize diagnostically important patterns and parts in lung tomograms. Quantitative analysis of histogram characteristics adds the results of visual analysis of processed image while the specification of nodule nature. A tutorial mode is available to support the physician's decision. Technology is realized on the base of personal computer and frame-grabber. It supports real-time image processing. Numerous scientific experiments confirmed the developed technology improves diagnostic efficacy.

Introduction

Nowadays lung cancer is one of the most often reason of mortality in all economically developed countries. Rising demands to early stage recognition of this disease and the necessity of high accuracy diagnosis call for the development of new technologies to improve this medical task solution. New computer-aided technology is designed to increase the effectiveness of lung tomogram analysis for specific diagnosis of solitary pulmonary nodules.

Materials and methods

The analysis of small (up to 3 cm in diameter) nodules was carried out. Linear lung tomograms with such objects were digitized with 100 mkm steps to get 512x512x8 bits matrix. Such digitized picture permitted to reach an acceptable space resolution of lung and nodule anatomic structures that were necessary for further analysis. Several investigations were then established to create technology.

a) Analysis of distortions of imaging system with test objects was carried out. It permitted to create algorithms for amplitude and aperture correction of distortions to avoid ambiguous information in the analyzed images.

b) The methods were developed for extraction of diagnostically important objects in linear lung tomograms. We used several models of background picture. For this purpose we applied several sets of background realizations. These models were to estimate necessary characteristics of useful signal and to find substantial methods of image processing [1].

c) X-ray morphological comparison was used to identify specific patterns on the processed images. Comparison helped one to select the most available models of background. These models were the best to create image processing methods that emphasized diagnostically important patterns without adding artifacts.

d) Recommendations were then developed for analysis of extracted patterns to diagnose the lesions [2]. These recommendation were used for tutorial mode development. It was designed to support the decision-making by physician.

e) The methods were found to step up processing rate to make real-time image processing available [1].

f) The methods for quantitative image analysis were proposed and tested to classify the objects of interest [2].

g) Test processing, image analysis and morphological verification of diagnostic decisions were carried out to calculate the effectiveness of developed technologies.

Technology realization

The technology was realized and tested on the workstation for radiologists [3]. It consisted of standard personal computer IBM PC/AT, TV-monitor for imaging, and input-output device. Drum-scanner or frame grabber with TV camera and medical light box were used to acquire up to 1024x1024x8 bits image matrix.

The software realization of technology accumulates the knowledge about optimal adaptive methods for tomogram processing and quantitative analysis of image characteristics to clarify important diagnostic features in lung tomograms.It provided the user with computer assistance and advice, supported a real-time image processing.

Special user interface was developed to do the work with computer convenient for physician who works as an expert in this system. It permits the user to interact with computer in habitual terms and does not require any special knowledge of used methods and programs. While user pushed one of the functional keys, one of the following tasks was fulfilled:

1.Input of the X-ray film.
The X-ray film was inputted and digitized by means of input-output device and stored in frame grabber. Noise correction was used to suppress amplitude and aperture image distortions that took place after input of the image (Fig.1.a).

2. Region of interest setting.
The physician determined some region of interest and oulined it with help of the cursor under the control of image displayed on TV monitor. In order to determine correct filter parameters the user had to restrict processing region rejecting objects with large signal variances (shadows of ribs in linear lung tomograms, etc.)

3. Selection of filter parameters.
To maintain optimal image processing the parameters of filtering had to be changed with the linear dimensions of the object of interest and with the number of gray levels in the processed image. When investigating solitary pulmonary nodule the user defined the nodule diameter by drawing it with cursor. It is enough for automatic calculation of necessary filter parameters.

4. Image filtration.
The image was processed by quasi optimal filter with parameters founded on the previous step. The processed image was displayed on TV monitor for visual analysis of extracted diagnostically important patterns (Fig.1.b).

5. Nodule outlining.

To calculate histogram statistics of nodule image the user had to outline the nodule boundary by means of cursor (Fig.1.c).

6. Histogram analysis.

The histogram of film densities inside the outlined nodule area was calculated and drawn on PC monitor (Fig.1.d). Some statistics of histogram were calculated and used for nodule classification [3]. The result of this classification was submitted to user. The physician took it into account among other diagnostic features making final decision.

7. Tutorial showing.

Data bases were in the allowance of physician that accumulated the knowledge of expert in image analysis and decision making. On user's request the computer gave examples of original and processed images and descriptions of the steps necessary for their analysis together with true diagnostic decisions.

Results

Analysis of post-processed images improves differentiation of malignant nodules from benign ones. For small (up to 3 cm) nodules total accuracy, sensitivity and specificity of diagnosis achieved 94.4, 97.2 and 91 percent respectively (data have been gotten after more then 600 cases). It is better that results of analysis of small nodules with CT technique.

The processing may be useful for correct diagnosis of small radiographically indeterminate pulmonary opacities and can add the data from conventional and computed tomography.

References

1. Belikova, T.P.: Synthesis of optimal linear filters for visualization of diagnostically important objects in medical radiological tasks. In book: Digital Optics in Medical Introscopy, Moscow: Science, 1992, 57-72 (in Russian).

2. Belikova, T.P.; Zavrazhina, I.N.; Yashunskaya, N.I.; Kogan, E.A.: Digital tomographic picture measurement in early differential diagnostics of lung cancer. Osterreichische Computer Gesellschaft. R. Oldenbourg Wien, 1991, 34-39.

3. Belikova, T.P.; Lashin, V.V.; Yaroslavsky, L.P.: Workstation for radiologist. In book: Proceedings of the Third International Seminar on Digital image processing in medicine, remote sensing and visualization of information. Latvia, Riga 1992, 99-101.

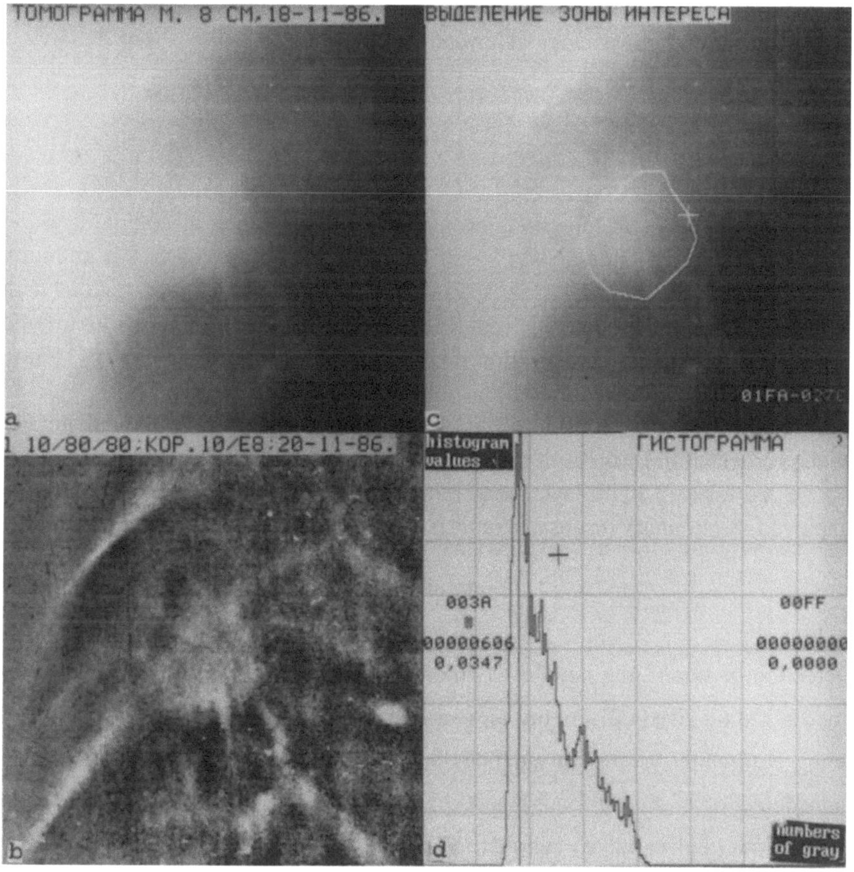

Fig.1. Malignant nodule.

a) The original tomogram.

b) The processed tomogram: The shape and edges of the nodule became more vivid. The processing emphasized details inside of the nodule. The state of the lung tissue around the nodule was well-distinct as well.

c) The nodule with outlined edges.

d) The histogram taken in the area of nodule has one of the shape that is specific for small malignant nodule.

A Neural Network System for Detecting Lung Nodules in Chest Radiograms

G. Coppini [1], R. Poli [2], R. Legittimo [2], R. De Dominicis [3], G. Valli [2]

1 - CNR - Institute of Clinical Physiology - Pisa

2 - Dept. of Electronic Engineering - University of Florence

3 - Dept. of Clinical Pathophysiology - University of Florence

Summary

A system for detecting small lung nodules is described. According to computer vision paradigms, the system is modular and based on a multi-scale approach. An attention focusing mechanism (operating at low spatial resolution) locates image regions with possible lesions. A second module analyzes (at full spatial resolution) the regions around the attention foci. Finally, a classification module estimates the probability of lesion presence.

Due to the complexity of the considered vision task, robustness of the adopted computational tools is crucial. For this reason the system modules were built using artificial neural networks. Feed-forward multi-layer nets trained with the back-propagation algorithm were employed. Preliminary results are reported.

Introduction

Detection of small lung nodules is among the most important and difficult tasks performed by radiologists [1,2]. For this reason, several researchers have faced the problem of building automatic systems to aid nodule finding in chest radiography [2,3,4,5,6]. Automatic detection of lung lesion is heavily hampered by several factors. In fact, one has to deal with low (and variable) contrast structures as well as with various sources of structural and non-structural noise. In particular, complexity and variability of lung parenchyma produces a troublesome background texture. Thus, extremely robust processing techniques must be developed. Moreover, the availability of *a priori* knowledge about the appearance of lesions is expected to be quite important. Unfortunately, the building of an efficient representation of prototypal shape of nodules is a challenging task.

The above considerations have prompted us to use a Computer Vision (CV) approach using Artificial Neural Networks (ANNs) to build an automatic system for detecting nodular lesions in chest radiograms. The basic project requirements have been:

a) nodules from 2-3 mm to 2-3 cm should be recognized, b) the overall system sensitivity should be high.

According to CV paradigms, the system we have developed is modular and operates on a multi-scale basis. To achieve high robustness we resorted to implement the various modules using neural nets. It is worth noting that other authors have reported on ANN-based systems for the detection of pulmonary nodules [5,6]. In this regard, it must be pointed out that ANNs exhibit several interesting features which are important in solving vision problems. For example, ANNs have the capability of storing and using efficiently the prototypal knowledge embedded in the training set.

The System

The adopted system architecture is an extension of a fully neural architecture we have recently developed for understanding discrete 3-D scenes [7]. Besides modularity, it exploits two basic mechanisms of biological inspiration: attention-focusing and multi-scale processing. The system, which is schematized in figure 1, includes three major blocks. The first one provides an attention-focusing mechanism by locating regions of interest (ROIs) with possible nodular lesions. In order to segment nodules out of anatomical background, the second module enhances the image in the ROI. A segmented ROI is built by assigning small intensity values to background pixels and large ones to nodule pixels. Finally, the third stage (classification module) processes the segmented ROIs and estimates the probability of nodule presence.

Attention-focusing allows a strong reduction of computation load as only small parts of the picture are processed by all modules. In addition, it is a light-weight task since it is performed by processing a low resolution version of the image. In this way, noise effects are smoothed out together with small, meaningless details. Low resolution images are obtained by Gaussian filtering followed by spatial decimation.

As concerns variability of background gray level, we have preprocessed the images with a Laplacian filter. Due to its derivative nature, Laplacian removes low frequency components which are responsible for average background variability. At the same time, it preserves detail contrast. Moreover, it must be noted that Laplacian operators and Gaussian kernels can be implemented with a single convolution with a Laplacian of Gaussian (LoG) mask.

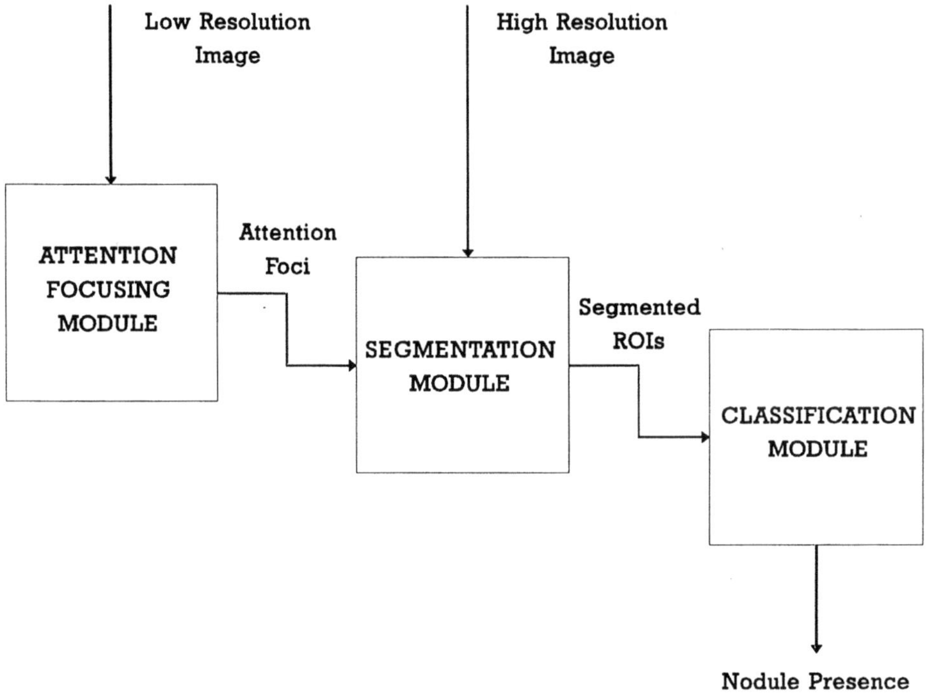

Fig. 1

Networks' Topology and Training

All modules depicted in figure 1 consist of feed-forward networks trained with the back-propagation algorithm [8].

The attention-focusing module includes an input layer with 19×19 units, three hidden layers with 13 , 9, and 3 units, and a single output neuron. This module operates by moving its input units as a filter mask on the image, the activation of the output neuron is high if a nodule-like structure is detected within the input mask.

The segmentation module is a block-connected net and comprises two main paths which join in a common layer with three units feeding an output neuron. The first path is devoted to gray-level processing and is made up of four layers with 19×19, 10, 6, and 3 units, respectively. Similarly to the focusing net, the input units are arranged as a mask which moves on each given ROI. Simultaneously, the second path processes the (binary coded) position of the central pixel of the mask. This path includes four layers with 12 (six for each coordinate), 6, 6, and 4 units respectively. This topology has been adopted to take the typical spatial arrangement of nodule gray levels into account. Knowledge about lesion shape is stored in the connection weights during the training phase. Afterwards, this net operates as a non-linear space-varying matched filter.

The classification module is a fully-connected network with four layers with 21×21, 4, 2, and 1 units respectively.

As to the training set, 32 chest radiograms were digitized by a laser scanner with a pixel size of 390 μm and twelve bits of density resolution. The acquisition of the lung region resulted in a 768×768 image. Subsequently, an expert radiologist was asked to mark the position of possible nodules and to segment them interactively. In this way, three sets with 4864, 8192, 1920 examples, respectively, were obtained and used to train the modules of the system.

As to LoG preprocessing, the original images are reduced to a resolution of

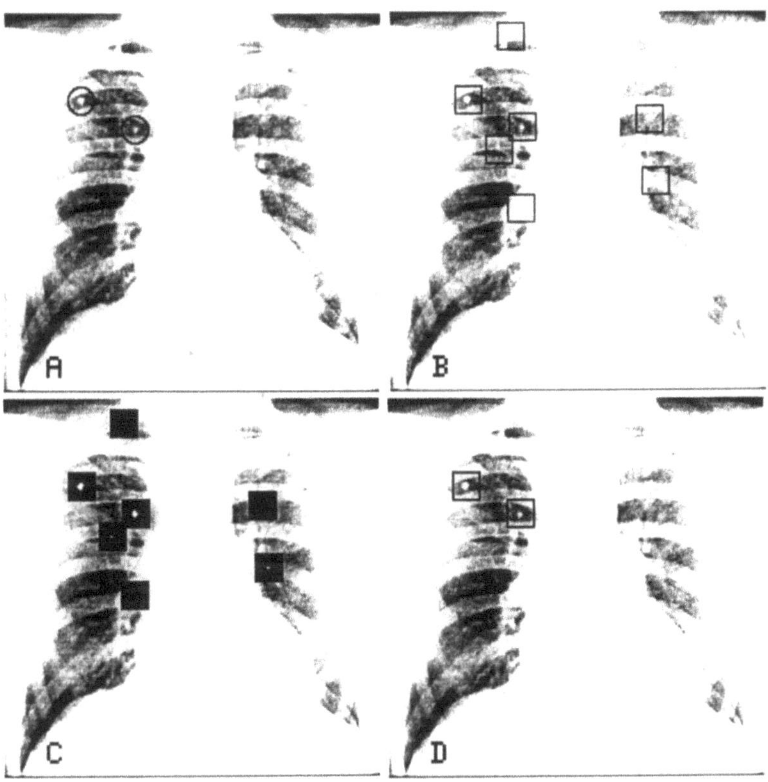

Fig. 2

256×256 images after LoG filtering with sigma of 8. These low-resolution pictures are utilized for the focusing process. The segmentation phase is carried out at full spatial resolution following LoG filtering with sigma of 3.

Results and Conclusions

Presently, we testing the system on a validation set. A typical processing example is illustrated in figure 2. Panel A shows a radiogram with two possible lesions marked with circles by the radiologist, in panel B the attention foci produced by the first module are indicated by squares, in panel C the segmented ROIs are superimposed on the original image, and, finally, in panel D the output of the classification module are indicated with circles. Presently, results are in good agreement with those observed by analyzing the performance of each module on the training set. In this case, we have computed the true positive (*TP*), and true negative (*TN*) rates for several values of the output thresholds. For example, the values: $TP \approx 0.99$, $TN \approx 0.98$ were observed for the focusing module, with a threshold value of about 0.8 (neuron activation is in the range [0,1]). Similarly, for the segmentation net (considered as a pixel classifier) with a threshold of 0.6 we obtained: $TP \approx 0.96$, $TN \approx 0.99$. Finally, for the classification module we had: $TP \approx 1.0$, $TN \approx 0.99$, with a threshold of 0.7.

References

1. J.H.M. Austin, B.M. Romney, L.S. Goldsmisth: Missed broncogenic carcinoma: radiographic findings in 27 patients with a potentially resectable lesion evident in retrospect. Radiology, 182: 115-122, 1992.

2. M.L. Giger, K. Doi, H. MacMahon: Image feature analysis and computer-aided diagnosis in digital radiography. 3. Automated detection of nodules in periferal lung fields. Med. Phys. 15: 158-165, 1988.

3. W.A. Lampeter, J.C. Wadtke: Computerized search of chest radiographs for nodules. Invest. Radiol. 21: 384-390, 1986.

4. H. Yoshimura, M.L. Giger, K. Doi, H. MacMahon, S.M. Montner: Computerized schema for detection of pulmonary nodules. Invest. Radiol 27: 124-129, 1992.

5 S. Garg. C.E. Floyd: The training of an artificial neural network for a detection task in medical images. Neural Networks 24: 2-3, 1991.

6 M.G. Penedo, D. Cabello, S. Barro, A. Mosquera, M.J. Carreira: A multilayered neural network for the recognition of lung nodules on digital chest radiographs. Proceedings CAR 91, 289-284, Springer Verlag, Berlin, 1991.

7 G. Coppini, R. Poli, M. Rucci, G. Valli: A neural network architecture for understanding discrete three-dimensional scenes in medical imaging. Computers and Biomedical Research 25: 568-585, 1992.

8 J.M. McClelland, D.E. Rumelhart *Parallel distributed processing.* MIT Press Cambridge, MA, 1986.

Analysis of Results in a Large Clinical Series of Computer-Aided Diagnosis in Chest Radiography

Katsumi Abe, MD[1], Kunio Doi, PhD[2], Heber MacMahon, MD[2],
Shigehiko Katsuragawa, PhD[1], Maryellen L. Giger, PhD[2], Toru Yanagisawa, MD, PhD[1]

[1]Department of Radiology, Iwate Medical University, Morioka, Iwate, Japan

[2]Kurt Rossmann Laboratories for Radiologic Image Research, Department of Radiology,

The University of Chicago, Chicago, Illinois, USA

Abstract

We are developing computer-aided diagnosis (CAD) schemes for chest radiography with which to alert radiologists to possible lesions, and thus improve diagnostic accuracy. Although preliminary results of individual CAD schemes have been encouraging, CAD is still at an early stage of its development. Therefore, we applied CAD programs for the automated detection of lung nodules, cardiomegaly, and interstitial infiltrates to 310 consecutive chest radiographs and analyzed their potential usefulness and limitations. CAD output was evaluated by radiologists and physicists for accuracy and technical problems, respectively. While approximately 70% of the CAD output were judged to be potentially acceptable, the number of false positives was relatively high. Technical problems included failure to detect subtle abnormalities and the occurrence of false positives caused by normal anatomical structures. We believe that CAD has the potential to be a valuable aid to radiologists in clinical practice if certain technical problems can be overcome and if optimal operating points can be defined for clinical use.

Introduction

Advantages of digital radiography include image processing, transmission, storage, and quantitative analysis of lesions. Many types of processing techniques for the detection of subtle lesions (1) as well as picture archiving and communication systems (2) are being developed. Results of quantitative analysis are being applied for computer-aided diagnosis (CAD) (3,4). Since 1985, at the University of Chicago, CAD schemes have been under development (5-10). The aim of these CAD programs is to alert the radiologist by indication of potential sites of lesions, and also to quantitate and classify detected abnormalities. Preliminary results of our individual CAD schemes for chest radiography have been encouraging (5-7, 9). However, these techniques are still at an early stage. In order to increase our understanding of the usefulness and limitations of the CAD schemes, we applied CAD to a large series of consecutive clinical cases.

Materials and Methods

A consecutive series of 310 posteroanterior (PA) chest radiographs was collected. Chest radiographs of infants and children were excluded. Abnormal findings were classified initially, and devices and technical artifacts were also noted by two chest radiologists independently. Table 1 summarizes the findings, and the occurrence of the most frequently seen devices and artifacts. Focal interstitial infiltrate was the most common abnormality. This high percentage reflects the inclusion of minimal focal interstitial abnormalities due to scarring in this category. To facilitate

Table 1. Abnormal findings in 310 consecutive chest radiographs as classified by two chest radiologists

Finding	No. of cases	Rate (%)
Normal	74	24
Interstitial infiltrate (focal)	114	37
Cardiomegaly	69	22
Calcified granuloma	59	19
Pleural effusion	54	17
Vascular access devices	49	16
ECG electrode snaps	40	13
Air space infiltrate	39	13
Interstitial infiltrate (diffuse)	37	12
Lung nodule	37	12
Surgical clips	25	8
Electrode cords	15	5
Safety pins for fixing surgical drains	9	3
Tracheostomy tubes	6	2
Indeterminate for cardiomegaly	7	2
Indeterminate for diffuse interstitial infiltrate	3	1
Indeterminate for focal interstitial infiltrate	3	1
Indeterminate for lung nodules	6	2

the evaluation of CAD results, these cases of focal interstitial infiltrate were distinguished from those with diffuse interstitial infiltrates. In addition, certain cases were categorized as "indeterminate" for a given abnormality due to inability of the radiologists to arrive at a reliable consensus, for reasons of obscuration of the relevant anatomy by other disease or technical problems. These cases were eliminated for purposes of evaluation of CAD results. Thus, 303 CAD results for detection of cardiomegaly, 270 for focal, 193 for diffuse interstitial infiltrate, and 304 for lung nodules were evaluated.

Then, these films were digitized, and CAD programs were applied. The CAD results were overlaid on the chest images on a CRT monitor, and then recorded by a video printer.

The chest CAD schemes used in this study included automated detection of lung nodules, analysis of interstitial infiltrates, and detection of cardiomegaly (5-7, 9). For the automated detection of lung nodules, from signal-enhanced and signal-suppressed images, a difference image is obtained to reduce the prominence of normal anatomic structures. Various feature-extraction techniques are applied in order to eliminate false-positive detections. Finally, suspected nodules are indicated by arrows as illustrated in Figure 1a (5, 9). For the analysis of interstitial infiltrates, several hundreds of square regions of interest (ROIs) are automatically selected from lung fields. ROIs over rib edges and artifacts are eliminated. Two texture measures are calculated. Normal and abnormal ROIs are classified based on comparison of these texture measures with those of a database, and the computer outputs are displayed as shown in Figure 1b (6). For the detection of cardiomegaly, the approximate location of the heart is determined initially. Lateral cardiac boundaries and diaphragm edges are detected. Detected cardiac edges are fitted to a model function. Lateral lung margins are also detected, and the smoothed outline is obtained. The cardiac outline together with distant points on both cardiac margin and ribcage edges are calculated and displayed as shown in Figure 1c (7).

The CAD results were evaluated subjectively and independently in terms of overall accuracy by two chest radiologists for detection of cardiomegaly and analysis of interstitial infiltrates using a five-point rating scale, with 5 corresponding to highly accurate, 4 to moderately accurate, 3 to marginally accurate, 2 to moderately inaccurate, and 1 to highly inaccurate. "Highly accurate" was used when the location of annotations of the CAD results corresponded almost perfectly to those of abnormal or normal findings that had been previously defined by the radiologists. "Highly inaccurate" was used when CAD results were clearly incorrect. Other intermediate rating scales were selected depending on the correspondence of the CAD output to the nature and location of important findings influencing the radiologic diagnosis.

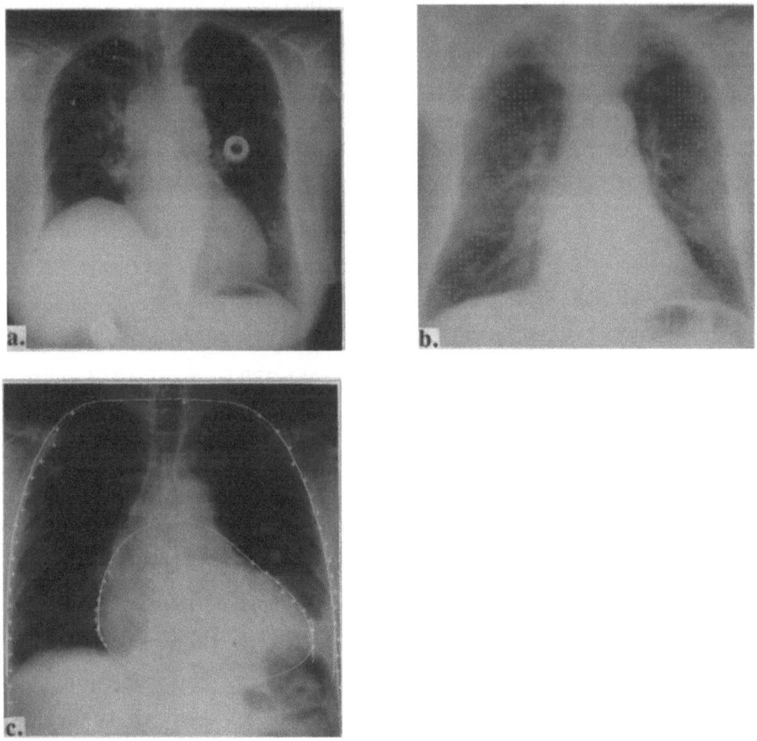

Figure 1. Chest radiograph with annotated CAD output for (a) multiple nodules, (b) diffuse interstitial infiltrates due to idiopathic pulmonary fibrosis, and (c) cardiomegaly. (a) CAD output indicates correct detections of four nodules (arrowheads in lung fields) and one false positive due to spine (arrowhead in mediastinum). (b) Pluses indicate normal texture patterns and other marks indicate abnormal texture patterns correctly. (c) CAD output indicates cardiac outline, most distant points on the right and left cardiac margin, rib edges, and most inner points of rib edges correctly.

The CAD output for cardiomegaly and interstitial infiltrates were also evaluated in terms of technical aspects under 7 categories. One radiologist and two physicists as a group evaluated these categories subjectively using a five-point rating scale-5 (definitely accurate detection), 4 (probably accurate detection), 3 (possibly accurate detection), 2 (probably inaccurate detection), and 1 (definitely inaccurate detection). Rating scales for technical evaluation were used in a way similar to those used in the evaluation of possible clinical usefulness.

Table 2. Five-point rating scale for evaluation of CAD output for lung nodules

Score	Normal	Solitary Nodule	Multiple Nodules
5	FP = 0	TP = 1, FP = 0	TP ≥ 2, FP = 0
4	FP ≤ 5	TP = 1, FP ≤ 5	TP ≥ 1, FP ≤ 5
3	FP ≤ 7	TP = 1, FP > 5	TP ≥ 1, FP > 5
2	FP ≤ 10	TP = 0, FP ≤ 5	TP = 0, FP ≤ 5
1	FP > 10	TP = 0, FP > 5	TP = 0, FP > 5

FP = false positives, TP = true positives.

The CAD results for detection of lung nodules was evaluated objectively using a five point rating scale by taking into account the numbers of false positives and true positives (Table 2). Based on previous results, we assumed that less than seven false positives for normal, and less than five for solitary and multiple nodules cases would not degrade radiologist's diagnosis by provoking excessive false positives (10). The detection of one true positive or more for multiple nodule cases was considered "acceptable" for the purpose of CAD in alerting to radiologists to potential lesions.

Results

The results of the evaluation in terms of possible clinical acceptability for individual CAD output are shown in Figure 2. For detection of cardiomegaly, 216 cases (71%) were in rating categories 3-5 (Fig 2a). For analysis of focal and diffuse interstitial infiltrates, 234 cases (87%) and 161 cases (83%), respectively, were in categories 3-5 (Fig 2b and 2c). For detection of lung nodules, 184 cases (61%) were in categories 3-5 (Fig 2d).

Table 3 shows the results from the evaluation of CAD output in terms of technical aspects for detection of cardiomegaly and analysis of interstitial infiltrates. The results are given in terms of accurate detection (categories 3 to 5). Notice that more than 70% of the CAD output were rated in these categories.

Discussion

Radiologists' diagnoses of normal and abnormal findings from radiographs are generally based on their own knowledge and experience that have been cultivated over a long period of time. However, since their judgements are subjective, interpretation of radiographs will inevitably involve missed diagnoses for various reasons. Austin et al investigated the nature of missed bronchogenic carcinoma (11). In their study, most missed lung cancers were visible in retrospect on previously obtained radiographs, and these were potentially resectable for cure at that time. Furthermore, Green emphasized the potential usefulness of computer-assisted detection of lesions for subtle lung nodules (12). The motivation of our developing of CAD schemes for detection of lung nodules based on these reasons.

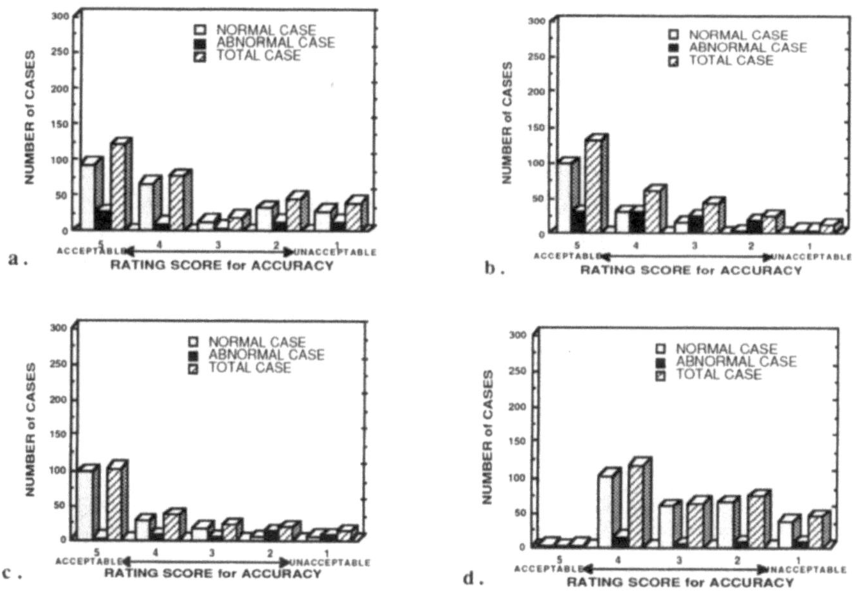

Figure 2. Distribution of rating scores for the evaluation of CAD output in terms of clinical acceptability for (a) detection of cardiomegaly, (b) analysis of focal, (c) diffuse interstitial infiltrate, and (d) detection of lung nodules.

Table 3. Results of CAD output: Technical aspects

Category	% Accurate detection
Detection of cardiomegaly	
Heart boundary detection	72
Curve fitting for heart shape	84
Lung top point	84
Rib cage edge detection	90
Curve fitting for rib cage shape	99
Thoracic maximum point	91
Analysis of interstitial infiltrates	
ROI location	91

MacMahon et al studied the nature and frequency of abnormal findings in chest radiographs (13). Pulmonary infiltrates were the most common abnormality and cardiac abnormalities were the second. In our study, focal interstitial infiltrate was the most common abnormality and cardiomegaly was the second. This high frequency ranking is one of the reasons for developing CAD schemes for the analysis of interstitial infiltrates and the detection of cardiomegaly.

Table 4.
Causes of inaccurate detection of cardiomegaly by the CAD scheme

Inaccurate heart boundaries	%
Inadequate detection of the apex of the heart	47
Incorrect detection of heart boundaries due to pulmonary vessels	24
Incorrect location of detected heart such as too high or too low	10
Other	<5
Incorrect detection of upper ribcage edge	
Detection of clavicles	35
Detection of artifacts	10
Detection of edges of scapula	8
Incorrect detection of top of lung	
Detection of cervical spine	33
Detection of thoracic spine	6
Difference in top location of each lung	28
Detection of mandible	6

In order to summarize the results of the clinical and technical evaluations, we used CAD scores of 3, 4, and 5 as "acceptably accurate" categories, and scores of 1 and 2 as "unacceptable or inaccurate" categories. Most of the CAD results for detection of cardiomegaly were in the "acceptably accurate" categories. However for abnormal cases, the percentage of "acceptably accurate" cases was only moderate, and the percentage of inaccurate detection of heart boundaries was relatively high (Table 3). Table 4 lists some of the causes of inaccurate detection of cardiomegaly. The results of CAD output for detection of interstitial infiltrates were acceptable for most normal cases, but for abnormal cases the frequency of unacceptable results was high in spite of the high percentage of acceptable cases with focal interstitial infiltrate. For subtle diffuse interstitial infiltrate, sensitivity was relatively low. On the other hand, if the infiltrates were very obvious, the CAD schemes tended to eliminate these findings during the process of removing of rib edges, because the edge gradient for the detected components of these infiltrates tend to be as large as those of rib edges. The causes of errors for detection of interstitial disease are listed in Table 5. Further investigations are needed to select appropriate parameters for thresholding between normal and abnormal texture measures, classification of diffuse and focal interstitial infiltrates, and complete elimination of rib edge. The locations of ROIs are critically important (Table 3) and development of an algorithm for accurate detection of ribcage edges is essential for various chest CAD schemes. For detection of lung nodules, the percentage of acceptable cases

Table 5. Causes of inaccurate analysis of interstitial infiltrates by the CAD scheme

Causes of false positives	%
Remaining rib edges	51
Pulmonary vessels	11
Edges of devices	5
Inaccurate analysis due to ROI problem	
Insufficient number of ROIs in upper lung	30
Improper location of ROIs due to incorrect detection of diaphragm edges	27
Improper location of ROIs due to incorrect detection of ribcage edges	21
Insufficient number of ROIs in perihilar region	7
Insufficient number of ROIs over the heart	3

Table 6. Causes (or location) of false negatives and false positives in the CAD scheme for nodule detection

Causes (or location) of false negatives	%
Overlap [illegible]	[illegible]
Low contrast	53
Small size	44
Large size	9
Irregular margin	44
Close to ribcage	15
Retrocardiac area	9
Retrodiaphragmatic area	9
Overlaps large pulmonary vessels	6
Artifacts such as improperly exposed films	6
Causes of false positives	
Ribs crossing	52
Rib and vessel crossing	22
Clavicle	5
Pulmonary vessel	4
Rib	3
Device	1

was only slightly larger than that of unacceptable cases, because of low sensitivity and high false positive rate. The causes of false negatives and false positives are listed in Table 6. The number of false positives was 6.4 per image. Apparently, further improvement of accuracy of the CAD technique both increasing the sensitivity for the detection of actual nodules and reducing the number of false positive detections.

We believe that CAD schemes may be incorporated into the clinical environment after certain improvements have been made. For instance, false negatives are common in mass lung cancer surveys (14). Interpretation by multi-readers is one procedures which can improve diagnostic accuracy in this situation, though, it is not easy to implement double reading in practice due to radiologists' time and resulting cost. Thus, CAD may be a solution if used as a "second reader" (15). In order to implement CAD in routine clinical studies, it would be necessary to improve the sensitivity and reduce false positives further. However, CAD has potential to be a practical tool in the future.

References

1. Abe K, Katsuragawa S, Sasaki Y, Yanagisawa T. A fully automated adaptive unsharp masking technique in digital chest radiograph. Invest Radiol 1992;27:64-70.
2. Dwyer III SJ, Templeton AW, Martin NL, et al. The cost of managing digital diagnostic images. Radiology 1982;144:313-318.
3. Doi K, MacMahon H, Katsuragawa S, et al. Utilization of digital image data for computer-aided diagnosis, in Mum SK, et al (ed): First International Conference on Image Management and Communication in Patient Care. IEEE Computer Society Press, Washington, DC, 1990; 128-135.
4. Doi K, Giger ML, MacMahon H, et al. Computer-aided diagnosis:development of automated schemes for quantitative analysis of radiographic images. Seminars in Ultrasound, CT, and MRI 1992;13:140-152.
5. Giger ML, Doi K, MacMahon H, Metz CE, Yin FF. Pulmonary nodules: computer-aided detection in digital chest images. RadioGraphics 1990;10:41-51.
6. Katsuragawa S, Doi K, MacMahon H, Nakamori N, Sasaki Y, Fennessy JJ. Quantitative computer-aided analysis of lung texture in chest radiographs. RadioGraphics 1990;10:257-269.
7. Nakamori N, Doi K, MacMahon H, Sasaki Y, Montner S. Effect of heart-size parameters computed from digital chest radiographs on detection of cardiomegaly: potential usefulness for computer-aided diagnosis. Invest Radiol 1991;26:546-550.
8. Giger ML, Nishikawa RM, Doi K, et al. Development of a "smart" workstation for use in mammography. Proc SPIE 1991;1445:101-103.
9. Yoshimura H, Giger ML, Doi K, MacMahon H, Montner SM. Computerized scheme for the detection of pulmonary nodules: a nonlinear filtering technique. Invest Radiol 1992;27:124-129.
10. MacMahon H, Doi K, Chan HP, Giger ML, Katsuragawa S, Nakamori N. Compute-aided diagnosis in chest radiology. J Thorac Imaging 1990;5:67-76.
11. Austin JHM, Romney BM, Goldsmith LS. Missed bronchogenic carcinoma: radiographic findings in 27 patients with a potentially resectable lesion evident in retrospect. Radiology 1992;182:115-122.
12. Green RE. Missed lung nodules: lost opportunities for cancer cure. Radiology 1992;182:8-9.
13. MacMahon H, Montner SM, Doi K, Liu KJM. The nature and subtlety of abnormal findings in chest radiographs. Med Phys 1991;18:206-210.
14. Matsumoto T, Yoshimura H, Giger ML, et al. Potential usefulness of computerized nodule detection in screening programs for lung cancer. Invest Radiol 1992;27:471-475.
15. Matsumoto T, Doi K, Nakamura H, Nakanishi T. Potential usefulness of computer-aided diagnosis (CAD) in a mass survey for lung cancer using photofluorographic films [in Japanese; abstract in English]. Nippon Act Radiol 1992;52:500-502.

Neural Network Approach for the Computer-Aided Diagnosis of Coronary Artery Diseases in Myocardial SPECT Bull's-Eye Images

Hiroshi Fujita [1] Tetsuro Katafuchi [2] Mitsuo Shinoda [1]
Toshiisa Uehara [3] Takeshi Hara [1] Tsunehiko Nishimura [3]

[1] Department of Electronics and Computer Engineering, Gifu University
Yanagido, Gifu 501-11, Japan (e-mail: fujita@fjt.info.gifu-u.ac.jp)

[2] Department of Radiology, National Cardiovascular Center
Suita, Osaka 565, Japan

[3] Osaka University Medical School
Suita, Osaka 565, Japan

ABSTRACT

We have developed a computerized scheme that can aid in the radiologist's diagnosis in the detection and classification of coronary artery diseases in ^{201}Tl myocardial SPECT (single photon emission computed tomography) bull's-eye images by use of artificial neural networks. The multi-layer feed-forward neural networks used with a back-propagation algorithm have 41-256 input units (pattern : compressed images), 50 or 100 units in a single hidden layer, and eight output units (diagnosis : one normal and seven different types of abnormalities). The neural networks consisting of two major networks for "EXTENT" and "SEVERITY" images were trained using pairs of training input data (bull's-eye image) and desired output data ("correct" diagnosis). The results show that the recognition performance of our neural-network-based system is comparable to that of two year experienced radiologists. Our study suggests that the neural network approach is useful for the computer-aided diagnosis of coronary artery diseases in myocardial SPECT bull's-eye images.

INTRODUCTION

Thallium-201 (^{201}Tl) myocardial SPECT imaging [1] has been reported to offer major improvements over planar imaging in the diagnosis of coronary artery disease. However, to overcome the difficulties of the interpretation of the myocardial SPECT images, a polar-map display, called a bull's-eye image, has been developed to characterize the three-dimensional images of the left ventricle in two dimensions [1]. Even with this technique, many problems have been indicated. Also, the number of experienced radiologists in this field is substantially few. The development of a computer-aided diagnostic system or expert system, therefore, is considered to be helpful for the diagnosis of bull's-eye images.

The purpose of this paper is to develop a computerized system, which can aid radiologist's diagnosis in the detection and classification of coronary artery diseases in ^{201}Tl SPECT bull's-eye images, by employing artificial neural networks. One of the advantages of the neural network approach is its powerful ability to analyze the physician's complicated decision-making or pattern-recognizing process in diagnosis without any need to write a special computer program. As a pilot study, we recently investigated the applicability of the neural network technique in the computer-aided diagnosis of coronary artery diseases only when the bull's-eye "EXTENT" images were used for analysis [2], and also studied on the effects of image processing and neuro parameters on the system performance [3]. In this paper, we present the results of recognition performance when EXTENT and "SEVERITY" images were used for analysis with multi neural networks.

MATERIALS AND METHODS

Thirty-six planar images of 64 × 64 matrix and 64 gray levels were obtained (30 sec/view) with a gamma camera (Shimadzu LFOV dual head) and these data were transferred to a data processing system (Shimadzu SCINTIPAC-2400) at the Department of Radiology, National Cardiovascular Center. This system produces three different types of bull's-eye images at the same time, i.e., "PIXEL CT", "EXTENT", and "SEVERITY" images, which respectively represent the original bull's-eye image, the image simply showing the extent area of disease relative to the averaged normal case, and the image showing the severity of the disease within the extent area. In the present study, we used both of EXTENT and SEVERITY images. Actually, when physicians interpret the bull's-eye images, they first look at the EXTENT image, and then look at the SEVERITY image carefully.

Coronary artery territories in the bull's-eye display are illustrated in **Fig.** 1, where the regions of three main coronary arteries, left anterior descending coronary artery (LAD), left circumflex coronary artery (LCX), and right coronary artery (RCA), are segmented [4]. It should be noted that this figure shows approximate territories and many variations, overlaps and exceptions in each territory can exist, so if one utilizes an AI rule-based expert system it may not be easy to construct a system. The coronary artery diseases can therefore be classified into seven different types due to the existence of single-, double-, and triple-vessel diseases. We collected a total of 74 bull's-eye images. All of them had been examined by coronary angiography, in which a coronary artery of more than 75% stenosis was diagnosed as "disease" according to the criteria of the American Heart Association (AHA). These results were employed as a gold standard or "correct diagnosis".

We employed a personal neuro-computer system (Neuro-07, NEC), which consists of a personal computer (PC-9801 VX21, NEC), a neuro-engine board (PC-98XL-02, NEC), and a neuro-software package ("Michi-Zane", NEC). The neural network software written in C language is based upon a feed-forward layered model with an input layer, one to three middle or hidden layer(s), and an output layer.

A preprocessing of the image data is required due to the limitation of memory capacity of the neuro-engine board. This is also important to save the computation time. Therefore, all of the bull's-eye images studied were compressed to produce the images of 16 × 16 matrices by averaging the

Figure 1
Coronary artery territories in the bull's-eye image.
LAD: left anterior descending coronary artery
LCX: left circumflex coronary artery
RCA: right coronary artery

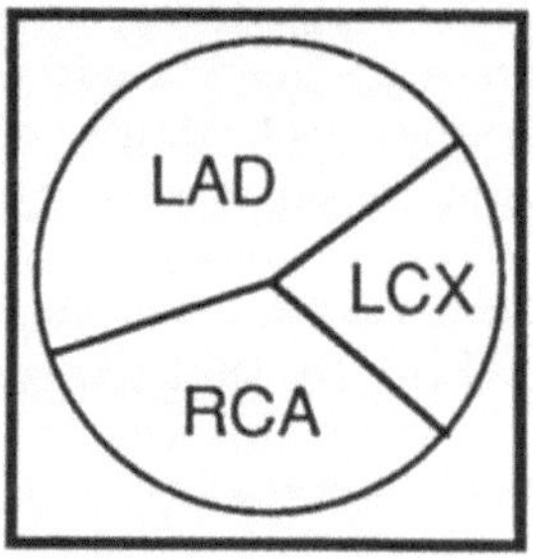

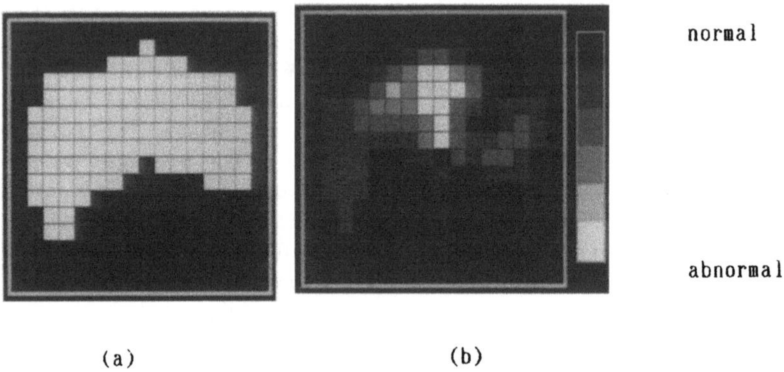

Figure 2 Preprocessed bull's-eye images in the case of LAD+LCX double vessel disease. (a) EXTENT image of a 16×16-matrix size with binary gray levels. (b) SEVERITY image of a 16×16-matrix size with 6 gray levels.

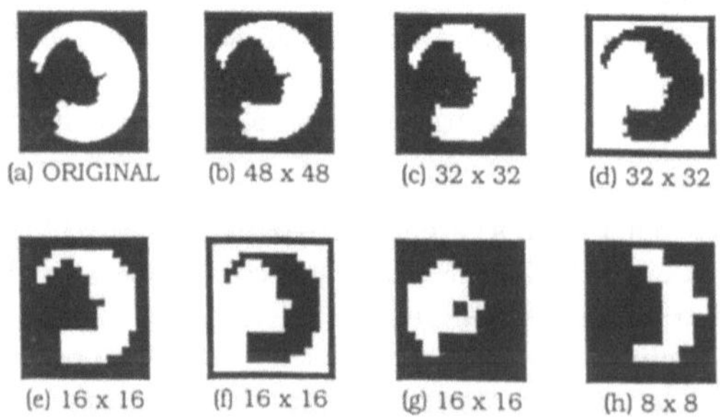

Figure 3 Effects of matrix size and pre-processing (inversion or background blackening) of EXTENT images. Inverse-processed 16×16 matrix is considered enough by the overall judgement in terms of the recognition rate, training time and data volume [3].

neighbouring pixel values and also to produce binary-gray-level images for the EXTENT and six-gray-level images for the SEVERITY. As an example, preprocessed images are shown in **Fig. 2**, which is a case of LAD+LCX double-vessel disease. The effects of the matrix size and preprocessing of the EXTENT images on the system performance are explained in **Fig. 3**. These compressed images were employed as an input to the neural network. The number of neurons in the output layer was fixed to eight units corresponding to the eight different types of diagnoses including normal. The neural network was trained using pairs of training input images (compressed images) and the desired output data ("correct diagnosis" based on the gold standard).

The overall flow of our system, called a "BULLsNET", for the recognition (testing) process is shown in **Fig. 4**, which includes the multi neural networks. In the case that the confidence level (CL) of the "extent neuro" is lower than 0.9, the "severity neuro" is performed, in which the each part of the vessel regions based upon the territories in **Fig. 1** is examined by LAD, LCX and RCA sub-neuro networks, then the output result from the severity neuro is used as diagnosis. The confidence level was determined from the weight values in the output layer of the network. On the other hand, in the case that the confidence level is equal or larger than 0.9, the output result from the extent neuro is simply used as diagnosis.

Table 1 Recognition rate for three different combinations of image data, only when the extent neuro was employed [2].

	Case A	Case B	Case C	Average
NN	69 %	75 %	88 %	77 %
I	56 %	81 %	69 %	69 %
II	75 %	75 %	88 %	79 %
III	75 %	88 %	88 %	83 %

NN: Neural network, I: three-month RI-experienced resident
II: two-year RI-experienced radiologist, III: ten-year RI-experienced radiologist

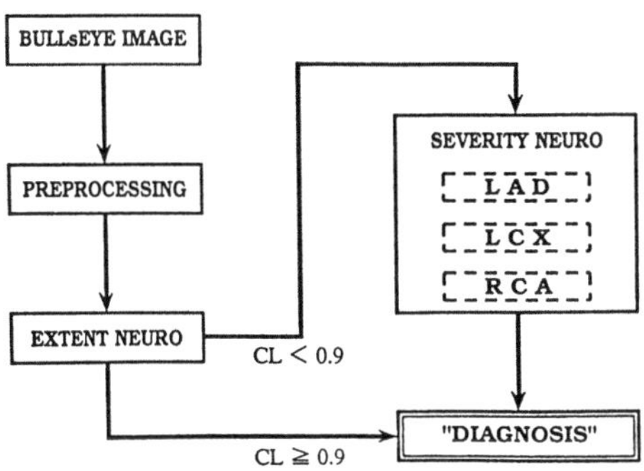

Figure 4
"BULLsNET" system for computer-aided diagnosis of SPECT bull's-eye images in nuclear cardiology.

RESULTS AND DISCUSSION

We varied combinations of image data for the training and testing (recognition) processes in the neural network and made three different combinations, cases A, B, and C, in which all images were chosen at random from a data base of 74 images.

In the case that only the extent neuro was performed, the following results were obtained. After the training procedure, all 58 images used for training were recognized correctly in the recognition process. The recognition rates (percentage of correct recognition) determined by the neural network (NN) for image data never used in the training process are listed in **Table 1**, together with those by one resident (I) and two radiologists (II and III) for comparison. This table demonstrates that the recognition rate depends on the image combinations, which can be explained from two different view points. One is that image data used for the training may be insufficient to recognize image data for testing. The other is that the inclination of image data in terms of their category in the training process and the degree of difficulty in diagnosis in the recognition process may cause variances in the recognition rate. These effects may be decreased by increasing the image data for training as well as for recognition. By comparing the averaged results, the performance of the neural network is better than that of the resident, and comparable to that of the two-year experienced radiologist, but worse than that of the ten-year experienced radiologist. All results shown above were obtained using the network that involves one hidden layer with 100 hidden units and 200 training iterations. Under these conditions the pure computation time for training was approximately 23 minutes. However, the time for training is not so important, because the user at the hospital may simply utilize the results obtained from the training process. On the other hand, the recognition of one image data in the testing process, including the preprocessing procedure, was performed in "real time".

It is worthwhile to include the SEVERITY image for analysis, as shown in **Fig. 4**, because they can help to differentiate lesions from artifacts. Actually, in the case of the physicians, we observed that the recognition rate with both of EXTENT and SEVERITY images results in 6–10% higher relative to that only with EXTENT images. The recognition rate determined by the neural networks using both images when the confidence level from the extent neuro is lower than 0.9 was 85%. It is considered to be comparable to that by the two-year experienced radiologist. The numbers of pixels of input image data (input units) for each region were 61 (LAD), 41 (LCX), and 47 (RCA), respectively. The number of training iterations was 150 for each region with 50 units in the hidden layer. The percentage of the case that the confidence level of the extent neuro was smaller than 0.9, that is, the severity neuro was necessary for analysis, was 35%.

Differing from AI expert systems in which complicate large programming is required for formation of rules based on capturing the knowledge of one or more experts, the neural network systems are able to form those rules by training procedure using pairs of training input data and desired output data (strength of neural networks). However, understanding how to effectively execute this "training" is the key point for the neural network system; in general, one has to collect enough patterns to train the network (weakness of neural networks). On the other hand, a rule-based expert system might be a better approach in the case of difficulty in preparing enough patterns to train it. Therefore, the process of learning or training can be an advantage of the neural network approach as well as a disadvantage. Moreover, hybrid systems, called "expert networks", that include both neural networks and rule-based expert systems, may be useful for further complicated applications [5].

CONCLUSION

The approach with artificial neural networks for a computer-aided diagnostic system of coronary artery disease in nuclear cardiology appears to show considerable promise, as shown in this study, when both of EXTENT and SEVERITY bull's-eye images are employed. The recognition performance of our improved system (BULLsNET) is comparable to that of the two-year experienced radiologist. However, in order to improve our system, it is required to increase the number of image data for training and testing processes. In addition, other clinical information, such as sex, temperature, and electrocardiogram data, has to be included in the overall analysis.

ACKNOWLEDGMENTS

We would like to thank Mr. Y. Horio of Gifu University (presently, Toyota Co.) and NEC Information Technologies for help with the technical assistance. We also are indebted to the radiologists and residents who participated in the decision performance studies.

REFERENCES

1.GUIBERTEAU MJ (ed.): Nuclear cardiovascular imaging: current clinical practice. New York: Churchill Livingstone 1990.

2.FUJITA H, KATAFUCHI T, UEHARA T, NISHIMURA T: Application of artificial neural network to computer-aided diagnosis of coronary artery disease in myocardial SPECT bull's-eye images. J. Nucl. Med. **33** (1992) 272-276.

3.SHINODA M, FUJITA H, KATAFUCHI T, et al: Development of a computer-aided diagnostic system for myocardial SPECT images: Effects of image processing and neuro parameters.
Med. Imag. and Inform. Sci. **10** (1993) 38-45.

4.GARCIA EV, VAN TRAIN K, MADDAHI J, et al: Quantification of rotational thallium-201 myocardial tomography. J. Nucl. Med. **26** (1985) 17-26.

5.EBERHART RC, DOBBINS RW: Systems considerations (Chap.3). In:EBERHART RC, DOBBINS RW, eds. Neural network PC tools: a practical guide. San Diego: Academic Press 1990, 59-79.

Measuring Edge Blur in Mammographic Lesions

E. Claridge[1], J. H. Richter[2], P. Stonelake[3]

[1] School of Computer Science, The University of Birmingham, Birmingham B15 2TT, U.K.
[2] Bull AG, European Projects, Max-Planck-Str. 17, Erkrath, Germany.
[3] Department of Surgery, The University of Birmingham, Birmingham B15 2TT, U.K.

Introduction

One of the diagnostic signs used by radiologists to assess whether mammographic lesions are benign or malignant is the sharpness or blur of the lesion periphery, the "edge blur". The blurring of lesion edges in mammograms (on radiographic film and in digital images) has a number of causes. Some blur is a result of processes occurring during image formation, such as the divergence of x-ray beam; x-ray scatter in the imaged tissues; light scatter from an intensifying screen; and light scatter in the process of image digitization (either from a light box or a light beam). Such blur is an unwanted artifact, distorting the image data; it can be partially eliminated by techniques based on modelling of the imaging process (e.g. [1]). Alternatively the amount of blur can be estimated from images of x-ray opaque objects with sharp edges, such as blocks or wedges.

A small amount of apparent blurring at the lesion periphery results because x-ray attenuation is dependent on tissue thickness. An example, in figure 1, illustrates how a lesion with a perfectly sharp periphery can produce an image with an apparently blurred edge. As the lesion thickness decreases towards the periphery the amount of x-ray photons absorbed by it decreases proportionally leading to a gradual rather than a sharp fall-off in brightness in the resulting film.

Whilst visual observers have no problems in assessing whether an edge is blurred or sharp, their assessment of the degree of blur is unreliable, even if they have only to compare the relative degree of blur of two edges which they can see simultaneously [2]. The comparisons get more difficult when the amount of blur increases and also when the contrast of the edge decreases. The increase in blur and the decrease in contrast also decrease the accuracy of edge localisation [2]. Blur which is of diagnostic interest arises because the periphery of malignant circumscribed lesions is thought to be poorly defined through the gradual spread of malignant tissue away from the centre. In such lesions a definite boundary between the lesion and the surrounding tissue is hard to find, because the malignant tissue is only slightly more dense than glandular tissue.

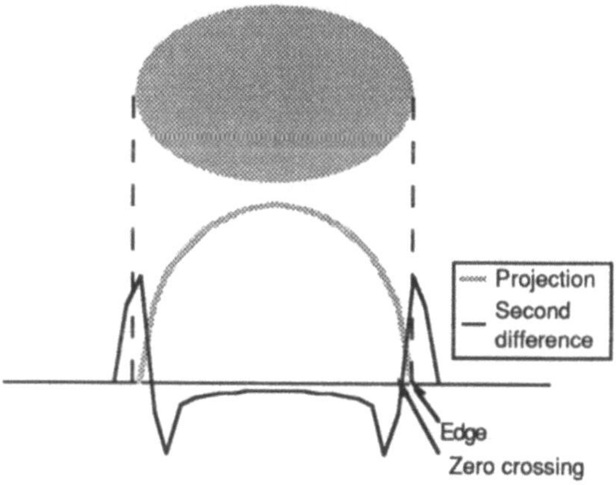

Figure 1. Profile of the projection of a solid ellipse and second derivative of the profile. Zero crossing of the second derivative does not coincide with location of the ellipse's edge.

Computer methods are of interest in measuring blur more reliably than is possible by humans because of limitations in human visual perception. As well as defining algorithms which permit objective measurements to be made computers enable compensation for imaging blur and for for shape-generated blur.

Objectives of work

The assertion that blurred lesions are malignant is based on observation and experience. It must be appreciated that established observation methods are subjective and in some cases may involve errors because of the way in which the human visual system works. The objective of this work is to develop methods of accurate measurement of edge blur by computer. At least these measurements are objective; they are not dependant on perception and they are directly related to the signal coming from the tissue.

To be able to make comparisons between different lesions, the blur measure of choice should have a natural progression. To be able to compare computer measurement with the observations of radiologists the blur measurement should ideally be expressed as a single value which captures just the property of blur as it is intuitively assessed

The overall primary objective of this work is to see whether blur, when accurately measured, may serve to distinguish between malignant and benign lesions. Secondly the work will seek to compare the computer measure of edge blur with the assessment of blur made by radiologists. Finally we will seek to use the computer to help to assess the location of the lesion's edge. This is very important if the lesion is to be surgically removed. Again there may be differences between assessments of the "real" edge of the lesion made by a computer and by a radiologist. Thus the work may have application in diagnosis, research and eventually in therapy. In this paper edge blur assessments and edge placements, by a radiologist and by computer, are compared for a set of confirmed malignant lesions.

Materials and methods

This study examined mammograms of 10 patients with biopsy-confirmed malignant lesions. For each patient two mammograms (lateral and AP) of a diseased breast were obtained. A consultant radiologist was asked to assess following features in each mammogram:

(1) The degree of blur for the entire periphery of the lesion

(2) The degree of blur for four sectors of 90° of the lesion

(3) The location of the lesion's edge

The degree of blur was expressed by a number from 1 (no blur) to 5 (severe blur). The subdivision into sectors and the location of the edge were drawn by the radiologist on a transparent film overlay firmly attached to the mammogram. Figure 2(b) shows a mammogram together with the overlay.

Computer methods of blur measurement

Mammograms both with and without the overlay were placed on a light box, photographed with a Canon still-video camera and digitised using a Neotech board. The resulting digital images have a linear spatial resolution of 0.3 mm/pixel and brightness resolution of 8 bits per pixel (256 grey levels). The digital images of the lesions with and without the overlay were aligned and the alignment was tested by subtracting the two images. As well as confirming the correct spatial alignment, the difference images showed clearly the location of the lesion outline marked by the radiologist. This outline served to calculate the centre of gravity for the lesion and was subsequently used as a centre of the polar coordinate transform applied to the lesion.

The blur measure was derived for each radial profile of the lesion (i.e. a column of the polar coordinate transform) using a method developed by Richter and Claridge [3]. In that method the step edge is taken to be an edge without any blur. The value of the parameter sigma (σ) in the Gaussian blurring function

$$G(\sigma) = \frac{1}{\sqrt{2\pi}\,\sigma} \exp[-\frac{1}{2} (\frac{x-\mu}{\sigma})^2]$$

which blurs the step edge so that it matches an edge being evaluated, is taken to be the blur measure for this edge. The parameter sigma has the desired property that it evaluates blur independently of the edge contrast. For a given edge profile the blur parameter is evaluated by de-convolving it with a step profile of the same contrast. The result of the de-convolution is a profile with an approximate Gaussian shape from which the parameter sigma is estimated.

Computer methods of edge location

For each radial profile two possible edge locations were calculated. The first one is placed at the zero-crossing of the profile convolved with a LoG operator [4]. This is a common practice in detecting edges of solid objects in images where pixel values register the amount of light (photons) reflected from various surfaces. The nature of mammograms is different in that pixel values correspond to the amount of x-ray photons being transmitted through the matter which is partially x-ray transparent. Assumptions underlying the application of the zero-crossing method are not correct for this kind of image and will lead to incorrect results. For example, the zero-crossing method locates edges incorrectly in the profile in figure 1.

For this reason a second edge location was calculated which depends both on the location of the zero crossing and the degree of blur. It is placed at distance 2σ from the zero crossing, which corresponds to the location of the cut-off line yielding 99% of the area of the normalised Gaussian convolved with the impulse function. Intuitively this is the location where the blurred edge just merges into the background. Figure 2(c) shows the zero crossing and the edge location for the lesion in figure 2(a).

Estimation of accuracy for computer methods

The accuracy of this method was estimated for artificial lesions with a known degree of blur. The error in the numerical calculations, assessed by performing the calculations on artificial lesions on a plain background, was below 1% [5]. The error associated with the entire method, assessed by performing calculations on artificial lesions embedded in a mammographic background, did not exceed 10% [3].

Comparison of results between radiologists and computer

The blur parameter was calculated for each of the 256 radial profiles of the lesion. The overall blur measure was calculated by taking the average of the individual measures. The blur measure for a sector was the average of individual values taken over that sector. In order to perform comparison between the blur parameter derived by the computer method and blur assessment provided by the radiologist the computer derived data was scaled to fit the range between 1 and 5. Measurements for several sectors had to be adjusted for lesions which were either very close to the edge of film or very close to the pectoral muscle. In these cases the average sector blur was calculated from radial profiles which were not upset by the film edge or the muscle.

Results

The degree of blur indicated by the radiologist for the sector data was in the range between 2 and 5. 45% of the blur values were the same for the radiologist and the computer evaluation; 85% differed by 1 or less. The discrepancy between the measurements was greater for higher values of blur. The overall blur measure for malignant lesions studied was 3.8 (standard deviation 0.7, variance 0.4).

Two possible reasons for the discrepancies were identified. One is related to inability of the algorithm to deal correctly with profiles showing not one but a number of edges (a "staircase"). Further work will be needed to decide on criteria of calculating blur measure in these cases. The second may be related to perception of edge sharpness. Where the lesion was either flanked by a dark streak or there was a bright ridge near its periphery the radiologist consistently reported blur values lower than those of computer; it is likely that apparent sharpening of the edge took place. Blur values reported by the radiologist tended to be higher than those of computer where contrast between the lesion and its background was low (normally for denser breasts). This problem needs to be researched further, possibly by using simulated images with artificial lesions of known blur.

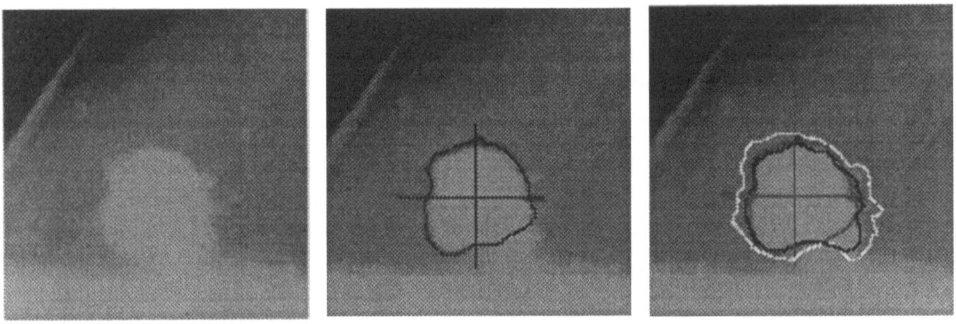

(a) (b) (c)

Figure 2. (a) A part of a mammogram. (b) Mammogram together with overlay drawn by the radiologist. (c) Mammogram with superimposed zero-crossing (black line) and the computer derived edge location (white line).

Discussion and work in progress

On the basis of this study we can conclude that there is a good agreement between the assessment of the edge blur between the radiologists and the computer derived measurements. On one hand this is pleasing in that the chosen method of blur assessment seems to reflect well human perception of blur. In this respect the method developed has potential for being included in a future computer-based screening system. On the other hand the results are not in agreement with data published by Watt [2]. However, Watt used untrained people from the general population in his study, rather than trained radiologists. It must be noted, however, that our preliminary study only involves assessment by a single radiologist.

At the time of going to press only a set of malignant lesions has been examined. Work is in progress to evaluate a similar set of known benign lesions and compare their blur measures. Results obtained to date on a very small sample of benign lesions suggest that there may be a difference of sigma about 2. It is expected that these results will be available for verbal presentation at the conference. In the near future we also plan to develop better models of sharp lesion profiles based on values obtained from physical models and from benign lesions and to develop computer methods of evaluation of correctness of the blur measure for a profile.

The definition of the of edge location yielded interesting results. The computer assessment of the edge location by the zero-crossings are in agreement with the delineations of the overlays provided by the radiologist. However, edge location calculated by adding spread estimated from the blur factor goes beyond the zero crossing. We intend to extend our investigation by attempting to remove imaging blur, so as to leave only physiological blur. The suspicion is that the extent of the lesion is underestimated by the radiologist because the human visual system detects zero-crossings whereas malignant tissue spreads further. Further plans involve the study of correlation between cell spread (assessed by cytological methods) and perceived and measured blur in mammograms.

Acknowledgements

The authors wish to thank Dr J. McMinn, a Consultant Radiologist from Dudley Road Hospital, Birmingham, for her work on assessing the mammograms; and to Mr M.J.R. Lee, a Consultant Surgeon, for his helpful comments and suggestions.

References

1. Highnam, R.P.; Brady, J.M.; Shepstone, B.J. Model-based enhancement of mammographic images. Paper submitted to IEEE Medical Imaging, April 1992. Corresponding author R. Highnam, Robotics Research Group, Department of Engineering Science, Oxford University, Parks Road, Oxford OX1 3PJ, U.K.
2. Watt, R.J.; Morgan, M.J. The recognition and Representation of Edge Blur: Evidence for Spatial Primitives in Human Vision. Vision Research 23 (1983) 1465-1477.
3. Richter, J.H.; Claridge, E. Extraction and measurement of lesion edge blur in mammograms by computer image analysis, Computer Assisted Radiology'91 (Lemke, H.U., Rhodes, M.L., Jaffe, C.C. and Felix, R., Eds): Springer Verlag (1991) 301-307.
4. Marr, D.; Hildreth, E. Theory of edge detection. Proceedings of the Royal Society of London B207 (1980) 187-217.
5. Richter J.H.; E Claridge Extraction of quantitative blur measures for circumscribed lesions in mammograms. Medical Informatics 16 (1991) 229-240: Taylor & Francis.

Analysis of the Requirements for Quantitative Evaluation of Mammograms by Computer

J.H. Richter[1] and E. Claridge[2]

1 Bull AG, European Projects, Max-Planck-Str. 17, 40683 Erkrath, Germany.
2 School of Computer Science, The University of Birmingham, Birmingham, B15 2TT, U.K.

General introduction

Breast cancer is the most common form of cancer in women today. The best long term outcomes are achieved with treatment that follows early detection. The results of breast screening programmes from a number of countries show that mammography can be used successfully for early cancer detection. A screening radiologist has to assess as many as 110 mammograms per hour [1], which is very tiring and can lead to decrease in diagnostic performance. The Forrest Report [1] recognises that "if mammography is to remain the standard method of early detection, further work will be needed on computer processing, storage and recall of mammographic images".

Screening and symptomatic diagnosis of mammograms

It is widely recognised that the interpretation of mammograms is very difficult and requires considerable skill and experience. A typical textbook recommends the following systematic approach to mammogram analysis. First the radiographic quality of the mammogram should be checked to assess film and image quality. Then the symmetry of both breasts is assessed and then a systematic search made for mass lesions and calcifications. Opacities or poorly defined masses of irregular contour and the presence of clustered microcalcifications are the two main direct signs of malignancy. Finally an examination of the skin line, lymphatic nodes and other detail is made to search for indirect signs of malignancy such as dilated ducts, architectural distortions, asymmetries or developing densities [2, 3].

There are two kinds of mammographic examinations. The objective of mammographic *screening* is to detect the presence of any abnormal features in a large number of mammograms, most of which are normal. Thus the screening radiologist has to take a binary decision: yes or no. At the screening stage it is essential that sensitivity (the percentage of abnormal mammograms classified as abnormal) is 100% so that no abnormal mammogram is classified as normal; that specificity (the percentage of normal mammograms classified as normal) should be as high as possible to keep to the minimum the number of further examinations and to prevent psychological stress in wrongly diagnosed patients.

The objective of *symptomatic diagnosis* is to perform careful examination of mammograms with abnormal appearance to decide whether these features are of benign or malignant nature. If malignancy is suspected the diagnostic mammograms are used to plan biopsy or surgery and so need to provide information about the location and the extent of the abnormal area. The number of mammograms that the radiologist sees is much smaller than in screening but each mammogram needs to undergo very careful examination. At this stage both sensitivity and specificity should ideally be 100% as an incorrect diagnosis of malignancy on the basis of the mammogram leads to a surgical examination.

Image formation

Before considering computer analysis of mammograms we will briefly examine the relationship between the breast tissue and its digital image. Fully digital mammography systems are not very common. In traditional X-ray mammography a beam of photons passes through breast tissue, where some photons are absorbed and some are scattered; the degree of absorption depends on the tissue density, its thickness and on the x-ray energy (figure 1a). The remaining photons are absorbed by an intensifying screen which in response emits visible light (figure 1b). The light photons cause darkening of the film (figure 1c and 1d). To obtain a digital image suitable for computer analysis the film is usually either scanned with a light beam or placed on a lightbox and exposed to a video camera; the resulting analogue signals are converted into integer values within a range dependent on the digital storage available.

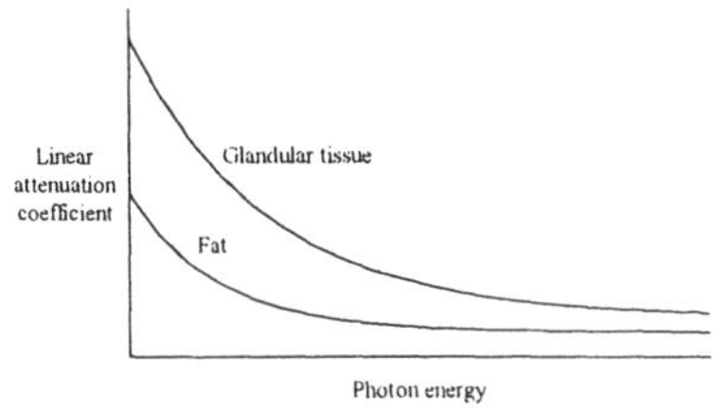

Figure 1a: Relationship between photon energy of x-ray beam and linear attenuation coefficient for different types of tissue (from [4]).

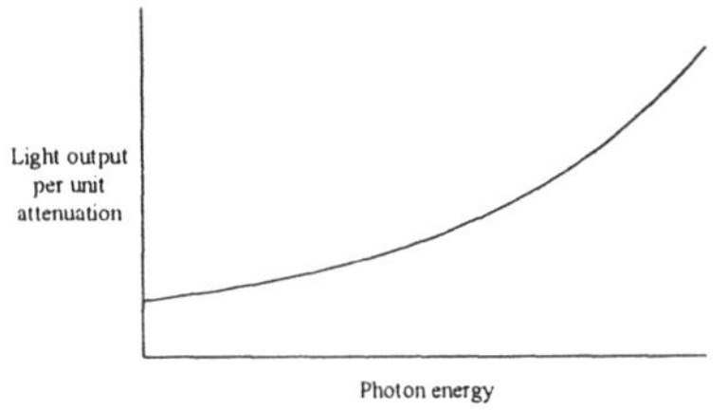

Figure 1b: Variation with photon energy of the light intensity emitted per unit x-ray attenuation for a typical intensifying screen (from [5]).

Figure 1c: Relationship between photon energy absorbed by an intensifying screen and film density.

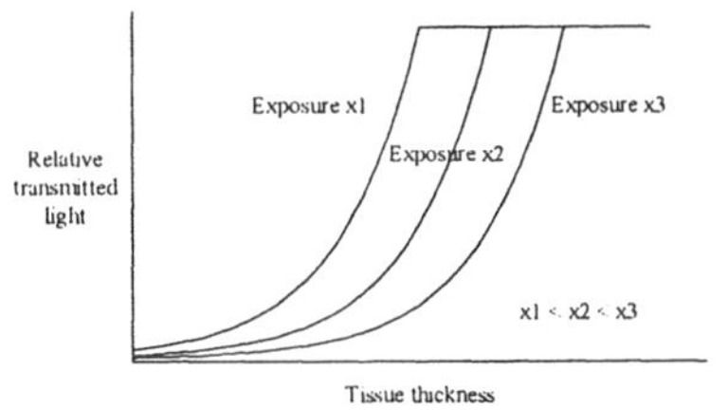

Figure 1d: Typical relationship between thickness of tissue and relative amount of light transmitted through a developed x-ray film for different exposure times (from [6]).

The subject of potential computer analysis is tissue density, the medium on which it is recorded is a digital image and it can be seen from the diagrams that the relationship between the two is not a simple one.

Computer assisted diagnosis

Computer systems developed so far cannot surpass radiologists in detection of subtle signs of disease. Fully computerised screening remains a long term research objective. In the field of computer-assisted screening research is carried out on image enhancement; on prompting (for human viewing); and on the detection of individual abnormalities. The main potential of computer systems is presently seen to be in the field of diagnosis where "objective" (usually quantitative) descriptions of abnormalities are derived from mammograms and used for classification.

This paper explores the objectivity of "objective" computer-derived descriptions and examines their usefulness as tools to the radiologist for whom they are intended. Potential problems with image enhancement are also discussed. The paper concludes with a number of recommendations for the qualitative evaluation of mammograms by computer. It does not consider further the problem of fully computerised screening; the detection of abnormalities or prompting tools.

Object depiction and geometric measurements

In this section we examine the problems associated with the derivation of quantitative descriptions of objects such as microcalcifications and lesions. Assuming that a suspicious area has been identified in a mammogram, either by the radiologist or by a computer, before any measurements can be taken the object has to be accurately de-lineated.

Microcalcifications are small dense objects and the problems are related to inadequate spatial resolution and excessive noise. Much attention has been devoted to these objects and the problems have largely been overcome. For a good treatment of the objective depiction and measurement of microcalcifications, see, for example Krassemeijer's work [7].

Most of the methods used for lesion delineation are based on classical signal-analysis approaches where the edge of the object is placed at the zero-crossing of a LoG-transformed image (i.e. at the centre of an intensity ramp between the lesion and background). This group of methods was devised for finding the edge locations of solid objects whose edges might have been blurred by distance, by recording or by the digitisation process. However, these methods are not suitable for localisation of edges in projected images where most of the matter being imaged is x-ray transparent to some degree. As a consequence, edges of relatively large objects, such as for example lesions, are misplaced and their size and extent are under-estimated (see [8]).

Brightness-related measurements

This area of assessment causes the greatest difficulty and raises the greatest concern for the validity of currently used methods. One of the "measurements" frequently performed on digitised mammograms is of the density of a lesion or of a microcalcification. This is considered to have diagnostic importance [9, 10, 11, 12, 13]. As mammograms are not routinely calibrated with respect to optical density (and even less so with respect to digitised grey level values), we must ask "what is being measured?" In the section "Image formation" above we listed a number of processes which affect grey level values in a digitised image. Most of these processes introduce non-linear transformations to the signal in its subsequent forms. Some transformations, e.g. field non-uniformity, can be modelled easily for a given machine and given x-ray energy; others, such as those introduced during film development, may vary between individual batches of films and cannot be modelled in advance. Thus readings obtained from a digital image characterise values of *pixels* but not necessarily the properties of *breast tissues*. Any quantitative comparisons may show differences between images and not between breast tissues.

The imaging process may affect both absolute values and the relationships between the values. Lesion density relative to parenchyma is considered to have diagnostic importance and is sometimes evaluated from digitised mammograms [14]. Consider, for example, a case where a film was over-exposed, with all other factors being equal; the relative density of a lesion will have decreased due to the non-linear response of the film to high-levels of exposure.

The density of the breast tissue also affects the relative lesion density; if the same lesion were embedded in a fatty breast and in a dense breast the image of the latter would indicate a smaller relative density. This problem is not limited just to digital imaging and is familiar to radiologists. A related problem is that of breast compression. It has been found that compression differences yielding only 1 cm difference in tissue thickness can affect the appearance of mammographic features to such a degree that different diagnoses can be made for the same breast [15]. It is not clear yet whether these two problems can be solved through quantitative evaluation and processing by computer. In any case, researchers in computer-based image analysis have been aware of this problem so that this measure is not used indiscriminately.

Image enhancement

Most of the work involving applications of computers to mammographic images concentrates on mammographic feature detection and evaluation. Comparatively less research is done on enhancement of mammographic images intended for subsequent viewing by a radiologist. Enhancement can take a number of forms. Contrast manipulation via histogram operations, such as the logarithmic transform, can be used to invert the exponential effect caused by the x-ray absorption [16, 17]. It can also be used to improve contrast by, for example, increasing the dynamic range of grey levels [18]. Noise removal operations can enhance the clarity of mammographic detail by decreasing the high frequency noise introduced by x-ray scatter, film processing and digitisation [19]. Edge sharpening operations, usually implemented by convolution, increase contrast at edge locations and therefore facilitate detection of edges that may have been blurred by x-ray scatter, or optical scatter during digitisation. Visual cueing is carried out by selective enhancement of features of interest, e.g. linear structures [20], bright speckles [21], or circular blobs [22] to indicate areas in the image that may require especially careful examination. Where the primary goal of image enhancement is to draw the attention of the viewer to the diagnostic features that otherwise might be camouflaged, an original mammogram is usually viewed alongside the enhanced one. This allows the radiologist to *detect* the feature in the enhanced image and *examine* it in detail in the original image. The cases where *only* the enhanced image is intended for viewing introduce a risk of mis-interpretation. For example, in a sharpened image a lesion's edge may appear less blurred and thus lead to incorrect diagnosis; enhanced noise may appear like microcalcifications; a histogram-equalised image may be not understandable to a radiologist who does not have training in image processing.

A notable example of image enhancement technique which does not suffer from similar problems is that proposed by Highnam [19]. Through modelling of physical processes involved in image formation it produces images free from degrading factors such as x-ray scatter and beam hardening and corrects for exposure errors.

Reporting

With all technical aspects appropriately considered, care must be taken about the way results are communicated to radiologists. If numbers or ratios produced by computer evaluation are expressed in purely technical language and bear no relation to terms that radiologists use themselves in describing their findings on mammograms, they are likely to be rejected. For example, in describing the shapes of individual calcifications one may use expressions such as "compactness = 0.9", "eccentricity = 6.2" and "elongatedness = 0.1". The meaning of these expressions is not immediately apparent to a non-technical person. Corresponding terms, such as "rounded", "branched" or "oblong" might convey the same information but in terms which are clearly understood by radiologists [16]. This transformation of quantitative terms into qualitative terms cannot be done in an *ad hoc* way and is best developed in collaboration with radiologists and through controlled experiments [8]. Results produced in this way have the benefit that the measurements are performed objectively but expressed in terms that are understandable and acceptable to radiologists.

Conclusions

The analysis of mammograms is a very complex procedure. Even if the focus of work is on the analysis of particular features, like lesions and microcalcifications, it is necessary to understand and take into account both the entire process of image formation and the perceptual processes involved in visual examination of mammograms.

One way to achieve this for the image formation process is through precise modelling of the entire process. A large body of knowledge exists, acquired by medical physicists, about the interactions between x-ray energy and matter and about properties of imaging systems and media; it ought to be used more frequently by medical image processing workers. The understanding of the image formation process will help to develop algorithms which remove factors dependent on imaging conditions so that analysis can be performed on tissue data and not on the image data.

Any techniques developed should be evaluated on model data for which correct results are known so that results obtained by processing can be compared against the model results. This opens the way to eliminate errors and weaknesses in the techniques and also may help define error bounds within which the techniques operate correctly.

If computer analysis is to deliver reliable, repeatable and comparable measurements then digital and digitised mammograms must contain means of spatial, brightness and possibly spectral calibration, such as rulers, step wedges and gratings.

Studying the perceptual processes involved in visual diagnosis can bring a number of benefits. Computer image analysis can concentrate on tasks for which the human visual system is known to have weaknesses, such as edge localisation, blur and size assessment. On the other hand, computer methods can learn from those tasks in which the human visual system is particularly efficient, such as segmentation and feature detection. The strengths of human and computer analysis can be combined in hybrid systems, sometimes by using computers to prompt suspicious areas for human evaluation, sometimes by letting radiologists point to areas detected by visual examination so they can be measured by computer.

Based on these considerations we conclude that quantitative evaluation of mammograms by computer be supported by the following endeavours:

- The modelling of breast tissue to understand how it affects a beam of radiation (including scatter) depending on tissue composition, thickness, compression, etc.
- The modelling of the imaging process, using the parameters of the imaging system geometry, radiation beam, etc., to derive a digital image which is independent of the imaging process and reflects just properties of the imaged tissue. This cannot be done without calibration and we recommend that mammography systems are equipped with calibration tools.
- The use of purpose-designed signal processing algorithms which are based on the above models.
- The utilisation of studies of the human visual perception to identify areas of difficulty in visual assessment and to improve segmentation and recognition algorithms.
- The careful design of feature descriptions which are quantitative and precise but also capture (and can be translated into) qualitative descriptors intuitively used by radiologists.

With these requirements met we believe that robust computer analysis of mammograms is possible and likely to be acceptable to radiologists.

References

1. Breast Cancer Screening: Report to the Health Ministers of England, Wales, Scotland & Northern Ireland by a working group chaired by Professor Sir Patric Forrest. Department of Health and Social Security, London, Her Majesty's Stationery Office (1986).
2. Gravelle, I.H.: Diagnostic Imaging in Breast Cancer. In: Breast Cancer - Clinics in Oncology (Baum, M. ed.). Sanders (1982).
3. Sickles, E.A.: Mammographic Features of "Early" Breast Cancer. American Journal of Roentgenology 143 (1984) 461-464.

4. Johns, P.C.; Yaffe, M.J.: Coherent scatter in diagnostic radiation. Medical Physics 10 (1983) 664-668.
5. Moores, B.M.: Films and Screens. In: Physical Aspects of Medical Imaging (Moores, B.M.; Parker, R.P.; Pullam, B.R. eds.). Wiley (1981) 253-265.
6. Highnam, R.P.; Brady, J.M.; Shepstone, B.J.: Computing the Scatter Component of Mammographic Images. Paper subitted to IEEE Medical Imaging, June 1992. Corresponding author: R. Highnam, Robotics Research Group, Department of Engineering Science, Oxford University, Parks Road, Oxford, OX1 3PJ, UK.
7. Krassemeijer, N.: A stochastic method for automated detection of microcalcifications in digital mammograms. XIIth Conference on Inf. Proc. in Med. Imaging, 7-12 July 1991, Wye College, Kent, UK. Corresponding author: N. Krassemeijer, Department of Radiology, St. Radboud University Hospital Nijmegen, The Netherlands.
8. Claridge, E.; Richter, J.H.: Measuring edge blur in mammographic lesions. In: Computer Assisted Radiology (Lemke, H.U.; Rhodes, M.L.; Jaffe, C.C.; Felix, R. eds.). Springer Verlag (1993) xxx-xxx.
9. Tabar, L.; Dean, P.B.: Teaching Atlas of Mammography. Stuttgart: Thieme (1985).
10. Smith, K.T.; Wagner, S.L.; Guenther, R.B.; Solomon, D.C.: The diagnosis of breast cancer in mammograms by evaluation of density patterns. Radiology 125 (1977) 383-386.
11. Brzakovic, D.; Luo, X.M.; Brzakovic, P.: An approach to automated detection of tumors in mammograms. IEEE Transactions on Medical Imaging 9(3) (1990) 233-241.
12. Spiesberger, W.: Mammogram inspection by computer. IEEE Transaction on Biomedical Engineering. BME26 (1979) 213-219.
13. Gilles, R.; Bouvet-Lefebvre, F.; Masselot, J.; Kahn, E.: Characterisation of Benign Clustered Microcalcifications with a New Image Analysis Method. In: Computer Assisted Radiology (Lemke, H.U.; Rhodes, M.L.; Jaffe, C.C.; Felix, R. eds.). Springer Verlag (1991) 295-300.
14. Lai, S.-M.; Li, X.; Bischof, W.F.: On Techniques for Detecting Circumscribed Masses in Mammograms. IEEE Transactions on Medical Imaging 8(4) (1989) 377-386.
15. Highnam, R.P.; Shepstone, B.J.; Brady, J.M.: Differential compression mammography. Submitted to The British Journal of Radiology, May 1992. Corresponding author: see [6].
16. Magnin, I.E.; El Alaoui, M.; Bremond, A.: Automatic microcalcification pattern recognition from x-ray mammographies. SPIE Science and Engineering of Medical Imaging 1137 (1989) 170-175.
17. Karssemeijer, N.; van Erning, L.: Iso-precision scaling of digitized mammograms to facilitate image analysis. Proceedings Medical Imaging V, San Jose, 1991. Corresponding author: see [7].
18. Richter, J.H; Claridge, E.: Towards better perception of diagnostic features in mammograms. Paper read at the workshop on "Computer-based methods in mammographic screening", organized jointly by the British Pattern Recognition Association (BPRA), the Institute of Physical Sciences in Medicine (IPSM) and the British Institute of Radiology (BIR), January 23rd, 1989, London, UK.
19. Highnam, R.P.; Brady, J.M.; Shepstone, B.J.: Model based enhancement of mammographic images. Paper subitted to IEEE Medical Imaging, April 1992. Corresponding author: see [6].
20. Ellison, T.P: Detection of stellate lesion in mammographic images. MSc Thesis. University of Manchester, Department of Medical Biophysics, Stopford Building, Oxford Road, Manchester, M13 9PT, UK. (1988).
21. Astley, S.M.; Taylor, C.J.: Combining cues for mammographic abnormalities. Proceedings of the BMVC. (1990) 253-258.
22. Kundel, H.L.; Nodine C.F.; Krupinski, E.A.: Computer-displayed eye position as a visual aid go pulmonary nodule interpretation. Investigative Radiology 25 (1990) 890-896.

Computer Diagnosis of Breast Cancer by Mammogram Processing

Hidefumi KOBATAKE*, Hua-Ron JIN*, Yukiharu YOSHINAGA* and Shigeru NAWANO**

* Faculty of Technology, Tokyo University of Agriculture & Technology
Koganei, Tokyo 184, Japan

**Department of Radiology, National Cancer Center Hospital East
Kashiwa, Chiba 277, Japan

Summary
This paper presents an automatic diagnostic system for breast cancer which is based on the processing of mammogram. The proposed system is composed of two subsystems. One is for tumor processing and the other is for microcalcifications. Two-dimensional adaptive filter called "Iris Filter" is developed to detect tumor candidates. New shape parameters and spicule detection method are introduced, which give the clue for benignancy/malignancy decision.The detection method of microcalcifications are based on the morphological analysis. Multi-structural elements are adopted. Experiments to test the performance of the proposed system have been made, whose experimental results show the possibility of its practical application to the breast cancer screening.

1. Introduction

Mass screening of breast cancer has become popular in advanced countries, and the number of mammograms to be diagnosed is increasing year by year. The development of automated screening system of breast cancer is urgently required. For reliable diagnosis of breast cancer, the identification of both malignant tumors and clustered microcalcifications is essential. Microcalcifications can be recognized as local maxima in terms of film density. To detect them, therefore, difference image approach has been adopted as a fundamental processing. That is, the difference between the original and the spatially smoothed images yields an image in which the variation in background density are largely removed. And thresholding is applied to the difference image to detect microcalcifications [1]-[3].

Identification of tumors is not easy from the view point of image processing. It is because the discrimination between the mass lesion and the mammary grands is difficult. Texture features have been adopted to characterize mass lesions [4], [5]. Deviations from the architectural symmetry of normal right and left breasts are also used to detect potential masses [6], [7]. Template matching method using circular templates was also used to detect mass lesions [8]. There have been many researches on automatic diagnosis of breast cancer as described above. However, it is required to develop a reliable diagnostic system of breast cancer such that a higher true-positive rate and a lower false-positive rate are attained at a time.

The purpose of our research is to develop a reliable system for automated diagnosis of breast cancer by CR (Computed Radiography) image processing. The proposed system basically consists of two subsystems. One of them is used to detect malignant tumors and the other is for detecting the presence of microcalcifications on mammogram. By combining the results of these two subsystems, a final decision is made. Experiments to test the performance of the proposed system are made, whose results show its effectiveness.

2. Tumor Detection System

The tumor detection subsystem consists of two processing steps. The first processing step is to detect candidates of malignant tumors. The second one is their discrimination between malignant tumors and other normal tissues. The details of their processing are as follows.

2.1 The First Processing Step

Malignant tumors are considered to be local high density areas on mammogram whose shapes are rounded as a zeroth-order approximation. However, they sometimes have very weak contrast in terms of a film density against their neighborhood, and they tend to be missed by simple spatial differentiation. We propose a novel filter called "*iris filter* "to detect candidates of mass lesions reliably ever if their contrast is weak [9].
Fig. 1 shows the region of support of the proposed *iris filter*. The output of the filter, x (i, j), is given as

$$x(i,j) = \frac{1}{N} \sum_{(k,l)\in R} \cos\theta(k,l)$$

where, (i, j) is the position of the pixel of interest, and N is the number of pixels in the region of support R of the *iris filter*. The variable θ (k, l) is defined as the angle between the gradient vector at (k, l) and the straight line connecting pixels at (i, j) and (k, l). The term cosθ (k, l) is a measure of convergence of the gradient vector upon the pixel of interest. If the pixel of interest is the pixel near the center of a mass lesion, the majority of the gradient vectors around it points to the pixel of interest,which means most of θ(k, l)'s are very small, that is, cosθ(k, l) = 1.0.
The shape of a malignant tumor can be approximated by a circle. Therefore, the output of the *iris filter* reflects the pattern of film density distribution. If the filter output is high, it means that the film density pattern around the pixel of interest is that of local high density area whose shape is approximately rounded.
The tumor has a great variety in sizes. To detect them reliably, the region of support of the iris filter must be adapted to the size of the local high density area. Such adaptation is introduced. The detail is given in the paper [9]. The *iris filter* is insensitive to long and slender shadows. It is sensitive only to rounded local high density regions. The areas whose outputs of the *iris filter* are higher than the experimentally determined threshold are decided as tumor candidates. They are discriminated between malignant tumors and other normal tissues by the second processing.

2.1 The Second Processing Step

Shape analysis is performed for tumor candidates to discriminate between malignant tumors and other normal tissues. Shape parameters adopted in the system are called *elongation* and *spreadness*. If a tumor candidate is rounded in shape, these parameters take small values. Then the square root of the squared sum of the two parameters is used to detect rounded candidates. Malignant tumor has a rough contour and sometimes spicules which are essential features differentiating between malignant and benign tumors. Two parameters *roughness* and *circularity* which are used to identify malignant tumors are defined as follows [10].

$$\text{Roughness} : R_f = \frac{1}{A(X)} \sum_{\lambda=1}^{N} \frac{PS(\lambda, B)}{\lambda}$$

$$\text{Circularity} : C_f = \frac{PS(\lambda, B)}{A(X)}$$

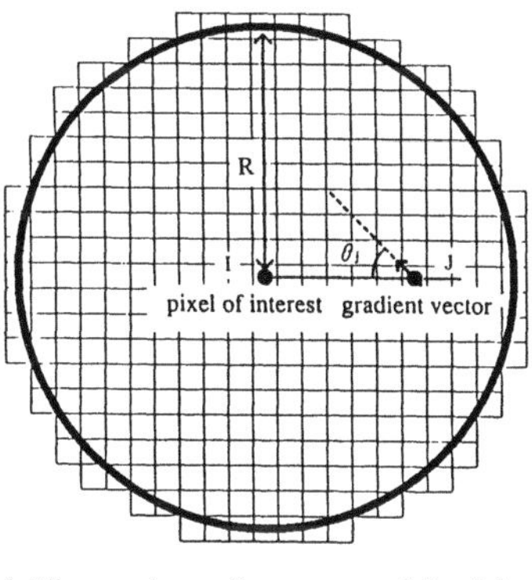

Fig.1 The region of support of the iris filter.

where,

A(X) : the area of the candidate,

PS(λ, B) : the pattern spectrum for the scale λ,

B : the basic structuring element,

λ : the scale,

and

N : maximum of the scale λ at which the pattern spectrum is not zero.

Spicules are detected as follows. First, morphological skeleton is obtained for the neighboring area of the candidate. Then Hough transform is applied to detect lines whose extrapolated lines intersect each other near the center of the candidate. If more than two lines which meet such conditions are detected, they are decided as spicules.

Feature parameters described above are combined to detect malignant tumors.

3. Microcalcification Detection System

The computerized scheme adopted in this subsystem detects microcalcifications using morphologic filters on the basis of their size, shape, and density. Morphologic filtering has been used to detect microcalcifications. In almost all applications of morphologic filtering, only a single structuring element with a disk shape or a square shape and very simple morphologic operations such as erosion have been used. We have developed a new effective morphologic filter, which is an extension of top-hat transformation with multiple structuring elements. It makes it possible to detect geometrical structures considered to be microcalcifications on the bases of their size, shape and density [11].

For the detection of microcalcifications on mammogram, two problems have to be solved. The first problem is how to combine basic operators to realize efficient filter for extracting microcalcifications. The second one is how to select the structuring element. It should be not only effective to detect microcalcifications but also insensitive to background noises such as mammary glands and mammary ducts. To solve the first problem, we evaluated the performance of several basic morphologic filters. Experimental results showed that Meyer's top-hat transform is effective to detect microcalcifications. Its performance can be controlled by the shape of the structuring element and its size. If we use a disc as the structuring element, mammary glands and ducts cause false microcalcifications. To reduce those false positive peaks, we adopted multiple structuring elements as shown in Fig.2. They are 16 primitive half lines. The morphologic filter adopted in our system is defined as follows.

$$
\begin{aligned}
P_\lambda &= f(i,j) - \max_{i \in (1,\cdots 16)} \{(f \ominus \lambda B_i) \oplus \lambda B_i\} \\
&= f(i,j) - \max_{i \in (1,\cdots 16)} \{(f_{\lambda B_i}\}
\end{aligned}
$$

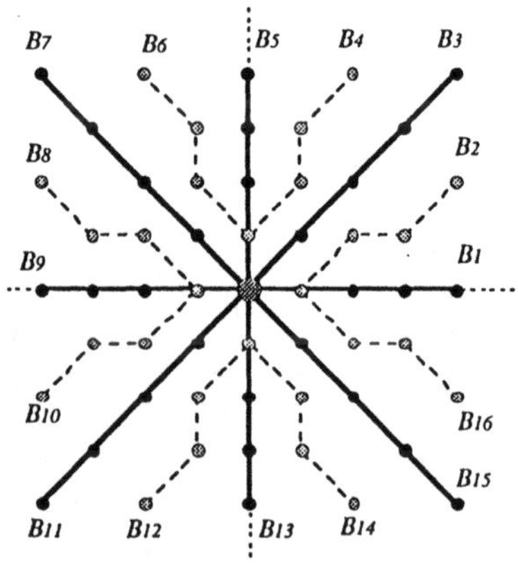

Fig.2 Multiple structuring elements.

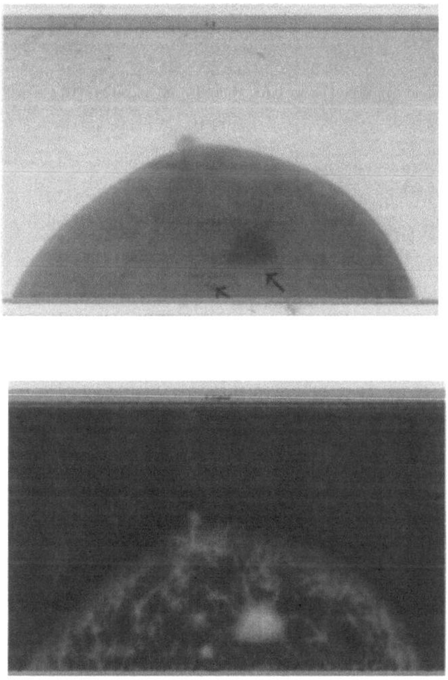

(b) Output of the iris filter.

Fig.3 An example of tumor enhancement.
Arrows indicate cancerous tumors.

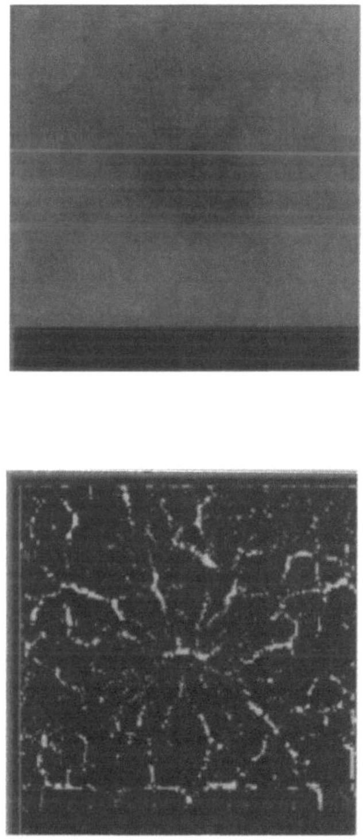

(b) Morphologic skeleton.

Fig.4 An example of spicule detection.

where, f is an image including microcalcifications. B_i's and λ are multiple structuring elements as shown in Fig.2 and a scale parameter, respectively.
To improve the detection sensitivity and to reduce the number of false positive detections, we introduced a morphologic gradient. True microcalcifications have usually edges much sharper than those of false calcifications. It means that they can be discriminated by the value of gradient. Let M_{grad} (i, j) be morphologic gradient of f at (i, j). If the value of P_λ is greater than a threshold and M_{grad} is also greater than a threshold, it means that suspicious microcalcification is extracted.
In order to detect clusters of microcalcifications, greyscale multi-scale closing and opening operators are used.

4. Software System

The automated diagnostic system for breast cancer works on the work station under UNIX operating system. The software system is composed of many software modules and it is easy to improve them and/or to add supplementary processing modules. The manipulation of the system is performed using a mouse. Every intermediate processing results can be accessible by pressing one of menu buttons on the CRT display, which makes it possible to evaluate system performance easily and to improve processing algorithms.

5. Experimental Results

5.1 Tumor Detection

Fig.3 shows an example of tumor detection. The original image and the output of the *iris filter* are given in the parts (a) and (b) of Fig.3, respectively. This mammogram shows two cancerous masses, one of which is large and the other is small (about 5mm in diameter). They are remarkably enhanced by applying the *adaptive iris filter* as shown in the part (b) of Fig.3. Final com-

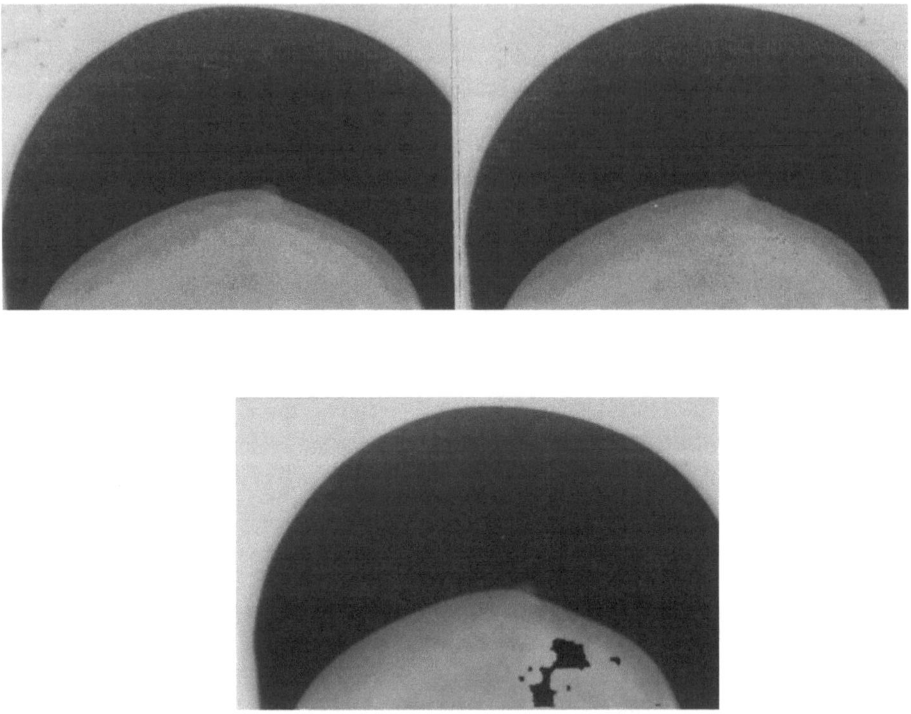

(c) Detected areas of clustered microcalcifications.

Fig.5 Detection of microcalcifications.

puter output indicates the two masses as malignant tumors correctly. No false positive detections are indicated for this example. Currently, with use of 50 mammograms (12 malignant and 38 normal mammograms), the automated diagnostic system achieved a true-positive rate of 100% and a false-positive rate of 24%. The average number of false-positive detections per image is only 0.6.

Fig.4 shows an example of morphologic skeletons. Spicules are identified as skeletons whose extrapolated ones intersect each other at around the center of the tumor. Such skeletons that meet those conditions can be detected by Hough transform.

5.2 Microcalcification Detection

Fig.5 shows an example of the detection of microcalcifications. Fig.5(a) shows a mammogram with a cluster of microcalcifications. Detected clustered microcalcifications after application of morphologic filtering techniques are shown in Fig.5(b). The clustered area is detected as shown in Fig.5(c).

The subsystem for microcalcification detection works very well. With use of 8 mammograms with clusters of microcalcifications, the computer scheme attained a 100% true-positive detection rate.

6. Conclusions

A new automated diagnostic system of breast cancer is presented. It consists of two subsystems. One of them is for the detection of mass lesions. A unique filter called *iris filter* is proposed to detect tumors. New shape parameters useful for differentiation between malignancy and benignancy are proposed. The other subsystem is for the detection of microcalcifications. Its processing techniques are based on mathematical morphology.

The performance of the proposed system was evaluated with use of CR images. Experimental results showed the effectiveness of the proposed system. The performance of the system is currently promising, and a large scale experiments are now being performed using a lot of CR images.

The Authors want to express great thanks to Dr. Yukio Tateno, Dr. Tatsuya Yamada, Dr. Kunio Odagiri and their colleagues for providing some parts of mammograms used in the experiments and for their helpful advices and discussions. This research was supported in part by the Grant-in-Aid for Cancer Research (3-45) from the Ministry of Health and Welfare, Japan.

References

[1] W. Siesberger, "Mammogram Inspection by Computer," IEEE Trans. on BME, Vol.26, No.4, pp.213-219, 1979.
[2] H. P. Chan, K. Doi, et al., "Image Feature Analysis and Computer-Aided Diagnosis in Digital Radiography," MEDICAL PHYSICS, Vol.14, No.4, pp.538-548, 1987.
[3] R.M. Nishikawa, M.L. Giger, K. Doi and R.A. Schmidt,"Automated Detection of Microcalcifications on Mammograms," Supplement of RADIOLOGY, Vol.177, No.11, pp.288, 1990.
[4] C. Kimme, B.J. O'Loughlin and J. Sklansky, "Automatic Detection of Suspicious Abnormalities in Breast Radiographs," In : A. Klinger, K.S. Fu and T.L. Kunii, eds., Data Structures, Computer Graphics and Pattern Recognition, Academic Press, New York, pp.427-447, 1975.
[5] W. Hand, J.L. Semmlow, L.V. Ackerman and E.S. Alcorn, "Computer Screening of Xeromammograms: A Technique for Defining Suspicious Areas of the Breast," Comput. Biomed. Res., Vol.12, pp.445-460, 1979.
[6] M.L. Giger, F.F. Yin, K. Doi, C.E. Metz, R.A. Schmidt and C.J. Vybomy, " Investigation of Methods for the Computerized Detection and Analysis of Mammographic Masses," Proc. SPIE, pp.183-184, 1990.
[7] F.F.Yin, M.L. Giger, K. Doi, C.E. Metz, C.J. Vyborny and R.A. Schmidt, " Computerized Detection of Masses in Digital Mammmograms: Analysis of Bilateral Subtraction Images," Med. Phys., Vol.18, pp.955-963, 1991.
[8] S.M. Lai, X. Li and W.F. Bischof, " On Techniques for Detecting Circumscribed Masses in Mammograms," IEEE Trans. Med. Imaging, Vol.8, pp.377-386, 1989.
[9] K. Matsumoto, H.R. Jin and H. Kobatake, "Detection Method of Malignant Tumors in DR Images - Iris Filter -, Trans. of the Inst. of Elec., Inf., and Comm. Eng., Vol.J75-D-II, No.3, pp.663-670, 1992.
[10] H.R. Jin and H. Kobatake, "Automated Detection of Malignant Tumors on Mammograms, Trans. of the Inst. of Elec., Inf., and Comm. Eng., Vol.J76-D-II, No.2, pp.288-295, 1992.
[11] H.R. Jin and H. Kobatake, "Extraction of Microcalcifications on Mammograms Using Morphological Filters with Multiple Structuring Elements, Trans. of the Inst. of Elec., Inf., and Comm. Eng., Vol.J75-D-II, No.7, pp.1170-1176, 1992.

Computer-Assisted Diagnosis of Bone Tumors

E.PELIKAN, K.BOHNDORF, T.TOLXDORFF, D.ZARRINNAM, B.WEIN

Institut für Medizinische Informatik und Biometie
Klinikum Aachen,
Pauwelsstraße 30
5100 Aachen

Summary

The main problems in radiological assessment of focal bone lesions are discussed. An approach to computer assisted diagnosis including a new systematic description of focal bone lesions is proposed. It is based on the MYCIN-formula in order to deal with the uncertainty and the vagueness associated with the assignment of X-ray findings to the diagnoses. The dialog generator CARAT, which is a computer based system to aid in diagnosing focal bone lesions is presented as an implementation of this conception.

Introduction

Focal bone lesions make up a very complex part of the differential-diagnostic spectrum. A very subtle and systematic assessment is needed in order to determine the tumor type and its biological behavior. The assessment depends on a fictitious similar appearance of different bone lesions. The way physicians and medical students deal with X-rays does not reflect the demands for systematic image examination. A consequent and systematic scheme of assessment is early terminated in favour to own associations. In consequence important details are overlooked and the case is missclassified. Moreover rare diagnoses are often out of scope of the examinators judgement.

Two main problems in the radiological assessment of focal bone lesions have been identified:

- the lack of knowledge concerning the relevant radiological features
- the lack of knowledge concerning dependencies between radiological features and diagnoses

To solve these problems and to enhance diagnostic quality a computer assisted dialog is proposed.

An approach to computer assisted diagnosis

A first linguistic description of primary bone tumors has been introduced by Lodwick [1] in 1966. Taking up this work an extended systematic description of focal bone lesions (Tab. 1) has been developed by our group [2].

1) Localization in the skeletal system ?
· (11 localizations)

2) Localization in the long bones ?
· (7 localizations)

3) Site in the bone ?
· central · excentric · unclear

4) Pattern of bone change ?
· solitary, well defined lytic lesion
· solitary, well defined increased density
· solitary mixed leasion (lytic/blastic)
· multicentric lucencies (with densities)
· multicentric densities (with lucencies)
· moth-eaten bone destruction
· permeative bone destruction
· increased density of the whole bone
· unclear

5) Margins: Tumor-Bone transition zone
· sharply delineated · no clearly deinable edge
· both kind of margin patterns · unclear

6) Sclerotic margins ?
· around the whole lesion
· partly to be seen
· not existing · unclear

7) No margin definable ?
(the whole bone is involved)
· true · not true · unclear

8) Cortex ?
· normal · thinned endostally
· destroyed (partly/totally)
· substituted by "neocortex"
· thinned periostally
· unclear

9) Periosteal Reaction ?
· not existing · solid (smooth)
· solid (lobulated,trabeculated)
· solid (interrupted)
· lamellar (uni, multi)
· spiculated, complexe reaction
· unclear

10) Expansion of bone ?
· not existing · <= 1cm · > 1cm · unklar

11) Soft tissue involvement ?
· not existing · existing · unclear

12) Matrix of the lesion?
· not existing · bony
· cartilage · ground glass (= fibrous)
· unclear

13) Intralesional fracture ?
· yes · no · unclear

14) Bowing of the shaft ?
· yes · no · unclear

15) Multiple Lesions ?
· yes · no · unclear

16) Involvement of the joint surface ?
· yes · no · unclear

17) Age of patient ?
· <= 10 · 11-20 · 21-30 · 31-40 · 41-50
· 51-60 · 61-70 · >= 70 · unclear

Tab. 1: Questions to describe focal bone lesions

It consists of 17 central questions and related answers. A systematic and complete examination of all relevant medical features can be reached using this description, which is available as a questionnaire.

The relation between the 45 implemented diagnoses (Tab. 2) and their radiological features is represented by the MYCIN-formula [3,4]. Due to the rarity of focal bone lesions no reliable statistical data are available. The MYCIN-formula has been chosen in order to deal with the uncertainty and the vagueness associated with the assignment of X-ray findings to the diagnoses.

Adamantinoma of the long b.	Fibrosarcoma	Paget disease
Brown Tumor	intraosseus Ganglion	Neurogenic Tumor
Chondroblastoma	Hemangioma	hypertr. Osteoarthropathy
periosteal Chondroma	Hemangiosarcoma	Osteoblastoma
Chondromyxoidfibroma	benign fibrous Histiocytoma	Osteochondroma
Chondrosarcoma	malignant fibr. Histiozytoma	Osteoid Osteoma
Chordoma	Bone infarct	Osteoma
periosteal Desmoid	aneurysmal Bone Cyst	acute Osteomyelitis
fibrous Dysplasia	simple Bone Cyst	chronic Osteomyelitis
Enchondroma	Bone Island	Osteopoikilosis
eosinophilic Eranuloma	Lipoma	central Osteosarcoma
Epidermoid	malignant Lymphoma	pigmented villon. Synovitis
Ewing sarcoma	Mastocytosis	multiple Myeloma
non-ossifying Fibroma	Melorheostosis	Giant Cell Tumor
ossifying Fibroma	Metastases	

Tab. 2: Diagnosis which are implemented in CARAT

Any combination of question and possible answer represents a "symptom". The questionaire from Tab. 1 yields 89 symptoms. The expert assigns a certainty-factor (CF) to each combination between symptom X and diagnosis Y:

-1 symptom X excludes diagnosis Y

0 symptom X does not promote diagnosis Y

+1 symptom X leads to diagnosis Y

The value for CF can be chosen from the interval [-1,1] and is not limited to one of the discret values shown above. In this case 89 * 45 = 4005 certainty-factors are needed to build the knowledge base For each question 45 CF-values are calculated using the formula

$$z = \begin{cases} m + n - m \cdot n & m > 0, n > 0 \\ m + n + m \cdot n & m < 0, n < 0 \\ \dfrac{m + n}{1 - \min(|m|,|n|)} & sonst \end{cases} \qquad (1)$$

Because only one answer can be given for each question $1.83325*10^{11}$ different combinations of symptoms are possible. This provides sufficient discrimination ability. The output of this algorithm is an ordered list of diagnoses. The expert will use this list as a guideline for his final judgement.

The dialog generator CARAT

Based on the conception just introduced the dialog generator CARAT (Computer Assisted Radiology Interactive Dialog Generator) has been developed. Figure 1 shows a screenshot of the dialog-window. In the left part 10 diagnoses with highest CFs are shown. The remaining diagnoses can be reached using the scroll arrows. In addition a single diagnosis can be traced during the whole dialog using the "Focus"-switch. If the focus-option is activated only the most important 8 diagnoses will be shown. The space below is used to display both a numerical value and a bargraph for the diagnosis in trace.

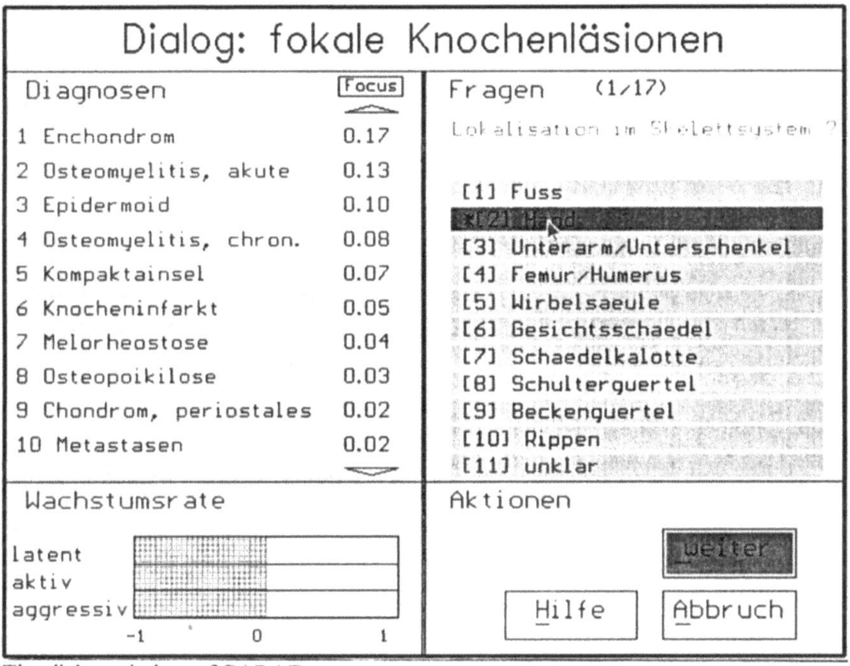

Fig. 1: The dialog window of CARAT

The growth rate, which was modified by our group is shown in the left lower part of the dialog window. Because only the questions 4 -12 are related to the growth rate the bars in the display show two kinds of appearance. If they are solid the current question is directly related to the growth rate, otherwise they are hashed.

In the right part of the screen the current question and the possible answers are displayed. If the user activates one answer a new list of diagnosis is calculated. The effect is shown immedeately. A reselection of the same answer or pushing the "weiter"-button leads to the next question. A step backwards to the preceeding question is also possible. In this case the prior selected answer is highlighted as a reminder.

A main component of the dialog generator is the help system. Help can be obtained at any time during the whole dialog in both graphical and written way for questions and diagnoses as well (Fig. 2).

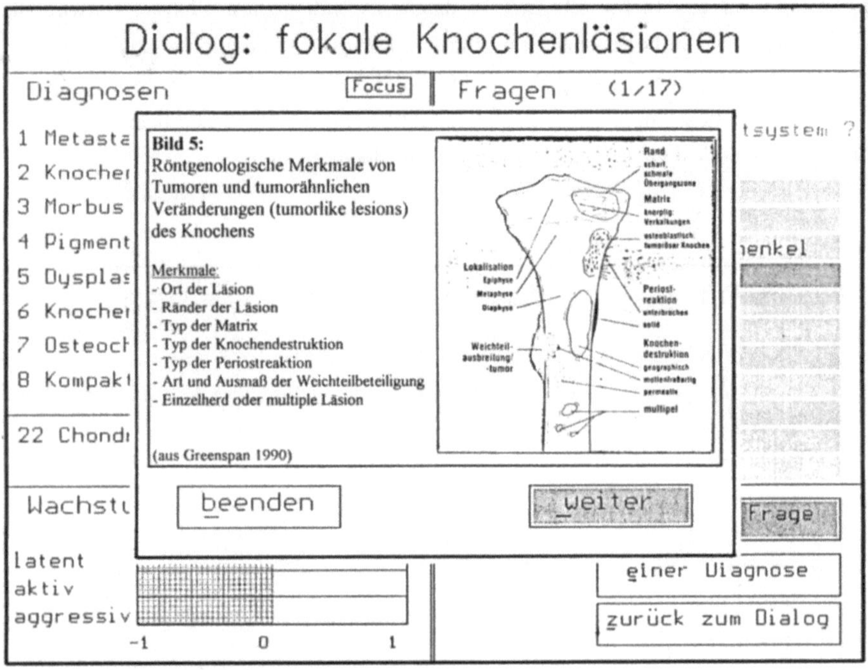

Fig.2: The Help system of CARAT

At the end of the dialog a report is shown. This report includes the complete sorted list of diagnoses, the values for the growth rate, a logging of all selected answers and a description of the underlaying database. For documentation purposes the report can be sent to the default printer.

CARAT 3.4 was implemented in C on a MS-DOS PC (286 and higher). Only a VGA graphic adapter and 1 MB harddisk space are required to run CARAT 3.4 including the help system. Windows 3.1 is supported but not required. A UNIX/X11 version is in preparation.

<u>References</u>

1. Lodwick, G.S.:Computer aided diagnosis in Radiology. A Research plan. Invest. Radiol. 1 (1966) 72.
2. Bohndorf, K.; Tolddorff, T.; Pelikan, E.; Zarrinnam, D.; Günther, R.W.: Neue Ansätze zur computerassistierten Diagnose von fokalen Knochenläsionen. Radiologe 32 (1992) 416-422.
3. Shortliffe, L.H.: Computer based medical Consultations: MYCIN. New York, Elsevier, 1976
4. Richter, M.M.: Prinzipien der Künstlichen Intelligenz, Stuttgart, Teubner, 1989

Classification of Medical Image Objects based on Fuzzy Sets

Thomas P. EISELE [1], Hans-Martin KLEIN [2], Walter AMELING [1], Rolf W. GÜNTHER [2]

[1] Rogowski-Institut for Electrical Engineering, RWTH Aachen, Schinkelstraße 2, D-5100 Aachen, Germany
[2] Clinic of Diagnostic Radiology, RWTH Aachen

Summary

Classification of medical image objects is based on association of a given object with one of several classes (i. e. diagnoses). We introduce a new scheme for the classification of medical image objects based on the theory of *fuzzy sets*. Each object does no longer belong entirely to only one class but rather to all classes with differing *membership values*.

The contribution of this paper is twofold. First, we extend the known fuzzy k-nearest-neighbor-algorithm to include a reject class without making any assumptions on class distributions. Second, we accelerate the classification phase by off-line classification of grid points and subsequent interpolation of the membership values.

We applied fuzzy classifiers to the diagnosis of focal liver lesions using dynamic CT-scanning. They showed considerably more robustness against outliers than their "hard" counterparts and yielded a good measure for the confidence of the classification.

Introduction

Computer-based classification of medical image data can simplify the difficult interpretation of images and provide additional information to the physician. It can be divided into three phases: *feature extraction*, *training* of the classifier, and the actual *classification* of unknown cases. In order to be appropriate in clinical practice, it has to cope with the following two requirements:

- During the *training phase*, untypical cases (due to imaging errors, abnormal physiological parameters of the patient, or untypical symptoms) have to be considered with a lower weighing factor, so that the knowledge acquired during the training contains the typical characteristics of the classes.
- The result of a *classification* of unknown cases has to contain information on how reliable the computer-based diagnosis is, i. e. how typical the classified case is for his class.

Both requirements are not met by conventional classifiers. Thus we base them on an extended class definition: the theory of the *fuzzy sets* [9]. One case does no longer belong entirely to one class but rather to all classes with differing *membership values*.

Image Acquisition

The most common focal liver lesions (Hemangioma, other benign tumours, and malignant tumours) can be differentiated only in rare cases in non-contrasted images. In order to provide a reliable diagnosis, a dynamic bolus CT scan has to be carried out, i. e. after quick injection of 100ml of contrast agent, a time series in a constant slide position is taken, thus recording the dynamic enhancement pattern of the lesion [5]. *Fig. 1* shows an image of a liver carcinoma after injection.

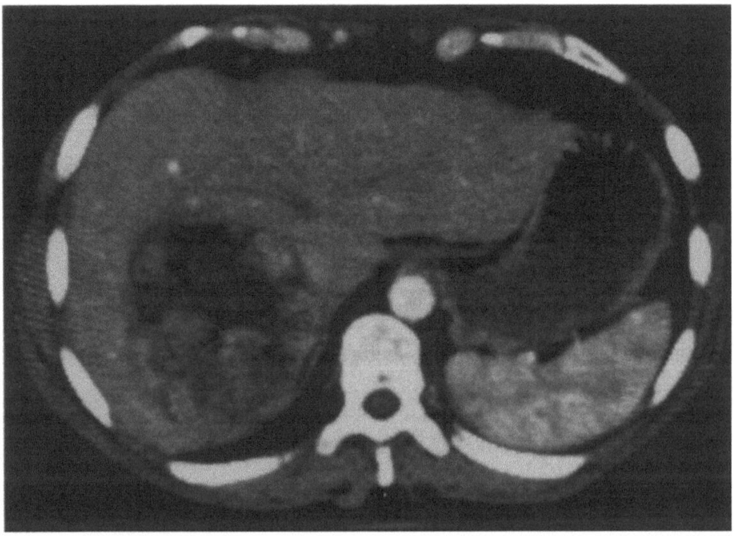

Fig. 1: Liver carcinoma after injection of contrast agent

We had dynamic bolus CT scans of 154 liver lesions at our disposal, namely, 60 hemangioma, 28 benign tumours, and 66 malignant tumours. Their diagnosis was verified by sonographical examinations and — in all cases except hemangioma — by percutaneous biopsy.

Three-dimensional Feature Extraction

We put the single slices obtained from the CT scanner together to a virtual 3D data volume with the time as third space axis, in order to register simultaneously grey level interactions both inside one image and between consecutive instances. Thus we are able to apply three dimensional texture analysis techniques. We gain various features out of the histograms of the lesion in the original pictures and in spatial and time gradient pictures (centre of gravity, deviation left and right of it, difference to healthy tissue, etc.) at several instances. Then we apply a principal component analysis to the resulting 55 features and yield three canonical features [2].

Training and Classification

Since no information on the probability distribution of the feature vectors and the *a priori* probabilities of the classes is available, we use non-parametric classification schemes. The two main classifiers of this category are the *threshold logic unit (TLU)* [7], which we use in the generalised form of the *locally trained piecewise linear classifier (LTPLC)* [8], and the *k-nearest-neighbor classifier (k-NN)* [1]. Fuzzy versions of the TLU and the k-NN-classifier are already mentioned in the literature [3,4]. In this paper we develop a fuzzy LTPLC.

The locally trained piecewise linear classifier: We fuzzified the LTPLC by incorporating the fuzzy TLU technique of [3], which significantly improved its convergence properties, and by allowing *fuzzy segments*, thus considerably reducing the number of not classifiable cases.

The *training phase* of the LTPLC consists of several steps. It first splits each class of the training set in feature space into several clusters by means of the k-means procedure. Then for each pair of clusters out of different classes a separating hyperplane is found by the TLU technique. The total of the hyperplanes divides the feature space into non-overlapping segments. Now each segment is assigned to the class to which the contained training samples belong. This is not possible if (i) the segment contains no training sample, or (ii) the training samples in this segment belong to different classes.

The fuzzy version of this classifier differs in two items:

- A *fuzzy membership* to each class is assigned to each training sample. Samples with low membership to their class have less influence on the determination of the hyperplanes [3].
- If a segment contains training samples of more than one class, we consider this case as *fuzzy segment* and assign it to that class to which the majority of the samples belong.

The *classification phase* is identical for both the "crisp" and the fuzzy LTPLC: For every unknown case the segment in which it lies is found. Then the case is assigned to the corresponding class. If it is located in a segment that is occupied by no class it is marked as *not classifiable*. One drawback of the LTPLC is that it is not able to produce fuzzy membership values as classification result.

The k-nearest-neighbor classifier: The "crisp" version of the k-NN classifier works without training phase. It simply looks in the training set for the k nearest neighbors in feature space for every unknown case and assigns to this case the class to which the majority of the neighbors belong. The fuzzy version [4] first assigns membership values to each element of the training set with respect to each class. Then, during the classification phase, each of the k nearest neighbors is weighted with its membership values and its distance to the actual, unknown case. As a result, the fuzzy k-NN classifier yields membership values to every known class, i. e. a *fuzzy classification*. *Fig. 2* shows the maximum membership value for every grid point in the 2D feature space depicted in the base of the graph.

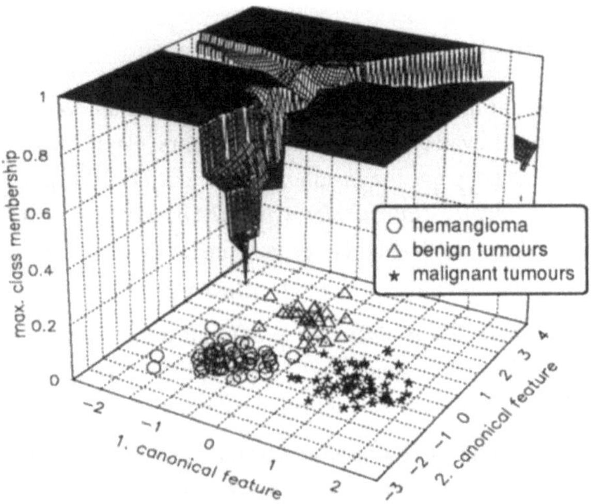

Fig. 2: Maximum membership values above the feature space

It can be clearly seen that the maximum membership values are near one not only where the training samples of the classes cluster, but also far behind the classes where no training samples exist. Thus, feature vectors that do not belong to any of the known classes may get full membership to one class, a result that is very unsatisfactory.

To overcome this drawback, we take the k nearest neighbors *out of each class* and consider only their membership values and their distances to the actual case. Feature vectors which lie far outside any known class have similar distances to all known classes and thus get similar membership values to each class. This indicates that the actual case does not belong to any of the known classes, i. e. it is assigned to the *reject class*. The corresponding graph is shown in *Fig. 3*.

One drawback of all k-NN classifiers is the high amount of computation time needed to classify an unknown case. This is particularly due to the calculation of the distances to all training samples. The fuzzy version can be accelerated by pre-classifying a certain amount of grid points in feature space. For classification of a new case only the nearest grid points have to be determined and then the membership values can be interpolated. This is *not* possible using the "crisp" k-NN classifier because the membership values he produces are not smooth. Acceleration factors of 2 to 10 are achievable, depending on the number of nearest neighbors and the size of the training set. Using only 500 grid points for each feature coordinate, the interpolation error was less than 1% for 98.7% of all tested feature vectors. The remaining 1.3% are due to the steps of the membership values visible in Fig. 2. Using our fuzzy k-NN classifier with reject class, these steps are eliminated.

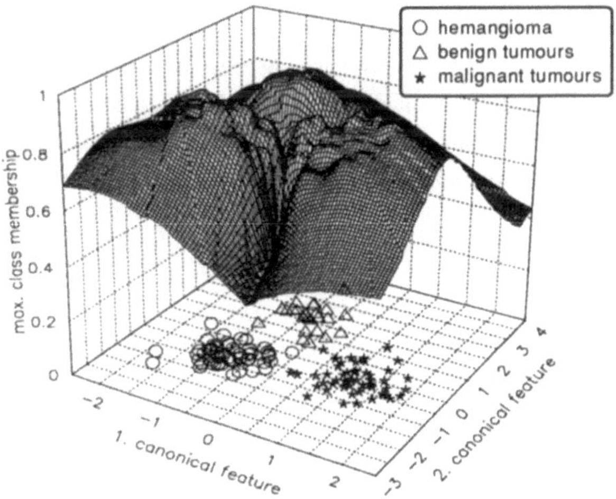

Fig. 3: Fuzzy k-NN with reject class

Results

We tested the different classifiers with the above mentioned material, which was not preselected, i. e. all dynamic bolus CT scans of liver lesions taken during clinical examinations between 1987 and 1991 and meeting certain conditions (completeness of the series, verified diagnosis, etc.) have been used, even small lesions (< 1cm) and series with respiration artefacts. For testing we used the *leave-one-out method*, i. e. the classifier is trained with all cases except the first and then the first is classified. Next it is trained with all cases except the second, and so on until all cases are classified [6]. This method is one of the least biased, but most expensive testing algorithms for classifiers. The results are shown in *Tab. 1*.

	crisp	fuzzy
LTPLC	75,3	85,1
k-NN (k = 4)	89,6	89,6
k-NN (k = 4) with reject class	89,0	89,0

Tab. 1: Percentage of correct classifications

The crisp LTPLC is sensitive against outliers in the training set, which results in a rather poor classification. Applying the fuzzy training method, a significant improvement can be stated. K-NN classifiers are more robust, even the crisp version, so that fuzzy techniques cannot reduce the amount of wrong classifications significantly. However, since fuzzy k-NN classifiers produce fuzzy membership values, the simple counting of true and false classifications (the maximum membership value belongs to the true or wrong class) does not reflect the amount of information contained in this result. A wrong classification due to similar membership values to several classes is much less severe because the physician notices that he cannot rely on this result. Furthermore, the inclusion of a reject class and acceleration using interpolation techniques are not possible using the crisp methods. The graphs in *Fig. 2* and *Fig. 3* give a much better idea of what fuzzy k-NN classifiers are able to achieve.

Acknowledgements

We wish to thank Dr. K. C. Klose for carrying out the CT examinations and Mrs. M. Brenner for rating and selecting the image data.

References

[1] Cover, Thomas M.; Hart, P. E.: *Nearest neighbor pattern classification,* IEEE Transactions on Information Theory, Vol. IT-13, No. 1, pp. 21-27, Jan. 1967

[2] Eisele, T.; Klose, K. C.; Klein, H. M.; Ameling, W.; Günther, R. W.: *Die Verwendung von Fuzzy-Klassifikationsverfahren für die Diagnostik fokaler Leberläsionen,* Biomedizinische Technik, Vol. 37, pp. 186-187, Schiele & Schön, Berlin 1992

[3] Keller, James M.; Hunt, Douglas J.: *Incorporating fuzzy membership functions into the perceptron algorithm,* IEEE Transactions on Pattern Analysis and Machine Intelligence, Vol. PAMI-7, No. 6, pp. 693-699, Nov. 1985

[4] Keller, James M.; Gray, Michael; Givens jr., James A.: *A fuzzy k-nearest neighbor algorithm,* IEEE Transactions on Systems, Man, and Cybernetics, Vol. SMC-15, No. 4, pp. 580-585, July/Aug. 1985

[5] Kurtz, B.: *Computertomographie herdförmiger Lebererkrankungen,* Röntgenpraxis, Vol. 42, pp. 297-304, 1989

[6] Lachenbruch, P. A.; Mickey, R. M.: *Estimation of error rates in discriminant analysis,* Technometrics, Vol. 10, No. 1, pp. 1-11, 1968

[7] Nilsson, Nils J.: *Learning machines,* McGraw Hill 1965

[8] Sklansky, Jack; Michelotti, Leo: *Locally trained piecewise linear classifiers,* IEEE Transactions on Pattern Analysis and Machine Intelligence, Vol. PAMI-2, No. 2, pp. 101-111, March 1980

[9] Zadeh, L. A.: *Fuzzy sets,* Information and Control, Vol. 8, No. 3, pp. 338-353, June 1965

Knowledge Based Systems

Using Expert Systems (ES) and Artificial Neural Systems (ANS) to Evaluate Serial Coronary Arteriograms in the POSCH Study

John M. Long and the POSCH Group
University of Minnesota.
Minneapolis, MN 55455

Summary
The evaluation of clinical data with a significant visual component such as coronary arteriograms are often considered unsuitable for artificial intelligence technology. The Program on the Surgical Control of the Hyperlipidemis (POSCH) tested expert systems (ES) and artificial neural systems (ANS) (back propagation) as a way of partially automating the clinical evaluation of a pair of serial coronary arteriograms taken three or more years apart. The manual reading of the films was still necessary, but can be done by a trained technician since it involves limited clinical interpretation. The clinical judgment aspects were incorporated into the ES and ANS.

The ES and ANS were compared to a manual interpretation performed by a panel of radiologists and cardiologists. Various sub-panels were compared to each other. Multiple linear regression (MLR) was also tested. All methods worked but the ANS was found unsuitable for these data.

The POSCH AI Project
Medicine relies on the manual evaluation by experts of large amounts of longitudinal data. These evaluations are usually performed by clinical specialists because of the clinical judgment required. They are tedious, expensive, logistically difficult and time consuming. The POSCH study has shown that expert systems (ES) can play a key role in reducing the time and expense of analyzing clinical data.[1]

Previous results are reinforced by showing that it is possible to use artificial intelligence technology including ES and artificial neural systems (ANS) to at least partially automate the evaluation of data even when some of the data are visual in nature. The poor performance of expert systems in tasks requiring perceptual skills does not rule out their application in studies requiring the evaluation of perceptual data if the perceptual data are first appropriately encoded. Often, the encoding can be done by a technician. In this case, the standard knowledge engineering process had to be modified. Domain experts, cardiologists and radiologists in the case reported here, are not accustomed to working on such problems in the absence of the standard visual information.

The system discussed involved the evaluation of serial coronary angiograms that come from the Program on the Surgical Control of the Hyperlipidemias (POSCH), a long term randomized clinical trial with 838 post-myocardial infarction patients. POSCH was designed to test the effect of reduced cholesterol in such patients.[2] Coronary angiograms were obtained at a patient's entry into POSCH, then at three, five and either seven or ten years into the study.

Following current standard practice, POSCH used highly skilled experts to examine pairs of angiograms taken from the same patient several years apart. The experts determined the progression or regression of atherosclerotic disease. They tried to distinguish between those changes caused by atherosclerotic disease and by other causes such as the formation of new clots and the dissolution of old clots. Other complicating factors included the variable visibility of blood flow in arterial segments and changes in the diameter of arteries resulting from recanalization, grafting or development of collaterals.

The process relies heavily on the highly subjective interpretation of visual data which can produce a large amount of observer variability. To reduce observer variability, POSCH used reading teams consisting of two people selected from the POSCH Arteriography Review Panel, a group of eight expert readers. The evaluations followed a well defined protocol previously documented in the literature.[3]

The evaluation of each film pair took an average of 20 to 30 minutes depending on the length of the angiograms and the quality of the film. The POSCH study attempted to collect four angiograms on each patient for a maximum number of 3352 (4 x 838) angiograms. There were 5028 possible pair combinations that needed to be evaluated. These numbers were substantially reduced by deaths or for medical or other reasons. Even so, manual processing of the pairs of angiograms was tedious. Because of this, the panel only evaluated baseline to follow-up comparisons. The two member panels met 12 times a year for two to four days, carrying out about 16 evaluations each day. They read a total of 28 days per year, producing about 450 evaluations, a few of which were repeated as part of the POSCH quality surveillance program. By rotating reading teams, any one panel member read four times a year. The process took about ten years!

POSCH wanted to make intervisit comparisons such as: 3 year to 5 year, 3 year to 10 year, etc., but the size of the manual task made it prohibitive. Because of our previous success with ETA, POSCH was encouraged again to investigate automating the process using artificial intelligence technology.[1]

We investigated expert systems and the self learning artificial neural systems. The expert system developed was called ESCA for expert system for coronary arteries. The more conventional statistical method, multiple linear regression (MLR) was also tested for a comparisons.

ESCA

ESCA's domain knowledge is organized in an inference network modeled after that used in AGNESS.[4] The generalized knowledge representation and expert defined inference methods of AGNESS made the development of ESCA simple and elegant. ESCA was developed using 50 cases previously evaluated by the panel; and subsequently tested on 200 additional cases. Knowledge engineering continued until ESCA reached a level of performance in the global assessment of change in atherosclerotic disease that was slightly better than that of the POSCH panel members. ESCA can determine the critical last step in the evaluations. Thus, by using specially trained technicians to identify and size lesions, POSCH could partially automate the evaluation of film pair combinations.

The knowledge engineering process used to develop ESCA has been published.[3] The evaluation process has also been published but is briefly described here in order for you to understand the comparisons made. Since the experts evaluated the change in each individual artery and then compared the various changes to arrive at a global assessment, ESCA does this also. The experts appeared to obtain a "gestalt" of blood flow, something the expert system was not able to imitate. In the prototype system used, a theory was developed by the collaborating domain expert based on the accumulation of evidence. It considers individual changes observed in each of 23 arterial segments and assigns points or weights to each change.

The rules used by ESCA allocate points that ultimately translated into the scale used by the experts of much better, better, slightly better, no change, slightly worse, worse and much worse. The global assessment is coded as +3 for "much better" to -3 for "much worse". The scale is shown in Figure 1. The evidence of change in the individual coronary artery segments were merged, to obtain the global assessment. Twenty major arterial segments were routinely evaluated and used as input to ESCA. In addition, three additional stenoses, either in already graded segments or other arterial segments were examined and used when the individual anatomy or special situations make these other areas important to the evaluations. The "no change" category was further divided into -0 and +0 where the expert attempted to provide some indication of the direction of the change even if it was very small. The two evaluations were coded as an exact agreement when the sub-panel and the other evaluation reached the same conclusion on the scale and as a close agreement when they used adjacent codes on the scale. These "points" are always integers corresponding to an assessment of severity of change. Since ESCA uses an inference network, the rules are represented naturally using an expert defined inference method.

Figure 1

-3	-2	-1	-0	+0	+1	+2	+3
much worse	worse	slightly worse	no change		slightly better	better	much better

The scale used to record the final global assessment

Artificial neural systems

One ANS algorithm, back-propagation, was investigated. One of the expectations of an ANS is that, once the network is trained with the back-propagation algorithm, it could generalize to a test set.

The neural system was trained on a training sample of 125 cases drawn from the set of 200 validated POSCH cases used previously to develop ESCA. The network was then tested on the remaining 75 cases. The independent variables were the stenoses observed in the 23 arterial segments from Film A and the 23 from Film B. Film A and Film B are taken several years apart, although their time sequence was masked.

The dependent variable was the expert's assessment of change on the 8-point scale from "much worse" to "much better". No change is coded as +0 or -0 for slightly worse, giving 8 categories as shown as Figure 1. During training, the independent variables were fed as inputs to the network and the network was trained to predict the dependent variable on the output. During testing, the independent variables were fed as inputs and the output of the network was compared to the assessment made by the human experts.

Units are arranged in layers in the back-propagation model. There is an input layer, an output layer, and either none or one or more hidden layers. Several network configurations with zero, one, two or three units were tested. The best results were obtained using three hidden layers. For more details see reference.[5]

Multiple linear regression (MLR) was also tried using the same 125 cases to build the equation and the remaining 75 cases to test it. Results from all three methods and the human experts are compared and summarized in Table I. The Kappa statistic adjusts for chance agreement.[6] Naturally, one would expect good results from the training set for either back-propagation or multiple linear regression. The appropriate comparison with ESCA and the human experts would be the test set results. For "exact agreement," the ANS does about as well as the human expert panel (39% vs. 38%) but not as well as ESCA (39% vs. 46%). MLR is comparable to ESCA. The "close agreement" rate for the MLR (93%) is comparable to ESCA and the panel, however the ANS (84%) does not do as well.

Table I

	Exact Agreement	Kappa	Close Agreement	Kappa
Panel vs panel	38%	0.23	92%	0.86
ESCA vs panel	46%	0.36	93%	0.87
ANS train vs panel	57%	0.48	98%	0.96
ANS test vs panel	39%	0.27	84%	0.70
MLR train vs panel	57%	0.49	98%	0.97
MLR test vs panel	44%	0.33	93%	0.87

Agreement rates between various panels and systems

POSCH found significant difficulties with the back-propagation method. It used an excessive amount of computer time and required considerable experimentation with various setups. The main problem was our inability in determining how one variable influences another. Perhaps it is more suitable for use with non-symbolic data. In any event, POSCH would need considerably more insight before the neural network method would be useful. MLR was also somewhat obscure with regard to cause and effect.

Based on the POSCH experience, expert systems seem to be a valid way to automate the analysis of clinical data that require expert clinical judgment. They are useful when the evaluations are repetitive and hence tedious for the clinical experts. POSCH experts were willing to help develop these knowledge based systems because they could relieve them of these tedious tasks.

POSCH results using ESCA also establishes the feasibility of separating the examination of the descriptive findings of the visual data from the evaluation of the conceptual data. Therefore, when conducting evaluations of large amounts of visual data, clinicians need not dismiss the possibility of using an expert system.

There are other potential benefits from using an expert system. For example, they can be used to edit the database and otherwise to manage the quality of the data. The logistics, tedium, and cost of carrying out large experimental trials over long periods of time can be ameliorated by using expert systems. Difficulties associated with maintaining the consistency of the evaluations of the data over time can be controlled. The classical time shift bias may be reduced. Changes that do occur, made by either the support staff or by participating experts, are more visible and can be documented. As appropriate, they can be compensated for or corrected. Such uses alone may provide sufficient motivation for the development of an expert system.

Acknowledgments

This work is supported by NHLBI Grant HL15265 and the Microelectronics and Information Sciences Center of the University of Minnesota.

References

1. Long JM, Slagle JR, Leon AS, Wick MW, Matts JP, Karnegis JN, Bissett JK, Sawin HS and Stevenson JP: An example of expert systems applied to clinical trials: Analysis of serial graded exercise ECG test data. Controlled Clinical Trials 8:136-145, 1987.

2. Campos CT, Matts JP, Fitch LL, Speech JC, Long JM, Buchwald H, and the POSCH Group: Lipoprotein modification achieved by partial ileal bypass: Five-year results of the program on the surgical control of the hyperlipidemia. Surgery 102:424-432, 1987.

3. Long JM, Irani EA, Hunter DW, Slagle JR, Matts JP, Casteneda W, Pearce M, Bissett J, Sawm H, Edmister A, Weber F, Anplaty K, Sanmarco M and the POSCH Group: Using a symbiotic man/machine approach to evaluate visual clinical research data. Journal of Medical Systems 12:327-339, 1988.

4. Slagle JR, Wick MR, Poliac MO: AGNESS, A Generalized Network-based Expert System Shell. Proceedings of the Fifth National Conference on Artificial Intelligence, AAAI 5:996-1002, 1986.

5. Irani EA, Matts JP, Long JM, Slagle JR and the POSCH Group: Using artificial neural nets for statistical discovery: Observations after using back-propagation, expert systems and multiple linear regression on clinical trial data. Complex Systems 3:295-311, 1989.

6. Fleiss JL: Statistical Methods for Rates and Proportions, 2nd Edition. John Wiley and Sons, New York, p. 217, 1981.

A Knowledge-based Multi-media Database System for Skeletal Radiology

R.K. Taira[1], A.F. Cardenas[2], W. W. Chu[2] C.M. Breant[3], T. Hall[1],

[1]Department of Radiological Sciences

[2]Computer Science Department

[3]Department of Surgery

University of California, Los Angeles

Los Angeles, CA 90024-1721

Abstract

This paper describes our research toward the development a knowledge-based multimedia database system for bone age research. The prototype system is designed to allow users to query for information based on patient demographic profile, patient medical history, evolution constraints of the developing hand, image content, and fuzzy high-level terms and concepts. The major highlights of this work include: 1) object-oriented database management system, 2) graphical user interface for query formulation, 3) object-oriented data model which supports both traditional "Where-type" predicates and new object evolutionary and temporal predicates, and 4) cooperative query answering capabilities for intelligent query modification in the event of a null solution to the original query.

Introduction

This paper describes the current research at UCLA toward the development of intelligent scientific database systems for medicine. The project is known as KMeD (Knowledge-based Multimedia Medical Distributed database system). The project goal is to provide database tools to assist researchers in answering high-level queries which involve correlating patient profiles (genetic profiles, psycho-social profiles) with various image features across both normal and abnormal human development. We have initially concentrated our efforts on pediatric bone-age research. We introduce a dynamic object-oriented temporal evolutionary data model for handling high-level medical imaging queries.

System Architecture

Figure 1 shows the overall multilayer architecture of our system [Carde87]. It consists of four layers: (1) a front-end layer (FEL), (2) a cooperative query answering layer (CQAL), (3) a canonical/distribution layer (CDL), and (4) an autonomous database layer (ADL).

Front End Layer - This layer is the primary interface between an application and the other layers of the system. A high-level domain-independent query language PICQUERY$^+$ has been designed, and a subset of it is being implemented for the KMeD project [Carde93]. Queries are friendly graphical interface. Both interfaces allow the user to specify high level operations including query definition, data analysis methods, and visualization methods for results. Using PICQUERY+ we are able to conveniently express, for example, the following high level medical imaging queries: *"Find an image of the proximal phalanx of the fifth finger for patient John Smith and obtain the length of the major axis of this bone."* and *"Show in a movie loop all images from patient cases which demonstrate the fusion phases of the 3rd distal phalanx ephiphysis and tubular bone."* See [Carde93] for further details.

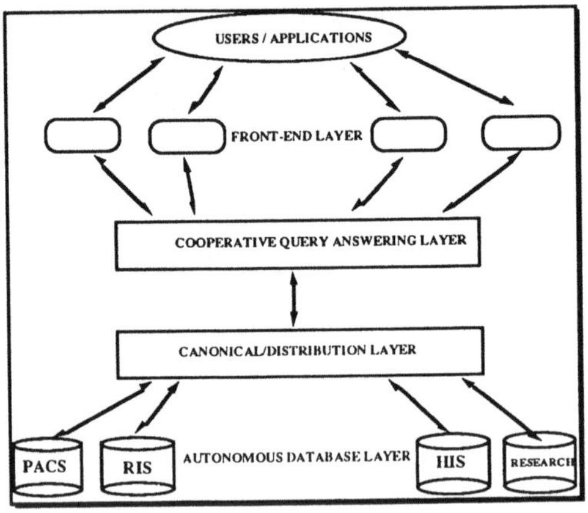

Fig. 1 - *KMeD multilayer architecture.*

Cooperative query answering layer - Research and clinical practice often involve queries which include fuzzy and imprecise descriptions of patients, image features, and disease processes. Traditional database management systems accept precisely specified queries and provide only exact answers, thus requiring users to understand the problem domain and database schema. They return limited or null information if exact answers are not available. The CQAL layer is designed to bridge the knowledge domains of the medical researcher and the low level understanding of the data held by traditional DBMS's. It provides approximate, summary and conceptual answers to the original query (within a certain semantic distance to the exact answer) [Chu91]. Furthermore, information conceptually related to but not explicitly asked by the query is provided as part of the query solution. We use a structured approach representing domain knowledge in type abstraction (concept) hierarchies. We then derive approximate answers using generalization and specialization relations from within a given concept hierarchy and association relations between different hierarchies. For example, consider the query: *"Retrieve all hand images demonstrating a fracture of the 3rd metacarpal of the hand for 12 year-old Korean boys"*. If no answer is available, then an approximate answer is found by relaxing various key concepts including image object (3rd metacarpal relaxed to any metacarpal), patient age (12 year old can be relaxed to pre-teen category), patient sex, and patient race (Korean can be relaxed to Oriental). A more general query can be formulated which relaxes some or all of these constraints. We have constructed the type abstraction hierarchies for several medical subject areas including the bone structure of the hand, disease classes of the hand, ethnic groups, and age groups. Futher details can be found in [Chu92].

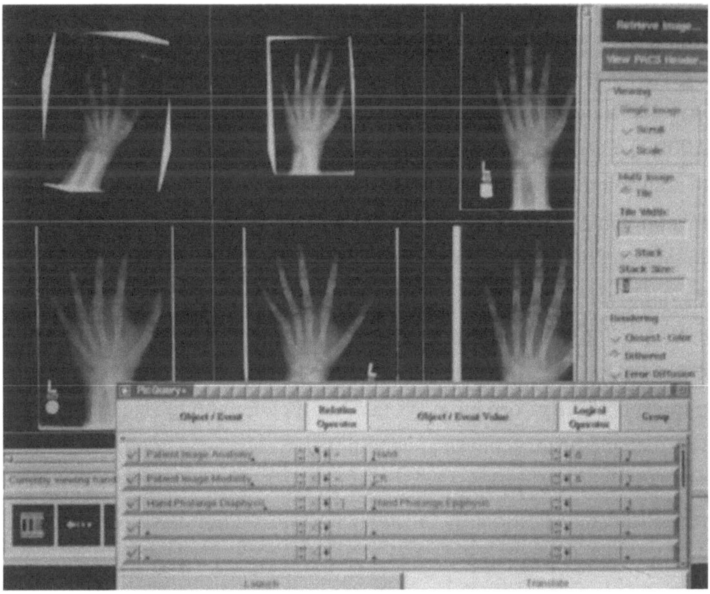

Fig 2 - *PICQUERY+ user interface for query specification.*

Canonical/distribution layer - This layer is the central interface to the federation of database systems. It manages both the aggregation of the local database views into a collected conceptual database description and the decomposition of the global view back into local databases views. Thus we provide powerful constructs by which a conceptual object's attributes can be a sum of features from multiple underlying heterogeneous data models. The implementation of this layer has been facilitated by the use of high-level graphical CASE data integration tools (Integration Works, Data Integration Solutions, Corp.). The integration tools, which execute in an object-oriented environment, includes a graphical global schema designer, communication utilities to access remote database server hosts, utilities to generate scripting control language to navigate to required data, a design consultant dictionary, data translators, and software version control utilities [Carde93].

Autonomous Database Layer - This is essentially the federation of autonomous databases, consisting of commercial database management systems running within their own data models and data manipulation languages. Currently, we are targeting the integration of the PACS (Sun based, UNIX, Sybase DBMS) and RIS (VAX based, VMS, Mumps DBMS) systems within the Department of Radiological Sciences, the hospital-wide HIS (IBM based, MVS, IMS DBMS), and our research object-oriented system (Sun, UNIX, Gemstone DBMS).

Patients referred to the UCLA Pediatric Radiology Department because of trauma will be the main population from which the normal development group will be formed. These patients are referred for evaluation for conditions unrelated to growth and development or chronic illness. In addition to the trauma patients, the Pediatric Radiology Section performs 7,000 to 10,000 skeletal examinations per year in the Medical Plaza and Center for Health Sciences. In addition to normals, the database will also include populations of abnormals (i.e. patients with a known pathology that affects growth and skeletal maturity). Hand radiographs for these patients will be obtained using a Fuji FCR-7000 computed radiography system. The digital image data from this system will be acquired by the departmental PACS system and stored in the PACS digital optical archives. All patients undergoing a bone age study will be asked fill out a questionnaire designed specifically to study skeletal development. The questionnaire will provide information about overall patient body size, patient genetic profile, maturity level, medical history, physical condition, psycho social condition, dietary habits, and sleep habits. Data from the questionnaire will be entered into an RIS terminal by a pediatric outpatient receptionist. Additionally, all relevant information stored in the PACS, RIS, and HIS systems will be made available.

Computer vision software has been developed to automatically extract bone features (lengths and widths of phalanges and epiphyses) from digital projectional hand images [Pietk90, Pietk91]. A new algorithm is currently under development which uses both local operators (density, gradients, etc.) as well as anatomical models to guide the segmentation procedure. It uses a hierarchical searching approach relying on the detection of a sequence of expected reference objects. Structural models of the hand at each human development stage will be used to assist segmentation of objects which are overlapping and/or are partially fused.

Temporal Evolutionary Data Model

Accurate data models are of great interest in medical databases due to the complexity of object features and object relationships. We have developed an object oriented temporal evolutionary data model (TEDM) for medical objects [Chu92]. Temporal evolutionary data modeling is important because structures in the human body are not static and often change their characteristics and/or existence over time. For example, at birth, only a limited number of bones are present. As we mature, microscopic growth centers **evolve** into new bones. In the wrist area, the eight carpal bones normally appear in roughly four stages, two bones forming during each stage. In the fingers, cartilage, the precursor to hard bones, begins to undergo chemical **transformations**. The epiphysis, a structure between the phylange bones of children, begins to **fuse** with the tubular phylanges at a certain period during skeletal maturation. Exceptional genetic conditions can cause some bones to undergo a **fission** process, splitting into multiple bones. We use evolutionary networks for modeling various object transformation processes and provide inheritance rules between objects that exist in various object evolutionary domains. Further details can be found in [Chu92].

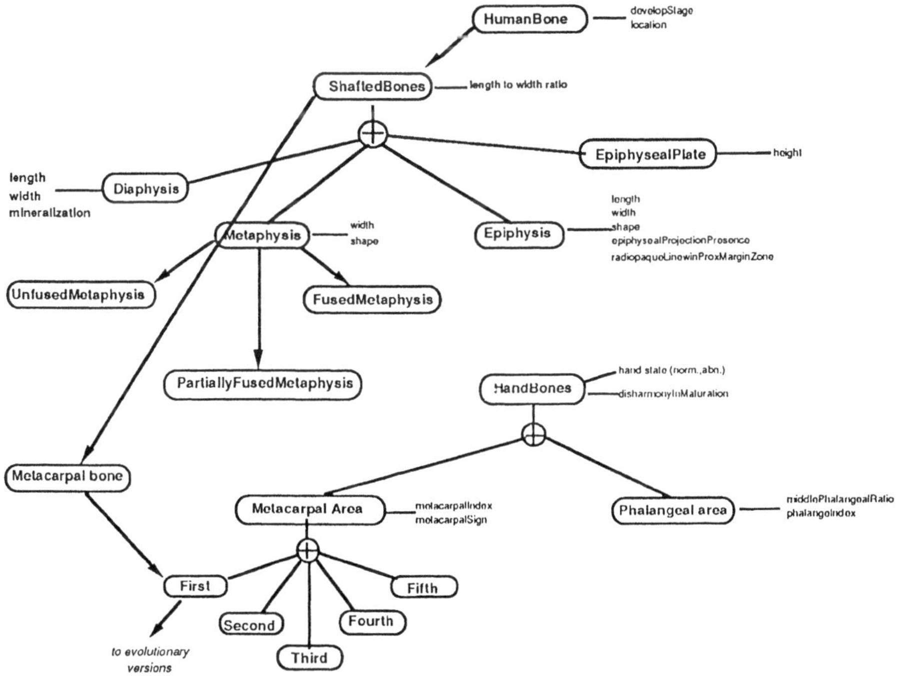

Fig. 3 - *TEDM model of evolving bones and structures in the human hand. Thick arrows represent is-a relationships. Thin lines represent is-part-of relationships.*

Query Processing Steps

High level queries are processed as follows: (1) the radiologist first thinks of a high-level research or clinical question to ask of the scientific database. A typical high-level query would be: "Find all cases of oriental males who are age pre-school who have stubby 3rd and 4th metacarpals. (2) the radiologist uses a graphical interface (FEL) to a high-level nonprocedural language (PICQUERY+) to convey the query to the computer system [Fig. 2]. (3) the PICQUERY+ entries are then translated into a lower level query representation, and initially bypasses the CQAL on its way to the CDL. The CDL manager uses a global system directory to decompose the query into more fundamental queries and dispatches these queries to the autonomous database layer. (4) the returned results are processed by a results interpreter within the CDL. If a solution to the query cannot be determined, the system will prompt the user to relax part of the query (i.e. expand the scope of the query to include more data candidates). The CQAL then uses hierarchical data models of the object and concepts involved to modify the original query to include a large scope of candidate objects. (5) the modified query is returned to the CDL. Steps 4 and 5 may be iterated a number of times until a satisfactory query result is reached. (6) the query results are returned to the FEL and they populate the PICQUERY+ user defined screens.

Conclusion

Research into the development of intelligent database systems which provide interoperability of medical databases and knowledge bases (anatomical models, disease classification models, etc.) will be crucial for research involving complex multidisciplinary collaboration. These systems will provide researchers with a convenient framework for which to share important research data toward, for example, the construction of morphological and physiological models of normality. This paper describes our efforts toward the development of a knowledge-based multimedia database system for bone age research. Each layer of the KMeD system is currently under development. We expect to fully process queries of moderate difficulty by the end of Summer 1993. Full clinical integration is expected in early 1994.

Acknowledgment

This work was supported in part by Public Health Service Grant No. PO1 CA 51198, awarded by the National Cancer Institute, Department of Health and Human Services and by grant number IRI 9116849 awarded by the National Science Foundation.

References

[Carde87] A.F. Cardenas, "Heterogeneous Distributed Database Management: The HD-DBMS", *Proceedings of the IEEE*, **75**(5):588-600, May 1987

[Carde93a] A.F. Cardenas, R.K. Taira, "Integration and Interoperability of Multimedia Distributed Database Systems", Proc. of the Uniform Executive Conference, San Francisco, CA, March 1993.

[Carde93] A.F. Cardenas, I.T. Ieong, R.K. Taira, R. Barker, C.M. Breant, "The Knowledge-Based Object-Oriented PICQUERY+ Language", *IEEE Transactions on Knowledge and Data Engineering*, August 1993.

[Chu91] W.W. Chu, Q. Chen, and R. Lee, "Cooperative Query Answering via Type Abstraction Hierarchy", in Cooperating Knowledge Based Systems, S.M. Deen (eds), Elsevier Science Publishing Co., Inc., 1991.

[Chu92] W.W. Chu, I.T. Ieong, R.K. Taira, C.M. Breant, "A Temporal Evolutionary Object-Oriented Data Model and Its Query Language for Medical Image Management", *Proceedings of the Very Large Database (VLDB) Conference*, 1992.

[Pietk90] E. Pietka, L. Kaabi, H.K. Huang, M.L. Kuo, "Feature Extraction for Bone Age Determination", *Radiology*, **177**, 1990.

[Pietk91] E. Pietka, M.F. McNitt-Gray, M.L.Kuo, H.K. Huang, "Computer-Assisted Phalangeal Analysis in Skeletal Age Assessment", *IEEE Transactions on Medical Imaging*, **10**(4):616-620, 1991.

Computer Assisted Radiation Therapy

Adopting the Shereouse Gratis Dose Calculation Code to Biological Treatment Planning

Mack, A. ; Dörner, K.-J. ; Kloos, U. * ; Nüsslin, F.

Department of Medical Physics
Radiological University-Hospital Tübingen
* Computer Graphics - Lab University Tübingen

Introduction

Striving for uncomplicated control of cancer is the main intention in radiation therapy. During the planning process the dose distribution within target volume and critical organs has to run through several criteria of optimization. Computer aided treatment planning based on CT-slices often ends up with the problem of comparing several treatment plans.
Useful tools for the comparison of treatment plans are isodose plots. Another possibility of representing the data are dose-volume-histogrames (DVH). Although DVHs tend to simplify the presentation of the data, they have a complex structure. The purpose of this paper is to develop a procedure to condense the dose distribution data to a single value.

Modelling Tissue Reaction

The major problem in modelling tissue reaction to radiation are the insufficient data. Clinical and experimental data are not easy to compare. The dose response relationship follows a sigmoidal curve. Therefore a variety of functions are capable to describe these data. Even in simplified cases the evaluation of the needed parameters is difficult. For this reason a phenomenological description has been chosen. The tissue response is taken to be a function a single variable only. The result occurs as a description given by a Gauss-graph. At a specific dose TD the tolerance of an organ exposed to radiation is reached. The tolerance doses of a set of individuals are settled around a mean. This unsharpness is described by the spreadparameter σ. The probability P for tissue complications up to a certain dose is calculated by integrating the probability density function from zero up to the applicated dose D:

$$P(D,V) = \int_0^{D'} \sigma^{-1} \, (2\cdot\pi)^{-\frac{1}{2}} \, e^{-\frac{1}{2}\left(\frac{D-TD(V)}{\sigma(V)}\right)^2} \, dD \qquad (1)$$

Isoeffect Curves

A substitution transformes the integral into the tabulated Error-function. The upper limit of this integral is zero for D = TD when the tolerance dose is reached. At this value the integral can be splitted. Using an analytical approach the following formular for the complication probability of

$$P(D;V) = P(TD;V) \pm ½ \sqrt{1 - e^{-0.6225 \left(\frac{D - TD(V)}{\sigma(V)} \right)^2}} \qquad (2)$$

normal tissue can be derived. The tolerance dose depends on the volume which is approximated by

$$TD(V) = TD(V_0) \cdot (^V/_{V_0})^{\Phi} \qquad (3)$$

All the parameters needed for calculatating P(D,V) are listed in the table, all required data for the isoeffect calculation are known. This is the tool for converting an isodose matrix into an isoeffect matrix.

Histograme reduction

The purpose of histograme reduction accounts for the finding of hypothetical homogenious irradiation with the same biological effect as the irradiation proposed originally. Therefore a closer look has to be taken at the plot of the complication-probability P at dose D and volume V. In this plot isoboles represent lines with the same P but different D and V. The basic idea of this algorithm is to transform the dose of each voxel along the isobole into an equivalent dose received by a reference volume. The equation reads:

$$D_{i_0} = \frac{D_k - TD_k(V_k)}{\sigma_k(V_k)} \sigma(V_0) + TD(V_0) \qquad (4)$$

This way the effects of irradiations in different volume elements with different doses are added up into a reference volume. The resulting complication probability P(D,V) is calculated with the formular (2) given above.

Sherouse Gratis Code

The Sherouse Gratis Code is an assemble of routines for treatment planning. Since the source course code is available, Gratis is a suitable tool for research purposes. The code is modularely constructed and the graphical presentation of parameters and results is sophisticated. The modules are:

IMEX. A tool for reading CT-data, contour definition and rendering of regions of interest within the CT-Slice.

XVSIM. A program for simulation set up parameters of the treatment machine. The xvsim consists of nine windows:

1. **Image Selection** for selection of CT-Slice and contours.
2. **Image Panel** for grey-scale and point manipulations.
3. **Virtual Simulator.** This is the steering control unit for the simulation of the treatment. Here the radiation field is declared and modified by blocks and filters etc. A lot of parameters for the visualization are controlled.
4. **Unit Control** to modify parameters like gantry angle, collimator angle, field width, table position, and moving the isocentrum.
5. **Unit Y** brings up a simplified model of the treatment machine and shows gantry, table and collimator settings from the top.
6. **Unit Z** acts like Unit Y, but gives a view from the front.
7. **BEV.** The Beams Eye View shows the patient contour, and the beam geometry from the source viewing along the central axis of the current radiation field.
8. **XVIEW.** A parallel projection is given of the beam geometry and patient contour along the x-direction.
9. **YVIEW.** Is the analogon to XVIEW in y-direction.

XPLACE-GRID. This module accounts for handling the dose calculation matrix. The dimensions of the matrix are independent in x, y and z - direction. The limits are visualized with the patient structure and with selected radiation fields.

XPLANDISP is the tool for displaying the dose distribution. Three orthogonal projections are supported. In the displayed picture patient contours, isodose lines and the demarcation of the radiation field are shown.

Modifications

Using the programming language "C" a modul has been programmed to calculate the biological effect: The module IMEX was used to link every voxel to a specific organ. The dose matrix calculated by Gratis delivers a dose value for each voxel. The tolerance dose of each voxel depending on the organ are calculated with the help of tabulated values of the tolerance doses. With formular (2) dose and distance to the tolerance are used to convert the dose matrix into a biological effect matrix. The resulting matrix of biological effect has been visualized by the Gratis modul XPLANDISP instead of displaying the dose matrix.

Additionally equation (4) has been programmed to project the dose of each voxel to the entire patient volume. The doses were summed and by using formular (2) a total effect of the irradiation has been received. This "over all complication probability" is displayed.

Example (calculated by OSS Philips)
The target volume should be the chest wall. Two treatment plans are proposed. The first way of realization is an approach with two photon fields. The second proposal is an electron arc therapy. The electron arc is divided up into three subarcs. The electron energy in each segment is selected depending on the thickness of the chest wall. The isodose curves and dose volume histogrames are shown in the fig. 2. To keep the example as simple as possible only one plane is taken into account. The two plans are compared. The dose in each volume element is projected to the volume of the lung. The projected $D_i(V_0)$ has the same effectiveness as if the original irradiation technique had been chosen. By summing up these $D_i(V_0)$ the probability $P(\Sigma D_{i0};V_0)$ is estimated.
The used values are TD = 25 Gy, σ = 9 Gy, Φ = -0.929 and V_{slice} = 46.625 ccm. In this example the complication probability resulting from the electron arc therapy is P_{el} = 16%. The irradiation with photons comes up with P_{ph} = 25%.

Discussion
The Sherouse Gratis Dose Calculation Code has a lot of advantages not least because of its open and modular concept. The problem are the time consuming dose calculation and the lack of an inhomogenity correction. Gratis possesses an advanced graphical environment and gives the possibility to add and replace treatment planning modules.
The proposed modelling of tissue reaction to radiation has the following advantages: Without any problems the known parameters $TD_{5,5}(V)$ and $TD_{50,5}(V)$ can be used. Thus the needed parameters TD, σ and Φ are evaluated. A fit of new parameters to clinical data is not nessecary. The problem how to add complication probabilities of different organs is solved. All changes in fractionation and in beam quality can be taken into account. From a critical point of view the results should not be overemphasizsed. The loss of information by reducing the histograme to a single number might be significant.
Thus the resulting value of the algorithm should not be taken as a complication probability on an absolute scale. The values are useful for the ranking of treatment plans.

Prospective studies would be necessary to show how useful the algorithm is in the daily routine.

Table of Tolerance Doses

Organ	Volume /age	$TD_{50/5}$	$TD_{5/5}$	σ	NSD	β/α	$-\Phi$
Eye		70	50	12	400		
Tissue	100 cm^2	70	55	9	1730	10	0.1196
Breast	2400 cm^2	100	50	31	1650		
Rectum	100 cm^2	80	60	12	1500	5-10	0.1674
Colon	100 cm^2	65	50	9	1400	3-7	
Brain		70	60	6	1800		0.2320
Bladder		80	60	12	1750	12	
Skin	100 cm^2	70	55	9	1730	10	0.1196
Heart	big part	55	45	6	1500		0.3273
Testes		2	1	0.6	500		
Hypophyse		200	45	15	1400		
Bone	10 cm adult	100	60	24	1800	2	0.0676
Liver		40	25	9	1200		0.3716
Lung	whole	25	10	9	760	3	0.9286
Lymphatics		70	50	12	2200		
Stomach	100 cm^2	55	45	6	1200		0.2741
Oral cavity	50 cm^2	75	60	9	1750		
Kidney		25	20	3	900	2	0.7447
Ear		70	55	9	1750		
Esophagus	75 cm^2	75	60	9	2000		0.0705
Ovar		9	2	4	400		
Sigma and Rectum		80	60	12	1750		
Thyroid		150	45	64	1400		
Uterus		200	100	60	2400		
Vagina		100	90	6	2200		

References

1.) Kutcher, J.: **Histogram reduction method for calculating complication probabilities for three-dimensional treatment planning evaluations.** Int.J. Radiation Oncology Biol.Phys. Vol.21,(1991), 137-146.

2.) Lyman, T.: **Optimization af radiation therapy, III: A method of assessing complication probabilities from dose volume histograms.** Int.J.Radioation Oncology Biol.Phys. Vol.13,(1987),103-109.

3.) Sherouse, G., Bourland, J., Reynolds, K., McMurry, H., Mitchell, T., Chaney, E. **Virtual simulation in the clinical setting: some practical considerations.** Int. J. Radiol. Onc. Biol. Phys. 19(4),1059-1065 (1990)

4.) Wendhausen, H.: **Untersuchungen zur Biophysikalischen Dosisplanung bei der Strahlentherapie maligner Tumore** Habilitationsschrift Kiel 1978

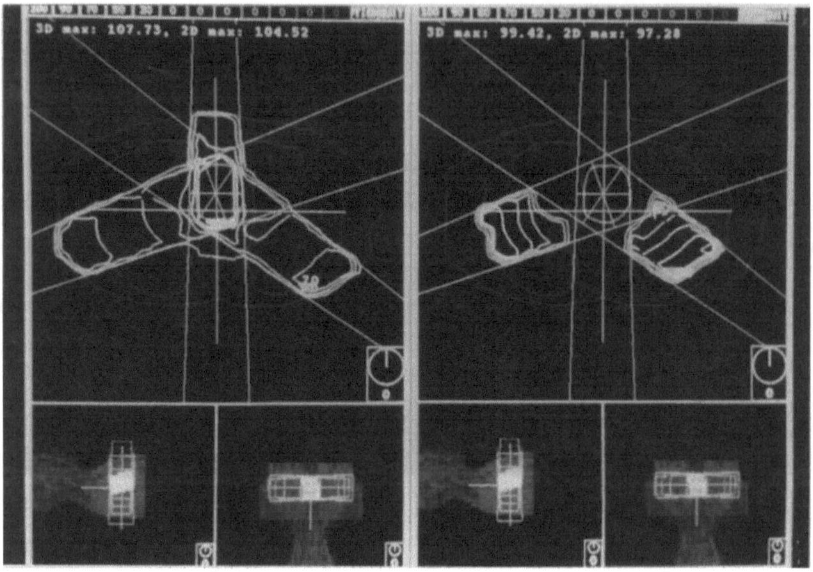

Fig. 1.) GRATIS-Output: Isodose- and Isoeffect-Curves

(see also in color on page 819)

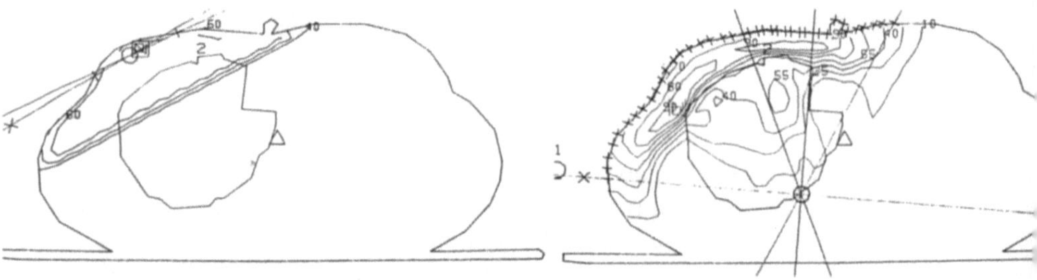

Fig. 2.) OSS-Output: Isodoseplot: Photon-irradiation and Electron Arc Therapy

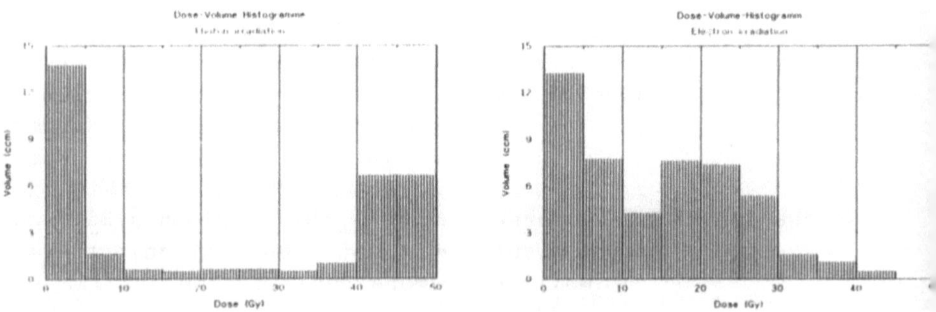

The corresponding Dose-Volume-Histogrammes

An Interactive 3D Visualization System for Dynamic Radiation Treatment Planning

U. Kloos[1], A. Mack[2], H.-H. Ehricke[1], F. Nüsslin[2] , W. Schlegel[3], W. Straßer[1]

[1]: University of Tuebingen, Computer Graphics Lab, FRG
[2]: Radiological University-Hospital, Dep. Medical Physics, Tuebingen, FRG
[3]: German Cancer Research Center, Institute for Radiology and Pathophysiology, Heidelberg, FRG

Summary

Technological progress in radiation treatment in recent years has led to an increasing complexity of treatment plans. Particularly in dynamic treatment planning a variety of parameters have to be adjusted, describing e.g. collimator leaf positions, gantry and table movements or modulation of the intensity profile. The paper will demonstrate that the design of an optimal treatment plan is substantially supported by the use of Computer Graphics methods.
We describe concept and implementation of a visualization system for dynamic treatment planning, allowing graphical definition of parameters as well as treatment simulation. The system is based on a user defined model of the treatment scene. This includes a geometric description of the treatment hardware components, their constraints in motion and of the individual patient-related objects (e.g. skin, target volume, organs at risk).

Introduction

The technological progess concerning radiation treatment planning has been followed by an increasing complexity of treatment plans. For example 3D conformal treatment techniques, subdivided into dynamic and multisegment treatment techniques, may consist of a large amount of coplanar and noncoplanar beams. For all beams a variety of parameters have to be adjusted, describing e.g. collimator leaf positions, gantry and table movements or modulation of the intensity profile.
This progess should also be taken into consideration concerning radiation treatment planning systems. The complexity of treatment plans constitutes great demands espacially for treatment design. A user of a treatment planning system needs planning tools allowing him to design a treatment plan in an intuitive way and within a reasonable amount in time.
Furthermore, treatment design is the first step in the optimization loop of a planning system. Due to the time consuming dose calculation algorithms it is important to find a treatment plan near to the optimum in the first iteration, thus having to run the dose calculation only few times. Therefore, the phase of treatment design is very important. In this context, similar to the problem of displaying dose distributions, Computer

Graphics methods may be very helpful. Using the potential of these methods it is possible to create a complete virtual treatment scene within the computer. With such a virtual scene the planning person is able to get to know exactly what happens inside the treatment room whenever he changes a treatment parameter. It is also possible to animate a dynamic treatment plan.
Schlegel demands a graphics system "... which faithfully reproduces the function and feel of a physical simulator" [4]. Mageras et al. described a radiation treatment system for a multisegment technique, which can be interpreted as a discretization form of a dynamic treatment technique, needing a simulation of a treatment plan directly at the treatment machine [3]. Both emphasize the need of a graphics system, allowing treatment simulation within the computer.
In this paper we describe concept and implementation of a visualization system for dynamic treatment planning, allowing graphical definition of parameters as well as complete treatment simulation. We will demonstrate that the design of an optimal treatment plan is substantially supported by the use of Computer Graphics methods.

Patient Modeling

The base data for patient structures consist of a stack of 2D images. For a 3D treatment planning system we need a 3D representation of the patient structures. Rendering 3D volume data directly by volume rendering techniques, is very time and memory consuming and can't be used interactively without special hardware in the short time.
Therefore, surface oriented reconstruction is prefered in the actual visualization system. After defining regions/volumes of interest the patient structures are represented in a polygonal form (contours). For a reconstruction, based on contours several methods are available. We decided to reconstruct the patient structures using a triangulation algorithm, ending up with triangle meshes for the internal representation of the patient related objects. Because triangles are primitives in graphic systems, the use of these geometric objects is directly supported by Computer Graphics systems.
A further advantage of triangle meshes is the existence of methods for direct triangulation of 3D data [2]. Structures which are easy to detect, such as skin or lung can be reconstructed by such algorithms avoiding time consuming 2D editing tasks.
We have the possibility for either manual or automatic segmentation of regions of interest. The set of contours are triangulated using an algorithm developed at our institute [6].

Machine Modeling

In contrary to the patient related objects, the treatment hardware components need to be modeled only once. In order to simulate movements of single components, we need a model that allows single movements in an easy way. Thus we decided to construct the machine related objects using the Denavit-Hartenberg notation, which is commonly used in robotics [1].

Each moveable component is defined as a single object with its own frame. The link between two of these objects is characterized by 4 parameters, of which only one is a variable. This parameter is the one being responsible for movements and can be defined as a function of time. The range of this parameter defines the motion scale, so it is easy to set constraints in motion. Furthermore, the parameters describe a transformation matrix defining an object frame in relation to its predecessor frame. Simulating a motion is easy to realize by the successive use of these matrices starting at the base of a movement unit.

The treatment hardware components are made up by two independed movement units, the gantry and the patient table. Figure 1 shows a schematic representation of the decomposition of both units into single objects, each characterized by one only (time depending) parameter. The table top is constructed as a single object, although it can be moved in lateral and longitudinal direction. Because our construction concept doesn't allow two variable parameters for one single object, we have to insert a virtual object. That means that the joint between the table column and the table top is subdivided into two joints, each of them responsible for one degree of motion.

For our visualization system we use the geometry of a LINAC SL-25 (Philips, Hamburg) treatment machine, which is used at our radiological university hospital.

To construct the virtual treatment scene we used the Inventor^{TM} (Silicon Graphics Inc., Mountain View) [7]. This is an object oriented toolkit for 3D graphics applications. Based on this toolkit it is quite easy to describe the single objects and the transformations needed to define the movement units.

The patient model is fixed to the last element of the table movement chain and thus moved whenever the table is set in motion.

Simulation

The simulation is based on two different kinds of representations both are shown in figure 2 .

One commonly used way of representation is the so called "Beams Eye View" (BEV). The camera position is fixed at the position of the source of radiation. The viewing direction is along the central beam axis and a perspective projection is used to account for the beam divergence. This is a very helpful tool for the interactive definition of beam direction and shape. The user has an immediate visual control of the radiation field and can avoid critical situations, such as organs at risk lying in front of the target volume. Every kind of interaction modifies the scene, keeping the camera position fixed. In this presentation the beam shape demarcated by the leaves of a multileaf-collimator can be adjusted.

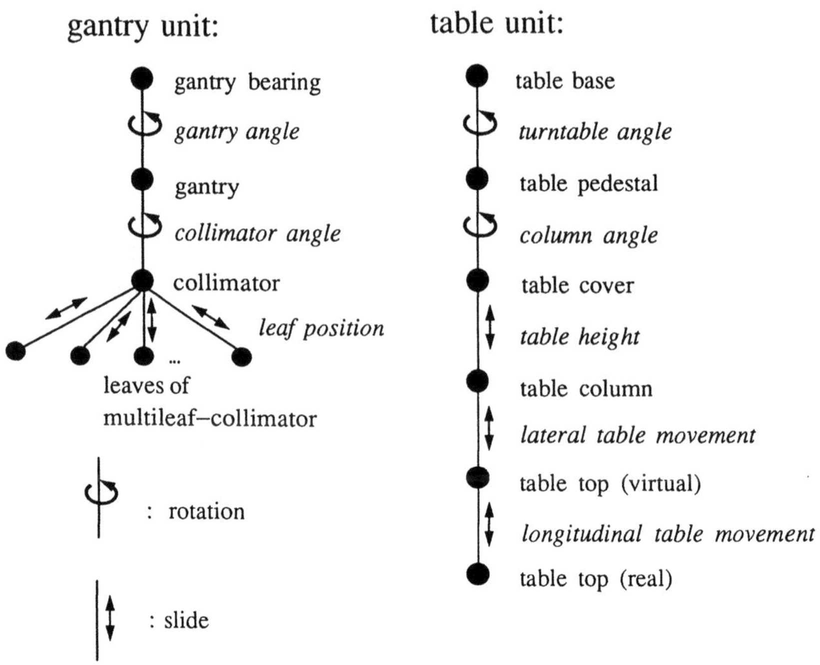

Figure 1: Decomposition of the gantry and table unit into single objects. Every joint between two objects is depending on only one paramter.

The BEV is helpful for defining single beams, but is not sufficient for simulating dynamic treatment plans. In order to approach this goal the so called "Observers Eye View" (OEV) is more suitable. In the OEV the user can modify the camera position and thus walk through the treatment scene, viewing the patient adjusted for treatment from any arbitrary position. In this representation it is possible to see what happens to the treatment hardware components and to the patient when treatment parameters are changed. This allows the user even in complex treatment techniques to avoid unwanted situations, for example the potential collision between patient and collimator.

Additionally it is possible to visualize the beam and to get an idea of those anatomic structures, the patient is exposed to. A further option is the mapping of the beam shape onto the patient's skin surface, so that it can be used for documentation purposes.

Treatment parameters can be set separately either by sliders or, as described above, with the BEV representation. It is also possible to load an external treatment plan and visualize the effect of its parameter settings.

For designing a dynamic treatment plan it is necessary to define a function of time for each parameter that should be changed during the treatment session. One way to

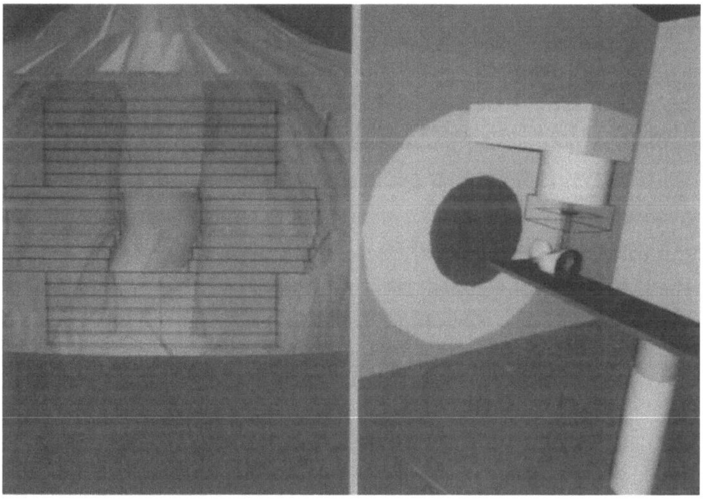

Figure 2: Graphical representation of the treatment scene. The Beams Eye View (left side) shows the multileaf-collimator and the patient structures, which are based on CT-images of an Alderson-phantom containing contours of skin, lung and an imaginary target. The Observers Eye View (right side) reveals the complete treatment scene, including gantry, table and patient structures.

(see also in color on page 819)

define this function is to select key values for each parameter and interpolate between them. Another possibility is the separate definition of each parameter function.
The visualization system has been implemented in C++, using OSF/Motif and SGI Inventor^{TM} for user interface development.
We also have a planning system from Chapel Hill, North Carolina [5], which is used for dose calculation and for visualization of the dose distribution.

Discussion

In the framework of the visualization system described in this paper it is possible to simulate a dynamic treatment plan. Using a flexible Observers Eye View and working with a virtual treatment scene the user gets a direct visual feedback of what happens while treatment paramters are changed. Critical situations, such as the collision between different treatment hardware components or between gantry unit and patient can be avoided.
By switching the beam on it is possible to watch the geometrical shape of the beam from any arbitrary position in order to check whether the target volume is completely

enclosed and whether organs at risk are affected or not. Matching the beam onto the patients skin surface can be used for documentation and for controlling the parameter transfer to the real treatment unit.
With the use of BEV and single parameter settings it is possible to use the system even for conventional treatment plan design.
Planning a dynamic treatment session requires new kinds of interaction tools, which overcome the limitation of explicit parameter definition. One possible solution would be the graphical interactive definition of a track on top of the patient surface to be followed by the treatment unit. The motion parameters are adjusted according to this prescription. The presented simulation system is a suitable framework for developing and testing such interaction tools.
Furthermore, it is necessary to integrate dose distributions into the virtual treatment scene. Geometric objects from the estimated dose data, like isodose surfaces or dose data matched to patient structure surfaces can be used within our system. The main handicap for the a use of the dose distribution visualization during interactive treatment design is the poor speed of dose calculation algorithms.
Further investigations are necessary to examine the integration of volume data for patient modeling.

References

[1] John J. Craig. *Introduction to Robotics, Mechanics & Control.* Addison-Wesley, 1986.

[2] W.E Lorensen and H.E. Cline. Marching cubes: A high resolution 3d surface construction algorithm. In *Proc. Siggraph*, pages 163–169. ACM, july 1987.

[3] G.S. Mageras, K.C. Podmaniczky, and R. Mohan. A model for computer-controlled delivery of 3-d conformal treatments. *Med. Phys.*, 19(4):945–953, Jul/Aug 1992.

[4] W. Schlegel. Computer assisted radiation therapy planning. In Pizer S.M. Höhne K.H., Fuchs H., editor, *3D Imaging in Medicine*, pages 399–410. NATO ASI Series, 1990.

[5] G.W. Sherouse, C.E. Mosher, K. Novins, J. Rosenman, and E.L. Chaney. Virtuals simulation: Concept and implementation. In I.A.D. Bruinvis, P.H. van der Giessen, H.J. van Kleffens, and F.W. Wittkaemper, editors, *9th International Conference on the Use of Computers in Radiation Therapy*, pages 433–436, 1987.

[6] R. Steinbrecher. *TURIST - ein oberflächenorientiertes System zur 3D-Rekonstruktion.* PhD thesis, Eberhards-Karls-Universität Tübingen, 1992.

[7] P.S. Strauss and R. Carey. An object-oriented 3d graphic toolkit. In *Computer Graphics Siggraph 92*, pages 341–349. ACM Siggraph, july 1992.

3D Visualization for Radiotherapy Treatment Planning

T. SCHIEMANN[1], B. DIPPOLD[1], R. SCHMIDT[2], A. POMMERT[1], M. RIEMER[1], R. SCHUBERT[1], U. TIEDE[1], K.H. HÖHNE[1]

[1] Institute of Mathematics and Computer Science in Medicine (IMDM)
[2] Department of Radiotherapy
University Hospital Eppendorf
Hamburg, Germany

e-mail: schiemann@imdm.uke.uni-hamburg.dbp.de

Abstract

3D radiotherapy treatment planning (RTP) requires visualization of complex scenes with many different objects (anatomical structures, isodose surfaces) in an easy and clear way. We have recently developed a new system for visualization of 3D medical data, which consists of a large set of functions for visualization and exploration of images based on a comprehensive data structure including all relevant knowledge. It is shown with different examples that this new approach is well suited for application in radiotherapy treatment planning.

Introduction

For 3D treatment planning systems in radiotherapy it is important to achieve a complete assessment of all aspects of the dose distribution [2, 5, 7]. For calculating dose distributions the density information is typically taken from a volume of coplanar CT-slices. These CT-slices can obviously be used very well to display the patient's anatomy together with the calculated dose data. Twodimensional display of patient anatomy with directions of beams and distribution of isodose lines is a necessary part of every state of the art treatment planning system. A generalization of this method is the display of arbitrarily positioned planes displaying the same matter. More sophisticated tools for true 3D display often suffer from the drawback of displaying surfaces only, or the system is too complex to handle.

Recently we have developed a new volume visualization tool (VOXEL-MAN/atlases) [3], which allows the visualization of pictorial and descriptive knowledge in a combined way. While its primary goal is the generation of 3D anatomical atlases, it turned out that it can be extended to various volume visualization tasks. It is the purpose of this paper to show that the concept can be applied for a substantial improvement for visualization and exploration in radiotherapy treatment planning.

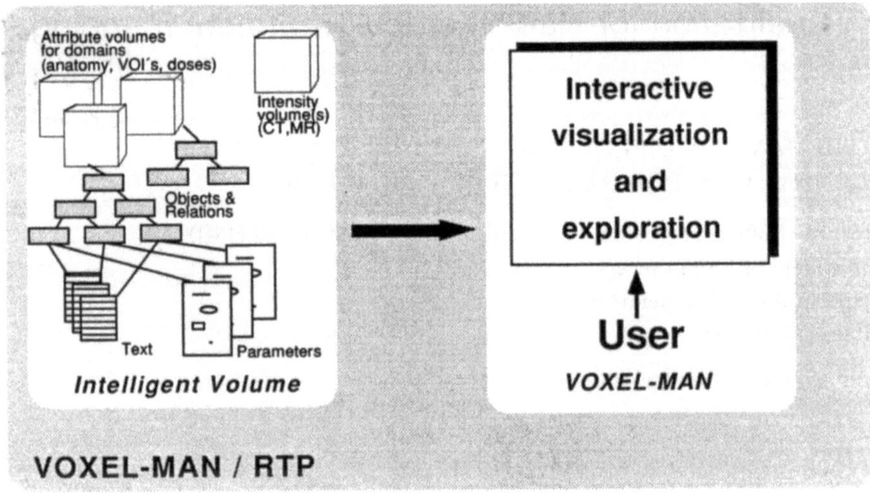

Fig. 1: Structure of VOXEL-MAN/RTP: Many visualization and exploration tools of VOXEL-MAN can be applied to the "Intelligent Volume" specific for radiotherapy treatment planning (RTP). The intelligent volume is a data structure, which includes spatial knowledge (attribute volumes) and descriptive knowledge (i.e. relations, text, or visualization parameters).

Method

The visualization system consists of a general volume visualization and exploration system and a new data structure, which we call "Intelligent Volume" [9]. Besides intensity volumes derived from imaging devices (CT,MRI), the intelligent volume consists of several attribute volumes for different domains of knowledge (anatomical structure, volumes of interest, and dose values) and a database providing descriptive knowledge about these domains (fig. 1). Generally, for every attribute volume the descriptive knowledge consists of a net of objects and relations (i.e. "The rectum belongs to the organs at risk"), a list of visualization parameters, and text information (i.e. "The dose distribution is created by four isocentric beams"). In particular the domains for radiotherapy are:

- *Anatomy:* Attributes are organs and structures, which can be segmented from image volumes acquired from the patient. In addition to the CT-volume needed for dose calculation these could also be obtained from MRI, MRA, or other modalities. The segmentation for single organs (i.e. bone, brain, liver, or kidney) can be performed easily and in a short time with our segmentation and volume editing system [4, 6].

- *Volumes of interest:* In this domain the attributes are the volumes of interest describing those parts of the anatomy, which are involved in the optimization of the treatment plan. The volumes of interest can also be defined by our segmentation system.

- *Radiation dose:* The dose volumes are delivered from an external dose calculation system [1, 8]. The dose values are considered as attributes in this domain. A particular dose volume can be modified during a session and several dose distributions can be handled at the same time for comparison.

The spatial knowledge (attribute volumes) has to be created for every single patient. The descriptive knowledge is stored in a generic database and can mainly be applied to any case, although modifications can be made if differences should occur. Once the intelligent volume has been filled with knowledge, its contents can be visualized and explored. The attribute volumes can be addressed separately for information on one domain only, or they can be used together for more complex questions (i.e. "Show those parts of the organs at risk, which get a dose above their tolerance level").

VOXEL-MAN has been implemented in the programming language C and is equipped with a comfortable OSF/MOTIF based user interface. It is running on UNIX workstations (DECstation 5000, DEC 3000 ("alpha"), and Sun SPARCstation), which are also used in modern imaging devices, and can be linked to many treatment planning systems. For such hardware the computing times for typical images range from 5–30 seconds (DECstation 5000).

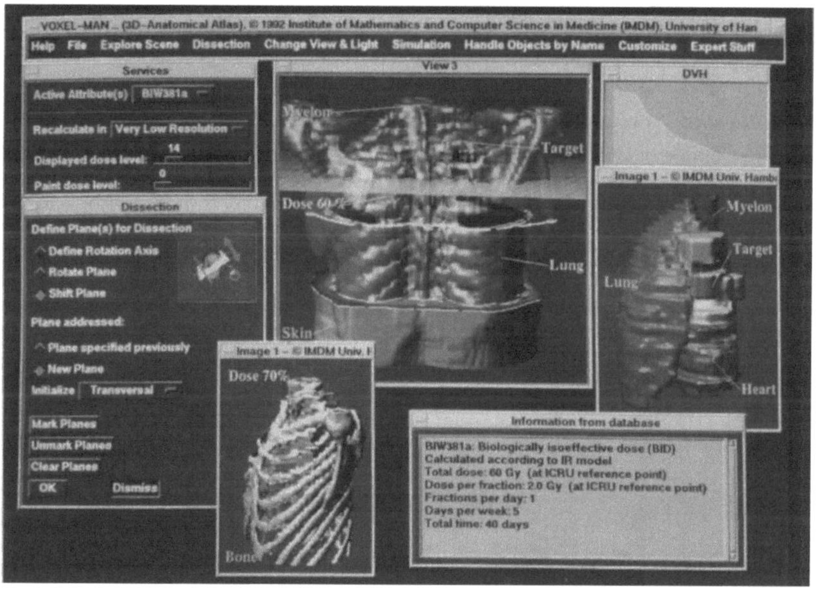

Fig. 2: View onto the screen during a typical session: Several 3D views can be explored simultaneously. Text output (relevant parameters of a selected dose distribution) has been obtained by pointing onto an isodose surface and a dose volume histogram for the lung has been calculated. The user has selected the dissection mode (active menu on the left) for unveiling structures of interest (large image at center).

(see also in color on page 820)

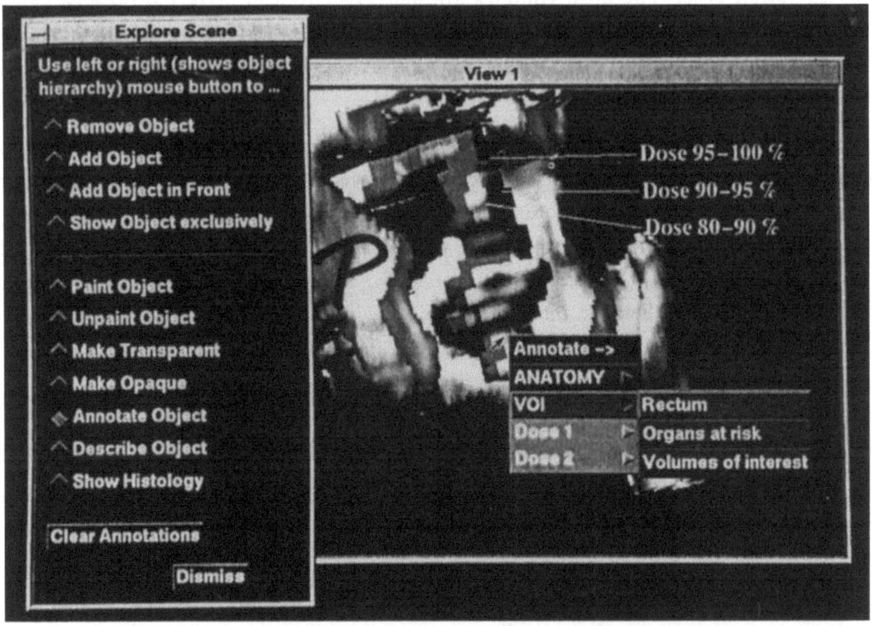

Fig. 3: Application of exploration tools selected from the menu on the left: After mouse click the list of available domains appears at the cursor position. The chosen function (in this case "annotate") is applied to the object, which has been selected from the hierarchy of objects available in this domain. (see also in color on page 820)

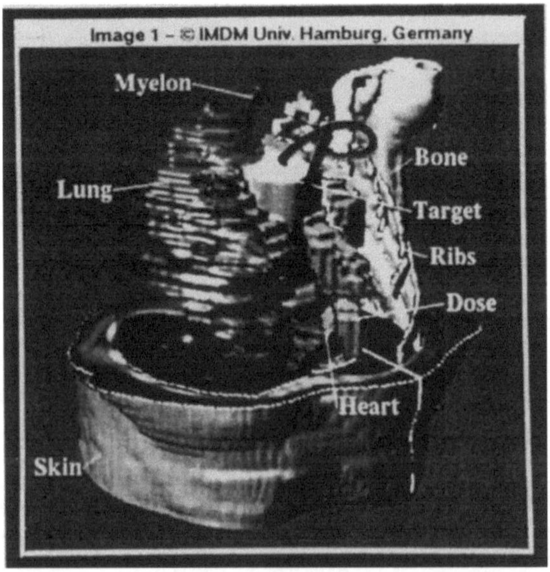

Fig. 4: Typical image after dissection and transparent rendering: The extent of an iso-surface of a dose distribution can be assessed. Skin and bone have partially been removed by dissection. The right lung is rendered transparent.

(see also in color on page 821)

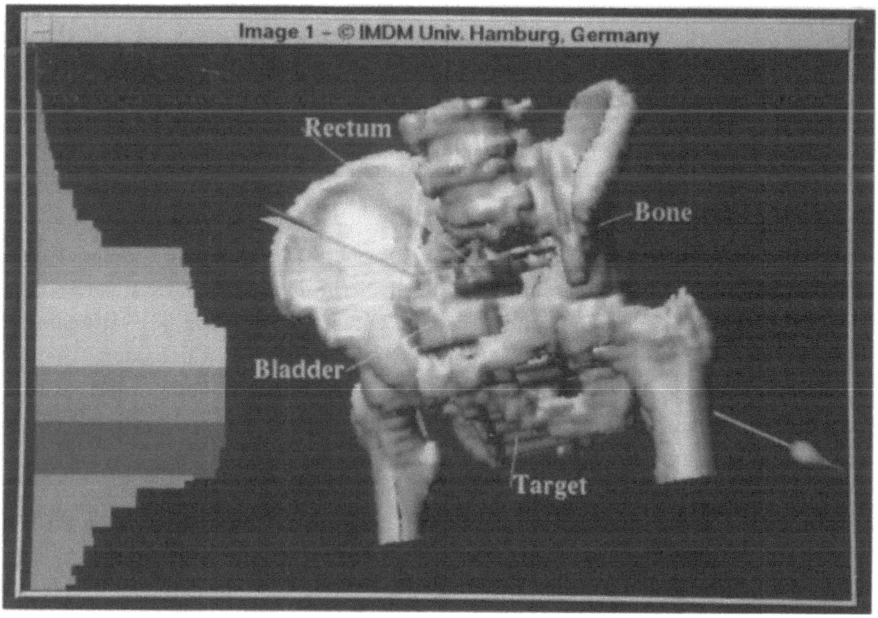

Fig. 5: Display of a line dose: A needle can be specified arbitrarily and the dose distribution along its path is shown as a histogram colored according to the volumes of interest.

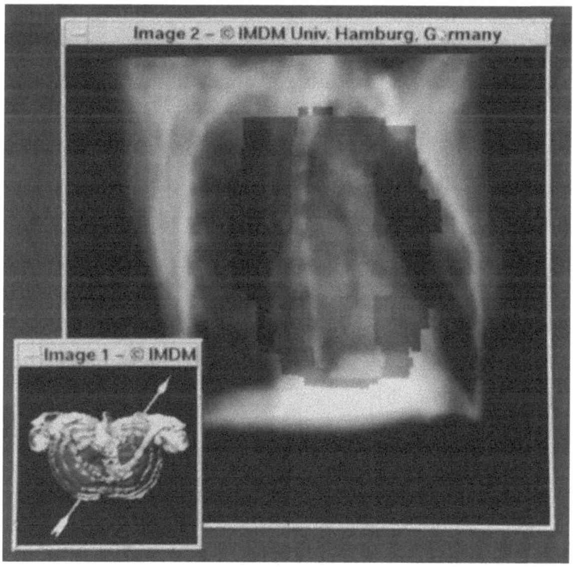

Fig. 6: Simulation of a radiograph. The regions of the target and a selected dose volume have been colored. The insert shows the direction of projection.

(see also in color on page 821 ,822)

Application

The application of the radiotherapy version of VOXEL-MAN, which we call VOXEL-MAN/RTP will be demonstrated with therapy plans for different carcinomas. Figure 2 shows a view of the system during a typical session. There are several 3D images, a dose volume histogram and text output from the database of the intelligent volume. All available functions can be controlled via menus, which can be selected from the menu bar at the top of the screen.

The functions grouped in the *Explore Scene* menu (fig. 3) are the most frequently used tools for exploration of an image and are applied in the pictorial context: After mouse click the list of available domains appears at the cursor position (fig. 3). Dragging the mouse to one of the domains causes a second list to be displayed, which shows the name of the object at this point for this domain and the membership of the object to higher level structures. Releasing the mouse button on an object of the list starts the selected action. Typical actions in a radiotherapy planning session are

- *Annotating* selected dose values
- *Painting* a selected dose range on an organ at risk
- *Adding* an isodose surface
- *Removing* an object, which obscures an isodose surface
- Making an object *transparent* in order to assess an isodose surface inside
- *Describing* a selected dose distribution with the description stored in the database

The system offers various functions to change the viewpoint and light parameters in an intuitive way. The displayed scene can be cut into different sectors and objects can be removed restricted to these regions providing effects similar to real dissection (fig. 4). As the system is based on a volume model, the whole knowledge is also available on the cut planes appearing after removal of an object. Unlike in reality, objects can also be added after they have been removed.

On every image an arbitrarily positioned needle can be specified for quickly assessing the dose distribution as a line dose along the path. The dose values are displayed in a fashion similar to a histogram, where the columns are colored according to the colors of the volumes of interest (fig. 5). For correlation with images obtained from a therapy simulator, the CT data can be used to reconstruct radiographs with objects of interest appearing as colored regions (fig. 6).

Conclusions

We have shown that our new approach for visualization and exploration of complex 3D medical data is very well suited to an application in the demanding field of radiotherapy treatment planning. According to his needs, a radiotherapist can choose from a large set of visualization tools, which includes all functionality described for other systems. As a substantial novelty, the knowledge available for any point of an image can be inquired in an intuitive way, thus the system can be used easily by any non experienced user.

As the system is not yet directly connected to a dose calculation system, it is currently in use for quality assurance using representative cases as references for clinical routine. A wide range of applications is also given by any kind of education, and there is no major obstacle to link the package to a suitable dose calculation platform.

Although our approach is very advantageous, computing times of up to 30 seconds (DECstation 5000) for typical images are not fully satisfying. Considering the fast increase in hardware technology, it can be expected that systems like the proposed one will be widely spread in the near future.

Acknowledgements

The authors are grateful to Drs. Krüll and Schwarz (Dept. of Radiotherapy) and all members of the IMDM who have supported this work.

References

1. Bauer-Kirpes, B., Schlegel, W., Boesecke, R., Lorenz, W.: Display of organs and isodoses as shaded 3-D objects for 3-D therapy planning. *Int. J. Radiat. Oncol. Biol. Phys. 13* (1987), 135–140.

2. Bourland, J., Camp, J., Robb, R.: Volume rendering: application in static field conformal radiosurgery. In Robb, R. A. (Ed.): *Visualization in Biomedical Computing II, Proc. SPIE 1808.* Chapel Hill, NC, 1992, 584–587.

3. Höhne, K. H., Bomans, M., Riemer, M., Schubert, R., Tiede, U., Lierse, W.: A 3D Anatomical Atlas Based on a Volume Model. *IEEE Comput. Graphics Appl. 12*, 4 (1992), 72–78.

4. Höhne, K. H., Hanson, W. A.: Interactive 3D-Segmentation of MRI and CT Volumes Using Morphological Operations. *J. Comput. Assist. Tomogr. 16*, 2 (1992), 285–294.

5. Rosenman, J., Sherouse, G., Fuchs, H., Pizer, S., Skinner, A., Mosher, C., Novins, K., Tepper, J.: Three-dimensional display techniques in radiation therapy treatment planning. *Int. J. Radiat. Oncol. Biol. Phys. 16* (1989), 263–269.

6. Schiemann, T., Bomans, M., Tiede, U., Höhne, K. H.: Interactive 3D-Segmentation. In Robb, R. A. (Ed.): *Visualization in Biomedical Computing II, Proc. SPIE 1808.* Chapel Hill, NC, 1992, 376–383.

7. Schlegel, W.: Computer Assisted Radiation Therapy Planning. In Höhne, K. H. et al. (Eds.): *3D-Imaging in Medicine: Algorithms, Systems, Applications*, Springer-Verlag, Berlin, 1990, 399–410.

8. Schmidt, R., Schiemann, T., Höhne, K. H., Hübener, K.-H.: Three-Dimensional Treatment Planning for Fast Neutrons. In Breit, A. (Ed.): *Advanced Radiation Therapy: Tumor Response Monitoring and Treatment Planning.* Springer-Verlag, Berlin, 1992, 643–648.

9. Schubert, R., Höhne, K. H., Pommert, A., Riemer, M., Schiemann, T., Tiede, U.: Spatial Knowledge Representation for Visualization of Human Anatomy and Function. In *Information Processing in Medical Imaging, Proc. IPMI'93*, Springer, Berlin, 1993. to be published.

VIRTUOS - A Program for VIRTUal radiotherapy Simulation

Rolf Bendl, Jürgen Pross, Mark Keller, Josef Bürkelbach and Wolfgang Schlegel

Deutsches Krebsforschungszentrum,
Abteilung für Biophysik und medizinische Strahlenphysik,
Im Neuenheimer Feld 280
6900 Heidelberg

Summary

VIRTUOS is the new user interface to the VOXELPLAN program for 3D radiotherapy planning of the DKFZ (German Cancer Research Center) Heidelberg. VIRTUOS is a virtual therapy simulator which was designed to support the therapist during the planning and evaluation of radiotherapy. Virtual therapy simulation means that the therapist can simulate and verify the complete radiation therapy on a graphic workstation before patient treatment. Clinical use has shown that the time needed for defining 3D radiotherapy plans can be reduced noticeably with VIRTUOS.

VIRTUOS offers a simple interface for information interchange with dose calculation and optimization programs, therefore it is a useful tool for other scientists, too, who want to evaluate their own algorithms and develop and examine new irradiation techniques.

Introduction

The aim of conformation radiotherapy is to fit the therapeutical dose distribution to the shape of the target volume in order to apply higher dose and this way improve tumor control probability and decrease side effects(6). Traditional 2D radiotherapy planning doesn't consider the three dimensional extension of target and organs at risk sufficiently. Also the dose distribution of non coplanar irradiation and other more sophisticated techniques cannot be calculated. Several algorithms for 3D dose calculation have been developped during the past few years. Unfortunately 3D planning is a complex and time consuming process because of the large number of possible treatment alternatives in each special case. In spite of its advantages, 3D radiotherapy planning hasn't asserted itself in clinical routine until now.

In the past many computer techniques have been proposed to assist the therapist in defining and evaluating 3D plans (back projection (4)(5), dose-volume-histograms(7), 3D display of dose distribution with target volume and organs at risk(6)). But so far there hasn't been a 3D treatment planning system available that combines these and other possible techniques in an appropriate user interface in order to assist the therapist in the best way during the planning phase.

For further research in this area there is no platform by which new software tools can be developped, tested and evaluated easily.

Therefore, on account of experiences made with earlier versions of the user interface of our VOXELPLAN program we have created a planning system called VIRTUOS (VIRTUal radiOtherapy Simulation) that supports the therapist in easy and quick definition and evaluation of 3D treatment plans (1)(3).

Methods

3D radiotherapy planning is based on a stack of CT-slices and a fully threedimensional model of the patient anatomy. This model includes all therapy relevant volumes like the target volume and the organs at risk.

During the planning phase the most time consuming task for the therapist is to define the direction of the irradiation fields determined by the angle of gantry and couch, the orientation, the size and shape of the irradiation field determined by the angle and size of the collimator. The shape of the irradiation field can be adjusted exactly to the shape of the tumor by means of a multi leaf collimator that consists of two rows of thin moveable leaves. So the therapist must define the position of the leaves.

Planning aids

Back projection - The CT-slices are overlayed with the extensions of the radiation fields(4)(5). If this information is presented to the therapist in three orthogonal planes, he can well imagine the position, direction and extension of the radiation fields. The consideration of all relevant anatomical structures - especially of those that are not in the planes of the presented CT-slices - is difficult.

Beam's Eye View - the therapist can look at the threedimensional model of the patient's anatomy from the position of the radiation source through the collimator's aperture(4). He can realize at once whether there are some organs at risk within the irradiation field respectively whether the target volume is included completely by the radiation field. The Beam's Eye View can also be used to determine size, orientation and shape of the radiation field.

Observer's Eye View - here the therapist can have a look at the threedimensional model of the patient's anatomy as well, but from arbitrary directions. In addition to the Beam's Eye View the threedimensional shapes of the beams are integrated. The therapist has got a supplementary guidance and he can see how the different beams of the plan interact with each other and where the best position for an additional beam might be.

Virtual therapy simulation - in the clinical routine irradiation direction, size and orientation are determined or checked with a therapy simulator. A therapy simulator is an x-ray unit with the same geometry as the irradiation device. From the position of the irradiation source a radiograph is made (corresponding to the Beam's Eye View) which is used to adjust and verify the irradiation parameters. If these possibilities are already available to the therapist within the planning program this is called **virtual** therapy simulation (8). The planning software should deliver the same control possibilities the therapist has got at the linear accelerator. The therapist should be able to define the parameters in the same way as at the accelerator. The best way to support this is to create a graphical user interface where for example parameters can be defined by means of scales or other graphical aids.

Display of dose distribution - the definition of a radiotherapy plan is an iterative process. Having defined a plan, the therapist has to verify the dose distribution and if he notices that it isn't sufficient he has to improve the plan. There are several aids for evaluating plans:

- Superimposing the dose information in form of isodose lines on the CT-slices. The therapist can check the dose distribution in the original patient data. But it is difficult to get a real impression of the entire spatial distribution.
- Integration of the isdodose lines in the 3D patient model. With the help of this technique, the therapist is able to gain an impression of the 3D geometry of the dose distribution in relation to the target volume and the organs at risk. At a glance he can recognize whether the target volume is surrounded completely by the therapeutic dose and whether the organs at risk are spared by doses higher than the tolerance dose(6).
- Dose-volume Histograms - It shows the volumes of the target or organs at risk in dependence of the dose those volumes get. Overdosages or underdosages can be identified very quickly(7).

Database with sample plans - Clinical evaluation has shown that for different patients with similiar tumor locations often the same kind of beam configuration is used. A database in which those plans can be stored and recalled will speed up treatment definition by experienced therapists and can be an excellent guidance for those who are not familiar with 3D radiotherapy planning.

Technical aspects

The efficiency of a planning system decisively depends on the speed of information presentation. In order to get a good impression of the spatial patient geometry the therapist should be able to rotate the patient model in realtime. This has much more importance than a realistic presentation with shaded surfaces or volume rendering. The therapist should be able as well to page quickly between the single CT-slices in order to check the extension of the beams or of the dose distribution.

Results

VIRTUOS is an interactive program that is easy to use. It has got a userfriendly human/computer interface.

The interface is menu- and dialog box driven and all interactions can be performed with the mouse. Special importance was attached to a clear arrangement of all therapy relevant information.

In the standard planning mode the user interface consists of three windows, containing the transversal, sagittal and frontal sections through the target point. A fourth window shows the Beam's Eye View. An additional window with the Observer's Eye View is available as well. Furthermore all relevant irradiation parameters are shown in a panel on the working surface. Each parameter label works like a button. If the button is pushed, a dialogbox pops up, in which the user can define the parameter by means of additional buttons, scales and list boxes. For example if the 'wedge' button is pushed, a list box appears containing all possible wedge angles and also a radio box in which the user can select the wedge orientation.

Device-dependent parameters like type of irradiation, energy, possbile wedges etc. are defined by a device file. According to the selected device the screen objects are configurated at run time.

CT-Windows - the predefined contours of target and organs at risk are overlayed to CT-images. In addition the cross sections of the CT-slices with the beam pyramides can be overlayed. So the therapist can easily judge the extension of the beams in the original patient data. In each CT-window reference lines show the position of the slices in the other two CT-windows. The user can move these lines with the mouse. As a result the slice which fits the new position is shown in the corresponding window. By means of buttons in a control panel the user can page through the CT-slices, zoom the pictures, and enlarge the windows. To avoid an information overflow in the CT-windows he can switch off the reference lines and the cross sections of the beams.

Beam's Eye View - with the mouse the patient model can be rotated in realtime around the axes of the irradiation device, so that the therapist can determine appropriate irradiation angles quickly and easily. The angles of gantry and couch are shown simultaneously on sliders in the parameter panel. The therapist can also drag the sliders in order to rotate the patient model. He can switch to different display techniques (contour lines, depth cueing, surface rendering, red-green stereo projection or pairs of stereo pictures) for better spatial resolution. He can also switch on and off the 3D objects or move a clipping plane for cutting them off and looking inside.

Graphical adjustment of the Collimator - Size and orientation of collimator or contours of irregular fields and the corresponding position of the leafs of multi leaf collimators can be adjusted automatically or by means of simple graphic edit functions.

Observer's Eye View - if the user rotates the patient model in the Beam's Eye View the corresponding beam in the Observer's Eye View is rotated synchronously.

Result mode - dose distributions can be loaded and superimposed in form of isodose lines on the CT-slices and integrated into the Observer's Eye View. Quantitative evaluation of dose distribution can be carried out by means of dose - volume histograms.

Database for sample plans - Sample plans for special treatment techniques or plans which have acchieved good results for special medical indications can be stored and retrieved easily. The database can be structured hierarchically. The hierarchy is presented in a listbox and the user can select an appropriate plan with a few mouse clicks. Each user can create his own database. At the moment we collect sample plans for the most frequent indications with the help of our clinical partners.

User configurable interface - the geometry of the user interface (size, position, zoomfactor etc. of the windows), colours, and visibility of the VOI's etc. can be stored in a setup file which is read automatically when the program starts. So each user can configure the interface according to his own demands.

Hardware independence - A further aim was to develop a system which doesn't need any special hardware, so that it can be run on a normal graphic workstation, in the meantime available in almost every clinic. So the system was written in C, using OSF/Motif toolkit and ISO PHIGS for 3D output. At the moment it runs on VAXStations under VMS and on an IBM RS6000 under UNIX. Installations on other hardware are in preparation.

Discussion

Meanwhile the program has been in clinical use for about 2 years. It has shown that VIRTUOS is easy to use and that the time needed for defining 3D radiotherapy plans can be reduced noticeably(3).

But it has also shown that therapists are in need of for additional support in 3D radiotherapy planning, above all in defining more sophisticated treatment techniques.

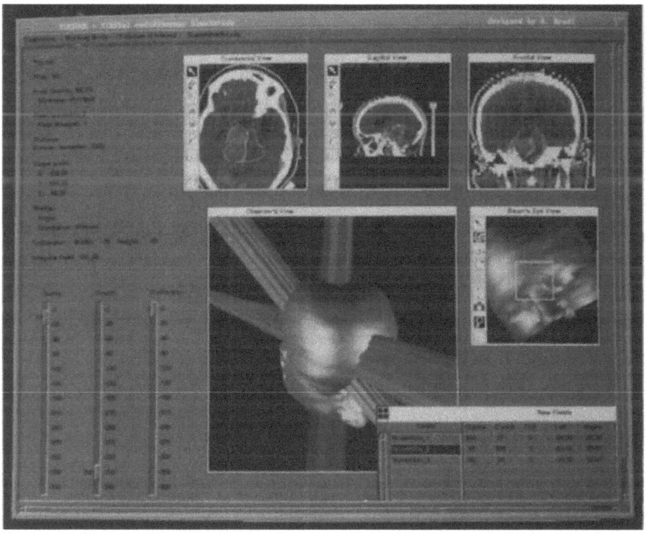

Pic.1:
The VIRTUOS user interface.

In the CT-Windows three orthogonal cross sections through the target point are shown.

In the Observer's Eye View the geometry of all fields of the current plan is visible.

In the Beam's Eye View the yellow rectangle represents the aperture of the collimator, the irregular contour of the current selected field is presented in blue.

Pic. 2:

The Beam's Eye View. The thin yellow lines show the position of the leafs of our multi leaf collimator according to the definition of the irregular field contour (blue).

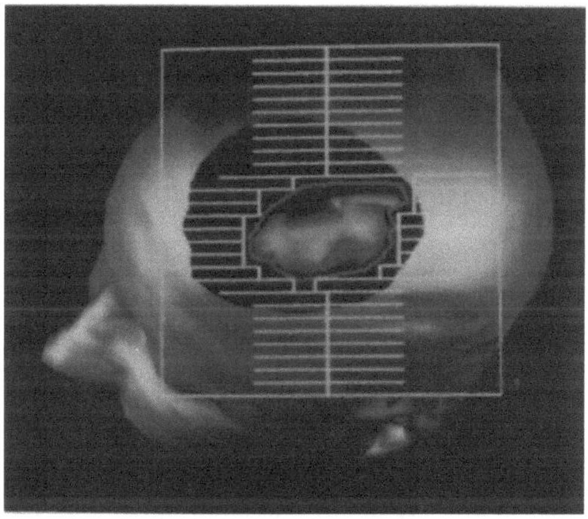

Pic. 3,4:

Target volume (red), eyes (violet), optical nerv (light blue), pituitary gland (dark blue). The yellow ribbons represents the 80% isodose. The left part of the tumor isn't completely surrounded by the 80% isodose, because this part is close to the brain steem (green).

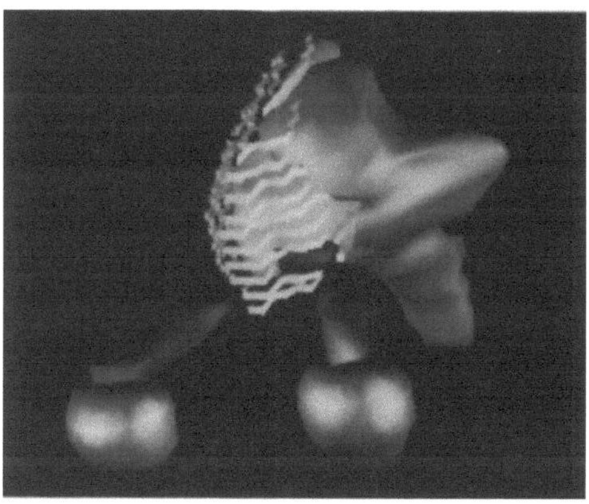

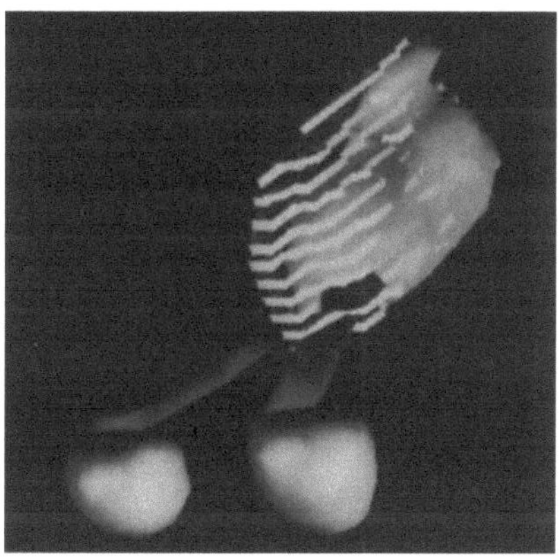

(see also in color on page 822, 823)

Further work should also be done concerning a comprehensive collision control. At the moment it is still possible that a user defines impossible gantry and couch positions, so that the therapists have to keep the possible values in mind. We plan an additional window which will show the accelerator. It should be rotated synchronously to the representation of the Beam's Eye View and should allow optical as well as automatical collision control .

Another issue is the qualitative and quantitative comparision of different plans. We intend to enlarge the result mode, so that the dose distribution of several plans can be evaluated in parallel.

Biological parameters should be integrated in the verification procedure, too.

Aknowledgement

This work was supported by Tumorzentrum Heidelberg Mannheim and the EC in the frame of the covira project.

References

1. Bendl R:
Entwicklung und Implementation einer Benutzerschnittstelle zur virtuellen Strahlentherapiesimulation.
Technical Report CVR 2/91, Institut für Radiologie und Pathophysiologie, Deutsches Krebsforschungszentrum, Heidelberg (1991)
2. Boesecke R, Doll J, Schlegel W, Lorenz, WJ: A multileaf collimator.
Strahlenther Onkol 164:151(1988)
3. Gademann G, Schlegel W, Bürkelbach J, Laier C, Behrens S, Brieger S, Wannenmacher M:
Dreidimensionale Bestrahlungsplanung. Untersuchungen zur klinischen Integration
to be published in Strahlenther Onkol(1992)
4. Goitein M, Abrams M, Rowell D, Pollari H, Wiles J:
Multi dimensional treatment planning: II. Beam's eye view, backprojection and projection through CT-sections. Int. J. Radiation Oncology Biol. Phys.,9:789-797 (1983)
5. Mohan R, Brewster LJ:
Intersection of shaped radiation beams with arbitrary image sections
Computer Methods and Programs in Biomedicine 24:161-168 (1987)
6. Schlegel W: Computer Assisted Radiation Therapy Planning.
In: 3D Imaging in Medicine (Höhne KH, ed.) NATO Series, Vol F 60, pp 399 - 410.
Springer Berlin Heidelberg (1990)
7. Schlegel W: 3D-Rekonstruktionen in der Bestrahlungsplanung
Radiologie 31:457-466 (1991)
8. Sherouse GW, Mosher CE: User interface issues in radiotherapy CAD software.
In: Bruinvis IAD et al (eds) The use of computers in radiation therapy
North Holland, Amsterdam,p,429 (1987)

Automatic Location of the Eyes in CT Images of the Head

Kostis Kaggelides, Peter J. Elliott and *Robert B. Fisher.

IBM UK Scientific Centre, Winchester, UK.

*Artificial Intelligence Dept, University of Edinburgh, UK.

Summary

A technique for locating the eyes in Computed Tomography (CT) brain scan data is described. The objective is to automatically localise the eyes for protection during radiotherapy planning. The image feature that is exploited is the circularity of the eyes. After data preprocessing to remove unwanted features in the images, signature analysis is performed to locate areas of interest. By applying the Canny edge detector to these areas, data is further reduced to the significant edge fragments. The Hough Transform is then applied to estimate radii and centres of the CT sections through the eyes. The Converging Squares algorithm is used as an efficient and robust method to search the Hough Transform parameter space. The results are processed by the hypothesis generation stage which clusters them according to the x,y,z coordinates of the suggested centres. The ISODATA algorithm is used for clustering. The hypotheses are assessed and sorted and the most valid hypothesis is selected and refined using a second Hough Transform. After the rejection of the invalid members of the hypothesis cluster, an ellipsoid is fitted to the new cluster centre and the results are drawn on the data. The method is fast and robust. The method was tested using five different data sets and it performed well on all of them.

Motivation and Problem Description

In radiotherapy treatment planning the objective is to plan a treatment which will deliver a high dose of radiation to the target volume, whilst restricting exposure to sensitive tissues (the organs at risk) to avoid undesirable side effects. The eyes are high on the list of organs at risk which have to be taken into consideration during the treatment planning stage.

This paper describes an algorithm which automatically identifies and locates the eyes in a Computed Tomography (CT) data set of the brain. The specialised human operator is thus relieved from the tedious and time-consuming process of manual segmentation, and the whole process of therapy planning is accelerated.

The system consists of a sequence of individual processes. The major feature exploited for the location of the eye throughout the system is the approximate circularity of the eye sections in every slice (following the assumption that the eye is approximately spherical and consequently any planar section is a circle).

Background Removal

The first step is to remove unwanted features in the background of the CT images e.g. the head supporting pads and the stereotactic frame. The existence of these objects is undesirable and the objective of this process is to swiftly remove them. A number of assumptions are made, namely that the objects are not connected to the head, they are the only features that are unchanged in shape and location throughout all the slices, and that they have a high intensity value. A simple algorithm is constructed which ANDs together all slices after thresholding to identify such objects. All pixels connected to these objects are then deleted. The method works well with our data.

Finding the Bounding Box

Speed is an important requirement for this system. In the case of the CT images only a third of the pixels contain valuable information and the rest are just background. The problem consequently is to calculate the bounding box of the head. Having done this, we can locate small regions within the image where we expect the eyes to be.

Signature analysis [3] is used for the calculation of the bounding box. The vertical and horizontal signatures of the system are essentially the number of pixels set at each row and column respectively. After the background removal process we expect to find only one object of significant dimensions present in the image, i.e. the head. Using the assumption that the patient's head has standard orientation upwards, the eyes are expected to be at the top of the bounding box. Hence, we can locate the eyes within two smaller subframes, usually 50x50 pixels.

Edge Detection

The size of the data to be processed is further reduced by the application of an edge detector, to delineate significant intensity changes between different regions within the image. The Canny operator [2] is used because of its optimal performance with respect to noise and localisation. Additionally, the Canny operator gives a unique response to a single step. A sigma value of 1.0 was selected for the Canny smoothing parameter, giving the desired level of detail in the area around the eye.

Circle finding

As mentioned above, the feature exploited for the localisation of the eyes is the approximate circularity of the eye sections. For this, the Hough Transform (HT) [1] is used, because it is known to perform well with incomplete or corrupted data. In general there are several inherent problems in HT applications, namely storage requirements and time complexity. Localising the feature by finding the bounding box reduces the space and time costs. Each pixel (of the edge detection output image) votes in a three dimensional data structure as the accumulator for all the possible models (circles in this case). Each cell in the accumulator represents a possible model (circle) that fits the data. Each coordinate of this cell represents a model parameter (x, y location of the circle's centre and radius r). In our algorithm the voting stage of the Hough Transform is viewed as a heuristic evaluation of the validity of the possible models, and the search procedure must interpret this evaluation correctly. The Converging Squares algorithm [4] is used for this. This efficient method of search considers peak density as well as peak value and is therefore a good choice of method for searching the accumulator in the case of approximately circular models. It proved to be superior to the maximum selection method, and is computationally efficient.

Hypothesis Generation

So far all the processes have been applied to 2D data. In this step the results are combined for the whole 3D dataset. The input to the process is the set of suggested circle centres (x, y, r together with the slice number and the number of votes that elected them).

These results are grouped in different hypotheses by using the efficient unsupervised clustering method known as ISODATA[5] in x, y, z space, where z is the slice number. The Euclidian space in which the ISODATA clusters the results is appropriate for centre location hypothesis generation, so the radius is excluded from the ISODATA input. The number of votes is also excluded because it is not very reliable information and possible errors are amplified. The different clusters represent different hypotheses for the eye location. The hypotheses are assessed and sorted using as a criterion the total number of votes for each cluster, and the leading hypothesis is selected.

Hypothesis Refinement

The ISODATA algorithm tries to minimise the total cluster dispersion in order to reach convergence, which may lead to the addition of spurious members in an otherwise valid hypothesis. So we need to refine the hypothesis using more high level knowledge. This is implemented as a second Hough Transform, this time between the members of the selected cluster, and for spheres. Each member of the selected cluster represents a section of a hypothetical sphere that contains the eye, with a horizontal plane. Each member votes for all the possible z locations and radius r of this hypothetical sphere. The radius varies in a known range within which we expect the eye to lie. Finally, the selected sphere is used as a filter to reject inconsistent parts of our hypothesis. After this stage, our estimate of the position of the eye centre is accurate.

Shape Fitting

We expect to miss the first and last slices that contain the eyes, where the eye segment is not circular because the eye itself is not exactly a sphere. To improve the results we use an ellipsoid model of the eye which is drawn from the calculated centre. This is not used in earlier stages in the process, because of the extra computational complexity involved in using an ellipsoid instead of a sphere.

Results and Conclusion

The system outlined above was implemented in C++. It takes approximately one minute on an IBM RISC System/6000 model 530H workstation to locate the eyes. It has been tested on five different datasets and locates the eyes with satisfactory accuracy in all of them. The results of all five data sets can be seen in Figure 1. The circles drawn there are the cross-section of the estimated ellipsoid with each slice. The x,y axis localisation of the eyes is very accurate. The z dimension of the ellipsoid fitted is such that the whole eye is included in it. The primary goal is to protect the eyes from radiation, so to ensure that the eye is completely contained within the ellipsoid found is important.

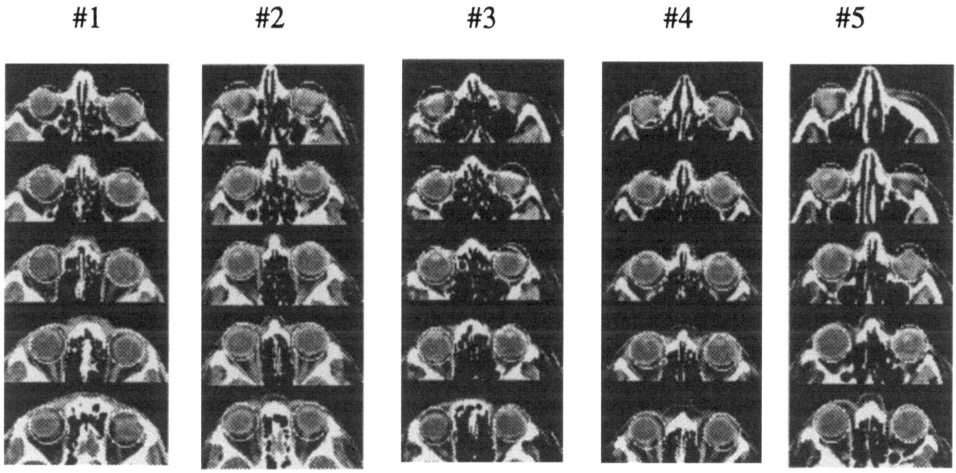

Figure 1: Results of the system applied to the five data sets

There are no data dependent parameters determining the operation of the system. All the parameters hardcoded in the program are model dependent (e.g. range of eye radii). This is the system's knowledge about the eyes as a general geometric shape. Minimal assumptions are also made about the nature of the data (e.g. one major object present, patient's head orientated towards the top side, etc.). These assumptions are used for the significant relaxation of the computational task to be performed.

The system is fast and robust. It is therefore a reliable alternative solution to manual or interactive segmentation, and can save valuable time for the human operator.

Acknowledgement

This work was partly funded by COVIRA (COmputer VIsion in RAdiology), project A2003 of the AIM (Advanced Informatics in Medicine) programme of the European Commission.

Participants in the COVIRA consortium are: Philips Medical Systems, Best (NL) (prime contractor) and Madrid (E); Philips Corporate Research, Hamburg (D); Siemens AG, Erlangen (D) and Munich (D); IBM UK Scientific Centre, Winchester (UK); Gregorio Maranon General Hospital, Madrid (E); University of Tuebingen, Neuroradiology and Theoretical Astrophysics (D); German Cancer Research Centre, Heidelberg (D); University of Leuven, Neurosurgery, Radiology and Electrical Engineering (B); University of Utrecht, Neurosurgery and Computer Vision (NL); Royal Marsden Hospital/Institute of Cancer Research, Sutton (UK); National Hospital for Neurology and Neurosurgery, London (UK); Foundation of Research and Technology, Crete (GR); University of Sheffield (UK); University of Genoa (I); University of Aachen (D); University of Hamburg (D); Federal Institute of Technology, Zurich (CH).

References

[1] C. M. Brown. Inherent bias and noise in the Hough transform. *IEEE Transactions on Pattern Analysis and Machine Intelligence*, PAMI-5(5):493–505, September 1983.

[2] J. Canny. A computational approach to edge detection. *IEEE Transactions on Pattern Analysis and Machine Intelligence*, PAMI-8:679–698, 1986.

[3] R.M. Haralick and L.G. Shapiro. *Computer and Robot Vision*, pages 80–83. Addison-Wesley, 1992.

[4] L. O'Gorman and A.C. Sanderson. The converging squares algorithm: An efficient method for locating peaks in multidimensions. *IEEE Transactions on Pattern Analysis and Machine Intelligence*, PAMI-6(3):280–288, May 1984.

[5] C.W. Therrien. *Decision Estimation and Classification : An Introduction to Pattern Recognition and Related Topics*. John Wiley and Sons, 1989.

3D Data Segmentation and Mesh Generation of Pathological Brain Tissue Based upon 3D MRI Data, in View of Acoustic and Thermal Modeling

P. FREY, B. SARTER, P. PEBAY, M. GAUTHERIE [(1)]
M. DESGEORGES, Ch. DEROSIER, F. HOR [(2)]

(1) Biomedical Thermology Laboratory, Faculty of Medicine
11, rue Humann, 67085 Strasbourg - France

(2) Val de Grâce Hospital, Dpt. of Radiology and Neurosurgery
75005 Paris - France

Abstract
Within the framework of a R&D project on ultrasound thermotherapy of brain tumors, studies are in progress to develop 3D numerical modeling of acoustic power deposition and temperature patterns in view of clinical treatment planning. Patient vs. tumor characterization is based upon 3D voxel data acquired from MRI examinations performed according to specific protocols which provide morphologic and angiographic data. Numerical simulations use the finite element method to solve the partial derivative equations describing (i) the interactions between ultrasound and brain structures and (ii) the processes of heat generation and heat transfer.

I. INTRODUCTION

A. Project of integrated ultrasound hyperthermia system

Our project is to design and develop a new scan-focussed ultrasound technology (multi-element matrix of transducers with feed-back control of phase and amplitude) and a 3D numerical thermal model for personalized treatment planning. It takes advantage from recent progress in ultrasound technologies, 3D MRI data acquisition, image processing and numerical simulation.

B. Ultrasound hyperthermia treatment planning

For brain tumors, which require highly selective and accurate heating of tumor tissue, treatment planning assumes all the more importance than there is no efficient method for non-invasive mapping of deep-tissue temperature. Our approach of treatment planning concurrently integrates data acquisition, data processing, thermal modeling and optimization and control of the heating system, and optimization of the heating system (the ultrasound transducer and the conditions of its use), in order to comply with the therapeutic requirements [1].

Treatment planning can only be effective if numerical modeling is based upon patient data, and takes the biophysical heterogeneity and variability of the brain tissue structures into

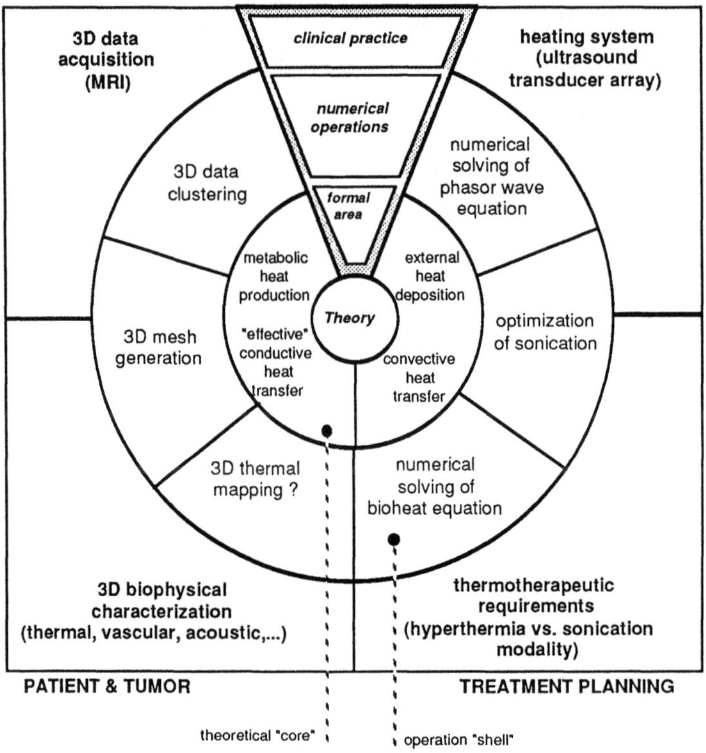

Fig. 1: Operating clinical treatment planning

account. This is the only acceptable approach to represent (i) the interactions of ultrasound with tissue, and (ii) the heat transfer phenomena including the vascular convective processes, in a significant way (fig 1).

II. NUMERICAL PROCESSING

Data acquisition uses magnetic resonance imaging (MRI) exclusively, which avoids the difficulties involved with recombining data from different acquisition techniques. Specific protocols of direct 3D acquisition and angiography make it possible to represent the tumor area and its vascularization with a matrix of voxels.

Data processing is performed prior to thermal modeling, in order to properly format the set of voxels in view of numerical simulations. It consists essentially in data clustering and

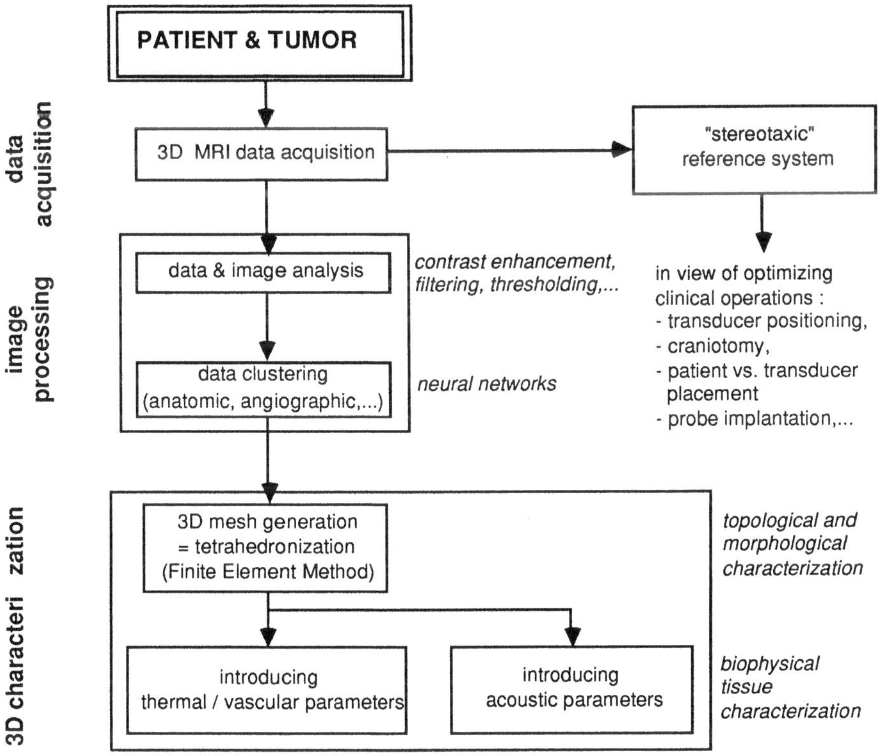

Fig 2 : 3D characterization of the regions of interest

mesh of the regions of interest in order that the finite element method (FEM) can be used to solve the partial derivative equations (bioheat transfer and phasor-wave) (fig 2).

A. Data segmentation

The algorithm for MRI data segmentation has to be designed so as to (i) minimize loss of spatial and spectral resolution, (ii) conform to the intrinsic topology of the voxel data set, and (iii) allow to develop automatic procedures.
Considering the complexity of the cerebral medium and the variability of pathological circumstances in terms of tumor characteristics, current image processing methods are not convenient to process 3D matrix of voxels, at least in their present state of development.
An adaptative neural network has appeared to be a method of choice. It permits a semi-automatic extraction of the regions of interest by proceeding with (i) a supervised learning of pre-classified examples representing the different "classes" to be extracted ("network

configuration step"), and (ii) a classification of voxels with error estimation based upon qualitative and quantitative criteria. The supervised learning uses a gradient error back-propagation algorithm in order to estimate the different weights of the network nodes. This operation is repeated as long as necessary according to an iterative process in order to optimize the "learning rate", and eliminate redundant examples [2] .

B. Fully automatic 3D mesh generation

The use of the finite element method (FEM) as solver of the bioheat transfer and phasor-wave equations in numerical simulations requires first the mesh of the domain of interest, in order to define a mathematical partition of this domain and the relevant mesh nodes. This processing step is of utmost importance since the quality of the numerical solution will closely depend upon the quality of the mesh [3]. The method should (i) preserve the intrinsic topology of the regions of interest (sets of voxels), (ii) allow the generation of tetrahedrons with a size of the order of magnitude of the discontinuities found in cerebral structures, (iii) permit proper characterization by the relevant biophysical parameters including the vascular ones, and (iv) allow to develop a fully automatic mesh generator.
A critical review of the current methods of mesh generation has shown that none of them is convenient to generate a mesh from 3D voxel data. A mesh generator was therefore developed, in which tetrahedronization uses a divide-and-conquer method, which provides small elements on the boundary of the domain of interest [4]. Voxels of the domain are subdivided according to an automatic procedure, with respect to the topology. Two sets of voxels are identified : (a) voxels totally included into the domain, and (b) boundary voxels.

III. RESULTS

Specific algorithms and software programs which comply with all the above requirements were developed on a Silicon Graphics workstation. Furthermore, a graphic environment was developed to permit easy and interactive design of complex image processing operators.
Two techniques (octree-based and voxel subdivision methods) have been combined to achieve fully automatic mesh generation. The resulting algorithm is robust for the tetrahedonization of complex domains in the cerebral medium (fig.3 shows the result of mesh generation of a brain tumor).
Each occurence of a polyhedron, a face, an edge and a vertex is inserted into the data structure using a hash table, which allows no double element storage. CPU time is O(n) for tetrahedra computation and insertion (fig 4).

ACKNOWLEDGMENT

The authors wish to express their gratitude to Pr. P.L. George and J. Henry (INRIA, Paris-Rocquencourt), for their scientific cooperation to the project. This work was made possible by research program funding from the Ligue Nationale Française contre le Cancer, the Fédération Nationale des Centres de Lutte contre le Cancer, the Association pour la Recherche sur le Cancer (ARC, Paris), and the Conseil Régional d'Alsace (Strasbourg).

REFERENCES

1. M. Gautherie, P. Frey, Toward clinical treatment planning for ultrasound hyperthermia, *Ultrasonics*, vol. 30, pp 135-137, 1992.
2. E. Engler, M. Gautherie, Method of image segmentation using a neural network. Application to MRI of brain tumors, *Inn. Tech. Biol. Med*, vol 13, pp. 115-129, 1992.
3. P.L. George, Automatic mesh generation, Masson (Paris), 1991.
4. P. Frey, M. Gautherie, Automatic 3D mesh generation in a set of voxels, *Inn.Tech.Biol.Med.*, vol 12, pp 428-442, 1991.

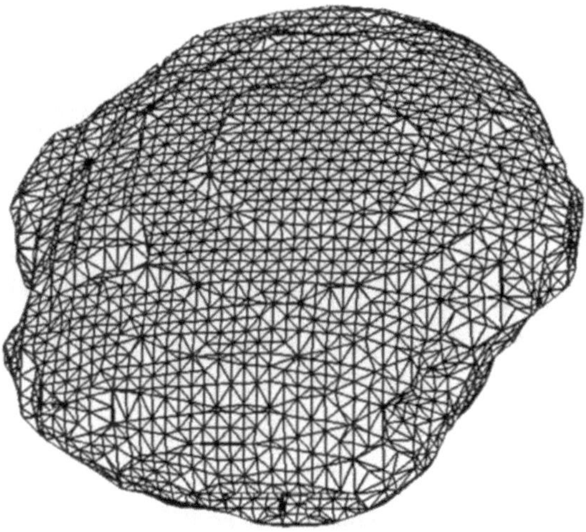

Figure 3 : mesh of a brain tumor (glioma). 3D MRI data : 1.5 T Signa imaging system (General Electric), domain of interest: iso-surface reconstruction (d=85). 34696 boundary mesh elements , 83242 faces, 61731 edges, 13284 vertices were generated in 47.27 sec.

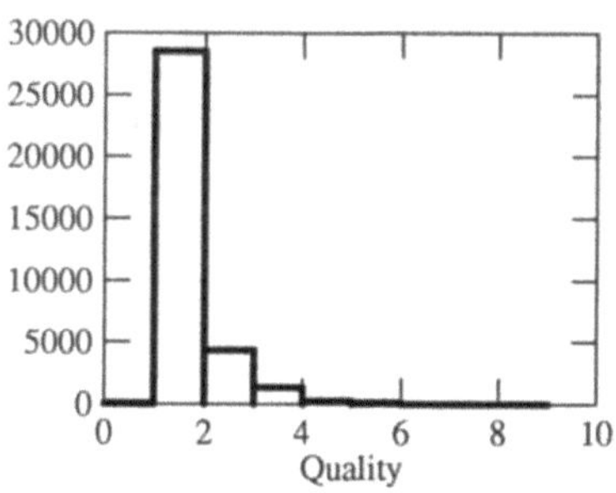

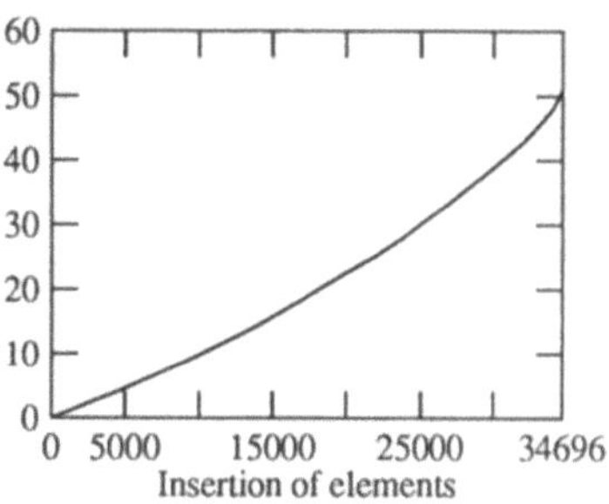

Figure 4 : Quality and CPU time requirement for the mesh of the boundary.

3D Numerical Modeling of Acoustic Power Deposition in Pathological Brain Tissue in View of Planning Ultrasound Thermotherapy and Surgery

Ph.PEBAY, P.FREY, M.GAUTHERIE, Louis Pasteur University, Strasbourg

Key-words : Ultrasound hyperthermia - ultrasound surgery - ultrasound transducers - acoustic modeling - brain tumors.

This work is part of a project on ultrasound thermotherapy and surgery dedicated to the treatment of brain tumors. This project aims at developing (i) ultrasound technologies for external and interstitial applications, and (ii) 3D numerical models for personalized treatment planning.

The proper design of ultrasound applicators as well as the optimization and control of the ultrasound system require numerical simulations of the 3D acoustic power deposition field within brain tissue.

A careful review of literature has allowed to collect specific data on the acoustic properties of normal and pathological brain tissue.

The finite element method has appeared to be appropriate to solve the phasor wave equations representing the interactions between ultrasound and biological structures.

3D numerical simulations use conveniently processed 3D MRI voxel data. Data processing has included data clustering and the generation of a 3D homogeneous mesh*.

In order to evaluate the respective contributions of the different interation phenomena separately, the simulations algorithms and relevant software programs were developed by (i) considering first absorption only, and then absorption, reflection and diffraction, and (ii) taking or not non-linear propagation into account.

These various simulations were made, first for single focused external transducers, and then for a variety of multi-element, electronically-controlled, scan-focused transducers designed for external and interstitial tumor treatment.

---

* see abstract by FREY et al., 3D DATA SEGMENTATION AND MESH GENERATION OF PATHOLOGICAL BRAIN TISSUE BASED ON 3D MRI DATA, IN VIEW OF ACOUSTIC AND THERMAL MODELING.

Aiding 3D Hyperthermia Planning by the Use of Combined Modelling and Visualization of Different Medical Image Data Sets

MARCUS RUDOLPH, ULF JACKISCH

Technical University of Berlin
Department of Computer Graphics
Sekr. FR 3-3
Franklinstrasse 28/29
D-10587 Berlin
e-mail: marc58@cs.tu-berlin.de, jackisch@cs.tu-berlin.de

Abstract

In today's increasingly complex medical environment, referring physicians are confronted with a huge amount of semantically different medical image information about the same region, with respect to the patient. This is due to the use of various pieces of hardware and software found in the modern clinic. If this information is geometrically correlated, then a joint processing of these different information classes becomes an important aid in diagnosis, therapy and surgery planning. This implies the need to have a joint representation of information as a basis. A "general model" is required as a development base for all the related applications. One task within the COMED (Cooperative Medicine) project covers the research area in the topic of combined modelling and visualization.

The COMED project is cooperating with the Hyperthermia Ward of the UKRV Hospital (Universitätsklinikum Rudolf Virchow) in Berlin. In the area of hyperthermia planning, the need exists to generate image relevant information of different semantics. The semantics of the information in our case, consists of CT/MRI datasets, E-field distribution, electric parameters such as conductivity and dielectricity constants, isotherms and perfusion distribution, among others. Bringing together the different semantical information implies the use of advanced representation and visualization methods. The presentation of such complex information to the physicians requires 2D and 3D visualization techniques and the possibility of combinations of partial types of the semantical information.

Fundamental concepts and first results will be discussed in this paper.

Introduction

The design and prototypical implementation of medical workstation software and specialized applications to support image based diagnosis and therapy planning in a fibre optic networking environment was one of the main goals of the MEDAP project (1990 - 1991) [1] and the succeeding COMED project (still in progress). Both projects receive their grant from the German company DETECON, a subsidiary company of the German TELECOM. In the framework of these projects a comprehensive and user-friendly tool was developed, with which one can generate every possible cross section within a medical image cube and visualize three-dimensional threshold defined objects [2]. After finishing this diagnosis orientated software, it is now intended to develop an application which covers the needs of the three-dimensional hyperthermia planning.

Hyperthermia

With the application of modern hyperthermia techniques in cancer therapy, it is intended to increase the vulnerability of the target volume in respect to radiation or cytostatics. This effect will be achieved by heating the malignant tumour to a therapeutic temperature greater or equal to 41° C. Thus, a hyperthermia application is just one part of combined oncological concepts, which include different types of cancer therapy methods. There are different application modes of hyperthermia, such as local, interstitial and regional hyperthermia [3], which are subjects of world-wide research and development activities. Each of these application modes is suited for special tumour classes or locations.

In the following, only the regional hyperthermia application mode practised by the working group of Dr. Wust in the hyperthermia ward of the UKRV Hospital, will be considered. This group has been researching and working since 1989 in the field of regional hyperthermia with Sigma BSD 1000/2000 RF-applicator systems [3] which work with phased arrays of antennas. After three years of research and development activities the group stated at the "6th European BSD Users-Conference", that their clinical experience indicates, that the effectiveness of Hyperthermia must be improved. Further, the group introduced their concepts to increase the clinical usefulness of hyperthermia. These concepts include the use of patient specific 3D modelling of the SAR distribution, the development of conformal antenna networks in suitable applicators, an adequate quality control and an online control for verification [3].

The most important point of the groups experience is the fact, that one has to treat the hyperthermia planning as a 3D problem, which is influenced by different categories of parameters. On one side there are the adjustments and properties of the hyperthermia device and the antennas, and on the other side there are patient specific variables, such as position, anatomy and perfusion conditions. In general, the planning of a hyperthermia treatment is on principle, an optimization problem, which has to be solved to achieve an optimal temperature distribution.

Because of the complexity of the variety of distinct parameters, which can partially influence each other or may correlate, a convenient and comprehensive computer based graphical application package is required, which provides feedback to settings and manipulations of the 3D hyperthermia planning system. Thus, the therapist would be supported in overviewing all system parameters and to assess his therapy plans.

The most important parameters or image relevant information types, which will become relevant during the process of hyperthermia therapy planning are listed as follows:

- CT Images [anatomical information]
- MRI Images [additional soft tissue information]
- MRI Images [blood perfusion information]

- 3D Grids [grids of tetrahedrons, which are derived from anatomical regions, which were segmented by the therapist]
- E-Field Distributions [electrical fields, which are the result of the energy input]
- SAR Distributions [absorbed power by the tissue]
- Perfusion Distributions [blood flow rate, which will cause a cooling effect if it is too high]
- Dielectrical Distributions [material constant]
- Temperature Distributions [contours of regions with the same temperature]
- Positions of the Antennas [placement of the antennas in 3D space]
- Position of the Patient [within the device]

Modelling

Due to the number of parameters or information types, which should be displayed solely or simultaneously in user defined combinations, it is necessary to integrate all these distinct image relevant information in a n-dimensional model. The central role of such a model in the context of a communication infrastructure, which connects medical workstations of distinct computational power and image acquisition devices, is illustrated in Fig. 1.

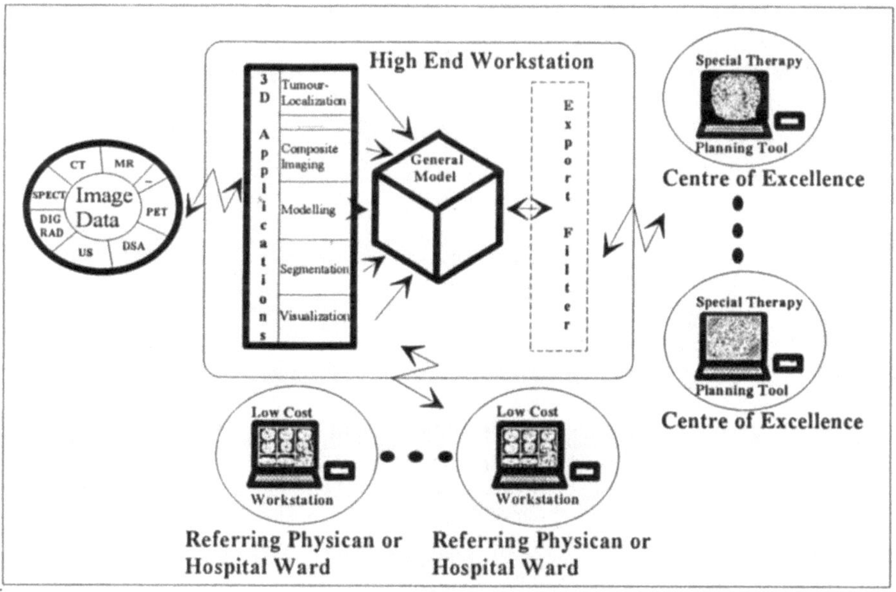

Fig. 1. The Generalized Model

The basic idea of such a n-dimensional model is quite old and was first introduced by S. Stiehl in 1980. He called it the "XCE Model" (eXtended Cell Enumeration) [4]. The advantages of this approach are on one hand its simplicity and on the other hand the simple and direct data access. Especially now, due to the availability of huge RAM capacities and decreasing RAM prices, it is possible to keep all the data in memory. Hence, one has to calculate the access offset of a selected voxel once and can peek thereafter at all information of this voxel with the same offset. The disadvantage of this approach is, that all the information which should be integrated in this model have to be correlated geometrically first. This implies the need of writing import filters for each information class, which is not defined in the space coordinates of the model. In the case of the hyperthermia information types, only the image data from the CT and MRI scanning devices require a geometric correlation, because the other information types are derivative information, which inherit the geometry of their ancestors.

Visualization

The first step in this work is the design and development of a 2D multidimensional visualizing tool, which allows the user to inspect distinct user defined combinations of relevant information classes in all three orthogonal directions.

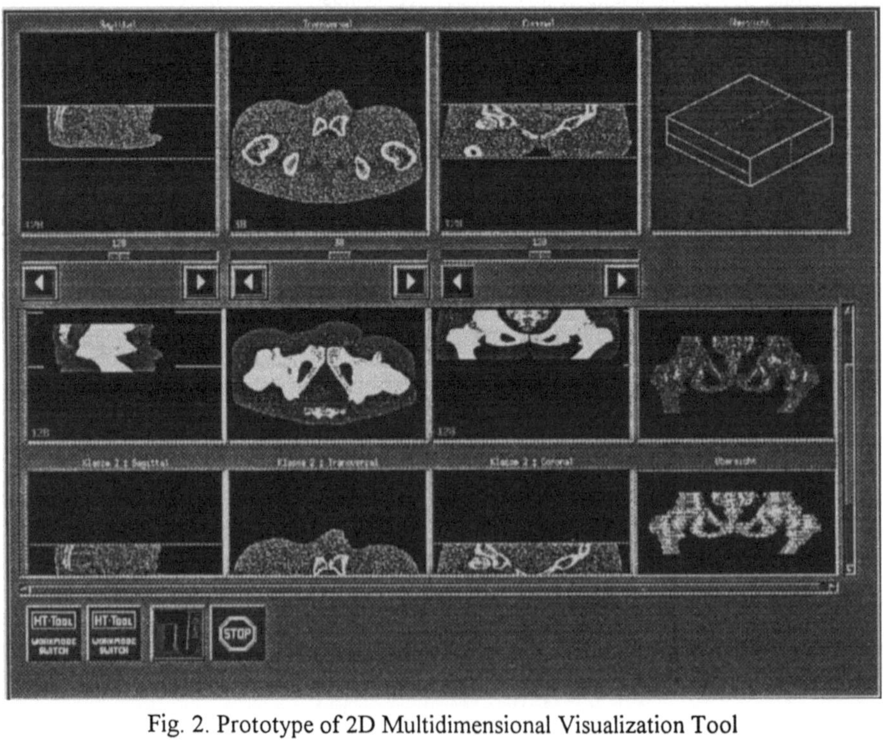

Fig. 2. Prototype of 2D Multidimensional Visualization Tool

Fig. 2. shows a screen shot of the first prototype of this tool, where the screen is separated vertically in a sagittal, a transversal, a coronal and an overview image column. Horizontally, the screen is split into rows, which represent different combinations of image information. The top row is reserved for the original scanner information, which is the basic anatomical information of the patient and which is the source for the segmentation of object contours, from which the 3D tetrahedron grid is calculated. The number and the individual semantics of the lower horizontal rows should be user defined and the user is allowed to scroll up and down the horizontal rows. In the top right window a navigation guide displays the recently adjusted orthogonal slice positions. One point of interest and special research would be the graphical presentation of merged image information with distinct semantics. In the frame of this 2D multidimensional visualization tool, 3D visualization only serves for the navigation aid.

The second step in this work is the design and implementation of a 3D application, which is dedicated for the graphical representation of the patient, the hyperthermia device and the antennas. This tool should provide a convenient user interaction for the interactive placement of these parameters.

User Interaction

It is intended to integrate and to assess alternatively the practicability of the two 3D input devices "GEOBALL" and "LOGITECH 3D Mouse" in respect to a friendly and convenient user interface. Both input devices can handle six degrees of freedom and do away with complicated and time consuming interactions that are involved with the 2D mouse. In the frame of this goal it is very important to create an intuitive user feedback.

Hardware and Software Environment

Due to the goals of portability and flexibility, the application software, which is and which will be implemented within this project is based on the X11R5 standard and uses the OSF motif tool kit to provide a standardized "look and feel". At this time, the development hardware configuration consists of a DEC 3000 AXP Model 500 system with an ALPHA processor architecture, running a OSF/1 operating system. The system is equipped with 256 MB of RAM and 4 GB hard disk space.

Final Remarks

This paper introduces an approach to aid hyperthermia planning with graphical orientated applications. It is intended to use and to assess a simple multidimensional voxel model to integrate geometric correlated information with distinct semantics. This voxel model should be the basis for further applications, such as 2D and 3D composite imaging. With the help of innovative 3D input devices the application control by the user will be eased. It is necessary to investigate how difficult the importing and exporting of information, into and from, the multidimensional data cube will be and which problems may occur.

Acknowledgement

The work reported in this presentation was done within the COMED project which is one of the BERKOM projects and is funded by the DETECON GmbH (Deutsche Telepost Consulting GmbH, Bonn) and by the Technical University of Berlin.

References

[1] H. Ricke, J. Kanzow, "BERKOM Breitbandkommunikation im Glasfasernetz", R. v. Decker Verlag, Heidelberg, 1991.

[2] U. Jackisch, M. Rudolph, "A highly interactive 3D visualization tool for sets of medical images in the environment of communicating low and high end workstations", in Proceedings of ISCAMI'91, London, December 13-15th, 1991, in Press.

[3] P. Wust, "Hyperthermia in Clinical Oncology", Abstract-Band zur 6. Europäischen BSD-Benutzer-Konferenz, Oktober 1992, Berlin.

[4] H. S. Stiehl, "Automatische Verarbeitung und Analyse von kranialen Computer-Tomogrammen", Technische Universität Berlin, Fachbereich Informatik, Dissertation, 1980.

Distortion in the MRI Stereotactic Planning of Gamma Knife Radiosurgery

Stefan M. Zechowy, Steven L. Fritz, Steven R. Roys

Department of Diagnostic Radiology
University of Maryland School of Medicine
Baltimore, Maryland, USA

Summary

In stereotactic localization using MRI, distortion of the image is a main factor contributing to errors of measurement. Although correction algorithms for distortion due to magnetic field incongruities are well known, there may still be residual distortion in the image after application of these corrections. As such, a linear scaling of the MR image may have to be applied before localization. In this study, axial and coronal MR images from Gamma Knife radiosurgery treatments were analyzed for the amount and possible effect of this residual distortion. It was found that there was measurable distortion in the images, and that the pattern of distortion suggested that the scaling factors were significantly different along the two axes in each imaging plane. The pixel sizes in the axial plane were 1.17 mm (.0044) along the Y-axis and 1.19 mm (.0038) along the X-axis. The pixel sizes in the coronal plane were 1.20 mm (.0055) along the Z-axis and 1.16 mm (.0031) along the X-axis. It was concluded that in order to reduce errors in localization, linear scaling factors must be applied in each imaging plane along each axis for every patient.

Introduction

In any stereotactic localization procedure, minimization of spatial errors is of primary importance. Particularly with MRI, image distortion is a main contributor to errors in localization[1-3,6-8,11-12]. The effects of MRI distortion are well documented, and can be attributed to any number of incongruities in the imaging process, including static field inhomogeneities, nonlinear gradient fields, and magnetic susceptibility in the patients themselves[18-11]. Algorithms for correction of these distortions to within acceptable limits have been developed, usually based upon the mapping of the magnetic field inhomogeneities[9-11] or the fitting of known point distortions to an algebraic equation[7,8]. The purpose of this study was to document any patterns of residual distortion in MR images after standard correction algorithms have been applied, and to assess the possible effects on stereotactic localization for Gamma Knife radiosurgery. More specifically, the intent was to study the distortion in axial and coronal images as it related to differences in the physical pixel sizes along the two orthogonal axes in the image. The accuracy of localization in the stereotactic planning for actual radiosurgery has been quantified in another study[3], but much significance was placed on human error in locating pixels on the image. We have found that even with software correcting for this error, there remained a significant difference in the pixel sizes along the two axes in the axial and coronal MRI scans of Gamma Knife patients. In the treatment of these patients, corrections were made when one axis seemed overly magnified or shrunken, but even then the corrections assumed that the same scaling factor should be applied to both axes. No systematic algorithm was applied to correct for the disproportionate axes even though the distortions occurred on a consistent basis and followed such a regular pattern. During the study, other reasons for possible errors in spatial localization were found and methods for their correction were proposed.

Materials and Methods

Fifty-six patients were scanned and treated with the Gamma Knife at the University of Maryland hospital over a period of seven months. A Leksell Stereotactic Coordinate Frame (Elekta Instruments; Tucker, Georgia) was used to provide the necessary scaling and measurement factors during the treatment. The frame was a hollow cube made of low atomic number, lightweight aluminum bars, and was specified to measure 180 mm on each side. The vertical bars which made up the main body of the frame were 120 mm apart. Diagonal bars connected the corners of the cube and provided a means of measuring the Z-coordinate of a scan taken perpendicular to the axis of the frame (and the bore of the magnet). All of the scans were performed using a 1.5 T Signa MRI Scanner (General Electric Medical Systems; Milwaukee, Wisconsin) which was calibrated every two weeks according to the manufacturer's guidelines.

An image series of each patient was taken and included coronal, axial, and occasionally sagittal scans. The sagittal scans were not used in this study since they did not contain the necessary fiducial marks to obtain the measurements needed. The goal was to obtain a sample of one axial and one coronal image for each patient in a systematic manner. Therefore the middle image in the first series encountered for each plane was taken, except when the image did not contain all of the necessary fiducial marks. In those cases, the next suitable image found was used. Some patients had no usable images, some had one, and some had both axial and coronal usable images. In all, 33 axial images and 45 coronal images were collected for the study. Some of the images used gadolinium as a paramagnetic contrast, but the effect on the measurement of the fiducial marks was not considered, since the frame was sufficiently far away from the intracranial contrast material.

The axial slices were 3 mm thick except for six which were 5 mm and one which was 6 mm. The coronal slices were 5 mm thick, except three which were 3 mm. The field of view for twenty-five of the axial images was 280 mm, and 300 mm for the other eight. All of the coronal images had a field of view of 300 mm. Each image was acquired in one of two ways: 1) Gradient echo, 20 flip, 15 ms TE, 225 ms TR; 2) Spin echo, 90 flip, 10 ms TE, 500 ms TR. All images were acquired in a 256 x 256 (phase x frequency) matrix.

The scans were stored and analyzed on a Columbia Scientific system (Columbia Scientific Inc.; Columbia, Maryland) using its proprietary software enhanced by the researchers for the purposes of this study. Each image had nine fiducial marks, labeled A through G, in a counter-clockwise fashion, starting from the upper left mark (Figure 1). The two outer marks on each side (A and C, D and F, G and I) corresponded to the straight bars of the frame while the middle marks (B, E, H) corresponded to the diagonal bars. A square region of interest selector was approximately centered over each fiducial mark and the pixels within the box that were above a certain threshold intensity were considered to be within the fiducial mark. The threshold was dynamically set to the median of the intensity value of all the pixels within the centroid box. The pixel coordinates of each mark were computed by calculating the centroid (geometric center) of the pixels that composed the mark. The location of the fiducial mark was thus computed to sub-pixel accuracy.

Initially, repeated measurements of the same marks were made to assure to consistency of this method. In order to obtain and compare the pixel sizes along the two axes, the two vertical pixel distances (AC and DF) and the horizontal pixel distance (GI) were calculated. Distances AC and DF were averaged into one measurement. The known physical distance between the bars (120 mm) was then divided by the two pixel distances to get the average pixel size along the vertical and horizontal axes. In axial scans, the vertical and horizontal pixel sizes corresponded to the Y and X axes, respectively; in the coronal scans they corresponded to the Z and X axes. The difference between the vertical and horizontal pixel sizes was computed for each image, and the mean and standard deviation of the two sets of differences (axial images and coronal images) were calculated.

The statistical significance of the difference was determined using a paired t test, with a null hypothesis that there was no difference between the pixel sizes in each axis for either imaging plane. The images that were analyzed on the computer screen had been magnified 4 times by the system software. Thus, the images were actually 1024 x 1024 pixels when measured. To obtain the pixel sizes in the original image, the computed sizes were multiplied by 4. Finally, as some axial images were taken with a 280 mm field of view, those pixel sizes were multiplied by a factor (300 mm/280 mm) for proper comparison.

Results

	mean (mm)	standard deviation (mm)
Pixel Size along X-axis	1.19	.0038
Pixel Size along Y-Axis	1.17	.0044
Difference in Pixel Sizes	0.02	.0059

Table 1: Axial plane results

	mean (mm)	standard deviation (mm)
Pixel Size along X-axis	1.16	.0031
Pixel Size along Z-axis	1.20	.0055
Difference in Pixel Sizes	0.04	.0069

Table 2: Coronal plane results

The nominal pixel size computed by the scanner for images with a 300 mm field of view was 1.17 mm. The data presented in Tables 1 and 2 have been corrected for differences in magnification and field of view according to the procedure described in Materials and Methods. There were 33 axial images and 45 coronal images studied. The difference in pixel sizes between each axis was significant for both axial and coronal imaging planes ($P < .001$). Interestingly, there was no correlation between any of the pixel size measurements or their differences, and the number of days since the last calibration of the scanner.

Discussion

In one of the early papers on the accuracy of stereotactic localization, Wyper et al.[12] stated that geometric scaling must be done when using MRI in order to correct for any distortion found in the image. Elekta Instruments, the manufacturer of the head frame used in this study, recommended averaging the image distance of the frame on both sides along one axis, and calculating the pixel size based upon the known length of the frame, regardless of the nominal pixel size reported by the MRI scanner. Since no mean pixel size matched the nominal pixel size within a standard deviation, such precautions are well justified. To account for tilt of the frame, the manufacturer's recommendation was to average the two diagonal marker measurements and use the average scale factor over the entire image. One potential source of systematic error is the fact that these calculated pixel sizes correspond to the orthogonal axes only if the frame is aligned with the scanning axes. There are two responses to this: First, during the scanning procedure, test images were taken to align the scanning plane with the frame. Second, the consistency of the results, plus the fact that the X-axis pixel size was different in the two planes, showed that random tilt could not be responsible for the regular pattern of different pixel sizes along the two axes in both imaging planes.

To illustrate the possible effects of the distortion found, consider the consequences of scaling along only one axis when attempting to localize a region of interest in the coronal plane. The mean pixel size in the Z-axis in the coronal plane was 1.20 mm. The mean pixel size in the X-axis was 1.16 mm, a 3.4 % difference. If the Z-axis scale were used for both axes, an area that is actually 30 mm, for example, to the left or right of center would be measured as if it were 28.98 mm from center. The main finding of this study was that if, in addition to the scanner's corrections, linear scaling is to be done before stereotactic localization, it must be done in both axes in each imaging plane, for each patient. The data clearly indicate that each pixel is not square-shaped, as was assumed in the Leksell manual, but rectangular, and the rectangularity varied between imaging planes.

The MRI scanner, as was noted, was calibrated regularly every two weeks, and any distortion due to the variability between calibrations was not a factor, since there was no correlation between the measurements and the number of days since the last calibration. Some of the random variability may have been due to any one of the following:

1) the fiducial marks varied in their placement in the field of view, and any non-linear distortions would have affected the precise pixel location;
2) the tilt of the frame could have affected the actual physical distance measured in the imaging plane, as well as introduce one axis' scale factor to another axis;
3) while the distance between the marking bars on the frame was assumed to be 120 mm, as calibrated by the manufacturer, no variability in this calibration was assumed. Any discrepancy in this distance could have produced variable results;
4) If the exact position of the marking bars do not correspond to the phase x frequency scanning matrix, the fiducial marks in the image may vary slightly, producing an offset in their centroid measurements.

Conclusions

When performing stereotactic localization every effort should be made to correct those factors which are correctable, leaving any distortion and error to those which are not. Therefore, the following conclusions were made:

1) In stereotactic localization using MRI, a scaling factor should be obtained for each plane in both axes, for each patient.
2) The fiducial marks should be located using centroid methods, reducing any arbitrary or inherent pixel error, as suggested by Kondziolka, et al.[3].
3) Any frame tilt that is recognizable through discrepancies in the diagonal fiducial marks, or from simple linear calculations should be corrected with rotation matrices.

References:

1. Ehricke H-H, Schad LR. MRA-Guided Stereotactic Radiation Treatment Planning for Cerebral Angiomas. Computerized Medical Imaging and Graphics 16:65-71, 1992.

2. Ehricke H-H, Schad LR, Gademann G, Wowra B, Engenhart R, Lorenz WJ. Use of MR Angiography for Stereotactic Planning. J Comp Assist Tomography 16:35-40, 1992.

3. Kondziolka D, Dempsey PK, Lundsford LD, Kestle JRW, Dolan EJ, Kanal E, Tasker RR. A Comparison between Magnetic Resonance Imaging and Computed Tomography for Stereotactic Coordinate Determination. Neurosurgery 30:402-407, 1992.

4. Leksell DG. Stereotactic Radiosurgery: Present Status and Future Trends. Neurological Research 9:60-68, 1987.

5. Leksell L, Leksell D, Schwebel J. Stereotaxis and nuclear magnetic resonance. J Neurology, Neurosurgery, and Psychiatry 48:14-18, 1985.

6. Rousseau J, Clarysse P, Blond S, Gibon D, Vasseur C, Marchandise X. Validation of a New Method for Stereotactic Localization using MR Imaging. J of Comp Assist Tomography 15:291- 296, 1991.

7. Schad LR, Ehricke H-H, Wowra B, Layer G, Engenhart R, Kauczor H-U, Zabel H-J, Brix B, Lorenz WJ. Correction of Spatial Distortion in Magnetic Resonance Angiography for Radiosurgical Treatment Planning of Cerebral Arteriovenous Malformations. Mag Res Imaging 10:609-621, 1992.

8. Schad L, Lott S, Schmitt F, Sturm V, Lorenz W. Correction of Spatial Distortion in MR Imaging: A Prerequisite for Accurate Stereotaxy. J Comp Assist Tomography 11:499-505, 1987.

9. Schmitt F. Correction of Geometrical Distortions in MR-images. In: Lembke HU, Rhodes ML, Jaffee CC, Felix R, eds. Proceedings of the International Symposium CAR '85--computer assisted radiology, Berlin: Springer; 15-23, 1985.

10. Sekihara K, Kuroda M, Kohno H. Image restoration from non-uniform magnetic field influence for direct Fourier NMR imaging. Phys in Med and Biol 29:15-24, 1984.

11. Weis J, Budinsky L. Simulation of the Influence of Magnetic Field Inhomogeneity and Distortion Correction in MR Imaging. Magnetic Resonance Imaging 8:483-489, 1990.

12. Wyper DJ, Turner JW, Patterson J, Condon BR, Grossart K, Jenkins A, Hadley DM, Rowan JO. Accuracy of stereotaxic localization using MRI and CT. Journal of Neurology, Neurosurgery, and Psychiatry 49:1445-1448, 1986.

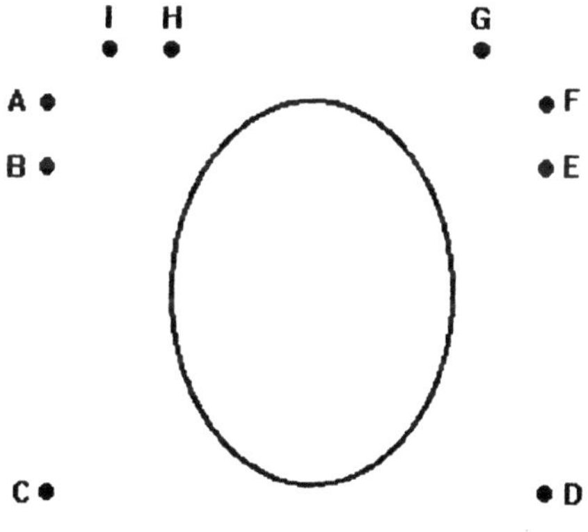

Figure 1. Schematic diagram of stereotactic coordinate fiducial markers.

Computer Assisted Surgical Planning

Reporting with 3D Solid Model: Another Concept for Radiologists

Ishtiaq Kasem, Kenji Matsuzaki Hiromu Nishitani, Fumio Shichijo

Dept. of Radiology, *Dept. of Neurosurgery, Medical School, Tokushima University, Tokushima, Japan

Summary

The advantages in introducing 3D reporting in the radiology department was assessed and the difficulties in overcoming the shortcomings are discussed. Ten clinical cases were selected and the reporting procedure was thoroughly monitored. The following article discusses the results and the factors presently hindering the concept of 3D reporting and the progress made so far to the solution of these problems.

Materials and Methods

A Toshiba TCT60 CT scanner was used to scan images of 5mm thickness. The images were then transferred on floppy disks to an HP Pascal Workstation which had a Pascal based Endoplan software with the capability to select the ROI semiautomatically. This provided the physician to have more control over the maneuver than conventional totally automatic ROI selection methods. Each slice was individually tagged and modified for the correct ROI. The images then underwent an image processing algorithm as devised by Endoplan which took few minutes to half an hour depending on the number of slices involved. But it was greatly enhanced in its speed with the help of transputers. The image was then displayed on the second graphical monitor of the system. At this juncture we had a clinical session with the neuro-radiologists to assess the model and to obtain their views and their expectations. After both the parties had agreed on the perfection of the model, Computer Numeric Coded (CNC) data was generated through a built in module of the the Endoplan and the data was subsequently transferred to a DOS based PC computer connected to a five axis milling machine to mill a model from a block of polyurethane. An average of 3 models were milled in each case depending on the neurosurgeons desire of observing the lesion from different cut sections.

Clinical Cases

Ten cases were referred to us by the neurosurgical department. The pathologies of all of them were confined to the cranial region since it was our desire to observe the capability of the

PATIENT NO.	AGE	SEX	CHIEF COMPLAINTS	DIAGNOSIS
1	51	F	headache	Meningioma
2	70	F	headache	Rt. Tentorial Meningioma
3	45	F	lt. exophthamus, double vision	Lt. Orbital Osteoma
4	61	M	rt. deafness, loss of conciousness	Rt. Acoustic Neurinoma
5	62	F	tinnitus	Rt. Glomus Jugular Tumor
6	58	M	total ophthalmoplegia	Lt. Temporal Lobe Abscess
7	54	F	rt. visual disturbance	Rt. IC Aneurysm
8	59	M	vertigo, lt auditory disturbance	Epidermoid
9	30	F	telescopic vision, blindness rt. eye	Meningioma
10	44	F	tinnitus	Rt. Glomus Jugular Tumor

Table 1: Clinical cases

DIAGNOSIS	SLICE #	TIME-HRS
Meningioma	21	6
Rt. Tentorial Meningioma	32	4.5
Lt. Orbital Osteoma	36	4
Rt. Acoustic Neurinoma	29	4
Rt. Glomus Jugular Tumor	22	8
Lt. Temporal Lobe Abscess	22	3
Rt. IC Aneurysm	23	3.5
Epidermoid	25	4
Meningioma	34	4
Rt. Glomus Jugular Tumor	20	6

Table 2: Comparison with Slice Numbers and Time

system to recreate the model of the intracranial lesions. The final report consisted of the solid models which were an average of three in all the cases. The models of the glomus jugular tumors were made by making one coronal and a sagittal section through the jugular foramen. This resulted in three individual models which pieced together like a jigsaw puzzle to create the whole model and provided a better perception for the surgeons to grasp the field of the lesion.

Results

The model making procedure could be devided into 3 distinct portions. The CT scan, the image processing on the ENDOPLAN and the milling of the model. The image processing consisted of a number of sequential modules like selecting the ROI, interpolating the selected contours and processing the 3D image followed by CNC data generation. Among these the contour selection offered the highest possibility in reducing the time factor. We calculated our performance during this period and sequentially estimated the time taken -- starting from the contour selection to the final CNC data generation. An average time of the three models in each case was taken. The slices of the CT scan ranged from 20 to 36 slices with a mean of 27. The mean time was 4.7 hrs. with a range of 3 to 8 hrs. But on plotting in a scattergram we found most of the models concentrated in the 3 to 4.5 hrs range. Only the intra-pyramidal tumors which were made in three pieces took more time but as we progressed with the second model the time was significantly reduced which was attributable to our experience and practice. The meningioma in the 6 hr. range was the first model we created which reflected the progress we made in curtailing time in subsequent models. Increase in number of slices prolonged the time required but it was easier to make solid bony contours which did offset the large number of slices in the osteoma case keeping the time within the reasonable range.

Discussion

From recent developments in the 3D aspect of imaging it is evident that 3D is proving to be more informative than conventional 2D images. CT and MRI among the modalities prove to be superior in producing 3D pictures in terms of ease and clarity. But when it comes to the form of using 3D in conventional reporting certain factors have to be overcome to accept 3D for reporting methods in daily cases.

Our experience with 10 different cases of varying pathology has shown that the most common hindrances in comparison to the conventional type of reporting are the time factor, cost factor, knowledge factor and the acceptability factor.

The first and foremost problem with 3D is the time consuming reconstruction procedures to create 3D image from 2D CT images. But the time factor depends on other subfactors like image processing algorithms, image slice numbers and 3D model reconstruction technique. We have noticed from our experiment that the slice numbers can be kept static for almost all types

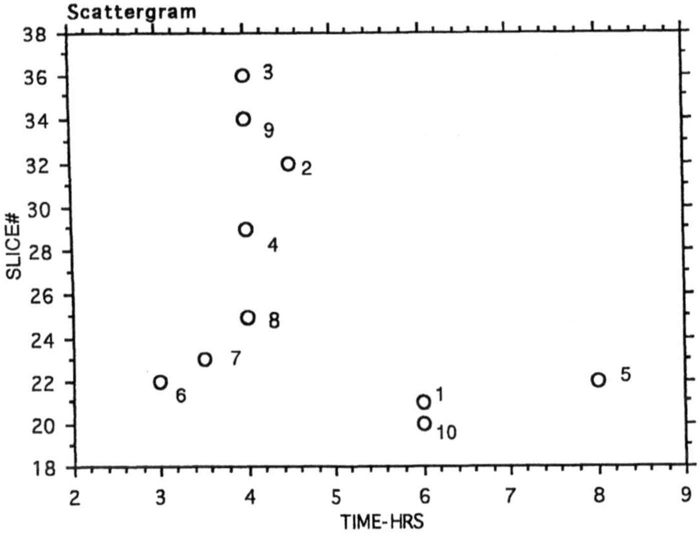

Fig 1: The scattergram shows that the time taken to produce the 3D model image on the graphical computer mainly concentrated in the 3 to 4.5 hours range. The two Jugular Tumors took the longest time as were made in three pieces. The first case of meningioma also took a longer period to create since it was the first case we encountered. The numerals in the chart are as follows: 1,2,9 Meningioma; 3 Orbital Osteoma; 4. Acoustic Neurinoma; 5,10 Glomus Jugular Tumor; 6 Temporal Lobe Abscess; 7 IC Aneurysm; 8 Epidermoid

	Mean	Std. Dev.	Std. Error	Minimum	Maximum	Variance	Range	Median
SLICE#	26.400	5.873	1.857	20.000	36.000	34.489	16.000	24.000
TIME/HRS	4.700	1.513	.478	3.000	8.000	2.289	5.000	4.000

Table 3: Descriptive statistical analysis of the data obtained from the ten clinical cases.

of intracranial lesions to within 25 to 30 slices. The imaging algorithm provides the user to select ROI semi-automatically and is the time consuming maneuver of the whole procedure. But judicious selection of ROI, increasing number of practices have turned out to reduce the time significantly. The slice thickness is inversely proportional to the time. As the slice becomes smaller the slice number increases and hence the time is prolonged in defining the ROI and processing the images for the final 3D outlook. A thickness constant of 5mm was kept in our experiment and it was acceptable in all our selected 10 cases. The model constructing procedure involves little interaction with the user and the process had a fixed time span depending on the scan numbers used in the image reconstruction. But wise ROI selection did make the final image distinct in clarity.

The initial resentment for any 3D procedure is probably due to the fear of inadequate knowledge of the whole process a radiologist encounters in creating a 3D model. But present day computers have reduced their complexity to a great extent to become user friendly. The final setback in our initial endeavor of introducing the 3D reporting was mainly from the human acceptability attitude. Our survey of two units of 14 doctors have shown that the time delay of accepting the new concept was prolonged among the senior doctors than the junior residents.

A comparative study was performed from the radiologists point of view of the 3D in comparison to the 2D conventional modalities which lead to the following inferences: i. better clarity of the lesion leading to a more accurate diagnosis, ii. quantitative analysis provides 3D measurements of the lesion preoperatively. From the surgeons point of view the models were a preoperative surgical tool which helped them to document all their maneuvers pre-operatively leading to a more successful and complete operative endeavor.

In conclusion, 3D models offer a versatile mode of interpreting radiological data and in its present form is possible to introduce it in radiological departments. With the advent of faster and lower priced computers and powerful softwares, leading to shorter time and plunging cost, the concept is becoming more acceptable gradually.

References

1. Udapa J.P., Herman G.T. *3D imaging in medicine,* CRC Press Inc., 1991:103-143

2. Parker J.A. *Image reconstruction in radiology,* CRC Press, 1990

An Interactive 3D-Atlas of Acetabular Fractures

CH. SEEBODE[1], R. SCHUBERT[1], A. POMMERT[1],
M. RIEMER[1], TH. SCHIEMANN[1], U. TIEDE[1], V. WENING[2], K.H. HÖHNE[1]

[1] Institute of Mathematics and Computer Science in Medicine (IMDM)
[2] Department of Traumatology
University Hospital Eppendorf
Hamburg, Germany

e-mail: seebode@imdm.uke.uni-hamburg.dbp.de

Abstract

Diagnosis of often complex-shaped acetabular fractures requires excellent spatial imagination by the traumatologist if he relies only on 2D-imaging techniques like radiographs or computed tomography. Linking these techniques with 3D-views of the fracture helps to understand how a three-dimensional situation appears in 2D. Based on our VOXEL-MAN visualization system, we developed a complete 3D-model to describe the anatomy and pathology of standard cases of acetabular fractures. This model allows to derive 3D-views at the user's will. Manipulation of the model like adding or removing objects, dissection, rotation, or zooming is feasible and allows an individual way of learning. Working on such a model bears the possibility to simulate radiographs or to reconstruct CT-slices which contain all the anatomical or pathological knowledge necessary to understand the images. The system proves to be powerful both in education and as clinical reference.

Introduction

Traumatic injury of the human pelvis is one of the challenging issues a traumatologist has to deal with. The three-dimensional shape of an acetabular fracture cannot be confined to a plane without loosing information. Tracking of fracture lines or dislocated fragments requires a good amount of spatial imagination by the traumatologist, when using 2D imaging techniques such as CT or plain film radiography in current diagnostic procedures. Nevertheless, the choice among a variety of surgical approaches to the pelvis and improved treatment by open reduction and fixation demands exact diagnosis.

A support of spatial imagination would facilitate the whole process of diagnosis. Any printed atlas offers a reasonable combination of static three-dimensional images and textual descriptions. A particular user remains in a passive position without the chance to adapt the content of the atlas to his requirements.

In order to overcome these drawbacks we want to build a 3D model which permits to learn anatomy and pathology of acetabular fractures providing:

- interactive 3D visualization and dissection of typical fracture cases from any viewing direction,
- support for the interpretation of classical radiological images by correlating simulated radiographs and cross-sectional images with the model.

According to the ASIF(Association for the Study of Internal Fixation) classification of acetabular fractures (modified after the original proposal by Judet and Letournel [4]) the atlas should contain at least one case representing each fracture class, thus covering the whole scope of possible surgical approaches to the acetabulum.

Method

Based on our experience with 3D visualization techniques [1, 5, 8, 9], we apply volume visualization using the generalized-voxel-model to realize 3D-views. The basic idea of all volume visualization methods is to describe a three-dimensional real-world object like the pelvis in terms of an image volume, which preserves the entire spatial information necessary for the reconstruction of 3D-views of the object. In the generalized-voxel-model each voxel "knows" to which object it belongs. We define an anatomical object as a subset of voxels sharing one object identification. This identification serves as the key addressing the attributes associated with the object in the knowledge base of the system. Attributes to objects are symbolic descriptions like names or plain text, or visualization parameters. Recombining the objects in relational descriptions (i.e. "right pubis is part of the right hip bone") forms higher-order objects. One voxel may represent different aspects of the same anatomy. For example, the pelvis may be described highlighting anatomical (i.e. "right femur") as well as classificational (i.e. "left posterior column") architecture. We call this combination of image data and symbolic descriptions an "Intelligent Volume" which serves as input to our VOXEL-MAN system [2, 7].

Entering the anatomical knowledge is done by segmentation and subsequent volume editing. This task is simple in principle, but strenuous in practice: selecting all voxels that make up a particular object and attach an identifier that the system interprets as an anatomical description such as "right femur" or "left pubis". Extracting bone from other tissues can be easily done by thresholding. Separating bony structures like the femur and the rest of the hip bone requires a more manual work. Only interactive visual control yields a good result. Our segmentation tool offers morphological processing aided by interactive visual control [3, 6].

The system runs on typical workstations like DEC station 5000. The software is written in C programming language in a UNIX environment. The user interface is based on the OSF/MOTIF standard.

Result

The actual version of the system contains two cases of acetabular fractures (bicolumnar fracture and posterior column fracture). Several months of work have been invested for segmentation and anatomical description of the data.

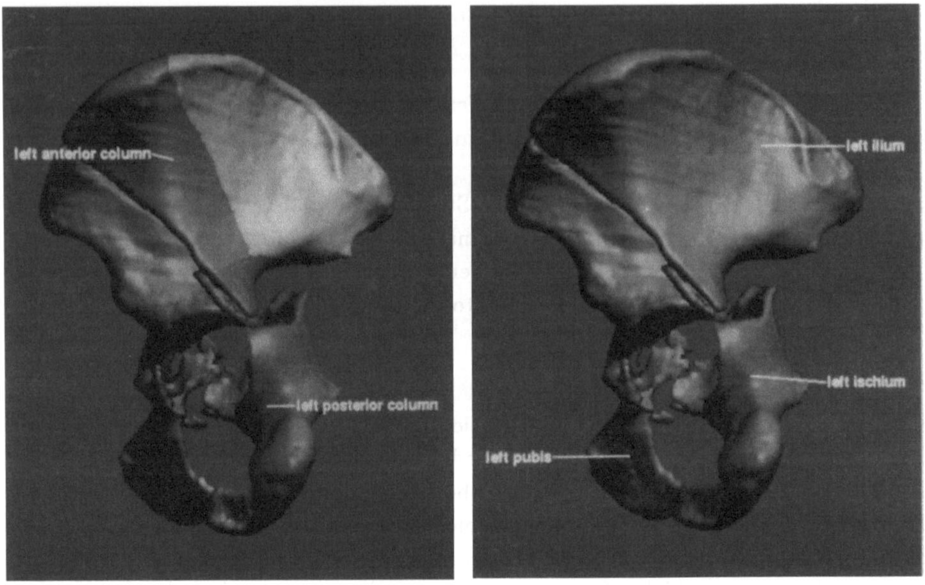

Fig. 1: Bicolumnar fracture of the left acetabulum. Only the left hip bone is shown, all other parts of the pelvis have been removed. The removed femur allows full insight into the fractured acetabulum. The damaged parts are identified considering the constituent bones of the hip bone (right) or the columnal architecture (left).

(see also in color on page 823)

The user communicates with the system through a sophisticated user interface. All actions are released by means of mouseclick on menues or pushbuttons. For exploration of the spatial relations between anatomical constituents of the pelvis and/or fracture fragments, initial 3D-views are computed by the system and the surgeon may modify them by rotation or zooming. In order to preserve orientation in a given view, all names of the displayed objects may be annotated at the cursor position. Moving a virtual lightsource enhances the impression of depth in the images. Especially the view from the pelvic cavity onto the inner face of the hip bone is no longer hindered. Any object that prevents the sight on an important detail may be removed, cut away, or made transparent. The cuts are done by positioning arbitrarily oriented planes through the objects. The fracture impact on the parts of the hip bone (whether pubis, ischium, ilium, or columns) is easily assessed, because these objects may be displayed, removed, or painted independently and fracture fragments may show their origin (see fig. 1).

Even though 3D imaging is increasingly applied in clinical work, the vast majority of the fracture cases is still assessed from radiographs or CT-slices. Our system lends itself perfectly to studying the appearance of fractures on CT and radiographs. As shown in fig. 2, a plane may be specified in the 3D-scene for which the corresponding CT-slice can be displayed with colour labels and annotations in both presentations. Thus the interpretation of CT can be thoroughly studied. In a similar fashion, radiographs may be generated from any direction and beam geometry. As a special feature, the intelligent volume ap-

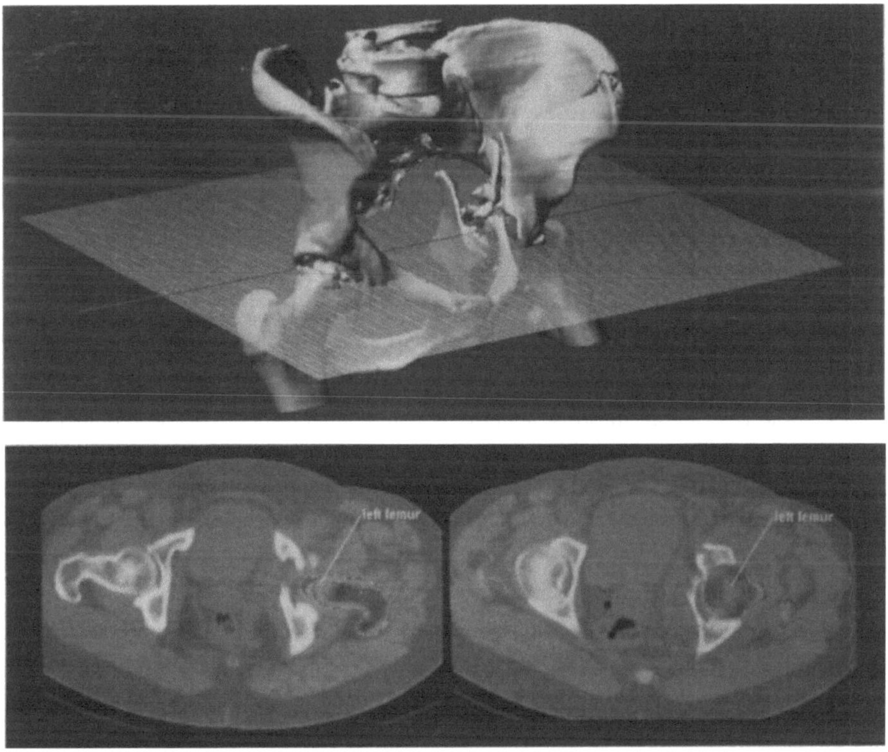

Fig. 2: Top: Specification of a CT plane within a 3D-view. Bottom: Corresponding CT-slices carrying the same colour labels as in the 3D-view. Annotations may be generated by the system via mouseclick.

(see also in color on page 824)

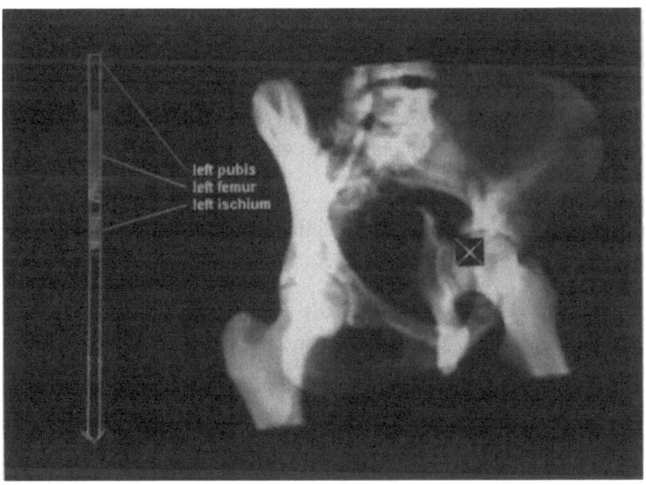

Fig. 3: A radiograph from the same viewing direction as in fig. 2 (top). The objects contributing to the intensity at cursor position are shown on the left.

proach delivers a way to figure out the contribution of different objects to a particular intensity value on the image, thus verifying or correcting a preliminary interpretation (see fig. 3).

Conclusions

First experience in educational and clinical trials show that learning pathology of acetabular fractures with our atlas proves to be an improvement compared to the more traditional way using printed atlases or textbooks as knowledge sources. Beside that, there was a general agreement that this kind of learning nourishes motivation, because all appearances of fracture cases (whether real perspective view, CT sequence, or radiograph) are easily accessible and animate the user to manipulate them as he desires.

There are some reasonable improvements that could be considered. The biomechanical aspects of acetabular fractures may be respected in various ways. Studying the stability of a fracture a surgeon may ask whether a fragment is stable or unstable. A corresponding object definition would use attributes like "stable fragment" or "unstable fragment". The answer would require nothing else then a simple mouseclick on the fragment in question. Biomechanical stress and strain could be included in the model as well.

Providing extended textual descriptions would reveal a more profound understanding of the pathology and surgery of acetabular fractures. An example of this could be the release of an explanation of a particular surgical treatment ordered by the user at any point of the atlas session.

Acknowledgements

We are grateful to all members of our department who have supported this work. The project is supported by the Deutsche Forschungsgemeinschaft (DFG).

References

[1] Höhne, K. H., Bomans, M., Pommert, A., Riemer, M., Schiers, C., Tiede, U., Wiebecke, G.: 3D-Visualization of Tomographic Volume Data Using the Generalized Voxel-Model. *Visual Comput. 6*, 1 (1990), 28–36.

[2] Höhne, K. H., Bomans, M., Riemer, M., Schubert, R., Tiede, U., Lierse, W.: A 3D Anatomical Atlas Based on a Volume Model. *IEEE Comput. Graphics Appl. 12*, 4 (1992), 72–78.

[3] Höhne, K. H., Hanson, W. A.: Interactive 3D-Segmentation of MRI and CT Volumes Using Morphological Operations. *J. Comput. Assist. Tomogr. 16*, 2 (1992), 285–294.

[4] Judet, R., Judet, J., LeTournel, E.: Fractures of the acetabulum: Classification and surgical approaches for open reduction. *J. Bone Joint Surg. 46A* (1964), 1615.

[5] Pommert, A., Höltje, W.-J., Holzknecht, N., Tiede, U., Höhne, K. H.: Accuracy of Images and Measurements in 3D Bone Imaging. In Lemke, H. U. et al. (Eds.): *Computer Assisted Radiology, Proc. CAR '91*, Springer-Verlag, Berlin, 1991, 209–215.

[6] Schiemann, T., Bomans, M., Tiede, U., Höhne, K. H.: Interactive 3D-Segmentation. In Robb, R. A. (Ed.): *Visualization in Biomedical Computing II, Proc. SPIE 1808*. Chapel Hill, NC, 1992, 376–383.

[7] Schubert, R., Höhne, K. H., Pommert, A., Riemer, M., Schiemann, T., Tiede, U.: Spatial Knowledge Representation for Visualization of Human Anatomy and Function. In *Information Processing in Medical Imaging, Proc. IPMI '93*, Springer-Verlag, Berlin, 1993.

[8] Schubert, R., Höltje, W.-J., Tiede, U., Höhne, K. H.: 3D-Darstellungen für die Kiefer- und Gesichtschirurgie. *Radiologe 31* (1991), 467–473.

[9] Tiede, U., Höhne, K. H., Bomans, M., Pommert, A., Riemer, M., Wiebecke, G.: Investigation of Medical 3D-Rendering Algorithms. *IEEE Comput. Graphics Appl. 10*, 2 (1990), 41–53.

CAD Techniques for the Preoperative Planning of Intertrochanteric Osteotomies

A.PEPINO(*), A.MOSCA(**), M.CESARELLI(*), F.DI SALLE(***), M.BRACALE(*)

* Department of Electronics, I University of Naples
** Orthopaedic Clinic, 2nd Faculty of Medicine and Surgery, I University of Naples
*** Radiology Clinic, 2nd Faculty of Medicine and Surgery, I University of Naples

Summary

Preoperative planning and computer assisted surgery techniques are two fields of applied research that are used in several medical specialities. Firstly, in orthopaedics these methods can be adopted in custom implants, internal prosthese design and osteotomy planning. These techniques all require the acquisition, reconstruction and analysis of a three dimensional representation of bones.

The authors report a method that first allows a solid model of the hip joint to be generated starting from CT data and then makes it possible to simulate the osteotomy on this model.

Introduction

Intertrochanteric osteotomies can be useful in the treatment of many disease processes in the proximal femur region. Some positive indications can be given for: malunion of fractures, pseudoarthroses of the femoral neck, congenital coxa vara and so forth.

Operative techniques generally require adequate preoperative planning, which is usually carried out on the basis of a number of X-rays (2D) reported on transparent sheets.

The osteotomies are simulated on these sheets via a manual cut, paste and draw procedure. Obviously this method is highly inaccurate and requires the operator to follow very strict protocols in the surgical intervention .

The final goal of our research is to provide the operator with a series of tools that will support preoperative planning of these osteotomies in a three-dimensional environment. In light of the above, such tools should be found in the field of 3D imaging and 3D CAD techniques.

Tools that effectively "render" a 3D image on a traditional 2D monitor have been considerably improved in recent times because of the considerable rise in computer performance accompanied by rapidly falling costs.

These innovations are now being implemented and will become standard tools in radiological diagnosis in the near future. However, they might be even more useful in fields of applications that are as yet totally unexplored, such as Preoperative Planning, Surgical Simulation or Computer Assisted Surgery.[1][2]

These keywords refer to all those techniques that make it possible to simulate, plan or assist surgical interventions on the basis of three-dimensional anatomic information. Of course, full simulation or planning is no simple task as this would require knowledge of all the possible events that might occur in the operating theatre. At this stage we will take into account only those aspects of the intervention that make a greater demand on calculations, computer modelling and so on.

Material and methods

We have seen that CAD techniques can be used in "Computer Assisted Surgery" applications. Naturally the information source is always a tomographic set of images but, in this case, a CAD solid model is generated from the volume-based model by means of appropriate calculations. Of course while the volume model contains all the available information, the solid model is relevant only to the requirements of surgical assistance.

A solid model usually consists of a 3D geometric structure with a set of secondary information on the material. Furthermore, two solid models can be joined to form a new solid or they can be subtracted from one another and so on, thus making it possible to build an infinite number of arbitrary solid models starting from a set of elementary solid models.

- Hardware:

Sparcstation II GX, 32 MB RAM, 700 MB Hard Disk, Tape Cartridge and CD ROM unit, connected via ethernet with several VAX systems equipped with 9-track tape units.

- Software:

Operating System:UNIX SUN OS 4.1.1

Tools:C and Fortran compilers. Developers Guide for design of user interface.

Libraries: Sun IP, Spider, AUTOCAD rel. 11 with AME (Advanced Modelling Extension) and ADS (Advanced Development System)

In our application, a solid model of the coxo-femoral structure is obtained from a set of 3D pictures coming from a CT CGR CE10000 and recorded on magnetic tape. This model is then used to simulate Intertrochanteric osteotomies using CAD techniques.

Our application consists of the following steps: [3]

1 Acquisition and decompression of image data
2 Contour detection
3 DXF Surface generation
4 Solid Model generation
5 Simulation of the surgical intervention.

The magnetic tape contains compressed data. In step 1 the computer program reads and decompresses it slice by slice.

After decompression, the slice is examined by a semi- automatic program to detect the boundaries of the anatomic structures involved. They are identified as several sets of X,Y vectors.

These contour sets (stacked contours) are exported in DXF format into AUTOCAD rel. 11. In the AUTOCAD environment the stacked contours are the basic frame for building the solid model.

This model can be displayed from an arbitrary point of view using conventional AUTOCAD commands. It is possible to make some arbitrary modifications on the model through conventional commands but, unfortunately, these are not easy to use. Some other C programs have been written to make these two operations more user-friendly,

If the cut plane intersects more than one solid, it is possible to select the solid to be cut. Consequently only that solid will be cut.

The cut operations can be performed by a blade perpendicular to the X,Y plane. Therefore, before any cut operations are carried out, a set of move operations should be performed so as to set the model in the proper position.

There is no predefined function in the AME environment that allows a solid to be "cut" arbitrarily. Fortunately the "cut" can be obtained indirectly by means of a number of logic operations (AND) between the source solid model and some temporary objects (e.g. cubes). The AND operations (SOLINT) between two solid entails the generation of a new solid equal to the common portion between them .

Through proper use of the move operations and by adequately changing the point of view, the simulation can be both effective and realistic.

Results

The method has been used on CT volumes recorded from the hip and femur region, consisting of 40 slices recorded with 1 mm thickness and an interslice gap ranging from 2 mm. (near the femoral head) to 8 mm (in the femoral stem zone).

The stacked contours of hip and femur structures were obtained through contour detection and DXF export programs A geometric solid model was obtained from the stacked contours model through the ADS programs "CON3DP" followed by "SOLUNION", as described above. Two separate models representing the original bones are obtained. The geometric measurements reported inside AUTOCAD are exactly the same as the ones in the real bones . As already said, we can manipulate these geometric models by means of the commands available in AUTOCAD11 + AME. Of course we are particularly interested in performing arbitrary cuts on the model followed by some movement (translations and rotations) of one or both models with respect to a reference joint with the model itself. This aims to reproduce some of the manoeuvres that will be carried out during the actual surgical intervention.

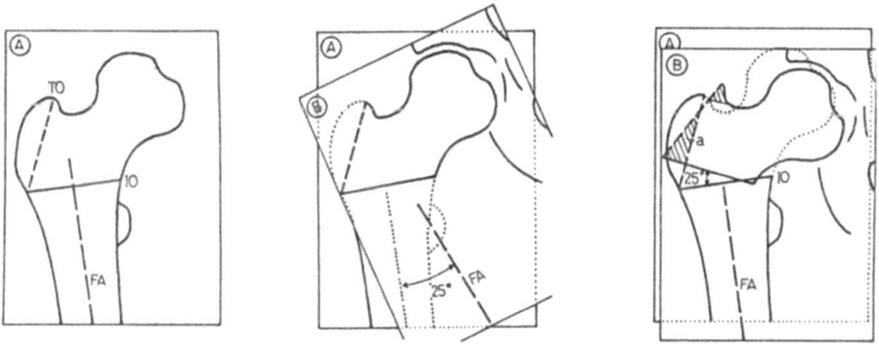

Fig. 1 Traditional procedure for the planning and drawing of a complex adduction (varus) osteotomy [4]

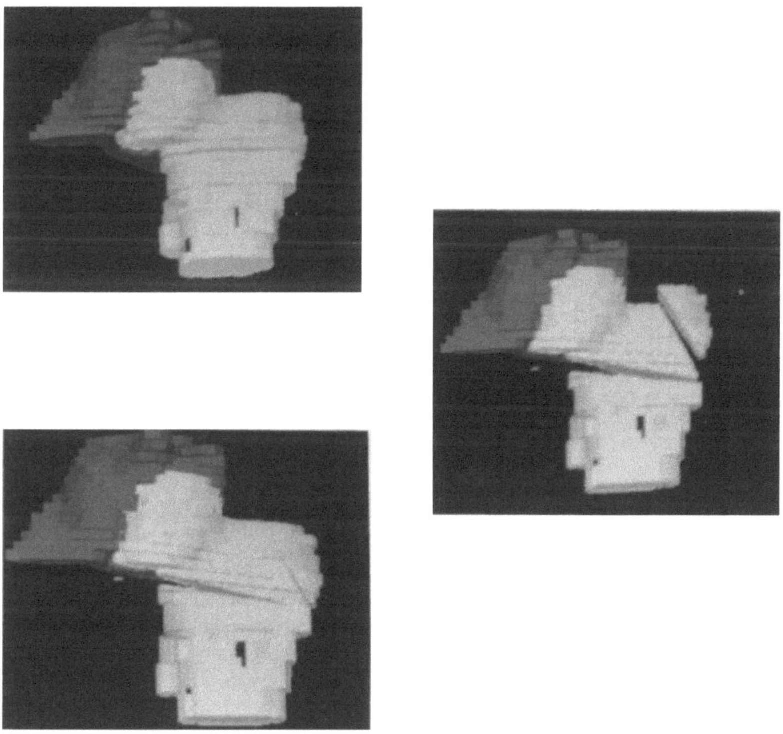

Fig. 2 An example of a CAD based procedure for the planning of an abduction (valgus) osteotomy.

For instance, fig.1 reports a traditional procedure for planning a resection of the femur. Fig.2 reports a similar procedure carried out on the solid model. The operator can of course check the result from any point of view by means of the AUTOCAD command "VPOINT"

Conclusion

By means of "volume rendering" techniques and 3D navigation systems, the "voxel-based" models allow the operator to obtain complete and detailed information on the muscle-skeletal structures. Unfortunately the "volume-based" models are unsuitable for individually manipulating the bone structures contained in the 3D image. In this case CAD-based models should be used. These models make it possible to simulate some of the surgical manoeuvres with realistic procedures.

At present an effective application of this method encounters certain problems:

a) The 3D TC image is usually recorded on magnetic media. This media should be standardized either in terms of hardware characteristics, or as software support (recording protocols). This would allow the user to access the information easily. Unfortunately at the moment these requirements are far from being satisfied as most manufacturers are not coordinated towards a common standardization goal.

b) The pointing and display devices are bi-dimensional. This forces the operator to make certain mental efforts to understand the 3D information, especially when wire-frame representations are displayed, in order to save time for display regeneration.

The next step in this work is to integrate our method with 3D pointing devices, such as "flying mouse" and 3D display devices as "private eye".

References

1. Murphy B.S. Hijewski K.P., Simon R.S., Chandler H.P., Griffin P.P., Reilly T.D., Penenberg B.L. & Landy M.M.:Computer Aided Simulation, Analysis and Design in Orthopaedic Surgery in the Orthopaedic Clinic of North America, Vol. 17 no.4 Oct 1986 637-649

2. Adams L.,Krybus W., Meyer-Ebrecht D.,Rueger R.: Computer Assisted Surgery in IEEE CG&A, May 1990 . 43-51.

3. A.Pepino, M.Cesarelli, F.di Salle, A. Mosca, M. Bracale:Some preliminary notes About the use CAD structures for the preoperative planning of intertrochanteric osteotomies. in supplement to "Functional Neurology" n.4,1992;175-178

4. M.E. Muller::Intertrochanteric Osteotomy:Indication,Preoperative Planning, Technique. The Intertrochanteric Osteotomy ed. by J.Schatzker 1984 Springer-Verlag 25-67

High Resolution 3D-CT and its Applications in Stereolithographic Computer Assisted Surgical and Implant Planning

Th. Fleiter, B. Erdtmann, C.D.Claussen

Departement of Diagnostic Radiology, Prof.Dr.C.D.Claussen, University of Tübingen, Germany

Introduction

Until now 3D-CT reformatting has been used as a kind of presentation tool, providing impressive images with usually limited clinical relevance. The main disadvantage of this technique is a missing transscription from the calculated image to a solid model, which can be used by surgeons for a precise preoperative planning. First steps to overcome this disadvantage were made with milling models based on CT-Data. But the spatial resolution and therefore the representation of fine details like nerve ducts was not sufficient. The application of stereolithographic 3D-modelling on CT- and MR-Data is actually the only way to preserve the high resolution of the original images into the resulting 3D solid model. In common these models are used as a 3D-reproduction of the scanned - usually - bone structure for surgical procedure planning. The purpose of our study was to evaluate the precision of High Resolution CT and Stereolithographic Modeling for preopertaive implant constructions in patients with skull defects and planned large tumor resections in the facial area. The intention was to combine a precise preoperative, computer asssisted resection planning with a precise preoperative implant construction. Therefore Stereolithographic models were not made of the whole CT-Data, but only of the planned implant. Because there is - up to now - no polymeric material availiable which can be implanted and stereolithographic exposed, the stereolithographic implant models were used as a positive for casting a ceramic material. The resulting ceramic implants are 1:1 copies of the calculated stereolithographic model and can be implanted after sterilization.

Method and Materials

The thin and fine bony structures of the middle face require High Resolution CT scanning. We usually use 1 mm slice thickness cobined with 1 mm table feed even in

Conventional and Spiral CT. All scans were performed on a fast scanning system (Siemens Somatom Plus S). For larger objects we use 2 mm slices with a 2 mm table feed. In Spiral mode we always use equal increment and table feed. Raw data of the Spiral-CT were reconstructed using both availiable standard reconstruction algorithms (360 ^{o} and 180 ^{o}). The scanning time for Spiral-CT `s was varied according to the length of the region of interest between 24 and 40 secs. More than one continous Spiral-CT was performed depending on the size of the object to acquire the whole volume of interest. All images and raw data were stored on optical disks and transferred via EtherNet to an Imaging System (Kontron Mipron). The images were 3D-reformatted and interactively manipulated to calculate the resection borders and to fill the resulting bony defects. Image mirroring and contour tracking were used to get the needed implant contours. The contours were converted from voxel to trigonometric data format and stereolithographic exposed. The 3D-solid model was used as a positve for casting the implantable ceramic material (IONOS Iconomer)[1]. In all cases a stereolithiographic model of the whole area of interest was made to verify the fitting of the calculated implant preoperative.

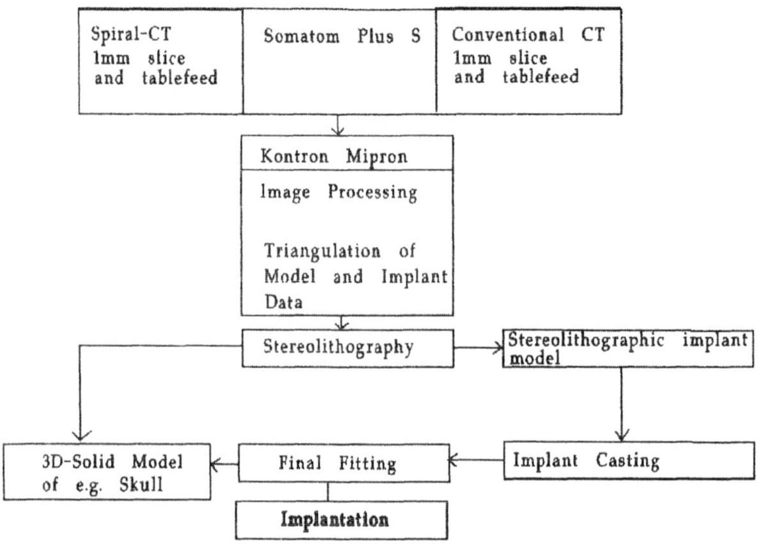

Fig. 1: System Configuration and Implant Production Process

Results

As this new technique requires High-Resolution-CT and reconstruction as well as the costintensive and timeconsuming stereolithographic exposure, only 5 patients could be examined until March 93. But there are already preliminary results according to the precision of the whole procedure: In all cases we found the 1mm spacing especially in the middle face area superior to the common used 2 mm slice thickness. Spiral-CT was always inferior according to the spatial resolution compared with High-Resolution conventional CT (1mm slice and tablefeed). But there is no significant difference between 3D-reconstructions calculated from Conventional and Spiral-CT. Because the Raw-Data of all scans were stored, we were able to reconstruct the Spiral-Scans with both recent availiable Spiral-CT algorithms (360° and 180°). There was no perceptable difference between the resulting 3D-Reconstructions and the 2D-images if the original scans were made with 1mm slice and tablefeed. The main advantage of the Spiral-Scans was a reduction of the needed examination time minimizing artefacts caused by patient movements. All stereolithographic reconstructions of the whole region of interest were of reliable precision and could be used for surgical treatment planning. All implants were calculated with overlapping borders and could be cut to the final fitting.

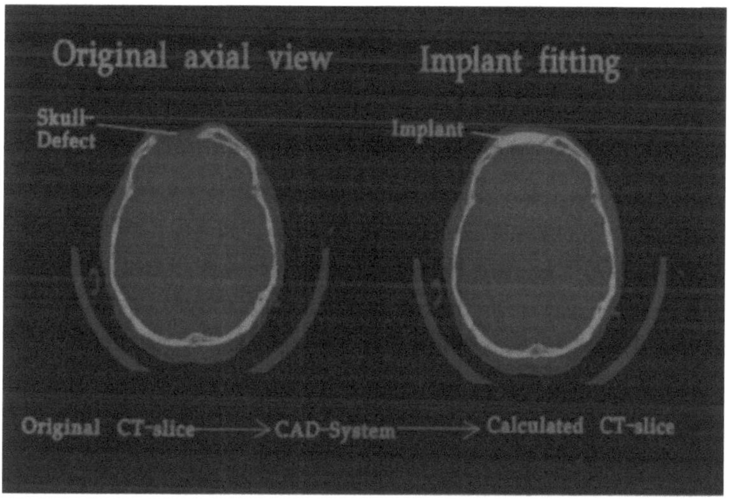

Fig. 2: Axial CT-Slice: Implant Fitting

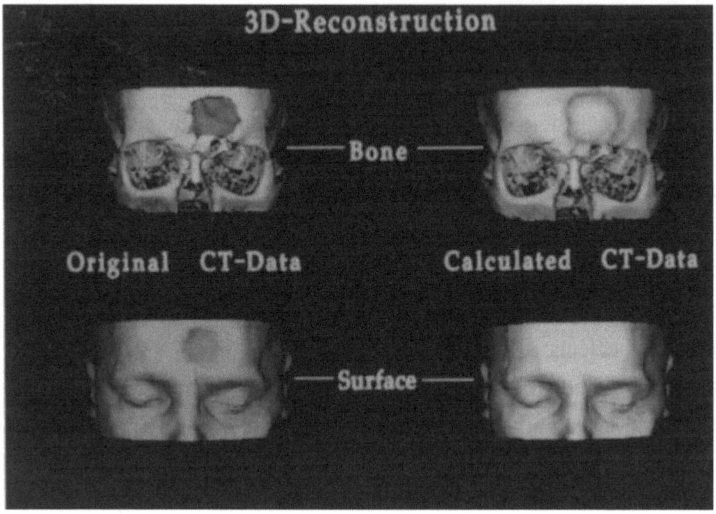

Fig. 3: 3D-CT: Implant Fitting

Stereolithography is a well established method for model manufacturing in technical purposes since years. But a reconstruction of complex structures like the human skull requires higher computer-capacities than actually availiable. The model production took up to 36 h. Nevertheless production time could be reduced by producing only the small implant models.

Discussion

The advantages of 3D-visualisations have been proven in serveral studies [2,3,4,5]. But in most cases the 3D-Reconstructions were used as a kind of presentation tool without significant diagnostic and therapeutic relevance. The 3D-image quality depends on several scan and reconstructions parameters [6]. Especially for Stereolithographic model manufacturing and precise preoperative implant construction High-resolution CT is a basic assumption. Stereolithographic model manufacturing seem to become the first application of 3D-scanning and visualisation techniques offering new therapeutic aspects [7]. The use of stereolithographic manufactured models and implants will improve the surgical preoperative procedure planning with the aim to shorten the operation-time and therefore for instance infection risks significantly. Even complex implants which were not producable with milling techniques are now possible. Of course, further studies have to be made to standarize the whole procedure and to improve the interactive, computer assisted resection planning.

References

1. L.M.Jonck, C.J.Grobbelaar, Ionos Bone Cement (Glass-Inomer): An Experimental and Clinical Evaluation in Joint Replacement. Clinical Materials 6 (1990):323-359

2. M.W.Vannier, J.L.Marsch, J.O.Warren, Three Dimensional CT reconstruction images for craniofacial surgical planning and evaluation. Radiology 150 (1984): 179-184

3. D.L. Burk, D.C.Mears, W.H.Kennedy, L.A.Cooperstein,D.L.Herbert, Three-dimensional computed tomography of acetabular fractures. Radiology 155 (1985):183-186

4. W.G.Wojcik,B.S.Edeiken-Monroe,J.H.Harris, Three-dimensional computed tomography in acute cervical spine trauma: a preliminary report. Skeletal Radiology 16 (1987):261-269

5. P.Soyer, A.Roche, M.Gad et al. Preoperative segmental localisation of hepatic metastases: Utility of three-dimensional CT during arterial portography. Radiology 180 (1991):653-658

6. D.R.Ney,E.K.Fishman, D.Magid, et al. Three-dimensional volumetric display of CT-Data: Effect of scan parameters upon image quality. J Comput Assist Tomogr 15(1991):875-885

7. H.M. Klein, W. Schneider, G. Alzen, E.D. Voy, R.W. Günther Pediatric craniofascial surgery: comparision of milling and stereolithographiy for 3 model manufactoring. Pediatr. Radiol. 22(1992)458-460

The Accurate 3D Registration of Stereotactic Radiographs and Non-Stereotactic CT Images Based on Volume Rendering

Louis Lemieux and Roger Jagoe
Epilepsy Research Group
Institute of Neurology
London, U.K.

We are currently developing a method for the accurate 3D registration of stereotactic radiographs with 'frameless' CT data based on volume rendering. The image acquisition geometry and the transformation between image and stereotactic space are found using the stereotactic markers in one or two radiographs. These parameters are used for perspective volumetric rendering of CT, to obtain radiograph-like images. In the proposed method, the orientation, position and scale of the CT in the rendering space are adjusted (using a standard minimisation routine) until a best match is found between the radiograph(s) and rendered CT. We present simulation results on the accuracy of the camera parameters as a function of the focal length and stereotactic marker localisation error. We find that sub-pixel localisation of the markers and two radiographs taken at right angles are necessary in order to achieve millimetre accuracy. We present an algorithm for localisation of fiducial markers in radiographs, and find it to be accurate to within 0.2 pixel (s.d. = 0.15). We present results of a set of manual registration experiments using CT images acquired under stereotactic control; the mean registration error is 1.1mm (s.d. = 1.67mm).

OPS - Operation Planning System for Neurosurgery

UDO JENDRYSIAK, BERNHARD NAFE, MICHAEL POLLECKER, HELMUT BRUNZLOW, ANDREAS ROESLER, JÖRG MICHAELIS

Klinikum der Johannes Gutenberg-Universität Mainz
Insitut für Medizinische Statistik und Dokumentation
Langenbeckstr. 1
6500 Mainz, Germany

Summary

The departments for neuroradiology and neurosurgery intend to use computer aided methods for individual planning of operations in the sella region. Essential structures like the optic nerve, arteries, aneurysms, the pituitary stalk, ventricles, and tumors should be segmented, reconstructed and visualized. The computed 3D view shall simulate the neurosurgeon's view through the operation microscope into the scene. This approach shall remedy the lack of available medical image analysis systems for neurosurgical intervention. The emphasis in this article will be on the segmentation algorithms and the impact of the system on daily neuroradiological work.

Requirements for Computer Aided Neurosurgery

The neuroradiologist provides the neurosurgeon with MR- and CT- image information needed for an intervention. The principle of neurosurgery is to achieve the greatest possible therapeutical effect by a minimal traumatization of the patient [Perneczky,91].

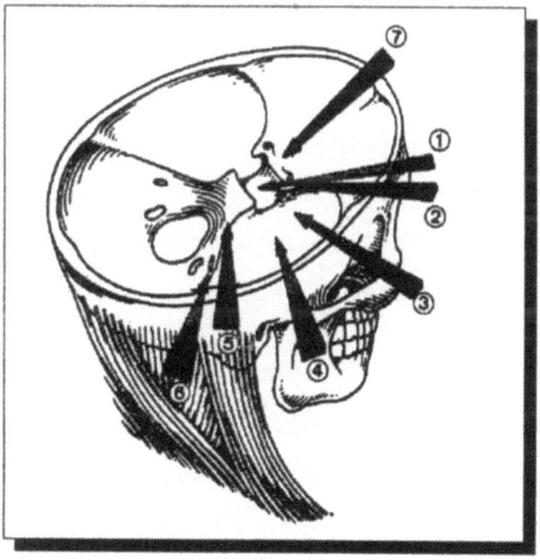

Fig. 1. Main access paths [Perneczky, 91]

The sella region is the target region for treatment in many cases - for aneurysms or tumors. While the neuroradiologist mentally reconstructs anatomical objects from the 2D slices, the neurosurgeon needs 3D visualization for optimal assistance. There are several main access paths for an operation in the sella region. The patient´s individual anatomy is the basis for the decision which path should be used. A threedimensional view of the sensitive structures should be adequate to the usual view offered by an endoscope.

Relevant Structures in the Sella Region:

A. carotis interna(3,4), A. cerebri anterior (6), Aa. cerebri posteriores, Aa. communicantes, A. basilaris, (circulus arteriosus Willisii), N. opticus with chiasma opticum (1,2), N. oculomotorius (5), infundibulum, ventricles, tumors, aneurysms, parts of the cranium (sella turcica).

Fig.2. The sella region seen by a microscope

Segmentation of Anatomic Structures in the Sella Region

The structures of the sella region are so variable in morphology and size that specialized segmentation algorithms have been developed. Traditional thresholding and region growing algorithms fail in most cases.

Fibre tracking - a Search Ray Algorithm

The nerves and vessels are thin, filiform structures, only a few voxel in diameter. They neither change diameter nor bending dramatically. The developed algorithm uses this knowledge to segment nerves and other filiform objects. First it tries to find the skeleton of the object in interest. It therefore calculates search rays in the most assumed directions and estimates the probability of true new skeleton points. Having finished that, it does region growing to get the whole object volume [Haberäcker, 87]. So we have to deal only with a few objects, we prefer interactevely predetermined starting points.

1. Determination of the Starting Direction

We expect the startpoint to be in a locally straight part of the nerve or at its end. This facilitates finding the correct starting direction and checking whether the starting point is in an object of the expected geometry.

First, an axis of a filiform structure is searched for from an interactively pre-determined starting point. With this, search rays are calculated in a hemispherical area from a known axis point which can be subsequently evaluated considering the points. Search rays are calculated for all 26 possible neighbour directions. They are all tested. Not more than three new search directions are allowed to get a high rating . Otherwise, the startpoint is inacceptable. In the case of three directions two of them must be in opposite directions. The third direction is omitted. In the case of two possible search directions, the angle between them must be in the range from 135 to 225 degrees. The search algorithm starts successively for both of them. If only one direction is in accordance with the criteria, the search starts only in this line.

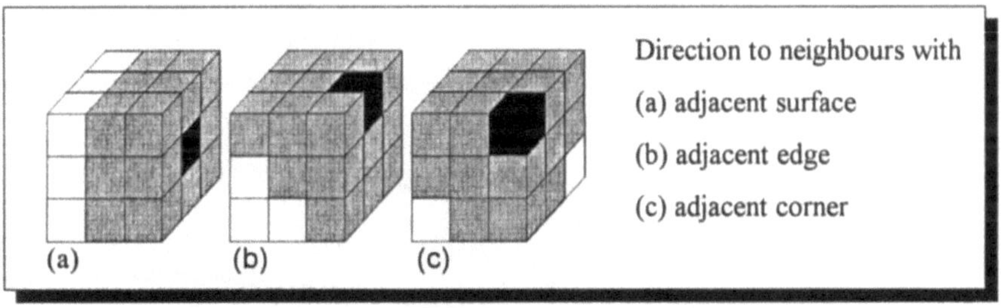

Fig. 3. Search directions in a 26 neighbourhood

2. Examination of Possible New Axis Points and New Directions

Following the object´s contour is an iterative process. It consists of:

- Calculating search rays in the assumed direction.
- Measuring criterias for each search ray.
- Estimating the criteria.
- Deciding for a new skeleton point and direction - or stop.

If a ramification is detected, the branch point is stored and the search algorithm starts for this point in two directions. Search rays are calculated only for a subset of 17 of the possible 26 neighbours of a voxel. The angle between the actual and a new direction must be smaller than 90 degrees.

3. Completing the Object Volume

Region growing is used to get the object volume from the skeleton points. By knowing inner skeleton points and the objects diameter, gaps in the contour can be closed.

Criteria for optimal search rays containing object skeleton voxels:

- The distance from the current point to the objects surface must be at a maximum.
- The sum of the absolute grey value differences of neighbouring points along the search ray must be at a minimum.
- The absolute grey level difference from the point of origin of the search ray to the mean grey value along the ray must be minimal.
- The difference between the tested and the current direction

The first voxel on the optimal search ray is used as the new starting point.

This is repeated until either the criteria cannot be satisfied by any possible search ray (e.g. at the end of the structure) or they can be satisfied by too many search rays of equal rating.

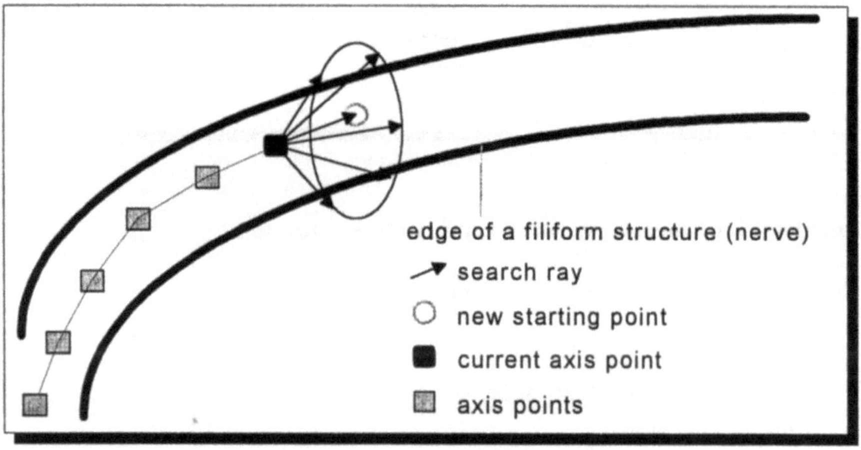

Fig. 4. Detecting an axis in a filiform structure using search rays

Contour detection is done by three different algorithms: a Laplace operator, Sobel operator, and a Non-contour-mask [Bässmann, Besslich 89].

The 3D mask detects voxels below a limit to be a contour voxel. Euclidic metric is used for distance measuring.

0	0	0	0	1	0	0	0	0
0	1	0	1	1	1	0	1	0
0	0	0	0	1	0	0	0	0
	z-1			z			z+1	

Fig. 5. 3D Non-contour-mask

Basic Modules of the Application

Programming is done in C on common Unix-Workstations (IBM RS/6000, Sparc2) with X-Windows/Motif. The system can be connected to scanners or medical imaging workstations. It reads the export data formats of NMR- and CT-scanners of the university hospital in Mainz. A graphical user interface helps to load examinations (image data) from CT and NMR scanners. Versatile 2D views are provided. Segmentation is done in a "point-and-click" way. Segmented objects have attributes like color or visibility. They can be viewed in 2D and 3D. Prepared 3D views can be used in the operation theatre as well.

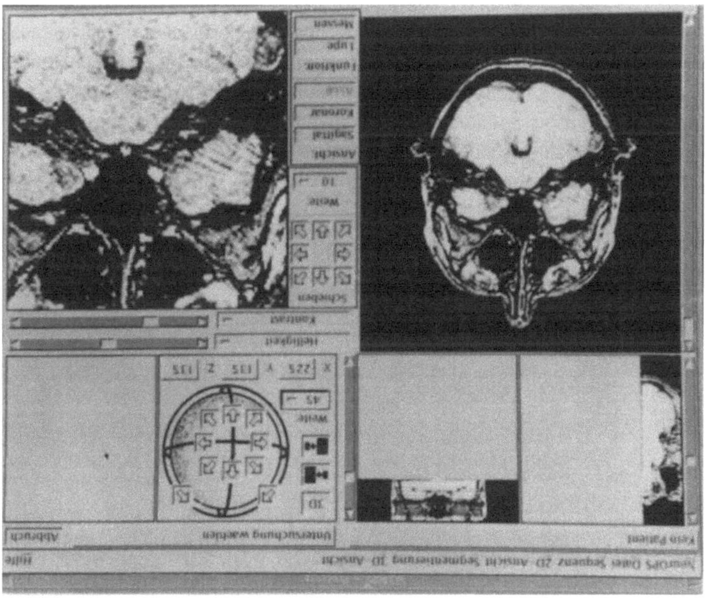

Fig. 6. User Interface of the Operation Planning System

References:

Bässmann, H. und Besslich, Ph.W.: *Konturorientierte Verfahren in der digitalen Bildverarbeitung.* Berlin; Heidelberg: Springer 1989

Haberäcker, Peter: *Digitale Bildverarbeitung. Grundlagen und Anwendungen.* München; Wien: Hanser Verlag 1987

A. Perneczky: Mikrochirurgie: *Optimale Therapie bei minimalem Trauma.* Münchener Medizinische Medizinische Wochenschrift 133 (1991) Nr. 6 S. 20-25. MMV Medizin Verlag GmbH München 1991.

Computer Planning of Stereotactic Iodine-125 Brachytherapy

S.W. Hughes, A.Sofat, N.Kitchen, R. Beany, J.E. Saunders, D.G.T. Thomas

Departments of Medical Physics and Radiotherapy, St. Thomas' Hospital, and the Gough-Cooper Department of Neurosurgery, The National Hospital for Neurology and Neurosurgery, London, UK.

Summary

A computer program is described for planning accurate positioning of catheters containing radioactive Iodine-125 seeds for the treatment of recurrent malignant brain tumours. A stereotactic reference frame is fitted to the patient's head and cross-sectional images acquired using a CT scanner. Images are transferred onto a computer and a three dimensional model of the head, tumour and frame constructed. A 3D cursor is used to mark catheter trajectories. The radiation dose distribution is calculated and displayed in two and three dimensions for various combinations of seed activity and position. The interactive 3D graphics and fast computation enables efficient optimization of plans.

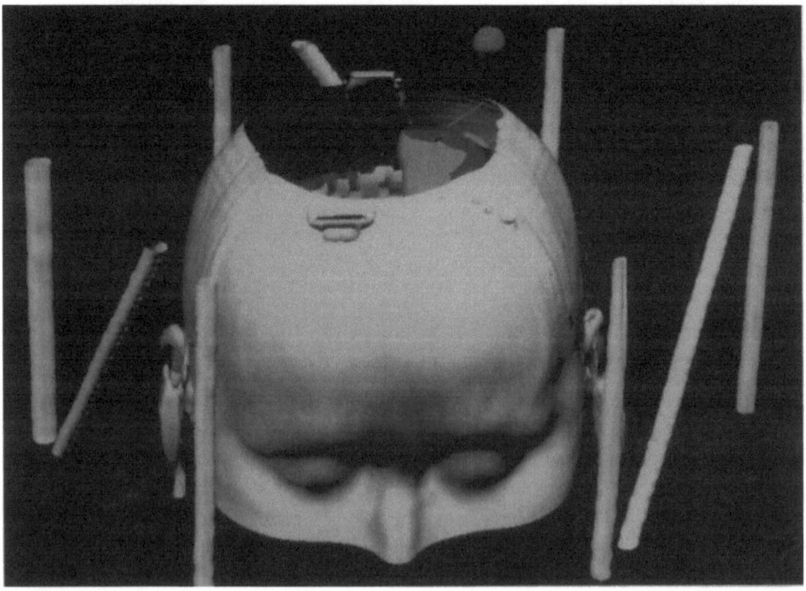

Fig. 1. Surface rendering of the head and stereotactic reference rods. The tumour (blue) and isodose (red) surface are visible near the top of the head. The green sphere is the 3D cursor used to mark the ends of the catheter trajectories (green lines). The surface rendering of the skin may be replaced by a surface rendering of the bone to show burr holes from previous surgery which may need to be taken into account when positioning the catheters.

(see also in color on page 824)

Introduction

Stereotactically implanted Iodine-125 seeds have been used for the treatment of aggressive malignant brain tumours over the last ten years or so [1,2]. A large number of calculations are required to produce a sufficiently detailed plan in the comparatively short time available before surgery; therefore we have made use of the latest developments in computing technology to facilitate planning such procedures.

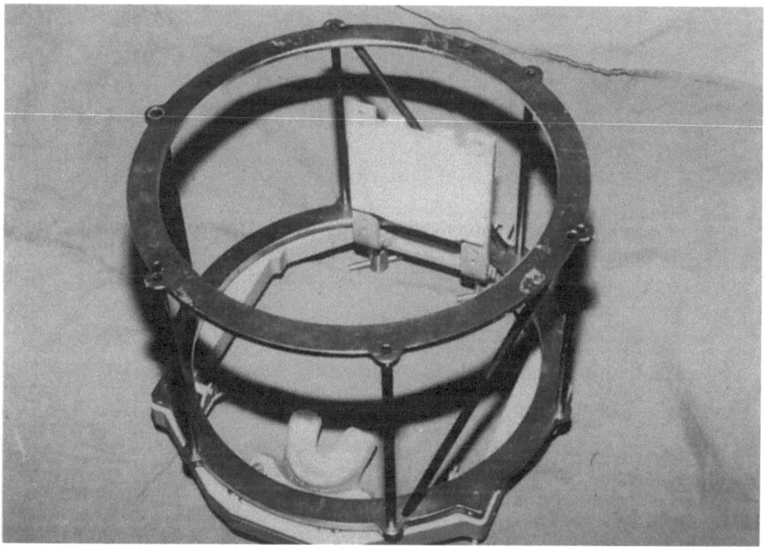

Fig. 2. Stereotactic frame. The dental tray that clips onto the patient's upper teeth, thus making the frame relocatable, is seen at the front. The rods (fiducials) serve as reference markers to convert points from CT to frame space. (N.B. The fiducials are shown surface rendered in fig. 1).

Method

The system has been developed on a Titan graphics supercomputer (Kubota Pacific Inc, Santa Clara, CA, USA) which has four 16 MFLOP processors, 64 Mbytes of RAM and 2.3 Gbytes of hard disc. The software was developed using Unix, C, X-windows and Dore (a 3D graphics library). A stereotactic Brown-Roberts-Wells (BRW) reference frame (Radionics Inc, Burlington, MA, USA) is fitted to the patient's head using a plastic dental tray that clips onto the upper teeth. This makes the frame relocatable [3] enabling the planning X-ray scan and surgery to be carried out on different days. The patient is scanned with the frame in place; sufficient images are acquired to encompass all of the tumour. The images are transferred onto the computer and the head, tumour, frame and any other relevant structures outlined on each CT image and a 3D model constructed.

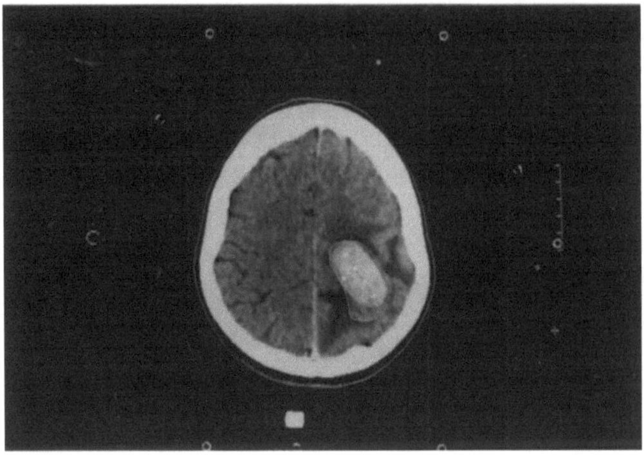

Fig. 3. CT image of head. The tumour is seen on the lower right of the head. The fiducial rods are seen in cross section around the head.

A 3D cursor is used to define catheter trajectories by marking catheter entry points on the side of the head and catheter tip points within the tumour. Several radioactive seeds may be positioned within each catheter if necessary. The radioactive dose distribution is calculated and displayed in two and three dimensions for various proposed combinations of seed activity and catheter position.

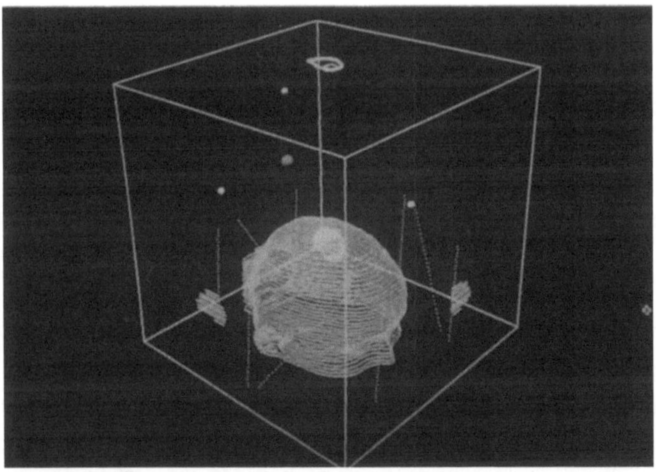

Fig. 4. 3D wire model of head, tumour and frame. The tumour outlines have been projected onto three faces of the reference box to aid in positioning the 3D cursor to mark the ends of the catheter trajectories.

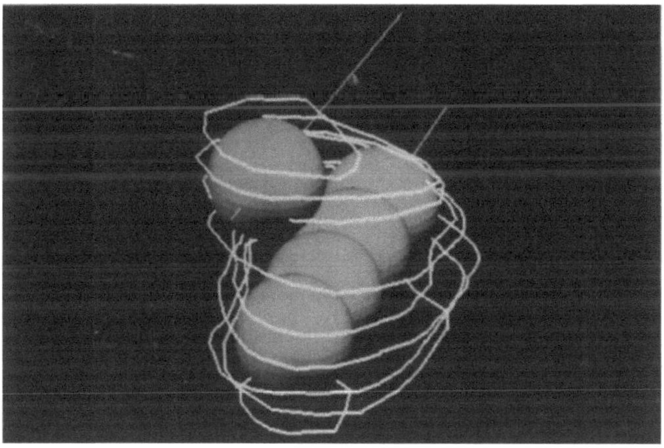

Fig. 5. 50 Gy per hour isodose spheres around seeds placed in the catheter

To aid in positioning the catheters, the 50 Gy per hour isodose surface around each seed is displayed. The surfaces are approximated by spheres. As the catheter trajectories are changed, the spheres maintain their relative positions. This technique enables a first approximation plan to be produced in just a few minutes. The plan may be fine tuned on the basis of the calculated dose distribution. A surface representation of the calculated dose distribution is generated by feeding the 3D matix of dose values into a surface finding routine. When this surface is plotted with a surface representation of the tumour, regions of over and under dosing are clearly indicated.

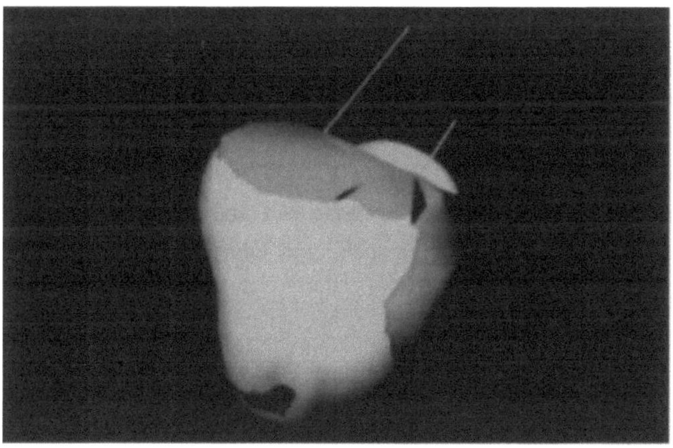

Fig. 6. 50 Gy per hour isodose surface (light) and tumour surface (dark)

Surface rendering of the skin (as seen in fig. 1.) is useful for ensuring that the catheters are not too close together, or passing through the ear lobes. Surface rendering of the bone reveals holes produced by previous surgery; the catheter tracks are routed so as to avoid these holes.

Surgery

When a suitable plan has been obtained the catheter insertion angles and depths are calculated. The relocatable frame is re-fitted to the patient in the operating room, and the catheters surgically implanted as planned. A reference box is placed on the base of the relocatable frame and orthogonal plain x-rays taken to check the positions of the catheters. The implanted catheters are afterloaded with the seeds the next day and remain in place for four to six days.

Discussion

The system described makes use of the latest advances in interactive 3D graphics technology and computational power. The interactive 3D graphics expedites the planning process by enabling the catheters and seeds to be positioned with greater ease. Plotting sensitive structures or regions in the 3D model enables the catheters to be routed away from these areas. The computational speed is invaluable for allowing a greater number of plans to be assessed in the limited time available.

References

1. Gutin, P.H.; Leibel, S.A.; Wara, W.M.; Choucair, A.; Levin, V.A.; Phillips, T.L.; Silver, P.; Silva, D.A.; Edwards, M.S.B.; Davis, R.L.; Weaver, K.A.; Lamb, S.: Recurrent malignant gliomas: survival following interstitial brachytherapy with high-activity iodine-125 sources. Journal of Neurosurgery 67 (1987) 864-873.

2. Rao, G.U.V.; Kan, P.T.; Howells, R.: Interstitial volume implants with I-125 seeds. International Journal of Radiation, Oncology, Biology, Physics 7 (1981) 431-438.

3. Thomas, D.G.T.; Gill, S.S.; Wilson, C.B.; Darling, J.L.; Parkins, C.S.: Use of a relocatable stereotactic frame to integrate PET and CT images: Application in human malignant brain tumours. Stereotactic and Functional Neurosurgery 54-55 (1990) 388-392.

Postersession

Implementation of PACS in Australia

B.L. CROWE,[1] M. de SILVA [2]

(1) Australian Institute of Health & Welfare, Canberra, Australia.

(2) JThe Children's Hospital, Sydney, Australia.

The Implementation of PACS projects continues to attract a great deal of interest in Australia from health departments, hospital administrators and radiologists. A project is being undertaken to install a PACS system in the new Childrens Hospital at Westmead, Sydney.

The hospital is designed as a 350 bed fully equiped hospital with 50 intensive care beds, theatres and emergency care. Radiological services are planned to include ultrasound, nuclear medicine, CT and MRI involving a total of twenty rooms. The PACS will be serviced by a comprehensive data base of images and reports with integration through a fibre optic backbone to a Clinical Information System and a Hospital Information System. The Imaging Department will have 11 plain X-ray rooms apart from 9 special service rooms, and in view of the wish to have a digitally based imaging system, computed radiography systems have been designed as an integral part of the operation of the hospital for both static and mobile x-ray procedures.

An overview document calling for expression of interest in the PACS project was issued in July 1992. Selected groups of suppliers will be asked to tender for the complete system during 1993/94 with a view to the system commencing operation in 1995/96. A pilot Telemedicine project will be conducted during 1993, linking the Childrens Hospital with three remote sites, to provide preliminary data on costs and performance.

An important part of the PACS project is evaluation which involves collection of data on the operating of the imaging department at the existing Childrens Hospital before it is moved to the new "filmless" site at Westmead in 1995/96. It is proposed to collect a wide range of cost and performance data during 1993/94 to enable comparative statements about cost efficiency to be made once the filmless imaging department becomes operational. Close contact is being maintained with the Hammersmith Hospital in UK which is engaged in a similar evaluation exercise. It is anticipated that the conduct of a comprehensive evaluation of the PACS project at the new Childrens Hospital will provide guidelines for other hospitals considering PACS in Australia.

A Large Clinical Image Archive System

Byrn Williamson, Jr., M.D.
Richard L. Morin, Ph.D.

Department of Diagnostic Radiology
Mayo Foundation
Rochester, Minnesota, and Jacksonville, Florida, U.S.A.

In January of 1989, Mayo Foundation began storing digital images on a central digital archive at the Rochester, Minnesota campus. Subsequently, a similar archive was established for digital images at the Mayo facility in Jacksonville, Florida. A two tier storage scheme is utilized: optical storage media provide long-term storage of images, and current images are stored on magnetic media on file servers. The combined optical archives in Rochester and Jacksonville currently contain more than 4 million images of which 350,000 images are available in an optical jukebox. Approximately 10,000 images are available for rapid access on file servers.

The operation of this PAC system is directed by an IBM AS/400 computer which controls the distribution of images, keeps track of the location of each set of images, and stores the diagnostic report of each examination. The archival process does not require operator intervention, and archiving and retrieving images do not interfere with the clinical workflow of individual imaging units.

Approximately 40 computers are incorporated into the PAC system including 15 viewing stations. Redundant components are used to provide high availability. The system accommodates images obtained from multiple imaging modalities and multiple vendors. Eleven MRI scanners and four CT scanners are currently connected to the system. The display program runs on multiple platforms utilizing DOS, OS/2, and UNIX operating systems.

Within the Rochester, Minnesota, campus and the Jacksonville, Florida, campus the PACS shares a 4 Mbps token ring Local Area Network with many non-radiology users. Rochester and Jacksonville are connected by a 1.5 Mbps land-based Wide Area Network utilizing commercial carriers.

Data Compression and Image Quality in Digital Coronary Angiography

U. Krauss, J. Beier, A. Wahle, H. Oswald, E. Fleck

German Heart Institute Berlin
Augustenburger Platz 1, D-W 1000 Berlin 65, Germany

Handling of digital coronary angiograms implies appropriate storage in digital archives, adequate image transfer between different in-house workplaces, and for external telemedicine applications. These tasks are restricted by the huge amount of data: one single angiographic examination results in approx. 2 GB of image data. Data compression techniques provide solutions to overcome this obvious bottleneck. The usage of compression techniques depends on various criteria: user requirements, technical and legal constraints. Compression algorithms can be devided into lossless and lossy methods. For lossless methods a compression factor (CF) of max. 3 can be achieved. The prominent criteria in coronary angiography is the image quality. To assess and quantify small vessels and stenoses, a remarkable loss of image quality cannot be accepted.

Using the well-known lossy JPEG image data compression, a study was performed to analyse the impact on image quality of routine angiograms and special phantoms. The question was, what extent of compression guarantiees a satisfying quality for clinical demands?

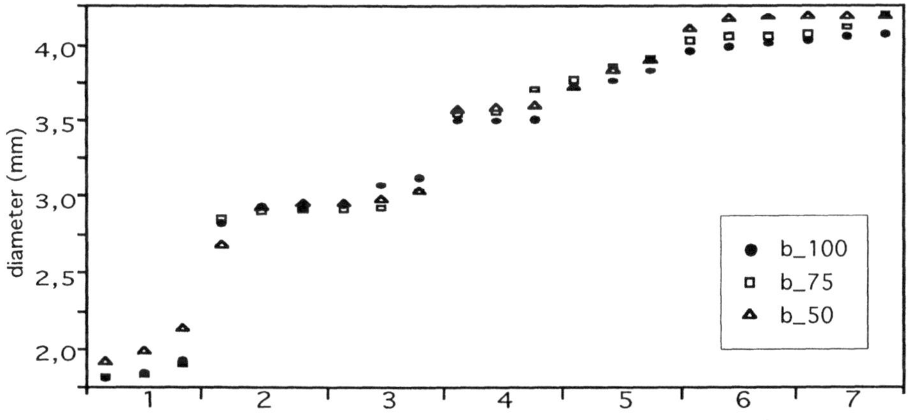

Fig. 1: Diameter evaluation of 7 vessel segments

For visual assessment, a JPEG quality setting < 75 per cent resulting in already perceptable changes cannot be accepted. For digital quantification of coronary arteries the most important factor is the evaluation of vessel diameters. Fig. 1 shows the evaluation of 7 vessel segments of an angiogram. A JPEG quality setting of 50 per cent leads to a significant difference in quantification parameters and it is therefore not acceptable for coronary angiography. Only a quality setting of 75 per cent is acceptable and leads to a CF of 12. Digitaly Coronary acquired images by a Philips DCI-System have an entropy of about 6 to 7. With a JPEG quality setting of 75 per cent CFs of 10 to 20 can be achieved.

MDIS: A PACS and Teleradiology Project - Present and Future Configuration

R.G. Leckie, M.D. CPT USA, M. Sheehy , LT USA, D. Lyche , CPT USAF , L. Cade, CPT USA , R. E. Detreville, CPT USA, G. Norton, MSgt. USAF , , F. Goeringer, M.S.C.E. COL USA.

USAMMA- MDIS PROJECT, FT. DETRICK, FREDRICK, MD. 21702-5000

The Medical Diagnostic Imaging Support System (MDIS) is a Picture Archiving and Communications System (PACS) and teleradiology project for the military. This presentation demonstrates the wide scope of the project, listing the multiple present and planned future site and their interactions. Sites are located both within the U.S. and abroad. These sites vary from portable /deployable units and small clinics to 1000 bed Medical Centers. For example , the proposed configuration for Tripler Army Medical Center in Hawaii will support part of Asia and the Pacific Basin - 50% of the world's surface area.

The MDIS project allows the military to provide rapid, expert radiologic support to our troops and their dependents even in obscure locations in times of peace and war.

A Modern Archive Concept for the Filmless Hospital

S. NISSEN-MEYER, U. FINK, M. PLEIER

Radiologische Klinik der Universität München
(Direktor: Prof. Dr. Dr. J. Lissner)
Klinikum Großhadern
Marchioninistr. 15
W - 8000 München 70

PACS has in the recent past been more and more frequently and controversially discussed in the literature, at workshops and at conferences. Should one buy PACS now or not? There seems to prevail a strong hesitance to take this step. One essential reason why seems to be that for the enormous amount of image data to be handled in the archive there have been to our knowledge no satisfactory solutions presented so far that aim for the filmless hospital. On the other hand, without realistic prospects for a functioning comprehensive digital archive, it is doubted by many that PACS will bring the promised benefits and also be cost effective.

Currently offered archive-components of PACS in Germany are all based on jukebox technologies with optical platters of some kind. Firstly, we argue that naive extrapolations of the costs of such systems to include the total conventional X-ray film archive lead to investments difficult to justify. Secondly, we present some simple arguments why such systems are insufficient to solve the radiological archive problem.

We draw attention to a an alternate, technologically mature and well tested archiving system. It employs a UNIX-based supercomputer controlling up to 4 robot towers respectively containing 220 25-GB 19 mm digital video tape cassettes. Each tower thus holds a total of 5.5 TB.

The employed HW- and SW-technologies plausibly allow for, and the vendor thus guarantees, unlimited data life times. These are achieved by regular automatic tape-error check- and correction-routines

Data transfer rates from the robots into the supercomputer are more than 7 times faster than with optical platters. This allows in particular for manageable migrations to newer technologies in case the tapes should become obsolete.

This system further allows for all data (old and new) pertaining to one patient to be written and held in contiguous areas on the tapes, thus strongly reducing robot arm movements when fetching old data for a patient when he/she is again newly admitted in the hospital.

System and medium costs are respectively of an order of magnitude lower than for optical disc jukeboxes per online Terabyte.

We present a rough conceptual framework for the organizational migration of a conventional analog archive to a digital one using this technology. It offers sufficient storage capacity for a long term scale with accompanying software database interfaces facilitating a very flexible integration into total PACS and hospital-communication environments. Based on rough cost estimates, this approach seems to move comprehensive PAC-Systems into the near future, pushed not only from the medical need, but also from the technological and economical point of view.

Integration of Picture Archiving Communications System (PACS) in a Nuclear Medicine Department

B. Bagni , S. Corcione, I. Bagni, F. Vita

Radiology and Nuclear Medicine Departments
S.Anna Hospital , Ferrara, Italy

A very large PACS system has been installed in S. Anna Hospital of Ferrara and is connected to three gamma cameras via optical fibre and ethernet local area network.The aim of our project is to improve diagnosis through better images management system and to obtain different pictures of the same patients in a unique visualisation system.The visualisation consoles are connected to a digital radiography, CT-Scanners, digital radioscopy, ultrasound scanners and nuclear medicine devices) via an optical fibre network and analog-to digital converters.
Diagnostic viewing stations are situated in remote sites and use high resolution B&W monitors (1K x 1.24K 8 bit depth) or colour monitors for nuclear medicine, in the PACS room a laser reader and a laser hard copy units are available for digitising and printing images coming from different peripheral devices. The comparison between productivity and diagnostic implications of the PACS and the Elscint network, ApexNet with ApexView systems for data exchange in the NM is related to the different modalities for frame grabbing and data transfer used by the two systems. At the moment only one-Way data transfer is possible since the NCR NEMA interfaces are not present in the AT&T system. A preliminary remark is possible today on the limitation represented by the monitors that simultaneously display only four low resolution images of compatible format, therefore compromising real time clinical evaluations. A new home-made low price consultation workstation using 80486 technology and C language is set-up and connected to the PACS network via a ethernet bridge. Our experience of 24 month shows evidentiate that the integration of whole diagnostic procedures currently used in an Imaging Department seems to be a unique opportunity for improving the diagnostic potential of the imaging modalities.

Real-Time Image Coding Using Vector Quantization-based Algorithm

Abdalla S.A.Mohamed

Department of Biophysics,Faculty of Medicine & Health Sciences,
United Arab Emirates University,
Al-Ain,P.O.Box17666,
United Arab Emirates.

Digital image coding using vector quantization [VQ] based techniques provides low-bit rates and high quality coded images ,at the expenses of intensive computational demands.The computational requirement which is due to the encoding search process had hindered application of VQ to real-time and high quality coding of color images.Reduction of the encoding search complexity through partitioning of a large code book by tree structured vector quantization [TSVQ] in the spatial domain is proposed in this paper.A. real-time TSVQ algorithm for encoding color is developed on Apollo workstation HP 9000/433.This algorithm contributes to the use of a free distortion measure and reduction of code vector dimension.The mean/quantized residual vector quantizer has been used for reducing computational complexity of the encoding search .Running at clock 33MHz,the proposed algorithm is capable of real-time processing of 512x492 pixels per frame with rate 30 frames/second.Application of this algorithm on a training image(lady face) shows the convergence with low distortion rate as shown in Fig.1(a-e).

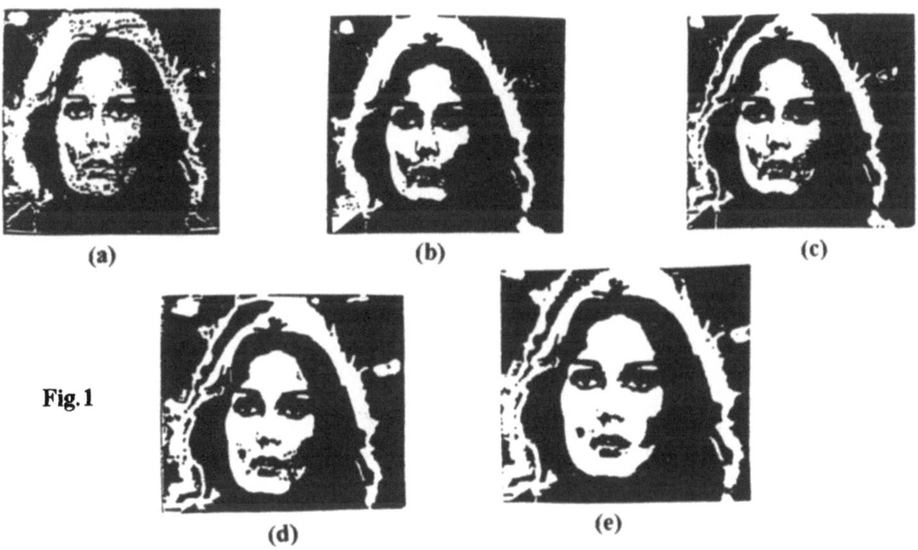

(a) (b) (c)

Fig.1

(d) (e)

An Object-oriented ACR-NEMA Data Dictionary in C++

S. L. Fritz, S. R. Roys, S. Munjal

University of Maryland School of Medicine,
Baltimore, Maryland, USA

Summary

An object-oriented data dictionary for the ACR-NEMA standard has been developed using the C++ object-oriented language. The dictionary consists of an object of class Dictionary which contains objects of class DictionaryEntry. Dictionary entry objects contain all the information in a single entry of the ACR-NEMA version 2.0 data dictionary. Each entry object has internal representation of the dictionary definitions and a set of public methods whereby that information may be accessed, for example for the purpose of parsing and displaying messages in an imaging workstation or construction ACR-NEMA images from proprietary format image data on a modality. The dictionary constructor reads a copy of the dictionary from hard disk and creates an instance of a dictionary object which contains dictionary entry objects in a form which provides for efficient lookup. The dictionary class library is intended to be part of a more general object-oriented class library for ACR-NEMA 3.0 (DICOM) applications.

Teleconferencing in a Medical Environment

Andreas Gehring

Technical University of Berlin
Department of Technical Computer Science
FR 3-3
Franklinstr. 28/29
D-1000 Berlin 10

Recent communication technologies and highspeed fibre networks open new fields for medical image communication. Applications like teleconferencing become feasible even in areas with large data volumes such as hospitals and other clinical environments. Medical workstations with integrated conferencing facilities can be used for Remote Expert Consultation (REC) or local meetings to save time and money.

Due to the complexity of such CSCW (Computer Supported Cooperative Work) applications problems arise which do not exist on local medical workstations. Mechanisms are needed to make it easy for the physician and the clinical personel to handle communication and security functions within a conference.

A teleconference has to be controlled on the technical and user level. On the one hand some restrictions are necessary to avoid confusing and intermixing of actions, on the other, the users should be as free in their work as possible. Thus, the design of conference user interfaces is essential for the usefulness of the system. The conference requirements must be supported in such a way that the doctors are not impeded in their work. A full integration of the application environment within the medical workstation is obligatory (hence all functions should work in the same way within a conference as they do locally).

A flexible client/server model is necessary on the software level, to make the conference system hardware and vendor independant.

The collecting of conference results and protocols is specific to the conference purpose. A ray therapy planning may produce different data than the finding of a diagnosis. The MEDAP/COMED project [1] is a possible approach to this problem. It combines the technology of Medical Workstations with the ability of Teleconferencing. Further studies in clinical environments are necessary to refine the requirements of physicians and radiologists for teleconference systems.

References

1. Heinz U.Lemke; Gabriele v. Voigt: The MEDAP Project: Medical Imaging and Cummunication Technology. BERKOM - Broadband Communication within the Optical Fibre Network (1992) 97-102

Data Transmission of Diagnostic Images in Clinical Routine

M. Daffertshofer, A. Schwartz.

Department of Neurology, University Heidelberg,
Klinikum Mannheim, Theodor-Kutzer-Ufer, Mannheim, FRG

Introduction:

A close cooperation in in- and out patient medical treatment requests swift and uncomplicated data-transmission of diagnostic images. Emergency diagnosis of therapeutical implications requier an image based decision and cannot be based on written results or phone calls only. Sending the images by mail or courier to the expert is expensive and time consuming; however today´s state of the art in computer processing enables electronic transmission of images. Most existing transmission systems need extended technical equipment, which incurs greater costs. Other systems only realize a transmission limitated to one building or need a fixed local network. For the highest flexibility, it is useful to transmit images via the commercial phone line, in which the teleradiology system gains highest mobility and economy [1,2,3].

Methods:

Hardware-configuration, system-requirements:

Data-input	Sender	Receiver
Video camera	Personal-Computer	Personal-Computer
Video recorder	Memory (Minimum of 640 Kb)	Memory (Minimum of 640 Kb)
Still video	Harddisk (Minimum 100 Mb)	Harddisk (Minimum 150 Mb)
Scanner	Display (Minimum VGA)	Display (Minimum VGA)
	Modem (Hayes compatible)	Modem (Hayes compatible)
Interface to scanning device (e.g. Computertomography)	public telephone line (analogue or even better ISDN)	public telephone line (analogue or even better ISDN)

Software:

To realize the transmission-project, software was developed, that fits IBM standard. Considering the requirement of mobility, the software is designed with the capability of a notebook computer. Concerning resolution and greyscales, its facilities are applied optimally. Screen resolution is 640x480 pixel with 16 greyscales for compatibility with a notebook and can can be increased with special graphic adapters to 1024x768 pixel with 256 greyscales. A maximum of 50 consecutive images can be summarized and transmitted, and a

UNIX-compatible compression guarantees swiftest transmission. Off-line treatment of the images can take place at the "calling" as well as at the "receiving" end: Control of brightness and contrast optimizes picture quality. Zoom-in is possible in any region of interest. By using high-performance modems (16800 Baud), the average transmission speed is 20 sec/image (Fig. 1).

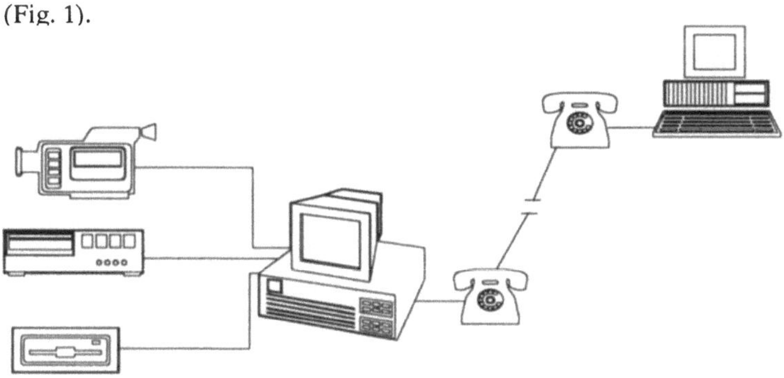

Fig 1:
Sketch of the system used.

Results:

In a clinical validation study over a 4 month period we tested the system during the neurological stand-by duty, which consists of a resident in the hospital and a chief resident at home. 84 investigations were transmitted from the hospital to the chief resident at home (73 CAT scans, 6 angiographies und 5 Duplex scans).

The residents described the additional time for carry out the data input and the transmission as tolerable and could be compensate for a faster process of diagnostic and therapeutical decision. The chief residents described the image quality as sufficient and the extra information as good. The receiver could make a diagnosis, that was confirmed after interpretation of the radiographic films in 64 CAT scans, 6 angiographies and 3 Duplex scans at the next day. In 12 CAT scans and 2 angiographies the transmitted data could help the chief resident in its decision to come to the hospital. A Diagnosis could not be made in 9 transmitted CAT scans and 2 transmitted Duplex scans. However, in these cases a severe cerebral lesion could be excluded. Moreover when looking to the radiographic films the next day only in 3 out of these 9 CAT scans a diagnosis could be made, while in 6 CAT scans the diagosis remains unclear also after interpreting the radiographic films and could only be made after additional examinations.

Discussion:

The presented system provides the transmission of diagnostic images with a resolution and an image quality which is sufficient for diagnostic procedure. Using this low cost solution, which can be easily integrated in already existing hardware, the on-call physician is linked to the scanning facility to provide emergency and extended coverage. Moreover by using the public telephone system the receiver is mobil. Thus so this system is useful for transmitting images from the clinic to the expert outside who only needs a standard portable personal computer with a modem and a telephone connection. Because of the variable data input, not only one sort of image can be transmitted but also nearly any material wich can be filmed and digitized.

References:

1.) Paakla T., J.Aalto, V.Kahara et al: Diagnostic performance of a teleradiology system in primary health care. Comput-Methods-Programs-Biomed. 1991; 36: 157-60.
2.) Odagiri K., H.Nakamae, T.Ohkoshi et al: Clinical evaluation of a teleradiology system utilizing personal computers and public telefon line. Nippon-Igaku-Hoshasen-Gakkai-Zasshi. 1991; 51: 1359-65.
3.) Dorhrmann P.J.: Low-cost teleradiolgy for Australia. Aust-N-Z-J-Surg. 1991; 61:115-7.

High-quality Digital Image Transmission with a High-speed Teleradiology System

V. Köchli[1], B. Kohlberger[2], D. Voellmy[1], G.C. McKinnon[1], G.K. v. Schulthess[1], and W.A. Fuchs[1]

[1]Department of Radiology, University Hospital, Zürich, Switzerland
[2]Department of Radiological Diagnostics, University Hospital, Freiburg i.Br., Germany

Summary:
One of the major drawbacks of current teleradiology systems is the relatively low data transmission rate of modem-based systems, resulting in image transmission times of several minutes up to hours for entire CT or MR examinations. Video image transmissions, on the other hand, result in significant quality reduction and loss of diagnostically relevant information. The objective for our teleradiology project therefore was to establish a high-speed digital image transfer system that would allow interactive sessions for discussing even large image series without loss of information.

High-speed Image Transmission:
The data transmission link we used is the MEGACOM self-dialling network of the Swiss Telephone company which is at the moment being integrated into the high-speed networks of the other european phone companies. These networks are based on fiber optics and satellite communication and offer a raw data transfer rate of 2 Megabits per second. A MEGACOM link was established as an Ethernet connection between two SUN workstations installed at the University Hospitals of Zürich, Switzerland, and Freiburg i.Br, Germany. Five imaging systems, 2 CT scanners and 3 MR units, by four different manufacturers were connected to the system. The Papyrus standard file format proposed by O. Ratib was chosen as a common image file format, into which all image data was converted before transmission. Conversion, display and transmission of the images was done by custom-made software. Several timing problems had to be solved before CT and MR images could finally be exchanged between the two hospitals over a distance of more than 700 km. The transmission time depended on the overall network load and was typically around 2 to 4 (max. 8) seconds for a 256 x 256 MR image with 16 Bits per pixel, which proved to be acceptable by the radiologists for interactive exchange and discussion of entire image series. At the same time, the full image quality is maintained through both the format conversion and the transmission. With its short transmission times and high image quality, our MEGACOM-based teleradiology system therefore promises to be a fast and cost effctive way for exchanging radiological images. Future perspectives include integration into PACS systems as well as use of the new ISDN telephone networks.

Functional Enhancement of Radiology Information Systems with Bar-code

R. Rechid, H. Dörner, A. Ullrich, R. Götzinger, R. Leppek, K.J. Klose
University Hospital Marburg
Department of Diagnostic Radiology
Baldingerstraße
D-3550 Marburg/Lahn

Since 1987, the Department of Diagnostic Radiology has been running the Radiology Information System (RIS) MEDOS. In order to optimise data entry, several bar-code readers were installed in June 1992. The bar-code *Code 3 of 9* was chosen because it is able to handle both numerical and alphanumerical data. The fields of application are:

1. **Patient label**: In order to access patient information more quickly on the RIS terminal, the bar-coded PID (patient identification number) was included on the patient label which is located on the film jacket.
2. **Post registration**: To improve post registration in ultrasound, MILLI ('Marburger Idee zur Leistungsquittierung mit Lichtgriffel') was developed. MILLI is a schematic one page overview of the body on which bar-code prints are assigned to organ related exam types.
3. **Film Tracking**: Since September 1990 a film tracking system has been used in our department (who is giving what to whom). The speed of data entry to the RIS has been increased by providing each individual with a bar-coded card (who is giving what) and by providing bar-coded destinations (to whom: wards, policlinics, physicians) which can be found either on a two pages list of destinations or on the film jacket labels which are printed after post registration.
4. **Reporting**: In order to generate faster report availability a mechanism was developed for radiologists to create standard reports without the use of dictation. MEDOS supports automated report creation by providing a tree structure of text blocks. Two trees have been implemented: the first is for chest reporting, the second is for intensive chest X-ray reporting. Bar-codes are used to select text blocks within a tree.

The application of bar-codes to support daily routine work in a radiology department has proven to be useful. The above list is only a sample of the possibilities of applications supported by bar-codes. Planned extensions are also presented in the poster.

Large Scale Feature Searches of Collections of Medical Imagery

M.W. Hedgcock Jr., M.D., W. Karshat, T.S. Levitt, Ph.D., Dmitry Vosky

Department of Veterans Affairs Medical Center/Radiology (114)
4150 Clement Street, San Francisco, Ca. 94121
U.S.A.
(415) 750-6901 FAX: (415) 750-6925

Summary

Large scale feature searches of accumulated collections of medical imagery are required for multiple purposes, including clinical studies, administrative planning, epidemiology, teaching, quality improvement, and research. To perform a feature search of large collections of medical imagery, one can either search text descriptors of the imagery in the collection (usually the interpretation), or (if the imagery is in digital format) the imagery itself. At our institution, text interpretations of medical imagery are all available in our VA Hospital Information System. These are downloaded daily into an off-line computer. The text descriptors of most medical imagery are usually formatted as free text, and so require a user friendly database search tool to make searches quick and easy for any user to design and execute. We are tailoring such a database search tool (Liveview), developed by one of the authors (Karshat). To further facilitate search construction, we are constructing (from our accumulated interpretation data) a dictionary of medical and radiological terms and synonyms. If the imagery database is digital, the imagery which the search discovers is easily retrieved from the computer archive. Computer assisted imagery searches from a clinical text database are far faster than manual searches, and allow automatic seaches to be performed frequently as part of the quality improvement process.

Rule-based Image Processing Environment Applied to Medical Imaging

Thomas Buck, Jürgen Fechter and Wolfgang Straßer

Universität Tübingen
Wilhelm Schickard Institut für Informatik
Graphisch–Interaktive Systeme
Auf der Morgenstelle 10, C-9
7400 Tübingen 1 – Germany

Data acquisition with medical imaging modalities (e.g. MRI, CT, US, and PET) has gained much importance in recent years, especially for diagnostic decision making purposes. To allow physicians to gain case-relevant information from these huge amounts of pictorial data, well designed image processing (IP) workstations are urgently required.

In this context, one of the most serious problems is the integration of such a system into the clinical routine. Since clinical users are usually not familiar with the field of IP, they are not likely to exploit the capabilities of the software, if we consider the great variety of IP algorithms with different parameters and their combinations. Embedded knowledge may be a source of assistance, taking over the role of the IP expert.

The objective of our project is to analyse the difficulties that arise with the use of an IP system by a IP non-expert, and to propose solutions for it. A rule–based system selects the algorithms that should appear in the pipeline of operators applied to an image. If the physician is not satisfied with the result, it is possible to input new information to the environment. This is done by using high–level descriptors for processing the images, allowing people who are not familiar with IP to express their desires in natural language like terms. The knowledge is represented in the form of facts and rules. Basically, the rules used represent explicit knowledge about image attributes, IP operators and their parameters, sequences of image operations (pipeline), anatomy and modality of image acquisition.

We developed an easy-to-understand language, named **Tseglan** (*T*übinger *seg*mentation *lan*guage), which is an important component of the user interface. For the implementation of the inference mechanism we have chosen CLIPS (C Language Integrated Production System), and for the IP part we used C for programming and X Windows as graphical user interface. The first prototype of our system has shown that interacting with the rule–based environment leads to advantages in speed and quality of the results, compared to directly manipulating the IP operators. We found that explicit representation of even trivial knowledge, although difficult to implement, is of great value.

A Topology-based Representation of Grey-Valued Images

S. Huang and Y. Bao

Department of Computer Graphics, Sekr. FR 3-3
Technical University of Berlin, Franklinstraße 28-29
D-1000 Berlin 10, Germany

In many medical image analysis tasks, e.g., image segmentation, anatomic/pathological object recognition, etc., the manual interference remains still, at least partially, unavoidable to achieve a meaningful result, because automatic algorithms are not always able to deal with the complex tasks / scene in images. Research effort has been increasingly given to development of an efficient and effective interaction scheme in performing image analysis. Interactive image analysis operations are very time-consuming and error-prone, because of the large data-volume (e.g. 512 x 512, 30-40 images per CT-scan). A high-level, hierarchical representation method for medical images are strongly demanded to improve the speed, the correctness and the quality of user interactions, and thus to render digital image analysis tools to be practically usable in medical diagnosis.

We represent in the paper a topological image representation. This representation is based on a topological description of the grey-value information carried by the image. Given an image, we build a pyramid structure. On each level of the pyramid, the domain of the image is partitioned into some subregions, each of which has a unitary property: the image, restricted on any subregion, has only one local maximum. Any two adjacent subregions are separated by the saddle point between them. If they are to be merged together, they are merged through the saddle point. For the subregions on the top level of the pyramid we use a relationship matrix to describe the merging properties. All the other subregions on the lower levels of the pyramid are first mapped onto the corresponding subregions on the upper level, and those subregions, which are corresponding to the same subregion on the upper level are then described by a relationship matrix associated to the subregion on the upper level.

Such a hierarchical description shows three main merits. First, for the user interaction, the user need only to treat subregions, rather than pixels. Second, the representation is hierarchical: the description for all the objects in the image is coarse-to-fine, which can be readily used to improve the efficiency of user-interaction. Third, the topological properties of an image is invariant against geometrical deformations and grey-value scaling. So the topological properties of objects, acquired through analysing some standard image examples, are better inertial estimation of the objects for the subsequent segmentation process.

Diffusion and Clustering Techniques for the Segmentation of Cranial MR-Images

I. C. CARLSEN, K. OTTENBERG, K. NEUMANN

Philips GmbH Forschungslaboratorien,
Forschungsabteilung Technische Systeme Hamburg,
P.O. Box 54 08 40, D-2000 Hamburg 54, FRG

Summary

In this paper we report on the segmentation of cranial MR images which has been set up as one of the approaches pursued within the framework of the European Community funded COVIRA project consortium. The approach combines anisotropic diffusion techniques [1] adapted to Magnetic Resonance (MR) signal structures with clustering for a purely grey-level based segmentation of MR images.

We have investigated clustering techniques [2] for MR image segmentation. The MR images to be clustered optionally can be smoothed via anisotropic diffusion processes prior to the clustering. For the cluster process we define the scatter for each pixel i with feature vector y_i and class k as:

$$S_i^k = (m_k - y_i)^T \Sigma^{-1} (m_k - y_i) + \alpha \sum_j \gamma_i \|m_k - m_j\|^2$$

where m_k is the cluster center of class k, Σ the covariance matrix, m_j the corresponding cluster center of a neighbor pixel and γ_j=1 for first and $1/\sqrt{2}$ for second-order neighborhood systems. The influence of the neighborhood term is controlled by α which is slowly increase from zero during the optimisation. After this a partial volume classification $\lambda_i(k)$ for every pixel is done by calculating

$$\lambda_i(k) = e^{-\beta s_i^k} / Z$$

where Z is a normalizing factor.

References

1. Nonlinear Anisotropic Filtering of MRI Data, G. Gerig, O. Kübler, R. Kikinis and F.A. Jolesz, IEEE Trans. Med. Imaging, June 1992 Vol. 11
2. Partial Volume Tissue Classification of Multichannel Magnetic Resonance Images-A Mixel Model, Hwan Soo Choi, David R. Haynor and Yongim Kim, IEEE, Trans. Med. Imaging 1991 Vol. 10 No 3.

Image Zoom Falsifies Vessel Diameters

T. Linderer, W. Wunderlich, F. Fischer, J. Nöring, R. Schröder.

Div. of Cardiology, Med. Klinik, Klinikum Steglitz, FU Berlin

Image zoom enhances visual perception of atherosclerotic lesions whereby the magnification required may vary. We studied the impact of varying image zoom between 1 and 6 fold on the quantitative assessment of vessel phantoms of 0.3 to 4.5 mm diameter, a range applicable e.g. to coronary arteries and to stenosis of larger vessels too. We used standard methods for computerized vessel analysis in digitized angiograms: optimum edge detection (optimized in the non-magnified image) and catheter calibration (catheter 2.5 mm ø). The figure shows the deviation of the measured from the true diameters for various zoom factors.

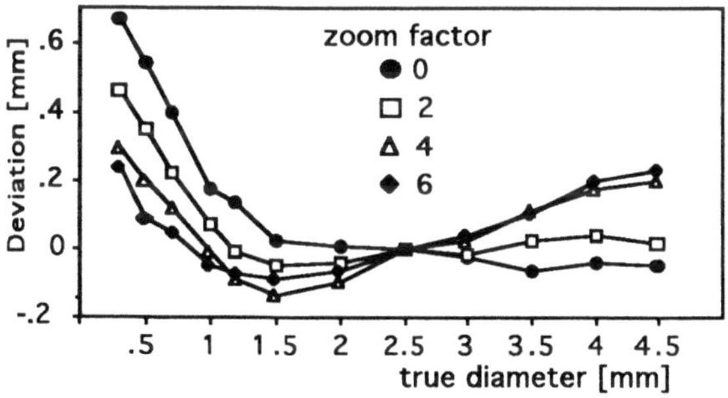

Results: The standard quantitative coronary arteriography provides accurate results in a range close to the calibration catheter only. Beyond, varying image zoom provides very different and diverging results. The position of the second derivative is zoom dependend and thus, any edge detection taking in the second derivative becomes zoom dependend.

Conclusion: Image zoom and edge detection have to be tuned very carefully. In long-term studies on progression/regression of atherosclerosis the zoom factor has to be kept constant regardless of visual perception since varying zoom may feign non-existent changes of vessel and lesion diameters.

Using „Declarative Domain-dependent and Independent Knowledge“ to Segment Cranial X-Ray CT Images

K.Natarajan, A.C.Hume, E.Claridge and J.A.Newell

School of Computer Science,
University of Birmingham, Edgbaston,
Birmingham, B15 2TT, U.K

When tackling the problem of grouping pixels and regions in Cranial X-Ray Computerised CT images it is possible to take either a procedural (algorithmic) or declarative approach. In general the procedural approach has been used to (i) describe the image in terms of meaningful image features (edges and regions[1])and (ii) to specify a sequence of image processing and analysis tasks to find specific anatomical objects. An example of the latter case being, to find ventricles,

-apply a Difference of Gaussians with specified scale space factors;
-connect edges to yield closed boundaries using parameter specific Euclidean geometry;
-choose a bounded region - calculate its feature vector (size,intensity, direction,adjacency);
-match feature vector with a pre-stored model.

This approach seems attractive for individual images or even comparable images sets but has limitations when dealing with the practical variability of cranial X-Ray CT scans. Alternatively it is possible to effectively use the declarative knowledge associated with brain anatomy, imaging modality[2]; and grouping concepts such as intensity attributes and n^{th} order spatial attributes; to suggest possible hypotheses and processing agendas for pixel and region grouping[3,4].

In the domain of Cranial X-Ray CT image segmentation, their is an abundance of information that can greatly constrain the task of grouping pixels and matching them to a set of modifiable models. Much of this information can be used directly at pixel grouping stages, such as the expected relative contrast between imaged neuro-anatomy; the spatial and directional relationships between neighbouring (or separated) structures; and relative sizes. These top-down constraints, can initially be used to suggest the anatomical structure that should be sought by determining which image attributes are the most discriminating for intra-structure and inter-structure grouping. Furthermore, through the use of meta-programming it is possible to dynamically adjust these constraints in the light of pixel merging, thus allowing essential feedback to be employed. With this approach, the choice and order of grouping, and matching precepts to be performed can be made dynamically via domain constraints and sensed data. The approach briefly discussed has been implemented using Quintus Prolog on a Sun SparcStation.

References:

1.Cawley M.G.;Natarajan K.;Newell J.A.: Data Representation for Subsequent Image Interpretation, in Medical Informatics. 16,No. 2, 1991.
2. Cawley M.G.; Natarajan K.;Newell J.A.: Automatic Generation of a Dynamic Anatomical Model for use in subsequent radiological image interpretation, CAR 91, Eds:Lemke,H.U, Rhodes,M.L., Jaffe, C.C. and Felix,R., Springer Verlag, pp 651-656, 1991.
3.Natarajan K.;Cawley M.G.; Newell J.A.: A KBS for the Automatic Interpretation of X-Ray CT scans, CAR 91, Eds:Lemke,H.U, Rhodes,M.L., Jaffe, C.C. and Felix,R., Springer Verlag, pp 618-623, 1991.
4.Natarajan K.; Cawley M.G.; Newell J.A. : A Knowledge Based System Paradigm for Automatic Interpretation of CT scans, in Medical Informatics, 16, No.2,1991.

Image Processing Requirements in PACS to Assist Diagnosis in a Multi-Modal Environment

N. SAEED

Picker International, Hirst Research Centre, East Lane, Wembley, Middlesex, HA9 7PP, U.K.

SUMMARY

The concept of an Automated Disease Diagnostic System (ADDS) in a multi–modal environment is becoming feasible with the advent of RIS, HIS, and PACS. It is anticipated that ADDS will be a neural network configuration which will make diagnosis based on rules adopted by expert radiologists, and will have the capability of interrogating RIS, HIS and PACS. However, for such a system to be feasible it is important to develop robust image segmentation, measurement, analysis and pattern recognition procedures that can be implemented on multi–modal scans. This agrees with the analogy of a radiologist examining scans and breaking the image into meaningful regions. The ideal location for the implementation of these techniques is in PACS as this system has fast access to a range of multi–modal images. The computing capabilities of PACS may need to be extended to cope with a multitude of time consuming tasks employed by the image segmentation, analysis and pattern recognition algorithms. In general, it is found that the image segmentation procedures developed for MR scans can quite easily be used on other modalities, however, the rules governing the initiation of these procedures may change. In the case of MR head imaging, image segmentation is based on a combination of edge detection, contour following, texture analysis, thresholding and region growing [1]. The choice of method that is finally employed to isolate a region from a scan is very much dependent on the MR sequence being employed [2]. Edge detection is used to highlight and segment the sulci. Contour following or thresholding is used to isolate the outer head boundary (OHB). Texture analysis and thresholding is used to segment the brain area. The lateral ventricles are isolated using thresholding or region growing procedures. Abnormal regions in a scan are identified by monitoring differences between the left and right hemispheres of the brain in the transverse or coronal plane. Knowledge of the MR sequence gives an indication of the likely T1 and T2 distribution for normal structures in the brain [3]. Here, structures that do not correspond to these known trends can be labelled as likely abnormal. Curvature measurements on the contour of isolated features can be used to identify useful anatomical markers. For example, the eyes, nose and ears can be identified by monitoring the contour of the outer head boundary.

[1] Saeed, N., **PhD Thesis**, University of Strathclyde, U.K., 1990.
[2] Saeed, N. and Young, I.R., **Comp. Asst. Radiol.**, pp 308, 1991.
[3] Bydder, G.M. and Steiner, R.E., **Neuroradiology**, vol23, pp231, 1982.

Fast Extraction of Line Properties by Tracking

Robert Kutka, Sebastian Stier

Siemens AG, Otto-Hahn-Ring 6, D-8000 Munich 83

Summary

A tracking algorithm is presented that extracts line features for use with a 3D reconstruction of blood vessels of the brain. The result is not only a segmented blood vessel tree but also a line intensity measure, and moreover directions, center lines and diameters are calculated.
The algorithm consists of four modules. The first module is a star-shaped prefilter whose result are directions and amplitudes of prospective line structures at each point of the input image. These features are input to a connectivity module that connects neighbouring pixels with similar properties by vectors. The next module tracks the lines along the vectors as long as particular continuity rules are fulfilled and calculates a line intensity measure. Based only on the directions and intensities the fourth module extracts structural information such as the line diameters and the center lines very fast. The only input parameter of the algorithm is the minimum contrast of the lines to be detected.
The line tracking will be compared with other algorithms. Its quality will be demonstrated on medical images.
This work is part of COVIRA (Computer Vision in Radiology), project A2003 of the AIM (Advanced Informatics in Medicine) programme of the European Commission.

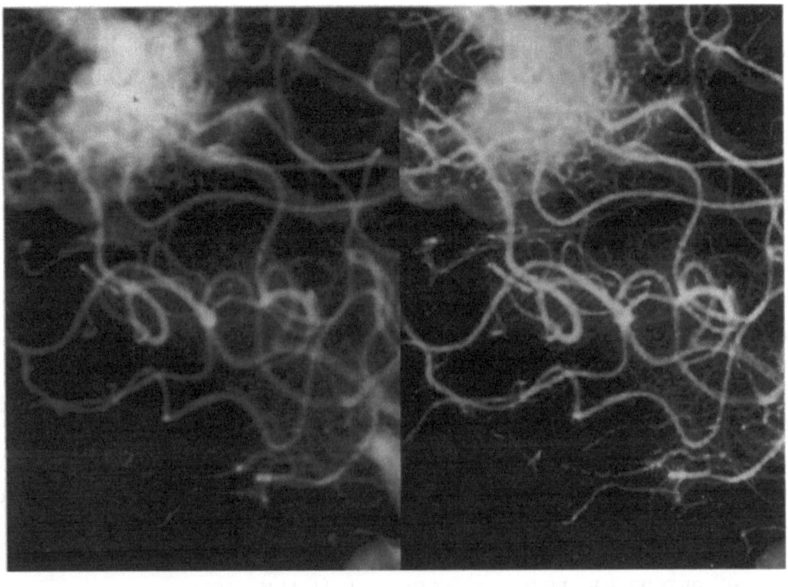

Fig.1: Comparison of an original DSA image (left) with the line intensity measure added to the original (right).

Two-Level Architecture for CAR Workstation

V. V. LASHIN, L. P. YAROSLAVSKY

Institute for Information Transmission Problems of Russian Academy of Science, Moscow, Russia

Two-level architecture of CAR workstation was suggested for computer-aided processing and analysis of medical images. It supports real-time processing and permits to design the verified technologies to improve decision making by radiologist. Low level consists of a set of basic algorithms for image processing, which are optimized by speed. This set is sufficient for decision of the most tasks concerned with image processing and analysis. High level supports friendly man-computer interface and decision making process. It uses technology database to process images automatically to improve medical diagnosis. This database consists of macrocommands, medical picture atlases and tutor for decision making. Suggested structure was realized on the base of IBM PC/AT computer. Software was tested on mammograms and lung tomograms and showed high results to improve preclinical cancer diagnosis.

A Software Platform for Fast Development of Medical Image Processing Modules

Lu J. Huang

Department of Radiological Sciences
School of Medicine
University of California, Los Angeles
Los Angeles, California 90024-1721
U.S.A.

The complexity of general-purpose graphics libraries, such as X Windows system, usually takes considerable amount of time for medical image processing scientists and programmers to master. Even after having obtained certain degree of proficiency in the use of these graphics libraries, medical scientists still often face certain difficulties, such as software modularity and extensibility in graphics programming environments; sharing of image data, which come from multiple image data sources and are shared among multiple processing modules. In this paper, we describe a solution to these problems, both in design and implmentation issues.

The fundamental approach we have taken is to design a programming environment which: (i). provides expressive and flexible image display capability so that the programmer can concentrate on algorithm and image processing module development; (ii). promotes image data sharing, which comes from multiple image data formats, shared among multiple image processing modules; (iii). facilitates multiple image data format file loading and header information extraction; (iv). reinforces software modularity in graphics programming environment; and optionally, (v). incorporates networking capability in a networked computing environment. Working under such a programming environment, programmers can produce codes more quickly and efficiently in a more modular fashion, thus increase productivity and code reusibility.

In our design of the graphics programming platform, there are several features which have been carefully designed and implemented. They are: (i) expressive and flexible display of images. To meet diverse image processing needs, the tools provided for image viewing and processing include multiple and single image display, with the multiple image display for viewing and the single image display for image processing. User can select an image by selecting from the multiple image display, which will display the image in a single window. Semantically, the selection means the image data displayed is pointed by a globally accesible pointer and is ready for manipulation. (ii). easy access to the image data buffer for image processing. Along with one image selection, all the loaded images are maintained in a list globally accessible so that the programmer can traverse through the list to maniputе all the images. (iii). convenient data import and export capability, which are important for many image processing applications. The processed data or a subset of the image can be saved into arrays or files for other use.

Such a programming environment has been seccussfully implemented using C language and Xlib/xview graphics libraries under Sun Solaris 1.x operating system running on Sun SPARC-series workstations. The implementation, named "Platform", is small yet powerful and robust, and has started gainning popularity among the medical scientists and programmers as well as radiologists in the Advanced Imaging Center and the Department of Radiological Sciences here at UCLA. The software modules built upon this platform include: PET-MR registration; Automatic MR Vascular Analysis; Interventional MRI, and for image viewing on UCLA PACS network computers.

An Interactive Visualization System for SPECT-Volumes

Clas Tegenfeldt[1], Jan–Erik Westlin[2], Björn Gudmundsson[1], Sten Nilsson[2]

[1] Image Processing Lab., Dept of EE, Linköping University, S–581 83 Linköping, Sweden, and
[2] Section of Nuclear Medicine, Dept of Oncology, Uppsala University Hospital, S–751 85 Uppsala, Sweden,

Abstract

A system for fast interactive visualization of volumes is presented. The system is implemented on a SUN Sparc work station under X–windows. Volumes can be viewed either by browsing through the stack of slices or by generating projections of the volumes using a number of different techniques. Surfaces of objects in the volume can be shaded by means of an illumination model (e.g. Phong–shading). Rotation sequences are generated by an algorithm that yields a speed–up factor of about 10 over traditional ray–casting methods. To further enhance the viewers perception of the third dimension, the algorithm also allows for fast generation of rotation sequences with shadows. In our implementation, rotation–sequences are simultaneously generated and displayed at a *sustained* rate of about 5 projections per second on a standard SUN Sparc IPC. The angle increment between successive projections is set by the user. Sequences can be saved for later replay at even higher animation rates (up to 25 frames per second). Other features of the system include variable thresholding and pseudo–colouring of intensity intervals interactively defined by the user.

The system is currently being used to visualize reconstructed SPECT–volumes from somatostatin receptor scintigraphy. We have earlier evaluated the value of 3D rendering (volume rendering plus transversal clipping) such volumes and we found that the high tumour to background ration in these images made anatomical orientation difficult without the aid of interactive volume rendering.

In an ongoing study the clinical value of the various features (e.g. fast rotation) of the interactive visualization system when applied to somatostatin receptor scintigraphy volumes is being evaluated.

Tissue Segmentation by Multiparameter Full-Color Composite Techniques: Clinical Application in Magnetic Resonance Imaging of Adnexal Masses

H. Keith Brown, Ph.D., Todd R. Hazelton, M.D., Anna K. Parsons, M.D., James V. Fiorica, M.D., Claudia G. Berman, M.D., and Martin L. Silbiger, M.D.

Departments of Anatomy, Radiology, and Obstetrics and Gynecology
University of South Florida, College of Medicine, Tampa, USA

SUMMARY:

Ovarian neoplasms represent a significant cause of cancer-related mortality in women. Since most women present with advanced stage disease, substantial diagnostic advances must take place to achieve improvement in survival. Magnetic Resonance (MR) imaging has been shown to exhibit greater sensitivity and specificity than computed tomography or ultrasound for the detection of certain adnexal masses, and has the potential to become an important tool in the diagnosis and characterization of ovarian neoplasms (1). A simple algorithm for tissue segmentation by multiparameter color composite generation (2,3) was applied to a variety of adnexal mass cases to determine the feasibility of this application. The pathologic diversity of adnexal masses provides a useful model for tissue segmentation and tumor characterization (4,5).

OBJECTIVE:

To apply color-enhancing image processing techniques to multiparameter (T1-weighted, proton density, and T2-weighted) magnetic resonance (MR) images of adnexal masses for data summation and improved tumor diagnosis and characterization.

METHODS:

For this project, a PC-compatible computer with full-color (24-bit) graphics and image processing software was utilized to apply specific color masks to multiparameter MR image sets of patients with suspected adnexal masses. In a process analogous to NASA's LANDSAT system for generating composites from multispectral remote sensing data (6,7,8), full-color composite images were generated through arithmetic addition of corresponding pixel sets within color-masked T1-, proton density, and T2-weighted MR images using the Red-Green-Blue (RGB) color model. To assess the potential clinical utility of this method, color composite images were correlated with surgical and pathologic data.

RESULTS:

With this simple algorithm, it was possible to generate full-color composites which simultaneously display uniquely color-coded anatomic and pathologic tissue information within the context of near-natural appearing summary images. Comparison of 95% confidence intervals and three-dimensional scatter plots of mean RGB tissue data demonstrate segmentation of most tissue types into quantitatively distinct areas within the full-color spectrum.

CONCLUSION:

This technique represents an inexpensive, viable segmentation algorithm for displaying multiparameter MR image data in MR imaging of adnexal masses, and may have broader applications within the field of MR.

REFERENCES:

1. Scoutt LM, McCarthy SM. Imaging of ovarian masses: Magnetic resonance imaging. Clin Obstet Gynecol (1991) 34:443-451.
2. Brown HK, Hazelton TR, Silbiger ML. Generation of color composites for enhanced tissue differentiation in magnetic resonance imaging of the brain. Am J Anat (1991) 192:23-34.
3. Brown HK, Hazelton TR, Putnam LM, Silbiger ML. Tomographic abdominal anatomy with magnetic resonance color composite. Clinical Anatomy (1992) 5:169-184.
4. Averette HE, Donato DM. Ovarian carcinoma: Advances in diagnosis, staging and treatment. Cancer (1990) 65:703-708.
5. Brown HK, Hazelton TR, Fiorica JV, Parsons AK, Clarke LP, Silbiger ML. Composite and classified color display in MR imaging of the female pelvis. Magn Reson Imaging (1992) 10:143-154. [Erratum, Magn Reson Imaging (1992) 10:495.]
6. Schowengerdt RA. Techniques for image processing and classification in remote sensing. Orlando, Fla: Academic Press; 1983.
7. Vannier MW, Butterfield RL, Rickman DL, Jordan DM, Murphy WA, Biondetti PR. Multispectral magnetic resonance image analysis. CRC Crit Rev Biomed Eng (1987) 15:117-144.
8. Alfano B, Arturo B, Ciarmiello A, Salvatore M. Simultaneous display of multiple MR parameters with "quantitative magnetic color imaging." J Comput Assist Tomogr (1992) 16:634-640.

Breast Screening: Defining the Implementation Criteria for Computer Assisted Diagnosis

A.G. GALE[1], A.R.M. WILSON[2] & E.J. ROEBUCK[2]
[1]Applied Vision Research Unit, University of Derby, UK
[2]Nottingham Breast Screening Training Centre, City Hospital, Nottingham, UK

Since 1986 national breast screening has been introduced throughout the UK. This involves all women in the age range 50-64 years and generates some 1.5 million mammograms annually. Currently specialised radiologists visually examine each case for the presence of cancer; the reported detected incidence of which is 6.2 cases per 1000 women examined. The potential for computer aided diagnosis lies in the difficulty of the task for human observers to detect this disease at an early stage of presentation from amongst the very large number mammograms of normal women examined. Other existing research programmes are currently investigating machine detection of particular mammographic features, such as microcalcifications, and the performance of these approaches typically is compared to that of expert radiologists. We present data from the first two years of a national programme for screening radiologists in the UK. One outcome of which is the demonstration of the typical performance characteristics of radiologists so illustrating where computer aided diagnostic systems can contribute most as an aid to radiologists in breast screening.

The Royal College of Radiologists, in co-operation with the NHS breast screening programme, have introduced a self-assessment programme for all screening radiologists who are annually asked to interpret a set of mammographic cases of known outcome. For each case they have to report whether key mammographic features are present and also to classify the diagnostic category of the case. Data are analyzed and results fed back to participating individuals so as to help them improve their own future performance.
This information is important for computer aided approaches as the results demonstrate; the wide range of inter-radiologist variability, which key mammographic features are difficult for radiologists (as a group or as individuals) to either detect or appropriately interpret, and particular mammographic cases are identified which are generally prone to misinterpretation. These data provide the background against which machine vision approaches can be implemented in this speciality and it is argued that for effective utilisation plasticity is required so as to complement the skills of individual radiologists.

Is Experienced „Eye Balling“ of Coronary Angiograms Obsolete in the Transition Period to Digital Quantitative Coronary Angiography?

Georg M. Stiel, Ludmilla S.G. Stiel and Christoph A. Nienaber

University Hospital Eppendorf, Department of Internal Medicine, Division of Cardiology, Hamburg, Germany

Computer-assisted image analysis systems for quantitative coronary angiography [QCA] require added investments in terms of both time for the analyses and money for additional equipment. Their utilization for routine work appears justified only by a clear superiority to usual visual-manual ("eye balling") quantitative coronary analysis [VQA].

To assess the accuracy of VQA and QCA for the assessment of coronary artery disease, coronary arteries of 24 excised human hearts were filled with a radiopaque acrylate after postmortem angiography and prepared by cutting-grinding technique to preserve coronary morphology. VQA of cine angiograms was performed by 3 experienced angiographers. The area stenosis was assessed by the same formula for 30 lesions by VQA and QCA and compared with the histological reference [HQA].

Results: Excellent linear correlation was found for both methods with $r = 0.96$, $p < 0.001$. SEE was 6.23% for VQA and 5.89% for QCA when compared to HQA. The mean values of inter- and intraobserver variability for % area stenosis estimate were 6±1% and 14±2% resp. using VQA and 5±1% and 10±2% resp. using QCA.

Conclusions: 1) QCA has no significant advantage over VQA for routine assessment of stenosis severity, but appears to reveal higher reproducibility. 2) The quality of quantitative coronary analysis is dependent mainly on the experience of the angiographer and the quality of the index angiogram. 3) At the moment VQA by experienced observers remains the method of first choice for routine clinical angiographic analysis. 4) The adventage of QCA cannot be utilized at its best as long as digital quantitative coronary angiography and ventriculography is not established for routine diagnostic and therapeutic work.

AWOS: Angiographic Workstation for Digital Quantitative Coronary Angiography

Georg M. Stiel, Karl Barth, Uwe Towara,* Birgit Eicker*, Christa Vogel**
and Christoph A. Nienaber
University Hospital Eppendorf, Department of Internal Medicine, Division of Cardiology, Hamburg and *Siemens AG, Medical Engineering Division, Erlangen, Germany

The angiographic workstation (AWOS) is an extended analysis system for digital quantitative coronary angiography. It is based on the SIEMENS SMI 5-image-processor architecture with a fast MULTIMUX-bus. A 32 Bit SUN SPARC II workstation with a 760 MBytes winchester system disk, a 3 GBytes fast image disk and a 128 MBytes scene memory are connected via ETHERNET to POLYTRON and via streamer tape to HICOR imaging system.

AWOS offers the possibility to display and to process digital angiograms from digital imaging systems, and 35 mm cine images digitized by an ARRIPRO-35 projektor with a TV-camera. All standard calibration and evaluation methods for quantitative coronary angiography and ventriculography are implemented.

AWOS is operated by a (menu-in) windows technique. Analysis results are displayed on the workstation monitor; a standardized VHS/S-VHS (CCIR-Standard) and hardcopy printouts are also available. At present the digital archive is based on 500 MB streamer tape.

Digital angiograms (HICOR) and the corresponding digitized cine angiograms (CINE) of 45 contrast filled phantoms for vessels of 0.1 to 5.0 mm were quantitatively evaluated by AWOS; the results were correlated with the true diameters of the phantom stenoses.

Results:	HICOR		CINE	
Zoom	no	2x	no	2x
Diameter (mm)	0.8 - 5.0	0.5 - 5.0	0.8 - 5.0	0.5 - 5.0
r	0.997*	0.998*	0.998*	0.998*
SEE	0.075	0.068	0.075	0.073

* $p < 0.0001$

With reference diameters < 0.8 mm , or < 0.5 mm at a 2 fold Zoom, no correlation with the angiographically measured diameter was found due to systematic overestimation of the true diameter by the imaging system.

Conclusion: AWOS is a workstation with on-line digital archiving of images. In concert with high definition digital angiocardiography equipment it represents a configuration of a digital angiographic system to realize the fully digital interventional catheterization laboratory today with all options for future developments.

A Framework Concept for Medical Work Station Design

R. BRILL and M. STAEMMLER

Fraunhofer Institut für Biomedizinische Technik - IBMT
Ensheimerstraße 48
D6670-St. Ingbert (Germany)

Introduction
Current medical practise deals with a variety of multi-modal information (MR, ultrasound, CT and X-Ray images, EEG and ECG signals, patient records etc.). Handling this data involves separate software tools and thus it is in no way consistent or easy to learn. Some solutions for handling medical image data are emerging on today's market. However, a consistent integration of non image data into these environments is difficult to achive.

Methods
The SUNRISE (St. Ingbert Ultrasound, Nuclear Magnetic Resonance and X-Ray Intelligent Software Environment) framework concept remedies these deficiencies in a twofold approach for users and programmers respectively:

- It provides a consistent Graphical User Interface for all modalities featuring (i) a common look and feel for all applications, (ii) real world coordinates (e.g. "m", "h:m:s", "Hz", ...) instead of pixel or voxel oriented presentations (iii) keeping track of user action and data processing history, (iv) user specific access control, and (v) user specific system configuration during runtime.
- Application development is supported by (i) a unified data representation based on real world units for n-dimensional data sets without any restrictions in dimensionality, (ii) inherent data storage and retrieval functions, (iii) a modular approach for building dedicated processing environments, (iv) full debugging support including error tracing, and heap debugging, (v) and a minimum code overhead (typically about 15 lines of code for an image filter skeleton).

Reqirements
SUNRISE requires a UNIX® workstation running X-Windows, and OSF/Motif or OPENLOOK for the graphical user interface. It operates stand-alone or within a heterogenious network based on the NFS standard allowing distributed data storage.

Results
The SUNRISE framework concept has been used successfully to build an evaluation environment for processing of MR-^{31}P-spectra. Currently it is beeing used as the integration platform for a neurological workstation beeing developed in the EC-AIM-project SAMMIE.

Developing an Improved Radiology Workstation User Interface: A Photographic Survey of Current Approaches

Vikas Bhushan, M.D., Henry H.-L. Shih, M.D., and Franklin N. Tessler, M.D., C.M.

Department of Radiological Sciences, CHS B2-220
University of California Los Angeles School of Medicine
10833 Le Conte Avenue
Los Angeles, CA 90024-1721 USA
Internet: p13721@mccs.ucla.edu, voice +1-310-825-7258, fax +1-310-206-5725

Most PACS/IMACS development effort has been directed to hardware-related issues such as networks, databases and storage. Unfortunately, little attention has been focused on user interface design for radiology workstations, which in part has limited system acceptance by radiologists. The next generation PACS/IMACS will need to incorporate a refined, high-performance user interface optimized for efficient primary diagnosis, with enhancements for quantitative analysis and simplifications for remote clinical viewing.

We undertook a limited photographic survey of academic and commercially-developed user interfaces for radiology workstations, examining input devices, workstation software and CRT display size and layout.

Input devices ranged from conventional keyboards, cursor keys, mice, and trackballs to more exotic vertical dial panels (2-8 dials), membrane touch keys (surrounding monitors), custom key pads and combination control panels. Display layout ranged from one to eight monitors, generally in stacked rows, with varying luminance, and with resolutions of 640 x 480 pixels to 2480 x 2000 pixels each. Workstation software was very heterogeneous, but shared certain features common to graphical user interfaces, including windows (tiled or overlapping), pull-down menus and/or image manipulation icons, scrollable list boxes, check boxes, sliders and pictorial (thumbnail) indexes. Basic functions included selecting patients, studies and images, choosing display layout, and adjusting contrast. Additional features included zoom, pan, color, cine, quantitation tools, multiplanar reconstruction, text and voice annotation, and varying levels of RIS/HIS integration. Each workstation used a unique, locally developed variation of the patient list/film jacket/light box metaphor, influenced by operating system, software toolkits, available display space and developers perception of needs.

As hardware-related problems are solved, user interface design becomes an increasingly important and limiting issue in PACS/IMACS success. Our *ad hoc* photographic survey shows widely disparate attempts at solving this problem, none of which individually appears adequate. Systematic refinement of interface design may be facilitated by modular, standards based development, sharing of experience, and research into novel input devices, navigation techniques, eye/hand movement optimization and comparative user interface performance in realistic clinical environments.

Physician's Work Stations under Multi-Platformed Computer Environment

Ken'ichiro Kajiwara, M.D.

Department of Medical Informatics Kurume University Hospital Kurume 830, Japan

The physicians's Work Station environment handling all kind of database including patient's history records, medical images and dictated voices was designed under UNIX/Ethernet platform. This platform is perfectly constructed with no machine dependent, OS dependent but file-format dependent system. The descriptive text data were handled with SQL language and images by JPEG/JFIF format. This system can accept all kind of computers under UNIX, X, Macintosh, MS-DOS and Windows. They can handle all data as if their originals which stored in UNIX server with their own GUI environments. This environment is achieved using NFS protocol under TCP/IP. Such system must become the basic component of hospital information system.

Virtual Reality in Medical Environments

MANFRED KRAUSS

Technische Universität Berlin
Institut für Technische Informatik
Sekr. FR 3-3, Franklinstr. 28/29
D 10587 Berlin
email: manfredo@cs.tu-berlin.de

Modern imaging modalities of Computer Assisted Radiology like CT or MRI scanners deliver large datasets containing consecutive image slices. These datasets, representing the anatomy of the patient, are very useful for the physicians in finding the diagnosis. The three dimensional reconstruction of these datasets takes place in the minds of the physicians by examination of slice by slice. To deliver this amount of such complex spatial information in a "natural way" which supports human perception, advanced interaction and visualisation techniques should be used.

A first approach within the C O M E D (Co-operative Medicine) project is M e d + V R (Medicine and Virtual Reality) a medical application using "virtual reality techniques". Med+VR offers the possibility of visualising precalculated stereoscopic images using a three dimensional output device. The precalculated images could be, for instance, a sequence of stereoscopic MRI images which are created out of consecutive MRI slices during a 3D reconstruction. The benefit of a 3D reconstruction is that the spatial information of the whole dataset is included in the sequence of the generated images. Upon creation of the precalculated set of images, it is for example, possible to change the viewpoint of the observer or to drive a cutting plane through the dataset. After calculation of the stereoscopic image sequence, the user can visualise the image sequence using the Med+VR application.

Med+VR offers the possibility of navigating interactively through the set of precalculated stereoscopic images using a three dimensional input device. The navigation through the stereoscopic images is like viewing an interactive film. Furthermore the user has the possibility to modify interactively the grey values of the shown images, which is known as "interactive windowing".

At present "virtual reality techniques" look promising for use in 3D medical applications like radiation therapy treatment planning, hyperthermia planning, surgical simulations and for teaching, but the realisation is still under development.

Computed Radiography: First Experiences in Intravenous Urograms

C.HUNDT, U.FINK, P.KOHZ, M.SCHÄTZL

Department of Radiology / Klinikum Großhadern
University of Munich / Germany

Summary:

Computed radiography offers new opportunities to optimize the x-ray quality in radiodiagnostic techniques.

This prospective study is a quality comparison of conventional screen-film and computed radiography in intravenous urograms. From September to December 1992 one hundred and twenty intravenous urograms were performed at the Department of Radiology in the Klinikum Großhadern (University, Munich). The x-rays were made in computed (5-minute-compression) and conventional (10-minute-compression) technique.

We used a FCR AC-I (Fuji) image processing system with HeNe-laserlight highly sensitive imaging plates.

Three radiologists assessed different anatomic structures (renal parenchym, renal pelvis, psoas shadow and spine) and found equal results in both techniques. The diagnostic sufficiency was evaluated by receiver operating characteristics (ROC) analysis. Results indicated no difference between digital and conventional screen-film urograms.

It has been shown that computed radiography can be used as a radiographic method in intravenous urograms. The highly sensitive imaging plates and computer assisted image processing allow a continuous film quality with automatic contrast adjustment. This leads to a reduced number of repeated films.

In the future the possibility of computed post-processing of stored information offers better and more differentiated diagnostic statements. The imaging plate system can be integrated into the PACS (Picture Archiving and Communication System).

Computed Radiography and Image Management Systems: Solutions for the Intensive Care Unit

Ralph Schaetzing, Ph.D.

Eastman Kodak Company, Health Sciences Division,
Rochester, New York, 14650

Diagnostic imaging using conventional screen/film systems in the Intensive Care Unit (ICU) poses clinical, technical, and operational challenges for radiographers, ICU physicians, and radiologists. For example, the difficulty of controlling exposure at the bedside can lead to image consistency problems (and, thus, to repeat exposures). Radiologists and ICU physicians wishing to consult on a patient's images must meet where the (original) film images are. Carrying cassettes between the ICU and the film processor can be burdensome and time-consuming. In addition, the identification of patient with cassette can be prone to errors.

Storage Phosphor Radiography (SPR), with its wide x-ray exposure latitude, is an ideal image acquisition system when exposures are highly variable and/or difficult to control. By using small, distributed SPR readers (near the ICU) in an 'open' network of workstations and laser printers, along with a unique electronic patient identification system, it is possible to address these ICU problems effectively. The separation of image capture from image display provided by such a system allows the images to be manipulated electronically (image processed for consistent, optimized diagnostic viewing), transmitted to multiple nearby or remote locations (minimal cassette transport, easier consultation), and stored. A number of such novel systems have already been installed around the world and are proving their usefulness in the ICU.

Applications areas other than the ICU can also be addressed. Through the use of a modular, 'open' architecture, and adherence to industry standards, image management systems can be configured to meet customers' varied needs. These can range from fairly simple networked systems consisting of one image input device with remote printing capability, to more complex systems that involve multiple acquisition modalities, diagnosis from high-resolution workstations, distributed printing, and optical disk archiving for long-term storage and retrieval.

The (ICU) application-specific combination of storage phosphor radiography and a networked image management system has demonstrated that improvements in image management can translate into improvements in patient care.

Storage Phosphor Digital Radiography Applied to the Study of Soft Tissue Masses

M. MICELI, M. LIPPARINI, G. SAGUATTI, S. SARTONI GALLONI

Department of Radiology I
Maggiore Hospital, Bologna, Italy

Summary

102 patients with soft tissue mass underwent both storage phosphor digital radiography (DR) [1,2] and ultrasound (US) examination, after a conventional radiography (CR) examination generally performed in other Radiology Departments. DR system is connected to a high resolution (1K) workstation for image viewing and processing (gamma correction and spatial filtering). CT and/or MR examination was performed for a more accurate assesment of the lesions. 70 patients underwent biopsy. DR was shown to be superior than CR in the assesment of the different pathologies observed (lipomas, cysts, angiomas, sarcomas etc). DR resulted more informative than CR for the depiction on the same film of an analogic-like and a xerographic-like image, and for the postprocessing possibilities on the workstation. The tumors were visualized directly either for different density or as "space occupying mass". The easy detection of the fascias permitted to evaluate their eventual dislocation or disappearance. The depiction of both cutaneos-subcutaneos and skeletal lesions at the same time resulted very useful. US-DR integration gave data sometime conclusive for the diagnostic investigation of some benign masses (lipomas, cysts etc).

References

1. Tateno Y, Iinuma T, Takano M: Computed Radiography. Tokyo: Springer-Verlag 1987.

2. Wegrin SA, Piraino DW, Richmond BJ, Schluchter MD, Muetani M, Freed HA, Meziane MA, Belhobek GA: Comparison of digital and conventional radiography: an observer performance study. Radiology 175 (1990) 225-228.

Digital Magnification Mammography

Michael T. Nelson, M.D., Eugene Elvecrog, M.D., Mary Lechner, M.D.
Park Nicollet Medical Center
Department of Radiology
5000 W. 39th Street
Minneapolis, MN 55416

Summary

A CCD camera consisting of a phosphor screen, fiber optic reducer and charge coupled device. With an imaging area of 24.6 mm x 24.6 mm with 1024 x 1024 pixels was used to examine breast masses less than 2 cm found on film screen mammography.

Abstract

Results comparing film screen magnification and digital laser images will be compared on phantom (ACR) images. Patient breast mass comparison will also be analyzed with film screen and direct digital imaging. The new Mammoscan (by Fisher Imaging, Denver, Colorado, USA) digital images not only help define breast masses, but help decrease the time required for a stereotactic breast biospy. The Mammoscan imager and Mammoscan imaging workstation will be discussed including the host 486/33 mHz computer and 24" 1280 x 1024 pixel video monitor.

Conclusion

This new digital imaging modality (Mammoscan ™) should improve breast diagnosis as well as decrease the biopsy time for stereotactic breast biopsy (Mammotest ™).

Achieving Adequate Resolution for Storage Phosphor Plate Mammography Using Existing Equipment

Matthew Freedman, M.D., M.B.A., S. K. Mun, Ph.D.,
Einar Pe, R.T.R., S.-C. Benedict Lo, Ph.D., Martha Nelson, M.D.
Department of Radiology, Georgetown University, Washington, D.C., U.S.A.

ABSTRACT:

The application of storage phosphor radiographic (SR) imagers to breast imaging has been limited by a perceived lack of adequate resolution of these systems when compared to conventional screen-film (SF) mammography. The actual resolution needed to be achieved by SR mammographic systems has remained undefined. Methods used here of direct geometric magnification of the initial exposure combined with image processing to change the degree of film blackness, contrast scale and edge enhancement are shown to provide greater visibility of all structures present in the breast imaging phantom used for American College of Radiology accreditation when compared to conventional screen-film mammography. At a simulated 5.5 x 6.5 lp/mm resolution, it is possible to obtain two images from a single exposure that together result in a perfect score on the accreditation phantom using the same exposure factors as for a conventional non-magnified grid mammographic view.

Our current film screen system typically demonstrates 4 or 5 of the fibers, 4 masses and 3 1/2 groups of calcifications. The SR images demonstrate all 6 fibers, all 5 masses, and all 5 groups of simulated microcalcifications.

The images in figures 1 a and b and 2 demonstrate digital images of the breast phantom obtained on two different commercial Storage Phosphor imaging plate systems (Fuji AC-1 (Miyanodai) and AGFA ADC (Antwerp)). The two images in Figure 1 (Fuji AC-1) resulted from a single exposure, but with two different image processing settings. On these two images, all objects in the phantom can be detected. Figure 2, obtained on the AGFA ADC SR imager demonstrates 5 fibers, 4 masses and 3 1/2 groups of microcalcifications; fewer objects than with the Fuji system, but with fewer artifacts.

It is possible that the image processing used to render visible all of the objects present on the ACR approved phantom might also result in images of the breast that are so over processed that they are unreadable. The breast biopsy radiographs in figure 3 suggest that the processing used might not interfere with interpretation of images of the whole breast. In figure 4, a small fibroadenoma with microcalcification is seen better with SR magnification mammography (a) than with magnification SF mammography (b).

These results show that using direct geometric magnification to simulate a plate with a resolution of theoretical 6.7 lp/mm (actual 5.5 x 6.5 lp/mm) is sufficient to demonstrate all features of the ACR accreditation phantom. Because this exceeds the detection rate possible with existing SF systems, existing 5 lp/mm storage phosphor systems should need only slight improvements in resolution to surpass conventional SF mammography.

Do Differences in Brightness Influence the Quality of Digitized Film Images?

A.Gahleitner[1],J.Kettenbach[1] ,E.Smutny[2] , A.Pinz[2], Ch.Herold[1]
1) Dept.of Radiology and Ludwig-Boltzmann-Institut, University Hospital,Vienna (Univ.Prof.Dr.H.Pokieser)
2) Dept.of Pattern Recognition and Image-Processing,Inst. f.Automation, Technical University of Vienna (Univ.Prof.Kropatsch)

Purpose: The digitization of radiologic images is increasingly used in documentation, modern radiologic teaching and education. The brightness of light-boxes, the degree of exposure of radiologic films, the surrounding light and different apertures of a CCD-Camera could significantly influence the gray scale-distribution and consequently the quality of digitized radiologic images. The purpose of this study was to assess the influence of these factors on image quality.

Materials and Methods: A frame-adjustable light-box, a single-chip b/w CCD-Camera with adjustable aperture,a personal computer with an additional Frame-Grabber card, image processing software and a densitometer were used. Radiologic films were digitized at different brightness-levels of the light-box,different surrounding light and alternated apertures of the CCD-camera. The gray-scale distribution of digitized images was compared to radiologic films with established degrees of exposure.

Results: Insufficient framing of the light-box, high surrounding light intensity and a large aperture of the CCD-camera,each lead individually to a reduced gray-scale range and consequently to decreased contrast-resolution.

Conclusion: The results establish guidelines to obtain high-quality digitized images.

Segmentation of MR Brain Images for Multimodality Fusion and Gray/White Matter Volumetry

Y. HUANG, U. KNORR, G. SCHLAUG, R.J. SEITZ, H. STEINMETZ

Department of Neurology , Heinrich-Heine-University
Moorenstr. 5
D-4000 Duesseldorf 1

Segmentation of MR images of the brain into white and gray matter is a prerequisite for brain morphometry and PET/MR image fusion [1, 2]. In combining 3-D oblique cut, active contour, edge detection and region growing, we have developed a semi-automatic segmentation method for 3D FLASH MR datasets with minimal user interaction. This approach consists of 4 steps:

1.) The user defines an active contour within a selected region of interest (ROI). The external energy function is defined by an expanding force proportional to the reciprocal value of the Euclidean distance from the center of the ROI. This function takes on an infinite value in pixel positions with sobel gradients above 95% of the maximal gradient (hard constraint).

2.) Brain tissue is separated by identifying and stripping away pixels representing scalp, skull and meninges, employing sobel edge detection and region growing. Two constraints are applied: Gray value must lie within a predefined range, and sobel gradient must be smaller than a limit. Steps 1 and 2 can be applied to any oblique cut from 3-D volume data. After these operations have been performed, segmented slices are written back to the 3-D volume data.

3.) An anisotropic edge preserving diffusion filter is applied to reduce noise.

4.) A histogram of all brain pixels is used to determine thresholds for segmentation of cerebro-spinal fluid (CSF), gray and white matter. Local maxima are identified and Gaussian distributions are fitted to the histogram. Each pixel is identified as either CSF, gray or white matter using a minimum probability of error decision rule. The described technique provides the basis for accurate surface matching of high resolution MR and PET scans. In addition, it allows the selective volumetry of whole brain, gray and white matter, necessary for the morphometric quantification of atrophic brain disease.

References:

(1) Huang Y., Steinmetz H., Knorr U. and Herzog H.: "3D Matching of Structural (CT, MR) and Functional (PET) Brain Image Data". Proceeding of the International Symposium Computer Assisted Radiology CAR '91 (229-233), Springer Verlag, Berlin, 1991.

(2) Knorr U, Huang Y, Schlaug G, Seitz R J, Steinmetz H: "High resolution PET images through Redistribution", Proceeding of the International Symposium Computer Assisted Radiology CAR '93 (in press), Springer Verlag, Berlin, 1993.

Algorithms for Contrast Enhanced Functional Cerebral MR Imaging

W. Wlodarczyk, J. C. Böck, B. Sander, R. Felix

Strahlenklinik und Poliklinik, Universitätsklinikum Rudolf Virchov

Augustenburger Platz 1, W-1000 Berlin 65, Germany

Maps of regional cerebral blood flow (rCBF) and volume (rCBV) yield complementary information about cerebral pathology. MR mapping of rCBF and rCBV is based on tracking the first cerebral passage of the paramagnetic contrast agent. Fast susceptibility (T_2^*) weighted gradient echo sequences ($TR/TE/\alpha = 25ms/21ms/10°$) were used in 1.5 T Siemens Magnetom SP. Time resolution of 0.8 s/image was achieved by applying keyhole technique. Precontrast base rawdata with larger matrix (64×128, overs.) was combined with smaller matrix (32×64, overs.) rawdata of the dynamic scan tracking an intravenously injected bolus of gadolinium-DTPA (0.1 mmol/kg b.w.). Time-dependent curves of susceptibility induced signal intensity changes were converted into the time-dependent concentration curves on a voxel-by-voxel basis. Zero value level of each curve was referenced to corresponding mean value of noise before bolus apearance. The recirculation part of curves was cutoff. After some analytical preconsiderations for choice of appropriate start parameters, model function was fitted to so prepared concentration curves using Levenberg-Marquardt algorithm for nonlinear modelling. Two model functions known from the indicator dilution theory were used and compared. Gamma-variate function as a model has three parameters but unfortunately can not be delayed. This required a time-consuming adaptive temporal shift of concentration curves in order to automatically fit the fixed model function. The lagged normal density function is a convolution of a normal density function with an exponential function and has four parameters. One of them represents the horizontal shift, thus exposing the delay of curve to the fitting process. After fitting the model function to measured data, tracer kinetic analysis was used. Area under the fitted curve was interpreted as rCBV. Mean transit time (MTT) was obtained by integrating time-weighted fitted curve and dividing it by rCBV. Finally, rCBF was calculated as a ratio of rCBV and MTT. Generating maps of rCBV and rCBF functional information of brain could be extracted.

Clinical Application of Echoplanar (EPI) MR Imaging

P. Reimer, F. Schmitt, R. Ladebeck, R. Grote, EJ. Rummeny, T. Lange, J. Hüsemann

Institute for Clinical Radiology, Westfälische Wilhelms-Universität, Münster and Siemens AG, Erlangen, FRG.

Purpose

The purpose of our study was to develop a strategy for optimized EPI in patients with abdominal disease to assess the clinical applicability of echoplanar MR imaging (EPI).

Materials and Methods

The study consisted of three parts: 1. In-vitro evaluation of potential bowel contrast agents for EPI, 2. In-vivo investigation of a bowel contrast agent in volunteers, and 3. Application of EPI with a bowel contrast agent for reduction of susceptibility artifacts and bowel marking in patients with abdominal disease. MR imaging was performed utilizing conventional and echoplanar techniques (1.0 Tesla , Siemens AG). Fat-suppressed EPI (Siemens AG, Erlangen) was performed utilizing a single-shot spin-echo technique. K-space was covered by a sinusoidal "resonant" read gradient together with a constant phase gradient.

Results

Phantom studies of the bowel contrast agents investigated showed high signal intensity on T2-weighted echoplanar images while signal intensity was intermediate on conventional T2-weighted MR images such as Flash or Turbo-Flash. In-vivo studies demonstrated that susceptibility artifacts from air-tissue interfaces (pre: 59% vs. post: 13% with significant susceptibility artifacts) were reduced on post-contrast images. Differentiation of abdominal organs from adjoining bowel was improved on post-contrast images (pre: 44% vs. post: 80% with sufficient bowel enhancement).

Conclusion

The oral administration of aqueous bowel contrast agents for bowel lumen marking at EPI is advantageous because susceptibility artifacts from air-tissue interfaces are reduced and bowel lumen signal is enhanced permitting differentiation of abdominal organs from bowel. EP-MR imaging offers the whole pulse sequence spectrum of conventional MR imaging with eliminated motion artifacts in just 20 - 30 % of conventional imaging time.

Accurate Alignment and Reslicing of PET Images

Y. HUANG, U. KNORR, H. STEINMETZ, R.J. SEITZ

Department of Neurology , Heinrich-Heine-University
Moorenstr. 5
D-4000 Duesseldorf 1

Accurate image alignment is a prerequisite for pixel-by-pixel comparison of PET data within subjects. Most current approaches utilize combinations of head immobilization techniques or external markers. However, their accuracy is limited by increased examination times and instability of markers.

We have further elaborated a retrospective technique of spatial intermodal image alignment using the MR-derived brain surface to determine transformation parameters for alignment and reslicing of PET images.

First, the MR-derived scalp surface is fitted to the PET transmission scan of the same subject. This is achieved by using a least square surface fitting algorithm which was originally designed for individual PET/MR integration [1].

The fitted MR images are then segmented, as described in an accompanying paper [2]. Using the same fitting algorithm, PET images of regional cerebral blood flow (rCBF) of the same individual are matched on the segmented MR images of his/her brain.

Finally, the PET images are resliced using the best fit transformation parameters. We employed a simulated PET camera to assess errors in alignment and quantitation. Computer simulations showed that alignment errors are usually less than the width of an MR image voxel (1.0 mm). Global errors in quantitation due to reslicing are smaller than 2%.

Our method cannot only align and reslice PET images, thereby allowing to calculate accurate pixel-by-pixel subtractions of consecutive PET scans of the same individual in activation studies. It also solves a more difficult problem, that is, to determine a proper attenuation correction by reslicing the rCBF PET images according to the spatial orientation of the PET transmission scan.

References :

(1) Huang Y., Steinmetz H., Knorr U. and Herzog H.: "3D Matching of Structural (CT, MR) and Functional (PET) Brain Image Data", Proceeding of the International Symposium Computer Assisted Radiology CAR '91 (229-233), Springer Verlag, Berlin, 1991.

(2) Huang Y., Knorr U., Seitz R.J., Steinmetz H.: "Segmentation of MR Brain Images for Multimodality Fusion and Gray/White Matter Volumetry", Proceeding of the International Symposium Computer Assisted Radiology CAR '93 (in press), Springer Verlag, Berlin, 1993.

Hardware of the System of Archiving and Storing Ultrasound Images of Thyroid Gland in Children Irradiated after Chernobyl Accident.

Victor K. Ivanov, Peter C. McBride, Alexan K. Sahakyan, Anatoly F. Tsyb

Medical Radiological Research Center RAMS of Russia

There is no doubt that the problem of examination of children exposed to radiation as a result of the Chernobyl accident for the thyroid disorders has the highest priority within the programmes of the accident aftermath abolition.

The main reason for that is the significant radioactive iodine release at the time of accident and consequent exposure of the thyroid, as iodine accumulates in this organ.

Therefore, the problem of flexible management of evaluation of thyroid state in children inhabiting contaminated areas rises.

Among a variety of non-invasive methods of the thyroid examination ultrasonic one is most acceptable. Paper [1] presents in details the system of collecting and archiving the ultrasonic images.

One of the major problems of the thyroid state evaluation is three-dimension representation of the organ as well as the possibility of comparing the cross-sections obtained for the same patient in a number of successive examinations with the main goal to search for changings in time.

This requires a number of technical problems to be solved, among them: detection of the ultrasonic scanner handler position with respect to the reference point, degree of the organ compression.

At present a number of devices, for example ne manufactured by Aloka company, are available at the market, which permit to register the position of the scanner handler, however, these devices are not convenient for usage. Authors of the present paper have developed two devices permitting to detect the spatial position of the ultrasonic scanner handler which could be discussed.

Fig. 1 presents the scheme of device permitting to detect spatial coordinates of the ultrasonic scanner handler, which must be fixed on the head of the patient. Fig. 2 presents the alternative version of the device, the major difference is: the latter must be fixed on the bed of the patient. In both cases physician should set the reference point and then examine the patient for disorders and volume of the organ. Demerit of the device presented in Fig. 1 is the necessity of fixing the device on the head of the patient before examination. However, this device is more precise when compared with that presented in Fig. 2, because it seems impossible to position the head of the patient the same way in a number of successive examinations and slight bias will certainly occur. Therefore, for the large scale studies involving big cohorts to be examined (e.g. in district hospitals) device presented in Fig. 2 is of preference. For the regional level, one presented in Fig. 1 is preferable.

At present the possibility of producing the experimental systems for detecting the spatial coordinates of the ultrasonic scanner is being considered.

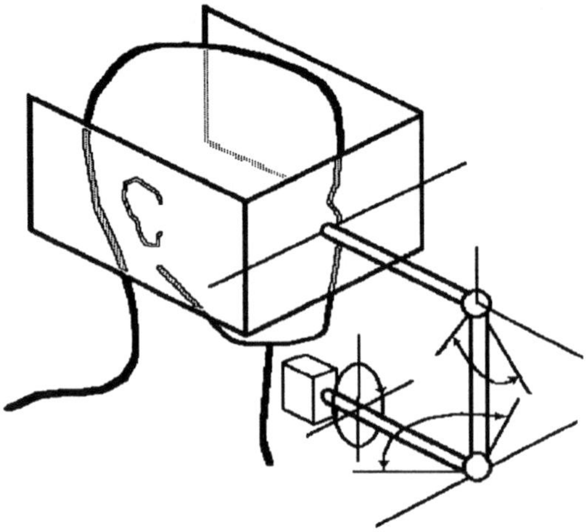

Fig.1 Principal scheme of the device for detection of the spatial coordinates of the ultrasonic scanner, which is fixed on the head of the patient.

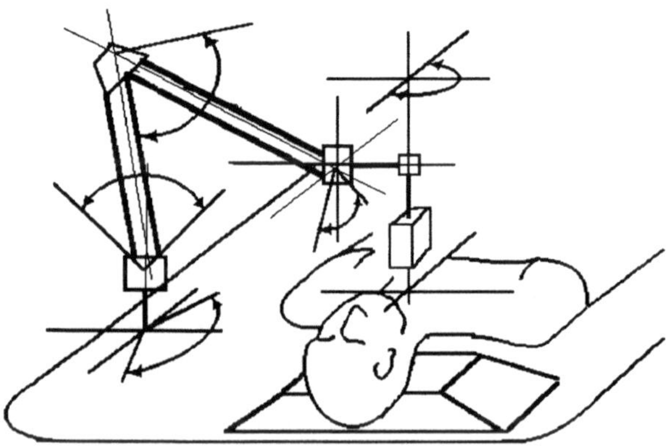

Fig.2 Principal scheme of the device for detection of the spatial coordinates of the ultrasonic scanner, which is fixed on the bed of the patient.

ACKNOWLEDGMENT

The authors acknowledge the financial support of QUEST Automation Ltd. company.

REFERENCES

1.V.K.Ivanov,A.K.Sahakyan,S.A.Ajrapetov,A.F.Tsyb//
Processing and Archiving of Ultrasound Images of Thyroid in Children Irradiated as a Result of the Chernobyl Accident, S/CAR'92, pp.272-277.

Automatic Detection of Ventricle Contours in Angiograms

J. BEIER, B. NIGBUR, S. LEMPERT, H. OSWALD, E. FLECK

Deutsches Herzzentrum Berlin, Abt. Digitale Bildverarbeitung,
Augustenburger Platz 1, W-1000 Berlin 65, Germany

The analysis of heart performance is based on the calculation of left and right ventricle volume, wall motion and derived functional values. These parameters are determined using the ventricle contours on non-substracted coronary angiograms. For reproducible measurements, the contours have to be entered as accurately as possible. A new algorithm for automatic contour detection provides an exact, fast and comfortable method for this task.

The inferior quality of ventricle angiograms (poor contrast, gaps in dye concentration) hampers fully automated detection algorithms. After entering three start points (anterior/inferior valve and apex) a circular ROI is calculated enclosing these three points. In this ROI a threshold is applied to receive a binary image. Radial from the ROI center, base points on the ventricle border are determined. Some constraints reject points which result from disturbances. The remaining points serve as base points in the following edge detection: A spline is interpolated using start and base points.To both sides of this spline a search region is defined. Within this matrix, a gradient and a dynamic search algorithm for the ventricle contour are applied. The user can easily correct the calculated contour by entering new points on a contour part. Based on this work, a frame-to-frame analysis through the entire cardiac cycle was realized. The current contour serves as a model for the dynamic search algorithm in the subsequent image.

Results: In more than 50% of cases no correction is necessary at all and in 25% only one slight correction has to be done. The algorithm works extremely fast: <2 seconds for the whole contour detection (SUN Sparc 2); corrections are also carried out instantly. This new approach is fully integrated in our cardiac quantification system for ventricle volumes and wall motion analysis. The contour detection and the volume calculation were validated using 60 post-mortem phantoms: 30 left and 30 right ventricles in standard RAO and LAO projection.

Accurate Quantitative 3-D Reconstruction and Calibration on Biplane Angiograms

A. Wahle, E. Wellnhofer, I. Mugaragu, J. Beier, H. Oswald, E. Fleck

German Heart Institute Berlin
Augustenburger Platz 1, D-W 1000 Berlin 65, Germany

3-D reconstruction of thoracic vessels aids in clinical management of patients with any kind of heart disease. Two simultaneous - but not necessarily orthogonal - angiographic views give 2-D projections of the 3-D space. Based upon known imaging geometry a 3-D model of complex vessel structures for visualization and quantification can be reconstructed [1-3]:

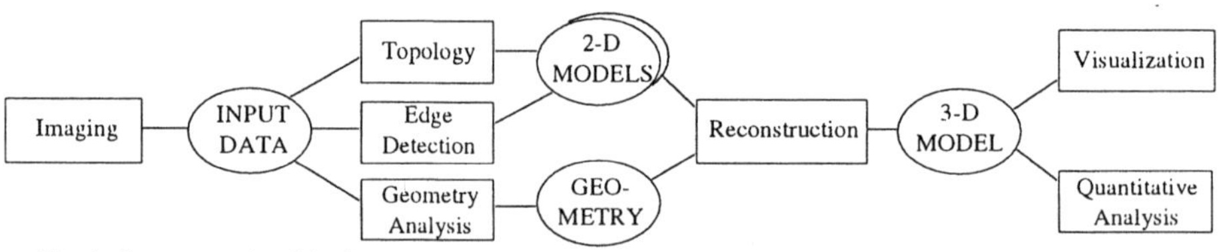

Fig. 1: Reconstruction Pipeline

The quantification quality strongly depends on the accuracy and completeness of the imaging parameters. To calculate missing and to correct distorted parameters a new geometry approximation has been developed [2]. The initial geometry is based on the geometric parameters of the X-ray device. This status is improved using several corresponding vessel points in both projections. The reconstruction errors of these points are analysed and correction coefficients are calculated. This is carried out in an iterative process.

To achieve accurate measurements in absolute dimensions a new calibration procedure was realized. The size of the image intensifiers can be determined by lead markers mounted on the entrance fields. Since continuous phantom measurements are impracticable and catheter radius determinations are prone to gross errors, an object of known dimensions and sufficient size visible in both projections was searched for. Markers on a catheter or the maximum diameter of the loop of a pigtail catheter were used as reference objects. The relation of actual to reconstructed object size provides additional correction coefficients for the approximation process.

The biplane calibration allows a reliable evaluation without phantom measurements or any assumptions. Common problems such as parallaxes do not occur. In addition with further corrections (image distortion, pointspreading) an accurate quantitative 3-D reconstruction is performed. The algorithm was validated using different phantoms of known sizes.

[1] Wahle A, Oswald H, Schulze G-A, Beier J, Fleck E. 3-D Reconstruction, Modelling and Viewing of Coronary Vessels. Computer Assisted Radiology, CAR'91, Berlin, Springer Verlag 1991;669-676

[2] Wahle A, Oswald H, Wellnhofer E, Beier J, Schulze G-A, Fleck E. Accuracy in Three Dimensional Reconstruction of Coronary Vessel Systems from Biplane Angiographic Views. Computer Applications to Assist Radiology, S/CAR'92, Baltimore, Symposia Foundation 1992;695-699

[3] Wahle A, Oswald H, Schulze-Neick I, Schneider M, Hausdorf G, Alexi V, Weng YG, Hetzer R, Lange PE, Fleck E. 3-D Reconstruction on Biplane Angiograms of Coronary Arteries in Pediatric Cardiology. Computers in Cardiology 1992, Durham, IEEE 1992;19-22

The Basic Method of Analytic Angiogram Calibration and its Error Regarding Developments

W. WUNDERLICH, F. FISCHER, T. LINDERER, J. NÖRING, R.SCHRÖDER

Div. of Cardiology, Med. Klinik, Klinikum Steglitz, Freie Universität Berlin, Berlin

Accurate dimensional measurements in angiograms are essential for interventional procedures. For transforming measurements in biplane angiograms to their absolute anatomic equivalents basically two calibration approaches are available: reference object calibration, using the ratio of known objekt size to its apparent pixel size and analytic calibration, using the rotational center of the X-ray system ("isocenter") for calculation of the exact radiological magnification. Isocenter calibration (ICC) avoids calibration errors, caused by the contour detection and the differential magnification of the scaling device and therefore yields more accurate measurements. Wollschläger [1] and Kirkeeide [2] provided ICC methods (the subsidiary-plane- and the minimum-distance-method), regarding geometrical inaccuracies of the X-ray system. Both approaches yield to different and complex computational formulas and therefore the acceptance of ICC is low yet. For this reason, we derived the basic formulas of ICC for an ideal accurate X-ray system. We simulated geometrical inaccuracies of the X-ray system on a Kontron Cardio 500 and compared the radiological magnification factors of the basic formulas with those of the error regarding developments.

Results: Following geometrical considerations, radiological magnification in ICC can be computed as the ratio of similar planes. When geometrical inaccuracies of the X-ray system vanish, both error regarding developments become equivalent and yield to the basic formulas. However, all three approaches provide nearly identical results, even if the geometrical inaccuracies increase.

Conclusion: Basic ICC is simply to derive and performs equivalently as its error regarding developments. The impact of geometrical inaccuracies of carefully installed X-ray systems on accurate radiological magnification computation seems to be lower, than expected.

References

1. Wollschläger, H.; Lee, P.; Zeiher, A.M.; Solzbach, U.; Bonzel, T.; Just, H.: Improvement of quantitative coronary angiography by calculation of exact magnification factors. Comput. Cardiol. (1985) 483-486.

2. Büchi, M.; Hess, O.M.; Kirkeeide, R.L.; Suter, T.; Muser, M.; Osenberg, H.P.; Niederer, P.; Anliker, M.; Gould, K.L.; Krayenbühl, H.P.: Validation of a new automatic system for biplane quantitative coronary arteriography. Int. J. Card. Im. 5 (1990) 93-103.

Quantification of Aortic Valve Insuffiency in Angiograms

KABEL N., BEIER J., OSWALD H., FLECK E.

Deutsches Herzzentrum Berlin, Abt. Dig. Bildverarbeitung,

Augustenburger Platz 1, W–1000 Berlin 65, Germany

Aortic valve insufficiency (AVI) is a disease of the aortic valve characterized by regurgitating blood (streaming backwards) from the aorta into the left ventricle (LV). The significance of this disease is expressed by the regurgitation fraction (RGF), defined as the regurgitation volume divided by the total volume ejected into the aorta.

Conventionally the severity of AVI is visually assessed by physicians watching the angiographic sequence. AVI is classified into 5 degrees ranging from 0 (without AVI) to class 4 (severe AVI). The visual diagnosis is based on the density ratio of LV and aorta, and the time required to clear the LV from dye. Visual grading is strongly inter– and intra–observer variable and cannot provide absolute regurgitation volumes. Computer densitometric analysis of digital subtraction angiography offers the potential to quantify AVI with high accuracy and reproducibility.

During examination a catheter is placed in the LV and contrast medium is injected. A number of pre–contrast images is acquired, covering at least one cardiac cycle. The imaging continues until the injected contrast medium has cleared the ventricle. All end diastolic (ED) and end systolic (ES) images after dye injection are selected. The LV contours on these images are determined using an automatic detection method. Two pre–contrast images in ED and ES serve as subtraction masks to eliminate pulsatile effects. Using the selected images and the ventricle contours, a time density curve (TDC) of the LV region is calculated. By selecting appropiate points on the TDC the RGF factor is determined.

Digital densitometrical quantification of AVI is more reliable than visual grading and is a relevant adjunct to conventional methods in assessing AVI severity.

Automated Pixel Shifting in Digital Subtraction Angiography - an Application of Cepstral Filtering

K-H Englmeier[1], U Fink[2], T Hilbertz[2]

[1] GSF- Forschungszentrum für Umwelt und Gesundheit, Institut für Medizinische Informatik und Systemforschung, Ingolstädter Landstr. 1, 8042 Neuherberg, Germany

[2] Radiologische Klinik, Klinikum Großhadern, Ludwig- Maximilians-Universität, Marchioninistr. 15, 8000 München 70, Germany

Since the development of medical workstations as well as the the use of picture archiving and communication tools, image processing and analysis is not only performed with the processing unit of the imaging device. An example of an interactive image analysis system is the Digital Subtraction Angiography. In order to shift the corrresponding mask and contrast agent images of the time dependent image sequence to their correct position, specialized hardware is used at the processing unit of the system console. The absence of these hardware utilities at graphic workstations requires preprocessing tools which support an automatic pixel shift.
One method for the estimation of translational shifts is the cepstrum filtering method. The cepstrum - an anagram of the word spectrum - is a well known non linear filter used for the detection of the echo arrival times in seismic signals and the estimation of local stereoscopic depth [2] [3]. It is defined as the power spectrum of the logarithm of its power spectrum [1].
In order to calculate the horizontal and vertical disparitites of two corresponding images of the Digital Angiography, the cepstral filter is applied to a pair of image patches. The disparity terms occur as bright spots in the cepstrum. A known big enough shift is introduced to move the maximum of the real shift to a position where the local maximum is not overlayed by the cepstrum itself. After detection of this local maximum by thresholding methods, the real shift of the two subsignals can be calculated. The shift of the corresponding images to their correct positions lead to improved subtraction results. The use of the cepstral filtering method has the following advantages: It is a fast preprocessing method, which is well known from previous applications. It is robust against noise and signal deformations [2]. It enables the correct localisation of the images in order to improve the subtraction results as well the the absolute coincidence between pre- and postcontrast exposures in case of peripheral DSA with automated stepping.

[1] **B P Bogert, M L R Healy, J W Tukey**: The Quefrency of Time Series for Echoes: Cepstrum, Pseudo-Autocovariance, Cross-Cepstrum and Saphe Cracking, Proceedings of the Symposium on Time Series Analysis, New York Wiley, 209-243, 1963

[2] **K-O Ludwig, H Neumann, B Neumann**: Robust estimation of local stereoscopic depth, in: Robust computer vision: Quality of vision algorithms, eds: Förstner, uwiedel, Karlsruhe, Wichmannn, 290-312, 1992

[3] **Y Yeshurun, E L Schwartz**: Cepstral Filtering on a Columnar Image Architecture: Fast Algorithm for Binocular Stereo Segmentation, IEEE Transactions on Pattern nalysis and Machine Intelligence, 11/7, 759-767, 1989

Detectability of K-edge Energy Subtraction Coronary Angiography by Iodine Filter Method Using Synchrotron Radiation

M. Akisada[1], K. Umetani[2], T. Takeda[3], K. Ueda[2] and T. Nakajima[1]

[1] Tama Health Management Center, Tachikawa, Tokyo 190, Japan
[2] Central Research Laboratory, Hitachi, Ltd., Kokubunji, Tokyo 185, Japan
[3] Institute of Clinical Medicine, University of Tsukuba, Tsukuba, Ibaraki 305, Japan

Iodine K-edge subtraction techniques using monochromatized X-ray have been investigated at several synchrotron radiation facilities for transvenous coronary angiography. We developed an iodine-filter imaging system for real-time imaging. A combination of the iodine filter and a single energy monochromatized X-ray produces two-energy images for the iodine K-edge subtraction technique. An X-ray exposure system consists of the iodine filter and a monochromator. X-ray energies are switched by the filter chopping of the X-ray beam. A detector consists of an X-ray image intensifier and a video camera. An oscillating mirror is installed between the collimating lens and camera lens in an image distributor. Each image that corresponds to the iodine- and aluminum- (non-iodine-) filtered images is focused side by side on the photoconductive layer of the camera tube by the oscillating mirror. Detectability of contrast material is calculated to be 0.39 mg/ml. This value corresponds to a vessel diameter of 0.1 mmØ. However, detectability of the system is limited by a video camera's dynamic range, X-ray flux and a spatial resolution of detection system. In vivo imaging was performed in dogs. In the present system the detectable minimum size of coronary artery is 0.3-0.4 mmØ in transvenous injection of a contrast agent. The imaging system produced subtracted images of coronary arteries in dogs in the form of motion pictures.

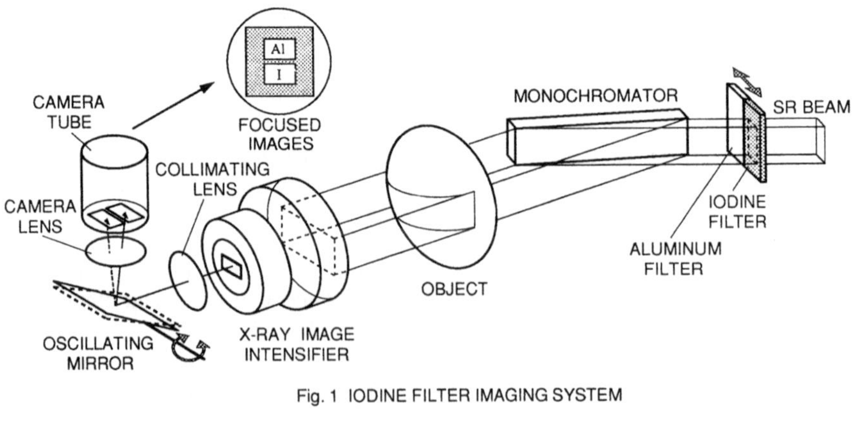

Fig. 1 IODINE FILTER IMAGING SYSTEM

Depth Separation in 10 Observers with a New Stereoscopic X-ray Acquisition System

T. MOLL*, C. PICARD**, G. FINET*, F. TURJMAN*, P DOUEK*, M. AMIEL*
*Hopital PRADEL 69, Bd PINEL 69003 LYON FRANCE fax (33) 72 35 72 91
**General Electric Medical Systems Europ BP 34 78533 BUC FRANCE

Introduction

Stereoscopic angiography provides a useful and efficient means to overcome the problem of depth separation and object overlapping on a single projection of a 3 dimensional object [1-3]. Although these advantages, stereoscopic Xray technique is still experimental, probably because of practical considerations. Technology of twin focal Xray tubes has solved inherent difficulties at the acquisition step of the radiological process but displaying stereoscopic images still remains difficult. The aim of this study is to assess in terms of depth detectability, the performance of a new stereoscopic angiographic system equiped with such stereoscopic videomonitor.

Method

The Xray tube was a twin focus angiographic tube. Focus size was 0.8 x 1 mm for both focus. Interfocal distance was 65 mm.

The light amplifier was a triple-field model with selectable field size of 160 , 230 or 320 mm. With a 230 mm field size and a 512^2 matrix, the pixel size was 450 micrometers.

A DSA system (DG 300 General Electric-CGR) was used for image acquisition, digitization and monoscopic display. Stereoscopic display was obtained by using a Liquid Crystal Stereoscopic Modulator (TEKTRONIX) placed in front of a black and white 1190 x 960 Video Monitor with a 120 Hz frame rate. Special polarizing glasses worn by the user provided the right and left eye views.

A "Plus Phantom" as those described by DOI [1] was constructed. It contains 8 steps, thickness ranging from 1.5 mm to 12 mm in 1.5 mm increment. The body of the phantom was made in methacrylate. 1 cm long copper wires of 0.9 mm diameter were placed on each side of the stepladder phantom so that to form on each step 8 plus signs. The location of the horizontal branch of each plus sign (horizontal or vertical) was randomly selected with an equal probability for the both configurations. 10 readers were involved in the reading protocol. 8 images of the Plus-Fantom were made. For each plus sign, they were asked to predict which branch (horizontal or vertical) was in the anterior plan.

Result

6 of the 10 readers were able to detect depth information from 12 to 1.5 mm

Conclusion

This experiment demonstrates that liquid crystal modulator and polarized glasses, coupled to classical DSA architecture is able to provide to radiologists an accurate mean to add three-dimensional information to radiographic projections. In addition, the absence of any wire connecting the passive polarized glasses suits this technology for application to stereofluoroscopy.

References

.1. DOI K. Detectability of Depth Information by Use of Magnification Stereoscopic Technique in Cerebral Angiography Radiology 1983 ; 146 : 91-95.
.2. HIGASHIDA Y. , HIRATA Y. , DOUDANOUKI S. , BUSSADA H. Depth Determination on Stereoscopic Digital Substraction Angiograms. Radiology 1988 ; 168 : 560-562.
.3. TAKAHASHI M. , OZAWA Y. Routine Biplane Cerebral Angiography with Stereoscopic Magnification. Radiology 1980 ; 136 :113-117.

Multimodality Imaging of Cerebral Infarction

H. POHJONEN[1], O. SIPILÄ[2], P. NIKKINEN[1], J. LAUNES[1], E. SALLI[2], X. YU[3], J. YLÄ-JÄÄSKI[3], T. KATILA[2], P. KARP[1]

[1]Helsinki University Central Hospital, 00290 Helsinki
[2]Helsinki University of Technology, 02150 Espoo
[3]Technical Research Centre of Finland, 02100 Espoo, Finland

We have developed a software that is clinically used to register, segment and analyze MR and blood perfusion SPET images of patients with cerebral infarction. These imaging modalities produce complementary information: the core ischemic infarct zone may be depicted by MRI, whereas in SPET both the core ischemic and the peri-infarct zones are detected, but not separated, however. The MR images are utilized for the segmentation of the core ischemic infarct zone and the total brain volume. Thereafter the peri-infarct zone can be segmented from the registered SPET data set. The volumes of these infarct zones are important in predicting the clinical outcome in brain infarction.

For the registration, we have used 6-8 skin markers containing coconut butter (50 μl) and ^{99m}Tc (12 μCi). The MR images are acquired with a Siemens 1.0 T device and the SPET images using a Picker rotating gamma camera and ^{99m}Tc-HM-PAO. The registration algorithm follows a noniterative least squares method using a singular value decomposition (SVD) of a 3x3 covariance matrix.

For preprocessing prior to segmentation, we have developed a one pass adaptive Gaussian filter to reduce noise. The filter operates faster and produces results comparable or even better than the standard iterative anisotropic diffusion method.

Two approaches have been utilized for the segmentation. The first approach is based on the Canny operator and hysteresis thresholding. The other is based on region growing: we have developed a method which localizes the borders more accurately than a mere region growing algorithm by combining it with edge detection. The accuracy of the segmentation results is validated visually and modified interactively, if necessary, by an experienced neurologist.

The results are visualized as 2D cross sections and 3D rendered images. Corresponding cross sections from the registered data sets can simultaneously be displayed for improved visualization of anatomical and functional details. The quantitative analysis includes the determination of the infarct volumes from the segmented data sets. These volumes may also be compared to the total brain volume. The count densities in the corresponding regions of interest in SPET images can also be compared.

Computer-assisted Tutorial System to Support the Diagnosis of Bone Lesions in the Instruction of Radiology

D.PRÜMER, E.PELIKAN, K.BOHNDORF, T.TOLXDORFF, B.WEIN, B.PRÜMER

Institut für Medizinische Informatik und Biometrie
Klinikum Aachen
Pauwelsstraße 30
W-5100 Aachen
Germany

In diagnosing very rare diseases using radiological images, such as in the diagnosis of bone tumors, there is often the need for additional support in form of tutorial systems. For the physician unexperienced with bone lesions this could mean access to an expert radiologist, books, journals and X-ray archives. We developed a computer-based tutorial system which combines all the aforementioned knowledge resources, becoming a powerful aid in diagnosing. It has been specifically developed for use in diagnosing focal bone lesions, which are relatively rare and have a similar appearance despite histological differences.

The intention of our tutorial system is to promote the diagnosing capability of the user. The theoretical background of this tutorial consists of a sample range of X-rays, grouped by topics to choose from. An illustrative questionnaire, formulated by an expert radiologist for each topic, is presented to the user during a tutorial session. In conjunction with the sample X-rays the user is asked to complete the presented questionnaire. The single diagnostic aspects determined with the help of the questionnaire are linked to a precise location depicted in the radiography. The diagnostic criterias discovered by the user are compared to the expert's analysis, providing the user with an immediate feedback. We have chosen this type of interactive communication deliberately in order to avoid the stress and strain of scoring a good hit-miss-ratio. The use of a form based questionnaire guarantees an analytical approach, which helps the user in making a complete finding. During a session the user can get support by samples of X-rays each showing an example for a single diagnostic aspect.

Our system provides as well an interface to the dialog generator CARAT, also developed by our group, which is an expert shell for the diagnosis of focal bone lesions. With the help of CARAT the user is supported in concluding a final diagnosis from the discovered diagnostic aspects. The tutorial system is supported by a relational database handling the X-rays and questionnaires to guarantee fast retrieval. Therefore it is easy to extend the tutorial in terms of new topics, as well as single aspects of titles already installed. Our system has been developed on a UNIX-workstation with X11 and OSF/Motif to provide an adequate graphical interface.

Computer Detection of Microcalcifications in Digital Mammograms

H. FUJITA[1] T. KIRITO[1] T. ENDO[2] K. HORITA[3]
T. MATSUBARA[1] M. IKEDA[2] C. KIDO[3] T. ISHIGAKI[2]

[1] Department of Electronics and Computer Engineering, Gifu University
Yanagido, Gifu 501-11, Japan (e-mail: fujita@fjt.info.gifu-u.ac.jp)

[2] Department of Radiology, Nagoya University School of Medicine
Showa-ku, Nagoya 466, Japan

[3] Department of Diagnostic Radiology, Aichi Cancer Center Hospital
Chikusa-ku, Nagoya 464, Japan

We have been developing an auto-analyzing system [1-4] for the detection of masses and microcalcifications in digital mammograms, which are obtained by an I.I.-TV based digital mammography system (1k × 1k), for application to mass screening of breast cancer. In this study, we investigate a new scheme for the detection of microcalcifications in finely-sampled mammograms; clinical screen-film mammograms are digitized using a laser scanner at a pixel size of 100 μm and 10-bit gray levels (2510 × 2000). The fully automated computer system consists of several parts such as region extraction of the breast, overall background-trend correction in the breast region, enhancement of the high frequency components, and the detection of microcalcifications and their cluster(s). Evaluation of the detection accuracy of the computer algorithm will be discussed.

References:

1. Kido, C.; Endo, T.; Horita, K.: Digital mammo-radiography (DMR) as mass screening for breast cancer.
Jpn. J. Cancer Chemother 15 (1988) 1665-1670.

2. Endo, T.; Kido, C.; Horita, K.: Mass screening of breast cancer with DMR.
Jpn. J. of Breast Cancer 5 (1990) 424-430.

3. Endo, T.; Kido, C.; Horita, K., et al.: Clinical evaluation of assistant diagnostic system for mammograms using the auto-analyzing method.
Radiation Medicine 10 (1992) 50-54.

4. Fujita, H.; Horita, K.; Endo, T., et al.: Neural network approach to classifications of benign and malignant tumors in mammograms.
Med. Imag. Tech. 10 (1992) 126-129.

Receiver Oriented Characteristics (ROC) Analysis of Pulmonary Nodule Detection in Digitized Chest X-ray Images Using Personal Computers

Y. Kurashita, T. Kitanosono, K. Suzuki, H. Konishi, M. Honda, T. Hishida

Showa University School of Medicine, Department of Radiology
1-5-8, Hatanodai, Shinagawa-ku, Tokyo, Japan

Summary
The apparatus used for PACS or CRT diagnosis are quite expensive and limits wide spread of such system. Recently, personal computers and commercially available image scanners are being used more frequently in limited uses such as Teleradiology, image filing and experiments. However, there have been few reports on diagnostic capabilities of images that are digitized and displayed through such equipments. Reciever oriented characteristic (ROC) analysis of digitized chest X-Ray images and conventional radiographs in detection of pulmonary nodules was carried out to evaluate the diagnostic ability of images that were digitized through personal computers and displayed on a commercially available CRT monitor.

Method
35 chest radiographs with pulmonary nodules and 35 radiographs without nodules were digitized with a film digitizer connected to a Macintosh personal computer and displayed on a 19 inch monitor with resolution of 72 dots per inch. The digitized images and the original radiographs were interpreted by six board certified radiologists and ROC analysis of pulmonary nodule detection was obtained.

Results
The area under the ROC curve was always greater in the conventional radiographs but the difference was not statistically significant ($t < 0.05$).There was a tendency for the digitiged images to produce false positive results.

Conclusion
In detection of pulmonary nodules, images obtained and displayed by commercially available hardware and software showed no statstically significant difference compared to conventional radiographs.However, there was a propencity for the digited images to produce false positive findings. This study indicates that images digitized by personal computers could be used to diagnose pulmonary nodules relatively safely. Further studies should be carried out to evaluate the digitized images in diagnosis of other more subtle conditions such as pulmonary interstitial diseases .

Workstation Design for Breast Screening

A.G. GALE & C.G. BLAIR-FORD
Applied Vision Research Unit, University of Derby, UK

Keywords: Breast Screening, Medical Workstations, Radiological Diagnostic Performance

Although digital mammography and associated computer aided diagnostic approaches are currently under investigation by many research groups it is unlikely that generally such approaches will be implemented for some considerable time. Currently in the UK all breast screening is performed using conventional Xray film.

The purpose of this research was to investigate the design of the dedicated mammographic multiviewers, employed in breast screening throughout the UK, in relationship both to established ergonomic principles and to official guidelines. The objective was to determine whether existing workstations were designed for maximum radiological efficiency and also to determine whether radiological screening performance could be bettered by improvements to these workstations, such modifications then having impact upon the design of future digital systems.

Design recommendations for multiviewers in the UK have been produced by the Department of Health and the NHS and these criteria were first compared to established ergonomic data which yielded some discrepancies. Two types of alternator were then selected for further detailed investigation and experienced radiologists were videotaped as they used them to report a series of cases. This yielded information on the inspection process itself as well as on the postures and movements of the radiologists whilst inspecting mammograms. These data were analyzed using standard ergonomic techniques. In addition structured interviews were carried out with other screening radiologists concerning both the desired, and the required, features which should be included in the multiviewer design. Several other video-based studies have been made of radiologists reporting standard sets of mammograms using the two viewers. It is concluded that existing guidelines could be improved and specific design recommendations are proposed which it is argued would aid the radiologists in their task so improving overall screening efficiency.

AUDIGON - An Expert System for MR-based Diagnosis of Osteoarthrosis of the Knee

D. Wetzel, P. Weierich, H. Niemann, B. Bühnemann

Bavarian Research Center For Knowledge-Based Systems (FORWISS)
Am Weichselgarten 7, D 8520 Erlangen, Germany, (*name*)@forwiss.uni-erlangen.de

K. Glückert, B. Kladny

Dept. of Orthopaedics, University of Erlangen
Rathsberger Straße 57, D-Erlangen, Germany, email: gluecker@forwiss.uni-erlangen.de

With MRI, esp. fast imaging sequences (FISP ...), different soft tissue components can be analysed in detail by their different signal and contrast behaviour. Early stages of cartilage degeneneration of the knee are detectable [1][3]. However, evaluations have shown to be too inaccurate and time consuming without adequate hardware and software support. Routine evaluation of MRI datasets therefore demands a knowledge based medical system which meets the following demands:

- Automated identification of articular cartilage using contrast optimized images
- Classification and quantification of the lesion including the generation of a report
- Display of all relevant information contained in the datasets
- Interactive handling by comfortable graphical user-interface
- Userfriendly maintainability with respect to changes of the medical knowledge and new imaging conditions
- Evaluation within acceptable time

A prototype of an expert system has been developed at the Bavarian Research Center for Knowledge Based Systems (FORWISS) [2][4]. Structural and procedural knowledge is represented within a semantic network, indistinct information by fuzzy logic [5]. Low level processing of the images (detection and segmentation of the articular cartilage) is achieved by polar transformation and dynamic programming algorithm. High level processing (analysis of the segmented cartilage and generation of the diagnosis) is carried out with fuzzy logic. Contour and texture features are utilized on different levels of abstraction. A graphic user interface allows the physicians not only to inspect the images and to generate diagnoses, but also provides a module to maintain the fuzzy rulebase.
The principle and medical application will be as well introduced as a comparision between arthroscopic findings and automatically generated diagnoses.

References:

[1] Glückert, K., Kladny, B., Hofmann, G., Willauschus, W., Blank-Schäl, A., Wirtz, P., MRI-Evaluation of early degenerative cartilage disease with a 3-D-gradient echo sequence. *in H.P. Higer, G. Bielke (Eds.): Tissue characterization in MRI* , Springer, Heidelberg-Berlin-New York 1990, pp. 185-191.
[2] Niemann, H.; Wetzel, D.; Weierich, P.; Sagerer, G.; Glückert , K. Methods of Artificial Intelligence in Medical Imaging, *in: proceedings of I.CO.GRAPHICS 1992, Milan, Italy Mondadori Informatica SPS - AICOGRAPHIC, Italy* 1992, pp. 253-260.
[3] Reiser, M. Magnetic resonance in cartilaginous lesions of the knee joint with three-dimensional gradient-echo imaging.*Skeletal Radiology*, vol. 17, pp. 465 - 471, 1988.
[4] Wetzel, D., Weierich, P., Niemann, H., Einsatz sematischer Netze in der automatischen Bildanalyse; *Proceedings of the MESSCOMP '92; Network GmbH., pp. 148 - 163; Wiesbaden,* September 1992
[5] Zimmermann, H.J., *Fuzzy Set Theory - and Its Applications*, 2nd ed., Kluwer Academic Publishers, Norwell, 1991.

Use of Computers in Orthopaedic Surgery

S.W. Hughes, R. Brueton, D. Reynolds, J. E. Saunders

Department of Medical Physics and Orthopaedics, St. Thomas' Hospital, London.

1. Quantification of Acetabular Fractures

A morphological classification of acetabular fractures has been devised by Judet and Letournel on the basis of plain radiograph AP pelvis and obturator oblique views. A method to analyse the displacement of a fracture has been developed from these three plain X-rays, but is not clinically useful. Measurement of the surface area of the articular cartilage of the acetabulum is of greater clinical relevance. We have devised a technique whereby the articular surface is outlined on the relevant CT slices and a triangulation algorithm used to join up the points to produce a three dimensional surface. The total surface area of the acetabular fragments is found by summing the areas of the individual triangles. Where possible, the fragment areas are compared with the area of the intact acetabulum on the other side. Fractures are assessed on the basis of the percentage of the acetabulum still in contact with the femoral head. The contact percentage is used to decide on the surgical intervention required.

2. Computer planning of Periacetabular Osteotomies.

The periacetabular osteotomie is a surgical technique that has been developed for the correction of malformed (dysplastic) hips. The procedure involves cutting around the acetabulum to detach it from the sacrum. The free floating acetabulum is then rotated about the anterior-posterior, lateral, vertical (x, y, z) axes in order to improve coverage of the femoral head. To plan this procedure, 2mm slice thickness CT scans are acquired from the acetabular roof to the femoral equator. The images are transferred onto a computer, the femoral head and articular surface of the acetabulum are outlined and a triangulation algorithm used to generate two surfaces. The centre of the femoral head is taken as the centre of rotation for the acetabular surface, and the acetabulum is rotated about the x, y, z axes on the computer display in order to improve acetabular coverage of the femoral head. The x, y and z axes divide the femoral head into quadrants; coverage is expressed as the percentage of each quadrant covered by articular surface of the acetabulum.

Pre- and Intraoperative Computer-Assisted Surgery: A Survey

C. SCHUH, K.-P. ADLASSNIG, C. CHIZZALI-BONFADIN, H. IMHOF

Department of Medical Computer Sciences
University of Vienna
Währinger Gürtel 18-20
A-1090 Vienna, Austria

A survey on current fields of pre- and intraoperative applications of computer-assisted surgery is presented.

Preoperative computer-assisted surgery aims at partially or totally simulating surgery operations by means of computer programs. Thus the surgeon has a tool in his/her hand to go through all steps of an operation before the intervention is done in the particular patient. By doing this, extended planning of the whole operation is possible to the benefit of the surgeon and, most important, to that of the patient. So far, computer-assisted surgery has been turned out to be most useful in planning craniofacial surgery to correct skeletal abnormalities and to aid in post-trauma operations. Orthopedic hip surgery, e.g., total hip replacement surgery, is a further important field of preoperative computer-assisted surgery simulation.

Intraoperative computer-assisted surgery has mainly been developed to aid in neurosurgery operations. It offers the possibility to gain an intraoperative view of the anatomical brain structure and may thus help to avoid or minimize neurological damages of patient's brain during operation.

Computer-based 3D visualisation of tomographic image sequences of CT and MR scans forms the foundation of all the above-mentioned approaches. In addition, surgery simulation needs software tools to cut, move, and paste anatomical blocks, whereas intraoperative computer assistance presupposes an accurate 3D positioning system. Both kinds of application demand high computer performance; intraoperative surgery definitely means real-time performance.

On the basis of the presented overview, it is anticipated that computer-assisted pre- and intraoperative surgery will routinely be applied most extensively in the coming future.

Automatic Localization and Display of MRI-Guided Needles in Interventional MRI

Lu J. Huang, Shantanu Sinha, Usha Sinha
Yoshimi Anzai, Dan Castro+ and Robert B. Lufkin

Department of Radiological Sciences
+Department of Surgery, Division of Head and Neck Surgery
School of Medicine
University of California, Los Angeles
Los Angeles, California 90024
U.S.A.

Because of its superior soft tissue contrast, MRI has become the imaging method of choice for the development of many interventional procedures. Such procedures, being developed at UCLA, include interventional therapies involving focal lesion destruction using laser, RF heating and ultrasound; MRI-guided fine-needle aspiration cytology; MRI-compatible deeply-embeded electrodes for accurate physiological monitoring of brain activity; among others.

In these MRI-guided interventional procedures, it is always very important to view the path and the tip of the guiding device, which is usually an MRI-compatible interventional needle. The fact that the needle is generally in an arbitrary orientation with respect to the normal MR imaging planes makes the detection of the entire needle path and tip difficult for routine scanning. We present a scheme which utilizes the 3D imaging capability of the MRI modality and the 3D image processing capability of computer workstations to automatically and time-efficiently localize the entire path and tip of the needle and display the oblique plane which cuts through the scanned object and contains the needle.

The scheme involves: (i) setting up a general-purpose interventional MRI environment; (ii) optimizing MRI protocols for fast MRI 3D data acquisition for needle detection; (iii) designing and implementing a computer algorithm for automatic needle path and tip detection and 3D display; and (iv) establishing communication channels between the local host computer and the MRI scanner host computer for image transfer, and the video projection device for the viewing of the needle path and tip by the surgeon and the radiologist inside the MRI scanner room.

We have successfully implemented this scheme in the Advanced Imaging Center at UCLA. In summary, the general-purpose interventional MRI environment consists of a GE MRI signa scanner networked with a Sun SparcII computer and a image video projector. Images can be transferred among the scanner, the Sun computer and the image projector. The 3D MRI protocols used were 3DFSPGR and FSE to time-efficiently generate contiguous-slices across a 3D volume of interest. A 3D image processing algorithm has been developed and implemented for automatic detection and display the path and the tip of the needle. The image containing the needle path and tip, which is in generally an oblique plane cutting the volume, was then automatically displayed on the the image projector display in the scanner room where the radiologist or the surgeon performs interventional prodecures. The 3D image display can be manipulated interactively to show the anatomy on different planes of the scanned volume. All the planes contain the needle path and tip. The scheme has been proven efficient and reliable. The scheme is being interfaced with various interventional procedures, such as MRI-guided biopsy and MRI-guided laser tumor ablation treatment to optimize the positioning of interventional devices, such as biopsy needle and laser optical fibres.

Appendix

Technical Requirements of Medical Diagnostic Databases, with Emphasis on the Field of Anatomic Pathology

Al M. Elsayed, M.D. and Seong K. Mun, Ph.D.,

Armed Forces Institute of Pathology and Department of Radiology, Georgetown University Medical Center, Washington, D.C., U.S.A.

Summary:
Display and interpretation of diagnostic images are pivotal for the performance of a medical consultation service. The goal of such a service is to enable the physician at the viewing site to render an accurate diagnosis. This is best served by presenting the images, identification data and patient's clinical information in a manner that follows the diagnostic decision making algorithms. Optimal database design can facilitate the process and enable post-diagnosis data for reference, educational and statistical purposes. The diversity of a data to be presented and operations to be performed impose a challenge for designing such databases.
Still-image files: Typical non-compressed still-images can approach 18MB each. For single patient presentation, typically between 8 and 36 images are required. However, in rare instances, up to 100 images could be needed.
Specialized files: These include live video with or without related sound tracks, pure sound track files and image overlay files.

Other data: These can range from modest discrete data stored in structured records to highly variable format data extracted from remotely located reference databases.

Procedures: These are programs that are registered with data types and specified tables. The procedures will manipulate various data types and decompress files for display or utilization. In some instances a specified procedure would be called upon to perform a sub-specialized task, such as making image overlays or simulating a focus-through effect by fading multiple images in sequence through the display buffer.

Conclusion: Database design has to facilitate performance of complex procedures utilizing conventional and non-conventional data types, and operational requirements. Databases with a mix of relational and object-oriented functional are most suitable for meeting diagnostic requirements.

Colour Figures

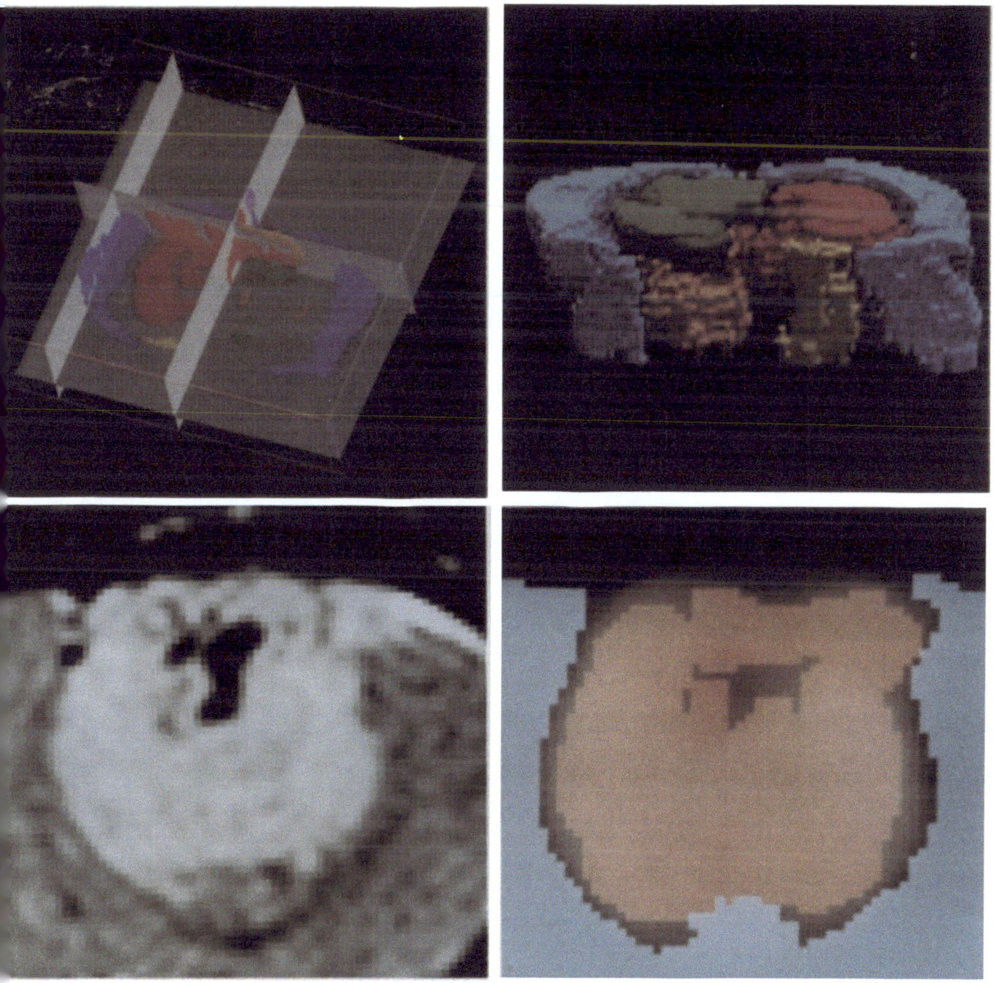

and et al., Fig.1, see p. 356

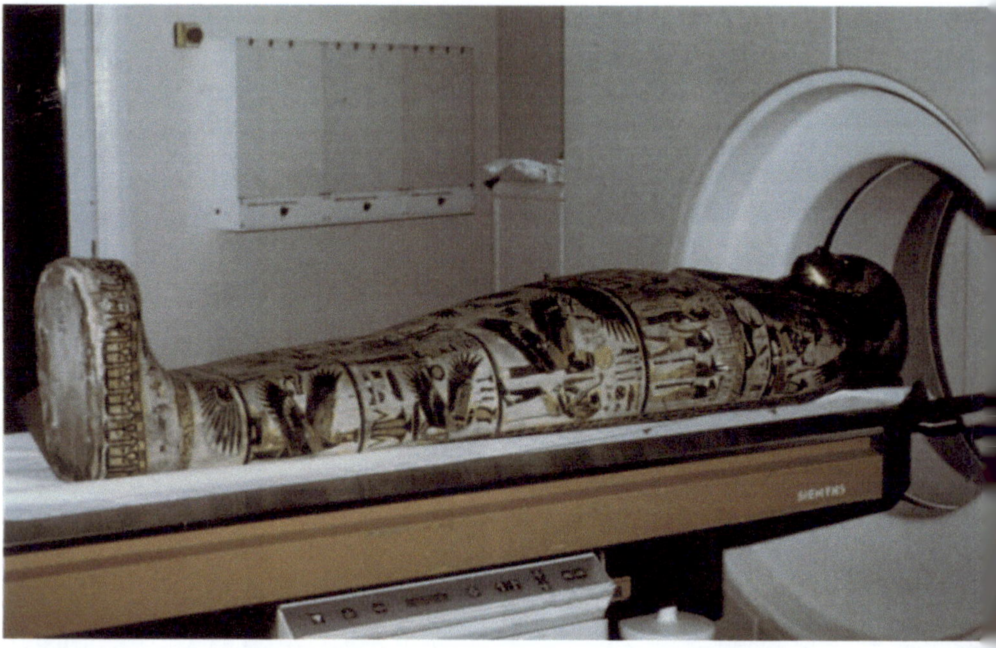

Hughes et al., Fig. 1, see p. 395

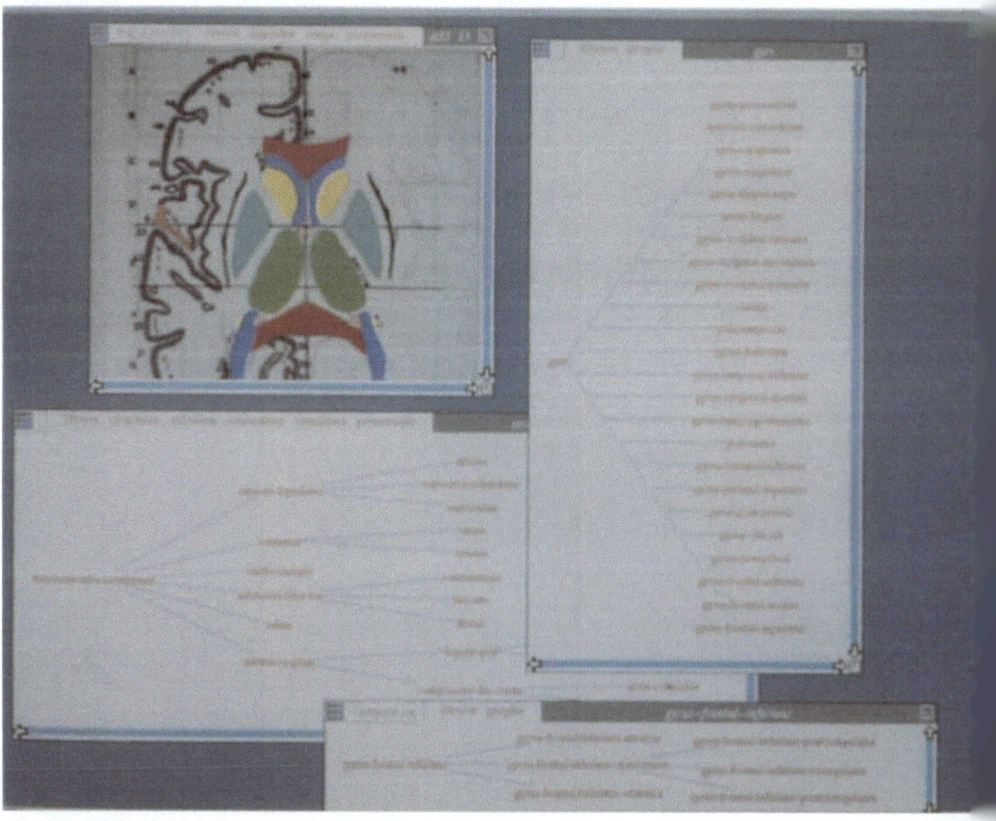

Montabord et al., Fig. 1, see p. 417

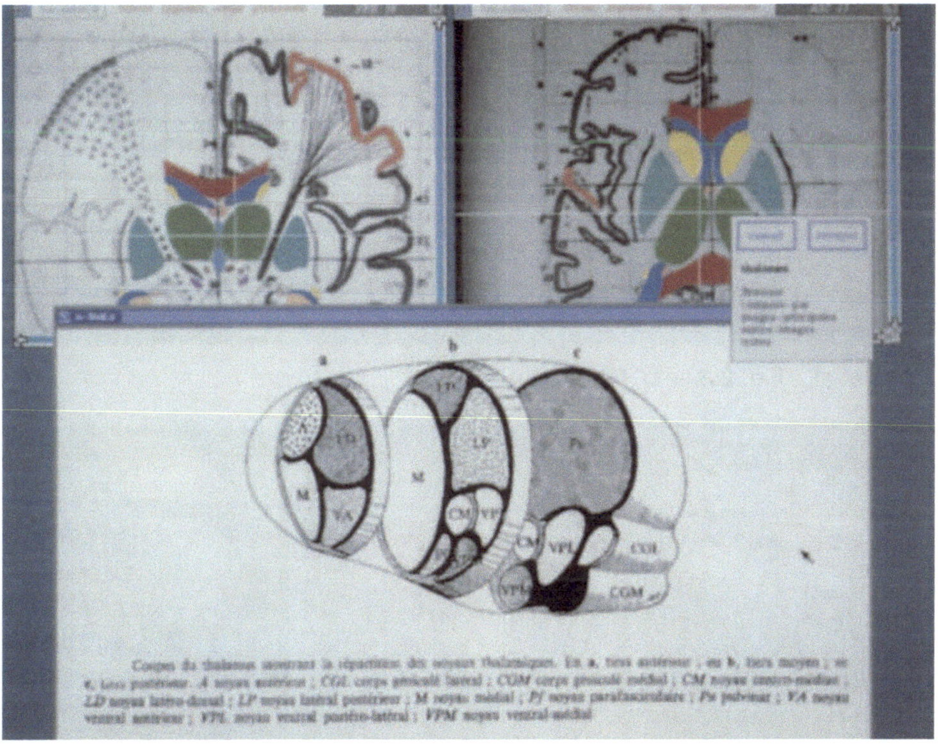

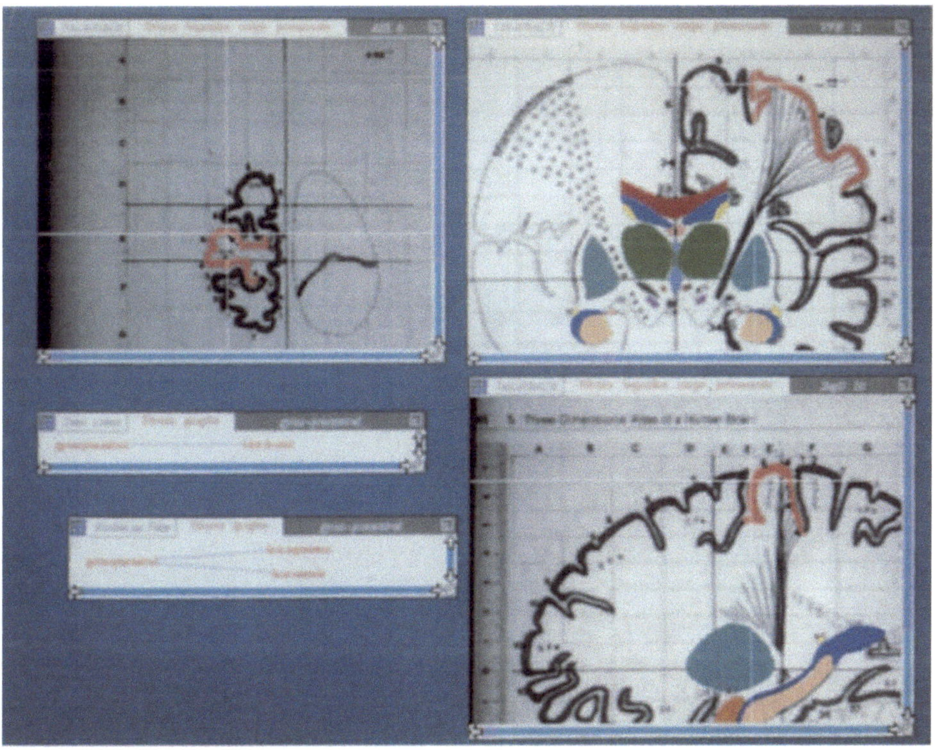

Montabord et al., Fig. 2, see p. 417

Stokking et al. , Fig. 2, see p. 425

Stokking et al., Fig. 3a-c, see p. 425

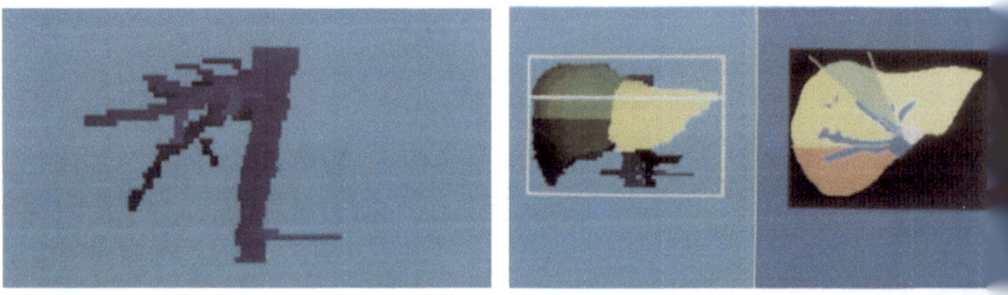

Stokking et al., Fig. 4a/b, see p. 425

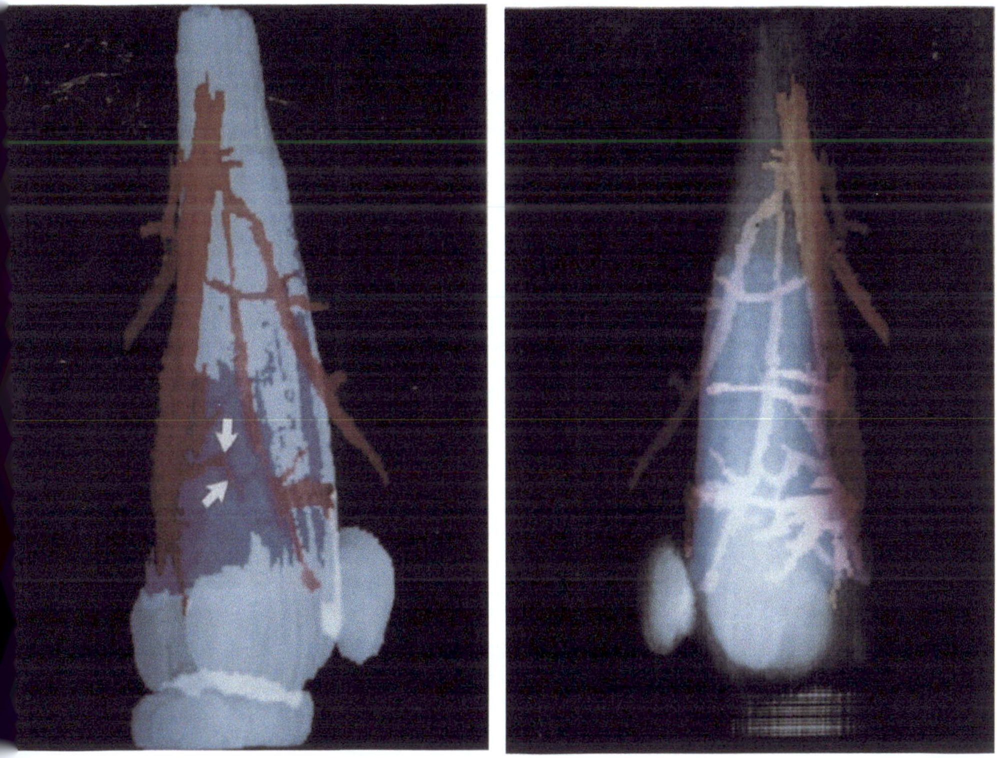

ang et al., Fig. 2A/B, see p. 473

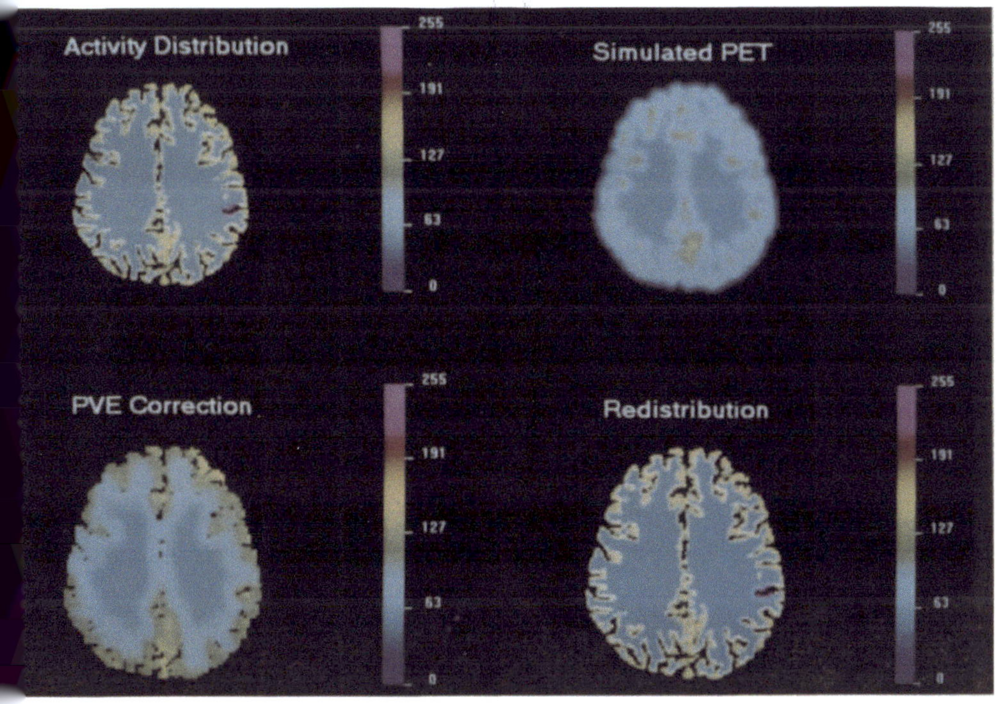

norr et al., Fig. 3, see p. 522

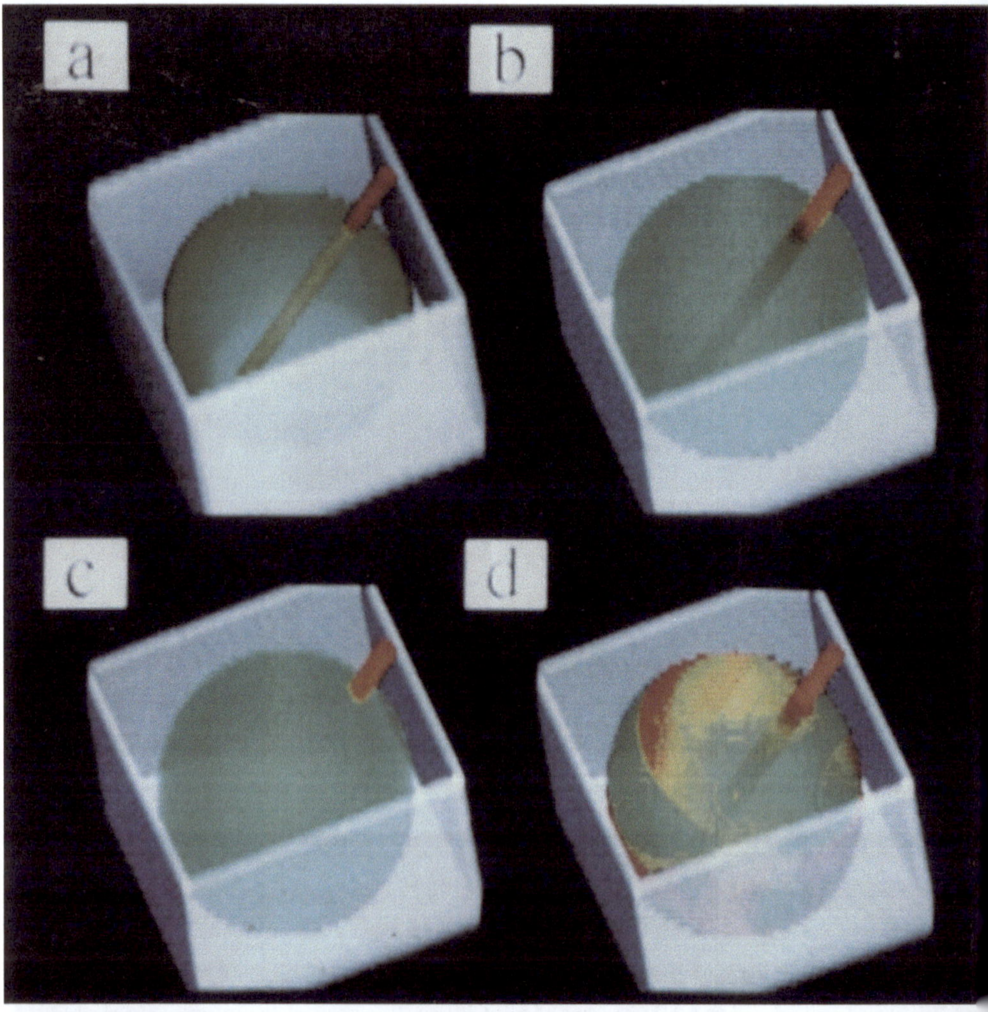

Ruff et al., Fig. 1a-d, see p. 578

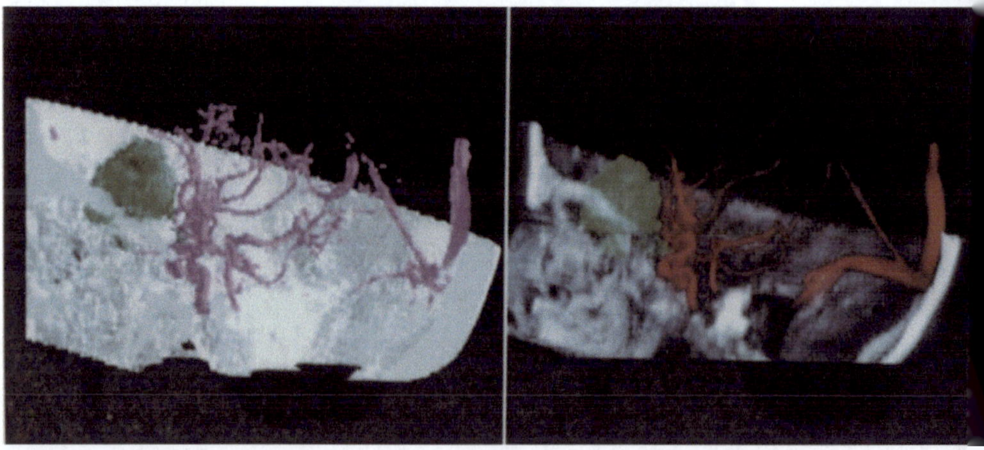

Ruff et al., Fig. 2, see p. 578

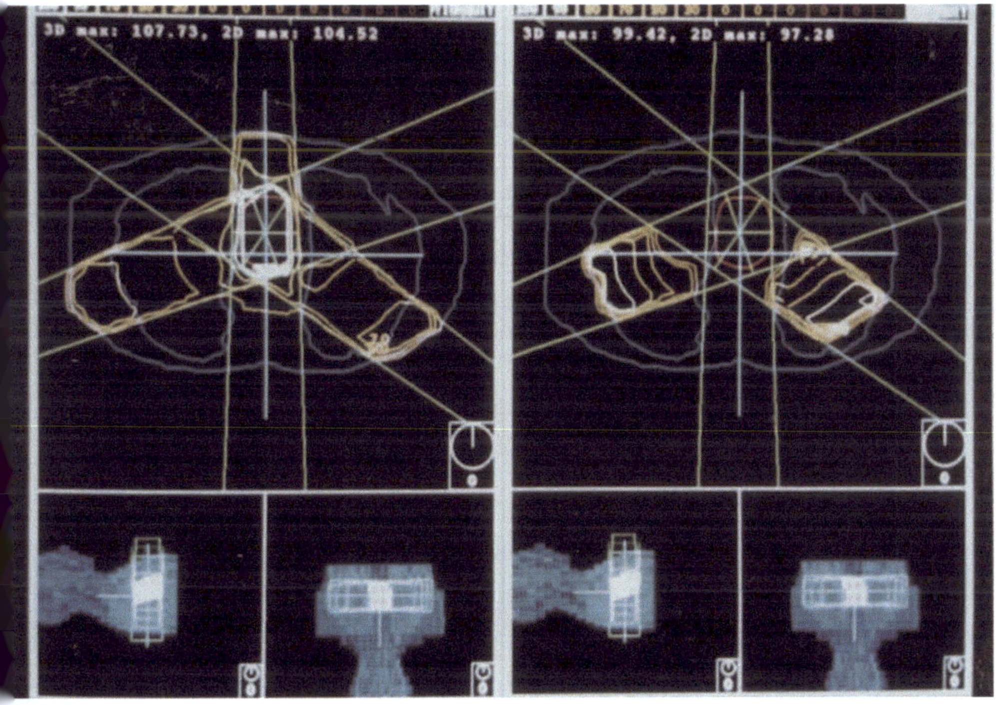

ack et al., Fig. 1, see p. 662

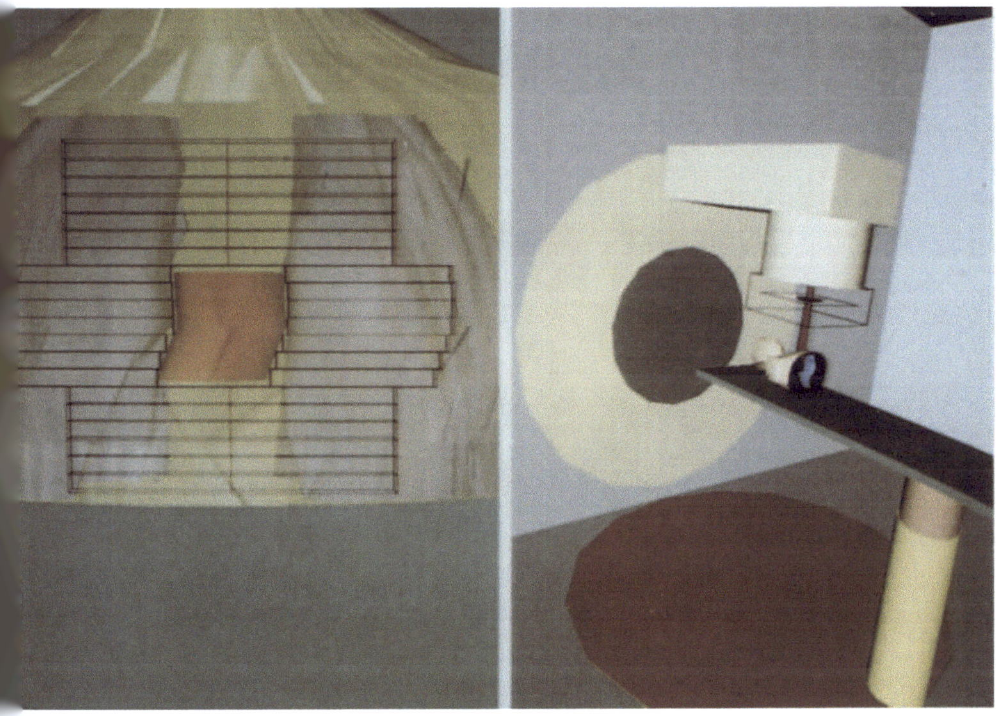

oos et al., Fig. 2, see p. 667

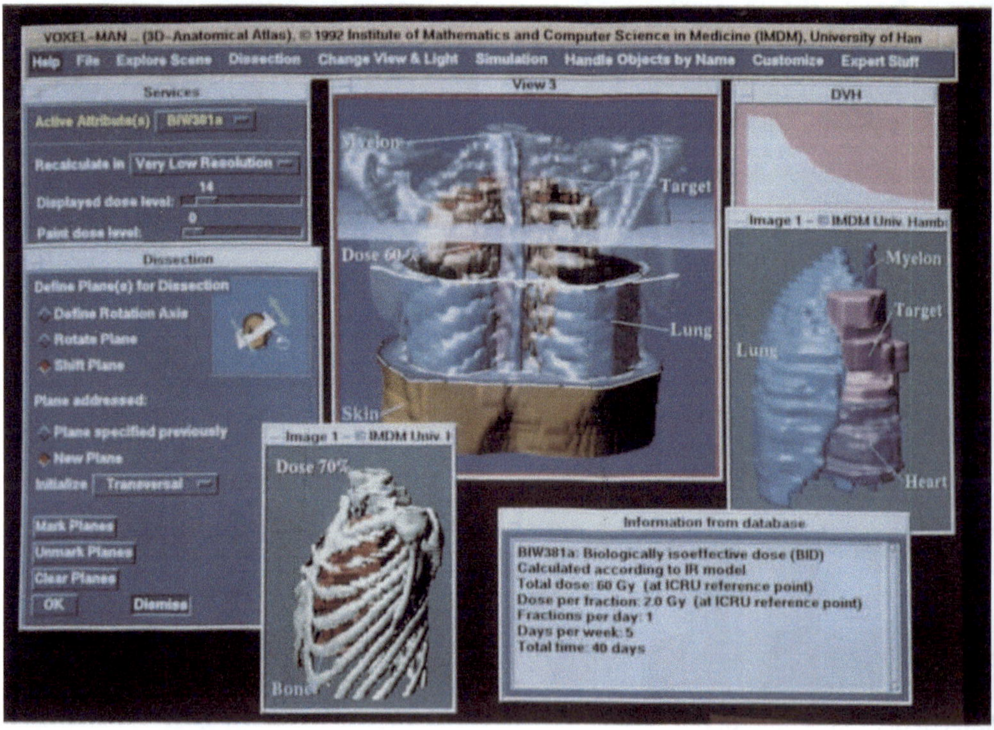

Schiemann et al., Fig. 2, see p. 671

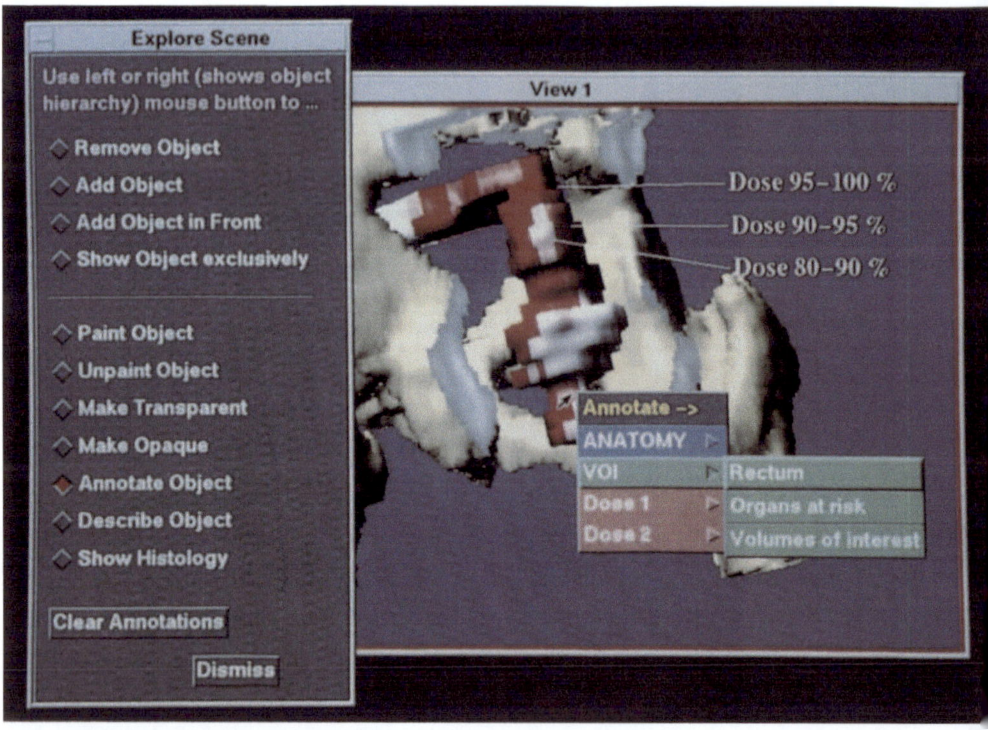

Schiemann et al., Fig. 3, see p. 672

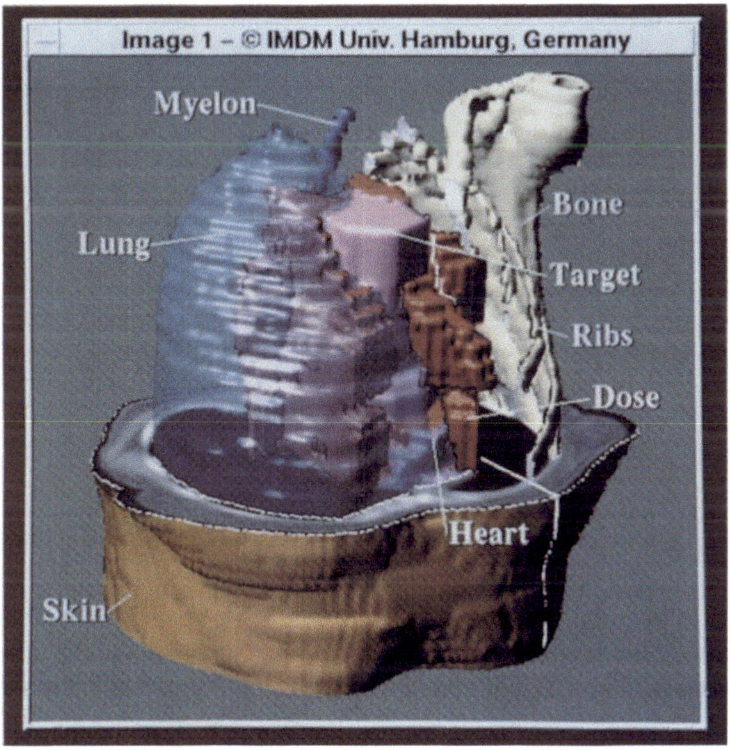

Schiemann et al., Fig. 4, see p. 672

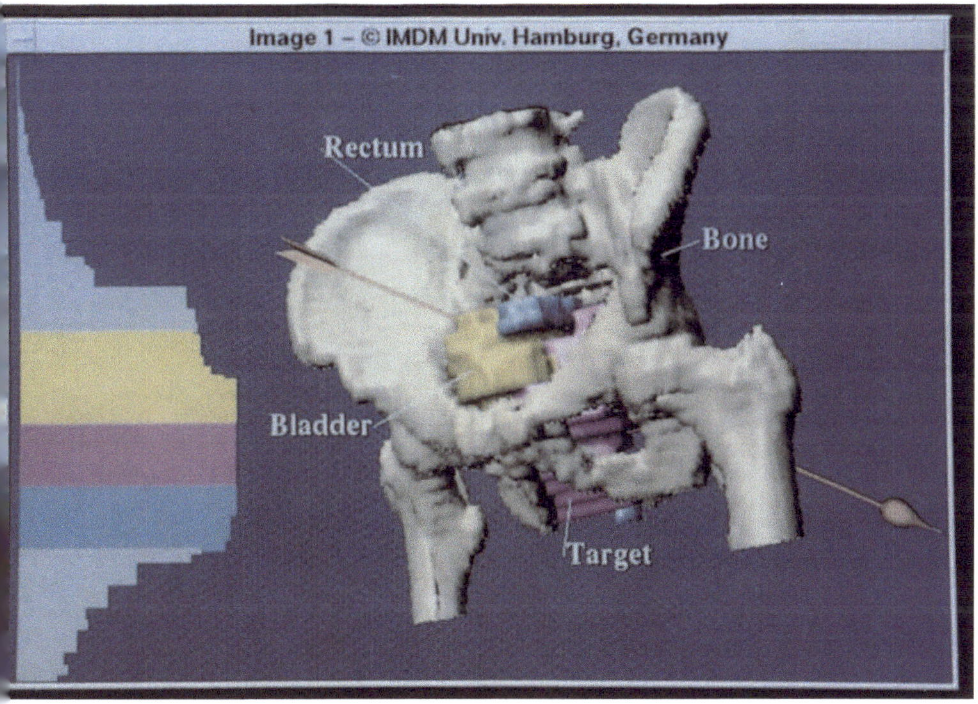

:hiemann et al., Fig. 5, see p. 673

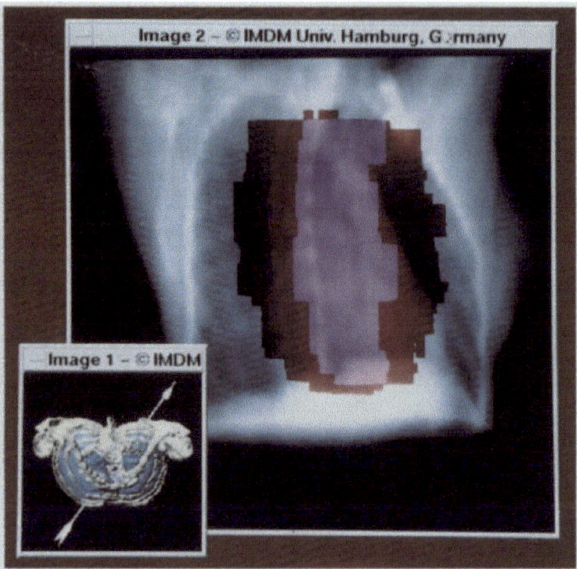

Schiemann et al., Fig. 6, see p. 673

Schiemann et al., Fig. 6, see p. 673

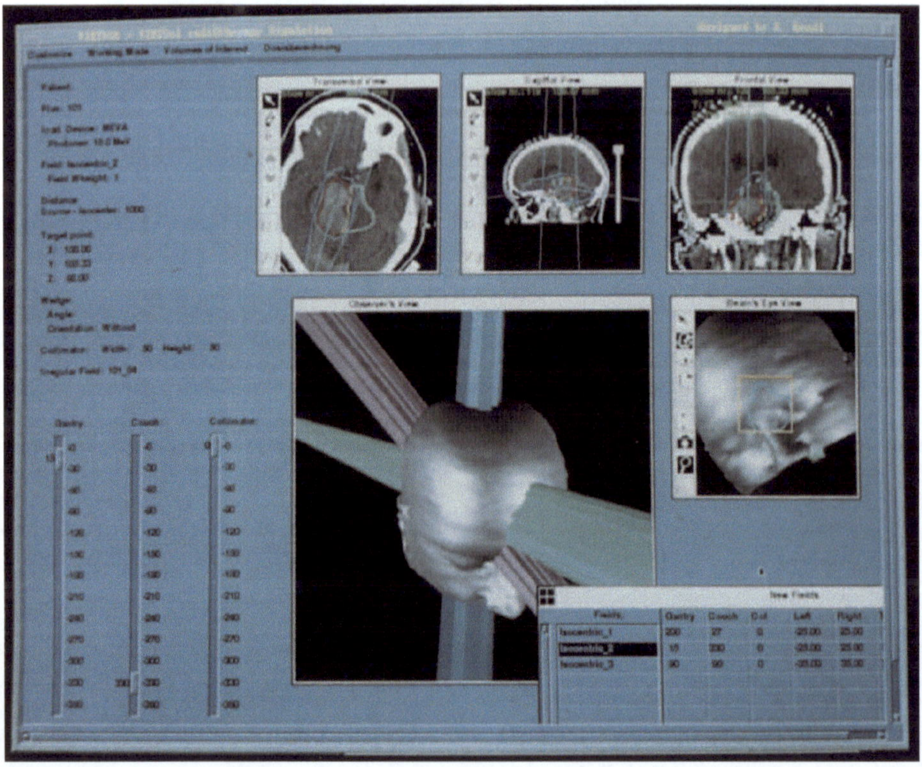

Bendl et al., Fig. 1, see p. 681

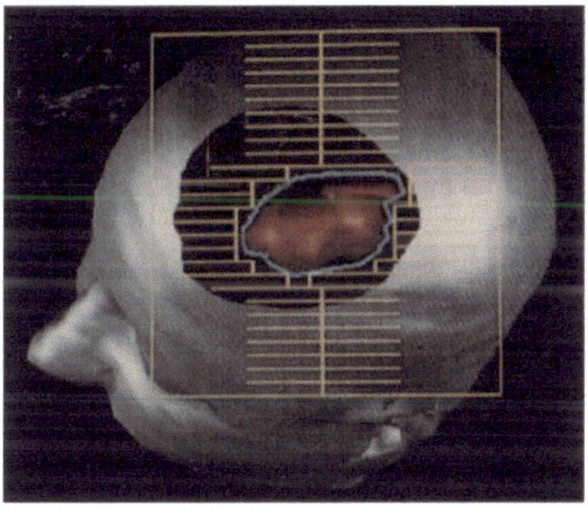

Bendl et al., Fig. 2, see p. 681

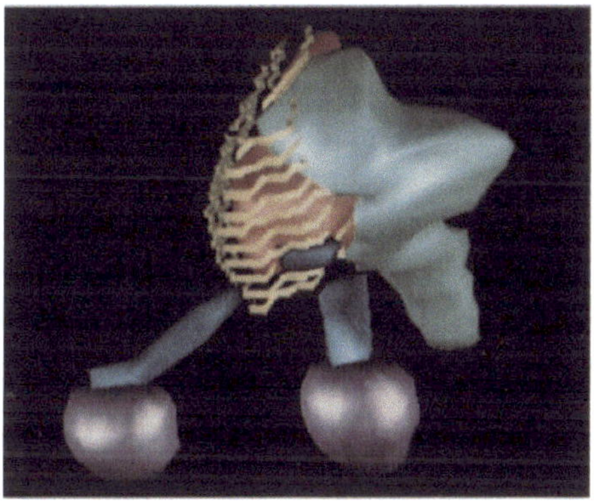

Bendl et al., Fig. 3, see p. 681

Bendl et al., Fig. 4, see p. 681

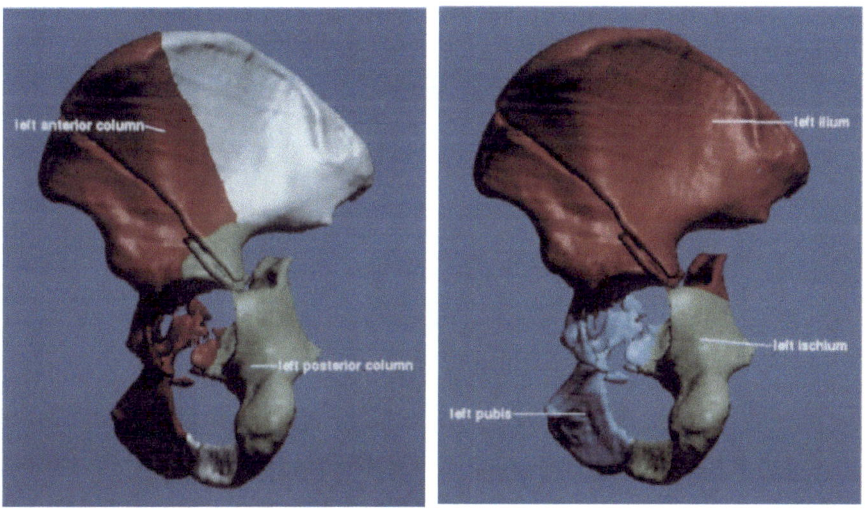

Seebode et al., Fig. 1, see p. 718

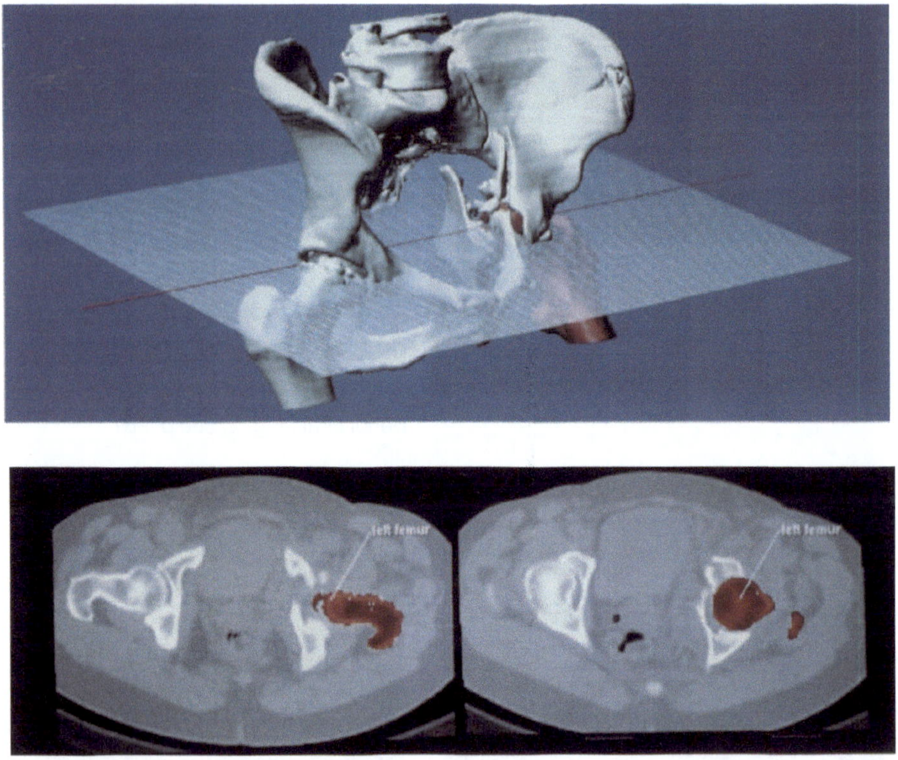

Seebode et al., Fig. 2, see p. 719

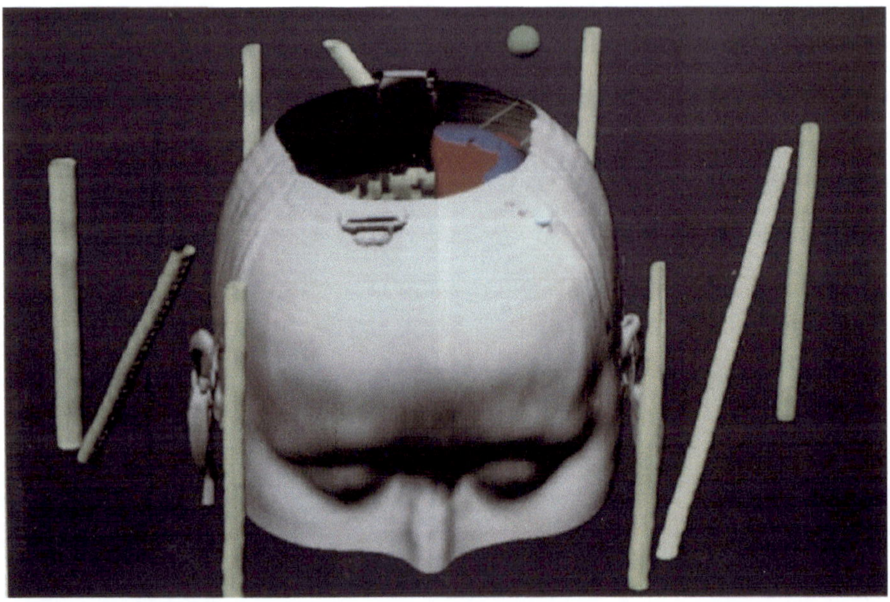

Hughes et al., Fig. 1, see p. 738

Glossary

This glossary has been compiled mainly from terms used in the proceedings of CAR. It is divided into medical, medical radiology, and technical terms, each of which will be extended in successive proceedings according to needs and feedback from readers.

Medical

Aneurysm - balloon-like dilation of a blood vessel or portion of the heart

Angina - chest pain caused by narrowing of arteries supplying the heart

Angiography - methods to visualize the blood vessels

Angioplasty - dilation of a narrowed blood vessel

Arteriosclerosis - disease of aging which results in narrowing of the blood vessels

Basal Ganglia - important neural structures at the base for analysis

Cerebrovascular - blood vessels that supply the brain

Congenital Disease - inherited disease

Contrast Media - substance used to increase the intensity difference between tissues. In radiography this is usually barium or iodine

Coronary Arteriography - imaging of the vessels that supply the cardiac muscle

Cyst - an abnormal fluid-filled mass arising from an organ, commonly the kidney or pancreas

Demyelinating - process in which the tissue sheath around the nerves is destroyed

Diastole - the period during which the cardiac muscle relaxes and the chamber fills with blood

DSA - digital subtraction angiography; an imaging method for demonstrating the new appearance of contrast media in blood vessels by subtracting a preinjection mask image from the image of a later, contrast filled vessel

ECG - electrocardiogram; electrical heart signals

Embolization - blockage of vessels by clot which has travelled from a downstream position within a vessel

Fibroma - abnormal mass of tissue but not cancerous

Hemangioma - abnormal tangle of blood vessels

Herniated Nucleus Pulposes - extrusion of tissue between the vertebra which presses on the spinal nerves and causes back pain

Histology - microscopic examination of organ tissues

Hybrid Subtraction - high and low kilovoltage technique subtracting one image from the other to remove representation of the soft tissues

Hypoechogenic - tissue that is free of echoes on ultrasound, usually clear fluid

Infarct - irreversible damage to tissue caused by lack of blood flow or oxygen

Ischemia - tissue suffering from lack of oxygen usually due to restricted blood flow

Laparotomy - surgical opening of the abdomen

Lymphoma - cancer of the lymph tissues

Malignancy - cancer

Mediastinum - organs in the central portion of the chest: primarily the heart and major vessels

Metabolism - cellular chemical activity required to maintain life

Metastasis - cancer which has spread to other organs from its original source

Monoclonal Antibody - biologically produced pure blood protein, capable of attaching to a specific target tissue

Myelography - imaging of the spinal cord

Myocardium - muscle of the heart wall

Neoplasm - a new and abnormal growth of tissue

Noninvasive Imaging - imaging achieved without the use of surgery or instruments that penetrate the skin. Examples are ultrasound or nuclear magnetic resonance

Nucleus Pulposes - the spongy tissue between the bony vertebra of the spine

Oncology - the study and knowledge concerning tumors

Percutaneous Transluminal Coronary Angioplasty - procedure by which a ballon catheter is expanded within a stenotic region of the coronary artery to dilate it and open blood flow to the myocardium

Peritoneum - internal contents of the abdomen containing bowel

PET - positron emission tomography; uses radioactively labelled compounds to selectively label target tissues. Tomographic image results from detection of the paired gamma rays from the positron decay

Retroperitoneum - structures behind the internal contents of the abdomen. This includes organs such as the pancreas, aorta, kidneys and muscular tissues

Scintigraphy - imaging by detecting gamma rays emitted from radioactively labelled tissues

SPET - single photon emission tomography; a tomographic technique using a rotating gamma ray detector to record, then reconstruct radioactive distribution patterns within the body

Supratentorial - above a tissue membrane that separates the cortex of the brain from the cerebellum. Lies in the lower aspects of the skull

Syringomyelia - abnormality in which a fluid cyst forms within the spinal cord

Systole - the period of contraction of the cardiac muscle in which blood is actively ejected

Therapeutic Radiology - treatment of cancer by targetting portions of the body with ionizing radiation

Thrombosis - clot formation in the arteries or veins resulting in obstruction

Tomography - slice imaging

Venogram - imaging of the veins which return blood to the heart

Ventriculography - display of the left ventricular cavity of the heart to determine its size and the motion of its muscular walls

Medical Radiology*

Associated Equipment - in a radiological installation, equipment other than that for the production and control of ionizing radiation, but essential for its application

Brachyradiotherapy - intracavitary, interstitial or superficial radiotherapy using one or more sealed radioactive sources

Cineradiography - indirect radiography of moving objects, usually in rapid series on cine-film

Computed Tomography (CT) - reconstructive tomography in which recording and processing is effected by a computing system

Dental Panoramic Radiography - direct radiography of a part of or the complete dentition by the use of an intra-oral X-ray tube

Direct Radiography - radiography in which the recording is effected at an image reception area

Direct Radioscopy - radioscopy in which the visible images are presented at the image reception area, or close to it, in the radiation beam

Direct Tomography - tomography in which recording of the image of a layer in the object is effected at the image reception area

Display - visual presentation of information

Electro-Optical X-Ray Image Intensifier - X-ray image intensifier incorporating an electro-optical vacuum device

* The medical radiology terminology has been extracted from the 'International Electrotechnical Commission IEC Standard', Publication 788, First Edition, 1984, Geneva, Switzerland.

Electromagnetic Radiation - propagation of emitted energy through space or through a material medium in the form of waves or in the form of kinetic energy of particles

Examination Room - in medical diagnostic radiology, room in which ionizing radiation is applied to the patient or measured on the patient, and which has the required structural means for radiation protection and for mechanical support

Fluorography - indirect radiography by means of a fluorescent screen

Fluoroscopy - technique of radioscopy by means of a fluorescent screen

Gamma Camera - equipment for scintigraphy, which produces an image by simultaneous detection of radiation emitted from the object

...graph, - in radiology, equipment designed to perform radiography preferably by means of a specific technique

Image Reception Area - in radiology, surface on which an X-ray pattern is received

Input Screen - in an electro-optical vacuum device, layer forming the imaging reception area

Intensifying Screen - layer of material used in direct radiography to convert the incident X-radiation or gamma radiation into radiation more suitable for the radiation-sensitive emulsion of the radiographic film

Irradiation - exposing of a living being or matter to radiation. In radiology, exposing of a living being or matter to ionizing radiation

Kymograph - in radiology, equipment designed to perform radiography preferably by means of a specific technique

Kymography - direct radiography in which an image of the displacements of the boundary of an object is obtained

Luminescence - phenomenon in which certain substances, when excited, emit light of wavelength characteristic of the substance

Medical Diagnostic Radiology - medical diagnosis using ionizing radiation

Medical Radiology - branch of radiology applied to human and veterinary medicine as well as to dentistry and chiropractic

Non-Screen Film - in direct radiography, radiographic film with relatively high sensitivity to the radiation emitted by fluorescence of an intensifying screen

Nuclear Medicine - use of unsealed radionuclides for diagnostic and therapeutic purposes

Operator - person utilizing equipment individually with or without the aid of an assistant, who controls some or all functions of the equipment in the operator's presence

Output Image - in an electro-optical vacuum device, light image generated in the output screen

Output Screen - in an electro-optical vacuum device, layer in which the electron pattern is converted into a light image

Patient - person or animal undergoing medical examination or treatment. For purposes of radiological protection, a person or animal is a patient only during the intentional application of ionizing radiation to that person or animal

Radiation - propagation of emitted energy through space or through a material medium in the form of waves or in the form of kinetic energy of particles

Radiography - technique for obtaining, recording, and optionally processing directly or after transfer, information contained in an X-ray pattern at an image reception area

Radiological - referring to ionizing radiation, its generation and application for scientific, medical and technical purposes

Radiological Equipment - equipment designed for use in radiology

Radiological Image - information obtained by using ionizing radiation presented as an image suitable for medical diagnosis

Radiological Installation - installed radiological equipment including all means for its intended operation

Radiology - science of ionizing radiation and its application

Radiotherapy - medical therapy essentially consisting of one or more treatments by ionizing radiation

Screen Film - in direct radiography, radiographic film with relatively high sensitivity to the radiation emitted by fluorescence of an intensifying screen

Stereoradiography - radiography in which an object is irradiated from two directions to produce a pair of radiograms which, when viewed by suitable optical means, give the illusion of a three-dimensional image

Teleradiotherapy - radiotherapy with a large radiation source to skin distance, usually not less than 50 cm

Tomograph - in radiology, equipment designed to perform radiography preferably by means of a specific technique

Tomography - radiography of one or more layers within an object

Transmission - in radiology, passage of ionizing radiation through, and emergence from, matter essentially in the direction of its incidence, without or after interaction with that matter

User - when used in an IEC standard on electro-medical equipment, organization or individual responsible for the use and maintenance of the equipment

Technical

Analog Signal - a signal in the form of a continnously varying physical quantity

Animation - presentation of a sequence of images on the display screen to create a sense of motion

ASCII - American Standard Code for Information Interchange. An 8 bit code for data representation and transfer

Attribute - a characteristic of a display entity, such as colour, intensity, detectability, etc.

Bandwidth - the difference in cycles per second or hertz between the high and low frequencies of a frequency band

Baud - a unit of signalling speed defining the number of discrete signal conditions

Bit-Rate - the speed at which bits are transmitted over a communication channel, usually expressed in bits per second

Calligraphic Display - a graphic display system for drawing vectors in a refreshed mode

Channel - a facility for a one-way transmission of data

Character - alpha-numerical symbol or other signs and symbols contained in a message

Circuit - a facility for a both-way transmission of data between two or more points

Computer Aided Design (CAD) - the application of computers to the design process

Computer Assisted Orthopaedics - computer based system to assist in orthopaedic surgical planning, e.g. by means of 3D-modelling and computer graphic display of patient specific anatomy and pathology. It may also refer to computer aided design (CAD) and computer aided manufacturing (CAM) methods of patient specific structure models and prothesis

Computer Assisted Radiation Therapy (CART) - radiation dose calculating models and procedures provided by a computerized dose planning system

Computer Assisted Radiology (CAR) - computer based system to assist in generating digital images, storing and transferring images, viewing, analysing and interpreting of images

Computer Assisted Surgical Planning (CASP) - computer based system to assist in general or stereotactic surgical planning

Computer Graphics (CG) - a set of computerized pictorial/graphical procedures for displaying 2D, 3D and 4D data

Computer Vision (CV) - models and procedures provided by a computerized system to support digital image segmentation, analysis and understanding

Data - encoded information that can be stored, manipulated and transmitted by digital devices

Data Communication - the transmission of data from one point to another by means of communication media

Data Tablet - a graphic input device which encodes X and Y data from a hand-held stylus on a tablet surface

Data Terminal Equipment - equipment attached to a data network to send or receive data or both

Digitiser - a device that codes images or shapes into digital data

Display Console - a hardware complex consisting of at least one display unit and usually an alpha-numeric keyboard, function keys and/or other input devices

DMA - Direct Memory Access. A technique for transferring data to and from a host memory without interrupting the host processing unit

Frame Buffer - a dedicated memory for a raster display unit used to store image data

GKS - Graphical Kernel System. A standard for programming computer graphic purposes

Graphic Input - a process to define graphical data for modelling or display purposes

Hardware - a generic term to include all equipment, both computer and communication, contained in a system

Hidden Lines, Hidden Surfaces - removal of line and surface segments obscured from view in a projected image of a 3D object

Highlight - a display characteristic, intended to draw the viewer's attention to a display entity. It is also used to describe a local maximum in rendering processes using illumination models

Image Transformation - algorithms applied to images to produce change in appearance

Light Pen - an hand-held stylus which detects light within a limited area on a refreshed display unit

Look-Up Table - a special memory array containing grey or colour values for pixel display

Man Computer Interaction (MCI) - procedures and devices optimized towards realizing a ‘natural’ interface between man and computer

Menu - a list of options shown on a display unit

Message - a sequence of characters used to convey information or data

Modelling - procedures and devices to represent 2D, 3D and 4D general or patient specific anatomic and clinical data in digital memories

Network - a system consisting of a number of points connected by communication channels

Picture Archiving and Communication Systems (PACS) - computer based system to assist in storage and archiving of images as well as soft and hard copy image generation

Pixel - an individual picture element of a digital image

Point - a coordinate position in 2D or 3D space. Also an addressable position on the display screen

Protocol - a formal procedure required for synchronizing data exchange, e.g. to initiate and maintain data communication

Raster - a rectangular matrix of pixels

Raster Scan - a technique for generating an image in a line-by-line sweep

Run-Length Encoding - a data compression technique based on describing the number of pixels along a scan line that have the same intensity value

Scaling - a transformation function to change the size of a display entity

Shading - computation of the intensities and colours of the surface pixels of 3D objects

Software - a generic term for computer programs and documentation

Spatial Resolution - the smallest distance between two pixels

Translation - movement of a display entity from one position to another

Vector - a line segment drawn between two points. It possesses the attributes of magnitude and direction

Window - a bounded area that contains a subset of the data to be shown on the display unit

List of Authors

MIX
Papier aus verantwortungsvollen Quellen
Paper from responsible sources
FSC® C105338

If you have any concerns about our products,
you can contact us on
ProductSafety@springernature.com

In case Publisher is established outside the EU,
the EU authorized representative is:
Springer Nature Customer Service Center GmbH
Europaplatz 3, 69115 Heidelberg, Germany

Printed by Libri Plureos GmbH
in Hamburg, Germany